DNA Nanotechnology for Cell Research

DNA Nanotechnology for Cell Research

From Bioanalysis to Biomedicine

Edited by Zhou Nie

WILEY VCH

Editor

Prof. Zhou Nie
Hunan University
College of Chemistry and Chemical Engineering
Lushan Road (S)
Yuelu District
Changsha 410082
China

Cover Image: © Vink Fan/Adobe Stock Photos

Library of Congress Card No.: applied for

British Library Cataloguing-in-Publication Data
A catalogue record for this book is available from the British Library.

Bibliographic information published by the Deutsche Nationalbibliothek
The Deutsche Nationalbibliothek lists this publication in the Deutsche Nationalbibliografie; detailed bibliographic data are available on the Internet at <http://dnb.d-nb.de>.

Print ISBN: 978-3-527-35173-2
ePDF ISBN: 978-3-527-84079-3
ePub ISBN: 978-3-527-84080-9
oBook ISBN: 978-3-527-84081-6

Typesetting Straive, Chennai, India
Printing CPI Group (UK) Ltd, Croydon, CR04YY
C9783527351732_060224

Contents

Preface

Nucleic acids, despite their primary role in genetic information storage and transmission, also function in synergy with other biomacromolecules to assemble intricate biological structures seen in nature, such as chromosomes and Holliday junctions. Beyond their structural capacities, these natural complexes are indispensable for many essential biological processes. This intricate intertwining of structure and function received a fresh perspective in 1982, when Nadrian C. Seeman introduced a pioneering concept that DNA could be harnessed as a programmable structural material. This groundbreaking idea has paved the way for substantial advancements in the field of DNA nanotechnology over the past four decades. The Watson-Crick base-pairing principle, the driving force behind the development of DNA nanotechnology, provides the foundation for the construction of static or dynamic nanostructures that offer nearly infinite programmability at the nanoscale and allow temporal control. DNA nanotechnology allows for precise spatial control in 3D space and introduces stimuli-responsive dynamic behaviors to an array of biological molecules. Recently, we have witnessed the development of DNA-based toolkits that have facilitated a higher-resolution exploration of cell structure, function, and behavior with improved specificity. Moreover, DNA nanotechnology also permits the precise control and intelligent programming of cellular functions, opening new possibilities for bioengineering and therapeutic applications.

This book, "DNA Nanotechnology for Cell Research: From Bioanalysis to Biomedicine," aims to fill a gap in current literature. Unlike previous books that focused mainly on the functionality of designer DNA nanostructures as well as their *in vitro* applications such as bioanalysis and biocomputing, this book presents a comprehensive overview of DNA nanotechnology and its diverse applications in cell research and engineering. We have invited active researchers with extensive expertise in the field of DNA nanotechnology to contribute their knowledge and insights to this volume. The coverage of this book ranges from the fundamental concepts of DNA nanotechnology to the most cutting-edge studies on its applications in fundamental cell research and associated biomedical fields, including (i) cellular recognition, (ii) cell imaging and intracellular sensing, (iii) regulation of cellular functions, and (iv) cell-targeted medical applications. We hope this book serves as an up-to-date, tutorial-style resource for researchers across diverse fields, from DNA nanotechnology, cell biology, chemical biology, and analytical chemistry, inspiring them to exploit the unique merits of DNA nanotechnology to unravel cellular mysteries and develop innovative therapeutic strategies.

We would like to express our sincere appreciation to all the contributing authors, reviewers, and editorial staff involved in this book. Their tremendous efforts, coupled with insightful perspectives, have significantly enriched the content of this volume. As we look to the future, we are confident that this collective endeavor will not only stimulate advancements but also inspire further exploration and innovation in this fascinating and ever-evolving field of research.

August 2023

Zhou Nie
Changsha, China

Part I

DNA Nanotechnology for Cellular Recognition (Cell SELEX, Cell Surface Engineering)

1

Developing DNA Aptamer Toolbox for Cell Research

Liang Yue[1], Shan Wang[1], and Weihong Tan[1,2,3]

[1] Hunan University, College of Chemistry and Chemical Engineering, College of Biology, Aptamer Engineering Center of Hunan Province, Molecular Science and Biomedicine Laboratory, State Key Laboratory of Chemo/Biosensing and Chemometrics, 2 Lushan Road, Changsha 410082, China
[2] Chinese Academy of Sciences, Hangzhou Institute of Medicine, The Cancer Hospital of the University of Chinese Academy of Sciences (Zhejiang Cancer Hospital), 150 Dongfang Street Xiasha, Hangzhou 310000, China
[3] Shanghai Jiao Tong University, Shanghai Jiao Tong University School of Medicine, Institute of Molecular Medicine, College of Chemistry and Chemical Engineering, Renji Hospital, 800 Dongchuan RD, Shanghai 200240, China

1.1 Cells and Their Complexity

Cells are the basic building blocks of life, initially discovered by Robert Hooke in 1665 [1, 2]. Their structural and functional complexity attracts significant research interest due to their intricate composition, abundance, and interactivity [3–5]. Even the simplest unicellular organism comprises various biomolecular components, ranging from nucleic acids, proteins, and lipids to carbohydrates, which are elaborately bonded together. Moreover, cells exhibit remarkable plasticity and can proliferate and differentiate to create distinct cell populations with radically different compositions and interactions, which enables them to implement specific and varied functions [6–8]. The organization of these cell populations yields a highly collaborative living system capable of carrying out all of life's biological processes, where cells execute their functions through molecular interactions with the external environment and other cells [9, 10]. Furthermore, cells are open systems that facilitate the exchange of information, energy, and matter between cells and their external environment [11–14]. The dissipation of energy and matter is a continuous process sustaining cellular functions, which drives the cells to operate far from equilibrium [15, 16]. The nonlinear interactions and reactions of the components further enhance the complexity of cellular systems, revealing extraordinary dynamic behaviors and functions adapting to ever-changing environmental cues [17–19].

To understand cells and their functions in their entirety, it is essential to investigate components that interact to develop and maintain high-order cellular structures and realize complex biochemical activities [20, 21]. Reliable and accurate molecular recognition tools are required to detect specific components from the

DNA Nanotechnology for Cell Research: From Bioanalysis to Biomedicine, First Edition. Edited by Zhou Nie.

cellular pool [22, 23]. In addition, suitable technologies and methodologies are needed to observe components' dynamics and analyze their nonlinear behaviors in biological processes [24–28]. This field has experienced substantial advancements propelled by breakthroughs in molecular recognition tools and analytical methods. Further, researchers have devoted themselves to developing new and innovative tools to observe, measure, and manipulate cellular components and systems at unprecedented levels of detail. These progresses broaden our understanding of cellular mechanisms and pave the way to develop novel therapies and treatments for diseases associated with cellular dysfunction [29–34].

Functional nucleic acids, as synthetic oligonucleotides with unique chemical-biological properties, sequence programmability, simplicity of modification, and facile functional manipulation, have demonstrated bioactivity, versatility, and *in vivo* bioavailability toward biological and biomedicine applications [35–39]. The rapid progress in DNA nanotechnology has offered remarkable opportunities for tackling the intricate nature of cellular systems [40]. In particular, introducing DNA aptamers with excellent molecular recognition ability into the modular design of molecular tools has opened new avenues in cell research, providing captivating prospects for handling complex molecular systems with precision and accuracy.

1.2 Features and Advantages of DNA Aptamers

DNA aptamers [41], termed "chemical antibodies", are short, single-stranded nucleic acid molecules (20–100 nucleotides) that can fold into complex 3D structures and bind with various targets (e.g. metal ions [42], small organic molecules [43], proteins [44], viruses [45], bacteria [46], cells [47], and even tissues [48]) with specificity and high affinity, Figure 1.1. Compared with natural antibodies, the core competency of DNA aptamers rests with the unique properties of synthetic nucleic acids [41], involving relatively low molecular weight, low-cost standardization synthesis, customized modification, programmable sequences and structures, sequence-dependent multifunctionality, etc. The size of DNA aptamers (10–30 kDa and ~3 nm in diameter) is much smaller than that of natural antibodies (~150 kDa and 10–15 nm in diameter), which endows DNA aptamers with fewer steric hindrance and unimpaired binding affinity in a confined environment, thus rendering a broader application [49]. For instance, Yang et al. reported a straightforward strategy for labeling and manipulating cell surface protein by utilizing DNA aptamers [50]. Specifically, an aptamer-based logic computing reaction is used to recognize selectively and covalently conjugate immune checkpoint antagonizing aptamers (e.g. D-aPDL1, D-Sgc8-PA, and D-Sgc8-99mTc) on the surface of cancer cells, improving the precision and robustness of immune checkpoint blockade therapy.

In contrast with the complex modifying process and unmanageable conjugation numbers and sites of natural antibodies [51], DNA aptamers are facile to be synthesized and modified as a consequence of the technological progress in phosphoramidite chemistry-based oligonucleotide synthesis over the past decades [52].

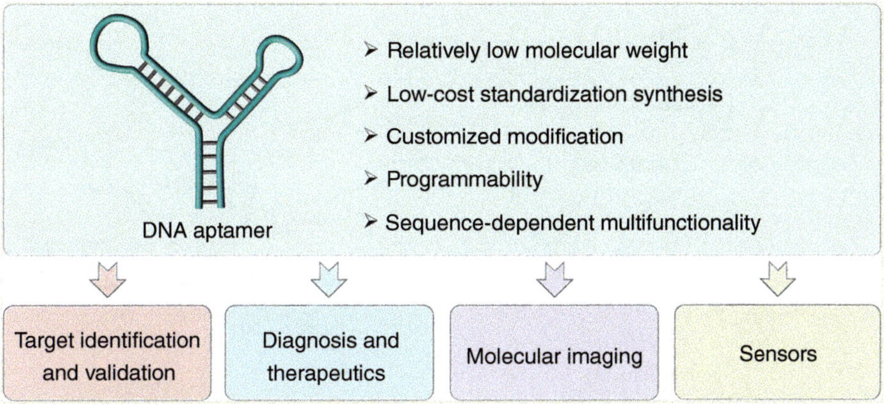

Figure 1.1 Schematic illustration of the DNA aptamer tools for various applications.

With recent advancements in automated modular synthesizers and phosphoramidite chemistry, DNA aptamers can now be conjugated with multiple functional groups at predetermined locations [53]. This approach provides many benefits for practical applications of DNA aptamers, including aptamer–drug conjugates (ApDCs) [54] and artificial nucleobase-expanded aptamers [55].

Furthermore, nucleic acids possess base sequence programmability and modular design characteristics, whereas natural antibodies currently do not. By integrating modular design and programmability into DNA aptamer systems, modular and programmable toolboxes can be created to meet future demands in advanced biological molecular tools and materials. Specifically, DNA aptamers with target recognition abilities can be regarded as functional modules that can be integrated, on demand, into nanostructures [56], nanodevices [57], and other functional systems [58]. These features enable researchers to engineer DNA aptamer-based systems with a high degree of flexibility and precision at the molecular level. Further, the highly predictable base-pairing and structural reconfiguration rules of nucleic acids provide a strong basis for designing programmable DNA structures and controlling DNA dynamics. Through their conjugation with specific functional components, DNA aptamers can respond to various environmental cues [59–61], such as pH, light, and temperature, and numerous biological targets (e.g. DNA/RNA, ATP, enzymes, and metal ions) leading to innovative nanodevices. With these superior merits, DNA aptamer-based systems have been considered promising candidates for various cellular applications.

1.3 On-demand Synthesis and Screening of DNA Aptamers

Highly accurate and pure DNA oligonucleotides are vital for DNA aptamer screening and their practical applications. Since its introduction by Caruthers in 1981 [62], phosphoramidite chemistry has emerged as the gold standard for

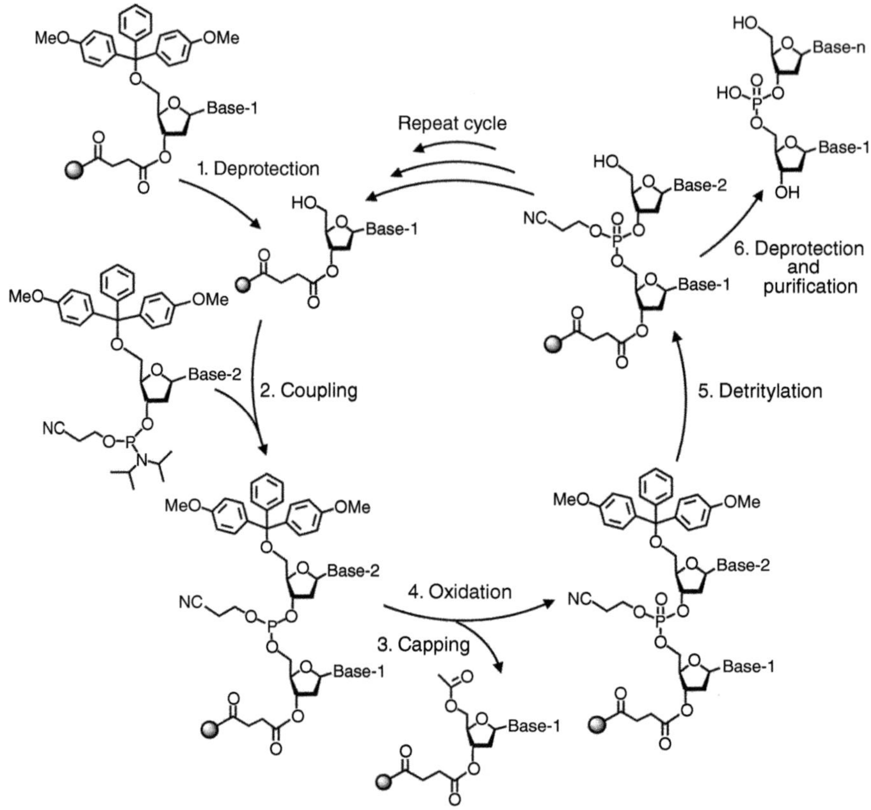

Figure 1.2 Schematic representation of solid-supported oligonucleotide synthesis.

the synthesis of short DNA oligonucleotides (<200 nucleotides). This innovative chemical method has revolutionized the field of DNA synthesis, which allows for the production of DNA sequences with high purity and chemical modifications customized for specific research or practical applications. To grow oligonucleotide chains using phosphoramidite chemistry via a solid-supported oligonucleotide synthesis, the following steps are typically involved, as depicted in Figure 1.2: (1) Deprotection of resin-supported nucleoside: The first nucleoside protected by a dimethoxytrityl (DMT) group is pre-attached to resin. Then, the 5′-DMT protecting group is removed by acid-catalyzed detritylation to expose the 5′-hydroxyl group as a reactive site for the addition of the next nucleotide; (2) Coupling: Taking tetrazole or its derivative as an activator, the diisopropylamino group of the incoming nucleoside phosphoramidite is protonated to enhance its reactivity as a leaving group. The 5′-hydroxyl group of the support-bound nucleoside attacks the phosphorus atom of the activated nucleoside phosphoramidite, resulting in a new phosphorus-oxygen bond; (3) Capping: To prevent the formation of any truncated sequences, a "capping" procedure is introduced after the coupling reaction, which effectively prevents unreacted 5′-hydroxyl groups from participating in subsequent coupling reactions. With all unreacted sites "capped," the oligonucleotide synthesis proceeds exclusively at full-length oligonucleotides; (4) Oxidation: The phosphite-triester (P^{III}) formed during the coupling step is susceptible to acid and needs to be converted into a stable form (P^{V}) before the implementation of next acidic detritylation step. This can be achieved through iodine oxidation in the presence of water and pyridine. The resulting phosphotriester is, essentially, the DNA backbone and is protected with a 2-cyanoethyl group from undesirable reactions during subsequent synthesis cycles; (5) Detritylation: After the oxidation step, the DMT-protecting group located at the 5′-end of the resin-bound DNA chain is removed for the following nucleotide phosphoramidite. Repeat steps 2 through 5 until the desired oligonucleotide sequence has been synthesized; (6) Deprotection and purification: Finally, all the remaining protecting groups (e.g. the protecting groups from the heterocyclic bases and phosphodiester backbone) are removed, and the oligonucleotide is purified using various methods, such as HPLC or PAGE.

The advancement of automated oligonucleotide synthesis technologies has significantly improved the synthetic length, speed, cost, and throughput, enabling high-throughput, cost-effective, and large-scale synthesis of DNA aptamers with high efficiency and precision. Because of the precise synthesis capability, the fine control over DNA aptamers is accessible, allowing for their incorporation with different chemical modifications where the position and quantity of modification can be manipulated to alter the properties and functions of DNA aptamers. The ability to engineer the structure and specificity of DNA aptamers through synthesis and chemical modification makes them a highly flexible and customizable tool for a diverse range of biological and biomedical research areas.

Relative to antibodies evolved from the natural immune response, aptamers are produced mainly by an *in vitro* evolution method, the systematic evolution of ligands by the exponential enrichment (SELEX) technology, developed by Szostak and Gold in 1990 [63, 64]. This technology involves iterative rounds of selection and amplification of specific nucleic acid sequences in a large oligo pool, through which various aptamers have evolved for targeting small molecules (e.g. organic dyes [65], ATP [66], cocaine [67], and mycotoxins [68]) or proteins (e.g. thrombin [69] and tumor biomarkers [70]), with specificity and high affinity. Nevertheless, the screened DNA aptamers through SELEX processes may undergo structural reconfiguration and function loss in a physiological environment, which limits their utilization in cell research.

In 2006, our group introduced an extension of the SELEX technology called cell-SELEX [47], aiming to select aptamers against living cells. Since then, the target has been developed into diverse living species [71, 72] (e.g. bacteria, viruses, and disease tissues). For the cell-SELEX technology, a large-capacity (10^{15}–10^{16}) oligonucleotide library needs to be engineered before the screening, and each oligonucleotide comprises random sequence regions in the middle and primer sequence-binding regions at both ends. A typical cell-SELEX process involves a six-step selection and amplification cycle: (i) Incubation of target cells with a library of single-stranded DNAs including a random domain; (ii) Collection of the target cell-bound oligonucleotides; (iii) Amplification of the collected oligonucleotides by PCR; (iv) Counter-selection step by using negative cells to reduce the common

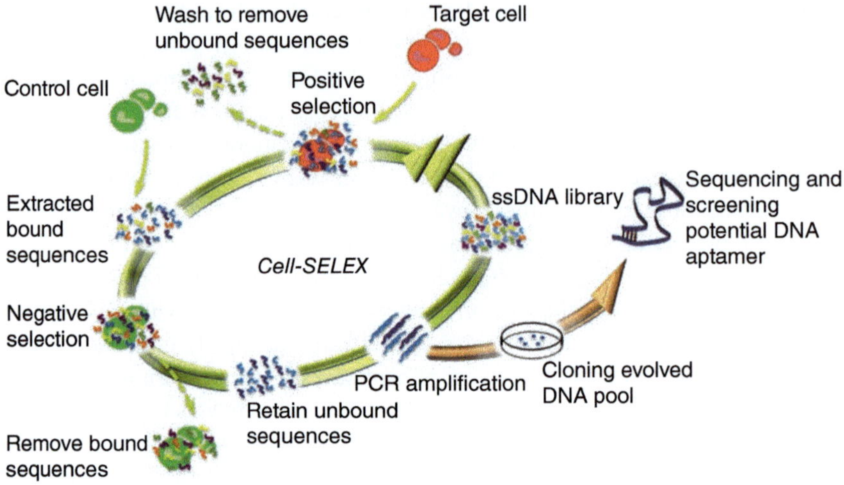

Figure 1.3 Schematic representation of DNA aptamer screening by the cell-SELEX process. Source: Reproduced with permission from Sefah et al. [72]. © 2010 Springer Nature.

sequences that bind with both cells; (v) Retainment of the unbound DNAs from counter-selection process; (vi) Amplification of the unbound DNAs to yield an enriched oligonucleotide pool for the following round selection. The process is iterated until the sequences specifically targeting the cell are highly enriched. The aptamer's affinity to the target molecules in a living cell can be significantly improved after 10–20 rounds of screening. Finally, corresponding pools are sorted and optimized for further cloning and sequencing, as shown in Figure 1.3.

Compared with the traditional SELEX strategy, cell-SELEX possesses several superiorities in the complex cellular system: (i) Massive biomolecules exist with different functional states and abundance in the cell. Aptamers produced by cell-SELEX can bind to these biomolecules naturally, conducive to understanding how these molecules interact and realize complex biochemical activities. (ii) It is unnecessary to know and purify target cells' molecular signatures in cell-SELEX since aptamers can target unknown molecules, and it enables the discovery of potential biomarkers. (iii) For cell research, aptamers generated from cell-SELEX are in their active and native states without structural unfolding in a cellular environment. (iv) Accompanied by the occurrence of disease in a living system, a fundamental change at the molecular level takes place, like the emergence of cancer with unique biomarkers. By analyzing the molecular difference between disease cells and normal cells, potential biomarkers and new molecular events can be targeted by cell-SELEX, thus distinguishing disease cells from normal cells. Cell-SELEX has successfully screened out hundreds of aptamers against more than 80 different cell lines, opening up new avenues for cellular applications [73, 74]. For example, Wu et al. reported a DNA aptamer, XQ-2d, which targets the transferrin receptor CD71 with high affinity and specificity [75]. CD71, a membrane glycoprotein and receptor of the iron-transferrin complex, is abundantly expressed in various cancer cells, including those found in the brain, liver, breast, lung, and colon. It is considered a significant independent prognostic marker and a promising therapeutic target for several cancers. Among 40 clinical samples of pancreatic adenocarcinoma investigated using Aptamer XQ-2d, CD71 is overexpressed in 82.5% of the samples. Furthermore, Aptamer XQ-2d has been used for the targeted delivery of doxorubicin (DOX) to pancreatic cancer cells. The selective delivery of loaded DOX into PDAC cell line PL45 results in a significant antiproliferation effect, implying the XQ-2d aptamer's great potential for targeted pancreatic cancer therapy.

With the advent of chemical synthesis and cell-SELEX technology, DNA aptamers have emerged as a powerful toolbox for the biological and biomedicine fields due to their unique properties, including high binding affinity, specificity, stability, and simplicity of synthesis. DNA aptamers have been widely applied for cell research to identify and validate new targets on the surface of cells [76], to investigate intracellular processes, or to control cell behaviors. Also, they have been employed as biosensors [41] to detect proteins, nucleic acids, or small molecules in real time or as therapeutic agents for targeted drug delivery [57] and gene regulation. Furthermore, DNA aptamers have assisted in the isolation and detection of rare cell populations as well as in cell imaging [56] and molecular diagnostics [77]. Overall, DNA aptamers constitute a versatile tool for cell research in favor of the elucidation of the complicated workings of cells. As research in DNA aptamers progresses, new and exciting applications are anticipated to emerge, which could significantly enhance their utility and impact on cell research. These advancements have the potential to revolutionize the field of biomedicine, resulting in improved patient outcomes.

1.4 Toward a Toolbox of DNA Aptamers for Cellular Applications

As life science continues to develop, there is a growing requirement for molecular tools that enable researchers to understand cellular compositions and precisely control their dynamic behaviors. Recent advances in molecular engineering in DNA aptamers have paved the way toward a DNA aptamer-based toolbox for various cellular applications [39, 78–80]. This toolbox consists of a standardized set of DNA aptamers that have been extensively validated for their specificity, affinity, protocols for synthesis, modification, integration, characterization, and utility in cellular applications. We believe the ongoing development of such a toolbox will unlock the full potential of DNA aptamers for cell research.

To develop and utilize a toolbox composed of a versatile set of DNA aptamers for cell research, here are the necessary protocols we can follow: (i) Identify the tasks and select appropriate aptamers. To reach research goals, we need to identify the target (e.g. biomolecules, viruses, cells, or tissues) and then select a suitable DNA aptamer with high affinity and specificity against the target. (ii) Verification of DNA aptamers: Once the DNA aptamer candidate is selected, it is crucial to verify its binding affinity and specificity by performing enzyme-linked oligonucleotide assay (ELONA) or surface plasmon resonance (SPR) analysis.

(iii) Aptamers modification: To meet the needs in the specific application scenario, it is necessary to modify them with appropriate functional groups. This process can be realized using various chemical or integrative approaches involving post-SELEX chemical modification, DNA nanotechnology, or supramolecular self-assembly. (iv) Application and optimization of DNA aptamer tools: After we have delivered DNA aptamer tools into cells for various applications, it is vital to evaluate their performance, optimize them based on feedback, and create new tools to fill any gaps in the toolbox when confronted with new tasks. By following these protocols and preparing for challenges, researchers can efficiently develop and utilize DNA aptamer tools in cell research.

In this context, we discuss recent advances in engineering DNA aptamers as targeting and regulating modules and integrating them with versatile modules for a multifunctional toolbox, which allows researchers to choose the desired aptamer tools and use them for specific applications without further screening and examination. We introduce the molecular engineering approaches of DNA aptamers, mainly categorized as chemical modification and DNA nanotechnology. We highlight some examples of DNA aptamers engineered for specific cellular applications, including biological regulation, targeted delivery of therapeutic agents, biosensing, cell imaging, and biomimicry.

1.4.1 Chemical Modifications via Solid-Supported Synthesis Strategy

DNA aptamers offer promising molecular recognition tools for cells. However, their full potential in life science has been limited by several challenges, such as poor chemical and biological stability and a narrow range of chemical diversity. To address these limitations, chemical modification of aptamers through a solid-supported synthesis strategy has been exploited to unleash their potential as molecular tools.

Solid-phase synthesis technology is a highly controllable and automated molecular synthesis method that can efficiently produce nucleic acids from individual phosphoramidite building blocks. This technology offers a sequence-predesigned DNA synthesis platform enabling the direct introduction of chemical groups into oligonucleotides in a controlled manner, which significantly enhances the aptamers' biostability and versatility [81–83]. Throughout the chemical synthesis process, aptamers can be modified at nucleotide components, including bases, sugars, or backbone groups, as shown in Figure 1.4. For instance, the backbone modifications on non-bridging oxygen of the canonical phosphodiester linkage with a methyl or ethyl group would provide uncharged backbones for minimizing the electrostatic repulsion between nucleic acid strands [84, 85]. The sugar group of the nucleotides can be modified by introducing a 2′-fluoro or 2′-*O*-methyl group and the locked nucleic acid [86]. It has been demonstrated that 2′-amino and 2′-fluoro substitutions can significantly increase the half-life of aptamers in human serum, extending from mere seconds (eight seconds) up to 86 hours. Furthermore, recent advances in nucleic acid chemistry have promoted the development of threose nucleic acid, peptide nucleic acid, and chirally inverted mirror-image nucleic acids (L-DNA) with unique structures different from natural nucleic acids [87, 88].

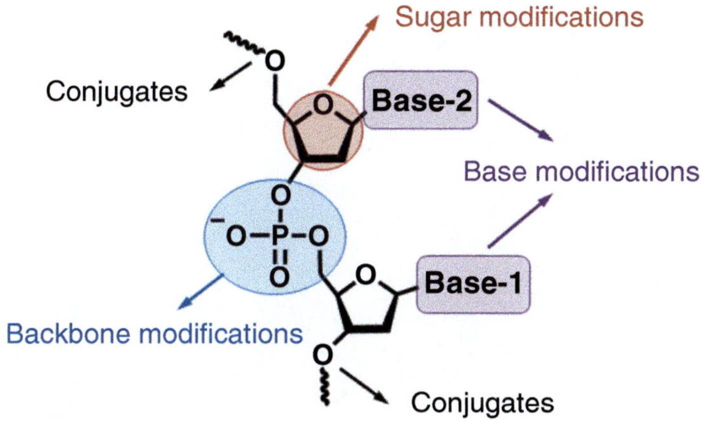

Figure 1.4 Schematic representation of common chemical modifications used in aptamers.

These artificial nucleic acids have also been explored as a promising category of nuclease-resistant and biostable aptamers that cannot be degraded by human blood and serum nucleases. As mentioned, these DNA aptamers can be readily synthesized by automated oligo synthesizer with commercially available materials, facilitating practical applications in diagnostics and therapeutics.

Meanwhile, this innovative approach enables the efficient synthesis of ApDCs using oligonucleotide synthesizers while preserving the biological activity of the included drugs. In 2014, Tan and coworkers developed the first ApDC phosphoramidite module, which incorporated an anticancer drug moiety and a photocleavable linker [53]. Upon the application of this module, automated and modular synthesis of ApDCs has been realized, allowing for the effective introduction of multiple drugs at predetermined positions. These ApDCs display specific targeting toward cancer cells and can be photoactivated to release drugs in a precise and controlled manner. Subsequently, various ApDCs have been exploited by using an automated DNA synthesizer. For example, Lv et al. reported a Sgc8-5FU ApDC by modifying DNA Aptamer Sgc8 with 5-fluorouracil (5-FU). The internalization and subsequent transportation pathways of the ApDC were investigated by the single-particle tracking (SPT) technique [89]. The results reveal that the ApDC predominantly entered the cells via caveolin-mediated endocytosis, similar to the pure DNA aptamer. Besides, Xuan et al. incorporated bioorthogonal chemistry with prodrug design to develop a novel aptamer prodrug conjugate (ApPdC) [90]. In the designed ApPdC, the hydrophilic DNA aptamer functions as a tumor-targeting subunit, whereas the hydrophobic prodrug promotes the self-assembly of ApPdC and acts as a free radical generator to enhance the efficacy of chemodynamic therapy.

Another approach involves the introduction of nucleotide analogs like Z:P, which mimic the structures and functions of natural nucleotides, into DNA aptamers to expand their chemical composition [91]. Artificial nucleotide bases promise to improve the diversity, properties, and functionalities of aptamers, making them more practical for various application scenarios. Our group proposed the concept of

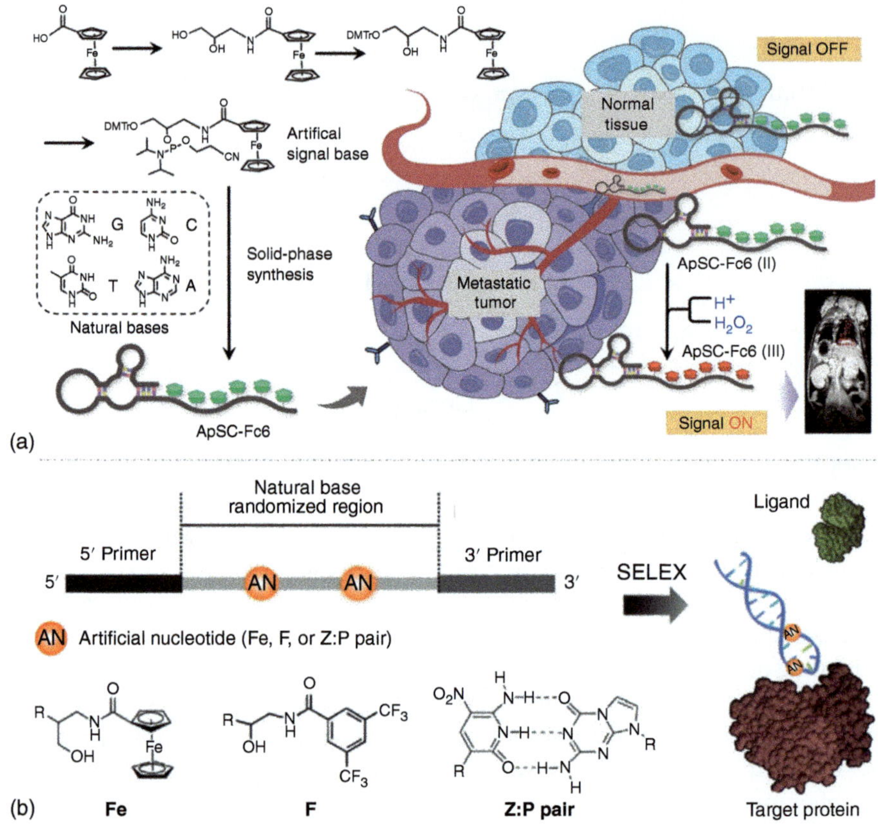

Figure 1.5 Schematic representation of (a) the logic-gated aptamer-signal base conjugate for targeted molecular imaging of metastatic cancer. Source: Reproduced with permission from Li et al. [92]. 2022 American Chemical Society. (b) The design of the artificial-nucleotide-expanded aptamer against integrin alpha3 for regulating cell activities. Source: Reproduced with permission from Tan et al. [93]. © 2018 John Wiley & Sons.

aptamer-signal base conjugates (ApSC) to develop AND-gated molecular tools for tumor-targeted molecular imaging [92]. As depicted in Figure 1.5a, the molecular probe consists of two segments: a DNA aptamer acting as the targeting ligand and tumor microenvironment-responsive artificial ferrocene bases for tumor cell imaging. The resulting ApSC demonstrates targeted accumulation in the tumor cells via aptamer-mediated molecular recognition. Meanwhile, the acidic pH and high levels of H_2O_2 in the tumor microenvironment trigger the *in situ* activation of the synthetic ferrocene bases through the Fenton-like reaction that leads to an amplified MRI signal over normal tissues.

The various synthesized nucleotides can be applied to enlarge the chemical space of the nucleic acid libraries in SELEX processes, thus increasing the diversity of DNA aptamers. Hence, an artificial-nucleotide-expanded SELEX approach was developed to evolve artificial nucleotide-based aptamers [93]. The nucleic acid library of cell-SELEX was extended with three artificial nucleotide bases: Fe base with a ferrocenyl group, F base with a trifluoromethyl group, and Z:P base pair (Figure 1.5b). And an artificial nucleotide-expanded aptamer named ZAP-1, targeting the integrin alpha3 (ITGA3), was screened through the artificial-nucleotide-expanded cell-SELEX process. The binding of aptamer ZAP-1 with ITGA3 can reduce the interaction between ITGA3 and its natural ligand, which inhibits the triple-negative breast cancer (TNBC) cells' adhesion and migration. We anticipate that the artificial-nucleotide-expanded aptamers evolved from this artificial nucleotide-assisted cell-SELEX approach would enable the elucidation of the molecular basis of biological activities and regulation of biological functions.

1.4.2 Chemical Modifications Through Covalent Conjugation

Covalent conjugation is a valuable technique that creates a solid covalent bond between the aptamer and a non-nucleotide moiety, like a fluorophore, drug, or nanomaterial (Figure 1.4). Covalent conjugation can significantly improve the performance of DNA aptamers *in vitro* and *in vivo*, offering a predicted advantage in various applications. To achieve covalent conjugation, different chemical reactions can be used, such as thiol-maleimide, click chemistry, or amine-reactive crosslinking, depending on the specific properties of the DNA aptamer and the intended applications.

Small-molecule drugs possess several advantages, such as easy storage and transportation, low immunogenicity, and oral administration [94]. However, their low specificity results in low efficacy, relatively high toxicity, and unavoidable side effects. Targeted delivery of small-molecule drugs has been proven to enhance drug efficacy significantly, thus reducing toxicity and side effects. With the efficient molecular recognition capability, DNA aptamers can serve as ideal targeting units when coupled with small-molecule drugs for targeted drug delivery. Upon the covalent conjugation approach, our group exploited the first ApDC in 2009 and then contributed much to the ApDC research [95]. Li et al. presented an Sgc8c-artesunate conjugate with much higher therapeutic efficacy than artesunate alone as a consequence of the DNA aptamer-induced accumulation of artesunate in target cells [96]. And its retention time in tumor cells is much longer than that in control cells, demonstrating the specific targeting capability of the Sgc8c-artesunate conjugate. In addition, Huang et al. investigated the impact of linkers on the properties of ApDC [97]. Combretastatin A4 is conjugated to the Sgc8c aptamer via three different linkers, namely disulfide, phosphodiester, and carbamate bonds, for the investigation of the drug release mechanism and anticancer efficacy, as shown in Figure 1.6. The results indicate that a nucleophilic attack of glutathione can cleave the phosphodiester bond, and the repeated cleavage of the linker endows the ApDC with higher anticancer efficacy. These imply that the design of the linker unit is also critical for the efficacy of ApDC.

TNBC is considered one of the most malignant cancers. Developing effective targeted TNBC therapy has been regarded as an essential research topic in TNBC treatment. He et al. developed an AS1411-triptolide-conjugate to treat TNBC, which reveals high specificity and cytotoxicity against the MDA-MB-231 cell line [98].

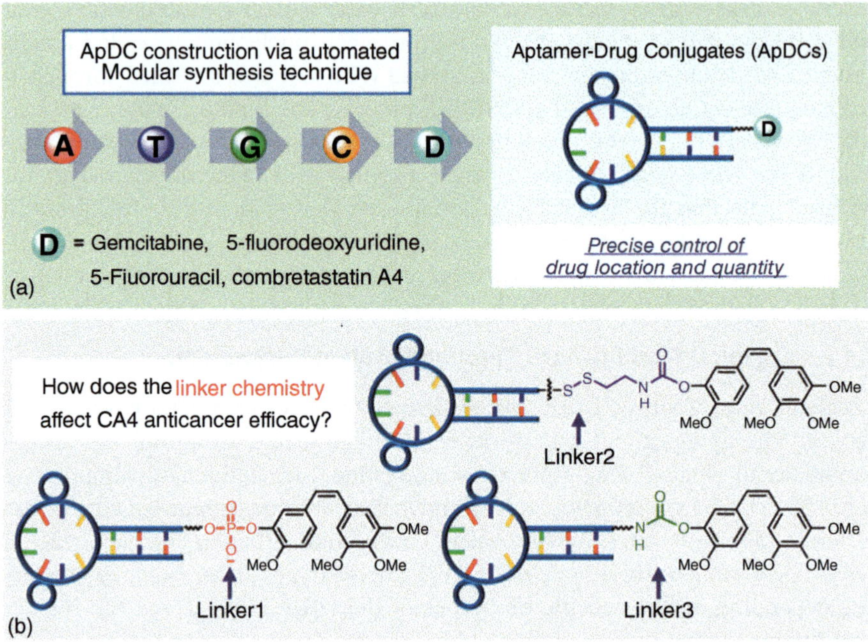

Figure 1.6 Schematic representation of (a) the construction of aptamer–drug conjugates through automated modular synthesis strategy. (b) Three different linkers: a phosphodiester bond (Linker 1), a disulfide bond (Linker 2), and a carbamate (Linker 3) involved in modifying the DNA aptamer. Source: Reproduced with permission from Huang et al. [97]. 2021 American Chemical Society.

More importantly, the AS1411-triptolide-conjugate has excellent *in vivo* anti-tumor efficacy for TNBC and negligible side effects on healthy organs. Compared with monotherapy using a single medication, polytherapy with multiple drugs can simultaneously target multiple mechanisms associated with tumor growth. This strategy can decrease individual drug dosages, enhance therapeutic efficacy against multiple targets, and overcome resistance mechanisms. However, there are challenges to creating effective combination therapies with an accurate tune of the drug ratio. To address these challenges, Zhou et al. developed cyclic bivalent ApDCs (cb-ApDCs) with a tunable drug ratio [99]. The resulting cb-ApDCs, remaining the specific recognition, are stable and can be quickly taken into cells. The drug ratio in cb-ApDCs can be easily controlled to strengthen the synergistic effect without complex chemistry. These cb-ApDCs offer a promising strategy for targeted anticancer therapy and hold the potential to stimulate the advancement of novel drug combinations for combinatorial cancer therapy.

In addition to small molecular drugs, various functional groups, such as fluorescent dyes, nanoparticles, enzymes, antibodies, and peptides, have been conjugated into DNA aptamers, which enriches their diversity and functionality. For example, by attaching fluorescent dyes to a DNA aptamer, the binding of the DNA aptamer with its target can be visualized and quantified in real time. This approach contributes to surveying the localization and dynamics of proteins in cells and

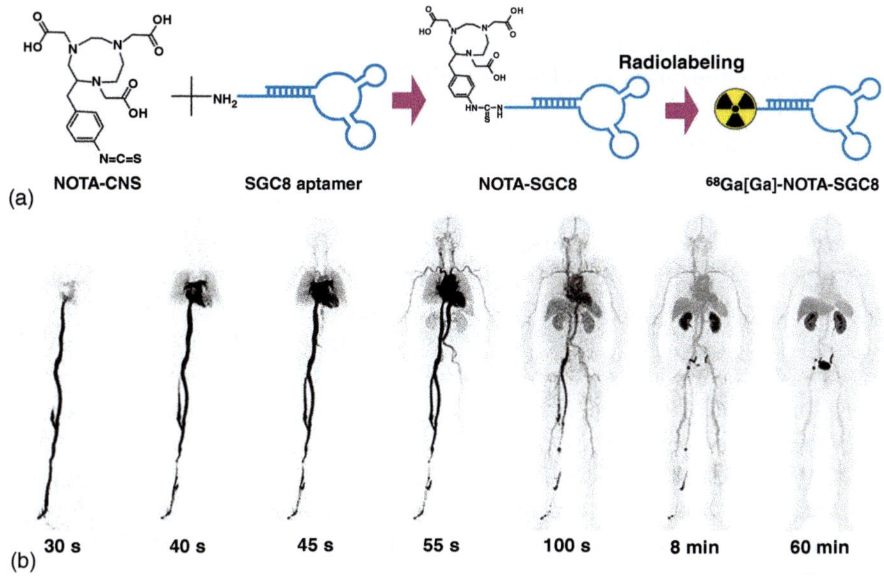

Figure 1.7 (a) Schematic illustration of chemical modification of Sgc-8 aptamer with gallium-68 radiolabel. (b) Whole-body dynamic imaging of the gallium-68 (^{68}Ga) radiolabeled aptamer-injected patient at different time points post administration. Source: Ding et al. [100]/American Association for the Advancement of Science/CC BY 4.0.

provides insights into cellular signaling pathways [60], significantly promoting our understanding of cellular processes and novel discoveries in cell biology. Also, fluorescent dye-conjugated DNA aptamers can be used to detect specific molecules in complex systems as biosensing devices.

Recently, Ding et al. developed a gallium-68 (^{68}Ga) radiolabeled aptamer (Figure 1.7a) to address the limited knowledge of the biosafety and metabolism patterns of DNA aptamers in the human body, which has impeded their clinical application in precision medicine [100]. The first-in-human pharmacokinetics study of the Sgc-8 aptamer was conducted using state-of-the-art total-body positron emission tomography (PET) technology. As shown in Figure 1.7b, the study captured dynamic distribution patterns of the aptamer in the human body, and reveals that the radiolabeled aptamer is safe for normal organs. The majority of the aptamer accumulated in the kidneys and was subsequently eliminated from the body through urine. Furthermore, a physiologically based pharmacokinetic model for the aptamer was developed, demonstrating its potential for predicting therapeutic responses and facilitating personalized treatment strategies.

1.4.3 Self-assembly Systems Based on Chemically Modified DNA Aptamers

Besides the simple covalent modification of aptamers, self-assembly systems based on chemically modified aptamers hold promise to enhance the cellular uptake, stability, and specificity of aptamers toward their targets. Such systems have been exploited for their potential in targeted drug delivery and imaging applications. The conjugation of highly hydrophilic DNA aptamers with hydrophobic small-molecule drugs results in the formation of amphiphilic ApDC. These amphiphilic conjugates can self-assemble into DNA aptamer-based micelle structures with a hydrophobic drug core and a hydrophilic DNA aptamer shield, suggesting a multivalent effect and an enhanced aptamer-target binding ability [101]. Such a supramolecular assembly could be a high-efficacy transport platform for cell imaging and drug delivery. Besides the multivalent effect induced by molecular self-assembly, Geng et al. reported a multivalent aptamer drug conjugate (ApMDC) [102] synthesized upon the coupling of a hydrophilic DNA aptamer with a hydrophobic monodendron anchored with four anticancer drugs (Figure 1.8). Amphiphilic ApMDC and its PEG-substituted analog synergistically assemble into nanomicelles, which display precise drug loading and tunable surface density of aptamers for optimal complementation between blood circulation and tumor-targeting ability. The released drug from the consequent degradation of nanomicelles induced by the acidic tumor microenvironment reinforces the immunogenic cell death of tumor cells. With these merits, ApMDC nanomicelles represent a robust platform for structure-function optimization of drug conjugates and nanomedicines.

Steadily increasing attention has been paid to the orchestration of supramolecular assemblies by using amphiphilic molecules due to their broad biological applications [103]. Upon the covalent coupling of hydrophilic DNA aptamer with hydrophobic lipids, the generated amphiphilic aptamer-lipid conjugates can self-assemble into DNA aptamer-based micelles, vesicles, and other complex assemblies, exhibiting enhanced target binding ability as a result of their multivalent effect, trim sizes, and increased cell permeability and carrier capacity. Therefore, DNA aptamer-based assemblies could be broadly applicable in nanobiotechnology, cell biology, and drug delivery systems. Our group pioneered the construction of DNA aptamer-based micelles. In 2009, they reported the development of TDO5 aptamer diacyllipid amphiphilic micelles [104] composed of a hydrophilic DNA aptamer head and a hydrophobic lipid tail. This study implies their potential as a targeted drug delivery system and a selective detection tool. One of the advantages of DNA aptamer-conjugated assemblies is their modularity and tunability. By altering assemblies' composition, size, and shape as well as the sequence and orientation of DNA aptamers, assemblies' properties can be fine-tuned to meet specific requirements. Recently, Li et al. presented a new strategy for constructing stable and specific aptamer-lipid micelles [105]. The DNA aptamer and lipid fragments were linked to a methacrylamide branch, which covalently crosslinks the aptamer-lipid micelle under photoirradiation, to increase the DNA aptamer's biostability.

Another advantage of DNA aptamer-conjugated assemblies is their biocompatibility and biodegradability. Unlike other nanoparticle-based systems, DNA aptamer-conjugate assemblies are usually non-toxic and metabolizable via normal metabolic pathways in the body, providing an ideal platform for various biomedical applications, including *in vivo* imaging and targeted drug delivery. For example, an aptamer-diacyllipid conjugate was synthesized by covalent coupling of an Sgc8 aptamer with a diacyllipid tail via a PEG linker [106]. Exosomes loaded

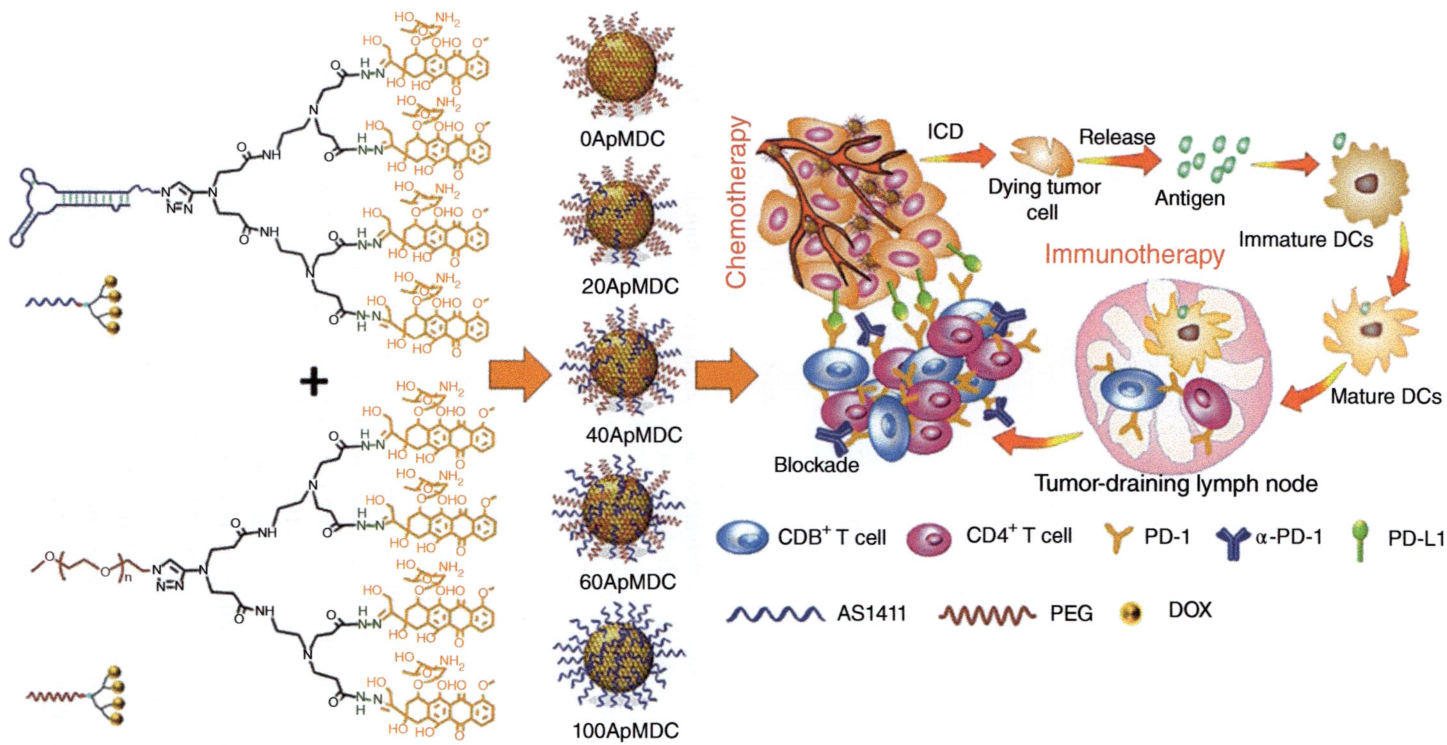

Figure 1.8 Schematic representation of the co-assemblies of a multivalent aptamer drug conjugate (ApMDC) and its PEG-substituted analog into nanomicelles for reinforcing the immunogenic cell death of tumor cells. Source: Reproduced with permission from Geng et al. [102]. © 2021 John Wiley & Sons.

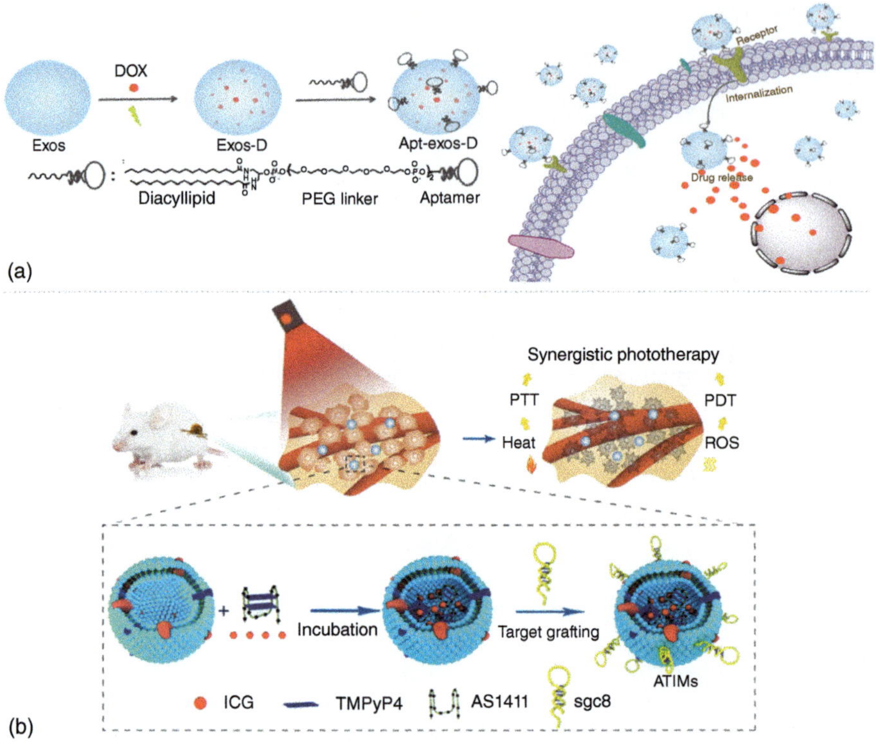

Figure 1.9 Schematic illustration of the design of (a) aptamer-functionalized exosomes (Apt-Exos) and (b) aptamer-cholesterol-modified vesicle for targeting delivery to cancer cells. Source: (a) Reproduced with permission from Zou et al. [106]. 2019 American Chemical Society. (b) Reproduced with permission from Luo et al. [107]. 2019 American Chemical Society.

with chemotherapeutic drugs were functionalized with this aptamer-diacyllipid conjugate, leading to aptamer-functionalized exosomes (Apt-Exos) for targeting delivery to cancer cells. Due to the natural delivery advantages of exosomes and specific molecular recognition properties of DNA aptamers, Apt-Exos can act as an efficient delivery tool for targeted cancer theranostics, Figure 1.9a.

Beyond the natural exosomes, biomimetic liposomes were also integrated with DNA aptamers for targeted therapeutics [108]. Liposomes self-assemble into vesicles with giant cavities mimicking the living cell membranes. They can be readily loaded with different kinds of cargo, from imaging reagents, drugs, and genes to biomolecule tools, aiming for various applications. However, the non-selectivity of liposome delivery is considered unfavorable for its practical applications. DNA aptamers' splendid molecular recognition properties make them ideal targeting modules to decorate with liposomes toward multifunctional target-specific delivery systems. Luo et al. exploited a new aptamer-cholesterol-modified vesicle loaded with therapeutic agents for cancer therapy, Figure 1.9b [107]. DNA aptamer-functionalized liposomes can precisely deliver the cargo into targeted cells. More promising applications of the DNA aptamer-based liposomes involve the delivery of miRNAs and CRISPR/Cas9 complex into specific cells to selectively manipulate cellular activities.

DNA aptamer-based assemblies represent a versatile and promising platform for various biomedical and biotechnological applications. They hold great potential for advancing our understanding of *in vivo* self-assembly principles and target delivery processes and improving the treatment of various diseases.

1.4.4 DNA Aptamers Engineered with Nanotechnology

DNA nanotechnology, where nucleic acids can be regarded as building blocks to orchestrate nanostructures and nanodevices with tunable sizes and shapes, significantly improves the capability to control molecular self-assembly [40]. Integrating DNA aptamers into nanotechnology has opened new possibilities for developing highly sensitive and specific biosensors, targeted drug delivery systems, and diagnostic molecular tools.

The major advantage of DNA aptamer-based nanotechnology lies in the molecular recognition ability of DNA aptamers. DNA aptamers screened by SELEX or cell-SELEX can bind to specific receptors, making them ideal probes for visualizing and detecting these receptors. This feature allows the advanced design of DNA aptamer-based nanodevices to detect and capture target molecules in complex biological samples precisely. In the study reported by Chen et al. [109], a DNA aptamer tool was developed for one-step fluorescence detection of antibody production and quality control. Trastuzumab, a humanized IgG1 antibody to the human epidermal growth factor 2 receptor, is selected as the model antibody drug. A DNA aptamer against trastuzumab is screened and identified by *in vitro* SELEX process. By using this DNA aptamer tool, the quality control and traceless purification of antibody drugs are demonstrated, which can support and accelerate the manufacture of antibody drugs.

Also, DNA aptamers can be engineered to undergo conformational changes to switch their molecular recognition abilities, which allows for the creation of "smart" nanomachines responding to specific environmental cues, such as pH changes or specific molecules. For example, Huang et al. designed a logic-gated nanodevice [110], as shown in Figure 1.10a. This nanodevice consists of a tetrahedron modified with a Sgc8 aptamer tail (Sgc8-CT) and a pH-responsive C-rich nucleic acid complementary with Sgc8 aptamer (i-motif/Sgc8-CT) to block the molecular recognition between Sgc8-CT and the target cell. Then the logic-gated DNA nanodevice is immobilized on the surface of nanovesicles filled with gold carbon dots (GCDs), forming a logic-gated nanovesicle capable of controlling the transportation of GCDs into the target cell. Initially, the C-rich domain of the nanodevice adopts a conformation interacting with the Sgc8 aptamer, which hinders its binding to target cells. Once the logic-gated nanovesicle is exposed to an acidic environment, the C-rich domain reconfigures into an i-motif structure, leading to its dissociation from Sgc8 aptamer and the recovery of Sgc8 aptamer's targeting ability. Thus, the nanovesicle can stimulate cargo delivery in the presence of an acidic environment and is a target

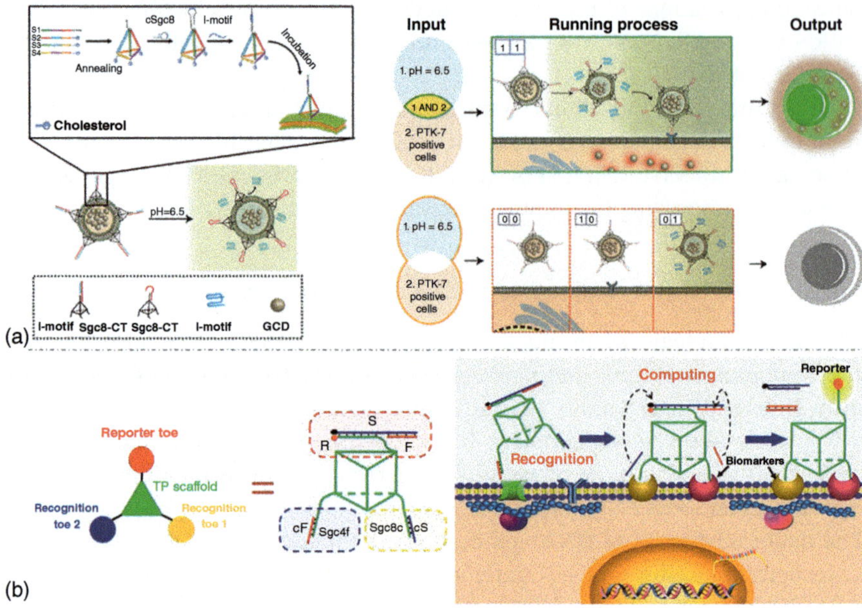

Figure 1.10 Schematic diagram of (a) the design of a logic-gated DNA nanodevice and its targeted GCD delivery process induced by an acidic environment and the overexpression of PTK-7. Source: Reproduced with permission from Huang et al. [110]. 2021 American Chemical Society. (b) A logic gate-guided DNA nanomachine for bispecific recognition and computing on cell surfaces. Source: Reproduced with permission from Peng et al. [111]. 2018 American Chemical Society.

biomarker for Sgc8-CT. As GCDs delivered into target cells, the intracellular redox status variation is monitored by the fluorescence changes of GCDs.

Furthermore, DNA aptamer-based nanotechnology has demonstrated enormous potential in DNA computing. Instead of traditional silicon-based computer chips, nucleic acids can be leveraged for implementing computing functions in living systems. Using DNA aptamers as molecular recognition elements, researchers have developed precise and efficient algorithms for solving logic problems in complex biological environments. Peng et al. fabricated a logic gate-guided DNA nanomachine for bispecific recognition and computing on cell surfaces [111], Figure 1.10b. A DNA triangular prism is decorated with two DNA aptamers, namely, Sgc8c and Sgc4f. These aptamers are designed to target two different overexpressed cancer biomarkers. Initially, their binding abilities are blocked by two specific single-stranded DNAs: cF (blocking Sgc4f) and cS (blocking Sgc8c). When two DNA aptamers interact with the respective biomarkers, two single-stranded DNAs, cF and cS, are released and tend to replace strand S from the DNA duplex structure (R/F/S) cooperatively. The coexistence of cF and cS turns the Boolean operator into an "AND" state, where the logic gate-guided nanomachine generates a true value. As a result, the released S strand triggers the fluorescence of the system, which indicates the simultaneous overexpression of both target biomarkers on the cancer cell, thus offering valuable information for cancer cell analysis.

In addition, the combination of DNA aptamers with nanotechnology provides exciting prospects for constructing complex nucleic acid-based dynamic networks [112–115]. These artificial networks aim at mimicking complex signaling dynamic behaviors and emerging functions observed in living systems. They are typically constructed using modular design principles, where a DNA aptamer is facilely integrated to generate a more extensive network. In the network, DNA aptamers can provide promising signal-recognizing and transmitting tools for the reception, processing, and feedback of biological signals. High binding affinity and specificity of DNA aptamers enable the network to function in the presence of low signal molecule concentration and substantial interference, which is essential for signal sensing, amplification, and processing as well as functional regulation in complex biological systems. He et al. presented a DNA-based signal transducer module that converts complex signal information into easy-to-read temperature output [116]. In this study, a switchable DNA G4 aptamer–Hemin complex (DGAH) was designed as a temperature-output DNA transducer. When DGAH is switched on to catalyze the oxidation of 3,3′,5,5′-tetramethylbenzidine (TMB) in the presence of hydrogen peroxide, the system's color changes from colorless to deep blue. Since the oxidized form of TMB exhibits strong and broad absorption of complementary colors across the yellow to near-infrared (NIR) regions, strong thermal conversion can be anticipated upon absorbing photons. Upon incorporating the temperature-output DNA transducer module into DNA reaction networks, the information encoded in nucleic acids can be successfully received, processed, amplified, and transduced into a high-sensitivity temperature output.

Integrating DNA aptamers into nanotechnology promotes the development of a versatile DNA aptamer-based toolbox. Combining controllable physicochemical properties and precise addressability of DNA nanotechnology with high binding specificity and affinity of DNA aptamers provides molecular recognition accessories for nanostructures and nanomaterials to target biomolecules, cells, or tissues with ultra-high sensitivity and specificity. These outstanding performances advance their applications in biosensing, bioimaging, targeted drug delivery, bioregulation, and biomimicry.

1.5 Summary and Outlook

Cells are highly complex systems whose structures and functionalities have been studied from many perspectives for decades. However, we are still far from a comprehensive understanding of their inner workings. In recent years, the advent of cell-SELEX technology has revolutionized the research field of DNA aptamers, making them valuable tools for molecular recognition and cell targeting. Also, the high affinity and specificity, programmable molecular structures, and facile chemical modification of DNA aptamers render them highly versatile tools for cellular applications. The development of the DNA aptamer toolbox has rapidly progressed through the incorporation with chemical modifications and advanced nanotechnology. These advancements have created aptamer-based molecular tools with exceptional versatility, such as ApDCs, aptamer-based molecular probes, aptamer-based nanodevices, and aptamer-based molecular computers, holding tremendous potential for numerous cellular applications, including sensing, imaging, targeted drug delivery, bioregulation, and biomimicry.

Despite the enormous potential and impressive advancements in DNA aptamer-based tools, their actual impact on biological and biomedical applications is yet to be fully realized. A few challenges remain in this emerging field, including complex DNA aptamer discovery strategies, limited understanding of the binding mechanisms between aptamers and bio-targets, low stability and efficacy of aptamer tools in biological research, and concerns regarding biosecurity in their applications. To push this emerging field, persistent efforts are required to address these fundamental gaps and challenges, such as the exploration of new aptamer screening methods and instruments, the investigation of the structural information of aptamer-target in complex physiological environments, and the development of the DNA aptamer toolbox to enhance the biological performance and biosecurity. With these efforts, it is conceivable that DNA aptamer tools will open up new ways for cell research, ultimately realizing clinical applications in the future.

Acknowledgments

This work was supported by the National Key Research and Development Project (2021YFA0909400) and the National Natural Science Foundation of China (NSFC 22101080 and 22207034).

References

1 Deatherage, F.E. (1975). Cells, the fundamental biological units. In: *Food for Life* (ed. F.E. Deatherage), 308–311. Boston: Springer. https://doi.org/10.1007/978-1-4684-0748-8_3.

2 Ellinger, I. and Ellinger, A. (2014). Smallest unit of life: cell biology. In: *Comparative Medicine* (ed. E. Jensen-Jarolim), 19, 19–33, 33. Vienna: Springer. https://doi.org/10.1007/978-3-7091-1559-6_2.

3 Alberts, B., Johnson, A., Lewis, J. et al. (2002). The chemical components of a cell. In: *Molecular Biology of the Cell*, 4e, 44–125. New York: Garland Science. Available from: https://www.ncbi.nlm.nih.gov/books/NBK26883/.

4 Wolkenhauer, O. and Muir, A. (2011). The complexity of cell-biological systems. In: *Philosophy of Complex Systems, Volume 10 in Handbook of the Philosophy of Science* (ed. D.M. Gabbay, P. Thagard, and J. Woods), 355–385. Elsevier. https://doi.org/10.1016/B978-0-444-52076-0.50013-4.

5 Barabási, A.L. and Oltvai, Z. (2004). Network biology: understanding the cell's functional organization. *Nature Reviews Genetics* 5: 101–113. https://doi.org/10.1038/nrg1272.

6 Trapnell, C., Williams, B.A., Pertea, G. et al. (2010). Transcript assembly and quantification by RNA-seq reveals unannotated transcripts and isoform switching during cell differentiation. *Nature Biotechnology* 28: 511–515. https://doi.org/10.1038/nbt.1621.

7 Yosef, N., Shalek, A.K., Gaublomme, J.T. et al. (2013). Dynamic regulatory network controlling TH17 cell differentiation. *Nature* 496: 461–468. https://doi.org/10.1038/nature 11981.

8 Whitfield, M., George, L., Grant, G. et al. (2006). Common markers of proliferation. *Nature Reviews Cancer* 6: 99–106. https://doi.org/10.1038/nrc1802.

9 Purvis, J.E. and Lahav, G. (2013). Encoding and decoding cellular information through signaling dynamics. *Cell* 152 (5): 945–956. https://doi.org/10.1016/j.cell.2013.02.005.

10 Strzyz, P. (2018). PrEView of cell–cell communication. *Nature Reviews Molecular Cell Biology* 19: 752–753. https://doi.org/10.1038/s41580-018-0073-3.

11 Kholodenko, B. (2006). Cell-signalling dynamics in time and space. *Nature Reviews Molecular Cell Biology* 7: 165–176. https://doi.org/10.1038/nrm1838.

12 Guimerà, R. and Nunes Amaral, L.A. (2005). Functional cartography of complex metabolic networks. *Nature* 433: 895–900. https://doi.org/10.1038/nature03288.

13 Jeong, H., Tombor, B., Albert, R. et al. (2000). The large-scale organization of metabolic networks. *Nature* 407: 651–654. https://doi.org/10.1038/35036627.

14 Hens, C., Harush, U., Haber, S. et al. (2019). Spatiotemporal signal propagation in complex networks. *Nature Physics* 15: 403–412. https://doi.org/10.1038/s41567-018-0409-0.

15 Goldbeter, A. (2017). Dissipative structures and biological rhythms. *Chaos: An Interdisciplinary Journal of Nonlinear Science* 27 (10): 104612. https://doi.org/10.1063/1. 4990783.

16 Andrews, L.B., Nielsen, A.A.K., and Voigt, C.A. (2018). Cellular checkpoint control using programmable sequential logic. *Science* 361 (6408): eaap8987. https://doi.org/10.1126/science.aap898.

17 Lezia, A., Csicsery, N., and Hasty, J. (2022). Design, mutate, screen: Multiplexed creation and arrayed screening of synchronized genetic clocks. *Cell Systems* 13 (5): 365–375. https://doi.org/10.1016/j.cels.2022.02.005.

18 Boya, P., Reggiori, F., and Codogno, P. (2013). Emerging regulation and functions of autophagy. *Nature Cell Biology* 15: 713–720. https://doi.org/10.1038/ncb2788.

19 Zhang, Y., Tian, Z., Ye, H. et al. (2022). Emerging functions of circular RNA in the regulation of adipocyte metabolism and obesity. *Cell Death Discovery* 8: 268. https://doi.org/10.1038/s41420-022-01062-w.

20 Ganser, L.R., Kelly, M.L., Herschlag, D. et al. (2019). The roles of structural dynamics in the cellular functions of RNAs. *Nature Reviews Molecular Cell Biology* 20: 474–489. https://doi.org/10.1038/s41580-019-0136-0.

21 Pantsar, T. (2019). The current understanding of KRAS protein structure and dynamics. *Computational and Structural Biotechnology Journal* 18 (2020): 189–198. https://doi.org/10.1016/j.csbj.2019.12.004.

22 Gorka, M., Swart, C., Siemiatkowska, B. et al. (2019). Protein complex identification and quantitative complexome by CN-PAGE. *Scientific Reports* 9: 11523. https://doi.org/10.1038/s41598-019-47829-7.

23 Alivisatos, P. (2004). The use of nanocrystals in biological detection. *Nature Biotechnology* 22: 47–52. https://doi.org/10.1038/nbt927.

24 Ko, J., Wilkovitsch, M., Oh, J. et al. (2022). Spatiotemporal multiplexed immunofluorescence imaging of living cells and tissues with bioorthogonal cycling of fluorescent probes. *Nature Biotechnology* 40: 1654–1662. https://doi.org/10.1038/s41587-022-01339-6.

25 Stehr, F., Stein, J., Bauer, J. et al. (2021). Tracking single particles for hours via continuous DNA-mediated fluorophore exchange. *Nature Communications* 12: 4432. https://doi.org/10.1038/s41467-021-24223-4.

26 Hou, S., Exell, J., and Welsher, K. (2020). Real-time 3D single molecule tracking. *Nature Communications* 11: 3607. https://doi.org/10.1038/s41467-020-17444-6.

27 Alon, U. (2007). Network motifs: Theory and experimental approaches. *Nature Reviews Genetics* 8: 450–461. https://doi.org/10.1038/nrg2102.

28 Lambiotte, R., Rosvall, M., and Scholtes, I. (2019). From networks to optimal higher-order models of complex systems. *Nature Physics* 15: 313–320. https://doi.org/10.1038/s41567-019-0459-y.

29 Ludwig, J. and Weinstein, J. (2005). Biomarkers in cancer staging, prognosis and treatment selection. *Nature Reviews Cancer* 5: 845–856. https://doi.org/10.1038/nrc1739.

30 Kwong, G.A., Ghosh, S., Gamboa, L. et al. (2021). Synthetic biomarkers: a twenty-first century path to early cancer detection. *Nature Reviews Cancer* 21: 655–668. https://doi.org/10.1038/s41568-021-00389-3.

31 Black, J.R.M. and McGranahan, N. (2021). Genetic and non-genetic clonal diversity in cancer evolution. *Nature Reviews Cancer* 21: 379–392. https://doi.org/10.1038/s41568-021-00336-2.

32 Weis, S. and Cheresh, D. (2011). Tumor angiogenesis: molecular pathways and therapeutic targets. *Nature Medicine* 17: 1359–1370. https://doi.org/10.1038/nm.2537.

33 Fares, J., Fares, M.Y., Khachfe, H.H. et al. (2020). Molecular principles of metastasis: a hallmark of cancer revisited. *Signal Transduction and Targeted Therapy* 5: 28. https://doi.org/10.1038/s41392-020-0134-x.

34 Komarova, N., Panova, O., Titov, A. et al. (2022). Aptamers targeting cardiac biomarkers as an analytical tool for the diagnostics of cardiovascular diseases: a review. *Biomedicines* 10 (5): 1085. https://doi.org/10.3390/biomedicines10051085.

35 Micura, R. and Höbartner, C. (2020). Fundamental studies of functional nucleic acids: aptamers, riboswitches, ribozymes and DNAzymes. *Chemical Society Reviews* 49 (20): 7331–7353. https://doi.org/10.1039/D0CS00617C.

36 Felsenfeld, G. and Miles, H.T. (1967). The physical and chemical properties of nucleic acids. *Annual Review of Biochemistry* 36: 407–448. https://doi.org/10.1146/annurev.bi.36.070167.002203.

37 Xu, W., He, W., Du, Z. et al. (2021). Functional nucleic acid nanomaterials: development, properties, and applications. *Angewandte Chemie International Edition* 60 (13): 6890–6918. https://doi.org/10.1002/anie.201909927.

38 Zhang, J., Lan, T., and Lu, Y. (2019). Molecular engineering of functional nucleic acid nanomaterials toward in vivo applications. *Advanced Healthcare Materials* 8 (6): e1801158. https://doi.org/10.1002/adhm.201801158.

39 Peng, T., Deng, Z., He, J. et al. (2020). Functional nucleic acids for cancer theranostics. *Coordination Chemistry Reviews* 403: 213080. https://doi.org/10.1016/j.ccr.2019.213080.

40 Seeman, N. and Sleiman, H. (2018). DNA nanotechnology. *Nature Reviews Materials* 3: 17068. https://doi.org/10.1038/natrevmats.2017.68.

41 Ku, T.H., Zhang, T., Luo, H. et al. (2015). Nucleic acid aptamers: an emerging tool for biotechnology and biomedical sensing. *Sensors* 15 (7): 16281–16313. https://doi.org/10.3390/s150716281.

42 Wang, H., Cheng, H., Wang, J. et al. (2016). Selection and characterization of DNA aptamers for the development of light-up biosensor to detect Cd(II). *Talanta* 154: 498–503. https://doi.org/10.1016/j.talanta.2016.04.005.

43 Baker, B.R., Lai, R.Y., Wood, M.S. et al. (2006). An electronic, aptamer-based small-molecule sensor for the rapid, label-free detection of cocaine in adulterated samples and biological fluids. *Journal of the American Chemical Society* 128 (10): 3138–3139. https://doi.org/10.1021/ja056957p.

44 Cai, H., Lee, T.M.H., and Hsing, I.M. (2006). Label-free protein recognition using an aptamer-based impedance measurement assay. *Sensors and Actuators, B: Chemical* 114 (1): 433–437. https://doi.org/10.1016/j.snb.2005.06.017.

45 Labib, M., Zamay, A.S., Muharemagic, D. et al. (2012). Aptamer-based viability impedimetric sensor for viruses. *Analytical Chemistry* 84 (4): 1813–1816. https://doi.org/10.1021/ac203412m.

46 Dunn, M., Jimenez, R., and Chaput, J. (2017). Analysis of aptamer discovery and technology. *Nature Reviews Chemistry* 1: 0076. https://doi.org/10.1038/s41570-017-0076.

47 Shangguan, D., Li, Y., Tang, Z. et al. (2006). Aptamers evolved from live cells as effective molecular probes for cancer study. *Proceedings of the National Academy of Sciences of the United States of America* 103 (32): 11838–11843. https://doi.org/10.1073/pnas.0602615103.

48 Huang, Z.X., Xie, Q., Guo, Q.P. et al. (2017). DNA aptamer selected for specific recognition of prostate cancer cells and clinical tissues. *Chinese Chemical Letters* 28 (6): 1252–1257. https://doi.org/10.1016/j.cclet.2017.01.002.

49 Mao, X., Liu, M., Yan, L. et al. (2020). Programming biomimetically confined aptamers with DNA frameworks. *ACS Nano* 14 (7): 8776–8783. https://doi.org/10.1021/acsnano.0c03362.

50 Yang, Y., Xu, J., Sun, Y. et al. (2021). Aptamer-based logic computing reaction on living cells to enable non-antibody immune checkpoint blockade therapy. *Journal of the American Chemical Society* 143 (22): 8391–8401. https://doi.org/10.1021/jacs.1c02016.

51 Park, J., Lee, S., Kim, Y. et al. (2021). Methods to generate site-specific conjugates of antibody and protein. *Bioorganic and Medicinal Chemistry* 30: 115946. https://doi.org/10.1016/j.bmc.2020.115946.

52 Brazier, J. (2023). Chemical synthesis of oligonucelotide sequences: phosphoramidite chemistry. In: *DNA Manipulation and Analysis* (ed. G. Scarlett), 185–193. New York: Humana. https://doi.org/10.1007/978-1-0716-3004-4_14.

53 Wang, R.W., Zhu, G., Mei, L. et al. (2014). Automated modular synthesis of aptamer–drug conjugates for targeted drug delivery. *Journal of the American Chemical Society* 136 (7): 2731–2734. https://doi.org/10.1021/ja4117395.

54 Wen, J., Tao, W., Hao, S. et al. (2016). A unique aptamer-drug conjugate for targeted therapy of multiple myeloma. *Leukemia* 30: 987–991. https://doi.org/10.1038/leu.2015.216.

55 Hampton, T. (2008). Researchers create artificial DNA bases. *Journal of the American Medical Association* 299 (11): 1251–1251. https://doi.org/10.1001/jama.299.11.1251.

56 Meng, H.M., Liu, H., Kuai, H. et al. (2016). Aptamer-integrated DNA nanostructures for biosensing, bioimaging and cancer therapy. *Chemical Society Reviews* 45 (9): 2583–2602. https://doi.org/10.1039/C5CS00645G.

57 Walia, S., Chandrasekaran, A.R., Chakraborty, B. et al. (2021). Aptamer-programmed DNA nanodevices for advanced, targeted cancer theranostics. *ACS Applied Bio Materials* 4 (7): 5392–5404. https://doi.org/10.1021/acsabm.1c00413.

58 Zhao, S., Tian, R., Wu, J. et al. (2021). A DNA origami-based aptamer nanoarray for potent and reversible anticoagulation in hemodialysis. *Nature Communications* 12: 358. https://doi.org/10.1038/s41467-020-20638-7.

59 Li, L., Jiang, Y., Cui, C. et al. (2018). Modulating aptamer specificity with pH-responsive DNA bonds. *Journal of the American Chemical Society* 140 (41): 13335–13339. https://doi.org/10.1021/jacs.8b08047.

60 Xie, S., Du, Y., Zhang, Y. et al. (2020). Aptamer-based optical manipulation of protein subcellular localization in cells. *Natture Communications* 11: 1347. https://doi.org/10.1038/s41467-020-15113-2.

61 Goda, T. and Miyahara, Y. (2011). Thermo-responsive molecular switches for ATP using hairpin DNA aptamers. *Biosensors and Bioelectronics* 26 (9): 3949–3952. https://doi.org/10.1016/j.bios.2011.02.041.

62 Beaucage, S.L. and Caruthers, M.H. (1981). Deoxynucleoside phosphoramidites—a new class of key intermediates for deoxypolynucleotide synthesis. *Tetrahedron Lett.* 22: 1859–1862. https://doi.org/10.1016/S0040-4039(01)90461-7.

63 Tuerk, C. and Gold, L. (1990). Systematic evolution of ligands by exponential enrichment: RNA ligands to bacteriophage T4 DNA polymerase. *Science* 249 (4968): 505–510. https://doi.org/10.1126/science.2200121.

64 Ellington, A. and Szostak, J. (1990). *In vitro* selection of RNA molecules that bind specific ligands. *Nature* 346: 818–822. https://doi.org/10.1038/346818a0.

65 Yu, H., Alkhamis, O., Canoura, J. et al. (2021). Advances and challenges in small-molecule DNA aptamer isolation, characterization, and sensor development. *Angewandte Chemie International Edition* 60 (31): 16800–16823. https://doi.org/10.1002/anie.202008663.

66 Huizenga, D.E. and Szostak, J.W. (1995). A DNA aptamer that binds adenosine and ATP. *Biochemistry* 34 (2): 656–665. https://doi.org/10.1021/bi00002a033.

67 Stojanovic, M.N., de Prada, P., and Landry, D.W. Fluorescent sensors based on aptamer self-assembly. *Journal of the American Chemical Society* 122 (46): 11547–11548. https://doi.org/10.1021/ja0022223.

68 Guo, X., Wen, F., Zheng, N. et al. (2020). Aptamer-based biosensor for detection of mycotoxins. *Frontiers in Chemistry* 8: 195. https://doi.org/10.3389/fchem.2020.00195.

69 Krauss, I.R., Merlino, A., Giancola, C. et al. (2011). Thrombin–aptamer recognition: a revealed ambiguity. *Nucleic Acids Research* 39 (17): 7858–7867. https://doi.org/10.1093/nar/gkr522.

70 Hongjie, X., Jianhua, Y., Cai, S. et al. (2019). Cancer protein biomarker discovery based on nucleic acid aptamers. *International Journal of Biological Macromolecules* 132: 190–202. https://doi.org/10.1016/j.ijbiomac.2019.3.165.

71 Fang, X. and Tan, W. (2010). Aptamers generated from cell-SELEX for molecular medicine: a chemical biology approach. *Accounts of Chemical Research* 43 (1): 48–57. https://doi.org/10.1021/ar900101s.

72 Sefah, K., Shangguan, D., Xiong, X. et al. (2010). Development of DNA aptamers using Cell-SELEX. *Nature Protocols* 5: 1169–1185. https://doi.org/10.1038/nprot.2010.66.

73 Dua, P., Kim, S., and Lee, D.K. (2011). Nucleic acid aptamers targeting cell-surface proteins. *Methods* 54 (2): 215–225. https://doi.org/10.1016/j.ymeth.2011.02.002.

74 Jin, C., Qiu, L., Li, J. et al. (2016). Cancer biomarker discovery using DNA aptamers. *The Analyst* 141 (2): 461–466. https://doi.org/10.1039/C5AN01918D.

75 Wu, X., Liu, H., Han, D. et al. (2019). Elucidation and structural modeling of CD71 as a molecular target for cell-specific aptamer binding. *Journal of the American Chemical Society* 141 (27): 10760–10769. https://doi.org/10.1021/jacs.9b03720.

76 Hou, Z., Meyer, S., Propson, N. et al. (2015). Characterization and target identification of a DNA aptamer that labels pluripotent stem cells. *Cell Research* 25: 390–393. https://doi.org/10.1038/cr.2015.7.

77 Zhou, W., Huang, P.J., Ding, J. et al. (2014). Aptamer-based biosensors for biomedical diagnostics. *Analyst* 139 (11): 2627–2640. https://doi.org/10.1039/C4AN00132J.

78 Akki, S.U. and Werth, C.J. (2018). Critical review: DNA aptasensors, are they ready for monitoring organic pollutants in natural and treated water sources? *Environmental Science & Technology* 52 (16): 8989–9007. https://doi.org/10.1021/acs.est.8b00558.

79 Huang, Z., Qiu, L., Zhang, T. et al. (2021). Integrating DNA nanotechnology with aptamers for biological and biomedical applications. *Matter* 4 (2): 461–489. https://doi.org/10.1016/j.matt.2020.11.002.

80 Bashir, A., Yang, Q., Wang, J. et al. (2021). Machine learning guided aptamer refinement and discovery. *Nature Communications* 12: 2366. https://doi.org/10.1038/s41467-021-22555-9.

81 Elskens, J.P., Elskens, J.M., and Madder, A. (2020). Chemical modification of aptamers for increased binding affinity in diagnostic applications: current status and future prospects. *International Journal of Molecular Sciences* 21 (12): 4522. https://doi.org/10.3390/ijms21124522.

82 Thirunavukarasu, D., Chen, T., Liu, Z. et al. (2017). Selection of 2′-fluoro-modified aptamers with optimized properties. *Journal of the American Chemical Society* 139 (8): 2892–2895. https://doi.org/10.1021/jacs.6b13132.

83 Gao, S., Zheng, X., Jiao, B. et al. (2016). Post-SELEX optimization of aptamers. *Analytical and bioanalytical chemistry* 408 (17): 4567–4573. https://doi.org/10.1007/s00216-016-9556-2.

84 Odeh, F., Nsairat, H., Alshaer, W. et al. (2019). Aptamers chemistry: chemical modifications and conjugation strategies. *Molecules* 25 (1): 3. https://doi.org/10.3390/molecules25010003.

85 Arangundy-Franklin, S., Taylor, A.I., Porebski, B.T. et al. (2019). A synthetic genetic polymer with an uncharged backbone chemistry based on alkyl phosphonate nucleic acids. *Nature chemistry* 11 (6): 533–542. https://doi.org/10.1038/s41557-019-0255-4.

86 Thiviyanathan, V. and Gorenstein, D.G. (2012). Aptamers and the next generation of diagnostic reagents. *Proteomics Clinical Applications* 6 (11–12): 563–573. https://doi.org/10.1002/prca.201200042.

87 Li, X., Li, Z., and Yu, H. (2020). Selection of threose nucleic acid aptamers to block PD-1/PD-L1 interaction for cancer immunotherapy. *Chemical communications* 56 (93): 14653–14656. https://doi.org/10.1039/d0cc06032a.

88 Chen, J., Chen, M., and Zhu, T.F. (2022). Directed evolution and selection of biostable L-DNA aptamers with a mirror-image DNA polymerase. *Nature Biotechnology* 40: 1601–1609. https://doi.org/10.1038/s41587-022-01337-8.

89 Lv, C., Yang, C., Ding, D. et al. (2019). Endocytic pathways and intracellular transport of aptamer-drug conjugates in live cells monitored by single-particle tracking. *Analytical chemistry* 91 (21): 13818–13823. https://doi.org/10.1021/acs.analchem.9b03281.

90 Xuan, W., Xia, Y., Li, T. et al. (2020). Molecular self-assembly of bioorthogonal aptamer-prodrug conjugate micelles for hydrogen peroxide and pH-independent cancer chemodynamic therapy. *Journal of the American Chemical Society* 142 (2): 937–944. https://doi.org/10.1021/jacs.9b10755.

91 Sefah, K., Yang, Z., Bradley, K.M. et al. (2014). In vitro selection with artificial expanded genetic information systems. *Proceedings of the National Academy of Sciences of the United States of America* 111 (4): 1449–1454. https://doi.org/10.1073/pnas.1311778111.

92 Li, Y., Li, T., Chen, H. et al. (2022). Engineering AND-gate aptamer-signal base conjugates for targeted magnetic resonance molecular imaging of metastatic cancer. *ACS applied materials & interfaces* 14 (15): 17032–17041. https://doi.org/10.1021/acsami.1c24048.

93 Tan, J., Zhao, M., Wang, J. et al. (2019). Regulation of protein activity and cellular functions mediated by molecularly evolved nucleic acids. *Angewandte Chemie International Edition* 58 (6): 1621–1625. https://doi.org/10.1002/anie.201809010.

94 Robert, A., Benoit-Vical, F., Liu, Y. et al. (2019). Small molecules: the past or the future in drug innovation? In: *Essential Metals in Medicine: Therapeutic Use and Toxicity of Metal Ions in the Clinic* (ed. L. Peggy), 17–48. Boston: De Gruyter. https://doi.org/10.1515/9783110527872-002.

95 Huang, Y.F., Shangguan, D., Liu, H. et al. (2009). Molecular assembly of an aptamer-drug conjugate for targeted drug delivery to tumor cells. *ChemBioChem* 10 (5): 862–868. https://doi.org/10.1002/cbic.200800805.

96 Li, Y., Peng, Y., Tan, Y. et al. (2021). A new paradigm for artesunate anticancer function: considerably enhancing the cytotoxicity via conjugating artesunate with aptamer. *Signal Transduction and Targeted Therapy* 6: 327. https://doi.org/10.1038/s41392-021-00671-8.

97 Huang, Z., Wang, D., Long, C.Y. et al. (2021). Regulating the anticancer efficacy of Sgc8–Combretastatin A4 conjugates: a case of recognizing the significance of linker chemistry for the design of aptamer-based targeted drug delivery strategies. *Journal of the American Chemical Society* 143 (23): 8559–8564. https://doi.org/10.1021/jacs.1c03013.

98 He, J., Peng, T., Peng, Y. et al. (2020). Molecularly engineering triptolide with aptamers for high specificity and cytotoxicity for triple-negative breast cancer. *Journal of the American Chemical Society* 142 (6): 2699–2703. https://doi.org/10.1021/jacs.9b10510.

99 Zhou, F., Wang, P., Peng, Y. et al. (2019). Molecular engineering-based aptamer–drug conjugates with accurate tunability of drug ratios for drug combination targeted cancer therapy. *Angewandte Chemie International Edition* 58 (34): 11661–11665. https://doi.org/10.1002/anie.201903807.

100 Ding, D., Zhao, H., Wei, D. et al. (2023). The first-in-human whole-body dynamic pharmacokinetics study of aptamer. *Research* 6: 0126. https://doi.org/10.34133/research.0126.

101 Xuan, W., Peng, Y., Deng, Z. et al. (2018). A basic insight into aptamer-drug conjugates (ApDCs). *Biomaterials* 182: 216–226. https://doi.org/10.1016/j.biomaterials.2018.08.021.

102 Geng, Z., Wang, L., Liu, K. et al. (2021). Enhancing anti-PD-1 immunotherapy by nanomicelles self-assembled from multivalent aptamer drug conjugates. *Angewandte Chemie International Edition* 60 (28): 15459–15465. https://doi.org/10.1002/anie.202102631.

103 Xie, S., Ai, L., Cui, C. et al. (2021). Functional aptamer-embedded nanomaterials for diagnostics and therapeutics. *ACS Applied Materials & Interfaces* 13 (8): 9542–9560. https://doi.org/10.1021/acsami.0c19562.

104 Wu, Y., Sefah, K., Liu, H. et al. (2010). DNA aptamer-micelle as an efficient detection/delivery vehicle toward cancer cells. *Proceedings of the National Academy of Sciences of the United States of America* 107 (1): 5–10. https://doi.org/10.1073/pnas.0909611107.

105 Li, X., Figg, C.A., Wang, R. et al. (2018). Crosslinked aptamer-lipid micelles for excellent stability and specificity in target-cell recognition. *Angewandte Chemie International Edition* 57 (36): 11589–11593. https://doi.org/10.1002/anie.201804682.

106 Zou, J., Shi, M., Liu, X. et al. (2019). Aptamer-functionalized exosomes: Elucidating the cellular uptake mechanism and the potential for cancer-targeted chemotherapy. *Analytical chemistry* 91 (3): 2425–2430. https://doi.org/10.1021/acs.analchem.8b05204.

107 Luo, C., Hu, X., and Peng, R. (2019). Biomimetic carriers based on giant membrane vesicles for targeted drug delivery and photodynamic/photothermal synergistic therapy. *ACS applied materials & interfaces* 11 (47): 43811–43819. https://doi.org/10.1021/acsami.9b11223.

108 Moosavian, S.A. and Sahebkar, A. (2019). Aptamer-functionalized liposomes for targeted cancer therapy. *Cancer letters* 448: 144–154. https://doi.org/10.1016/j.canlet.2019.01.045.

109 Chen, K., Zhou, J., Shao, Z. et al. (2020). Aptamers as versatile molecular tools for antibody production monitoring and quality control. *Journal of the American Chemical Society* 142 (28): 12079–12086. https://doi.org/10.1021/jacs.9b13370.

110 Huang, H., Guo, Z., Zhang, C. et al. (2021). Logic-gated cell-derived nanovesicles via DNA-based smart recognition module. *ACS applied materials & interfaces* 13 (26): 30397–30403. https://doi.org/10.1021/acsami.1c07632.

111 Peng, R., Zheng, X., Lyu, Y. et al. (2018). Engineering a 3D DNA-logic gate nanomachine for bispecific recognition and computing on target cell surfaces. *Journal of the American Chemical Society* 140 (31): 9793–9796. https://doi.org/10.1021/jacs.8b04319.

112 Yue, L., Wang, S., Wulf, V. et al. (2019). Consecutive feedback-driven constitutional dynamic networks. *Proceedings of the National Academy of Sciences of the United States of America* 116 (8): 2843–2848. https://doi.org/10.1073/pnas.1816670116.

113 Wang, S., Yue, L., Wulf, V. et al. (2020). Dissipative constitutional dynamic networks for tunable transient responses and catalytic functions. *Journal of the American Chemical Society* 142 (41): 17480–17488. https://doi.org/10.1021/jacs.0c06977.

114 Yue, L., Wang, S., Zhou, Z. et al. (2020). Nucleic acid based constitutional dynamic networks: from basic principles to applications. *Journal of the American Chemical Society* 142 (52): 21577–21594. https://doi.org/10.1021/jacs.0c09891.

115 Wang, D., Yang, Y., Chen, F. et al. (2022). Network topology-directed design of molecular CPU for cell-like dynamic information processing. *Science advances* 8 (32): eabq0917. https://doi.org/10.1126/sciadv.abq0917.

116 He, L., Chen, F., Zhang, D. et al. (2020). Transducing complex biomolecular interactions by temperature-output artificial DNA signaling networks. *Journal of the American Chemical Society* 142 (33): 14234–14239. https://doi.org/10.1021/jacs.0c05453.

2

Bacterial Detection with Functional Nucleic Acids: *Escherichia coli* as a Case Study

Yash Patel and Yingfu Li

McMaster University, Department of Biochemistry and Biomedical Sciences, 1280 Main Street West, Hamilton, ON L8S 4K1, Canada

2.1 General Introduction to Bacteria

Bacteria are prokaryotic single-celled microscopic organisms found virtually everywhere in nature, including soil, water, air, and even in other organisms [1]. They are one of the most abundant forms of life on our planet when measured by biomass, dwarfed only by plants [2]. Bacterial cells come in various sizes and shapes, from sphere to rod to even star-shape [3, 4]. Bacteria are key organisms that drive life on Earth [5, 6]. The cycling of five of the six most important elements required for life, hydrogen, carbon, oxygen, nitrogen, and sulfur, is performed by bacteria [6]. They break down pollutants, waste, and ingested food by recycling organic matter that can be used by other life [5]. There are estimates that the ratio between bacterial and human cells in the human body is anywhere between 1 : 1 and 10 : 1 [7]. Bacterial life is located throughout the human body and even plays a very important role in human health [7]. For example, the gut microbiota, which consists of the largest quantities of microorganisms and the largest diversity of species, is significant in supporting protection from pathogens, enhancing the immune system, digestion, and metabolism, controlling epithelial cell proliferation and differentiation, modifying insulin resistance and secretion, and influencing brain–gut communication [8]. Disturbance of the gut microbiome populations can cause various complications and may even require fecal microbiota transplantation, which is the transplantation of fecal matter from a donor to re-establish the microbiome diversity and population in the recipient [8, 9]. However, despite many beneficial roles, some bacterial species can also cause various human diseases. Out of the estimated 10,000 different bacteria species on Earth, approximately 1500 are known to be pathogenic to humans, and 1200 regularly cause diseases in humans [10, 11]. Common bacterial infections include tuberculosis, whooping cough, gonorrhea, and urinary tract infections (UTIs), just to list a few.

Since the discovery of penicillin by Alexander Fleming in 1929, antibiotics have been used to treat bacterial infections [12, 13]. The introduction of antibiotics was

DNA Nanotechnology for Cell Research: From Bioanalysis to Biomedicine, First Edition. Edited by Zhou Nie.

arguably one of the greatest medical breakthroughs of the 1900s that has allowed the possibility of many modern medical procedures such as cancer treatment, organ transplants, and open-heart surgery [13]. However, the widespread misuse of antibiotics, especially broad-spectrum antibiotics, has led to the rapid rise of antimicrobial resistance in bacteria [13, 14]. Resistance to antibiotics is associated with high morbidity and mortality in infected individuals [14]. The Centers for Disease Control and Prevention (CDC) estimates that at least 23,000 individuals die every year due to multidrug resistance (MDR) in bacterial pathogens, which causes $20 billion in direct healthcare costs [15]. By 2050, there are estimates that MDR will cause 10 million deaths per year, with up to $100 trillion in cost to the global economy [13, 15]. These dramatic increases in MDR combined with poor infection control practices pose a significant risk to the human population worldwide, as these drug-resistant bacteria can transfer drug-resistance genes to other bacteria through horizontal gene transfer [14]. Given the lack of success in the development of novel synthetic antibiotics for clinical use to combat MDR, better infection control and detection methods are required. This chapter aims to discuss bacterial detection, exemplified by the detection of *Escherichia coli* (*E. coli*).

2.2 *E. coli*

Escherichia coli, commonly known as *E. coli*, is a gram-negative, facultative anaerobic bacterium that is commonly found in the intestines and feces of warm-blooded animals while also being found in the environment in water, soil, and sediment [16, 17]. It was first isolated and described by Theodor Escherich in 1885, who was studying the role of microbes in digestion and intestinal disease [18]. *E. coli* is one of the first bacteria to colonize the human gut at birth [18]. While *E. coli* species only make up 0.1–5%, they play a very important role in the human gut microbiome [18]. *E. coli* helps deplete oxygen from the gastrointestinal surface, which allows for strict anaerobes to dominate the gut microbiota [18, 19]. *E. coli* also produces vitamin K and vitamin B12, which are essential molecules that are required by its mammalian hosts for survival [18, 19]. Most *E. coli* strains found in the human body are considered commensal, which include *E. coli strain crooks*, *E. coli HS*, *E. coli K12*, and H10407 [20]. However, pathogenic *E. coli* can cause severe human diseases and great economic harm [21].

Pathogenic *E. coli* infections usually cause enteric/diarrhoeal disease, UTIs, and sepsis/meningitis [22]. Pathogenic *E. coli* can be grouped into enteropathogenic *E. coli* (EPEC), enterohaemorrhagic *E. coli* (EHEC), *enterotoxigenic E. coli* (ETEC), *enteroaggregative E. coli* (EAEC), *enteroinvasive E. coli* (EIEC), *diffusely adherent E. coli* (DAEC), *adherent invasive E. coli* (AIEC), *uropathogenic E. coli* (UPEC), *meningitis-associated E. coli* (MNEC), *septicemia-associated E. coli* (SEPEC), and *avian pathogenic E. coli* (APEC) [20, 22]. EPEC, EHEC, ETEC, EAEC, EIEC, and DAEC are intestinal pathogens, while UPEC causes UTI, MNEC and SEPEC cause meningitis/sepsis [22]. One noteworthy example of EHEC is *E. coli* O157:H7, which is known to cause potentially fatal diarrhea, hemorrhagic colitis, and hemolytic-uremic syndrome (HUS) [23–25]. While *E. coli* is naturally susceptible to antibiotics, MDR has become an increasingly common trend in this species [26]. It is becoming increasingly important to ensure proper testing at hospitals and food facilities to ensure that infection from this bacterium is avoided. This chapter will aim to provide an overview of the current detection methods that can be used to detect *E. coli* as well as report the development of new tests based on functional nucleic acids (specifically DNAzymes and DNA aptamers), with an emphasis on our own work.

2.3 Conventional Methods for General *E. coli* Detection

Before using a detection method, it is often suggested to perform enrichment to increase bacterial concentration. The conventional testing methods for *E. coli* do not specify what type of *E. coli* is found [27]. The food/fecal matter to be tested is homogenized with PBS, added to lauryl tryptose (LST) broth or lactose broth, and incubated at 35 °C for 24 hours for ideal bacterial growth [27]. If there is evidence of gas production in the tubes, such as displacement of the medium or effervescence when the vial is agitated, the tubes are re-incubated for another 24 hours [27]. Examples of gas production in the tubes include displacement of the medium in the fermentation vial or effervescence when agitated. Any tubes that show signs of gas are re-incubated for another 24 hours and re-examined. This is a presumptive test for *E. coli*, which is further tested to get a confirmed *E. coli* reading using a liquid culture method or a solid agar plating method.

If the liquid culture of the sample in *E. coli* broth remains negative for gas production for 72 hours, the tested sample is not contaminated with *E. coli* [27]. However, if the liquid culture tests positive, the suspension is streaked on Levine's eosin-methylene blue (L-EMB) agar, incubated for 24 hours at 35 °C, and visually examined for suspicious colonies [27]. These colonies are transferred to plate count agar (PCA) slants and incubated for 24 hours at 35 °C followed by gram staining to confirm that colonies are indeed gram-negative [27]. The gram-negative cultures are then tested using IMViC (Indole, Methyl red, Voges–Proskauer, Citrate) assay [27]. *E. coli* tests positive or negative for indole, negative for Voges–Proskauer, positive for methyl red, and negative for citrate [27]. If the test results align, the sample is contaminated with *E. coli*.

For the alternative solid medium method, violet, red bile agar (VRBA) is prepared and streaked with the LST suspension [27]. 4-Methyl-umbelliferyll-β-D-gucuronide (MUG) mixed with additional VRBA is added on top of the agar plating, and the plate is inverted for incubation at 35 °C for 24 hours [27]. In the presence of longwave (365 nm) UV light, if blue fluorescence is observed around the colonies, then *E. coli* is present in the initial sample [27]. 95% of *E. coli* contains the β-glucuronidase (GUD) enzyme that cleaves the MUG to release 4-methylumbelliferone (MU), which emits a blue fluorescence under longwave UV light [27, 28]. Another method to perform the MUG is the LST-MUG method [27, 28]. For this test, the sample is prepped in the same manner as done for the previous tests, but instead of incubating in an

LST broth, the sample is incubated in an LST-MUG assay. One noteworthy example of an *E. coli* serotype that does not cleave MUG to MU is *E. coli* O157:H7, an EHEC serotype that commonly causes severe disease in humans [23–27]. These conventional methods are labor-intensive and time-consuming, pointing to the need for novel detection systems.

2.3.1 Sorbitol MacConkey Agar

One of the oldest currently utilized detection methods for *E. coli* O157:H7 is sorbitol MacConkey agar (SMAC) [29, 30]. This agar consists of bile salts, sorbitol (a carbohydrate), and an indicator. The agar is plated, spiked with a sample, and left to incubate over 24 hours at around 36 °C [29–31]. In normal laboratory conditions, *E. coli* O157:H7 does not ferment sorbitol [31]. *E. coli* O157:H7 colonies will appear as colorless colonies, while other enterobacteria show up as pink colonies. This detection method has a sensitivity of 100%, a specificity of 85%, and an accuracy of 86% [31]. However, there are other bacteria or *E. coli* strains, such as *Escherichia hermanii* and *E. coli* O157:H16, that also do not cause the fermentation of sorbitol and can result in false positives of the test [30]. There have been enrichment broths that can enhance the detection of *EHEC* by providing nutrients that allow specific organisms to produce more colonies [29]. Some examples of these nutrients and chemicals include peptone, vancomycin, cefixime, cefsulodin, and potassium tellurite. This is the preferred method utilized by public health agencies around the world. Current estimates for testing turnaround time for this is approximately seven days.

2.3.2 PCR

Polymerase chain reaction (PCR) is another common technique for the detection of *E. coli* in contaminated fecal and food samples [29]. While PCR can also be done without enrichment of the sample, the United States Food and Drug Administration (FDA) requires enrichment to improve detection [32]. PCR is a much faster technique than traditional culturing methods. Takahashi et al. described a real-time PCR technique that can detect 51 different strains of *E. coli*, including four serotypes of *E. coli* O157:H7 [33]. TaqMan PCR probes were designed for *E. coli* based on the alignment of 22 different uidA sequences. These uidA genes encode for the β-D-glucuronidase enzyme, which cleaves MUG to create MU. While *E. coli* O157:H7 does not contain the enzyme, it still has the uidA gene [33]. Following traditional qPCR with 40 cycles of amplifications, Takahashi et al. were able to detect *E. coli* in samples with concentrations as low as 10 CFU/ml with a seven-hour enrichment in TSB culture or 10^3 CFU/ml without an enrichment step [33].

Another PCR method for the detection of *E. coli* is the BioFire FilmArray Gastrointestinal (GI) panel [34]. This panel can detect up to 22 targets, including EAEC, EPEC, ETEC, EHEC (including *E. coli* O157:H7), and EIEC [34]. Each GI panel test consists of automated nucleic acid extraction, reverse transcription, amplification, and the analysis of the results within a one-hour runtime per specimen [34]. This GI panel multiplex PCR has a sensitivity of 100% for ETEC and EHEC and at least 94.5% sensitivity for EAEC, EPEC, and EIEC [34]. The specificity is greater than 97.1% [34].

Other developed PCR testing methods are specific for *E. coli* O157:H7. The FDA has a manual on qPCR-based *E. coli* O157:H7 detection that amplifies stx1, stx2, and uidA genes [32]. However, they still require culture confirmation on an agar plate such as SMAC if probable *E. coli* O157:H7 contamination is detected by qPCR [32]. Other researchers have described the use of specific forward and reverse primers to amplify the stx2 gene present in *E. coli* O157:H7, which can detect as low as 100 colony-forming units (CFU) [35]. Another research group has been able to detect as low as 2 CFU/g of soil by amplifying a wider variety of genes present in *E. coli* O157:H7 [29]. While qPCR is a faster technique, it requires trained personnel and expensive equipment to properly work as well as the standard culturing techniques that are utilized after. There are also concerns about false positives and negatives due to the lower sensitivity and specificity of the technique. The requirement of lysing the cells also adds another layer of complexity. While some protocols have been recently developed to obtain same-day detection of *E. coli* O157:H7, it still does not solve the other concerns associated with qPCR-based detection of the pathogen such as its inability to differentiate among active infections, asymptomatic carriers, and past infections as nucleic acids may remain in human bodies for several months [36, 37].

2.4 Biosensors for *E. coli* Detection

While the above-reported techniques are commonly utilized, this section will focus on novel detection methods that are currently in research, such as biosensors. One of the major problems with the development of a rapid *E. coli* test is the lengthy enrichment process and the post-processing [29]. Also, colony counting, PCR, ELISA, and other conventional methods are labor-intensive, expensive, and cannot be used on-site due to the requirements of specialized equipment [38, 39]. Biosensors are in development to provide a detection method that eliminates the requirement of some or most of pre- and post-processing steps.

2.4.1 Protein Biosensors for *E. coli* Detection

Antibodies can be generated to recognize a specific strain of a bacterium, such as *E. coli* O157:H7. They can be further explored to create biosensors for the detection of the bacteria of interest using various signal transduction mechanisms; a comprehensive review on this topic has been written by Bai et al. [39]. To provide a specific example, gold electrodes on glass slides and anti-*E. coli* O157:H7 IgG were used to create the biosensor to detect *E. coli* O157:H7 based on electrochemical impedance spectroscopy (EIS) [38], a technique that is increasing in popularity due to it being label-free, simple to use, nondestructive, and having low detection limits, thus representing an attractive design for the development of an inexpensive, portable, and rapid detection platform. Upon incubation between the sensor and the contaminated sample, electrochemical measurements can be taken using a potentiostat and a redox probe. This biosensor has a clinically relevant limit of detection (LOD) of 48 CFU/ml [38]. This biosensor is highly selective, sensitive, and easily reproducible. As it can also be self-assembled, this makes it an attractive design in the development of fast, portable, and inexpensive detection platforms.

2.4.2 Functional Nucleic Acid-Based Sensors for *E. coli* Detection

Functional nucleic acids refer to nucleic acid molecules with catalytic properties, binding properties, or both [40, 41]. Functional nucleic acids include aptamers and nucleic acid enzymes (DNAzymes and ribozymes). Functional nucleic acid-based biosensors, particularly those made of DNAzymes and DNA aptamers, offer interesting new methods of detection that provide many benefits over conventional biosensors [42, 43]. DNA aptamers and DNAzymes with high affinity and specificity for virtually any target can now be routinely isolated from synthetic nucleic acid libraries. They are versatile, easy to synthesize, and also have a long shelf life. In this section, we will discuss several examples where DNAzymes and DNA aptamers were used as the key components for the design of biosensors for the detection of *E. coli*.

2.4.2.1 DNAzyme-Based Biosensors for *E. coli* Detection

Many RNA-cleaving fluorogenic DNAzymes (RFDs) have been developed by our group for the recognition of a multitude of bacteria [36]. These DNAzymes can be further developed into biosensors for bacterial detection in either a point of care (POC) or laboratory setting. While various reporting systems may be used, the three commonly used systems are fluorogenic, colorimetric, and electrochemical systems [36].

Fluorogenic DNAzyme systems utilize a fluorescent probe and a quencher that surround a ribonucleotide as the cleavage site (Figure 2.1) [36]. When the DNAzyme system is exposed to the target bacteria, the DNAzyme cleaves at this site, which results in the separation of the quencher from the fluorophore, accompanied by the generation of a strong fluorescence signal [36]. Furthermore, sandwiching the cleavage site with the quencher and fluorophore molecules can significantly decrease nonspecific cleavage of the ribonucleotide site by RNases in biological samples, thus improving the probe stability and specificity [44].

A specific RFD, named RFD-EC1, was described by our group in 2011 for specific recognition of *E. coli* [45]. Since then, this DNAzyme has been incorporated into various fluorogenic detection systems for the rapid, sensitive, and specific detection of *E. coli* [46–50]. For example, we reported a graphene-DNAzyme-based fluorescent biosensor for *E. coli* detection [48]. The DNAzyme is adsorbed to the graphene using π–π stacking interactions. The graphene quenches the fluorescence of RFD-EC1. Once *E. coli* is added to the sample, the target from *E. coli* that activates the DNAzyme competes with the graphene to bind to the DNAzyme. The released DNAzyme self-cleaves at its RNA linkage and produces a detectable fluorescence

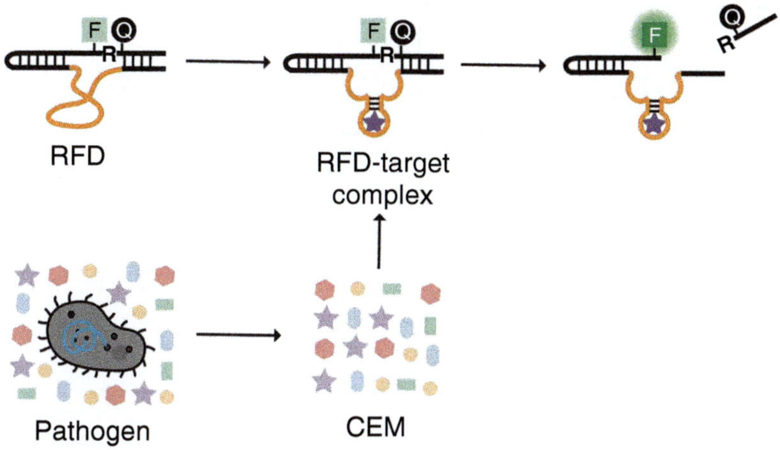

Figure 2.1 Cleavage and reporting mechanism of RNA-cleaving fluorogenic DNAzyme (RFD) system for the detection of a bacterial pathogen. In the presence of the crude extracellular mixture (CEM) of the target pathogen, the DNAzyme cleaves at the ribonucleotide site, which separates the quencher molecule from the fluorophore, allowing for the generation of a detectable signal. Source: Reproduced with permission from Chang et al. [36]. 2021 American Chemical Society.

signal. This biosensor was able to detect *E. coli* specifically at a concentration as low as 10^5 CFU/ml without a pre-enrichment step. With an additional pre-enrichment step, the sensor was able to detect one single CFU [48].

In another example, RFD-EC1 was used to set up a paper-based detection system that can be used by untrained personnel at sources of contamination or for at-home testing, which involves wax-based printing of microzones followed by ink-jet printing of an ink loaded with RFD-EC1 (Figure 2.2) [47]. A nitrocellulose paper backed with a thin plastic layer was used in conjunction with Whatman #1 filter

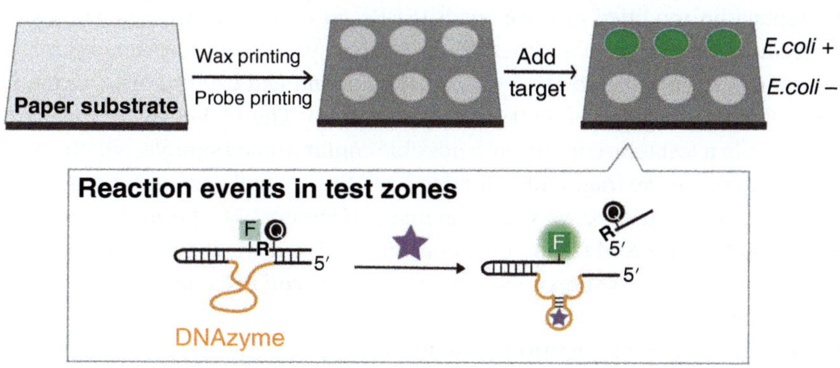

Figure 2.2 Paper-based biosensor that can combine the cleavage reaction and the fluorescence detection on one platform for *E. coli* detection. Wax containing the DNAzyme was printed on nitrocellulose paper, which could be used to detect a signal in the presence of *E. coli*. Source: Reproduced with permission from Chang et al. [36]. 2021 American Chemical Society.

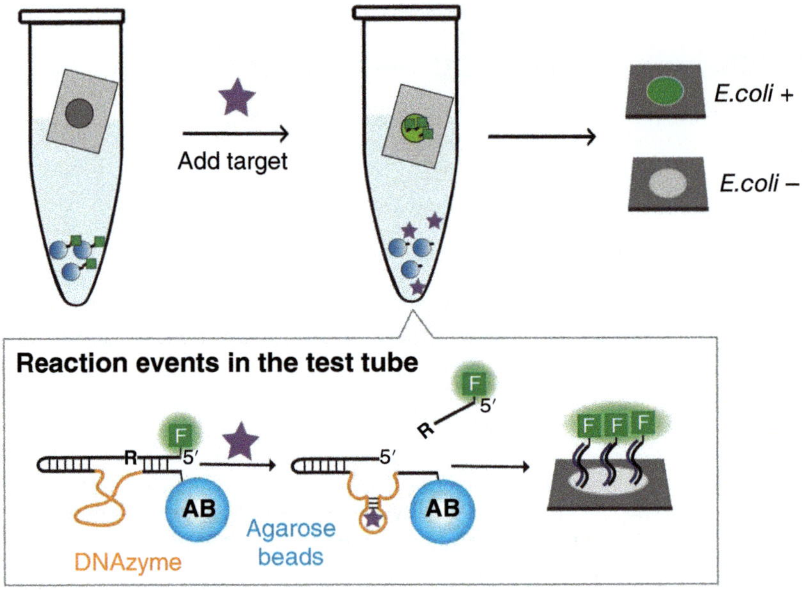

Figure 2.3 Surface-to-surface signal enrichment-based paper biosensor for *E. coli* detection. The EC1 DNAzyme is attached to agarose beads, and a nitrocellulose paper printed with a capture probe can capture the cleaved fragments. Source: Reproduced with permission from Chang et al. [36]. 2021 American Chemical Society.

paper for its low background fluorescence and its ability to prevent leaching of aqueous solutions. This sensor was found to be stable for six months at ambient temperature.

A different version of the paper-based system uses a surface-to-surface product enrichment (S2SPE) approach (Figure 2.3) [49]. In this method, the DNAzyme is modified with a 5′-FAM fluorophore and a 3′-biotin component. The biotin can bond with streptavidin-modified agarose beads (SABs) to create an EC1-SAB conjugate. A nitrocellulose paper strip, referred to as P^CDNA, printed with a wax microzone that contained a biotinylated DNA oligonucleotide, named capture DNA sequence 1 (CDS1), hybridizes with the EC1 cleavage fragment. The EC1-SAB are incubated with P^CDNA in a test tube containing a possible contaminated sample, which causes the release of a cleavage fragment to bind to a relatively small area of P^CDNA microzone, which can be retrieved, washed, and imaged (Figure 2.3). The advantage of this method is that the cleaved fragment is concentrated in a small zone, which allows for a stronger signal. This sensor was able to detect *E. coli* at a concentration as low as 100 CFU/ml [49].

The detection systems mentioned above use fluorogenic DNAzymes as their detection platform. However, RFD-EC1 has also been developed into electrochemical systems [36, 51, 52]. This system used the DNAzyme modified to release electroactive DNA barcodes in the presence of *E. coli*, an electrochemical differential chip (e-DChip) with two star-shaped channels that can capture and release the electroactive DNA barcodes, and three-dimensional nanostructured electrochemical

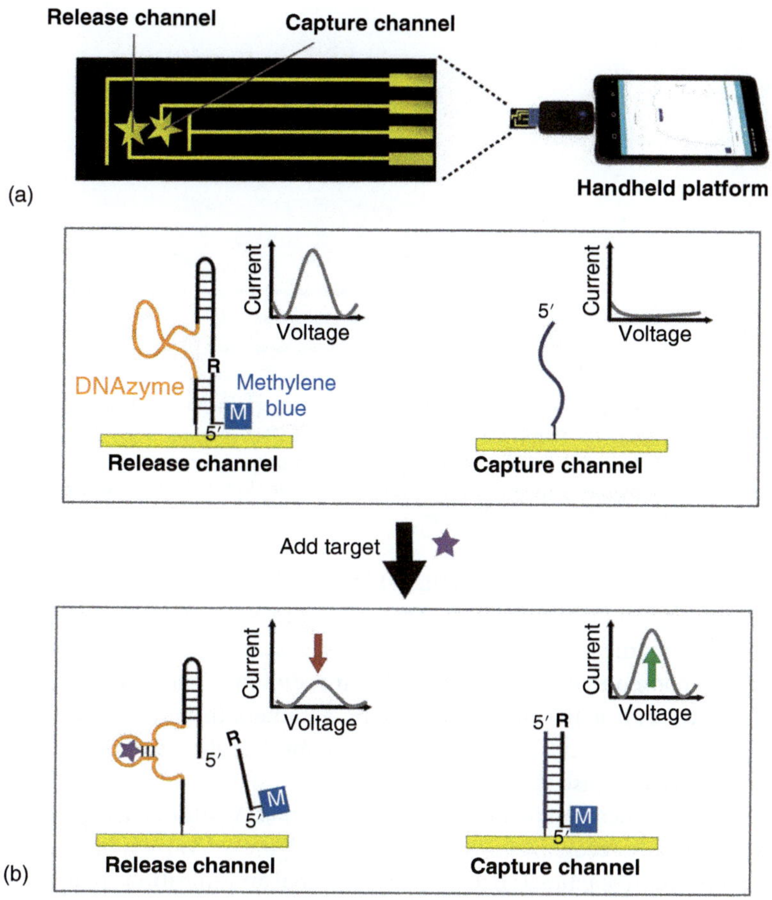

Figure 2.4 DNAzyme-based electrochemical biosensor for the detection of *E. coli*. (a) Sensor Design: the sensor is designed to have two channels for the release and capture of redox barcodes. (b) The working principle of the sensor. The release channel initially has a higher current; however, in the presence of the pathogen, the DNAzyme cleaves and the cleaved fragment binds to the capture channel, which causes it to have an increase of current passing through. Source: Reproduced with permission from Chang et al. [36]. 2021 American Chemical Society.

channels (electrodes) to increase the capture and hybridization efficiency of DNA barcodes (Figure 2.4). The integration of these three components enables bacterial detection to be performed with minimal user intervention and without the need for external reagents or labels. The release-diffuse-capture approach that this system utilizes significantly reduces background signal and thus increases the detection sensitivity [51, 52].

2.4.2.2 Aptamer-Based Sensors for *E. coli* Detection

Several aptamers have also been developed for the recognition of *E. coli*, particularly *E. coli* O157:H7. For example, Kim et al. used a whole bacterial cell-SELEX process

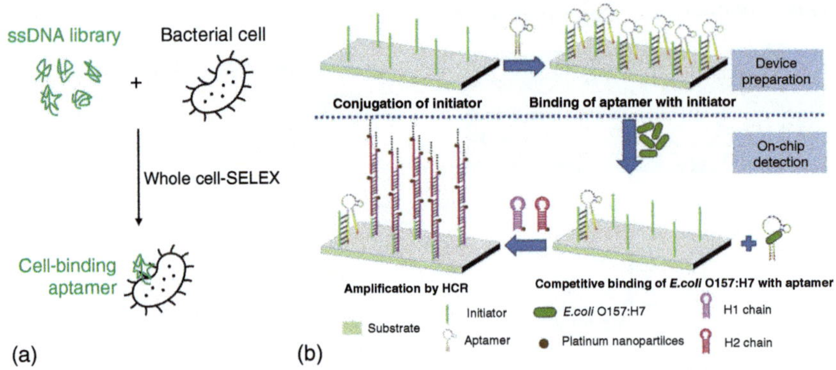

Figure 2.5 An example of aptasensor for *E. coli* detection. (a) Cell-binding aptamer SELEX. (b) Aptamer-HCA-based assay for *E. coli* O157:H7. The initiator DNA is immobilized on the substrate surface, followed by the binding of the aptamer. Binding of *E. coli* O157:H7 frees up the initiator, which participates in HCA to generate a readout signal. Source: Reproduced with permission from Li et al. [53]. 2020 Elsevier.

to isolate *E. coli* cell-binding DNA aptamers (Figure 2.5a) [54]. Several aptamers were derived from a synthetic DNA library that are highly specific to *E. coli*. While the targets for their aptamers have not been identified, researchers hypothesize that they likely bind to components of the outer cellular membrane such as LPS, outer membrane proteins, and/or flagella. With specific aptamers in hand, researchers can proceed to create different types of biosensors for *E. coli* detection. In this section, we will selectively discuss a few examples.

Li et al. developed a microfluidic eye-based aptasensor (EA-sensor) using a DNA aptamer that specifically recognized *E. coli* O157:H7 (Figure 2.5b) [53]. The sensor used hybridization chain reaction (HCR) to achieve signal amplification. To prepare the sensor, the initiator was immobilized on the surface of a reaction well. This was followed by the addition of aptamer, which would hybridize with the surface-bound initiator. *E. coli* O157:H7 cells would compete with the initiator for the binding to the aptamer, which resulted in the departure of the aptamer from the initiator (as the aptamer prefers *E. coli* as the binding partner). Platinum nanoparticles (PtNPs) labeled H1 and H2 were then added to the system to initiate HCA. PtNPs would then catalyze H_2O_2 to produce O_2, which oxidizes a dye to produce a detectable signal in a microfluidic channel. This enabled fast detection of the pathogen. This biosensor achieved the detection of *E. coli* O157:H7 within 75 minutes of exposure. The LOD for the reaction was 250 CFU/ml in a regular buffered sample and 400 CFU/ml in a milk sample [53].

Diaz-Amya et al. reported the use of DNA aptamers that were covalently conjugated to 4-aminothiophenol–gold nanoparticle complexes for the sensitive and specific detection of *E. coli* O157:H7 [55]. This group then performed surface-enhanced Raman spectroscopy (SERS) analysis for detection and quantification. To begin, gold (Au) nanoparticles were modified with 4-aminothiophenol (4-ATP). The *E. coli* O157:H7 specific aptamer was modified with a thiol functionalization that could be chemically activated with a disulfide bond reduction using

Tris(2-carboxyethyl) phosphine hydrochloride, TECP. The aptamer is functionalized with the GNP and washed with BSA to block unbound amine groups. The aptamer system can be utilized to detect *E. coli* O157:H7 using Raman spectroscopy. The researchers reported that the LOD for the sensor was 10 CFU/ml.

Zhou et al. also used gold nanoparticles and Raman scattering with their aptamer-based sensor for detecting *E. coli* O157:H7 [56]. Zhou et al. conjugated their aptamer (Apt-1) with the signal molecule rhodamine B (RhB) to gold nanorods (GNRs), which prepared gold nanobones (NBs). Apt-2 was utilized as a capture molecule. Apt-2 capture probe was prepared by conjugating Apt-2 with streptavidin-conjugated magnetic nanoparticles (streptavidin-MNP). For detection, the capture probe is incubated with a sample containing the signal probe. The solution is analyzed using Raman spectroscopic analysis to determine the presence of *E. coli* O157:H7 in the sample. The researchers were able to achieve a LOD of 3 CFU/ml for the detection of the pathogen [56].

Jiang et al. reported an ultra-sensitive photoelectrochemical (PEC) aptamer biosensor for detecting *E. coli* O157:H7 [57]. The biosensing platform was fabricated based on non-metallic plasmonic tungsten oxide hydrate nanosheets ($WO_3{\cdot}H_2O$) that were coupled with nitrogen-doped graphene quantum dots (N-GQDs). By conjugation with an *E. coli* O157:H7 binding aptamer, the PEC aptasensor can achieve high sensitivity for the detection of the pathogen. The sample to be tested can be added to the aptamer sensor and washed with PBS. The PBS washing will remove any unbound bacterial cells. The aptasensor is detected for PEC detection as the capturing of *E. coli* O157:H7 on the aptasensor decreases the photocurrent signal of the electrodes. This is due to steric hindrance caused by the capture of the pathogen, which can be measured. The LOD for this aptasensor was 0.05 CFU/ml, while it had a linear detection range of $0.1–10^4$ CFU/ml [57].

2.5 Conclusion

In this chapter, we have discussed what *E. coli* is, what the conventional testing methods are, what their drawbacks are, and what novel detection methods are currently undergoing research. While there have already been plenty of specific and sensitive *E. coli* detection methods, these conventional methods do have many drawbacks that include a long test turnaround time, the need for specialized equipment, and the use of lengthy procedures, or lack of high accuracy. The new functional nucleic acid-based detecting methods in development are aiming to resolve all these issues to deliver a rapid, sensitive, and specific test that can also be used in point-of-care (POC) testing. We introduced a few novel functional nucleic acid-based detection methods as they have several advantages over traditional techniques, including, but not limited to, ease of synthesis, high specificity, high sensitivity, versatility, and long shelf life. While these functional nucleic acid techniques are still in their infancy, they do represent the next-generation methods that are not only attractive for the detection of *E. coli*, but also offer the versatility and adaptability to target many other bacterial pathogens for detection.

References

1 Rodriguez-Valera, F. (2011). Bacteria. In: *Encyclopedia of Astrobiology* (ed. M. Gargaud, R. Amils, J.C. Quintanilla, et al.), 137–139. Berlin: Springer.

2 Bar-On, Y.M., Phillips, R., and Milo, R. (2018). The biomass distribution on Earth. *PNAS* 115 (25): 6506–6511.

3 Shibai, A., Maeda, T., Kawada, M. et al. (2019). Complete genome sequences of three star-shaped bacteria, *Stella humosa*, *Stella vacuolata*, and *Stella* species ATCC 35155. *Microbiol. Resour. Announce.* 8 (32): e00719-19.

4 van Teeffelen, S. and Renner, L.D. (2018). Recent advances in understanding how rod-like bacteria stably maintain their cell shapes. *F1000Res* 7: 241.

5 Bell, T., Newman, J.A., Silverman, B.W. et al. (2005). The contribution of species richness and composition to bacterial services. *Nature* 436 (7054): 1157–1160.

6 Falkowski, P.G., Fenchel, T., and Delong, E.F. (2008). The microbial engines that drive Earth's biogeochemical cycles. *Science* 320 (5879): 1034–1039.

7 Sender, R., Fuchs, S., and Milo, R. (2016). Revised estimates for the number of human and bacteria cells in the body. *PLoS Biol.* 14 (8): e1002533.

8 Gomaa, E.Z. (2020). Human gut microbiota/microbiome in health and diseases: a review. *Antonie van Leeuwenhoek* 113 (12): 2019–2040.

9 Gupta, S., Allen-Vercoe, E., and Petrof, E.O. (2016). Fecal microbiota transplantation: in perspective. *Ther. Adv. Gastroenterol.* 9 (2): 229–239.

10 Bartlett, A., Padfield, D., Lear, L. et al. (2022). A comprehensive list of bacterial pathogens infecting humans. *Microbiol. Aust.* 168 (12): 001269.

11 Louca, S., Mazel, F., Doebeli, M., and Parfrey, L.W. (2019). A census-based estimate of Earth's bacterial and archaeal diversity. *PLoS Biol.* 17 (2): e3000106.

12 Fleming, A. (1929). On the antibacterial action of cultures of a *Penicillium*, with special reference to their use in the isolation of *B. influenzæ*. *Br. J. Exp. Pathol.* 10 (3): 226–236.

13 Hutchings, M.I., Truman, A.W., and Wilkinson, B. (2019). Antibiotics: past, present and future. *Curr. Opin. Microbiol.* 51: 72–80.

14 Frieri, M., Kumar, K., and Boutin, A. (2017). Antibiotic resistance. *J. Infect. Public Health* 10 (4): 369–378.

15 Akova, M. (2016). Epidemiology of antimicrobial resistance in bloodstream infections. *Virulence* 7 (3): 252–266.

16 Durso, L.M., Smith, D., and Hutkins, R.W. (2004). Measurements of fitness and competition in commensal *Escherichia coli* and *E. coli* O157:H7 strains. *Appl. Environ. Microbiol.* 70 (11): 6466–6472.

17 Tenaillon, O., Skurnik, D., Picard, B., and Denamur, E. (2010). The population genetics of commensal *Escherichia coli*. *Nat. Rev. Microbiol.* 8 (3): 207–217.

18 Martinson, J.N.V. and Walk, S.T. (2020). *Escherichia coli* residency in the gut of healthy human adults. *EcoSal Plus* 9 (1). https://doi.org/10.1128/ecosalplus.ESP-0003--2020.

19 Blount, Z.D. (2015). The unexhausted potential of *E. coli*. *eLife* 4: e05826.

20 Leimbach, A., Hacker, J., and Dobrindt, U. (2012). *E. coli* as an all-rounder: the thin line between commensalism and pathogenicity. In: *Between Pathogenicity*

and Commensalism, Current Topics in Microbiology and Immunology, vol. 358 (ed. U. Dobrindt, J.H. Hacker, and C. Svanborg), 3–32. Berlin: Springer.

21 Russo, T.A. and Johnson, J.R. (2003). Medical and economic impact of extraintestinal infections due to *Escherichia coli*: focus on an increasingly important endemic problem. *Microbes Infect.* 5 (5): 449–456.

22 Kaper, J.B., Nataro, J.P., and Mobley, H.L.T. (2004). Pathogenic *Escherichia coli*. *Nat. Rev. Microbiol.* 2 (2): 123–140.

23 Mead, P.S. and Griffin, P.M. (1998). *Escherichia coli* O157:H7. *Lancet* 352 (9135): 1207–1212.

24 Sun, H., Wang, M., Liu, Y. et al. (2022). Regulation of flagellar motility and biosynthesis in enterohemorrhagic *Escherichia coli* O157:H7. *Gut Microbes* 14 (1): 2110822.

25 Tarr, G.A.M., Stokowski, T., Shringi, S. et al. (2019). Contribution and interaction of Shiga toxin genes to *Escherichia coli* O157:H7 virulence. *Toxins* 11 (10): 607.

26 Poirel, L., Madec, J.Y., Lupo, A. et al. (2018). Antimicrobial resistance in *Escherichia coli*. *Microbiol. Spectr.* 6 (4). https://doi.org/10.1128/microbiolspec.ARBA-0026-2017.

27 Feng, P., Weagant, S. D., Grant, M. A., and Burkhardt, W. (2020). BAM Chapter 4: Enumeration of *Escherichia coli* and the Coliform Bacteria. Available from: https://www.fda.gov/food/laboratory-methods-food/bam-chapter-4-enumeration-escherichia-coli-and-coliform-bacteria

28 Moberg, L.J., Wagner, M.K., and Kellen, L.A. (1988). Fluorogenic assay for rapid detection of *Escherichia coli* in chilled and frozen foods: collaborative study. *J. Assoc. Off. Anal. Chem.* 71 (3): 589–602.

29 Deisingh, A.K. and Thompson, M. (2004). Strategies for the detection of *Escherichia coli* O157:H7 in foods. *J. Appl. Microbiol.* 96 (3): 419–429.

30 Padhye, N.V. and Doyle, M.P. (1992). *Escherichia coli* O157:H7: epidemiology, pathogenesis, and methods for detection in food. *J. Food Prot.* 55 (7): 555–565.

31 March, S.B. and Ratnam, S. (1986). Sorbitol-MacConkey medium for detection of *Escherichia coli* O157:H7 associated with hemorrhagic colitis. *J. Clin. Microbiol.* 23 (5): 869–872.

32 U.S Food and Drug Administration (2015). Testing methodologies for *E. coli* O157:H7 and *Salmonella* species in spent sprout irrigation water (or sprouts). Available from: https://www.fda.gov/media/94349/download.

33 Takahashi, H., Kimura, B., Tanaka, Y. et al. (2009). Real-time PCR and enrichment culture for sensitive detection and enumeration of *Escherichia coli*. *J. Microbiol. Methods* 79 (1): 124–127.

34 Buss, S.N., Leber, A., Chapin, K. et al. (2015). Multicenter evaluation of the BioFire FilmArray gastrointestinal panel for etiologic diagnosis of infectious gastroenteritis. *J. Clin. Microbiol.* 53 (3): 915–925.

35 Lv, X., Wang, L., Zhang, J. et al. (2020). Rapid and sensitive detection of VBNC *Escherichia coli* O157: H7 in beef by PMAxx and real-time LAMP. *Food Control* 115: 107292.

36 Chang, D., Zakaria, S., Esmaeili Samani, S. et al. (2021). Functional nucleic acids for pathogenic bacteria detection. *Acc. Chem. Res.* 54 (18): 3540–3549.

37 Garrido-Maestu, A., Azinheiro, S., Carvalho, J. et al. (2020). Optimized sample treatment, combined with real-time PCR, for same-day detection of *E. coli* O157 in ground beef and leafy greens. *Food Control* 108: 106790.

38 Lin, D., Pillai, R.G., Lee, W.E., and Jemere, A.B. (2019). An impedimetric biosensor for *E. coli* O157:H7 based on the use of self-assembled gold nanoparticles and protein G. *Microchim. Acta* 186 (3): 1–9.

39 Bai, Z., Xu, X., Wang, C. et al. (2022). A comprehensive review of detection methods for *Escherichia coli* O157:H7. *TrAC Trends Anal. Chem.* 152: 116646.

40 Qian, S., Chang, D., He, S., and Li, Y. (2022). Aptamers from random sequence space: accomplishments, gaps and future considerations. *Anal. Chim. Acta* 2022 (1196): 339511.

41 Morrison, D., Rothenbroker, M., and Li, Y. (2018). DNAzymes: selected for applications. *Small Methods* 2 (3): 1700319.

42 McConnell, E.M., Cozma, I., Morrison, D., and Li, Y. (2020). Biosensors made of synthetic functional nucleic acids toward better human health. *Anal. Chem.* 92 (1): 327–344.

43 McConnell, E.M., Cozma, I., Mou, Q. et al. (2021). Biosensing with DNAzymes. *Chem. Soc. Rev.* 50 (16): 8954–8994.

44 Chang, D., Chang, T., Salena, B., and Li, Y. (2020). An unintentional discovery of a fluorogenic DNA probe for ribonuclease I. *ChemBioChem* 21 (4): 464–468.

45 Ali, M.M., Aguirre, S.D., Lazim, H., and Li, Y. (2011). Fluorogenic DNAzyme probes as bacterial indicators. *Angew. Chem. Int. Ed.* 50 (16): 3751–3754.

46 Aguirre, S.D., Ali, M.M., Salena, B.J., and Li, Y. (2013). A sensitive DNA enzyme-based fluorescent assay for bacterial detection. *Biomolecules* 3 (3): 563–577.

47 Ali, M.M., Brown, C.L., Jahanshahi-Anbuhi, S. et al. (2017). A printed multicomponent paper sensor for bacterial detection. *Sci. Rep.* 7 (1): 12335.

48 Liu, M., Zhang, Q., Brennan, J.D., and Li, Y. (2018). Graphene-DNAzyme-based fluorescent biosensor for *Escherichia coli* detection. *MRS Commun.* 8 (3): 687–694.

49 Samani, S.E., Chang, D., McConnell, E.M. et al. (2020). Highly sensitive RNA-cleaving DNAzyme sensors from surface-to-surface product enrichment. *ChemBioChem* 21 (5): 632–637.

50 Zhou, Z., Brennan, J.D., and Li, Y. (2020). A multi-component all-DNA biosensing system controlled by a DNAzyme. *Angew. Chem. Int. Ed.* 59 (26): 10401–10405.

51 Pandey, R., Chang, D., Smieja, M. et al. (2021). Integrating programmable DNAzymes with electrical readout for rapid and culture-free bacterial detection using a handheld platform. *Nat. Chem.* 13 (9): 895–901.

52 Pandey, R., Lu, Y., Osman, E. et al. (2022). DNAzyme-immobilizing microgel magnetic beads enable rapid, specific, culture-free, and wash-free electrochemical quantification of bacteria in untreated urine. *ACS Sens.* 7 (4): 985–994.

53 Li, T., Ou, G., Chen, X. et al. (2020). Naked-eye based point-of-care detection of *E. coli* O157: H7 by a signal-amplified microfluidic aptasensor. *Anal. Chim. Acta* 1130: 20–28.

54 Kim, Y.S., Song, M.Y., Jurng, J., and Kim, B.C. (2013). Isolation and characterization of DNA aptamers against Escherichia coli using a bacterial cell–systematic evolution of ligands by exponential enrichment approach. *Anal. Biochem.* 436 (1): 22–28.

55 Díaz-Amaya, S., Lin, L.K., Deering, A.J., and Stanciu, L.A. (2019). Aptamer-based SERS biosensor for whole cell analytical detection of *E. coli* O157:H7. *Anal. Chim. Acta* 1081: 146–156.

56 Zhou, S., Lu, C., Li, Y. et al. (2020). Gold nanobones enhanced ultrasensitive surface-enhanced Raman scattering aptasensor for detecting *Escherichia coli* O157:H7. *ACS Sens.* 5 (2): 588–596.

57 Jiang, D., Yang, C., Fan, Y. et al. (2021). Ultra-sensitive photoelectrochemical aptamer biosensor for detecting *E. coli* O157:H7 based on nonmetallic plasmonic two-dimensional hydrated defective tungsten oxide nanosheets coupling with nitrogen-doped graphene quantum dots ($dWO_3 \bullet H_2O$@N-GQDs). *Biosens. Bioelectron.* 183: 113214.

3

From Ligand-Binding Aptamers to Molecular Switches

Sanshu Li[1], Xiaojun Zhang[2], Tao Luo[1], Xuejiao Liu[1], Tingting Zhai[3], and Hongzhou Gu[3]

[1]Huaqiao University, Medical School, 668 Jimei Road, Jimei District, Xiamen 361021, China
[2]Huaqiao University, Chemical Engineering Institute, 668 Jimei Road, Jimei District, Xiamen 361021, China
[3]Shanghai Jiao Tong University, Frontiers Science Center for Transformative Molecules, National Center for Translational Medicine, Department of Chemical Biology, School of Chemistry and Chemical Engineering, 800 Dongchun Road, Minhang District, Shanghai 200240, China

3.1 Aptamers Can Be Generated by SELEX

Aptamers are single-stranded DNA or RNA molecules that can fold into unique secondary or tertiary structures and bind to ligands with high specificity and affinity [1, 2]. They have been engineered as a variety of sensitive biosensors for food and environmental monitoring [3–7], protein detection [8–10], and chemical sensing [11–15]. In addition, they have been developed as nucleic acid drugs [9, 16]. For example, Macugen is an aptamer drug targeting vascular endothelial growth factor (VEGF) [9]. And the advancement of nanomaterial synthesis technology also enables the combination of aptamers and nanomaterials for tumor-targeted therapy [17]. Therefore, aptamers have broad applications in biotechnology and medicine.

Aptamers are commonly produced by the systematic evolution of ligands by exponential enrichment (SELEX), a strategy originally developed by Tuerk [1] and Ellington [2] to screen for oligonucleotides that specifically bind to T4 DNA polymerase and organic dye molecules, respectively. The basic steps of SELEX are to synthesize a large oligonucleotide sequence library (around 10^{15} molecules), incubate the library with a ligand, separate and purify oligonucleotides that can bind to the ligand, and then use the selected signal as templates for PCR amplification to rebuild the library. Through repeated cycles of binding, selection, and amplification, aptamers with ligand-binding affinity are enriched [1, 2, 18] (Figure 3.1).

DNA Nanotechnology for Cell Research: From Bioanalysis to Biomedicine, First Edition. Edited by Zhou Nie.

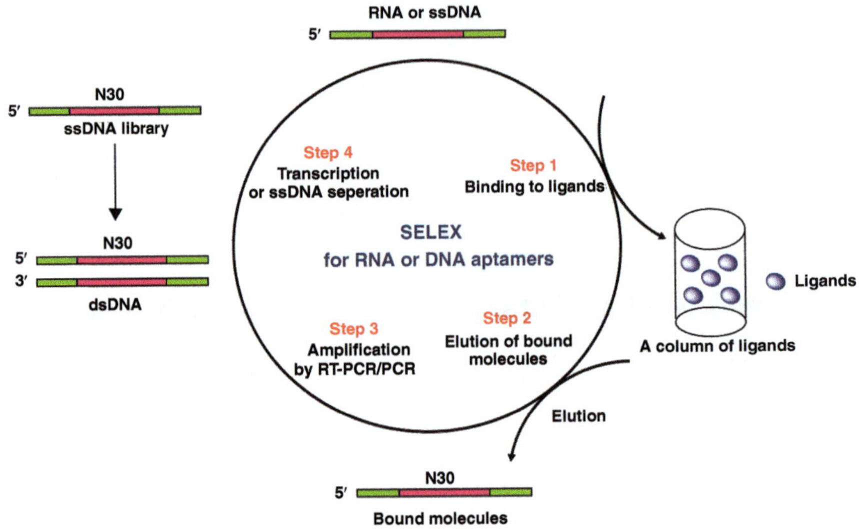

Figure 3.1 The SELEX cycle for the selection of aptamers that can bind to a ligand. Single-stranded (ss)DNA/RNA library contains primer-binding sites at the two ends and random nucleotides in the middle. Step 1: binding of ssDNA/RNA to ligands in the column. Step 2: Use buffer to elute the aptamer candidates from the column. Step 3: Amplification of the eluted ssDNA/RNA by PCR/RT-PCR amplification. Step 4: separation and purification of the PCR product by NaOH treatment (to digest the embedded ribonucleotide in the primer region) and by denaturing PAGE to obtain full-length ssDNA for next cycle selection. For the selection of RNA aptamers, a reverse transcription step needs to be included for signal amplification, and a corresponding transcription step is required for RNA generation.

3.2 Various Subtypes of SELEX Have Been Invented

So far, SELEX-based screening technology has produced thousands of aptamers targeting amino acids, proteins, metal ions, organic molecules, viruses, and even whole cells [16]. During the process, various subtypes including negative SELEX [18], counter-SELEX [19], capillary electrophoresis SELEX (CE-SELEX) [20], Cell SELEX [10], *in vivo* SELEX [21], and Capture-SELEX [22], have been invented.

Negative SELEX is helpful in eliminating nonspecific binding. For example, during the selection of DNA aptamers that can bind to dye molecules, including cibacron blue (CB), reactive green (GR), and reactive blue 4 (B4), after three rounds of selection, the majority of aptamer candidates can non-specifically bind to either the agarose matrix or the non-cognate dye molecules. In the fourth round of selection, a blank column with agarose beads and a non-cognate dye column were applied as negative SELEX to remove the non-specifically binding ssDNA. Eventually, mini-aptamers (25–27 nt) were evolved with high affinity (μM level) and high selectivity to GR [19].

Counter-SELEX uses a closely related analog (to the ligand) as an elution reagent to increase the stringency of the selection process. For example, during the selection of theophylline aptamer, bound RNAs were first eluted with caffeine, which differs from theophylline by a single methyl group at the N-7 position. Then the remaining RNAs were eluted with theophylline. This resulted in theophylline aptamers with 10,000-fold better binding affinity to caffeine [19]. Therefore, counter-SELEX is commonly applied to increase selection specificity.

CE-SELEX is a special selection approach, in which capillary electrophoresis is applied to separate ligand-binding sequences from unbound ones based on the binding-induced mobility shift that can be collected as a separate CE fraction. A successful example is to select for Immunoglobulin E (IgE) targeting aptamers. Only four CE-SELEX rounds were performed, each with 1 pM of IgE, which yielded aptamers with a K_D value of ~27 nM [20].

Cell SELEX uses a whole cell as a docking matrix to isolate aptamers that can recognize and bind to the surface proteins (potential biomarkers) of various cancer cells [10, 23, 24]. For example, a glioblastoma-derived cell line, U251, was used as the target by plating a monolayer of the cells to enable the separation of the interacted (bound) DNAs from a pool of ssDNA sequences. As a result, a DNA aptamer, GBI-10, was found to be able to interact with tenascin-C, an extracellular protein in the tumor matrix [10].

In vivo, SELEX was invented to select RNA-binding motifs using tumor-bearing mice and generate RNA molecules that are capable of specifically recognizing and binding to tumor cells *in vivo*. One of the selected RNA aptamers was found to target protein p68, an RNA helicase that is upregulated in colorectal cancer [21].

Capture-SELEX harnesses a short sequence to base-pair with the middle region of the aptamer candidates [22]. Ligand binding is expected to induce conformational changes, which break the base pairing and release the aptamers. Capture-SELEX is an approach that does not require immobilization of the ligand.

The multiple types of SELEX as well as other alternative approaches that are not mentioned here [25, 26] (refer to reviews [27, 28] for more details) allow us to discover many DNA/RNA aptamers targeting various molecules. Besides the artificial ones, aptamers have also been discovered and verified in nature, more specifically, in the 5′ untranslated region (UTR) of mRNA, which is always linked with an expression platform called riboswitches.

3.3 Riboswitches Are Natural RNA Aptamers Carrying Expression Platforms

Riboswitches are mRNAs that can bind to metabolites or other molecules to regulate gene expression [29–38]. Upon ligand binding, the conformation of riboswitch changes, resulting in the alternation of gene expression by blocking ribosomal binding sites (RBS), sequestering the rho-independent terminators, or even changing the splicing sites. So far, at least 56 classes of natural riboswitches have been identified and verified [39] (Figure 3.2). The classification is based on either the unique folded architecture or a special ligand.

The first riboswitch class was identified in 2002 by Dr. Breaker's lab [40]. It is called cobalamin riboswitch, which controls the cellular concentration of the

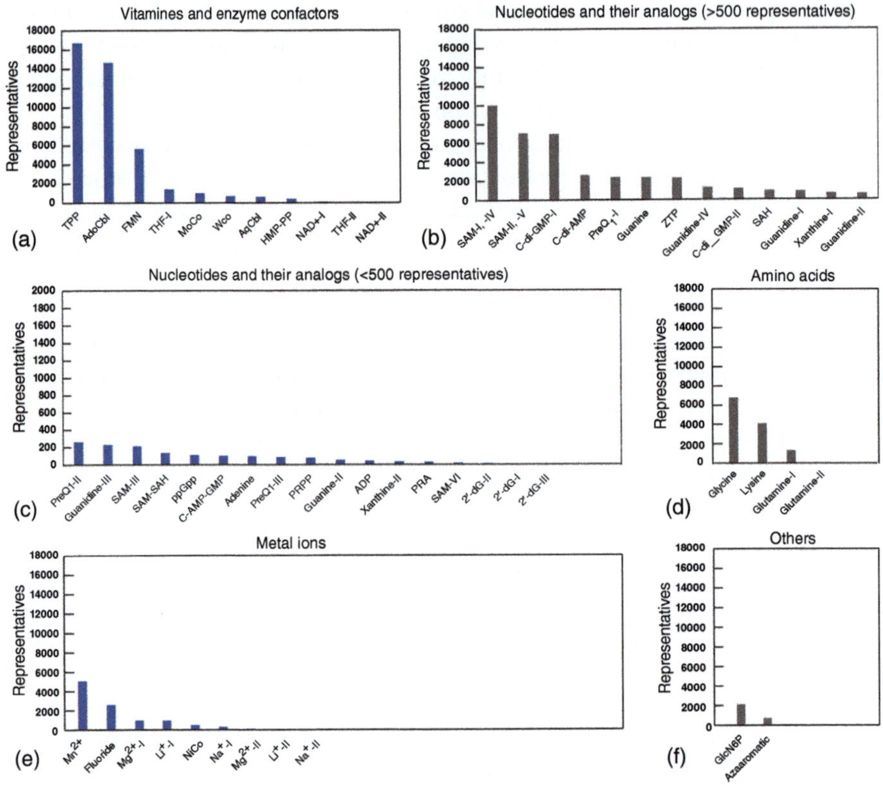

Figure 3.2 Riboswitch classes verified so far. (a) The classes that sense nucleotides and their analogs (>500 representatives for each class). (b) The classes that sense nucleotides and their analogs (<500 representatives for each class). (c) The classes that sense vitamins and enzyme cofactors. (d) The classes that sense metal ions. (e) The classes that sense amino acids. (f) The classes that sense other molecules. 56 classes have been verified by both experimental data and bioinformatics. They have been described in detail in a recent review paper. Source: Adapted from McCown et al. [29].

coenzyme Vitamin B_{12} by regulating the synthesis of the cobalamin transport protein BtuB. Soon after, the TPP (thiamine pyrophosphate) riboswitch was discovered to be involved in thiamine (vitamin B_1) biosynthesis by binding to thiamine or its pyrophosphate derivative without the need for any protein cofactor [32]. The mRNA-effector complex adopts a distinct structure that sequesters the ribosome-binding site and leads to a reduction in gene expression. Almost at the same time, the FMN (flavin mononucleotide) riboswitch was discovered in the 5′ UTR of prokaryotic mRNAs that encode for FMN biosynthesis and transport proteins [41]. Later, riboswitches of SAM [42], guanine [43], adenine [44], and lysine [45] were reported.

SAM riboswitch is a highly conserved RNA element that can bind to the coenzyme *S*-adenosyl-methionine (SAM) with high affinity and specificity [42]. Upon binding, the RNA undergoes structural reorganization, and the allosteric change regulates the expression of genes in *Bacillus subtilis*.

Guanine and adenine riboswitches are located in the 5′ UTR of numerous prokaryotic genes [43, 44], where they control the expression of proteins involved in purine salvage and biosynthesis. Guanine riboswitch was also found to be in tandem with a guanidine riboswitch to regulate the metabolism of both guanine and guanidine, which is a close analog of urea.

Lysine riboswitch is located in the 5′ UTR of the lysC gene. The lysine riboswitch in *B. subtilis* binds to L-lysine with an apparent K_D value of approximately 1 μM and exhibits a high level of molecular discrimination against closely related analogs, including D-lysine and ornithine [45].

In recent years, computer science (bioinformatics) has been applied to riboswitch discovery [46–48]. More than 100 riboswitch candidates were predicted from bacteria and metagenomes, many of which have been verified by experimental data [39] (Figures 3.2 and 3.3). Among them, the *glmS* riboswitch is a relatively special one due to the presence of a ribozyme in its expression platform [31]. Upon ligand binding, the conformational change results in the self-cleavage of the ribozyme.

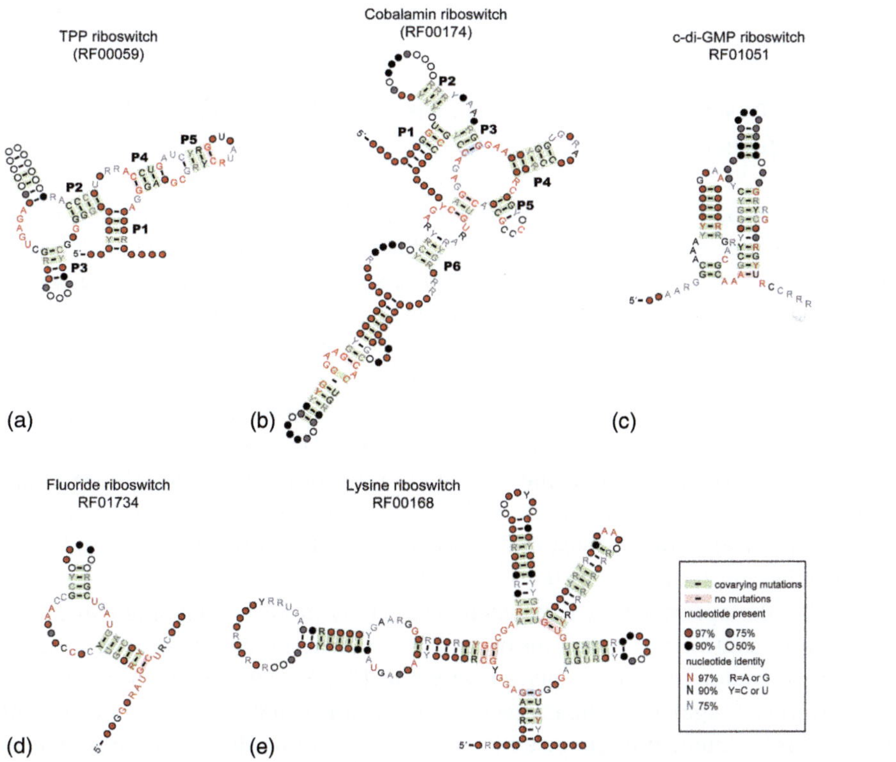

Figure 3.3 Consensus sequence and secondary structures of a few riboswitch classes. (a) TPP riboswitch. (b) AdoCbl riboswitch. (c) C-di-GMP riboswitch. (d) Fluoride riboswitch. (e) Lysine riboswitch. These riboswitch secondary structures are obtained from riboswitch sequences (in Stockholm format) downloaded through Rfam. Source: Adapted from Winkler and Breaker [49].

In nature, riboswitches are specifically evolved to recognize a wide range of molecules (ligands) and regulate gene expression in response to environmental changes [29, 39]. Riboswitch ligands include several common enzyme cofactors, such as vitamin TPP [32], adenosyl-cobalamin (Adocbl) [50], FMN [41], tetrahydrofolate (THF) [51], molybdenum cofactor (Moco) [52], tungsten cofactor (Tuco) [52], nicotinamide adenosine diphosphate (NAD^+) [53, 54], and the thiamin precursor HMP-PP [55] (Figure 3.2a). They can also be numerous ribonucleotide-like molecules, such as guanine [56], adenine [44], xanthine [57], $preQ_1$ [58], guanidine [59], c-di-GMP [60], ppGpp [61], SAM [42], ADP [62], phosphoribosyl pyrophosphate (PRPP) [63], and so on (Figure 3.2b,c). In addition, some amino acids, such as glycine, lysine, and glutamine, can also be riboswitch ligands (Figure 3.2d). Certain metal ions such as F^-, Mn^{2+}, Mg^{2+}, Ni^{2+}, Co^{2+}, Na^+, and Li^+ are also recognized by the corresponding riboswitches (Figure 3.2e). Other riboswitch ligands include glucosamine-6-phosphate (GlcN6P), which is used in cell-wall synthesis [31], and azaaromatic compounds [64] (Figure 3.2f). The abundance of riboswitches that sense the cofactors and numerous other ribonucleotide-related ligands supports the hypothesis that some modern riboswitch classes are descendants of ancient ribozymes that synthesize enzyme cofactors to promote early metabolic processes.

3.4 Riboswitches Use Various Mechanisms to Regulate Gene Expression

Riboswitches possess both the abilities of ligand binding and switching on/off the gene expression largely because they have evolved sophisticated structures to fulfill the goals [65–67] (Figure 3.3). These structures are usually composed of certain conserved nucleotides in the single-stranded loop and double-stranded stem regions to form specific ligand-binding pockets. Conserved ribonucleotides in the riboswitch are critical for ligand recognition and interaction. For example, the guanine riboswitch and the adenine riboswitch contain a distinctive U or C in the internal loop region to enable the specific recognition of guanine or adenine, respectively [44]. In addition, the size of riboswitches is variable. The minimal $PreQ_1$ riboswitch [58] and fluoride riboswitch [68] contain only about 40 and 60 nt, respectively, while cobalamin riboswitch, FMN riboswitch, Cobo, and WCo riboswitches are at the size of 100–300 nt [41, 52].

Usually, the aptamer in a riboswitch interacts with an overlapping expression platform to control the formation or availability of terminators, RBS, or alternative splicing sites to regulate gene expression [39, 67, 69, 70] (Figure 3.4). For transcriptional regulation, the expression platform is usually the rho-independent terminator or attenuator. Here, we take the lysine riboswitch as an example to explain its working mechanism [45]. The aptamer region of the lysine riboswitch resides immediately upstream of a putative transcription terminator. A small region of the aptamer (called anti-terminator) (Figure 3.4a) could potentially base-pair with the terminator to disrupt the stem of the latter. When the cellular concentration of lysine rises, most of the aptamers will be bound by lysine. Consequently,

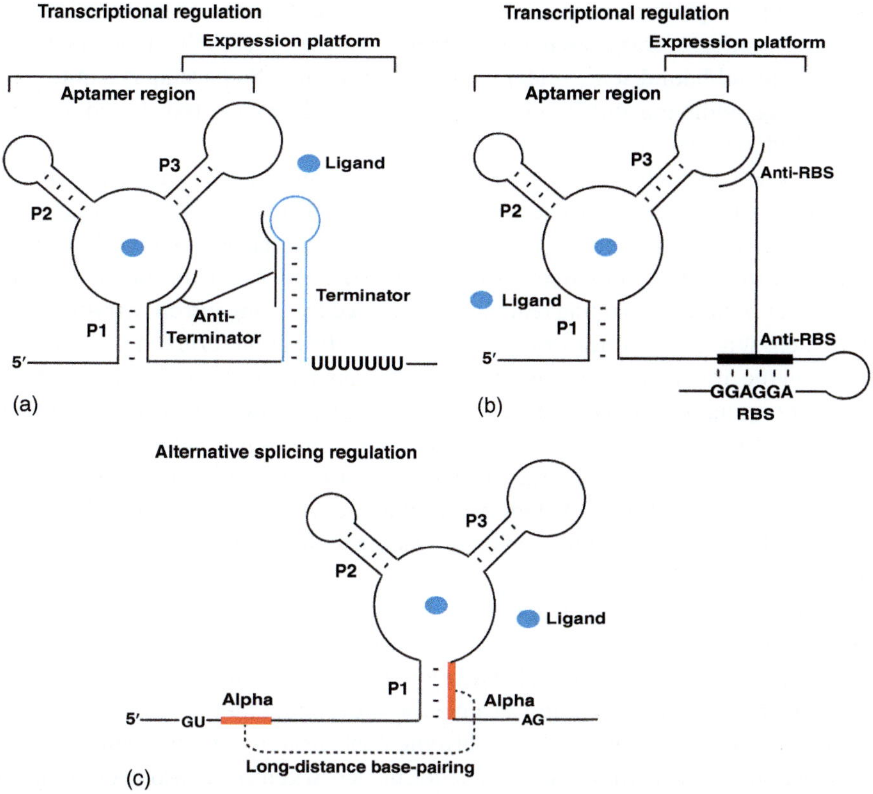

Figure 3.4 Riboswitch regulation of gene expression. (a) Transcriptional regulation. Riboswitches interact with the rho-independent terminator to regulate the formation of the terminator. Anti-terminator is part of the riboswitch expression platform that can base-pair to the stem region of the terminator to disrupt it. (b) Translational regulation. Riboswitches interact with the RBS to regulate its accessibility. Anti-RBS is part of the riboswitch expression platform that can base-pair to RBS. (c) Alternative splicing regulation. One of the TPP riboswitches in fungi uses the P1 portion (alpha′), which is close to the 3′ splicing site (AG) to interact with a region (alpha) that is close to the 5′ splicing site (GU) by base-pairing. Source: Adapted from Breaker [39] and Li and Breaker et al. [69].

the anti-terminator of the aptamer participates in the ligand-binding and does not base-pair with the terminator anymore, resulting in the termination of the transcription. However, when the concentration of cellular lysine gets low, most of the aptamers are free from ligand-binding. Consequently, the anti-terminator base pairs with part of the terminator to disrupt terminator formation, and the transcription could continue. The transcribed mRNA is later translated into proteins to synthesize more lysine molecules.

For translational regulation, a good example is the TPP riboswitch in *Escherichia coli* [32]. This riboswitch is located in the 5′ UTR of the mRNA-encoding enzymes involved in TPP (vitamin B1) biosynthesis. When the level of TPP is low, one part of the aptamer (anti-anti-RBS) could base-pair with the anti-RBS (anti-ribosome binding site), which exposes the RBS and makes it available for translation initiation, resulting in the production of proteins to synthesize thiamine pyrophosphate. As the concentration of thiamine pyrophosphate increases, most TPP aptamers would be in the ligand-bound state, under which the anti-anti-RBS participates in the ligand binding, while the anti-RBS base pairs to RBS to block the translation, leading to a reduction in gene expression (Figure 3.4b).

Interestingly, even the riboswitches that sense the same ligand can evolve with different regulation mechanisms in different microorganism species. For example, the TPP riboswitch in the fungal species *Neurospora crassa* harnesses a mechanism of alternative splicing regulation [69]. In the gene NCU01977 that encodes a predicted transporter protein, the coding region is interrupted by a long intron carrying a TPP riboswitch. When TPP concentration is low, most TPP aptamers are in the unbound state, and the alpha' region of the aptamer's P1 stem could pair to the alpha region, thereby bringing the otherwise distal 3′ and 5′ splice sites in proximity to favor their splicing and produce a contiguous ORF (open reading frame) (Figure 3.4c). However, in the TPP-bound state (as the TPP level increases), the alpha' region is involved in the formation of a stable P1 stem in the aptamer, which would expose the potential 5′ splicing site for alternative splicing, leading to the reduction of the ratio of the contiguous ORF (Figure 3.4c).

In addition, multiple riboswitches can be arranged in tandem to yield complex gene control outcomes [71, 72]. Tandem arrangements were originally encountered with the glycine riboswitches [73], wherein two glycine aptamers tandemly control glycine metabolism by regulating a single rho-independent terminator, resulting in far greater responsiveness to change glycine concentrations than a single glycine riboswitch does. SAM riboswitch and AdoCbl riboswitch have also been found to tandemly control gene expression by regulating the folding of two rho-independent terminators to sense *S*-adenosylmethionine (SAM) and coenzyme B_{12} (adenosylcobalamin), respectively [71]. Besides, guanine riboswitch and guanidine riboswitch can cooperate to regulate a single rho-independent terminator. Overall, riboswitches can form different logic gates including "AND," "OR," and "NOR" to meet the requirements of complex cellular regulation. Herein, we use SAM and AdoCbl tandem riboswitches to illustrate how this works (Figure 3.5). MetE and MetH are proteins that independently catalyze the formation of methionine

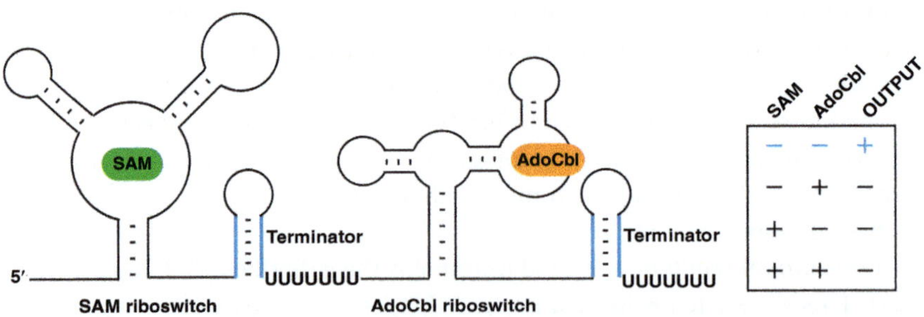

Figure 3.5 SAM and AdoCbl tandem riboswitches regulating the gene-producing MetE. Each riboswitch is followed by a rho-independent terminator to stop the transcription independently. The right panel is the output of gene expression based on the presence of SAM and/or AdoCbl. Source: Adapted from Li et al. [63] and Sudarsan et al. [71].

from homocysteine in certain bacteria. Methionine is further used by the protein MetK to produce SAM. The mRNAs that encode for these three proteins (MetE, MetH, and MetK) carry a SAM riboswitch, which can repress the pathway when SAM is sufficient [40, 42]. However, in some bacteria, the mRNA that encodes for MetE carries the SAM and AdoCbl tandem riboswitches [71], the arrangement of which forms a sophisticated regulation system. In bacteria, MetH usually produces methionine more efficiently. The turnover number of MetH is about 50-fold higher than that of MetE [74]. However, the activity of MetH is dependent on the coenzyme AdoCbl. When the cell has enough AdoCbl to support MetH activity, it would use MetH to synthesize SAM and turn off the MetE gene expression, since MetE is less efficient and cost-ineffective [74]. Only when both SAM and AdoCbl are absent in the cell, the gene expression of MetE will be turned on. To handle this complex regulation, the cell uses the tandem riboswitches to form a NOR logic gate [63, 71] to regulate the MetE gene expression (Figure 3.5). When either SAM or AdoCbl are sufficient, they would bind to an aptamer of the riboswitches and form a terminator to turn off the gene expression. Only when both SAM and AdoCbl are absent, the ligand-unbound aptamers would disrupt both terminators, and the MetE gene expression would be turned on (Figure 3.5).

3.5 Riboswitches Are Potential Drug Targets

Accumulating evidence indicates that riboswitches can be novel antimicrobial drug targets [75–78]. It has been found that the antimicrobial action of the thiamine analog pyrithiamine (PT) is mediated by TPP riboswitches in bacteria and fungi, in which PT is converted to pyrithiamine pyrophosphate (PTPP) to bind to the TPP riboswitches and shut down the thiamine production [77]. In addition, roseoflavin, a natural product of *Streptomyces davawensis*, is an analog of riboflavin. It can target FMN riboswitches of multiple bacterial species to achieve antimicrobial ability [75, 76]. Furthermore, the Breaker lab has identified several lysine analogs that bind to lysine riboswitches and inhibit *B. subtilis* growth, probably through the repression of lysine biosynthesis mediated by the lysine riboswitch [78]. Merck Research Laboratories have reported the discovery of ribocil, a highly selective molecule that targets bacterial riboflavin riboswitches in a phenotypic screen. Although ribocil is structurally distinct from the natural ligand FMN, it can bind specifically to the FMN riboswitch, which mediates the growth inhibition of bacteria [79]. It is believed that in the future, drugs targeting riboswitches will appear in the pharmaceutical market [80].

3.6 Fusing Aptamer with Expression Platform to Construct Artificial RNA Switches

In the early 1990s, the Breaker lab fused aptamers of ATP, FMN, and theophylline to the hammerhead ribozyme (HHR) [81, 82] by a nucleotide-bridge domain to create allosteric ribozymes (Figure 3.6), whose enzymatic activity can be controlled by the

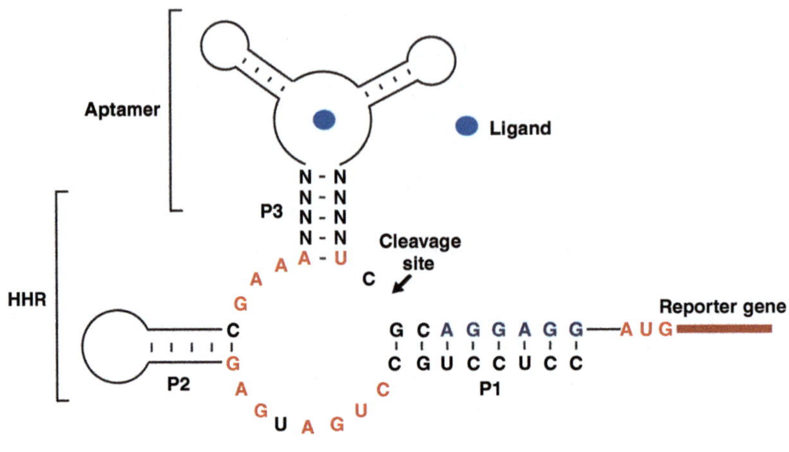

Figure 3.6 An example of combining aptamer with ribozyme to control gene expression. Tested ribozyme candidates can be HHR, twister, pistol, or HDV ribozymes. P3 is a bridge domain largely used to connect the aptamer and the HHR ribozyme. The RBS is embedded in the P1 stem.

aptamer ligand. Soon after, natural allosteric ribozymes were discovered. The representative glmS riboswitch [31] can bind to gluosamine-6-phophate and turn on the cleavage of the embedded ribozyme to down-regulate gene expression.

By using similar working strategies, a series of RNA switches have been made to operate as molecular sensors or as a new class of genetic control elements. For example, Win and Smolke fused an HHR at its stem II with a theophylline or a tetracycline aptamer to enable ligand binding to trigger the tertiary structural shift to induce or block self-cleavage of the HHR [83]. They integrated this construct into the 3′ UTR of the GFP mRNA and demonstrated that this riboswitch-like gene-regulatory platform performed well in *Saccharomyces cerevisiae* (*S. cerevisiae*) to control cell growth and monitor the metabolite [83]. In addition, Smolke's group also developed a general framework to assemble RNA devices, including a HHR and ligand-sensing aptamers to function as logic gates (AND, NOR, NAND, or OR gates), in which two inputs from the aptamers can coordinate together to turn ON or OFF the ribozyme activity to control cellular gene expression [84].

The theophylline aptamer was also successfully integrated into the guide RNA (gRNA) tetra-loop region for gene regulation in living cells [85] by Xu's group. In the absence of theophylline, the aptamer was in a loose state, leading to an unstable RNA structure with compromised activity of gRNA. When theophylline is present, the aptamer would form a tightly defined structure to facilitate the formation of an effective Cas9/gRNA complex, resulting in gene regulation through the CRISPR-Cas function.

Besides, expression platforms other than the HHR have been successfully explored. For example, the Hartig group and others have attached aptamers to the twister ribozyme [86] or antisense RNA and confirmed that the engineered RNA constructs can conduct gene regulation in bacteria [87, 88], yeast [89–91], and mammalian cells [92]. All these examples show that artificial riboswitches can regulate gene expression by affecting transcription elongation, translation initiation, or mRNA stability, just like natural riboswitches do.

3.7 Conclusions

In conclusion, aptamers are single-stranded DNA or RNA molecules that can bind to ligands with high specificity and affinity. They can be generated by SELEX-based approaches including negative SELEX, counter SELEX, capillary electrophoresis SELEX, Cell SELEX, *in vivo* SELEX, Capture-SELEX, Expression-SELEX, and so on. Aptamers have been widely engineered as biosensors, diagnostic tools, and therapeutic reagents to target amino acids, proteins, metal ions, organic molecules, viruses, and even whole cells. In nature, RNA aptamers exist as part of the riboswitches and overlap with expression platforms to control the formation or availability of terminators, RBS, or alternative splicing sites to regulate gene expression. Inspired by natural switches, artificial RNA switches have been engineered by fusing RNA aptamers with ribozymes, which enriches the RNA-based tools for biosensing.

Acknowledgments

This work was supported by the National Natural Science Foundation of China (Grants 81861138004, 82121002, 31770882, and 32271522), the Program of Shanghai Academic Research Leader (22XD1421500), the Project of Science and Technology of Quanzhou (2021C039R), the Natural Science Foundation of Xiamen City (3502Z20227194), and the Huaqiao University Research Funding Project (Z16Y0015).

References

1 Tuerk, C. and Gold, L. (1990). Systematic evolution of ligands by exponential enrichment: RNA ligands to bacteriophage T4 DNA polymerase. *Science* 249 (4968): 505–510.

2 Ellington, A.D. and Szostak, J.W. (1990). In vitro selection of RNA molecules that bind specific ligands. *Nature* 346 (6287): 818–822.

3 Spöring, M., Boneberg, R., and Hartig, J.S. (2020). Aptamer-mediated control of polyadenylation for gene expression regulation in mammalian cells. *ACS Synth. Biol.* 9 (11): 3008–3018.

4 Xiang, W., Lv, Q., Shi, H. et al. (2020). Aptamer-based biosensor for detecting carcinoembryonic antigen. *Talanta* 214: 120716.

5 MacKay, S., Wishart, D., Xing, J.Z. et al. (2014). Developing trends in aptamer-based biosensor devices and their applications. *IEEE Trans. Biomed. Circuits Syst.* 8 (1): 4–14.

6 Yi, J., Xiao, W., Li, G. et al. (2020). The research of aptamer biosensor technologies for detection of microorganism. *Appl. Microbiol. Biotechnol.* 104: 9877–9890.

7 Barbas, A.S. and White, R.R. (2009). The development and testing of aptamers for cancer. *Curr. Opin. Invest. Drugs* 10 (6): 572–578.

8 Liu, Y., Tuleouva, N., Ramanculov, E. et al. (2010). Aptamer-based electrochemical biosensor for interferon gamma detection. *Anal. Chem.* 82 (19): 8131–8136.

9 Ng, E.W., Shima, D.T., Calias, P. et al. (2006). Pegaptanib, a targeted anti-VEGF aptamer for ocular vascular disease. *Nat. Rev. Drug Discovery* 5 (2): 123–132.

10 Daniels, D.A., Chen, H., Hicke, B.J. et al. (2003). A tenascin-C aptamer identified by tumor cell SELEX: systematic evolution of ligands by exponential enrichment. *Proc. Natl. Acad. Sci.* 100 (26): 15416–15421.

11 Bonel, L., Vidal, J.C., Duato, P. et al. (2011). An electrochemical competitive biosensor for ochratoxin A based on a DNA biotinylated aptamer. *Biosens. Bioelectron.* 26 (7): 3254–3259.

12 Mayer, G. (2009). The chemical biology of aptamers. *Angew. Chem. Int. Ed.* 48 (15): 2672–2689.

13 Du, X., Cheng, X., Li, W. et al. (2021). Engineering allosteric ribozymes to detect thiamine pyrophosphate in whole blood. *Anal. Chem.* 93 (9): 4277–4284.

14 Li, H.Y., Yin, F.F., Li, X.Y. et al. (2021). Novel aptasensor-based assay of sonic hedgehog ligand for detection of portal vein invasion of hepatocellular carcinoma. *Biosens. Bioelectron.* 174: 112738.

15 Manna, S., Truong, J., and Hammond, M.C. (2021). Guanidine biosensors enable comparison of cellular turn-on kinetics of riboswitch-based biosensor and reporter. *ACS Synth. Biol.* 10 (3): 566–578.

16 Zhou, J. and Rossi, J. (2017). Aptamers as targeted therapeutics: current potential and challenges. *Nat. Rev. Drug Discovery* 16 (3): 181–202.

17 Panigaj, M., Johnson, M.B., Ke, W. et al. (2019). Aptamers as modular components of therapeutic nucleic acid nanotechnology. *ACS Nano* 13 (11): 12301–12321.

18 Ellington, A.D. and Szostak, J.W. (1992). Selection in vitro of single-stranded DNA molecules that fold into specific ligand-binding structures. *Nature* 355 (6363): 850–852.

19 Jenison, R.D., Gill, S.C., Pardi, A. et al. (1994). High-resolution molecular discrimination by RNA. *Science* 263 (5152): 1425–1429.

20 Mendonsa, S.D. and Bowser, M.T. (2004). In vitro evolution of functional DNA using capillary electrophoresis. *JACS* 126 (1): 20–21.

21 Mi, J., Liu, Y., Rabbani, Z.N. et al. (2010). In vivo selection of tumor-targeting RNA motifs. *Nat. Chem. Biol.* 6 (1): 22–24.

22 Nutiu, R. and Li, Y. (2005). In vitro selection of structure-switching signaling aptamers. *Angew. Chem. Int. Ed.* 44 (7): 1061–1065.

23 Fang, X. and Tan, W. (2010). Aptamers generated from cell-SELEX for molecular medicine: a chemical biology approach. *Acc. Chem. Res.* 43 (1): 48–57.

24 Kim, Y., Liu, C., and Tan, W. (2009). Aptamers generated by Cell SELEX for biomarker discovery. *Biomarkers Med.* 3 (2): 193–202.

25 Kaur, H., Shorie, M., and Sabherwal, P. (2020). Biolayer interferometry-SELEX for Shiga toxin antigenic-peptide aptamers & detection via chitosan-WSe2 aptasensor. *Biosens. Bioelectron.* 167: 112498.

26 Robertson, D.L. and Joyce, G.F. (1990). Selection in vitro of an RNA enzyme that specifically cleaves single-stranded DNA. *Nature* 344 (6265): 467–468.

27 Carmi, N., Shultz, L.A., and Breaker, R.R. (1996). In vitro selection of self-cleaving DNAs. *Chem. Biol.* 3 (12): 1039–1046.

28 Duan, Y., Zhang, C., Wang, Y. et al. (2022). Research progress of whole-cell-SELEX selection and the application of cell-targeting aptamer. *Mol. Biol. Rep.* 49 (8): 7979–7993.

29 McCown, P.J., Corbino, K.A., Stav, S. et al. (2017). Riboswitch diversity and distribution. *RNA* 23 (7): 995–1011.

30 Breaker, R.R. (2012). Riboswitches and the RNA world. *Cold Spring Harbor Perspect. Biol.* 4 (2): a003566.

31 Winkler, W.C., Nahvi, A., Roth, A. et al. (2004). Control of gene expression by a natural metabolite-responsive ribozyme. *Nature* 428 (6980): 281–286.

32 Winkler, W., Nahvi, A., and Breaker, R.R. (2002). Thiamine derivatives bind messenger RNAs directly to regulate bacterial gene expression. *Nature* 419 (6910): 952–956.

33 Xu, X., Egger, M., Chen, H. et al. (2021). Insights into xanthine riboswitch structure and metal ion-mediated ligand recognition. *Nucleic Acids Res.* 49 (12): 7139–7153.

34 Edwards, T.E., Klein, D.J., and Ferré-D'Amaré, A.R. (2007). Riboswitches: small-molecule recognition by gene regulatory RNAs. *Curr. Opin. Struct. Biol.* 17 (3): 273–279.

35 Huang, L., Liao, T.W., Wang, J. et al. (2020). Crystal structure and ligand-induced folding of the SAM/SAH riboswitch. *Nucleic Acids Res.* 48 (13): 7545–7556.

36 Ren, A., Rajashankar, K.R., and Patel, D.J. (2012). Fluoride ion encapsulation by Mg2+ ions and phosphates in a fluoride riboswitch. *Nature* 486 (7401): 85–89.

37 Shanahan, C.A., Gaffney, B.L., Jones, R.A. et al. (2011). Differential analogue binding by two classes of c-di-GMP riboswitches. *JACS* 133 (39): 15578–15592.

38 Crum, M., Ram-Mohan, N., and Meyer, M.M. (2019). Regulatory context drives conservation of glycine riboswitch aptamers. *PLoS Comput. Biol.* 15 (12): e1007564.

39 Breaker, R.R. (2022). The biochemical landscape of riboswitch ligands. *Biochemistry* 61 (3): 137–149.

40 Nahvi, A., Sudarsan, N., Ebert, M.S. et al. (2002). Genetic control by a metabolite binding mRNA. *Chem. Biol.* 9 (9): 1043–1049.

41 Winkler, W.C., Cohen-Chalamish, S., and Breaker, R.R. (2002). An mRNA structure that controls gene expression by binding FMN. *Proc. Natl. Acad. Sci.* 99 (25): 15908–15913.

42 Winkler, W.C., Nahvi, A., Sudarsan, N. et al. (2003). An mRNA structure that controls gene expression by binding S-adenosylmethionine. *Nat. Struct. Mol. Biol.* 10 (9): 701–707.

43 Mandal, M., Boese, B., Barrick, J.E. et al. (2003). Riboswitches control fundamental biochemical pathways in *Bacillus subtilis* and other bacteria. *Cell* 113 (5): 577–586.

44 Mandal, M. and Breaker, R.R. (2004). Adenine riboswitches and gene activation by disruption of a transcription terminator. *Nat. Struct. Mol. Biol.* 11 (1): 29–35.

45 Sudarsan, N., Wickiser, J.K., Nakamura, S. et al. (2003). An mRNA structure in bacteria that controls gene expression by binding lysine. *Genes Dev.* 17 (21): 2688–2697.

46 Weinberg, Z., Lünse, C.E., Corbino, K.A. et al. (2017). Detection of 224 candidate structured RNAs by comparative analysis of specific subsets of intergenic regions. *Nucleic Acids Res.* 45 (18): 10811–10823.

47 Weinberg, Z., Wang, J.X., Bogue, J. et al. (2010). Comparative genomics reveals 104 candidate structured RNAs from bacteria, archaea, and their metagenomes. *Genome Biol.* 11 (3): 1–17.

48 Weinberg, Z., Barrick, J.E., Yao, Z. et al. (2007). Identification of 22 candidate structured RNAs in bacteria using the CMfinder comparative genomics pipeline. *Nucleic Acids Res.* 35 (14): 4809–4819.

49 Kalvari, I., Nawrocki, E.P., Ontiveros-Palacios, N. et al. (2021). Rfam 14: expanded coverage of metagenomic, viral and microRNA families. *Nucleic Acids Res.* 49 (D1): D192–D200.

50 Winkler, W.C. and Breaker, R.R. (2003). Genetic control by metabolite-binding riboswitches. *ChemBioChem* 4 (10): 1024–1032.

51 Ames, T.D., Rodionov, D.A., Weinberg, Z. et al. (2010). A eubacterial riboswitch class that senses the coenzyme tetrahydrofolate. *Chem. Biol.* 17 (7): 681–685.

52 Regulski, E.E., Moy, R.H., Weinberg, Z. et al. (2008). A widespread riboswitch candidate that controls bacterial genes involved in molybdenum cofactor and tungsten cofactor metabolism. *Mol. Microbiol.* 68 (4): 918–932.

53 Panchapakesan, S.S., Corey, L., Malkowski, S.N. et al. (2021). A second riboswitch class for the enzyme cofactor NAD+. *RNA* 27 (1): 99–105.

54 Malkowski, S.N., Spencer, T.C., and Breaker, R.R. (2019). Evidence that the nadA motif is a bacterial riboswitch for the ubiquitous enzyme cofactor NAD+. *RNA* 25 (12): 1616–1627.

55 Atilho, R.M., Mirihana Arachchilage, G., Greenlee, E.B. et al. (2019). A bacterial riboswitch class for the thiamin precursor HMP-PP employs a terminator-embedded aptamer. *elife* 8: e45210.

56 Serganov, A., Yuan, Y.R., Pikovskaya, O. et al. (2004). Structural basis for discriminative regulation of gene expression by adenine-and guanine-sensing mRNAs. *Chem. Biol.* 11 (12): 1729–1741.

57 Yu, D. and Breaker, R.R. (2020). A bacterial riboswitch class senses xanthine and uric acid to regulate genes associated with purine oxidation. *RNA* 26 (8): 960–968.

58 McCown, P.J., Liang, J.J., Weinberg, Z. et al. (2014). Structural, functional, and taxonomic diversity of three preQ1 riboswitch classes. *Chem. Biol.* 21 (7): 880–889.

59 Nelson, J.W., Atilho, R.M., Sherlock, M.E. et al. (2017). Metabolism of free guanidine in bacteria is regulated by a widespread riboswitch class. *Mol. Cell* 65 (2): 220–230.

60 Furukawa, K., Gu, H., Sudarsan, N. et al. (2012). Identification of ligand analogues that control c-di-GMP riboswitches. *ACS Chem. Biol.* 7 (8): 1436–1443.

61 Sherlock, M.E., Sudarsan, N., and Breaker, R.R. (2018). Riboswitches for the alarmone ppGpp expand the collection of RNA-based signaling systems. *Proc. Natl. Acad. Sci.* 115 (23): 6052–6057.

62 Sherlock, M.E., Sadeeshkumar, H., and Breaker, R.R. (2018). Variant bacterial riboswitches associated with nucleotide hydrolase genes sense nucleoside diphosphates. *Biochemistry* 58 (5): 401–410.

63 Sherlock, M.E., Sudarsan, N., Stav, S. et al. (2018). Tandem riboswitches form a natural Boolean logic gate to control purine metabolism in bacteria. *elife* 7: e33908.

64 Li, S., Hwang, X.Y., Stav, S. et al. (2016). The yjdF riboswitch candidate regulates gene expression by binding diverse azaaromatic compounds. *RNA* 22 (4): 530–541.

65 Kavita, K. and Breaker, R.R. (2023). Discovering riboswitches: the past and the future. *Trends Biochem. Sci* 48 (2): 119–141.

66 Mandal, M. and Breaker, R.R. (2004). Gene regulation by riboswitches. *Nat. Rev. Mol. Cell Biol.* 5 (6): 451–463.

67 Breaker, R.R. (2018). Riboswitches and translation control. *Cold Spring Harbor Perspect. Biol.* 10 (11): a032797.

68 Baker, J.L., Sudarsan, N., Weinberg, Z. et al. (2012). Widespread genetic switches and toxicity resistance proteins for fluoride. *Science* 335 (6065): 233–235.

69 Li, S. and Breaker, R.R. (2013). Eukaryotic TPP riboswitch regulation of alternative splicing involving long-distance base pairing. *Nucleic Acids Res.* 41 (5): 3022–3031.

70 Cheah, M.T., Wachter, A., Sudarsan, N. et al. (2007). Control of alternative RNA splicing and gene expression by eukaryotic riboswitches. *Nature* 447 (7143): 497–500.

71 Sudarsan, N., Hammond, M.C., Block, K.F. et al. (2006). Tandem riboswitch architectures exhibit complex gene control functions. *Science* 314 (5797): 300–304.

72 Welz, R. and Breaker, R.R. (2007). Ligand binding and gene control characteristics of tandem riboswitches in Bacillus anthracis. *RNA* 13 (4): 573–582.

73 Mandal, M., Lee, M., Barrick, J.E. et al. (2004). A glycine-dependent riboswitch that uses cooperative binding to control gene expression. *Science* 306 (5694): 275–279.

74 González, J.C., Peariso, K., Penner-Hahn, J.E. et al. (1996). Cobalamin-independent methionine synthase from *Escherichia coli*: a zinc metalloenzyme. *Biochemistry* 35 (38): 12228–12234.

75 Lee, E.R., Blount, K.F., and Breaker, R.R. (2009). Roseoflavin is a natural antibacterial compound that binds to FMN riboswitches and regulates gene expression. *RNA Biol.* 6 (2): 187–194.

76 Pedrolli, D.B., Matern, A., Wang, J. et al. (2012). A highly specialized flavin mononucleotide riboswitch responds differently to similar ligands and confers roseoflavin resistance to *Streptomyces davawensis*. *Nucleic Acids Res.* 40 (17): 8662–8673.

77 Sudarsan, N., Cohen-Chalamish, S., Nakamura, S. et al. (2005). Thiamine pyrophosphate riboswitches are targets for the antimicrobial compound pyrithiamine. *Chem. Biol.* 12 (12): 1325–1335.

78 Blount, K.F., Wang, J.X., Lim, J. et al. (2007). Antibacterial lysine analogs that target lysine riboswitches. *Nat. Chem. Biol.* 3 (1): 44–49.

79 Howe, J.A., Wang, H., Fischmann, T.O. et al. (2015). Selective small-molecule inhibition of an RNA structural element. *Nature* 526 (7575): 672–677.

80 Breaker, R.R. (2009). Riboswitches: from ancient gene-control systems to modern drug targets. *Future Microbiol.* 4 (7): 771–773.

81 Soukup, G.A. and Breaker, R.R. (1999). Engineering precision RNA molecular switches. *Proc. Natl. Acad. Sci.* 96 (7): 3584–3589.

82 Tang, J. and Breaker, R.R. (1997). Rational design of allosteric ribozymes. *Chem. Biol.* 4 (6): 453–459.

83 Win, M.N. and Smolke, C.D. (2007). A modular and extensible RNA-based gene-regulatory platform for engineering cellular function. *Proc. Natl. Acad. Sci.* 104 (36): 14283–14288.

84 Win, M.N. and Smolke, C.D. (2008). Higher-order cellular information processing with synthetic RNA devices. *Science* 322 (5900): 456–460.

85 Liu, Y., Wang, Y., Lin, J. et al. (2021). Theophylline-induced synergic activation of guide RNA to control CRISPR/Cas9 function. *Chem. Commun.* 57 (44): 5418–5421.

86 Breaker, R.R. (2017). Mechanistic debris generated by twister ribozymes. *ACS Chem. Biol.* 12 (4): 886–891.

87 Rehm, C. and Hartig, J.S. (2014). In vivo screening for aptazyme-based bacterial riboswitches. In: *Artificial Riboswitches: Methods and Protocols*, vol. 1111 (ed. A. Ogawa), 237–249. Humana Press.

88 Felletti, M., Bieber, A., and Hartig, J.S. (2017). The 3′-untranslated region of mRNAs as a site for ribozyme cleavage-dependent processing and control in bacteria. *RNA Biol.* 14 (11): 1522–1533.

89 Klauser, B., Atanasov, J., Siewert, L.K. et al. (2015). Ribozyme-based aminoglycoside switches of gene expression engineered by genetic selection in *S. cerevisiae*. *ACS Synth. Biol.* 4 (5): 516–525.

90 Klauser, B., Rehm, C., Summerer, D. et al. (2015). Engineering of ribozyme-based aminoglycoside switches of gene expression by in vivo genetic selection in *Saccharomyces cerevisiae*. *Methods Enzymol.* 550: 301–320.

91 Bayer, T.S. and Smolke, C.D. (2005). Programmable ligand-controlled riboregulators of eukaryotic gene expression. *Nat. Biotechnol.* 23 (3): 337–343.

92 Ausländer, S., Ketzer, P., and Hartig, J.S. (2010). A ligand-dependent hammerhead ribozyme switch for controlling mammalian gene expression. *Mol. Biosyst.* 6 (5): 807–814.

4

DNA Nanotechnology-Based Microfluidics for Liquid Biopsy

Qi Niu[2], Shanqing Huang[2], Chaoyong Yang[1,2], and Lingling Wu[1]

[1] *Shanghai Jiao Tong University, Institute of Molecular Medicine, School of Medicine, Shanghai Key Laboratory for Nucleic Acid Chemistry and Nanomedicine, Renji Hospital, No. 160, Pujian Road, Shanghai 200127, China*
[2] *Xiamen University, College of Chemistry and Chemical Engineering, The MOE Key Laboratory of Spectrochemical Analysis and Instrumentation, Discipline of Intelligent Instrument and Equipment, No. 422, Siming South Road, Xiamen 361005, China*

4.1 Introduction

Cancer is a leading cause of death around the world, especially in developing countries. Liquid biopsy is an emerging technology for disease diagnosis and monitoring, which can detect and analyze tumor-derived circulating targets in body fluids, including circulating tumor cells (CTCs), extracellular vesicles (EVs), proteins, circulating free DNA (cf-DNA), and microRNAs (miRNAs) [1, 2]. The analysis of these biomarkers holds great significance for diagnosis, prognosis, and personalized treatment of tumors. Compared with traditional tissue biopsy, liquid biopsy exhibits many inherent advantages, ushering in a novel era of precision medicine [3]. First, circulating biomarkers can appear in the early stages of diseases. Thus, liquid biopsy enables early diagnosis for timely treatment [4, 5]. Second, without invasive surgical or puncture-based tissue sampling procedures, the convenient and noninvasive sampling from body fluids allows for real-time detection and longitudinal analysis of tumors, enabling treatment assessment, prognosis evaluation, and recurrence monitoring [6]. Third, liquid biopsy can provide comprehensive and dynamic information on both primary and metastatic tumors to guide precision medicine [7].

Despite promising prospects, liquid biopsy still faces some technical challenges in the isolation and detection of circulating targets. Firstly, circulating targets often exist at low concentrations in relatively complex body fluid matrices, requiring highly efficient enrichment techniques and strategies for high signal amplification [8, 9]. Secondly, considering the subsequent downstream analysis, gentle isolation and release methods are required for these fragile targets [10–13]. Additionally, the heterogeneity of circulating targets, such as CTCs and EVs, requires single-cell or single-particle analysis [12, 13]. To overcome these challenges, there is an

DNA Nanotechnology for Cell Research: From Bioanalysis to Biomedicine, First Edition. Edited by Zhou Nie.

urgent need to develop effective isolation, sensitive detection, and high-throughput heterogeneous analysis of circulating targets.

DNA, as the fundamental genetic material, is necessary for all forms of life [14]. Beyond this, according to the base-pairing rule, DNA can form numerous nanostructures and play a crucial role in liquid biopsy [15, 16]. Firstly, some DNA strands can be folded into unique secondary and tertiary structures that can perform recognition or catalytic functions for target detection [17, 18]. Secondly, the highly programmable design of DNA framework nanostructures can be used to construct precise and controllable isolation interfaces. This enhances the efficiency of target separation [19]. Third, the DNA framework nanostructure can be combined with multiple nucleic acid amplification strategies, greatly improving detection sensitivity [20, 21]. On the other hand, DNA possesses powerful encoding capability. DNA barcodes with reasonable design can transform individual circulating targets, as well as their genotype and protein phenotype information into nucleotide sequences [22]. Through high-throughput second-generation sequencing, it achieves high throughput parallel analysis of the heterogeneity of cells and EVs [23, 24].

Microfluidic techniques with the merits of precise manipulation of fluids, miniaturization, integration, and automation provide unprecedented opportunities for liquid biopsy [10, 25, 26]. Hereinto, DNA nanotechnology-based microfluidics combines the advantages of DNA nanostructures and microfluidic chips, opening new avenues for isolation, detection, and heterogeneous analysis of circulating targets. First, it can employ different DNA nanostructures to construct precise, controllable capture substrates within microchannels, enabling efficient separation and sensitive analysis of circulating targets. Second, the integration of DNA-based amplification strategies and picoliter-to-nanoliter reaction systems enabled by microfluidics offers rapid and efficient signal amplification reactions for sensitive detection. Furthermore, microfluidics can realize the isolation and manipulation of single-cells/single-vesicles, making single-cell/single-vesicle analysis possible [11, 27]. Coupled with DNA barcoding strategy, high-throughput analysis of CTCs and EVs at single-cell or single-particle level can be achieved for a better understanding of the mechanism underlying disease occurrence.

We herein focus on the two most important types of liquid biopsy targets: CTCs and EVs. CTCs are shed from solid tumor tissues and enter the bloodstream, potentially leading to the formation of metastases at distant sites. Because CTCs have cellular structure, components, and functions, they better represent cellular states and molecular features of derived tumor tissues. EVs are nanoscale lipid bilayer vesicles released almost by all cells, which exist in various biological fluids and participate in numerous physiological and pathological processes. CTCs or EVs-based liquid biopsy have important implications for early cancer diagnosis, treatment monitoring, prognosis evaluation, and personalized therapy. This chapter summarizes the recent advancements in DNA nanotechnology-based microfluidics for liquid biopsy. First, recent progress in DNA nanotechnology-based microfluidics for the isolation of circulating targets is summarized. Then, the use of DNA nanotechnology-based microfluidics for the release and detection of circulating targets is described. Third, the applications of DNA-assisted microfluidics for single cell/vesicle analysis are summarized. Finally, future perspectives and challenges in the field of DNA nanotechnology-based microfluidics for liquid biopsy are discussed.

4.2 DNA Nanotechnology-Based Microfluidics for Isolation of Circulating Targets

Considering the extremely low concentration of circulating targets in complex body liquids, efficient and specific isolation and enrichment strategies are urgently needed in liquid biopsy. Affinity isolation, a commonly employed technique in liquid biopsy, leverages highly specific interactions between recognition ligands (such as antibodies, peptides, or aptamers) immobilized on the capture interface and receptors present on circulating targets. The efficacy of affinity isolation depends on both interface affinity and mass transfer. DNA nanotechnology-based microfluidics, with their inherent capacity to regulate interface affinity and enhance mass transfer efficiency, has provided powerful platforms for efficient and specific isolation of circulating targets. DNA-based affinity interfaces can be mainly classified into two categories: aptamer interface and DNA framework-supported affinity interfaces.

4.2.1 Aptamer-Modified Micro/Nano-Substrate Microfluidic Chip

Aptamers, as single-stranded functional oligonucleotides, are evolved through an *in vitro* iterative process termed systematic evolution of ligands by exponential enrichment techniques (SELEX) [28]. They can form secondary and tertiary structures for specific recognition of targets with high affinity. Compared with other types of recognition ligands, aptamers have advantages such as small size, high affinity, good stability, and minimal batch variation. These attributes make them particularly advantageous in the field of liquid biopsy [29, 30]. In addition, aptamers exhibit the advantages of facile evolution and chemical synthesis, site-specific modification of functional moieties, and programmable structural design, facilitating the assembly of high-performance capture platforms. Since Tan's group first reported aptamer-based microfluidic platforms for tumor cell isolation in 2009, many methods have been reported in this field to improve the isolation efficiency of circulating targets [31, 32]. In addition, mounting studies indicated that nanostructured substrates integrated into microfluidic chips facilitate improved isolation efficiency [33, 34]. These substrates increase contact area for higher-density aptamer immobilization, thus enhancing affinity interactions. Moreover, they enhance local topographic interactions between the capture interface and nanoscale structures of cells such as cell microvilli [35, 36]. To date, a variety of aptamer-functionalized inorganic or organic micro/nano substrates have been engineered to isolate CTCs and EVs in microfluidic chips [37]. In addition, various multivalent aptamer-functionalized micro/nano-substrates have been cleverly designed to further enhance interface affinity through multivalent binding effects [10].

4.2.1.1 Inorganic Nanostructured "Hard Substrates"

Inorganic nanostructured substrates, also known as "hard substrates", can be fabricated by chemical synthesis, deposition, or etching and are widely used in microfluidics devices [38, 39]. Currently reported hard substrates include gold nanoparticle substrates, silicon or gallium nitride nanowire arrays, TiO_2 nanorods/nanofiber, and hydroxyapatite/chitosan nanofilms [37]. Sheng et al. developed a microfluidic device consisting of eight parallel channels with an array of over 59,000 isotropically etched micro-scale elliptical micropillars (Figure 4.1a) [40]. The aptamer-functionalized micropillar array enhanced the likelihood of interactions between tumor cells and the aptamers coated on the channel/pillar surfaces, yielding a capture efficiency of approximately 95% and a purity of around 81% with a relatively high throughput (>2 ml h^{-1}). Nano-substrates can exhibit enhanced local topographic interactions with cell surface nanoscale structures (such as microvilli), which are similar to the working principle of velcro [43]. Fang's group developed a NanoVelcro Chip that combined aptamer-coated silicon nanowire substrate (SiNWS) with overlaid PDMS chaotic mixer to achieve efficient capture of non-small cell lung cancer (NSCLC) CTCs from blood (>80% efficiency, Figure 4.1b) [41]. Building upon this foundation and considering the high degree of heterogeneity of CTCs in patients, Fang's group further functionalized the NanoVelcro Chip with rationally designed aptamer cocktails to capture heterogeneous CTCs in NSCLC patients (Figure 4.1c) [42]. This approach demonstrated that the overall composition of CTC subpopulations obtained by multiple combinations of aptamers could provide more comprehensive information for treatment monitoring.

4.2.1.2 Organic Nanostructured "Soft Substrates"

Organic nanostructured substrates, known as "soft substrates," which comprise hydrogels and organic polymers, can establish a soft contact interface. They can better match the soft nature of cells to maintain the viability of CTCs. Chitosan, poly(lactic-coglycolic acid) (PLGA), and other organic polymer materials have been used to construct aptamer-based nanointerfaces due to their excellent biological functions and easy chemical modification properties. Pei's group has developed a series of aptamer-based microfluidic chips with "soft substrate" to efficiently capture tumor cells. As shown in Figure 4.2a, they prepared a "soft substrate" on the surface of chitosan nanoparticles fabricated by electrospray [44]. Subsequent modification with aptamers as capture molecules enabled specific capture of viable rare CTCs from artificial white blood cell (WBC) samples [44]. In addition, the developed "soft substrates" facilitated *in situ* culture of the captured cells, enabling further purification and proliferation of the rare captured/target cells. Furthermore, their group developed a porous network nanostructure based on chitosan nanofibers to simulate the function of the extracellular matrix. This functional biointerface served to provide a multivalent aptamer interface and reduce nonspecific cell adhesion for efficient and specific CTC capture (Figure 4.2b) [45]. To achieve efficient isolation of different phenotypic CTCs, Pei's group developed a dual aptamer-modified PLGA nanofiber-based microfluidic to achieve efficient capture of epithelial and mesenchymal CTCs of ovarian cancer, with a capture efficiency of >88% for both types of CTCs (Figure 4.2c) [46]. The results indicated

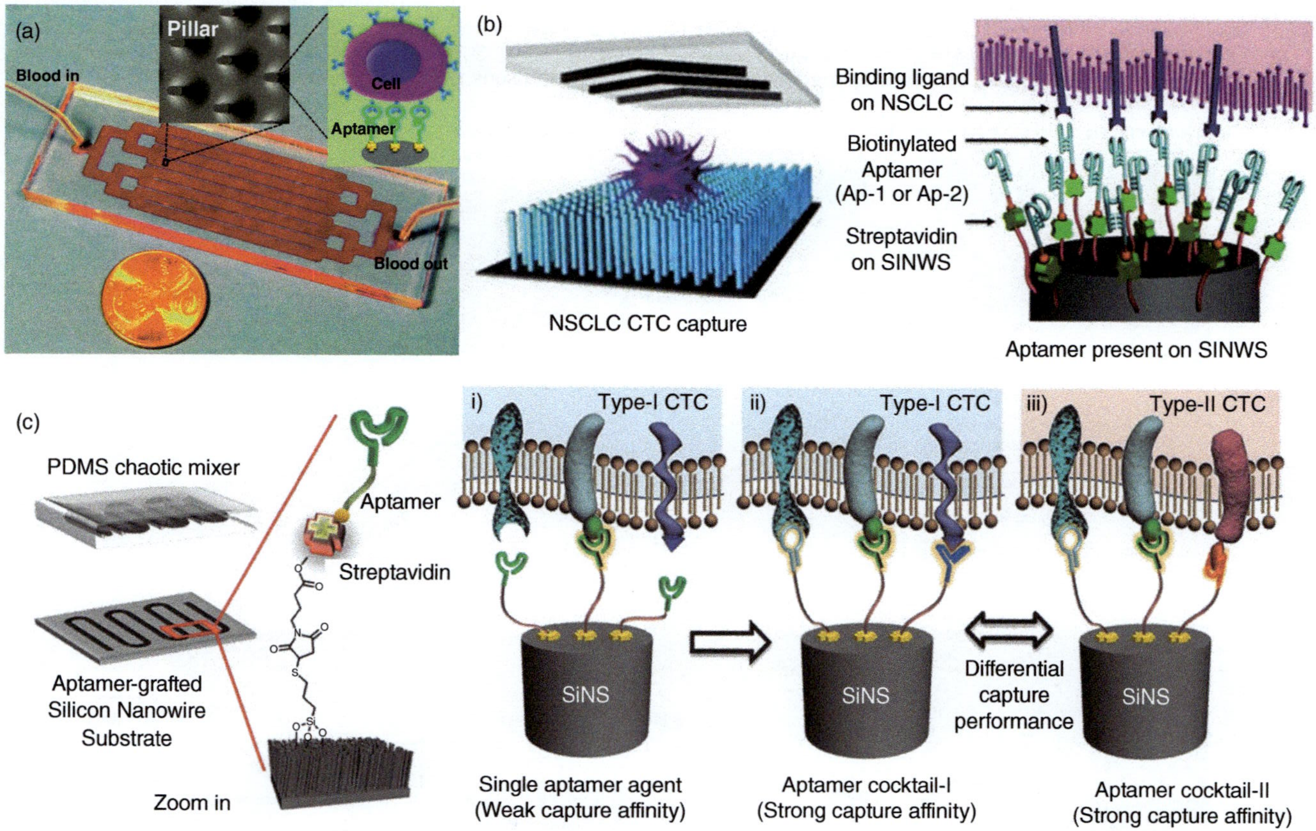

Figure 4.1 Aptamer-functionalized nanostructured "hard substrates." (a) Aptamer-functionalized micropillar array chip for tumor cell isolation. Source: Reproduced with permission from Shen et al. [40]. © 2012 American Chemical Society. (b) Aptamer-coated NanoVelcro Chip for isolating NSCLC CTCs. Source: Reproduced with permission from Shen et al. [41]. © 2013 John Wiley & Sons. (c) Aptamer cocktail-coated NanoVelcro Chip for enhanced and differential capture of CTCs. Source: Reproduced with permission from Zhao et al. [42]. © 2016 John Wiley & Sons.

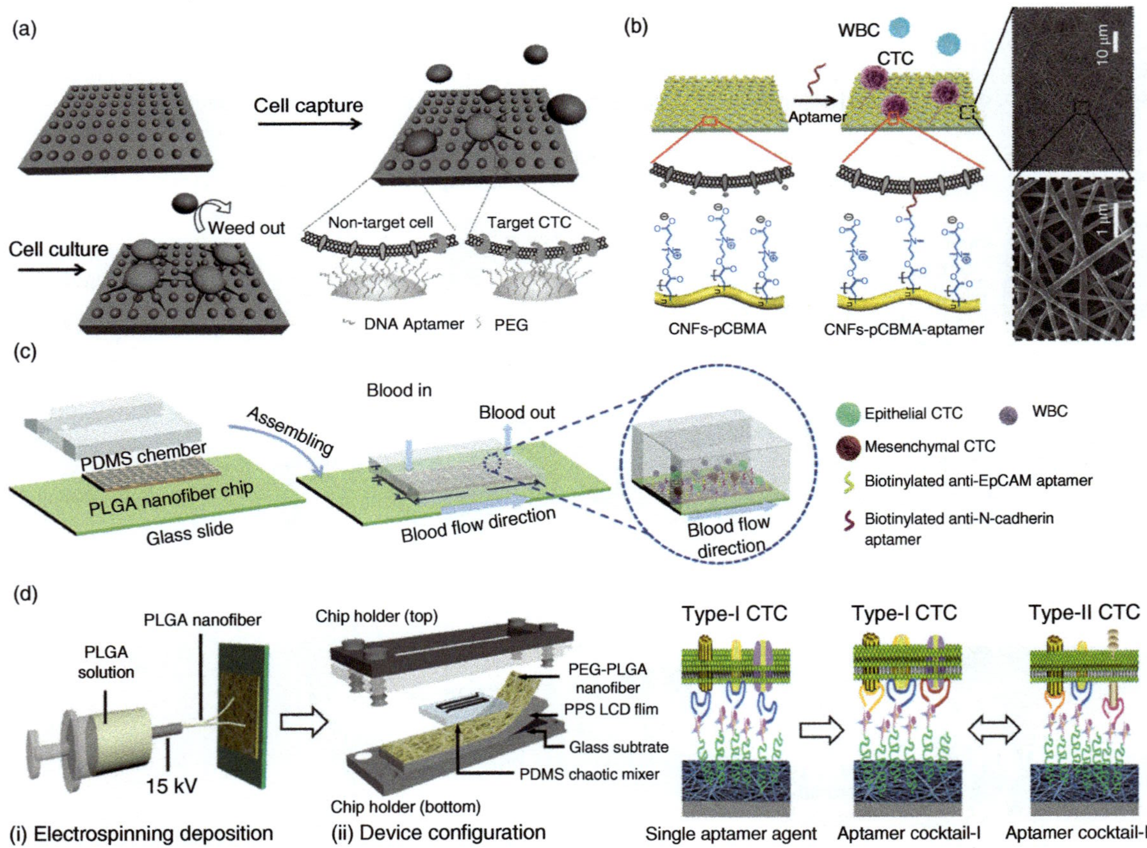

Figure 4.2 Aptamer-functionalized nanostructured "soft substrates." (a) The chitosan nanoparticle substrate for CTC isolation from nontarget cells. Source: Reproduced with permission from Sun et al. [44]. © 2015 John Wiley & Sons. (b) Chitosan nanofibers for specific capture of CTCs. Source: Reproduced with permission from Sun et al. [45]. © 2015 John Wiley & Sons. (c) The dual aptamer-modified PLGA nanofiber-based microfluidic device to capture CTCs of the different phenotypes. Source: Reproduced with permission from Wu et al. [46]. © 2013 Royal Society of Chemistry. (d) Aptamer-modified PPN microfluidic system to recognize rare subtypes of CTCs in lung cancer patients. Source: Reproduced with permission from Zhu et al. [47]. © 2020 American Chemical Society.

the potential for efficient capture of two phenotypic CTCs to guide the diagnosis, treatment, and prognosis of ovarian cancer patients. During the progression of tumors, CTCs undergo clonal evolution, leading to phenotype transformation. In order to dynamically monitor the clonal evolution of cancer, Zhu et al. developed an aptamer cocktail-modified PEG-PLGA-nanofiber (PPN) microfluidic system by depositing electrospun poly(ethylene glycol)-conjugated poly(lactic-coglycolic acid) (PEG-PLGA) nanofibers on glass slides (Figure 4.2d) [47]. This system exhibited excellent cell capture efficiency with high cell viability and could identify rare subtypes of CTCs in lung cancer patients and monitor the clonal evolution of solid tumors. However, the universal nanoscale interactions between cells and nanoscale surfaces can result in nonspecific cell adhesion on nanostructured substrates. In order to avoid nonspecific adsorption of WBCs or free proteins on the substrates, various antifouling molecules such as PEG, poly(methacrylic acid carboxyl betaine) (pCBMA), and bovine serum albumin (BSA) have been incorporated into the linker between aptamers and soft substrates or grafted onto the interface [33, 44, 45, 47]. These strategies significantly improved the purity of captured CTCs while minimizing interference with downstream analysis.

4.2.1.3 Multivalent Aptamer-Functionalized Micro/Nano-Substrates

Multivalent binding has been widely used to improve binding affinity by harnessing simultaneous interactions between multiple ligands on one entity and multiple receptors on another. The easily modifiable nature of aptamers allows for their dense assembly at the nanointerface for multivalent binding. Inspired by the self-adaptive skeleton and multiple suckers or tendrils of climbing plants, Yang's group reported a biomimetic multivalent aptamer nanoclimber (MANC)-functionalized microfluidic chip to detect minimal residual disease in the blood of patients with leukemia (Figure 4.3a) [48]. Through a free radical polymerization reaction, acrylamide co-polymerized with acrylate-poly T-sgc8 aptamer (targeting protein tyrosine kinase 7, PTK-7) and acrydite-poly T-TEG-biotin to form MANCs with flexible skeleton and multiple aptamers. The acrydite-poly T-TEG-biotin was introduced to anchor the MANC polymer chains on the chip substrate through streptavidin-biotin interaction. This device achieved a high capture efficiency of 92.2% for leukemia cells. Furthermore, Zhang et al. designed Y-shaped aptamer nanostructures with stronger recognition ability and stability, which were grafted onto a linear polymer chain through polymerization reactions. The polymer chain featured multiple divalent aptamer side chains to provide multi-site interactions, thereby resulting in a remarkable affinity enhancement (Figure 4.3b) [49]. Theoretically, this multivalent polymerization strategy held the potential to simultaneously assemble multiple types of aptamers, but the synthesis and purification process were relatively complex.

Nanoparticles can be modified with high-density DNA strands, termed spherical nucleic acids (SNAs) [52]. Sheng et al. developed a microfluidic device whose substrates were coated with aptamer nanospheres, and AuNPs served as scaffolds to assemble aptamer SNAs (Figure 4.3c) [50]. The binding affinity of AuNP-aptamers

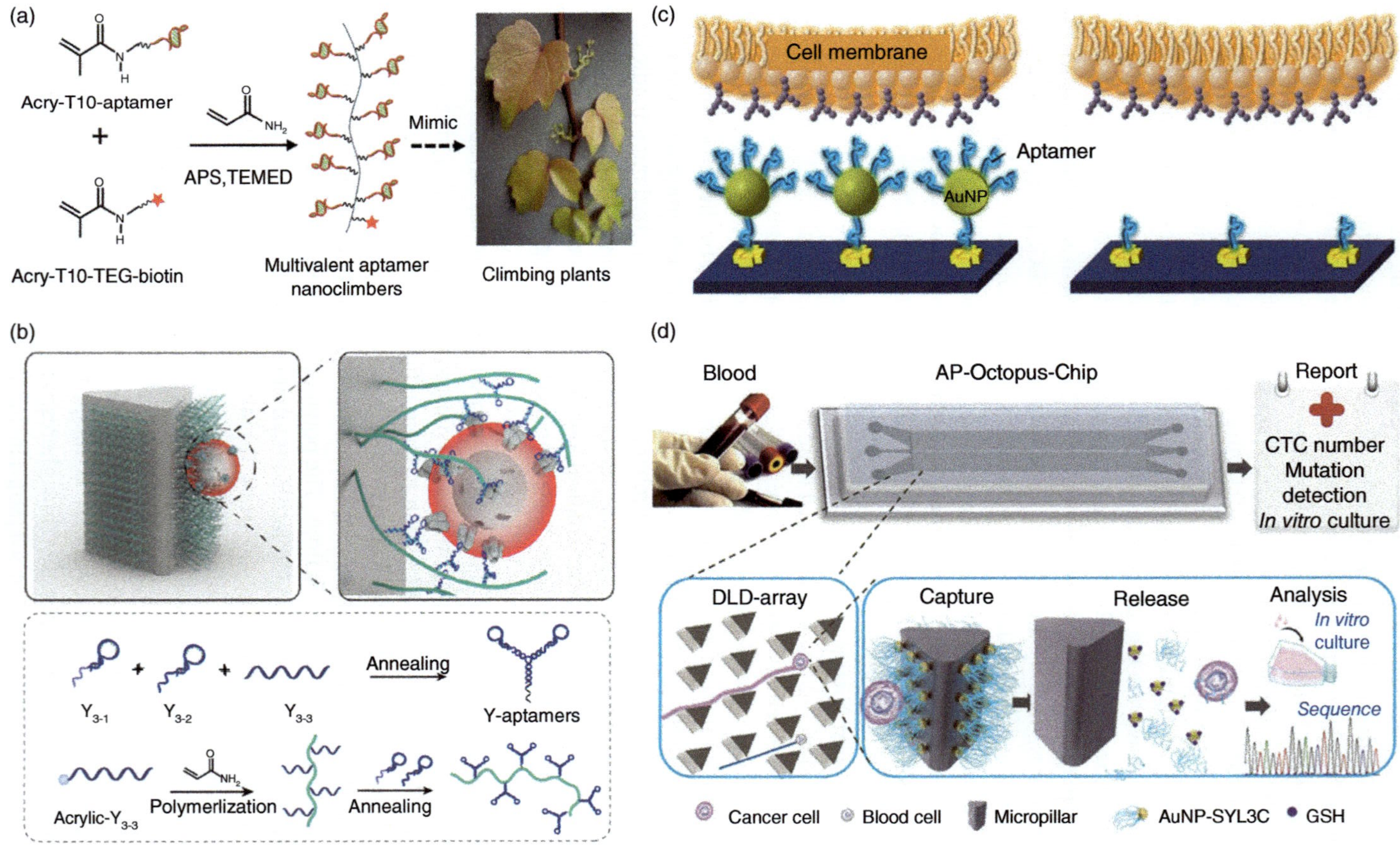

Figure 4.3 Multivalent aptamer-functionalized micro/nano-substrates. (a) Biomimetic multivalent aptamer nanoclimber constructed by free radical polymerization. Source: Reproduced with permission from Liu et al. [48]. © 2020 John Wiley & Sons. (b) Aptamer-based hierarchically multivalent antibodies mimic interfaced microfluidic chips for CTC isolation. Source: Reproduced with permission from Zhang et al. [49]. © 2021 American Chemical Society. (c) Multivalent DNA nanospheres to enhance the capture efficiency of CTCs in microfluidic devices. Source: Reproduced with permission from Sheng et al. [50]. © 2013 American Chemical Society. (d) Micropillar arrays in a DLD model and engineered with Apt-AuNPs substrate. Source: Reproduced with permission from Song et al. [51]. © 2020 John Wiley & Sons.

(up to 95 aptamers attached to each AuNP) was 39-fold higher than that of free aptamers, and the capture efficiency of tumor cells increased from 49% to 93%. Furthermore, Song et al. combined AuNP-aptamers with a deterministic lateral displacement (DLD)-patterned microfluidics chip to develop an aptamer-tailed octopus chip (AP-Octopus-Chip) [51]. The DLD chip was arranged with rotated triangular micropillars, with each row of micropillars having a fixed shift (Figure 4.3d). CTCs larger than the critical diameter (D_c = 13 μm) would continuously and frequently collide with the micropillars, while blood cells smaller than CTCs would migrate along the original flow direction, thereby minimizing the interaction with the micropillar array [53]. Therefore, this geometric microfluidic chip not only improved the capture efficiency of CTC by over 300% but also reduced the adsorption of background cells.

In addition, the utilization of biomimetic nanomaterials or interfaces has brought new opportunities for the capture and detection of circulating targets [54]. Xie's group developed a biomimetic microfluidic system for CTC isolation (Figure 4.4a) [55]. Aptamer-functionalized and leukocyte membrane-camouflaged magnetic nanoclusters were utilized to assemble the capture microarray with the help of a magnet. Capitalizing on the enhanced collision of magnetic arrays and the anti-adsorption attributes of leukocyte membranes, over 90% of rare tumor cells could be captured from whole blood with minimal nonspecific adsorption of leukocytes. The inherent lateral fluidity of the natural biomembrane interface enables multivalent ligand-receptor interactions to greatly enhance binding affinity for various cell activities. Inspired by this, Wu et al. engineered a DLD microfluidic chip with a fluidic multivalent nanointerface, which was decorated with aptamer-functionalized leukocyte membrane nanovesicles in the substrate and triangular micropillars (Figure 4.4b) [56]. The resultant fluidity-enhanced multivalent binding improved the interface affinity by 4 orders of magnitude, resulting in a sevenfold increase in capture efficiency. The employment of fluidity-enhanced multivalent binding strategies in the isolation of tumor EVs has also yielded promising results. Niu et al. developed an aptamer-functionalized fluid nanoporous microinterface (FluidporeFace) and decorated it with supported lipid bilayers (SLBs) on nanoporous herringbone microstructures (Figure 4.4c) [57]. At the microscale level, the herringbone micropattern promoted the mass transfer of tumor-derived EVs (T-EVs) to capture the surface. At the nanoscale level, nanoporousity could overcome boundary effects for close contact between T-EVs and the interface. At the molecular level, fluid SLBs enabled recognition molecules to be clustered at binding sites, fostering multivalent binding with an ~83-fold increase in affinity compared with nonfluid interfaces. Through synergistic enhancement of mass transfer, interface contact, and binding affinity, FluidporeFace has achieved a 1.4- to 1.6-fold increase in the capture efficiency of T-EVs and 1-2 orders of magnitude enhancement in the detection sensitivity, with a detection limit (LOD) of 10 T-EVs μl^{-1}. In summary, multivalent aptamer-functionalized micro/nano substrates in microfluidics significantly improve binding affinity through multivalent binding effects, enabling efficient isolation of circulating targets in liquid biopsy.

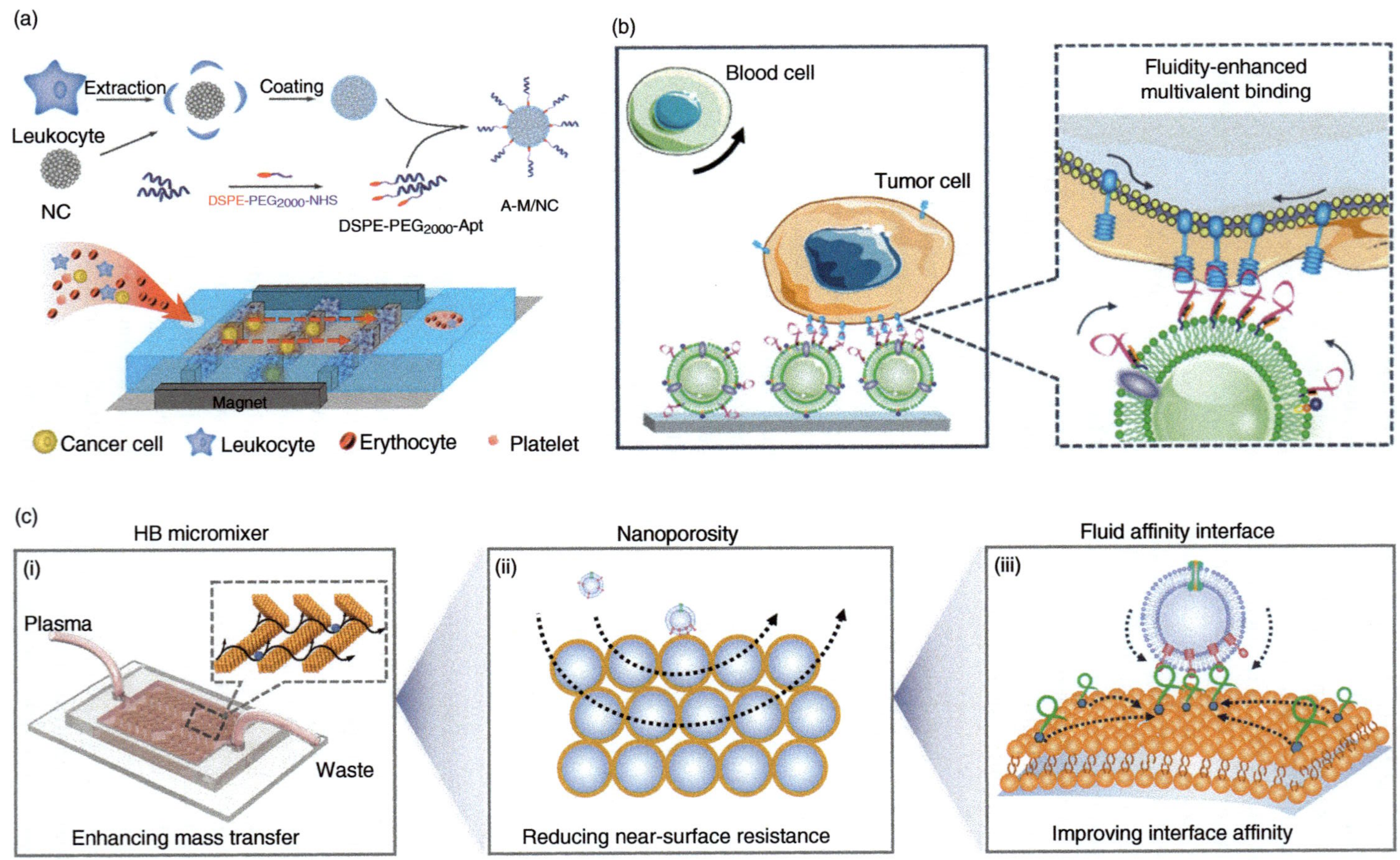

Figure 4.4 Aptamer-functionalized biomembrane interfaces. (a) Leukocyte membrane camouflaged and aptamer-functionalized biomimetic microfluidic system for specific and efficient capture of CTCs. Source: Reproduced with permission from Zhang et al. [55]. © 2019 American Chemical Society. (b) Fluid multivalent membrane nanointerface for efficient, high-selective, and gentle capture of CTCs. Source: Reproduced with permission from Wu et al. [56]. © 2020 American Chemical Society. (c) A fluid nanoporous microinterface for efficient isolation of T-EVs. Source: Reproduced with permission from Niu et al. [57]. © 2022 National Academy of Science.

4.2.2 DNA Framework-Supported Affinity Substrate Microfluidic Chip

In addition to two-dimensional chain structures, pioneering DNA nanotechnology enables the creation of various three-dimensional DNA-based nanostructures through precise and programmable nucleotide base pairing. For example, DNA framework structures of various shapes and sizes can serve as discrete scaffolds for site-specific chemical/biochemical functionalization of small molecules, proteins, and nanoparticles either at the edges or in the cavity. As a result, DNA nanostructures can serve as scaffolds for the attachment of recognition molecules (aptamers, antibodies, etc.), enabling the directional arrangement of recognition molecules. With this strategy, the loss of recognition ability caused by disordered orientation or aggregation can be reduced, leading to enhanced capture efficiency or detection sensitivity of the recognition interface. In addition, the reduction of assembly defects on the DNA frame structure assembly interface and the rigid spatial structure also reduces nonspecific adsorption of the interference.

4.2.2.1 DNA Framework-Supported Aptamer Substrates

Aptamer structures can be affected by interstrand entanglement and nonspecific interactions between the aptamer and the substrate. Therefore, aptamers were prone to orientation disorder at the microfluidic interface, which greatly restricted the interaction between the aptamers and the target antigens. To circumvent this problem, Zhang et al. developed a highly ordered recognition microfluidic interface, in which tetrahedral DNA nanostructures (TDNs), containing a pendant aptamer at the top vertex, were grafted onto the chip substrate as recognition elements (Figure 4.5a) [58]. The precise nano-scale and rigid TDN framework ensured aptamers with a highly ordered upright orientation, thereby augmenting capture efficiency by nearly 60%. Two types of EpCAM (epithelial cell adhesion molecule) aptamers, SJ3C2 and SYL3C, have the binding merits of high binding affinity and fast binding rate toward EpCAM proteins, respectively. By simultaneously assembling them into a TDN, Ling et al. developed a merit-complementary-heteromultivalent aptamers-functionalized microfluidic interface (Figure 4.5b) [59]. This strategy regulated the thermodynamics and kinetics of interface binding reactions. As a result, it enabled tight and rapid recognition of EpCAM, leading to high efficiency and throughput in capturing target cells. The rigid tetrahedral DNA scaffold with controllable size can precisely control the distance from the substrate, facilitating the creation of three-dimensional reaction space at the nanoscale by raising aptamers away from the surface. Therefore, this strategy significantly improves the reaction kinetics at the interface. Based on this, Zhu et al. constructed a three-dimensional nucleic acid nanoscale reaction interface based on a tetrahedral framework in microfluidics channels (Figure 4.5c) [60]. The highly ordered orientation of the surface aptamers and the nanoscale stereo reaction space significantly improved the reaction kinetics during molecular recognition, and the platform was applied to the "one-stop" assay of *Escherichia coli* O157: H7.

In addition to tetrahedral structures, other DNA frameworks, such as cubic or triangular prism DNA nanostructures, have been used to assemble multiple

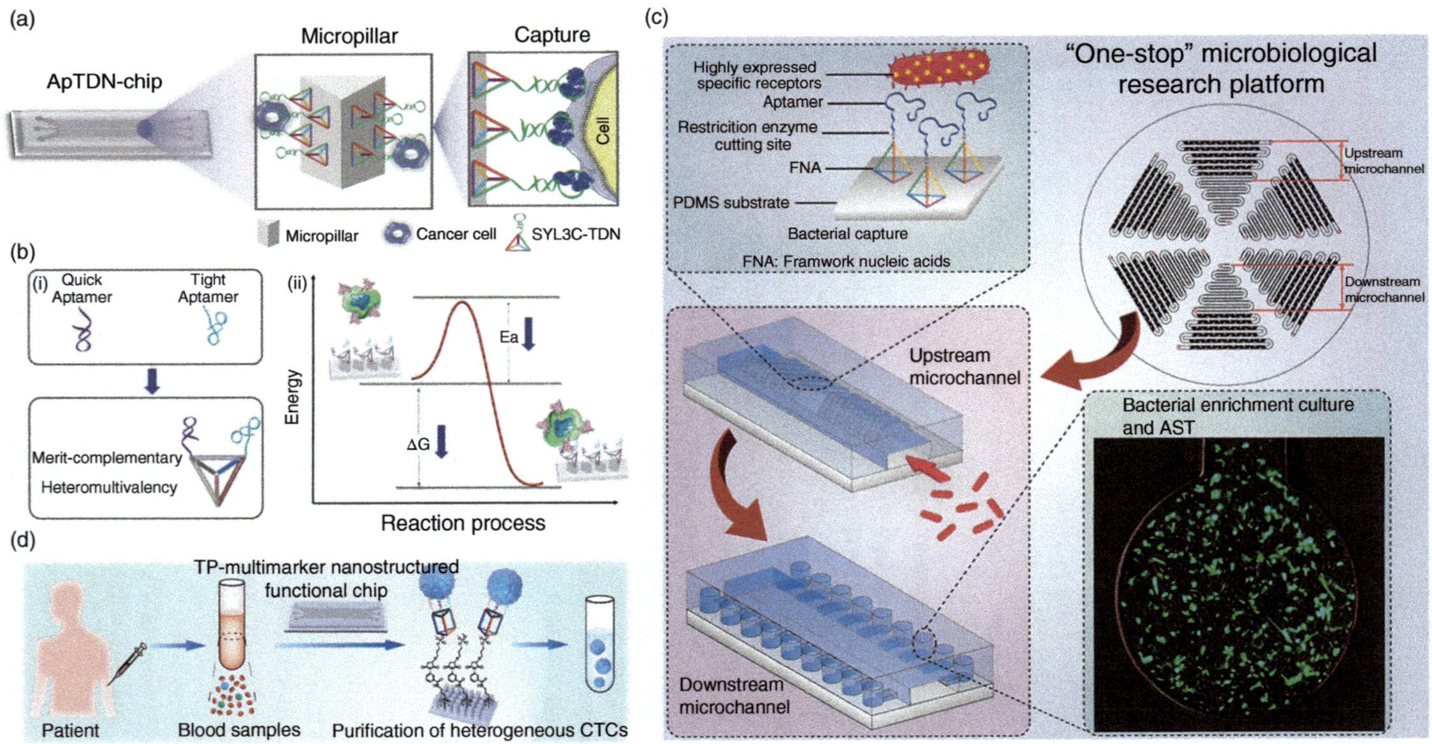

Figure 4.5 DNA framework-supported aptamer substrates. (a) Engineered aptamer-TDN as a nano-scaffold anchored on the microfluidic interface for efficient capture of CTCs. Source: Reproduced with permission from Zhang et al. [58]. © 2020 John Wiley & Sons. (b) A merit-complementary-heteromultivalent aptamer-functionalized TDN interface for high efficiency and throughput target cell capture. Source: Reproduced with permission from Ling et al. [59]. © 2022 American Chemical Society. (c) Tetrahedral framework in an integrated microfluidic platform for "one-stop" assay of *E. coli* O157: H7. Source: Reproduced with permission from Zhu et al. [60]. © 2021 Elsevier. (d) Heterovalent DNA framework recognition element-functionalized microfluidics for the purification of heterogeneous CTCs. Source: Reproduced with permission from Liu et al. [61]. © 2022 American Chemical Society.

aptamers to improve the affinity of microfluidic interfaces. These frameworks not only serve as rigid scaffolds for preparing highly ordered microfluidic nanostructures but also provide more recognition of molecular anchor points. Thus, it enables the synergistic effect of ordered orientation and multivalent binding. For example, Peng et al. reported a cubic DNA nanostructure (CDN) programming strategy to precisely control the orientation and valency of aptamers on the microfluidics interface (CDN-Apt-Chip) for enhanced capture of CTCs [62]. Due to the heterogeneous expression of protein biomarkers of CTCs, a single type of aptamer may not be able to capture CTCs with different phenotypes. Many methods based on the cocktail of recognition ligands can be employed to cooperatively achieve efficient capture of heterogeneous CTCs. Liu et al. utilized multimarker aptamer cocktail DNA nanostructures to functionalize microfluidic chips for the enrichment of heterogeneous CTCs (Figure 4.5d) [61]. Triangular prism (TP) DNA nanostructures were constructed as scaffolds, and each anchor point of their upper surfaces was anchored with three different types of aptamers targeting EpCAM, Her-2, and EGFR, respectively. As a result, the capture efficiency toward cell lines with both high and low expression of EpCAM was improved. The anisotropic recognition interfaces provided an effective strategy to enhance the interaction between heterogeneous circulating targets and ligands. In conclusion, the modular design platform based on DNA nanostructures allows for specific recognition of different cell phenotypes of CTC by changing the size of the scaffold unit and the type of aptamers, thus providing a deeper understanding of the mechanism of tumor progression.

4.2.2.2 DNA Framework-Supported Antibody Substrates

In traditional immunological assays, the sensitivity is often limited by the low efficiency of heterogeneous immune responses. This limitation is partly due to the uneven distribution and inconsistent spatial orientation of capture antibodies during the assembly process. These factors can lead to a decrease in the number of interface antibodies and even inactivation, which is unfavorable for rapid and highly sensitive detection of proteins. Antibodies can be directly or indirectly coupled to the TDNs through easy chemical modification. This enables the generation of nanostructured surfaces with good control over the spacing and direction of capture antibodies. As a result, the accessibility of target molecule recognition is enhanced, the binding energy barrier is reduced, and high sensitivity and specificity can be achieved. For example, tumor necrosis factor-α (TNF-α) antibodies could be easily and reversibly fixed to the surface of DNA nanostructures using DNA linking chains. This allowed for the conversion of the DNA recognition layer into a protein recognition layer, which significantly improved the sensitivity of target protein detection with LODs as low as 100 pg ml^{-1} (Figure 4.6a) [63]. Subsequently, Yuan et al. proposed a covalent coupling method to attach anti-immunoglobulin G (IgG) antibodies at the top vertex of TDN scaffolds. This direct and DNA-bridge-free antibody immobilization approach enabled ultrasensitive detection of IgG (LOD: 2.8 pg ml^{-1}, Figure 4.6b) [64]. It is well established that the combination of nanostructured substrates and affinity recognition elements can enhance the capture

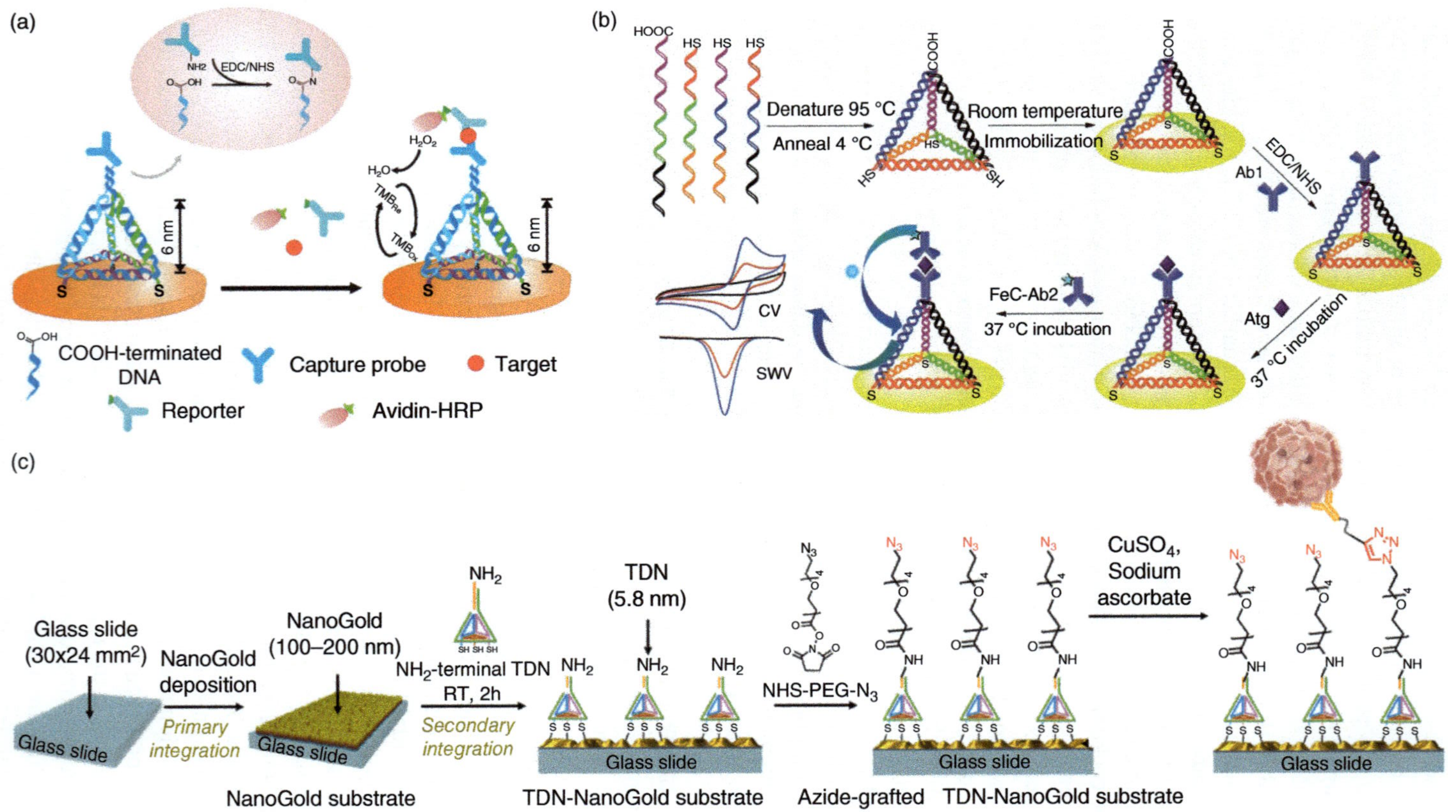

Figure 4.6 DNA framework-supported antibody substrate. (a) Antibodies-anchored-DNA framework for ultrasensitive electrochemical detection of PSA. Source: Reproduced with permission from Pei et al. [63]. © 2011 Royal Society of Chemistry. (b) Sandwich immunoassay of IgG using the DNA nanopyramids-modified Au electrode. Source: Reproduced with permission from Yuan et al. [64]. © 2014 Springer Nature. (c) A microfluidics chip with TDN-nanogold substrates for enhancing cell-capture performance. Source: Reproduced with permission from Sun et al. [65]. © 2023 Elsevier.

performance of the substrate. Using this strategy, Sun et al. proposed a top-down approach to manufacture antibody-modified TDN-nanogold substrates in microfluidics chips (Figure 4.6c) [65]. The resultant chips allowed for effective isolation and purification of CTCs of neck squamous cell carcinoma (HNSCC) with intact cell morphologies and molecular integrity. In summary, DNA framework-supported antibody substrates greatly increase the effective collision of target molecules, enabling highly sensitive detection of circulation targets. In addition, different recognition molecules (DNA, antibodies, nucleic acid aptamers, etc.) can be modified on DNA frameworks to develop a universal detection platform for liquid biopsy.

4.3 DNA Nanotechnology-Based Microfluidics for Release and Detection of Circulating Targets

4.3.1 Efficient Release

Genomics, transcriptomics, and proteomics analysis of circulating targets can provide comprehensive genetic and phenotypic information about derived tumor tissues. Therefore, there is an urgent need to develop efficient, specific, and gentle release strategies to release captured targets from the capture interface for downstream analysis. Compared with other affinity interfaces, DNA nanotechnology-based affinity interfaces can be controllably disassembled. In addition, the binding reaction between circulating targets and DNA nanotechnology-based microfluidic interfaces can be disrupted for target release using external stimuli, such as nuclease, thermal denaturation, and other stimuli. This part mainly introduces the release strategy of DNA nanointerfaces in microfluidic chips using aptamers as recognition ligands. There are generally two main types: aptamer degeneration and aptamer detachment from substrates.

4.3.1.1 Aptamer Conformation Alteration

The recognition capability and affinity of aptamers rely on their folded three-dimensional conformations. The conformation of the aptamers can be conveniently disrupted through nuclease degradation, hybridization inactivation, and thermal denaturation, which induces the loss of aptamer binding capability, thereby releasing the captured circulating targets. Aptamers, as functional nucleic acids, can be degraded by nucleases. For example, in NanoVelcro Chip, aptamers modified on SiNWS surface could be cleaved by benzonase nuclease, resulting in the specific release of CTCs (Figure 4.7a) [41]. This approach enabled the release of more than 85% of tumor cells with a viability of 78–83%, while less than 5% of adsorbed WBCs were non-specifically released. In addition, the DNA nanopolymers and DNA framework structures as scaffolds make DNA nuclease more accessible to aptamers and lead to an efficient release. This type of release strategy is commonly conducted under mild physiological conditions, which can maintain cell activity and integrity, making it compatible with downstream analysis. For example, using

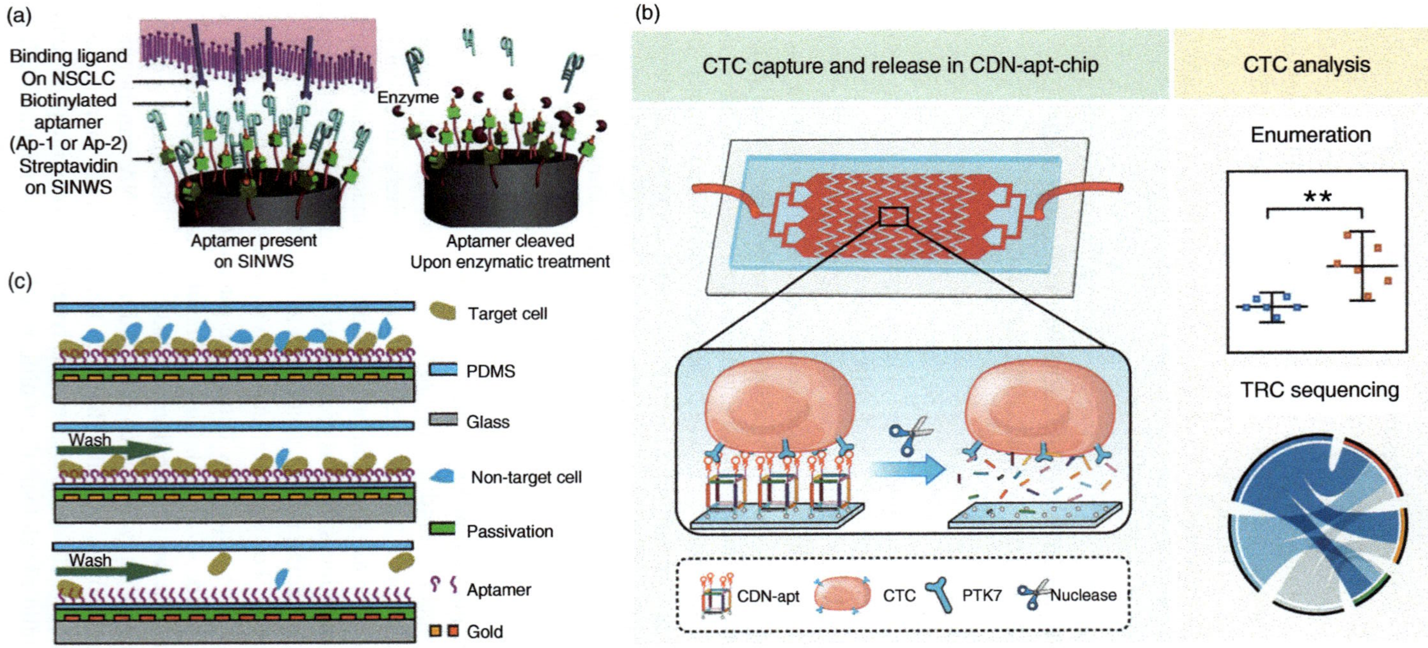

Figure 4.7 Aptamer denaturation for CTC release. (a) Nuclease degradation of aptamers for releasing NSCLC cells. Source: Reproduced with permission from Shen et al. [41]. © 2013 John Wiley & Sons. (b) Nuclease-mediated nondestructive release of CTCs for TCR sequencing. Source: Reproduced with permission from Peng et al. [62]. © 2022 American Chemical Society. (c) Temperature-mediated release of tumor cells. Source: Reproduced with permission from Zhu et al. [66]. © 2012 Royal Society of Chemistry.

CDNs as scaffolders, Zhu's group constructed aptamer interfaces (CDN–Apt) with ordered orientation and polyvalent states in microfluidic chips (Figure 4.7b) [62]. After nuclease degradation of CDN-APT, approximately 97.8% of captured cells were released, while maintaining a high cell viability of around 91.6%. Using this technique, they successfully isolated CTCs from the peripheral blood of T-cell leukemia patients, and the released cells were subjected to nuclease treatment for T-cell receptor sequencing, demonstrating their compatibility with downstream assays.

The folded conformations of aptamers are temperature-sensitive, enabling the release of captured targets in a temperature-controlled manner. Lin's group developed a microfluidic-based approach for temperature-mediated cell release (Figure 4.7c) [66]. After specific cell capture, the bottom temperature control device provided moderate elevated temperature (from 30 to 48 °C for two minutes) to reversibly disrupt the cell-aptamer interaction, enabling non-destructive, temperature-mediated cell release and recovery. In addition, the aptamer-functionalized surfaces were successfully regenerated and shown to be reusable. An increase in temperature will accelerate the conformational change of the aptamers, but higher temperature poses a greater risk of cell damage, so the trade-off between them exists. Alternatively, hybridization reaction between antisense oligonucleotides and aptamers provides a facile and mild strategy to block the binding of aptamer and attenuate CTCs binding on the interface, so that intact target cells can be nondestructively released from the substrate [11].

4.3.1.2 Detaching Aptamers from Capture Interface

Aptamers can be immobilized on substrates by different modification strategies. Disrupting the aptamer-substrate interactions via physical or chemical stimuli represents another type of release strategy. As for aptamers attached to substrates through hybridization reactions, competitive complementary sequences can also be applied to trigger chain displacement to release aptamers and captured targets from substrates. Zhou's group developed a programmable DNA-responsive microchip to capture and release CTCs by nucleic acid hybridization (Figure 4.8a) [67]. Upon the dissociation of aptamers using complementary sequences, ~86% of captured cells were released with a viability of around 81%. Functionalizing gold nanosubstrates with aptamers via Au-S bonds is a common interface modification method. The Au-S bond can be disrupted by electrochemical reduction or competitive ligand exchanges, providing excellent capture-and-release interfaces. For example, in the AP-Octopus Chip, the AuNP-aptamers-functionalized nanointerface was disrupted by the competitive binding of glutathione (GSH) to AuNPs, resulting in the release of CTCs with exceeding 80% efficiency and 96% viability (Figure 4.8b) [51]. The light-controlled release strategy can achieve precise spatiotemporal release, even at the level of single cells, by light-induced chemical bond changes or substrate disassembly. In an aptamer-modified PEG-PLGA-nanofiber (PPN) microfluidic system, a laser capture microscope (LCM) was performed on PEG-PLGA-nanofiber to achieve controlled single-CTC release (Figure 4.8c) [47]. Aptamer-based microfluidic technology offers controllable and gentle release of targets, which lay a solid

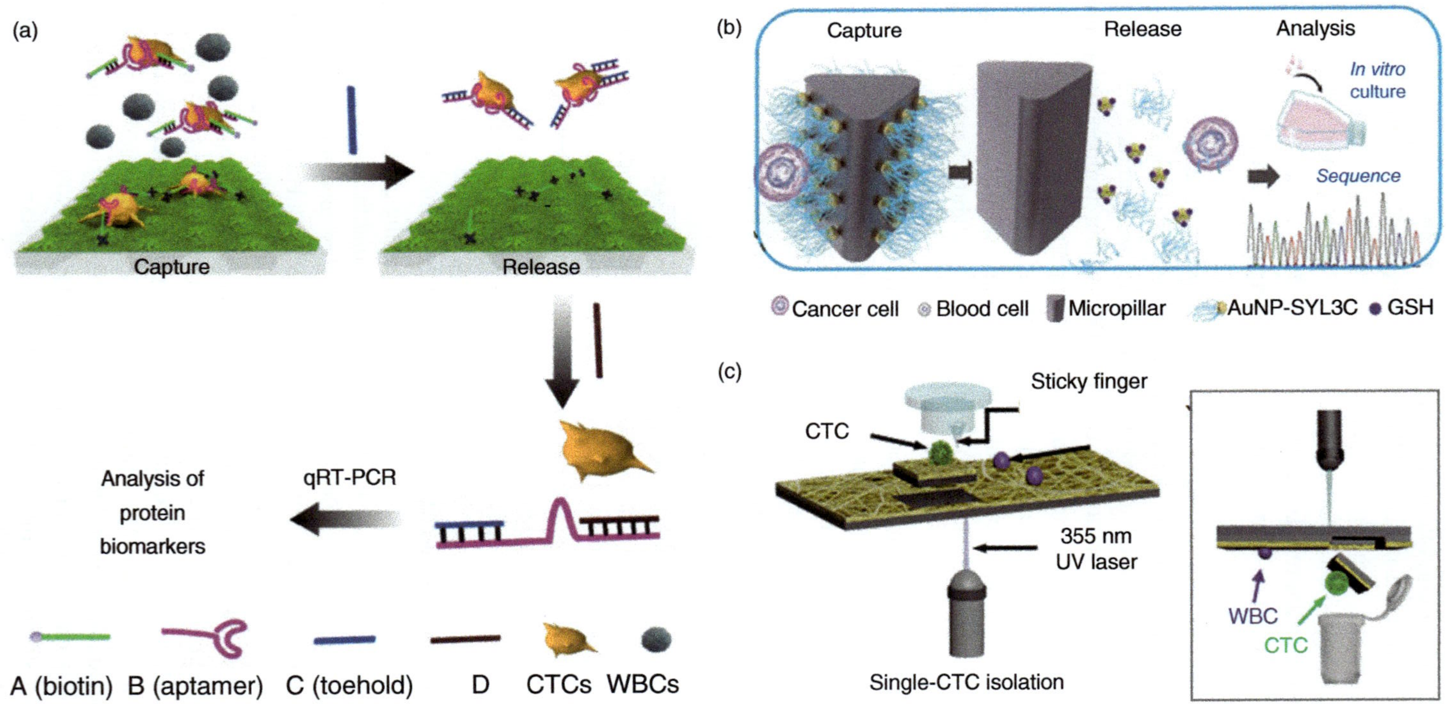

Figure 4.8 CTC release by detaching aptamers from capture interface. (a) Complementary sequence-activated aptamer detachment for CTC release. Source: Reproduced with permission from Guo et al. [67]. © 2018 Springer Nature. (b) CTC release by detaching aptamers from AuNPs via the competitive exchange of GSH. Source: Reproduced with permission from Song et al. [50]. © 2019 John Wiley & Sons. (c) Controllable release of single CTCs using LCM technique. Source: Reproduced with permission from Zhu et al. [47]. © 2020 American Chemical Society.

foundation for downstream analysis to provide more comprehensive biological information.

4.3.2 Analysis and Destruction of CTCs

4.3.2.1 CTC Detection and In-depth Analysis

The detection and enumeration of CTCs can be used for clinical diagnosis, prognosis, and therapy assessment. Following their capture on chips, some aptamer probes have been utilized to specifically label tumor cells, enabling the conversion of cell numbers into measurable signals. Liu et al. developed an aptamer-based "sandwich" microfluidic platform for the capture and detection of rare cells based on chemiluminescence (CL) analysis (Figure 4.9a) [68]. The aptamer-modified AuNPs were introduced to label the captured cells and catalyzed the decomposition of luminol to 3-aminophthalate in hydrogen peroxide solution, triggering a short and strong CL reaction at 425 nm. A satisfactory linearity of the calibration curve between the CL intensity and the number of target cells was observed with a LOD of 30 target cells in a 3 μl cell mixture. Yang's group designed a volumetric bar chart chip (V-Chip) that utilized aptamer-conjugated Pt nanoparticles (ACNPs) to label CTCs (Figure 4.9b) [69]. The production of oxygen in the catalytic reaction between ACNPs and H_2O_2 facilitated the movement of the ink along a vertical channel in the chamber, allowing for the visualization and quantification of CTC numbers. Thus, a linear correlation was observed between the distance moved by the ink and the number of cells in the sample, with LODs as low as a single cell.

CTCs have inherent heterogeneity in both genotype and phenotype. Compared with antibodies (12–15 nm in diameter), aptamers with smaller sizes (2–3 nm in diameter) allow for more accurate quantification of cell surface markers for the identification of CTC subpopulations. For phenotypic analysis of CTCs, Kelly's group has reported several aptamer-mediated microfluidics for CTC spatial sorting and profiling. CTCs with different phenotypes can be labeled with different numbers of aptamer-functionalized magnetic nanoparticles, resulting in different magnetic susceptibilities. Thus, CTC subpopulations can be spatially sorted by balancing magnetic force and fluid drag force. According to the expression levels of two different markers, EpCAM and HER2, CTCs were classified into 16 subgroups by a two-dimensional sorting approach (Figure 4.9c) [70]. The constant evolution of tumor cells with different combinations of genetic mutations makes cancer treatment difficult and ineffective. Reinholt et al. designed a two-intersecting orthogonal microchannel device, which enables the capture of cancer cells in the first aptamer-functionalized orthogonal microchannel and captures genomic DNA (gDNA) of lysis CTCs in the second aptamer-functionalized orthogonal microchannel (Figure 4.9d) [71]. Thus, this chip allowed for efficient retention and purification of gDNA in microchannels, enabling the amplification of specific genes for off-chip sequencing. The point mutation of the TP53 gene was successfully detected in CAOV-3 ovarian cancer cells. This approach offered a way to select cells of interest and monitor multiple genetic mutations. Overall, in-depth analysis of CTCs, including genomic, proteomic, and functional profiling, can fully reveal the

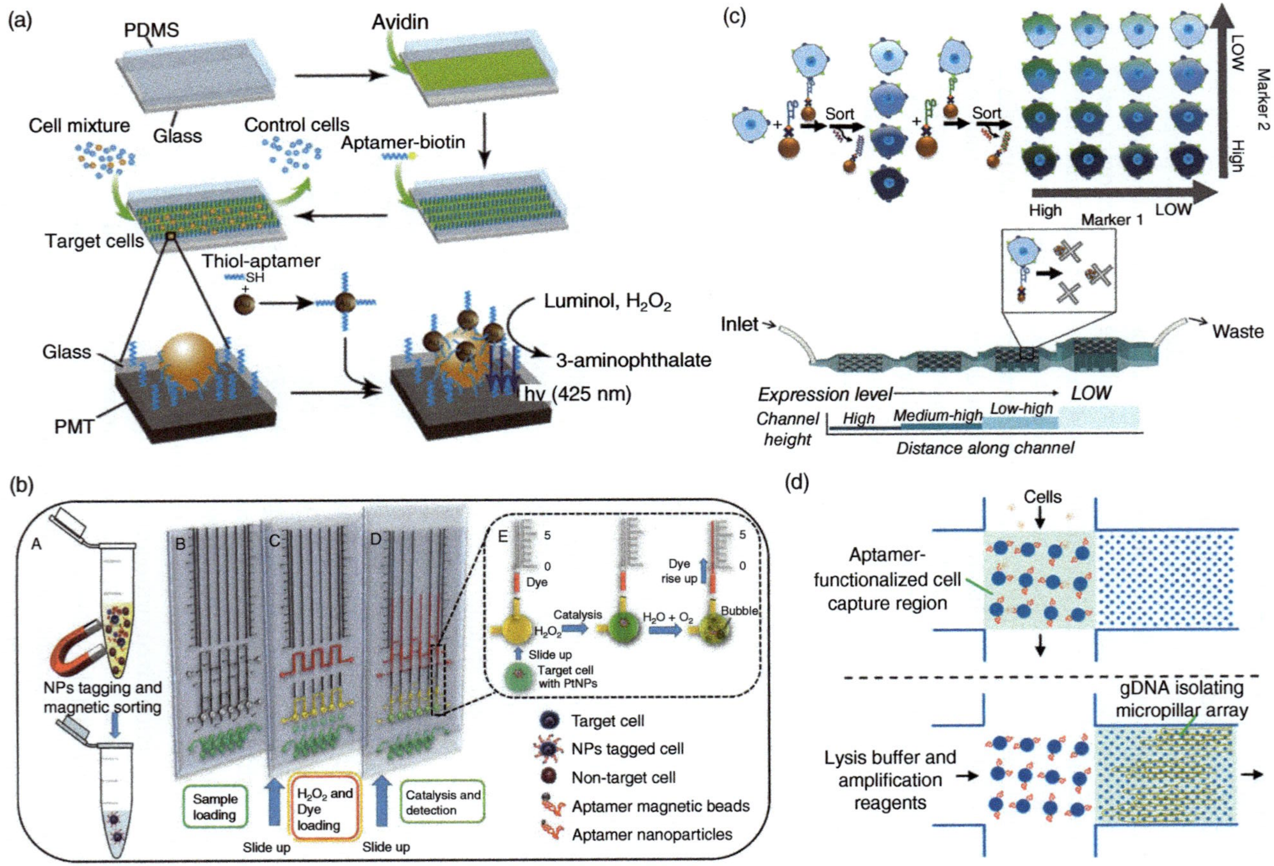

Figure 4.9 DNA nanotechnology-based microfluidics for CTC detection and in-depth analysis. (a) CTC detection on an aptamer-modified microfluidics with chemiluminescence analysis. Source: Reproduced with permission from Liu et al. [68]. © 2011 Elsevier. (b) Volumetric bar chart chip for visual quantitative detection of CTCs using aptamer-conjugated nanoparticles. Source: Reproduced with permission from Abate et al. [69]. © 2019 John Wiley & Sons. (c) Aptamer-functionalized magnetic nanoparticles and antisense-mediated two-dimensional microdevice for CTC subpopulation sorting according to the expression level of two surface markers. Source: Reproduced with permission from Labib et al. [70]. © 2016 American Chemical Society. (d) Microdevice consisting of two intersecting orthogonal micropillar arrayed microchannels for CTC capture and genetic mutation detection. Source: Reproduced with permission from Reinholt and Craighead [71]. © 2018 American Chemical Society.

molecular mechanisms of tumor heterogeneity, metastasis, and drug resistance, facilitating personalized therapy for patients.

4.3.2.2 CTC Destruction for Therapy

CTCs, shed from solid tumor tissues, are closely related to the malignant progression of cancers. Unfortunately, the existing drug delivery systems, whether systemic or localized, are inadequate for effectively killing CTCs. In general, injecting anticancer drugs through the systemic drug delivery system may cause unavoidable damage to normal tissue and make it difficult to effectively destroy CTCs. The localized drug delivery system cannot kill CTCs in the circulatory system. To address the above problem, Wang's group designed a CTC-triggered localized drug release strategy that combined photodynamic therapy for targeted CTC destruction (Figure 4.10a) [72]. In a two-part microfluidic device, the photosensitizer-labeled hairpin switch adaptor (HSA) probes could bind to CTCs from the peripheral blood, thereby tagging them as photosensitizer-labeled aptamers from Part I. Subsequently, when the CTCs flowed through transparent capillary microchannel (Part II), the photosensitizer was activated by light, generating $^1O_2$ and leading to cell apoptosis. In order to increase drug loading capacity, Wang's group further developed a "sense-and-treat" localized drug delivery system based on an HSA-anchored DNA device. DNA nanodevices can promote the internalization of anticancer drugs, increase drug loading capacity, and allow for synergistic chemotherapy and photodynamic therapy, thereby enhancing the killing ability of CTCs (Figure 4.10b) [73].

Different from the above strategy of directly killing CTCs, Zhang's group proposed a liquid biopsy-guided drug release system with integrated functions of cancer diagnosis and therapy. The system could control the amount of drug released according to the number of detected CTCs, offering personalized therapy. A recognition area (Area I) and a drug release area (Area II) were integrated into a magnetically controlled microfluidic chip. When the CTCs bound to the recognition region, the complementary chain (cDNAs) hybridized with the aptamer was released, triggering the release of anti-cancer drugs in the downstream region (Figure 4.10c) [74]. On this basis, they further constructed a three-dimensional magnetic-controlled chip (3DMC-system)-based extracorporeal circulation system for in vivo detection and real-time monitoring of CTCs (Figure 4.10d) [75]. By using the signal conversion strategy, this system could integrate detection and treatment modules to achieve CTC-response tumor treatment. Compared with in vitro detection, in vivo detection enabled the detection of more CTCs. Moreover, it allowed for the discovery of CTCs at the very early stage of tumors, even before tumor metastasis was observed in imaging. Therefore, this system enabled timely triggering of drug release to achieve more effective therapeutic effects. These CTC-based integrated diagnosis and treatment platforms greatly improved the efficacy of anti-tumor drugs, offering personalized therapy and maximum therapy efficacy.

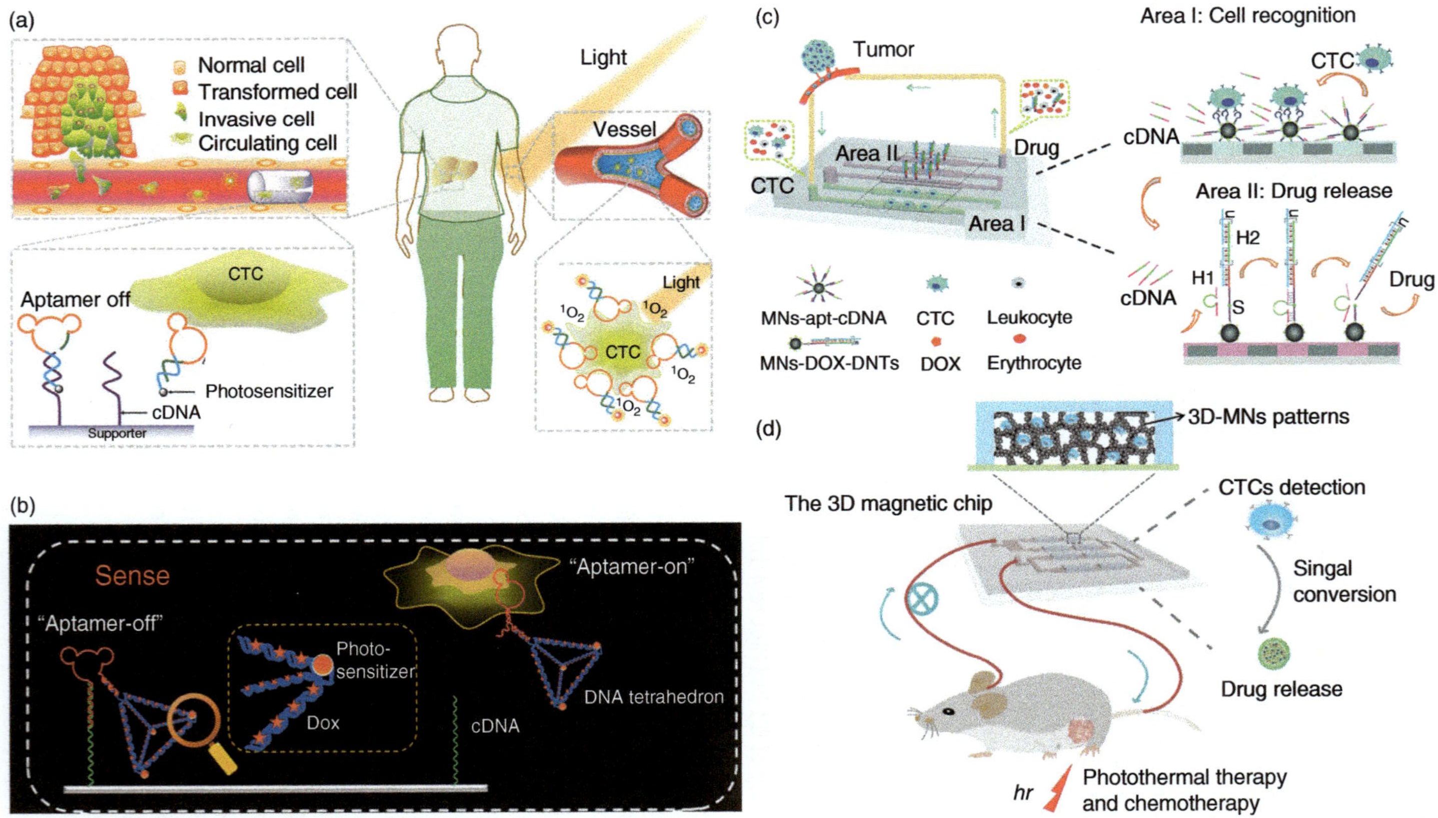

Figure 4.10 DNA nanotechnology-based microfluidics for CTC destruction and CTC-response tumor treatment. (a) CTC-triggered localized release of anticancer agent via a structure-switching aptamer. Source: Reproduced with permission from Chen et al. [72]. © 2016 Royal Society of Chemistry. (b) Aptamer-anchored DNA nanodevice for synergetic destruction of CTCs. Source: Reproduced with permission from Chen et al. [73]. © 2016 American Chemical Society. (c) A liquid biopsy-guided drug release system for cancer theragnostic based on a magnetically controlled microfluidic platform. Source: Reproduced with permission from Xu et al. [74]. © 2020 Royal Society of Chemistry. (d) 3DMC system for CTC *in vivo* detection and monitoring. Source: Reproduced with permission from Feng et al. [75]. © 2023 American Chemical Society.

4.3.3 Sensitive Detection of EVs

4.3.3.1 Signal Transduction for EV Detection

Aptamer-based detection strategies can transform the recognition events of EV-aptamers into physical and chemical signals, which can be interpreted as quantitative and molecular information on EVs. Electrochemical detection has many advantages, such as low cost, high sensitivity, portability, ease of use, and minimal sample requirement. Aptamer-based electrochemical detection strategies have been widely applied to detect EVs. For example, Zhou et al. developed an aptamer-based electrochemical platform for quantitative detection of exosomes (Figure 4.11c) [76]. The interaction between exosomes and CD63 aptamers was immobilized on gold electrode surfaces, which resulted in the displacement of the antisense strand, leading to a decrease in redox signal. Without the need for pre-treatment procedure, this aptasensor could detect as few as 1×10^6 exosomes per ml. The convenient chemical modification of the aptamers allowed for labeling of various organic dyes or fluorescent particles on aptamers. The aptamer-based fluorescent probes can directly label targets for fluorescence qualitative and quantitative detection. Sun's group employed two kinds of fluorescent dye-conjugated aptamers to label EVs, which underwent size-selective separation using λ-DNA-mediated viscoelastic microfluidics (Figure 4.11b) [77]. By measuring the intensity of labeled fluorescence, two-dimensional analysis could be performed on individual EVs according to the expression levels of HER2 and EpCAM. In addition, the strategy of responsive fluorescence signal induced by aptamer-target interaction could significantly reduce background noise and improve the sensitivity of the platform. For example, Ye's group proposed a rapid isolation chip for tumor-associated exosomes, which could achieve sensitive quantification using a signal switch strategy based on the interaction between cationic perylene tetracarboxylic acid diimide derivative (PTCDI) and aptamers (Figure 4.11c) [78]. By coupling the efficient enrichment capability of Tim4 beads and this superior fluorescence signal transduction strategy, the LOD of HepG2 exosomes on this platform reached 8.69×10^3 particles per ml, and the detection range spanned 6 orders of magnitude.

4.3.3.2 Signal Amplification

The low concentration of T-EVs in body fluids and the limited abundance of biomarkers unavoidably require signal amplification strategies to improve detection sensitivity. Naturally, compared with other recognition ligands, aptamer has higher programmability, enabling the integration with various signal amplification strategies. Taking advantage of its nucleic acid nature, many studies have designed or modified aptamer's sequence to trigger hybridization chain reaction (HCR) or rolling circle amplification (RCA) to amplify the signal from targets. For example, Zhang's group proposed a highly sensitive T-EV phenotype analysis strategy using size-coded microbeads. By employing these different-sized microbeads labeled with hairpin aptamer probes, specific capture of T-EVs with different phenotypes was achieved (Figure 4.12a) [79]. After flowing through a microfluidic chip containing spacer arrays, these size-coded microbeads were trapped in distinct chip

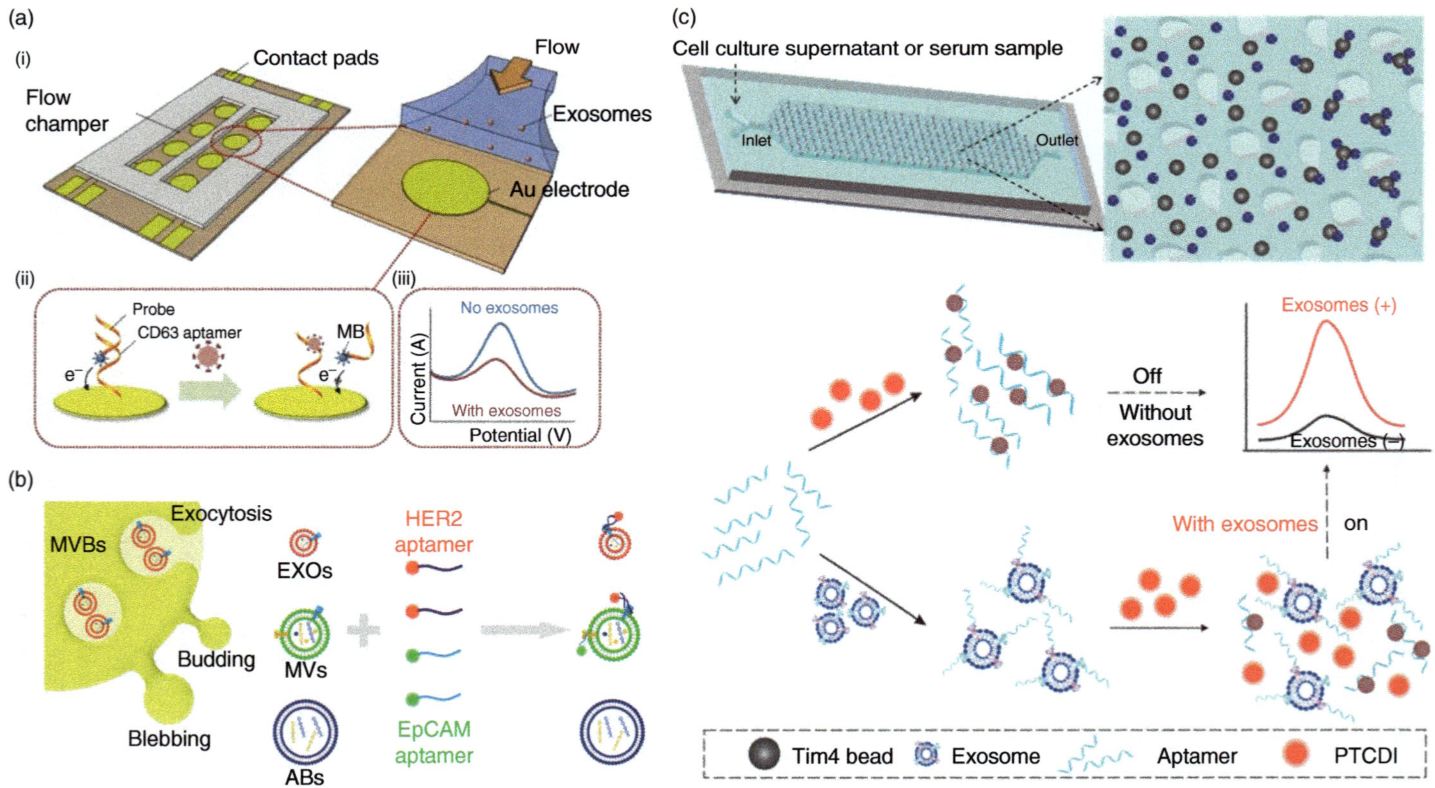

Figure 4.11 DNA nanotechnology-based signal transduction for EV detection. (a) An aptasensor for electrochemical detection of exosomes. Source: Reproduced with permission from Zhou et al. [76]. © 2016 Elsevier. (b) Aptamer-mediated fluorescence analysis of EVs. Source: Reproduced with permission from Liu et al. [77]. © 2019 American Chemical Society. (c) PTCDI–aptamer fluorescence signal transduction strategy for exosome detection. Source: Reproduced with permission from Zheng et al. [78]. © 2022 American Chemical Society.

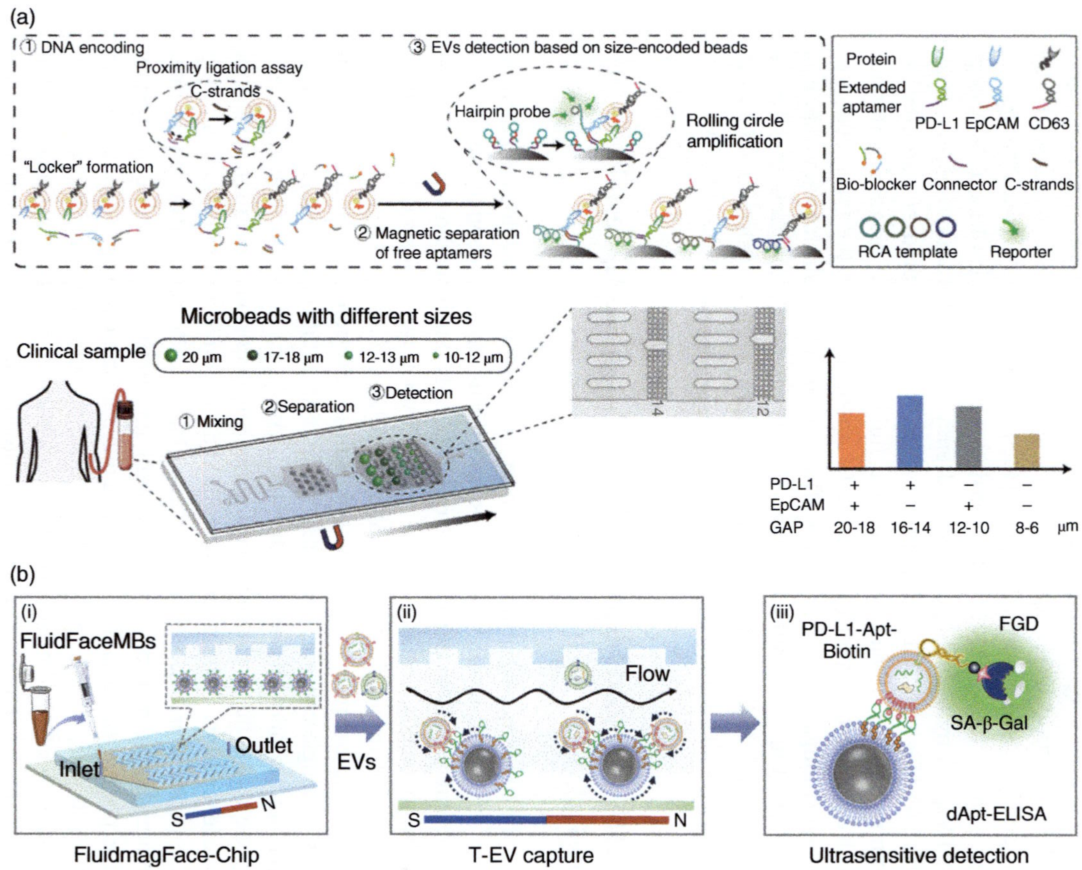

Figure 4.12 DNA nanotechnology-based signal amplification for EV detection. (a) Size-coded strategy for multiplex EV phenotyping by *in situ* RCA signal amplification. Source: Reproduced with permission from Wu et al. [79]. © 2022 American Chemical Society. (b) On-chip dual-aptamer-labeled sandwich enzyme-linked fluorescence amplification strategy for ultrasensitive detection of PD-L1 protein on T-EV. Source: Reproduced with permission from Niu et al. [80]. © 2023 John Wiley & Sons.

regions, generating location-specific signals based on different EV biomarkers. Subsequently, *in situ* RCA was adopted to amplify signals to analyze the levels of EV biomarkers on these microspheres. This strategy allowed fluorescence spectroscopy to quickly and simultaneously detect six different T-EV phenotypes, free of spectral overlap.

In addition to functional sequences, aptamers can also be coupled with other molecular tags for signal amplification. After binding to the protein markers on EVs, biotin-coupled aptamers can introduce enzymes to catalyze the reaction and produce amplified signals [57, 80]. For example, Wu's group has designed a microfluidic platform for sensitive detection of PD-L1 on T-EV surfaces (T-$EV_{PD\text{-}L1}$) by a dual-aptamer-based enzyme-linked immunosorbent assay (dApt-ELISA) (Figure 4.12b) [80]. In these chips, T-EVs could be captured by high-affinity EpCAM aptamer-immobilized fluid multivalent magnetic interface and labeled by the biotin-terminated PD-L1 aptamers. After the introduction of streptavidin-conjugated β-galactosidase, fluorescein-di-β-D-galactopyranoside was catalyzed, resulting in amplified fluorescence for quantitative detection of PD-L1 protein on T-EV. This approach allowed for ultra-sensitive detection of PD-L1 positive T-EVs with a LOD of 10 EVs μl^{-1} and a dynamic range of 10^4–10^7 EVs ml^{-1}.

4.4 DNA-Assisted Microfluidics for Single-Cell/Vesicle Analysis

Cells are the fundamental units of organism structure and physiological function. However, numerous traditional studies only focus on cell populations, which merely present a holistic portrayal of multiple cells. Such methods can conceal cellular heterogeneity, leading to the loss of important biological information. To solve this dilemma, single-cell analysis has been developed and is becoming an attractive technique in modern biomedical research [23]. Traditional cell isolations in single-cell analysis, such as limiting dilution, micromanipulation, and laser-capture microdissection, have been employed to obtain individual cells one by one but suffer from time-consuming and laborious operation, potential cell damage, and low throughput. Microfluidics, regarded as a miniaturized "chip lab," can perform precise single-cell manipulation with many advantages. First, its miniaturization features reduce reagent consumption, resulting in greatly reduced experimental costs. The micro-reaction system increased the concentration of target/template, which increased the reaction efficiency and shortened the analysis time. Second, the isolated reaction chamber effectively avoids external and intercellular contamination. Third, its integration and parallelization features allow a series of operations on parallel samples in a chip, realizing high-throughput single-cell analysis.

The emergence of DNA barcoding for single cell is a pivotal point for high-throughput single-cell sequencing. DNA barcodes mainly consist of a universal PCR primer site, a cell barcode, a molecular index, and an mRNA capture sequence of an oligo (deoxythymidine) (oligo(dT)). Various materials, such as hydrogels, resins, or magnetic materials, are widely used as carriers for DNA barcodes [81]. Cell barcodes on each bead are identical but different from other beads, serving as the distinctive barcode for the corresponding cell. The molecular indices are introduced to correct amplification bias and accurately quantify mRNA levels. With the help of microfluidic fluid control and microstructure design, the high-throughput pairing of individual cells and barcode beads can be achieved on the same microfluidic chip. In addition, a series of processes such as cell lysis, mRNA capture, reverse transcription (RT), and amplification can be integrated on the chip, thus empowering multiplexed processing of thousands of cells. Furthermore, barcoded recognition ligands for protein labeling, including antibodies and aptamers conjugated with DNA barcodes, further promoted single-cell multi-omics analysis [24]. Single-EV analysis was important for resolving EV heterogeneity, and the strategy based on single-cell barcode beads was also applied to single-EV analysis.

4.4.1 Single-Cell Analysis by Sequencing

Single-cell RNA sequencing (scRNA-seq) technology has become an important tool for the interrogation of gene expression patterns, facilitating the dissection of cellular heterogeneity [82]. As the field of scRNA-seq, Since the development of the first scRNA-seq method, multiple iterations and updates have been made to the scRNA-seq methods, with a core goal of improving sensitivity, transcript coverage, and cell throughput. Generally, scRNA-seq protocols follow several modular steps, including single-cell isolation and lysis, mRNA extraction, RT, complementary DNA (cDNA) synthesis, sequencing library construction, sequencing, and data analysis (Figure 4.13A) [81, 83]. Currently, several high-throughput scRNA-seq platforms have been reported to combine barcode beads (Figure 4.13B) with microfluidic technology, including droplet- and microwell-based platforms, etc. These platforms have been summarized in published reviews [82, 86]. Unfortunately, most of the existing high-throughput microfluidic platforms suffer from low pairing efficiency of single cells and single barcode beads due to the restriction of Poisson distribution. Under this circumstance, reagent waste and cell loss are inevitable during the process, which is unpractical for the analysis of rare cells, such as CTCs [87]. Here, we focus on the methods for single CTC sequencing that have been developed to tackle the scarcity of CTCs.

The application of current scRNA-seq in analyzing CTCs is often limited by cell scarcity and the severe contamination of blood cells. To overcome this challenge, Cheng et al. developed Hydro-seq that effectively captured rare CTCs (Figure 4.13C) [85]. In Hydro-seq, capture chambers equipped with control valves were designed to facilitate the pairing of single CTCs and single beads. After cell loading, a single CTC could be captured in a size-matched capture site while other smaller blood cells passed through. Washing steps were then performed by the valves on capture chambers to wash all unwanted cells away. After that, beads were loaded for pairing, followed by cell lysis and retrieval of mRNA-captured beads for subsequent library construction. In addition, the chamber array can be scaled up

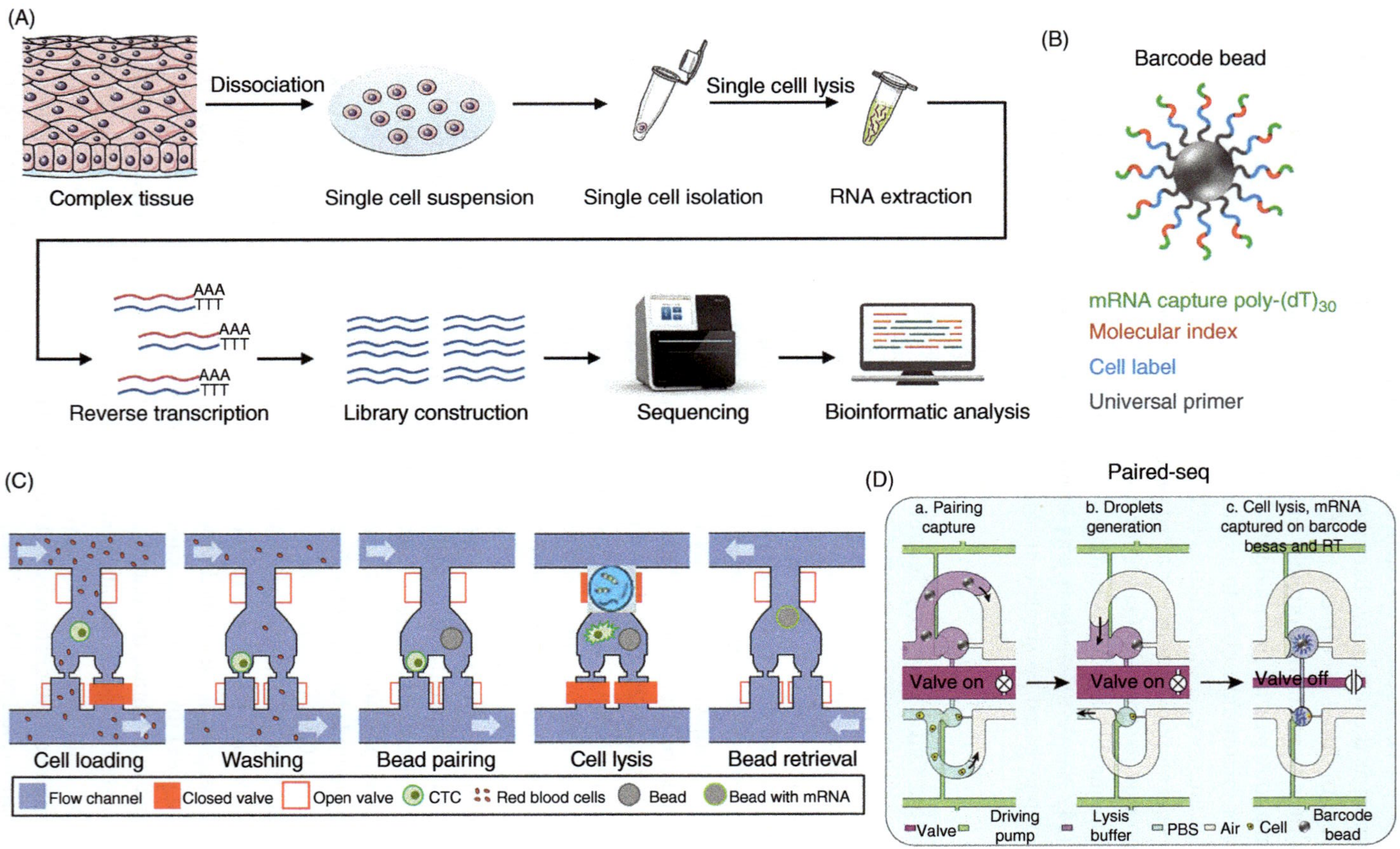

Figure 4.13 DNA-assisted microfluidics for single-cell analysis by sequencing. (A) Major steps of scRNA-seq workflows. Source: Reproduced with permission from Chen et al. [83]. © 2021 John Wiley & Sons. (B) Structure of barcode on the barcode beads. Source: Reproduced with permission from Zhang et al. [84]. © 2020 Springer Nature. (C) Operation workflow of Hydro-seq, integrating cell loading, bead pairing, cell lysis, and RT in microfluidic circuits with the assistance of valves. Source: Reproduced with permission from Cheng et al. [85]. © 2019 Springer Nature. (D) Schematic of the basic workflow for single cell and bead parallel manipulation on paired-seq chip. Source: Reproduced with permission from Zhang et al. [84]. © 2020 Springer Nature.

to thousands of chambers for high-throughput parallel analysis. However, the cell capture site and bead capture site in Hydro-seq cannot be completely separated, making it difficult to rule out contamination with free mRNA. To overcome this challenge, Yang's group developed a Paired-seq platform with high pairing efficiency and innovative cell-free RNA removal capability (Figure 4.13D) [84]. Based on the principle of paired differential flow resistance capture, the design facilitated the pairing of single cell/bead with 95% cell utilization efficiency. Besides, cells and beads were loaded and washed in independent channels, allowing efficient removal of cell-free RNAs. Efficient cell lysis, mRNA capture, and RT were performed in hundreds of picoliter droplets, allowing for highly parallel processing of tens to thousands of cells with high sequencing accuracy and sensitivity.

Transcriptome analysis of partial transcripts is insufficient to elucidate single-cell heterogeneity, and multi-omics analysis can better construct single-cell atlas. Proteins act as key executors of biological functions of organisms, and the dynamic changes of their abundance have important effects on various life processes. At the single-cell level, the joint analysis of proteome and transcriptome can help to resolve the relationship between gene expression and protein expression. Protein-recognition ligands (e.g. antibodies, aptamers) conjugated with DNA barcodes have enabled the utilization of high-throughput sequencing for single-cell proteome analysis. Furthermore, the introduction of amplification allows for the detection of low-abundance proteins, which might not be detected by conventional methods. The antibody-barcode sequence generally contains a poly A sequence, an antibody barcode, and a PCR handle. The poly(A) sequence can be captured by barcode beads and is compatible with scRNA-seq platforms. The antibody barcode is used for protein identification, and the PCR handle is used for subsequent PCR amplification. When antibody-barcode complexes are incubated with cells, the antibodies specifically bind to proteins on the cell surface. Then the cells are introduced into microfluidic platforms, which are compatible with scRNA-seq workflows. Through size separation, the amplified cDNAs and antibody-derived tags (ADTs) can be separated for library construction and analysis, respectively.

Based on Paired-seq, Yang's group developed a multi-Paired-seq platform, which integrated high-throughput scRNA-seq and protein barcoding techniques, enabling simultaneous analysis of mRNA and protein on a single cell (Figure 4.14a) [88]. The platform accurately classified three known breast cancer cells based on mRNA and protein expression profiles. Additionally, the dynamic expression and correlation between mRNA and protein in cells before and after drug treatment were identified. In addition to transcriptome, proteome can also be analyzed in conjunction with other omics, such as genome analysis. Demaree et al. proposed the Dab-seq platform for high-throughput joint profiling of single-cell DNA genomes and surface proteins (Figure 4.14b) [89]. Stained acute myeloid leukemia (AML) cells were paired and encapsulated with a barcode bead using a microfluidic device, followed by multiple PCRs to amplify genomic targets and capture antibody tags with a unique cell index. By sequencing, the platform enabled high-throughput analysis of 49 DNA targets and 23 hematopoietic markers in AML cells. It elucidated the complex interplay between DNA mutations and their effects on protein expression in cancers.

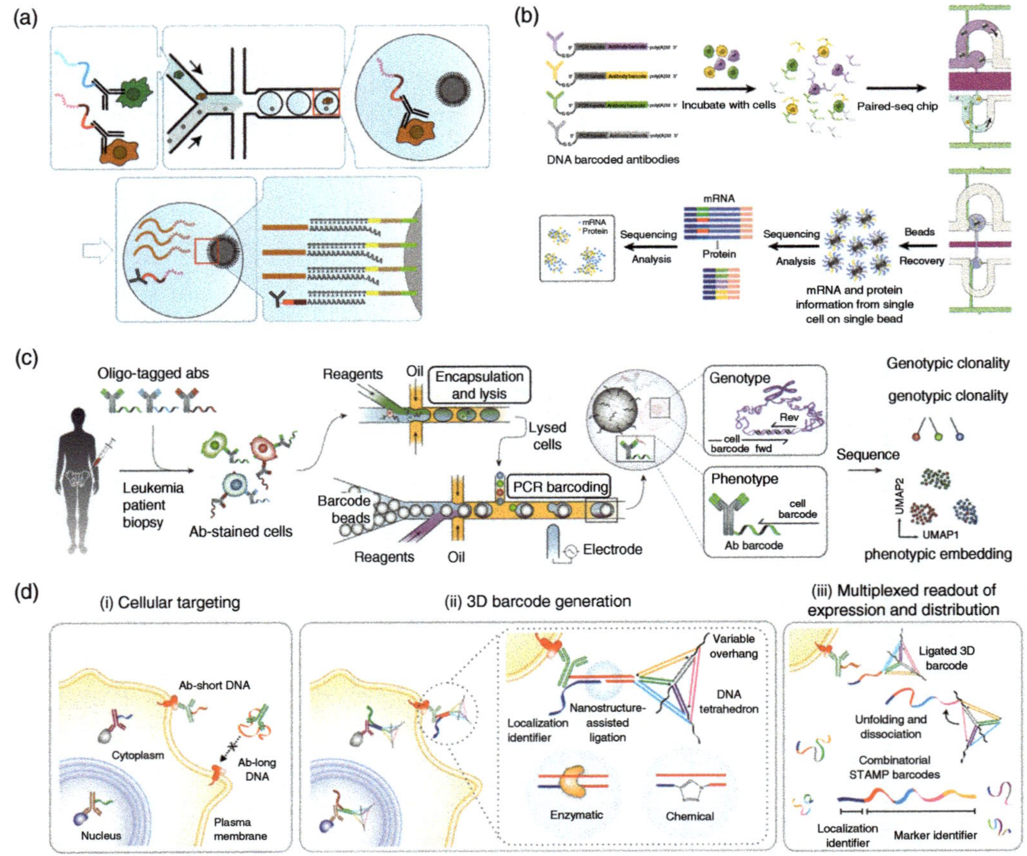

Figure 4.14 Antibody-barcode-based single-cell sequence for protein analysis and multi-omics analysis. (a) Workflow of multi-paired-seq for protein and transcriptome analysis. Source: Reproduced with permission from Xu et al. [88]. © 2022 American Chemical Society. (b) Joint profiling of DNA and proteins in single cells by DAb-seq platform. Source: Reproduced with permission from Demaree et al. [89]. © 2021 Springer Nature. (c) Operation workflow of STAMP assay, including cellular targeting, 3D barcode generation, and multiplexed readout. Source: Reproduced with permission from Sundah et al. [90]. © 2019 Springer Nature.

However, existing methods only achieve cellular-level analysis, which limits the resolution of more fine-grained regulatory networks in subcellular regions. To surmount this obstacle, Shao's group reported a barcoding approach for high-throughput multiplexed profiling of the subcellular expression and distribution of proteins in whole cells (Figure 4.14c) [90]. This barcoding approach combined DNA nanostructures with localization labels to generate multiplexed DNA sequence-topology assembly, defined as multiplexed profiling (STAMP) barcodes. The established sequencing methods ensured that a STAMP barcode could be analyzed by high throughput measurements. Combined with microfluidic devices, STAMP assay could accurately classify the molecular subtypes of breast cancer cells and identify the subcellular spatial markers of disease aggressiveness. Overall, the development of barcode strategies represents a significant advancement in the field, making high-throughput, and parallel single-cell analysis possible. Multi-omics analysis of single tumor cell offers multi-dimensional perspectives of disease-associated information and affords a comprehensive solution for studying tumor regulatory mechanisms and precision medicine.

4.4.2 Single-Cell Analysis by Spectroscopy

Precise single-cell phenotypic analysis requires extremely high resolution. Spectrograph-based analysis techniques, especially surface-enhanced Raman scattering (SERS) and mass spectrometry (MS), exhibit good resolution and multiplexing capability and are widely used in the analysis of single-cell phenotypes. For example, Cui's group achieved an on-chip protein profiling of individual CTCs by combining size-based microfluidic filter cell separation with multi-spectral orthogonal SERS analysis (Figure 4.15a) [91]. In order to achieve multiple phenotype analysis, three kinds of spectrally orthogonal SERS-aptamer probes were designed. After the binding of cocktail probes composite spectral features of surface protein expression levels were obtained for individual cells. This method facilitated the identification of cancer subtypes, holding great potential in clinical cancer diagnosis and precision medicine. However, the utilization of size-based microfluidics filters poses a challenge to the analysis of rare CTCs in patients. Gao et al. fabricated two types of SERS-aptamer nanotags and designed a new microfluidic chip with a lantern-like bypass structure to capture rare CTCs from whole blood (Figure 4.15b) [92]. This strategy enabled efficient capture and *in situ* heterogeneous phenotype analysis of CTCs from hepatocellular carcinoma patients at the single-cell level.

In recent years, inductively coupled plasma-mass spectrometry (ICP-MS) has been applied to the quantitative analysis of CTCs based on noble metal labeling. Inspired by the efficient preying mechanism of sea urchins, Wang's group designed a sea urchin-like nanoprobe for superefficient discerning and background-free phenotypic analysis of single CTC on microfluidics devices (Figure 4.15c) [93]. The sea urchin-like nanoprobe was constructed by conjugating Sgc8 and SYL3C aptamers onto AuNPs, and single CTC was separated and purified by the separation unit and hydrodynamic filtration purification device. Then, the amount of ^{197}Au isotope in the targeted cell was determined by ICP-MS, and the biomarker protein

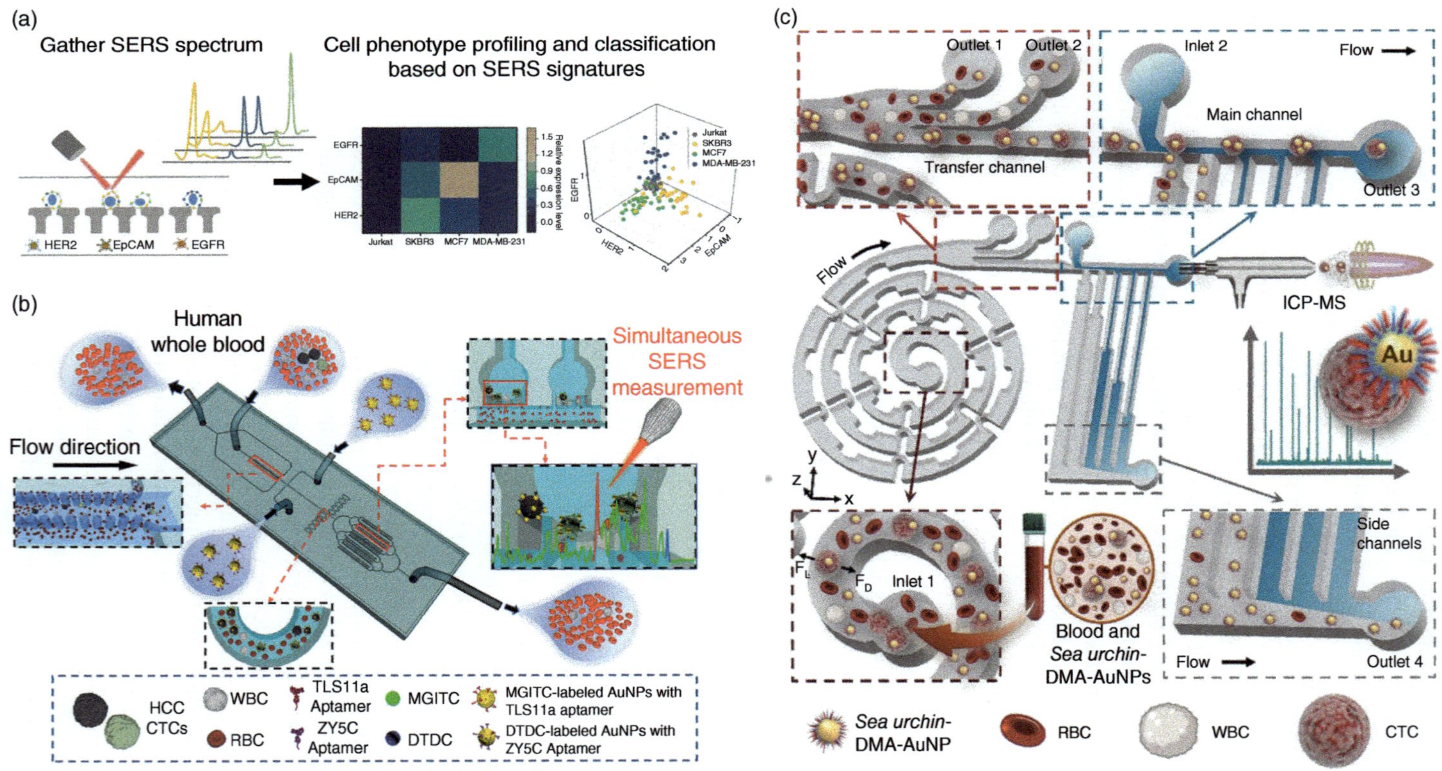

Figure 4.15 DNA-assisted microfluidics for single-cell analysis by spectroscopy. (a) Multiplex SERS nanovectors and multivariate analysis for *in situ* profiling of CTC phenotype in a microfluidic chip. Source: Reproduced with permission from Zhang et al. [91]. © 2018 John Wiley & Sons. (b) The workflow of SERS-aptamer-based microfluidic chip for hepatocellular carcinoma CTC capture, SERS measurement, and single-cell phenotype analysis. Source: Reproduced with permission from Gao et al. [92]. © 2021 Royal Society of Chemistry. (c) Superefficient discerning of single CTC analysis with ICP-MS. Source: Reproduced with permission from Zhang et al. [93]. © 2021 American Chemical Society.

of a single CTC was analyzed without the interference of background signals. This strategy enabled highly efficient and sensitive analysis of single-cell EpCAM or PTK7 phenotypes, providing favorable information for clinical diagnosis and tumor metastasis prediction. In summary, single-cell spectroscopy analysis can reliably identify cancer subtypes with spectral information, holding great potential for clinical cancer diagnosis and precision medicine.

4.4.3 Single-Vesicle Analysis

EVs are highly heterogeneous in terms of biomolecule composition and physical properties, necessitating the study of EVs at the single-particle level [94]. The integration of a DNA nanostructure-based EV detection strategy with a microfluidic platform for single-EV isolation enables the detection of single EV with high sensitivity. Li's group has developed several single-EV imaging analysis platforms, combined with different DNA signal amplification strategies for direct visualization and quantification of surface proteins and internal miRNAs of single EV [95, 96]. For example, they constructed an ultra-sensitive single-vesicle imaging assay to label the captured exosomes with an activatable aptamer probe (AAP), triggering and initiating the HCR response to amplifyfluorescence signals (Figure 4.16a) [95]. This strategy enabled direct visualization and quantification of tumor exosomes in 1 μl plasma samples at the single-vesicle level. This assay achieved a high signal-to-noise ratio imaging for quantitative analysis of exosomes and detection of surface protein expressions of individual exosomes. Wang's group has developed a fiber aptasensor based on hollow-core anti-resonant optical fibers (HcARFs) and a sandwich detection strategy to achieve the detection of single cancer exosome (Figure 4.16b) [97]. The utilization of HcARFs could improve Raman scattering by over three orders of magnitude compared with a planar setting. Consequently, multiplexed analysis of surface proteins at the single-EV level could be achieved, thereby facilitating accurate identification of the cellular origin of exosomes and differentiation of different types of cancers.

Optical sensing suffers from limited signal amplification, low multiplexing capability, and low throughput. Inspired by scRNA-seq methods and DNA-barcoded ligands for single-cell protein analysis, researchers have developed integrated sequencing and barcoding strategies for high-throughput and sensitive profiling of single EV. Wu et al. proposed a proximity-dependent barcoding assay (PBA) to profile surface proteins of single exosome using antibody-DNA barcode and next-generation sequencing technique [99]. As a high-throughput approach, this assay could simultaneously profile 38 surface proteins present on individual exosomes, offering important application values for the analysis of single-exosome level heterogeneity. Furthermore, Weissleder's group used droplet microfluidics to separate and encode single EV (Figure 4.16c) [98]. The antibody-DNA barcode was sequenced by a single-cell sequencing process to determine the protein composition of individual EVs. Thousands of EVs and possibly dozens of biomarkers of interest can be analyzed in one experiment, enabling the identification of rare EV subtypes with reasonable certainty. In summary, these single-EV analysis methods, facilitate

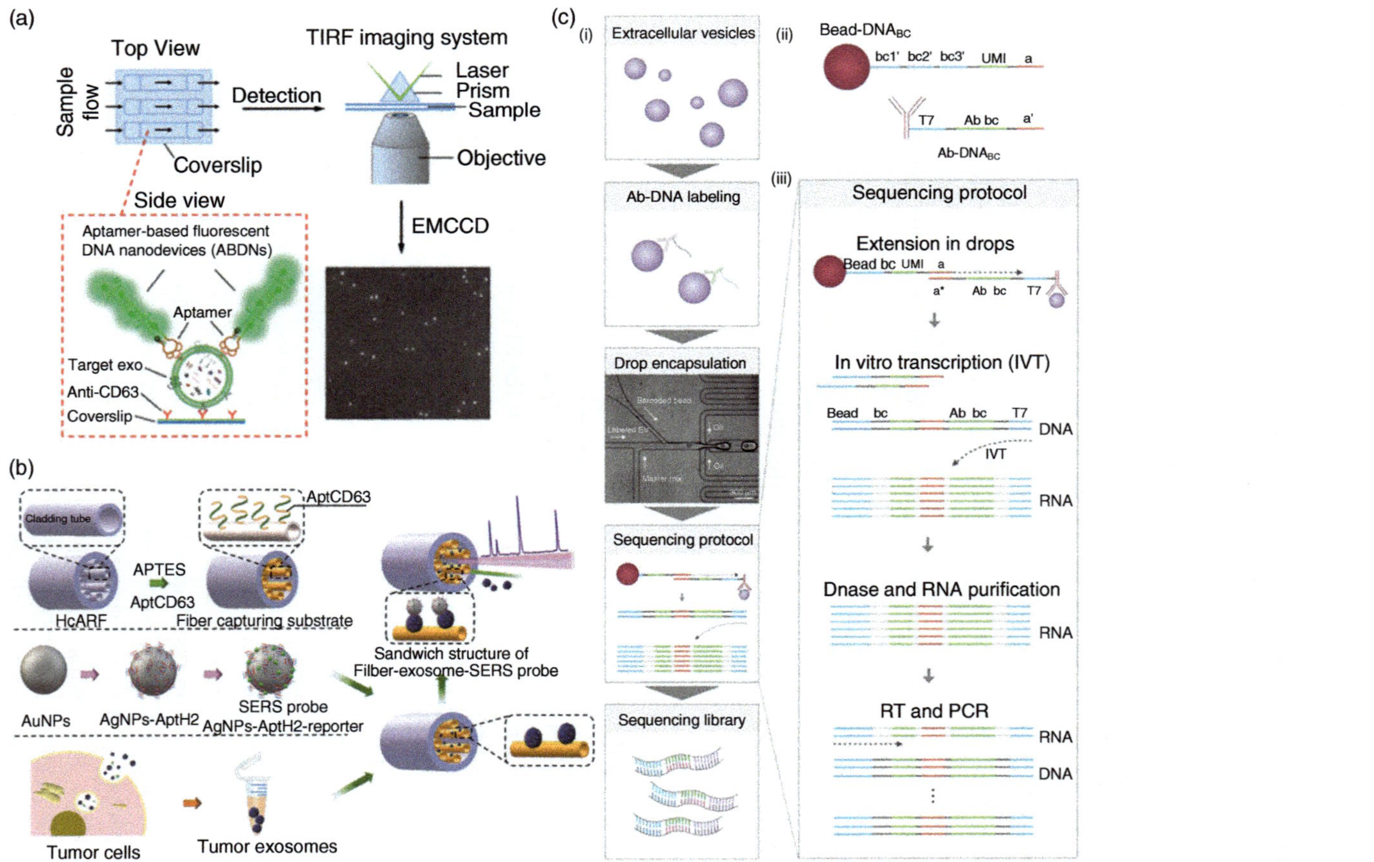

Figure 4.16 DNA-assisted microfluidics for single-vesicle analysis. (a) Aptamer-based DNA nanodevices for single-vesicle imaging and detection. Source: Reproduced with permission from He et al. [95]. © 2019 American Chemical Society. (b) Giant enhancement of Raman scattering by HcARF sensor for single cancer exosome detection. Source: Reproduced with permission from Xia et al. [97]. © 2023 American Chemical Society. (c) Antibody-based immunosequencing for multiplexed protein molecules measurement of single extracellular vesicles. Source: Reproduced with permission from Ko et al. [98]. © 2021 American Chemical Society.

the comparison of EV profiling, which contributes to a deeper understanding of the fundamental biology and heterogeneity of EVs and improves tumor diagnostic specificity.

4.5 Summary and Outlook

Liquid biopsy can provide real-time and comprehensive physiological and pathological information in a noninvasive sampling manner. It has become a powerful supplement to traditional tissue biopsy and is expected to promote the development of precision medicine. DNA nanotechnology harnesses the unique structure and functional advantages of DNA, providing a powerful tool for liquid biopsy. Thereinto, aptamers, as functional nucleic acids, can serve as recognition ligands for target isolation and labeling. Highly programmable DNA nanostructures can be used to engineer precise and controllable isolation interfaces to improve isolation efficiency. Meanwhile, with DNA-based amplification strategies, sensitive detection of rare circulating targets could be achieved. In addition, according to the intrinsic properties of DNA nanostructures, efficient release of captured targets can be achieved through simple external stimulation. Furthermore, the powerful encoding capability of DNA, combined with second-generation sequencing, can facilitate high-throughput parallel analyses of targets within blood circulation, especially CTCs and EVs. DNA nanotechnology-based microfluidics can further integrate the merits of precise manipulation of fluids, miniaturization, and integration of microfluidics into DNA nanotechnology, providing a more efficient and sensitive platform for liquid biopsy. Through precise microchannel design and fluid manipulation, the mass transfer efficiency between the targets and capture interfaces has been significantly increased, which achieves efficient isolation of circulating targets. The miniaturization feature of microfluidics increases reaction kinetics in picoliter-to-nanoliter reaction system for rapid and sensitive detection. Furthermore, the integration and flexible operation of chips enable precise isolation and manipulation of single-cell/single-vesicle, making single-cell/single-vesicle analysis possible. In this chapter, we discussed the recent advances in DNA nanotechnology-based microfluidics for liquid biopsy, mainly focused on their applications in the isolation and detection of circulating targets and the applications of DNA-assisted microfluidics for single-cell/vesicle analysis.

DNA nanotechnology-based microfluidics has demonstrated enormous advantages and practical potential in the field of liquid biopsy, but there are still some challenges. Firstly, the production cost of DNA nanostructures is still high, such as complex DNA origami, which may contain hundreds of single-stranded DNA. However, the high cost would hinder the high-throughput applications of DNA nanotechnology. The development of synthetic biology may offer a good opportunity to overcome this challenge. A new strategy of "single-strand origami" can form single-strand DNA nanostructures through self-folding, which has great potential in amplification, replication, and cloning. This single-strand DNA can be produced on a cost-effective scale by means of enzymatic and biological replication, greatly reducing the cost of synthesis of short-strand DNA. Secondly, DNA nanostructures suffer from unsatisfactory stability in complex samples, such as plasma and whole blood. Although the anti-nuclease ability of DNA nanostructures is significantly enhanced, compared with linear DNA sequences, it is still unsatisfactory for long-term monitoring and analysis. Introducing specific structural motifs or unnatural bases to the DNA nanostructures may be a promising method for improving their stability in body fluids. Thirdly, the various DNA nanostructures currently used for target detection are mostly fixed. These rigid structures can provide precisely controlled nanoscale reaction spaces for identifying molecules and targets but they lack the ability to autonomously approach, identify, and capture targets. Therefore, the development of intelligent dynamic DNA nanostructures, namely intelligent DNA molecular nanorobots, may be a promising direction for future development. For liquid biopsy, the intelligent DNA molecular nanobots would automatically identify circulating targets and then quickly assemble themselves to launch a "siege," facilitating the capture and signal amplification. Such technique leads the way for a revolution in early diagnosis of disease and therapeutic monitoring. It is believed that under continuous development, DNA nanotechnology-based microfluidics will be applied in clinical liquid biopsy in the foreseeable future, and this cutting-edge technique will promote the development of precision medicine and personalized medicine.

References

1 Crowley, E., Di Nicolantonio, F., Loupakis, F. et al. (2013). Liquid biopsy: monitoring cancer-genetics in the blood. *Nat. Rev. Clin. Oncol.* 10 (8): 472–484.

2 Ring, A., Nguyen-Strauli, B.D., Wicki, A. et al. (2023). Biology, vulnerabilities and clinical applications of circulating tumour cells. *Nat. Rev. Cancer* 23 (2): 95–111.

3 Diamantis, A., Magiorkinis, E., and Koutselini, H. (2009). Fine-needle aspiration (FNA) biopsy: historical aspects. *Folia Histochem. et Cytobiol.* 47 (2): 191–197.

4 Mellby, L.D., Nyberg, A.P., Johansen, J.S. et al. (2018). Serum biomarker signature-based liquid biopsy for diagnosis of early-stage pancreatic cancer. *J. Clin. Oncol.* 36 (28): 2887–2894.

5 Aro, K., Wei, F., Wong, D.T. et al. (2017). Saliva liquid biopsy for point-of-care applications. *Front. Public Health* 5: 77.

6 Lone, S.N., Nisar, S., Masoodi, T. et al. (2022). Liquid biopsy: a step closer to transform diagnosis, prognosis and future of cancer treatments. *Mol. Cancer* 21 (1): 79.

7 Maheswaran, S., Sequist, L.V., Nagrathet, S. et al. (2005). Detection of mutations in EGFR in circulating lung-cancer cells. *N. Engl. J. Med.* 359 (4): 366–377.

8 Tian, F., Liu, C., Lin, L. et al. (2019). Microfluidic analysis of circulating tumor cells and tumor-derived extracellular vesicles. *TrAC, Trends Anal. Chem.* 117: 128–145.

9 Poudineh, M., Sargent, E.H., Pantel, K. et al. (2018). Profiling circulating tumour cells and other biomarkers of invasive cancers. *Nat. Biomed. Eng.* 2 (2): 72–84.

10 Wu, L., Zhu, L., Huang, M. et al. (2019). Aptamer-based microfluidics for isolation, release and analysis of circulating tumor cells. *TrAC, Trends Anal. Chem.* 117: 69–77.

11 Wu, L., Wang, Y., Zhu, L. et al. (2020). Aptamer-based liquid biopsy. *ACS Appl. Bio Mater.* 3 (5): 2743–2764.

12 Green, B.J., Saberi Safaei, T., Mepham, A. et al. (2016). Beyond the capture of circulating tumor cells: next-generation devices and materials. *Angew. Chem. Int. Ed.* 55 (4): 1252–1265.

13 Song, Y., Tian, T., Shi, Y. et al. (2017). Enrichment and single-cell analysis of circulating tumor cells. *Chem. Sci.* 8 (3): 1736–1751.

14 Hu, Q., Li, H., Wang, L. et al. (2019). DNA nanotechnology-enabled drug delivery systems. *Chem. Rev.* 119 (10): 6459–6506.

15 Zhao, Y., Chen, F., Li, Q. et al. (2015). Isothermal amplification of nucleic acids. *Chem. Rev.* 115 (22): 12491–12545.

16 Bhatia, D., Wunder, C., and Johannes, L. (2021). Self-assembled, programmable DNA nanodevices for biological and biomedical applications. *ChemBioChem* 22 (5): 763–778.

17 Zuo, X., Xia, F., Xiao, Y. et al. (2010). Sensitive and selective amplified fluorescence DNA detection based on exonuclease III-aided target recycling. *J. Am. Chem. Soc.* 132: 1816–1818.

18 Allawi, H.T., Dahlberg, J.E., Olson, S. et al. (2004). Quantitation of microRNAs using a modified Invader assay. *RNA* 10 (7): 1153–1161.

19 Ye, D., Zuo, X., and Fan, C. (2018). DNA nanotechnology-enabled interfacial engineering for biosensor development. *Annu. Rev. Anal. Chem.* 11 (1): 171–195.

20 Dirks, R.M. and Pierce, N.A. (2004). Triggered amplification by hybridization chain reaction. *Proc. Natl. Acad. Sci . U. S. A.* 101 (43): 15275–15278.

21 Ali, M.M., Li, F., Zhang, Z. et al. (2014). Rolling circle amplification: a versatile tool for chemical biology, materials science and medicine. *Chem. Soc. Rev.* 43 (10): 3324–3341.

22 Jovic, D., Liang, X., Zeng, H. et al. (2022). Single-cell RNA sequencing technologies and applications: A brief overview. *Clin. Transl. Med.* 12 (3): e694.

23 Xue, R., Li, R., and Bai, F. (2015). Single cell sequencing: technique, application, and future development. *Sci. Bull.* 60 (1): 33–42.

24 Labib, M. and Kelley, S.O. (2020). Single-cell analysis targeting the proteome. *Nat. Rev. Chem.* 4 (3): 143–158.

25 Lin, B., Lei, Y., Wang, J. et al. (2021). Microfluidic-based exosome analysis for liquid biopsy. *Small Methods* 5 (3): e2001131.

26 Culbertson, C.T., Mickleburgh, T.G., Stewart-James, S.A. et al. (2014). Micro total analysis systems: fundamental advances and biological applications. *Anal. Chem.* 86 (1): 95–118.

27 Wu, L., Wang, Y., Xu, X. et al. (2021). Aptamer-based detection of circulating targets for precision medicine. *Chem. Rev.* 121 (19): 12035–12105.

28 Tan, W., Donovan, M.J., and Jiang, J. (2013). Aptamers from cell-based selection for bioanalytical applications. *Chem. Rev.* 113 (4): 2842–2862.

29 Cho, E.J., Lee, J.W., and Ellington, A.D. (2009). Applications of aptamers as sensors. *Annu. Rev. Anal. Chem.* 2: 241–264.

30 Ma, H., Liu, J., Ali, M.M. et al. (2015). Nucleic acid aptamers in cancer research, diagnosis and therapy. *Chem. Soc. Rev.* 44 (5): 1240–1256.

31 Phillips, J.A., Xu, Y., Xia, Z. et al. (2009). Enrichment of cancer cells using aptamers immobilized on a microfluidic channel. *Anal. Chem.* 81: 1033–1039.

32 Ye, X., Phillips, J.A., Yan, J. et al. (2009). Aptamer-based microfluidic device for enrichment, sorting, and detection of multiple cancer cells. *Anal. Chem.* 81: 7436–7442.

33 Wan, Y., Mahmood, M.A., Li, N. et al. (2012). Nanotextured substrates with immobilized aptamers for cancer cell isolation and cytology. *Cancer* 118 (4): 1145–1154.

34 Ding, P., Wang, Z., Wu, Z. et al. (2020). Natural biointerface based on cancer cell membranes for specific capture and release of circulating tumor cells. *ACS Appl. Mater. Interface* 12 (18): 20263–20270.

35 Fischer, K.E., Alemán, B.J., Tao, S.L. et al. (2009). Biomimetic nanowire coatings for next generation adhesive drug delivery systems. *Nano Lett.* 9: 716.

36 Bettinger, C.J., Langer, R., and Borenstein, J.T. (2009). Engineering substrate topography at the micro- and nanoscale to control cell function. *Angew. Chem. Int. Ed.* 48 (30): 5406–5415.

37 Dong, J., Chen, J.F., Smalley, M. et al. (2020). Nanostructured substrates for detection and characterization of circulating rare cells: from materials research to clinical applications. *Adv. Mater.* 32 (1): e1903663.

38 Wang, Y., Zhou, F., Liu, X. et al. (2013). Aptamer-modified micro/nanostructured surfaces: efficient capture of Ramos cells in serum environment. *ACS Appl. Mater. Interfaces* 5 (9): 3816–3823.

39 Zhang, Q., Wang, W., Huang, S. et al. (2020). Capture and selective release of multiple types of circulating tumor cells using smart DNAzyme probes. *Chem. Sci.* 11 (7): 1948–1956.

40 Sheng, W., Chen, T., Kamath, R. et al. (2012). Aptamer-enabled efficient isolation of cancer cells from whole blood using a microfluidic device. *Anal. Chem.* 84 (9): 4199–4206.

41 Shen, Q., Xu, L., Zhao, L. et al. (2013). Specific capture and release of circulating tumor cells using aptamer-modified nanosubstrates. *Adv. Mater.* 25 (16): 2368–2373.

42 Zhao, L., Tang, C., Xu, L. et al. (2016). Enhanced and differential capture of circulating tumor cells from lung cancer patients by microfluidic assays using aptamer cocktail. *Small* 12 (8): 1072–1081.

43 Chen, L., Liu, X., Su, B. et al. (2011). Aptamer-mediated efficient capture and release of T lymphocytes on nanostructured surfaces. *Adv. Mater.* 23 (38): 4376–4380.

44 Sun, N., Wang, J., Ji, L. et al. (2015). A cellular compatible chitosan nanoparticle surface for isolation and in situ culture of rare number CTCs. *Small* 11 (40): 5444–5451.

45 Sun, N., Liu, M., Wang, J. et al. (2016). Chitosan nanofibers for specific capture and nondestructive release of CTCs assisted by pCBMA brushes. *Small* 12 (36): 5090–5097.

46 Wu, Z., Pan, Y., Wang, Z. et al. (2021). A PLGA nanofiber microfluidic device for highly efficient isolation and release of different phenotypic circulating tumor cells based on dual aptamers. *J. Mater. Chem. B* 9 (9): 2212–2220.

47 Zhu, Y., Zou, C., Zhang, J. et al. (2020). Dynamically monitoring the clonal evolution of lung cancer based on the molecular characterization of circulating tumor cells using aptamer cocktail-modified nanosubstrates. *ACS Appl. Mater. Interfaces* 12 (5): 5671–5679.

48 Liu, Y., Zhang, H., Du, Y. et al. (2020). Highly sensitive minimal residual disease detection by biomimetic multivalent aptamer nanoclimber functionalized microfluidic chip. *Small* 16 (20).

49 Zhang, H., Yu, X., Liu, Y. et al. (2021). HUNTER-chip: bioinspired hierarchically aptamer structure-based circulating fetal cell isolation for non-invasive prenatal testing. *Anal. Chem.* 93 (19): 7235–7241.

50 Sheng, W., Tan, W., and Fan, Z. (2013). Multivalent DNA nanospheres for enhanced capture of cancer cells in microfluidic dDevices. *ACS Nano* 7 (8): 7067–7076.

51 Song, Y., Shi, Y., Huang, M. et al. (2019). Bioinspired engineering of a multivalent aptamer-functionalized nanointerface to enhance the capture and release of circulating tumor cells. *Angew. Chem. Int. Ed.* 58 (8): 2236–2240.

52 Cutler, J.I., Auyeung, E., and Mirkin, C.A. (2012). Spherical nucleic acids. *J. Am. Chem. Soc.* 134 (3): 1376–1391.

53 Ahmed, M.G., Abate, M.F., Song, Y. et al. (2017). Isolation, detection, and antigen-based profiling of circulating tumor cells using a size-dictated immunocapture chip. *Angew. Chem. Int. Ed.* 56 (36): 10681–10685.

54 Wu, J.C., Tseng, P.Y., Tsai, W.S. et al. (2013). Antibody conjugated supported lipid bilayer for capturing and purification of viable tumor cells in blood for subsequent cell culture. *Biomaterials* 34 (21): 5191–5199.

55 Zhang, F., Wu, L., Nie, W. et al. (2019). Biomimetic microfluidic system for fast and specific detection of circulating tumor cells. *Anal. Chem.* 91 (24): 15726–15731.

56 Wu, L., Ding, H., Qu, X. et al. (2020). Fluidic multivalent membrane nanointerface enables synergetic enrichment of circulating tumor cells with high efficiency and viability. *J. Am. Chem. Soc.* 142 (10): 4800–4806.

57 Niu, Q., Gao, J., Zhao, K. et al. (2022). Fluid nanoporous microinterface enables multiscale-enhanced affinity interaction for tumor-derived extracellular vesicle detection. *Proc. Natl. Acad. Sci. U.S.A.* 119 (44): e2213236119.

58 Zhang, J., Lin, B., Wu, L. et al. (2020). DNA nanolithography enables a highly ordered recognition interface in a microfluidic chip for the efficient capture and release of circulating tumor cells. *Angew. Chem. Int. Ed.* 59 (33): 14115–14119.

59 Ling, J., Liu, D., Zhang, J. et al. (2022). Thermodynamic and kinetic modulation of microfluidic interfaces by DNA nanoassembly mediated merit-complementary heteromultivalency. *ACS Nano* 16 (12): 20915–20921.

60 Zhu, F., Bian, X., Zhang, H. et al. (2021). Controllable design of a nano-bio aptasensing interface based on tetrahedral framework nucleic acids in an integrated microfluidic platform. *Biosens. Bioelectron.* 176: 112943.

61 Liu, Y., Lin, Z., Zheng, Z. et al. (2022). Accurate isolation of circulating tumor cells via a heterovalent DNA framework recognition element-functionalized microfluidic chip. *ACS Sens.* 7 (2): 666–673.

62 Peng, J., Liu, Y., Su, R. et al. (2022). DNA-programmed orientation-ordered multivalent microfluidic interface for liquid biopsy. *Anal. Chem.* 94 (24): 8766–8773.

63 Pei, H., Wan, Y., Li, J. et al. (2011). Regenerable electrochemical immunological sensing at DNA nanostructure-decorated gold surfaces. *Chem. Commun.* 47 (22): 6254–6256.

64 Yuan, L., Giovanni, M., Xie, J. et al. (2014). Ultrasensitive IgG quantification using DNA nano-pyramids. *NPG Asia Mater.* 6 (7): e112–e112.

65 Sun, N., Zhang, C., Wang, J. et al. (2023). Hierarchical integration of DNA nanostructures and NanoGold onto a microchip facilitates covalent chemistry-mediated purification of circulating tumor cells in head and neck squamous cell carcinoma. *Nano Today* 49: 101786.

66 Zhu, J., Nguyen, T., Pei, R. et al. (2012). Specific capture and temperature-mediated release of cells in an aptamer-based microfluidic device. *Lab Chip* 12 (18): 3504–3513.

67 Guo, S., Huang, H., Deng, X. et al. (2018). Programmable DNA-responsive microchip for the capture and release of circulating tumor cells by nucleic acid hybridization. *Nano Res.* 11 (5): 2592–2604.

68 Liu, W., Wei, H., Lin, Z. et al. (2011). Rare cell chemiluminescence detection based on aptamer-specific capture in microfluidic channels. *Biosens. Bioelectron.* 28 (1): 438–442.

69 Abate, M.F., Jia, S., Ahmed, M.G. et al. (2019). Visual quantitative detection of circulating tumor cells with single-cell sensitivity using a portable microfluidic device. *Small* 15 (14): e1804890.

70 Labib, M., Green, B., Mohamadi, R.M. et al. (2016). Aptamer and antisense-mediated two-dimensional solation of specific cancer cell subpopulations. *J. Am. Chem. Soc.* 138 (8): 2476–2479.

71 Reinholt, S.J. and Craighead, H.G. (2018). Microfluidic device for aptamer-based cancer cell capture and genetic mutation detection. *Anal. Chem.* 90 (4): 2601–2608.

72 Chen, N., Yang, X., Wang, Q. et al. (2016). Proof of concept for inhibiting metastasis: circulating tumor cell-triggered localized release of anticancer agent via a structure-switching aptamer. *Chem. Commun.* 52 (41): 6789–6792.

73 Chen, N., Qin, S., Yang, X. et al. (2016). "Sense-and-Treat" DNA nanodevice for synergetic destruction of circulating tumor cells. *ACS Appl. Mater. Interfaces* 8 (40): 26552–26558.

74 Xu, C.M., Tang, M., Feng, J. et al. (2020). A liquid biopsy-guided drug release system for cancer theranostics: integrating rapid circulating tumor cell detection and precision tumor therapy. *Lab Chip* 20 (8): 1418–1425.

75 Feng, J., Xu, C.M., Xia, H.F. et al. (2023). Three-dimensional magnetic chip with extracorporeal circulation for circulating tumor cell in vivo detection and tumor growth inhibition. *Anal. Chem.* 95 (22): 8735–8743.

76 Zhou, Q., Rahimian, A., Son, K. et al. (2016). Development of an aptasensor for electrochemical detection of exosomes. *Methods* 97: 88–93.

77 Liu, C., Zhao, J., Tian, F. et al. (2019). lambda-DNA- and aptamer-mediated sorting and analysis of extracellular vesicles. *J. Am. Chem. Soc.* 141 (9): 3817–3821.

78 Zheng, L., Wang, H., Zuo, P. et al. (2022). Rapid on-chip isolation of cancer-associated exosomes and combined analysis of exosomes and exosomal proteins. *Anal. Chem.* 94 (21): 7703–7712.

79 Wu, J., Lin, Z., Zou, Z. et al. (2022). Identifying the phenotypes of tumor-derived extracellular vesicles using size-coded affinity microbeads. *J. Am. Chem. Soc.* 144 (51): 23483–23491.

80 Niu, Q., Shu, Y., Chen, Y. et al. (2023). A fluid multivalent magnetic interface for high-performance isolation and proteomic profiling of tumor-derived extracellular vesicles. *Angew. Chem. Int. Ed.* 62 (21): e202215337.

81 Fan, H.C., Fu, G.K., and Fodor, S.P. (2015). Expression profiling. Combinatorial labeling of single cells for gene expression cytometry. *Science* 347 (6222): 1258367.

82 Klein, A.M., Mazutis, L., Akartuna, I. et al. (2015). Droplet barcoding for single-cell transcriptomics applied to embryonic stem cells. *Cell* 161 (5): 1187–1201.

83 Chen, Y., Song, J., Ruan, Q. et al. (2021). Single-cell sequencing methodologies: from transcriptome to multi-dimensional measurement. *Small Methods* 5 (6): e2100111.

84 Zhang, M., Zou, Y., Xu, X. et al. (2020). Highly parallel and efficient single cell mRNA sequencing with paired picoliter chambers. *Nat. Commun.* 11 (1): 2118.

85 Cheng, Y.H., Chen, Y.C., Lin, E. et al. (2019). Hydro-Seq enables contamination-free high-throughput single-cell RNA-sequencing for circulating tumor cells. *Nat. Commun.* 10 (1): 2163.

86 Gierahn, T.M., Wadsworth, M.H. 2nd, Hughes, T.K. et al. (2017). Seq-Well: portable, low-cost RNA sequencing of single cells at high throughput. *Nat. Methods* 14 (4): 395–398.

87 Wang, Y. and Navin, N.E. (2015). Advances and applications of single-cell sequencing technologies. *Mol. Cell* 58 (4): 598–609.

88 Xu, X., Zhang, M., Zhang, X. et al. (2022). Decoding expression dynamics of protein and transcriptome at the single-cell level in paired picoliter chambers. *Anal. Chem.* 94 (23): 8164–8173.

89 Demaree, B., Delley, C.L., Vasudevan, H.N. et al. (2021). Joint profiling of DNA and proteins in single cells to dissect genotype-phenotype associations in leukemia. *Nat. Commun.* 12 (1): 1583.

90 Sundah, N.R., Ho, N.R.Y., Lim, G.S. et al. (2019). Barcoded DNA nanostructures for the multiplexed profiling of subcellular protein distribution. *Nat. Biomed. Eng.* 3 (9): 684–694.

91 Zhang, Y., Wang, Z., Wu, L. et al. (2018). Combining multiplex SERS nanovectors and multivariate analysis for in situ profiling of circulating tumor cell phenotype using a microfluidic chip. *Small* 14 (20): e1704433.

92 Gao, R., Zhan, C., Wu, C. et al. (2021). Simultaneous single-cell phenotype analysis of hepatocellular carcinoma CTCs using a SERS-aptamer based microfluidic chip. *Lab Chip* 21 (20): 3888–3898.

93 Zhang, X., Wei, X., Men, X. et al. (2021). Dual-multivalent-aptamer-conjugated nanoprobes for superefficient discerning of single circulating tumor cells in a microfluidic chip with inductively coupled plasma mass spectrometry detection. *ACS Appl. Mater. Interfaces* 13 (36): 43668–43675.

94 Spitzberg, J.D., Ferguson, S., Yang, K.S. et al. (2023). Multiplexed analysis of EV reveals specific biomarker composition with diagnostic impact. *Nat. Commun.* 14 (1): 1239.

95 He, D., Ho, S.L., Chan, H.N. et al. (2019). Molecular-recognition-based DNA nanodevices for enhancing the direct visualization and quantification of single vesicles of tumor exosomes in plasma microsamples. *Anal. Chem.* 91 (4): 2768–2775.

96 He, D., Wang, H., Ho, S.L. et al. (2019). Total internal reflection-based single-vesicle in situ quantitative and stoichiometric analysis of tumor-derived exosomal microRNAs for diagnosis and treatment monitoring. *Theranostics* 9 (15): 4494–4507.

97 Xia, Z., Zhang, X., Yao, J. et al. (2023). Giant enhancement of Raman scattering by a hollow-core microstructured optical fiber allows single exosome probing. *ACS Sens.* 8 (4): 1799–1809.

98 Ko, J., Wang, Y., Sheng, K. et al. (2021). Sequencing-based protein analysis of single extracellular vesicles. *ACS Nano* 15 (3): 5631–5638.

99 Wu, D., Yan, J., Shen, X. et al. (2019). Profiling surface proteins on individual exosomes using a proximity barcoding assay. *Nat. Commun.* 10 (1): 3854.

5

Spatiotemporal-Controlled Cell Membrane Engineering Using DNA Nanotechnology

Wenxue Xie[1], Cong Ren[1], Minjie Lin[2], and Hang Xing[1]

[1]Hunan University, Institute of Chemical Biology and Nanomedicine, College of Chemistry and Chemical Engineering, State Key Laboratory of Chemo/Biosensing and Chemometrics, Hunan Provincial Key Laboratory of Biomacromolecular Chemical Biology, Changsha 410082, China
[2]Zhejiang University, Second Affiliated Hospital, School of Medicine, Department of Emergency Medicine, Hangzhou 310009, China

5.1 Background

The cell membrane plays a crucial role in various cellular processes such as adhesion, proliferation, differentiation, and apoptosis [1–3]. Acting as a bridge between the cell and the external environment, the cell membrane separates the cytoplasm from the outside world [4]. Cells undergo dynamic responses to external cues by interacting with physical or chemical signals through their membrane structure [5, 6]. This interaction leads to pronounced alterations in cellular shape and function, enabling effective communication with the extracellular microenvironment and facilitating the analysis of intricate interactions between cells and their surrounding milieu, facilitating homeostasis regulation, intercellular communication, and tumor metastasis [7, 8]. For instance, Vascular Endothelial Growth Factor (VEGF) promotes endothelial cell and neuronal precursor proliferation, with VEGF concentration gradients guiding vascular growth toward hypoxic regions [9, 10]. Transforming Growth Factor (TGF) stimulates fibroblast differentiation into myofibroblasts, crucial for wound healing and fibrosis development, when secreted or present at elevated levels. The interactions between cells primarily occur through specific and dynamic regulation of membrane proteins [11, 12]. For instance, tumor cells rapidly upregulate PD-L1 on the cell membrane in response to pro-inflammatory factors in the tumor microenvironment, inhibiting T-cell immune responses through PD-L1/PD1 binding [13, 14]. Therefore, understanding the biological functions of the cell membrane provides a solid foundation for unraveling complex biological questions and developing advanced therapeutic approaches.

Wenxue Xie and Cong Ren contributed equally to this work.

DNA Nanotechnology for Cell Research: From Bioanalysis to Biomedicine, First Edition. Edited by Zhou Nie.

To investigate the biological functions of the cell membrane, a common approach involves employing chemical and biological methods to introduce synthetic handles onto the cell surface. These incorporated chemical and biological molecules offer potential opportunities not only for exploring the characteristics of the cell membrane but also for actively participating in the regulation of cell behavior and expanding cellular functionality. Currently, various types of chemical and biological handles, such as lipids, glycans, proteins, and nucleic acids, have been successfully incorporated onto the surfaces of live cells as synthetic receptors. These achievements have opened up a wide range of applications, including tissue engineering, cell-based immunotherapy, and enhanced cell delivery. These studies have significantly enhanced our understanding of cellular physiological processes, facilitated the regulation of signal communication between cells and the environment, and ultimately reshaped cell behavior and function [15–20].

Among all reported synthetic receptors, nucleic acids, with their precise Watson–Crick base pairing roles, have emerged as a novel and promising molecular tool for modifying the cell surface. These nucleic acids possess distinct biological recognition properties, a flexible and designable structure, precise programmability, and other advantageous features [21–23]. Recent advancements in DNA nanotechnology have facilitated the development of various dynamic and responsive nanodevices, as well as intricate DNA nanostructures. Dynamic DNA nanodevices, including DNA walkers [24, 25], DNA plugs [26, 27], DNA tweezers [28, 29], and others, can undergo conformational changes in response to different stimuli through chain displacement reactions, aptamer recognition sequences, and specialized structural modules [30, 31]. Furthermore, more complex DNA nanostructures, such as DNA origami [32, 33], DNA Lego [34, 35], and DNA meshes [36, 37], have significantly improved addressability and structural control, enabling precise positioning of proteins and nanomaterials at the nanoscale within three-dimensional spaces. The powerful tools provided by DNA nanotechnology have opened up new avenues for applications in synthetic tissues, whole-cell biosensing, and programmable intercellular communication.

In this book chapter, we have discussed DNA nanostructures for cell surface engineering and their dynamic behaviors that coordinate cell activities. Unlike previous review articles that primarily focused on the engineering of the external surface of the cell membrane, our focus in this chapter is on the use of functional nucleic acids for spatial control in cell membrane engineering, encompassing both the outer and inner leaflets. We provide an overview of the chemistries used to install DNA onto the exterior cell membrane surface, including nonspecific and specific modification methods. Nonspecific modifications encompass covalent modification and hydrophobic insertion, while specific modifications involve genetic engineering, metabolic processes, nucleic acid aptamer methods, as well as antibody binding. Furthermore, we survey the utilization of synthetic functional DNA nanodevices for cell surface sensing and imaging, cell communication, and cell protection. Additionally, we highlight potential chemistries for immobilizing functional motifs onto the inner leaflets of live cell membranes, which could be used for incorporating functional DNAs. Moreover, we introduce the liposome fusion-based transport (LiFT) approach developed by our research group. The LiFT approach allows the labeling of both sides of the plasma membrane with distinct sets of bioorthogonal DNA strands, enabling the regulation of intercellular communication and the monitoring of intracellular metabolites (Figure 5.1).

5.2 DNA Modifications on the External Cell Membrane Surface

The engineering of live cell membranes using DNA presents a significant challenge due to the dynamic nature of the cell membrane and the need for modification methods that do not disrupt normal biological functions. Various chemistries have been developed to incorporate DNA onto live cell membranes, primarily focusing on the external membrane surface. In this section, we provide an in-depth analysis of DNA modification methods on the extracellular membrane surface, along with their respective applications. These methods are categorized into nonspecific and specific modification approaches. Furthermore, we explore the diverse applications of these modification methods, including sensing and imaging, cellular assembly and communication, and cell protection.

5.2.1 Strategies to Incorporate DNA onto External Cell Membrane Surface

An important aspect of incorporating DNA strands onto the external surface of the cell membrane is the establishment of a stable anchor as a chemical "bond" that connects DNA with the biomolecules on the membrane. Considering the electrostatic repulsion between negatively charged DNA and lipids present on the membrane, the stability of this "chemical bond" is essential to ensure reliable immobilization. Furthermore, it is imperative for the bond to be biocompatible to avoid interfering with the biological processes of live cells. In this context, we highlight synthetic approaches for DNA modification, distinguishing between nonspecific and specific modifications based on whether DNA strands are attached to specific target molecules on the external surface of the cell membrane.

5.2.1.1 Nonspecific DNA Modification Strategies

The most commonly employed nonspecific modification strategies involve the covalent conjugation of DNA strands onto native functional groups exposed on the cell membrane, such as primary amine or thiol groups. The presence of various membrane proteins on the cell membrane results in the exposure of lysine and cysteine residues containing amino or thiol groups. These active residues serve as readily available but random sites for covalent DNA coupling. For instance, DNA strands with terminal mercaptothiol groups can be functionalized with *N*-hydroxysuccinimide (NHS) activated esters, and the resulting NHS-DNA conjugate can react with exposed primary amine residues on the cell surface to form a covalent conjugation (Figure 5.2a) [40]. Bertozzi and coworkers utilized

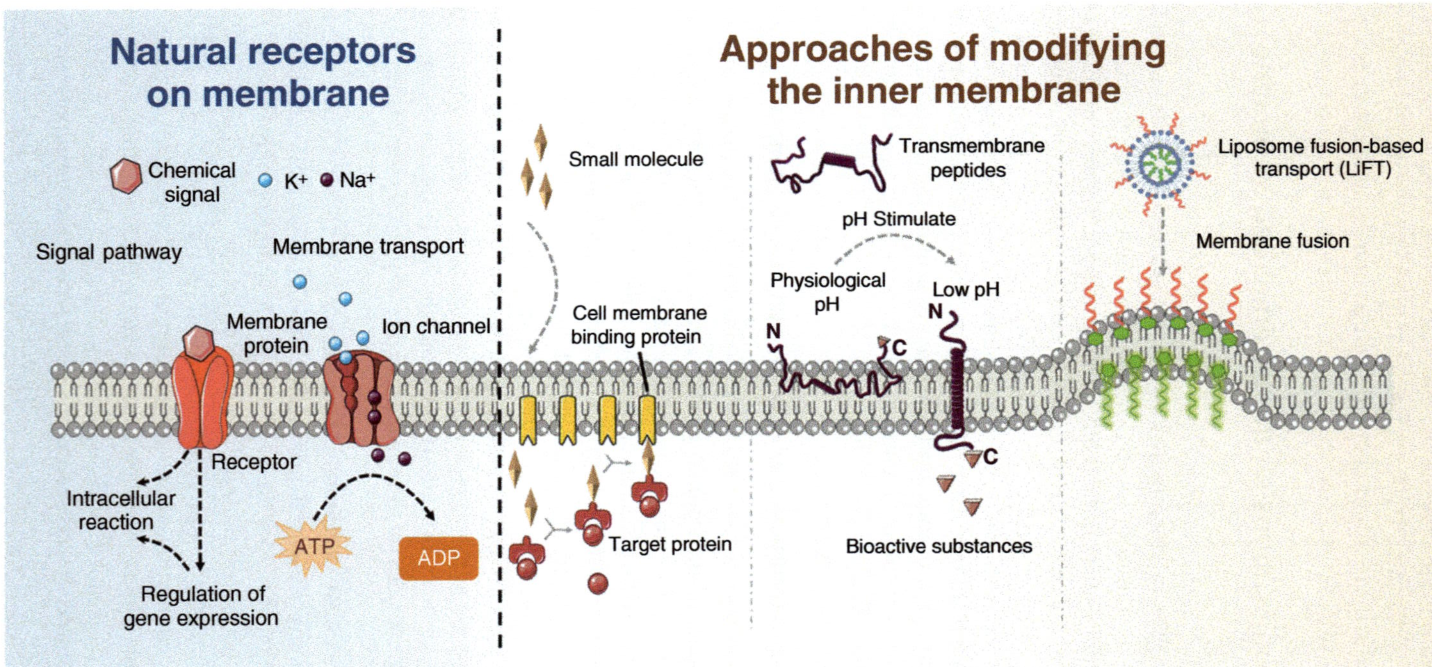

Figure 5.1 The left figure illustrates the involvement of the natural cell membrane in various physiological activities, such as signaling pathways [38] and membrane transport [39]. The right figure presents three potential approaches for modifying the inner side of the cell membrane, namely the small-molecule approach, the transmembrane peptide approach, and the LiFT approach.

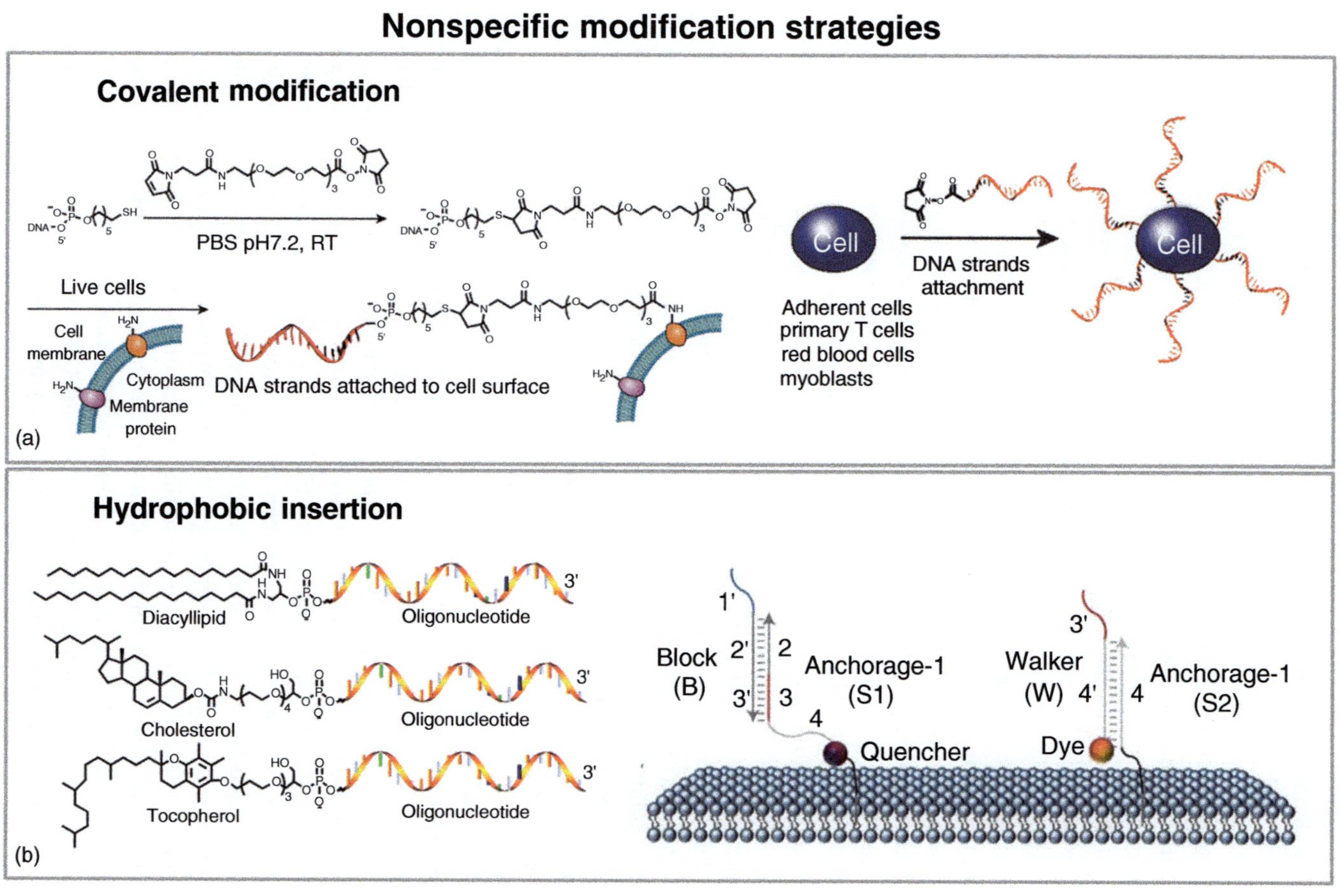

Figure 5.2 Nonspecific modification strategies to incorporate DNA onto the external surface of the cell membrane. (a) Covalent modification strategy. Thiolated single-stranded DNA was initially reacted with NHS-PEG-maleimide to form the NHS-DNA conjugate. Subsequently, the conjugate was incubated with live cell suspensions, allowing the NHS-DNA conjugate to attach to the exposed primary amine groups on the cell membrane [40]. Source: Reproduced with permission from the American Chemical Society. (b) Hydrophobic insertion strategy. DNA strands were modified at their termini with hydrophobic motifs, such as diacyl lipid, cholesterol, and tocopherol. These hydrophobic motifs were then inserted into the lipid bilayer, enabling the immobilization of DNA strands on the outer leaflet of the cell membrane [41]. Source: Reproduced with permission from Springer Nature.

NHS-DNA conjugates to directly modify the surfaces of different mammalian cells, including red blood cells and primary T cells. This method is rapid and efficient, enabling cell surface modification within one hour. Alongside amine-NHS coupling chemistry, the biotin-streptavidin interaction is another widely employed approach. In this method, surface-exposed free amine groups on the cell membrane are biotinylated using NHS-biotin, followed by binding of the biotinylated amines to streptavidin-DNA conjugates [42, 43]. Both approaches involve reactions with native primary amines on the cell membrane, which could potentially have undesired effects on cell surface receptors.

In addition to the covalent conjugation of functional DNA strands directly onto native protein residues, another commonly employed nonspecific cell surface modification approach is hydrophobic insertion. In this method, target DNA strands are modified at their termini with hydrophobic molecules. Since the cell membrane primarily consists of phospholipid bilayers with localized cholesterol, the terminal hydrophobic motifs on DNA can insert into the membrane, acting as anchors for immobilizing the target DNA strands. Various hydrophobic end-modifying groups have been used, including C18 chains, phosphatidyl, perfluorodecyl, cholesterol, and tocopherol [44, 45]. For instance, Darrell J. Irvine and coworkers developed amphiphilic oligonucleotides by covalently coupling hydrophobic dilauroyl tails with chemically stable RNA or DNA oligonucleotides. These amphiphilic oligonucleotides can hydrophobically insert into the surfaces of cancer cells, enabling direct labeling of tumor cells through hydrophobic insertion when injected into solid tumors. This approach has been employed for *in vivo* cell modification and localized immune therapy [46]. Tan and You's group developed DNA probes with terminal natural membrane components, such as phospholipids and cholesterol, which can modify live cell membranes as force interaction probes. These DNA probes can capture, accumulate, and transduce transient biological signals at the cell surface, generating detectable signals (Figure 5.2b) [41]. In another study, Sugimoto and coworkers used DNA probes with terminal cholesterol to modify cell membranes and monitor the dynamics of DNA hybridization on the cell membrane in real time [47]. Among the commonly used hydrophobic anchors, phospholipids such as 1,2-distearoyl-sn-glycero-3-phosphoethanolamine (DSPE) with two C18 hydrophobic carbon chains exhibit the strongest interactions and provide the most stable non-covalent DNA modification. However, an additional conjugation step is required to modify the DNA between dibenzocyclooctyne (DBCO)-containing DNA and azide-DSPE. It is important to note that different cells possess varying membrane lipid compositions, leading to different interactions between hydrophobic anchors and cell membranes, resulting in variations in binding affinity and dynamics. Therefore, careful optimization of hydrophobic interactions is necessary to achieve stable DNA modification without a rapid release from the membrane.

The aforementioned nonspecific approaches for DNA modification on the cell membrane are widely used due to their simplicity and efficiency. However, these approaches have inherent drawbacks and limitations. For instance, the hydrophobic insertion approach offers good anchoring efficiency, but it exhibits low binding affinity, which may result in the detachment of modified DNA from the cell membrane due to processes like endocytosis. On the other hand, covalent modification provides better chemical stability, but it relies on reactions with native primary amine groups at specific sites, leading to relatively lower anchoring efficiency. Moreover, covalent modification raises concerns regarding biocompatibility as it involves direct and random DNA modifications on native membrane proteins, potentially affecting the overall functionality of the cell membrane and cellular behavior. Additionally, nonspecific DNA modifications lack spatial control, preventing specific targeting of particular cell types or membrane proteins. Consequently, there is a pressing need for specific DNA modification strategies that enable precise engineering of cell membranes with high levels of control.

5.2.1.2 Specific DNA Modification Strategies

One of the major limitations of nonspecific DNA modification strategies on the cell membrane is their potential to modify any exposed amine or thiol groups, which can disrupt the natural membrane and protein functionality. Additionally, nonspecific modifications result in random positioning of DNA and cannot achieve cell-specific modifications. To address these limitations, specific DNA modification approaches have been developed, which selectively recognize and modify target handles on the cell membrane. These approaches involve the generation of bioorthogonal functional groups that are typically absent from the cell surface, as well as the recognition of specific surface receptors on the target cells. In this section, we highlight two specific DNA modification strategies. The first strategy is metabolic labeling, which entails introducing unnatural amino acids onto the cell membrane as bioorthogonal handles for sequential DNA coupling. The second strategy involves the use of specific targeting motifs, such as antibodies and nucleic acid aptamers, that selectively recognize specific receptors on the cell membrane, enabling receptor and cell selectivity in DNA modification.

Metabolic labeling is a specific modification strategy that has been widely employed, initially developed by Bertozzi and colleagues [48, 49]. This approach utilizes the inherent cellular mechanisms to incorporate tag groups onto biomolecules on the cell membrane using labeled metabolic precursors, such as unnatural amino acids. The introduced tags on the membrane can then serve as handles for sequential covalent coupling of other exogenous molecules, including DNA. Metabolic labeling offers significant advantages in cell-based therapeutics, live cell imaging, and cell tracking. A seminal example of metabolic labeling to introduce DNA onto the cell membrane was demonstrated by Bertozzi and coworkers, who treated cells with tetracarbonyl glucosamine-containing azide groups. The azide groups were then incorporated into cell surface glycans during glycosylation, providing a tag for subsequent coupling with alkyne-modified DNA strands through azide-alkyne click chemistry. These DNA strands were subsequently employed to program live cell adhesion using strands with complementary sticky ends. Additionally, phosphine-modified ssDNA was prepared using 5′-amine-modified ssDNA with phosphine pentafluorophenyl (PFP) ester, allowing Staudinger ligation for the attachment of phosphine-DNA to azide-modified cell membranes through amide bond formation (Figure 5.3a) [50]. In a similar vein, Zev J. Gartner and

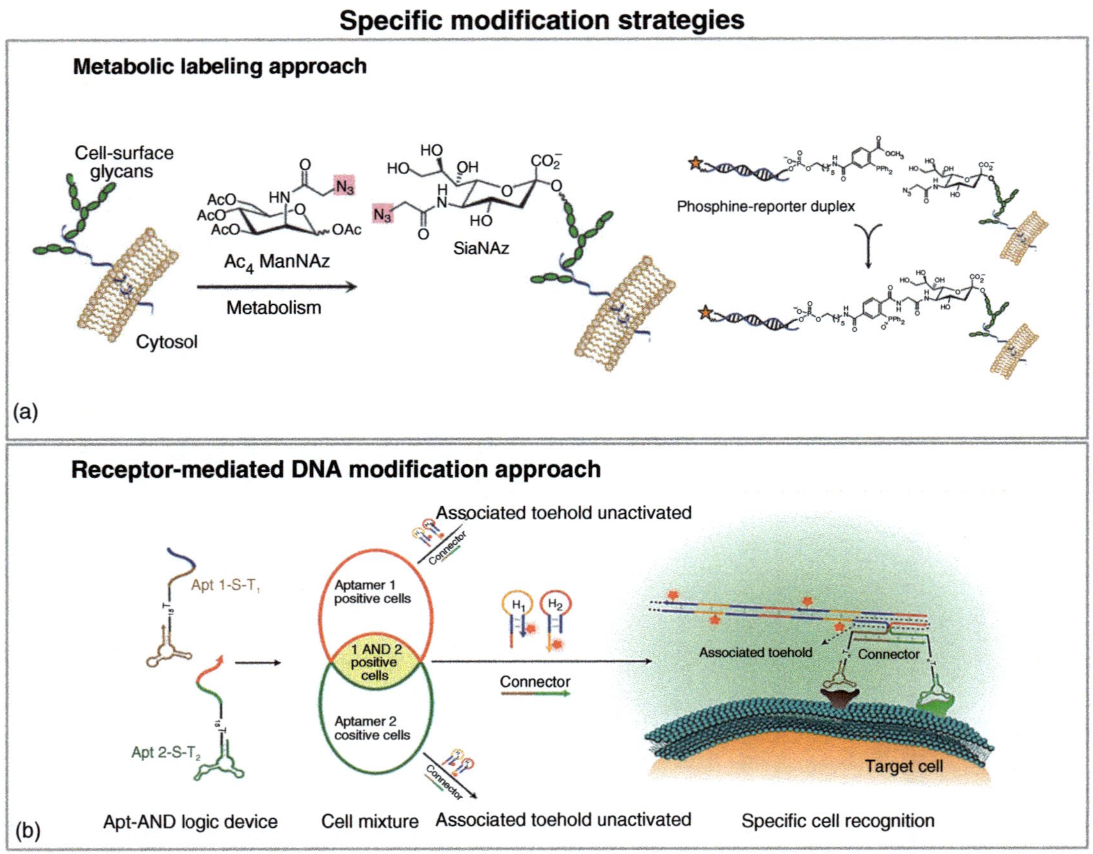

Figure 5.3 Strategies for specific DNA modification. (a) Metabolic labeling approach. Unnatural azide-modified monosaccharides are internalized by cells, metabolically processed, and incorporated into glycoconjugates. Phosphine-modified DNAs are then attached to the azide-modified cell membrane via Staudinger ligation [50]. Source: Reproduced with permission from John Wiley & Sons publication. (b) Receptor-mediated DNA modification approach. Functional DNA strands are introduced onto the cell surface through receptor-mediated modification. Recognition motifs, such as DNA aptamers, selectively recognize cell surface proteins and immobilize the strands onto the cell surface. The figure depicts a dual-aptamer-based AND logic device for target cell recognition and signal amplification [51]. Source: Reproduced with permission from Springer Nature.

colleagues utilized difluorocyclooctyne (DIFO)-modified DNAs to functionalize azido-treated cells. They employed this approach to regulate cell adhesion and membrane dynamics, investigate intercellular interactions, and control the modular bottom-up assembly of organized cellular structures. The use of metabolic labeling in this context relies on the incorporation of unnatural substrates referred to as bioorthogonal. These substrates lack natural reaction partners within the cellular environment, ensuring they do not interfere with normal cellular processes [52]. Moreover, the DNA conjugation process associated with metabolic labeling does not generate potentially harmful side reactions or byproducts that could be detrimental to the cells.

Unlike the metabolic labeling approach, which involves introducing exogenous bioorthogonal tags onto the cell membrane, the receptor-mediated DNA modification approach relies on the recognition of specific natural receptors present on the cell membrane [23]. Aptamers, which are biorecognition motifs, are commonly used for cell surface modification due to their accessibility similar to nonspecific modification strategies, as well as their ability to specifically target cell binding sites. Through the systematic evolution of ligands by exponential enrichment (SELEX) process from random oligonucleotide libraries, aptamers can be selected against cell membrane target substances with high specificity and affinity [53, 54]. Several DNA aptamers have been successfully screened, validated, and characterized for their ability to recognize specific cell surface receptors, including receptor tyrosine kinases (RTKs) such as MET, FGFR1, VEGFR2, and PDGFR [55], as well as integrins [56] and T cell receptors [57]. Aptamers do not require additional coupling modifications due to the designability of oligonucleotides and can be paired with other DNA sequences to extend from the cell surface. Han and Tan's group reported the use of dual DNA aptamers to recognize and modify cancer cells. In their study, aptamers sgc8c and TCO1 were used to target PTK7 on CEM cells and an unknown overexpressed protein, respectively. They designed a dual-aptamer-based AND logic device for target cell recognition and promoted rolling circle amplification of circular DNA templates, generating cancer cell-specific signals. They achieved single-step cancer cell identification and isolation with good sensitivity and accuracy (Figure 5.3b) [51]. Similarly, Yang and Tan's group used the TLS11a aptamer to target and modify HepeG2 cells, and they modified NK cells with an aptamer against PD-L1. The small size of aptamer-modified NK cells allowed them to penetrate solid tumors and recognize tumor cells for DNA-mediated cancer immunotherapy [58].

In comparison to nonspecific modification approaches, specific DNA modification methods typically exhibit lower anchoring efficiency, resulting in lower surface DNA density on the cell membrane. However, these specific DNA modification strategies offer the advantage of targeting specific cells and membrane receptors, thereby enabling potential spatiotemporal control of extracellular DNA membrane modification.

5.2.2 Applications of External Membrane-modified DNA

5.2.2.1 Sensing and Imaging on Cell Membrane Using DNA Probes

Functional DNA molecules, such as aptamers, molecular beacons, and DNAzymes, have been extensively employed in biosensing applications to monitor various biological targets in the extracellular microenvironment, such as metal ions, pH, and metabolic molecules. In this section, we provide an overview of the sensing and imaging applications utilizing functional DNA probes immobilized on the surface of live cell membranes [59].

The extracellular pH within the microenvironment is important for cellular physiological processes. Consequently, the modification of the extracellular leaflet of the cell membrane with functional DNA probes enables direct and accurate monitoring and detection of extracellular pH, providing valuable insights for biomedical research. Yang et al. developed a cell surface-anchored pH probe utilizing a hydrophobic-anchored DNA nanodevice. This DNA nanodevice consisted of a pH-sensitive i-motif-based pH indicator and a pH-insensitive fluorescent dye serving as an internal reference for detecting extracellular pH. The DNA-based pH probe demonstrated high sensitivity and reversibility to pH changes within the range of 6.0–8.0, enabling precise detection of extracellular pH even in three-dimensional tissue cultures [60]. Similarly, Nie et al. constructed a cell surface-anchored DNA probe with an i-motif structure that undergone a transition to a four-strand structure under low pH conditions, allowing for real-time monitoring of extracellular pH. The pH-sensitive i-motif sequence controls the opening and closing state of the tweezer, and the output is detected through a pair of FRET fluorescent groups, enabling sensitive detection of extracellular pH values [61].

In addition to extracellular protons, living cells also employ the secretion of signaling molecules to integrate, encode, and transmit information to their surrounding environment [62]. Exploring these secreted signaling molecules is crucial for understanding cell signaling and intercellular communication processes. One prominent example is the monitoring of extracellular adenosine triphosphate (ATP) [63–65]. ATP, as a vital extracellular signaling molecule, plays a role in various physiological and pathological processes, including cell growth, trauma response, tissue injury, inflammation, and more. However, accurately measuring ATP concentration with dynamic monitoring capability and high spatial resolution presents a challenge [66]. To address this challenge, Li et al. designed a cell surface-anchored, self-phosphorylating DNAzyme sensor (Figure 5.4a) [67]. In the presence of ATP, the substrate strand can transfer a phosphate group from ATP to its 5′ end, leading to self-phosphorylation. Consequently, the DNAzyme digests the substrate strand, generating a fluorescence signal. The membrane-anchored sensor demonstrated its utility as a powerful tool with high temporal resolution and ligand specificity for monitoring extracellular ATP [71]. Upon binding to adenine, the conformation of the fluorescent aptamer changes, resulting in a rapid increase in fluorescence signal within seconds. The sensor enables real-time monitoring of adenine release.

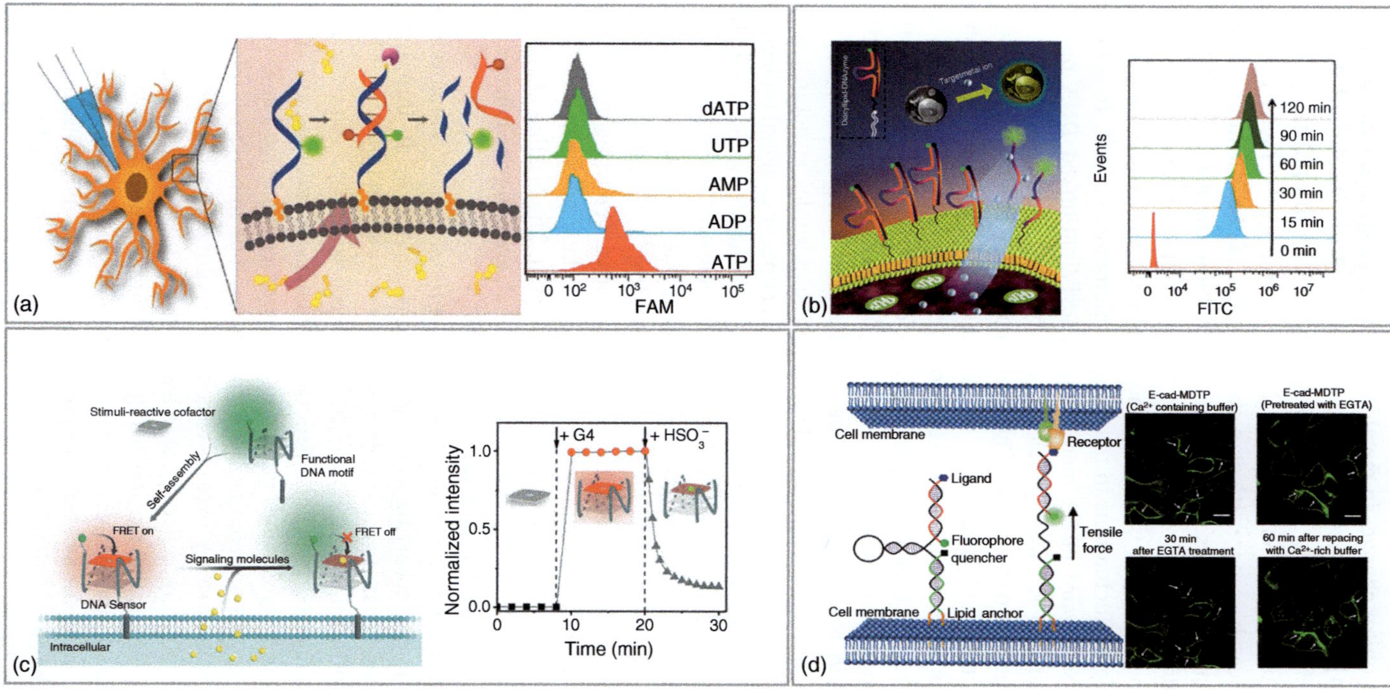

Figure 5.4 Sensing and imaging applications on a cell membrane using immobilized DNA probes on the outer leaflet. (a) Cell surface-anchored self-phosphorylating DNAzyme sensor enables fluorescence imaging of extracellular ATP transients occurring on a subsecond time scale [67]. Source: Reproduced with permission from the American Chemical Society. (b) Cell membrane-anchored Mg^{2+} DNAzyme probe for real-time monitoring of metal ions in the cellular microenvironment [68]. Source: Reproduced with permission from the American Chemical Society. (c) Stimuli-reactive cofactor-based DNA G4 sensor for monitoring SO_2 signaling molecules in the microenvironment [69]. Source: Reproduced with permission from John Wiley & Sons publication. (d) Membrane-anchored DNA tension probes for visualization of tensile forces at cell junctions [70]. Source: Reproduced with permission from the American Chemical Society.

Metal ions play a crucial role as biologically important metabolites involved in diverse cellular functions. DNA-based sensors immobilized on the cell surface offer the ability to monitor changes in metal ion concentrations in the extracellular environment [72–74]. For instance, Tan et al. anchored a Mg^{2+} specific DNAzyme sensor to the cell membrane by incorporating a modified hydrophobic group at the terminus. The presence of extracellular Mg^{2+}, serving as the metal cofactor for the DNAzyme catalytic core, triggers the cleavage of the substrate strand, resulting in an enhanced fluorescence signal and enabling the detection of Mg^{2+} concentration (Figure 5.4b) [68]. However, as the DNAzyme sensor exhibits irreversible behavior toward Mg^{2+}, the cell surface-anchored sensors cannot provide real-time monitoring of fluctuations in the cellular microenvironment. To achieve a reversible response toward target metal ions, diacyl lipid-linked DNA probes capable of responding to K^+ were developed for dynamic monitoring of K^+ in the extracellular microenvironment. These G-quadruplex-based DNA probes were immobilized on the cell membrane. In the presence of K^+, the DNA probes undergo a transition from an extended state to a contracted state, leading to an increase in the fluorescence signal. Importantly, the conformational change of the G-quadruplex DNA structure is reversible, allowing for the dynamic quantification of K^+ fluctuations in real time within the cellular microenvironment [75]. In addition to metal ions, Nie et al. assembled DNA G-quadruplex structures with stimuli-responsive cofactors capable of responding to endogenous signal molecules such as HSO_3^-. In this study, they synthesized sulfur dioxide derivatives as active cofactors by introducing a conjugated bridge in the structure of the benzothiazole moiety. The benzothiazole core structure was selected due to its substantial fluorescence enhancement when combined with G-quadruplexes, and its bridging double bond readily undergoes reactions with SO_2 derivatives through Michael addition. The assembled G-quadruplex structure with cofactors was immobilized on the cell membrane, resulting in a turn-on FRET signal. The binding of HSO_3^- to the structure leads to the turn-off of the FRET signal, enabling the detection of cell-released HSO_3^- signals (Figure 5.4c) [69].

Furthermore, cell membrane-immobilized DNA probes can be employed to detect mechanical forces occurring at the cell interface. You et al. introduced DNA-based membrane tension probes for visualizing forces at cell junctions. These probes, modified with lipids, can efficiently and securely anchor onto the cell membrane. When subjected to tension generated by adjacent cells, the unfolding of the probe results in a significant increase in fluorescence intensity. These DNA membrane tension probes are compatible with fluorescence microscopy and can be widely utilized for measuring forces between cells (Figure 5.4d) [70]. Moreover, You et al. presented a fluorescence lifetime-based method for precise quantification of intercellular molecular tension using membrane-anchored DNA tension probes. Compared to traditional fluorescence intensity-based DNA tension probes, a major advantage of lifetime-based measurements is their independence from concentration and the elimination of the need for additional reference fluorophores for calibration. The fluorescence tension probes were applied to quantify the correlation between tension in E-cadherin and N-cadherin proteins [76].

5.2.2.2 DNA-Mediated Cell Assembly and Communication

Multicellular systems rely on cell-to-cell interactions to coordinate intercellular signaling pathways and regulate cellular functions. Cell communication plays a crucial role in biological processes, coordinating intracellular biochemical pathways through transmembrane signaling to respond to changes in environmental conditions as well as regulating physiological and pathological behaviors through intercellular signaling pathways in multicellular organisms. Disruption of these intercellular interactions can lead to biological dysfunction, including metabolic disorders and diseases such as autoimmune diseases and cancer [77, 78]. Cell communication occurs through two main pathways. The first is contact-based communication, where neighboring cells interact through surface ligands and receptors. When a receptor recognizes a ligand on adjacent cells, conformational changes occur, initiating subsequent signaling processes within the cells. For instance, when T cells recognize antigens presented by antigen-presenting cells (APCs), an immunological synapse forms, involving T cell receptors, major histocompatibility complex (MHC), co-stimulatory receptors, adhesion molecules, intracellular kinases, and scaffold proteins, to activate T cell receptor signaling [79, 80]. The second form of cell communication is diffusion-based communication, where cells indirectly communicate by secreting diffusible proteins or small molecules. Localized sources of diffusible proteins can create concentration gradients, providing positional information to surrounding cells. For example, released interleukin-2 (IL-2) promotes the proliferation and activation of T cells [81]. Studying cell-to-cell interactions provides insights into fundamental multicellular behavior and holds great significance. Leveraging the unique advantages of DNA nanotechnology, DNA modification on the cell surface offers an attractive approach for controlling intercellular interactions and serves as a powerful tool for regulating cell assembly and communication.

One of the pioneering works exploring the use of DNA to program the assembly of mammalian cells was reported by Bertozzi and coworkers in 2009. They utilized cell membrane-anchored complementary DNA sequences to guide the assembly of cells, forming three-dimensional microtissues (Figure 5.5a) [82]. The specific cell-to-cell contacts and multi-cell structures could be deliberately controlled through DNA hybridization. Subsequently, Gartner and coworkers performed stepwise assembly of dissociated cells modified with single-stranded DNA onto surfaces modified with complementary DNA sequences. By carefully designing the DNA sequences and the order of introducing DNA-modified cells, they successfully achieved programmable construction of three-dimensional organ-like microtissues, allowing for more precise control over the overall cell assembly. These DNA-mediated cell assemblies could be enzymatically digested by DNase I after formation, providing modular tools for three-dimensional tissue structure cultivation (Figure 5.5b) [83]. In addition to DNA base pairing, the specific biorecognition abilities of DNA can also be utilized to control interactions between cells. For instance, Lu and coworkers reported a DNAzyme-based approach that employed metal ions to control intercellular interactions (Figure 5.5c) [84]. They used ion-selective DNAzymes to modify cell surfaces, enabling reversible cell assembly and disassembly based on

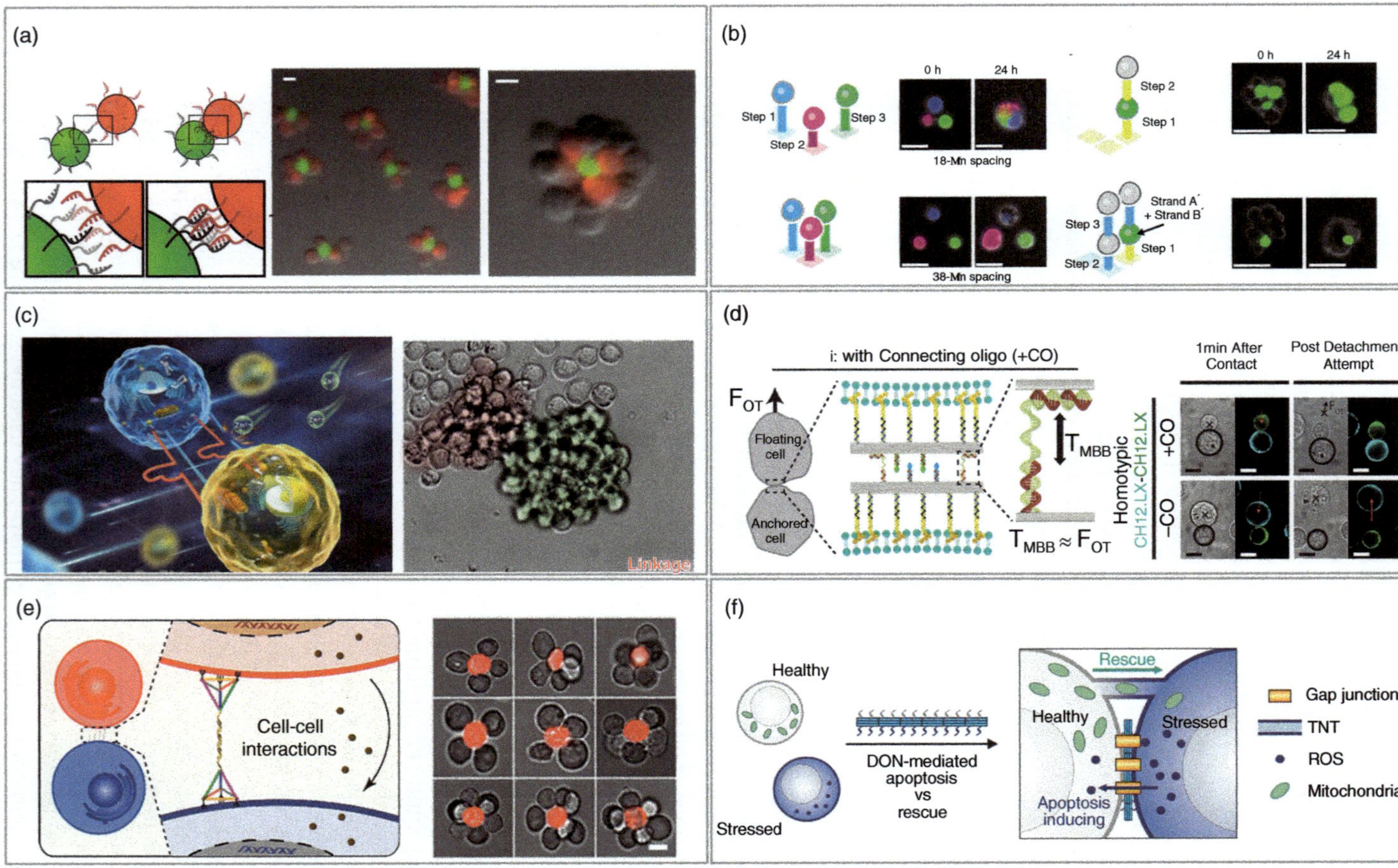

Figure 5.5 DNA-mediated cell assembly and cell communication. (a) Stable multicellular clusters are formed by cells bearing complementary cell surface oligonucleotides [82]. Source: Reproduced with permission from PNAS. (b) Stepwise surface-based DNA programming of cell assembly for the reconstruction of multicellular tissue size, shape, composition, and spatial heterogeneity, leading to programmed organoids [83]. Source: Reproduced with permission from Springer Nature. (c) Metal-dependent RNA-cleaving DNAzymes are employed to regulate cell–cell interactions in response to metal ions [84]. Source: Reproduced with permission from the American Chemical Society. (d) Engineering of cell functionalities and interactions using DNA origami nanostructures [85]. Source: Reproduced with permission from John Wiley & Sons publication. (e) Cell assembly is mediated by DNA tetrahedrons, and cell communication is facilitated through diffusion [84, 86]. Source: Reproduced with permission from the American Chemical Society. (f) Formation of gap junctions and control of contact-based cell communication using DNA origami. Source: Adapted from Li et al. [87].

DNA hybridization and metal ion-induced cleavage of the substrates. Furthermore, they applied the DNAzyme-engineered cell membrane to regulate T cell-cancer cell interactions. By combining MUC-1 aptamer and DNAzymes, the engineered T cells could recognize neighboring cancer cells, leading to an enhanced killing effect in zinc-deficient cancer cells [88]. This work provides a powerful tool to promote cancer cell apoptosis through dynamic regulation of cellular interactions. Yang and Tan's group employed DNA aptamers to modulate the interactions between natural killer (NK) cells and cancer cells as well as to obstruct the PD-1/PD-L1 checkpoint to facilitate adoptive immunotherapy [58].

More intricate DNA nanostructures, including DNA tetrahedrons, DNA origami, and DNA circuits, have been documented as means to govern cell assembly and communication. Carlos E. Castro and coworkers devised specific and reversible methodologies to functionalize the cell membrane utilizing DNA origami nanodevices. These DNA origami structures exhibit robust membrane functionalization across epithelial, mesenchymal, and non-adherent immune cells. By immobilizing the DNA origami on the membrane, they facilitated the formation of high-order DNA assemblies and programmed cell–cell adhesion among homotypic and heterotypic cells through sequence-specific DNA hybridization (Figure 5.5d) [85]. Tan et al. employed hydrophobic end-modified tetrahedral DNA probes to modify the cell surface, guiding intercellular interactions (Figure 5.5e) [86]. Multiple anchoring points on the DNA tetrahedron facilitated effective cell membrane modification via hydrophobic interactions. Building upon this platform, they demonstrated a dynamic simulation of membrane proteins and manipulation of cellular interaction responses to environmental changes. Furthermore, they employed membrane-anchored DNA tetrahedron linkers of varying lengths to systematically investigate the impact of the distance between APC cells and T cells on T cell activation. They discovered that T-cell activation can be regulated by adjusting protein recombination and mechanical force [89]. Additionally, Pei et al. proposed a method based on DNA feedback loops to modify the cell surface and achieve programmable regulation of the interaction between immune cells and cancer cells [90]. Similarly, Fan et al. utilized DNA origami nanotubes to establish biomimetic membrane channels for studying diverse types of cell–cell communication. They harnessed this system to organize T cells and cancer cells in predetermined ratios and geometries, thereby programming cell proliferation and apoptosis. The demonstrated apoptotic cells were able to induce apoptosis in adjacent cells through gap junctions, while healthy cells transferred functional mitochondria to stressed cells to rescue them (Figure 5.5f) [87].

The examples discussed in this section indicate that integrating DNA nanodevices onto the cell membrane offers a novel platform for controlling cell assembly and regulating the functions of living cells, thereby holding promise for future advancements in immunocancer therapies.

5.2.2.3 Cell Protection Using DNA Structures on Membrane

Organisms employ various biological defense mechanisms to safeguard and transmit genetic information and protect cellular components in adverse conditions.

One effective strategy involves the formation of robust and flexible armor. Plant and microbial cells, for instance, possess a resilient outer cell wall composed of cross-linked polysaccharides, enabling their survival and recovery even in harsh environments. In contrast, mammalian cells feature a vulnerable outermost layer, a phospholipid bilayer membrane with proteins, making them susceptible to environmental changes. Therefore, the development of membrane protection materials becomes crucial to advance the applications of mammalian cells. DNA has emerged as a promising alternative to peptides and polysaccharides for the modification of live cell membranes to serve as biological protection materials.

Wang and coworkers pioneered the construction of biomimetic cell walls using a synthetic framework on the membrane to shield live mammalian cells. The framework combines DNA and a polyelectrolyte complex comprising alginate and polylysine. The synthesis involves consecutive DNA hybridization chain reactions (HCR) and polyelectrolyte complexation performed directly on live cells. By operating under physiological conditions, the biomimetic cell walls maintain the biological activity of the mammalian cells. Importantly, the biomimetic cell wall not only protects the cells from physical attacks but also provides defense against biological attacks [91]. Li and coworkers demonstrated a programmable and biocompatible encapsulation method utilizing *in situ* DNA-directed polymerization to graft DNA polymers onto live cells. The process involves two steps: membrane-anchored initiator primers mediating rolling circle amplification reactions, followed by enzymatic polymerization to generate cross-assemblies of DNA polymers. These densely grafted DNA polymers form a protective coating on the cell surface. The DNA coating was successfully demonstrated on bacteria, yeast, and mammalian cells [92]. Additionally, Zhu and coworkers proposed an elastic biomineralization strategy for immune cells, achieved through a double aptamer-based hybrid chain reaction and calcium ion-driven assembly. This consensus recognition design enhances selectivity and effectively shields cells from environmental influences. The method can be utilized for the protection, separation, and identification of circulating tumor cells, preserving their natural state and providing complete genetic information for downstream analysis [93].

5.3 DNA Modifications on the Internal Cell Membrane Surface

As discussed earlier, most studies have primarily focused on DNA modifications occurring on the extracellular side of the cell membrane, with limited reports on engineering the intracellular side. However, the internal side of the cell membrane plays an even more critical role for cells in natural systems. The inner leaflet of the cell membrane directly interacts with the cytoplasm and is involved in cellular metabolic reactions and transmembrane signaling pathways. RTKs [94] and G protein-coupled receptors (GPCRs) [95], two main types of transmembrane signaling systems, undergo dimerization or conformational changes during signal transduction, generating intracellular signaling molecules that regulate cell behavior through their intracellular domains. Therefore, studying the biological processes occurring on the inner leaflet of the cell membrane is highly important but often challenging due to the lipid bilayer barrier. Inspired by nature, there has been significant interest in developing artificial transmembrane systems and synthetic approaches to engineer the inner leaflet of the cell membrane. Efforts have been focused on the internal side of the cell membrane to investigate the molecular mechanisms of natural transmembrane receptors, monitor quantitative changes in intracellular metabolites, and explore applications such as controlling protein translocation and developing new therapeutics.

In this section, we specifically discuss approaches that enable DNA modifications on the internal cell membrane surface, including transmembrane modification and direct engineering of the inner leaflet. Firstly, we introduce transmembrane modification approaches utilizing DNA origami nanostructures as well as internally modified single- and double-stranded DNAs. Secondly, we present several synthetic approaches for inner leaflet engineering, which may currently be applied to small molecule or peptide modifications but hold potential for DNA modification. Lastly, we highlight a LiFT approach developed by our group, enabling efficient modification of both sides of the cell membrane using two bioorthogonal functionalities.

5.3.1 Transmembrane Modification of DNA

To engineer the internal side of the live cell membrane using DNA, a straightforward approach involves utilizing DNA structures with hydrophobic groups modified in the middle region of the structure. Similar to the hydrophobic insertion approach used to anchor DNA sequences onto the external surface of cell membranes through terminal hydrophobic modification, researchers have demonstrated that by localizing hydrophobic anchors onto the internal position of the phosphate backbone, transmembrane modification of functional DNA structures can be achieved, with half of the structure inside and half outside the membrane.

In this section, we provide an overview of two different examples of transmembrane modification of functional DNA structures. The first example involves the direct insertion of internally modified single- or double-stranded DNAs, with cholesterol modifications, into the cell membrane through hydrophobic insertion. The second example entails the use of more complex DNA nanostructures, such as DNA origami, to mimic ion channel proteins on the cell membrane. We also discuss the relevant applications and current challenges associated with these approaches.

5.3.1.1 Transmembrane Modification Using Single-Stranded and Double-Stranded DNA

By incorporating cholesterol-modified single- or double-stranded DNA in the middle region of the backbone, the cholesterol molecules can act as hydrophobic anchors, inserting into the lipid bilayer and allowing the DNA sequences to adopt a transmembrane configuration, with half exposed to the intracellular environment and half to the extracellular environment. Liu and coworkers developed an artificial transmembrane signaling pathway based on the formation of an i-motif under low pH conditions and the dimerization of DNA-based artificial membrane receptors. This pathway activates fluorescence resonance energy transfer and G-quadruplex/heme-mediated fluorescence amplification in giant unilamellar vesicles (Figure 5.6a) [96]. The artificial transmembrane unit consists of dimeric structures, with each monomer featuring three functional modules: a pH-responsive split i-motif as the top signal recognition module, a cholesterol-labeled sequence as the transmembrane spanning connector module, and a split G-quadruplex as the bottom signal output module, which performs peroxidase-like catalytic reactions. The cholesterol insertion enables the DNA transmembrane tool to traverse the lipid bilayer, and the dimerization of the artificial receptor leads to the production of fluorescence signals in giant unilamellar vesicles. This study represents a significant advancement in the design of environmentally responsive artificial signal systems and offers opportunities to establish signaling networks in primitive cell colonies. The researchers further extended this strategy to develop other artificial transmembrane receptor signaling systems using split aptamers responsive to ATP and enzymes [100].

Liu and coworkers demonstrated the transmembrane signaling system in artificial cells. Additionally, Yang and coworkers employed similar DNA designs to immobilize dsDNA as a transmembrane connector for the detection of HGF-induced VEGF secretion on live mammalian cell membranes (Figure 5.6b) [97]. HGF-induced c-Met dimerization leads to VEGF secretion, which can be directly monitored by surface-immobilized dsDNA probes, providing insights into cell signal perturbations. This work enables transmembrane DNA modification in live cells, allowing for the exploration of related signal components in endogenous cell signaling. However, although DNA sequences are exposed on the intracellular side of the membrane, the functions of the inner surface DNAs have not been demonstrated. Furthermore, Göpfrich et al. reported that dsDNA with porphyrin modifications inserted into membranes can locally induce the formation of hydrophilic pores. They further constructed a hollow barrel structure comprising four double helices, which, upon insertion into the cell membrane, exhibited catalytic activity, promoting lipid transfer between membrane leaflets at a three-fold higher rate than protein enzymes [101].

One notable limitation of the internal cholesterol insertion strategy is the difficulty in controlling the orientations of immobilized transmembrane DNA strands, as there is minimal selectivity for DNA sequence during the hydrophobic insertion process. This approach typically results in transmembrane DNA structures with random orientations, with each strand having a 50% chance of its 5′-end or 3′-end facing outward.

5.3.1.2 Transmembrane Modification Using DNA Nanostructures

In addition to ssDNA and dsDNA, DNA nanostructures have also been explored for transmembrane modification on live cell membranes. Currently, research in this field mainly focuses on two aspects: one is to develop DNA nanopores with larger pore sizes to transport larger molecules, and another is to control the state

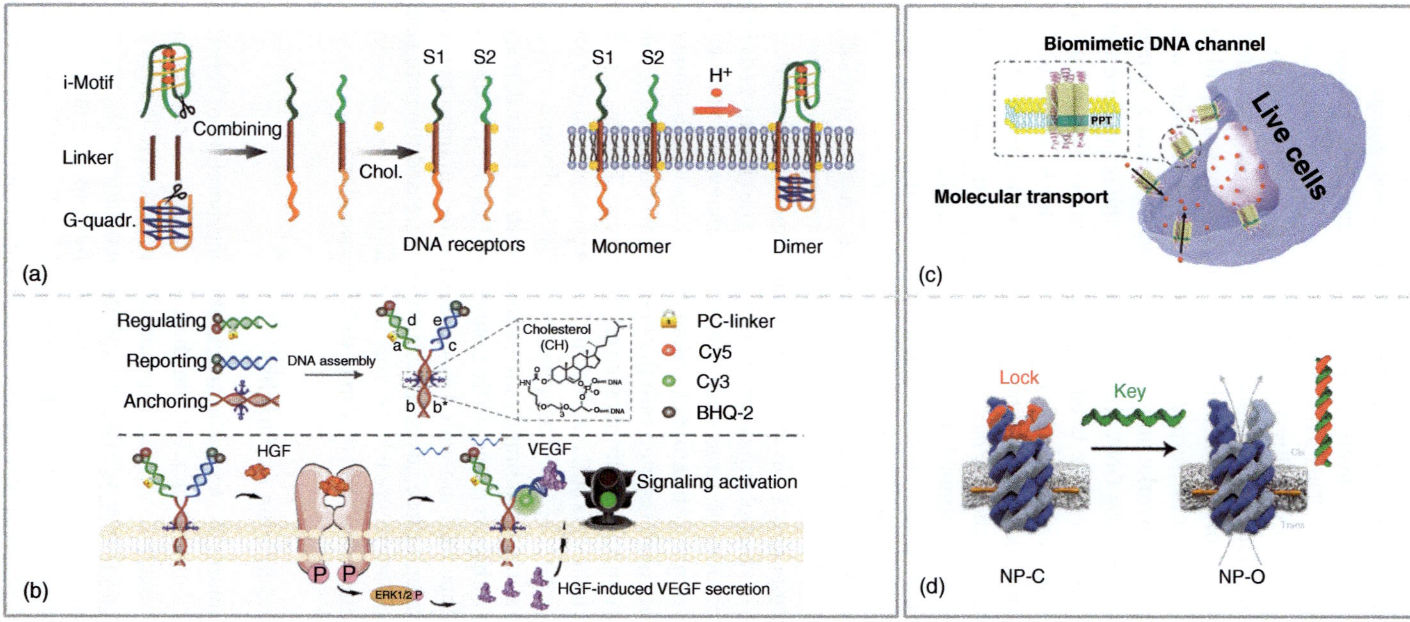

Figure 5.6 DNA modifications on the internal cell membrane surface. (a) A DNA-based artificial membrane receptor enabling transmembrane signal transduction in artificial cells through chemical input-mediated dimerization [96]. Source: Reproduced with permission from John Wiley & Sons publication. (b) Cell membrane-anchored dsDNA structure for detecting HGF-induced VEGF secretion in cellular signaling [97]. Source: Reproduced with permission from John Wiley & Sons publication. (c) Phosphorothioate-modified DNA bundle structures enable structural mimicry of biological channels for molecular transportation across live cell membranes [84, 98]. Source: Reproduced with permission from the American Chemical Society. (d) DNA origami-based nanopore structure with a nanomechanical and sequence-specific gate for regulating transmembrane flux [99]. Source: Reproduced with permission from Springer Nature.

(open and closed) of DNA-constructed nanopores to achieve stimuli-responsive transport of cargo molecule. Tan and coworkers have reported the development of a sulfur-containing phosphoric acid-modified DNA nanochannel, composed of six duplex DNA bundles, that can effectively mimic biological channels for molecular transport on live cell membranes. This nanochannel possesses a stable structure, a small hollow size (<2 nm), and the ability to interact with lipid molecules, enabling its stable insertion into the membrane. Importantly, it has been demonstrated that this transmembrane channel can facilitate the transport of both ions and anti-tumor drugs to neurons and cancer cells, respectively, within a specific time window (Figure 5.6c) [98]. Similarly, Howorka and coworkers have designed a switchable molecular composition valve using DNA strands to control the open-close state of DNA nanopores. The molecular valve is equipped with four cholesterol groups, which aid in inserting the negatively charged pore into the hydrophobic lipid membrane. This valve can selectively bind to specific ligands to regulate the states of the transmembrane channel, thereby facilitating the selective transport of organic small molecules and specific inorganic substances. This artificial nanopore exhibits notable selectivity to some extent and demonstrates structural and functional similarities to natural ion channels. Furthermore, the researchers have developed a temperature-controlled DNA nanofluidic valve in the form of a barrel-shaped nanofluidic valve composed of seven oligonucleotides. This valve remains in a closed state at ambient temperature but opens when the temperature exceeds 40 °C, enabling the transport of small molecules across the membrane (Figure 5.6d) [99].

To enable the construction of large-scale DNA nanochannels, Simmel and coworkers utilized DNA origami structures to create nanopores with a diameter of 4 nm. These nanochannels were inserted into the lipid bilayer through the interaction between streptavidin and the hydrophobic cell membrane. However, the pore size achieved in these structures is still insufficient for the passage of larger protein molecules or nanoparticles [102]. In subsequent work, Howorka and coworkers successfully constructed a transmembrane nanochannel with an inner diameter of 7.5 nm, enabling protein transport through DNA nanochannels [103]. Hatzakis and Kjems's group also developed a 9-nm-wide DNA nanochannel on liposomes using DNA origami [104]. Although these advancements have increased the pore size of DNA nanochannels, their selectivity in transporting molecules remains poor, and further optimization of their structural design is still necessary [105].

5.3.2 Inner Leaflet Modification Approaches

In addition to transmembrane DNA modifications, the incorporation of functional DNA and DNA nanostructures exclusively on the inner leaflets of the cell membrane has received limited attention. This is primarily due to the challenges posed by the lipid bilayer barriers, which restrict the penetration of large-scale, negatively charged DNA strands. Consequently, traditional nonspecific or specific modification approaches have encountered limitations in accessing the cell membrane's inner side, thereby rendering DNA modifications on the inner leaflet more challenging. In this review, we provide a comprehensive overview of existing approaches for inner leaflet modification. While most of these approaches have been primarily employed for proteins or small molecules and have not yet been extensively used for DNA modification, they hold the potential for delivering DNA onto the inner leaflets. Notably, we emphasize the LiFT approach developed by our research group, which represents the pioneering example of successful inner leaflet DNA modifications on the inner membrane surface.

5.3.2.1 Small-Molecule Ligand Targeting Approach

Small molecules can readily penetrate the cell membrane and traverse into the cytoplasm via concentration gradients. These small molecules, known as "chemical inducers of dimerization" (CIDs), can induce protein-protein interactions, including heterologous or homologous protein dimerization on the cell membrane, particularly on the inner side. For instance, rapamycin serves as an illustrative example. Upon entering the cytoplasm, rapamycin induces the translocation of cytoplasmic proteins onto the inner leaflet of the plasma membrane in living cells [106]. The fusion of one of the rapamycin binding domains (FRB) to the desired site on the target protein (POI), while the other rapamycin binding domain (FKBP) remains fused to the cytoplasm, establishes an anchor at the target site. Following the addition of rapamycin to the cell, the compound diffuses across the plasma membrane and initially binds to FKBP. Subsequently, the rapamycin-FKBP complex binds to FRB. Within seconds of rapamycin addition, the protein of interest (POI) forms a complex and translocates to the location of the FRB domain, namely the inner leaflet of the plasma membrane. If the POI possesses intrinsic enzymatic activity, it can carry out its function at the membrane, where the corresponding enzyme substrate resides. Rapamycin-based CIDs have provided a broad array of tools that rely on CID-dependent mechanisms and cover numerous cellular signaling modules, ranging from small G proteins, heterotrimeric G protein subunits, phosphoinositide metabolic enzymes, receptor and non-receptor tyrosine kinases, and serine/threonine protein kinases, to transmembrane receptors and signal adapter proteins without enzymatic function (Figure 5.7a) [107].

By employing CIDs as binder motifs, Tsukada et al. developed a small-molecule ligand modified with coumarin acid at one end, enabling direct targeting of the inner side of the cell membrane. The ligand forms a dimeric structure by binding to the target protein in the cytoplasm via the other end of the ligand, thereby guiding the protein to the inner side of the cell membrane. This modular approach offers programmable modification of intracellular target proteins on the inner side of the cell membrane by adjusting ligand polymerization (Figure 5.7b) [108]. An existing limitation of rapamycin and rapalogues lies in the challenge of controlling the translocation process, as rapamycin does not dissociate from its binding protein. Therefore, one approach is to design a photo-cleavable version of CIDs that can induce specific membrane modification of target proteins on the inner side of the cell membrane through light modulation. This enables controlled inner leaflet modification and regulation of transient signaling patterns [109].

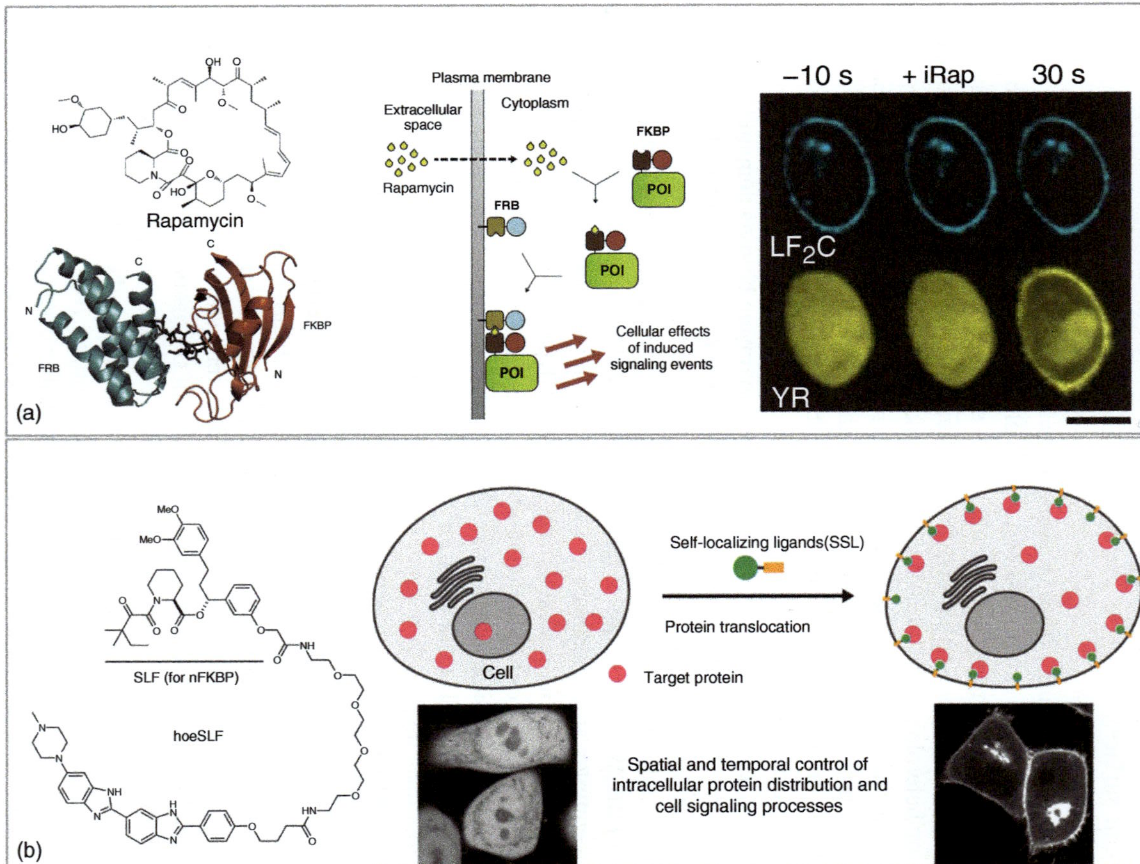

Figure 5.7 Small-molecule ligand targeting approach. (a) Structure of rapamycin and its complex with protein partners FKBP1 and FRB. Upon the addition of rapamycin to cells, the compound diffuses across the plasma membrane and initially binds to the fusion protein of interest (POI) with FKBP. Subsequently, the rapamycin-FKBP complex interacts with FRB, facilitating the translocation of the POI to the inner leaflet of the membrane. Source: Putyrski and Schultz [107]/John Wiley & Sons/CC BY 3.0. (b) Self-localizing ligands (SLLs) bind to their target proteins and facilitate their relocation from the cytoplasm to specific targeting sites on the inner leaflet of the cell membrane [108]. Source: Reproduced with permission from the American Chemical Society.

5.3.2.2 Cell-Penetrating Peptide Approach

Cell-penetrating peptides (CPPs) belong to the category of membrane-active peptides (MAPs) and antimicrobial peptides (AMPs). These peptides are typically unstructured in aqueous solution but fold into well-defined secondary structures near the membrane. They can penetrate the cell membrane without compromising its integrity. Furthermore, CPPs not only have the ability to traverse the cell membrane but also to transport bioactive substances, such as proteins, peptides, DNA, RNA, fluorescent labels, nanoparticles, viruses, liposomes, oligonucleotides, and drugs, either covalently or non-covalently, into cells. In addition to the use of CPPs to deliver bioactive substances intracellularly, they can also be used for modification of the inner leaflet of cell membranes [110].

An exemplary case of an amphiphilic cell transmembrane peptide suitable for inner leaflet modification is the pH-(low) insertion peptide (pHLIP), developed by Engelman and coworkers [111]. pHLIP is a pH-dependent CPP that exhibits unique characteristics. Its transmembrane domain contains two negatively charged aspartic acid residues, which are uncharged at near-neutral pH. However, when exposed to an acidic environment, the increased concentration of protons leads to the protonation of these charged residues at specific positions within the peptide. Consequently, pHLIP becomes more hydrophobic and adopts a helical conformation, enabling its insertion as a transmembrane helix in the lipid bilayer. The N-terminus of the inserted peptide remains in the extracellular space, while the C-terminus crosses the membrane and enters the cytoplasm. The lipid composition of the cell membrane plays a role in regulating the binding and insertion of pHLIP (Figure 5.8a) [112, 116].

In a study conducted by Yana K. Reshetnyak and coworkers, they investigated the cellular delivery of α-amanitin, a polar cell-impermeable toxin, and an RNA polymerase II inhibitor, using pHLIP as the carrier. Their findings revealed that pHLIP can deliver α-amanitin into cells in a pH-dependent manner, leading to cell death within 48 hours. The translocation capability of α-amanitin could be modulated by conjugating it to the C-terminus of pHLIP using linkers of varying hydrophobicities, which could be cleaved in the cytoplasm. The pH-selective antiproliferative activity of the pHLIP-amanitin conjugate was found to be four to five times higher at pH 6 compared to pH 7. The primary mechanism of amanitin delivery involves direct translocation (flip) across the membrane facilitated by pHLIP, followed by cleavage of the dithiol bond in the cytoplasm. The pHLIP-mediated cytoplasmic delivery of amanitin holds significant potential for its use as a potent pH-selective anticancer agent, particularly targeting highly proliferative cancer cells in low extracellular pH environments (Figure 5.8b) [113].

Currently, pHLIP is being integrated with DNA technology for cancer cell detection and the regulation of cell functions. Li and coworkers have developed a pHLIP and DNA-based probe that anchors to the cell membrane in the acidic environment of cancer cells. The pHLIP component ensures membrane anchoring, while DNA enables signal amplification, resulting in enhanced sensitivity and imaging of the tumor microenvironment (Figure 5.8c) [114]. Another study by Qiu and coworkers involves the development of an artificial signal pathway using DNA-pHLIP

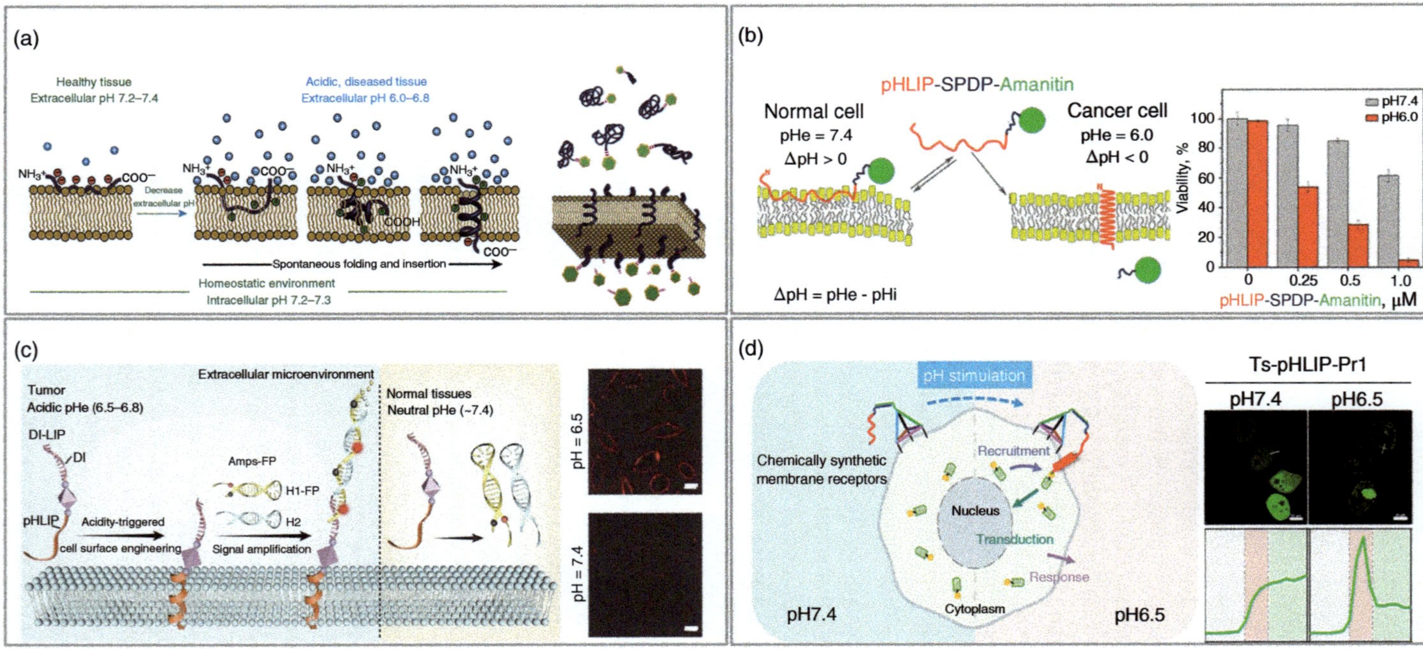

Figure 5.8 Cell-penetrating peptide approach. (a) Mechanism of pHLIP insertion into the cellular membrane. Source: Adapted from Scott et al. [112]. The pHLIP peptide inserts into the membrane, leaving its N-terminus in the extracellular space while the C-terminus translocates across the membrane into the cytoplasm. Source: Reproduced with permission from the American Chemical Society. (b) pHLIP-mediated delivery of a cell-impermeable drug across the lipid bilayer into the cytoplasm of cancer cells [113]. Source: Reproduced with permission from the American Chemical Society. (c) pHLIP-based detection of cancer cells. pHLIP enables controlled recruitment of a DNA signal amplifier to the specific acidic tumor microenvironment, enhancing detection sensitivity [114]. Source: Reproduced with permission from John Wiley & Sons publication. (d) Membrane-anchored DNA-pHLIP conjugate for sensing and transducing extracellular pH changes, triggering intracellular signaling processes to regulate cellular responses [115]. Source: Reproduced with permission from the American Chemical Society.

conjugates as membrane-anchored synthetic receptors. Upon sensing external pH stimuli, the conformational switch of pHLIP facilitates translocation across the cell membrane, exposing recruiting sites on the inner leaflet. This enables specific protein recruitment, such as PKC and ZAP70, at the membrane proximal end, subsequently triggering downstream signal cascades. Thus, this membrane-anchored artificial receptor can sense and transduce extracellular pH changes, initiating intracellular signal programs to regulate cell responses (Figure 5.8d) [115]. While pHLIP is a prominent example, other CPPs with transmembrane activity hold the potential for inner leaflet modifications. For instance, the trans-activator of transcription (TAT) protein derived from the HIV-1 virus facilitates membrane penetration and direct interaction with the cell nucleus [117]. However, CPPs have yet to be directly employed for immobilizing DNA onto the inner leaflet, likely due to the high positive charge and molecular weight of oligonucleotide strands. There is a need for novel CPP designs to overcome these challenges.

5.3.3 Liposome Fusion-Based Transport (LiFT) Approach

Membrane fusion is a natural process observed in various systems, including fusogenic extracellular vesicles (EVs), which facilitate the transfer of exogenous molecules between cells for cell–cell communication. Artificial lipid nanoparticles, such as liposomes, can also undergo membrane fusion to deliver exogenous molecules [118]. Fusogenic liposomes have lipid formulations that closely resemble the components of the cell membrane and exhibit strong interactions with the cell membrane, facilitating fusion between liposomes and the cell membrane. Liposomal fusion enables the transmembrane delivery of drugs or proteins, avoiding lysosomal degradation during the endocytosis pathway and ensuring the stability of delivered substances, particularly protein drugs [119]. Dutta et al. demonstrated that cationic liposomes, such as DOTAP, can induce fusion between ketone-modified liposomes and fibroblast cells without causing any changes in cell behavior or morphology, highlighting the noninvasive nature of the liposome fusion strategy [120].

Drawing inspiration from the fusion process between EVs and cells, Xing and coworkers developed a liposome fusion-based transport strategy known as LiFT, which enables precise directional control and efficient integration of DNA sequences into the inner and outer membranes of cells (Figure 5.9a) [121]. The high spatial control was achieved by fusing asymmetric fuso-SNA bearing two orthogonal DNA strands onto the cell membrane, maintaining the DNA-facing orientations. This represents the first successful demonstration of DNA modification onto the inner leaflet using an efficient nongenetic approach. In particular, the DNA modification of the inner leaflet may bring opportunities for the detection of metabolites of the cell membrane, transmembrane transport, and other applications. Additionally, Xing et al. applied the LiFT approach to anchor bioorthogonal catalysts onto both sides of the cell membrane with high orientation control, overcoming challenges associated with the lack of localization control for intracellular catalysts. The study revealed that the inner surface of the plasma membrane can serve as a reliable shelter for the

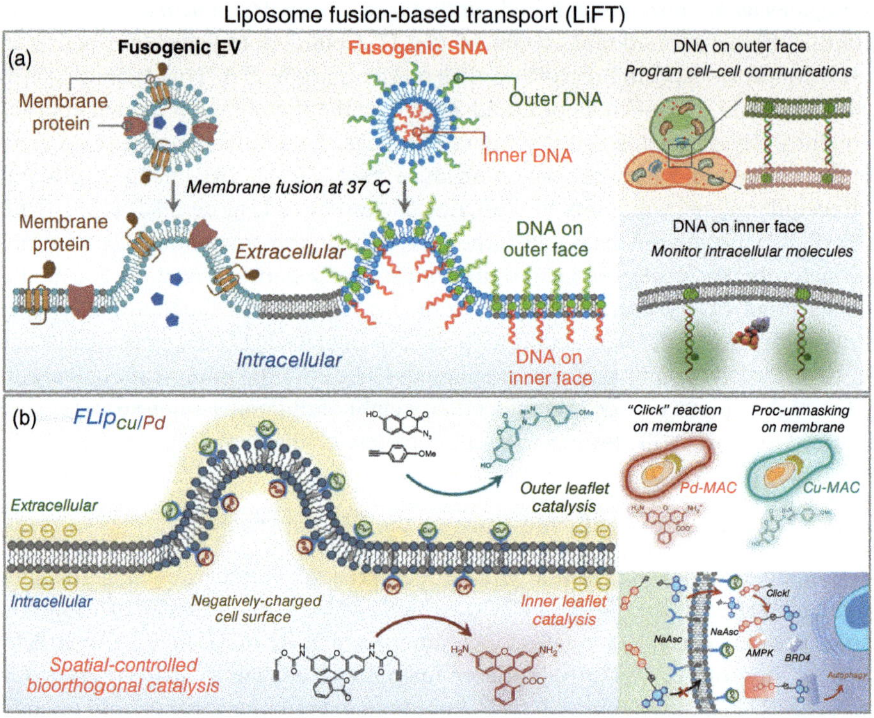

Figure 5.9 Liposome fusion-based transport strategy. (a) Spatially controlled membrane engineering using DNA strands, enabling the regulation of cellular communication and the detection of intracellular metabolites [121]. Source: Reproduced with permission from the Authors and John Wiley & Sons publication. (b) Anchoring of two bioorthogonal catalysts onto the cell membrane using the LiFT approach [122]. Source: Reproduced with permission from the Authors and AAAS open-access publication.

metal center, protecting the catalyst complex from deactivation and significantly reducing the amount of catalyst required for effective intracellular catalysis. Importantly, this approach enables the in-cellulo synthesis of a chimeric binder drug as an autophagic agent, opening up possibilities for the in-cellulo synthesis of non-permeable drugs from easily permeable substrates and showcasing potential as future therapeutic strategies (Figure 5.9b) [122]. Granted, there are currently very limited approaches to modifying DNA on the inner leaflet of cell membranes, and more convenient, efficient, and universal methods for achieving inner cell membrane modification of functional DNA need to be developed. These advancements would contribute to expanding the range of applications and enhancing the effectiveness of DNA modifications on the inner leaflet of the cell membrane.

5.4 Perspectives

As discussed in this chapter, significant progress has been made in the application of DNA molecules for cell surface engineering. DNA-based cell surface engineering offers promising opportunities for precise regulation of cell–cell interactions, manipulation of cell behavior, and various cell-based biomedical applications. In addition to conventional outer membrane surface engineering using DNA, we have also explored potential approaches for anchoring functional DNA on the inner leaflet of the cell membrane. The ability to modify the inner surface of the cell membrane with DNA allows for enhanced observation of cellular activities, transmembrane signaling, and the creation of novel functions not naturally present in cells. With the capability to engineer both sides of the cell membrane with high spatial control, cells engineered with different DNA molecules hold broad prospects for diverse applications.

However, despite the advancements in DNA-based cell surface engineering, there are still important challenges that require further investigation and resolution. One major issue is the stability of modified DNA on the membrane, given the dynamic nature of the cell membrane. For instance, hydrophobic insertion, while facilitating the entry of therapeutic molecules into the cell membrane without disrupting cellular physiology, may lead to the transfer of inserted molecules into the cytoplasm through flipping enzymes distributed on other organelle membranes. Moreover, the half-life of DNA on the cell membrane is relatively short, with reported retention times ranging from 4 to 12 hours. To address this concern, combining DNA with biopolymers has been proposed as an effective approach to enhance the stability and retention time of DNA on the cell membrane surface. Crosslinking modified DNA strands into meshes with diameters exceeding 500 nm, matching the size of lipid raft structures on membranes, could potentially compete with endocytosis mechanisms mediated by grid proteins and caveolae, which are associated with lipid raft internalization.

Moreover, in addition to stability, a key concern is that current modification strategies primarily involve a "one-time" approach, where the anchoring of DNA strands on the cell surface becomes diluted during cell proliferation and growth. This limitation significantly impedes the dynamic modification and regulation of cells. To overcome this challenge, a genetic engineering approach can be employed to express endogenous membrane proteins that can associate with DNA and shape membrane function. As a tentative initial example, the Nie group expressed a "click" type protein on the cell membrane surface and directly conjugated it to DNA by controlling the extracellular ion environment. This innovative approach enabled the detection of Mn^{2+} in the cellular environment, showcasing the potential of genetic engineering techniques for addressing this issue [123].

Additionally, the LiFT method developed by the Xing group holds promise for the selective modification of more complex DNA nanostructures, such as DNA origami, extending beyond ssDNA or dsDNA. However, the application of LiFT for DNA modifications has been mainly validated in a limited number of *in vitro* cell culture systems. To advance this field, it is crucial to optimize the lipid composition of membrane fusion liposomes to enhance their efficiency in the cellular environment. Furthermore, validating their efficacy *in vivo* models is also necessary to pave the way for the development of novel therapeutics. Collectively, we anticipate that the future of spatiotemporally controlled cell membrane engineering using DNA nanotechnology will thrive, enabling a wide range of biomedical applications.

Acknowledgments

This work is financially supported by NSFC (22374043), Hunan Provincial Science and Technology Department (2022SK2003, 2022JJ10007), and the Science and Technology Innovation Program of Hunan Province (2022RC3046).

References

1 Langecker, M., Arnaut, V., List, J. et al. (2014). DNA nanostructures interacting with lipid bilayer membranes. *Acc. Chem. Res.* 47: 1807–1815. https://doi.org/10.1021/ar500051r.

2 Hanczyc, M.M. and Monnard, P.A. (2017). Primordial membranes: more than simple container boundaries. *Curr. Opin. Chem. Biol.* 40: 78–86. https://doi.org/10.1016/j.cbpa.2017.07.009.

3 Shan, Y.P. and Wang, H.D. (2015). The structure and function of cell membranes examined by atomic force microscopy and single-molecule force spectroscopy. *Chem. Soc. Rev.* 44: 3617–3638. https://doi.org/10.1039/c4cs00508b.

4 Bretscher, M.S. (1973). Membrane structure: some general principles. *Science.* 181: 622–629. https://doi.org/10.1126/science.181.4100.622.

5 Mager, M.D., LaPointe, V., and Stevens, M.M. (2011). Exploring and exploiting chemistry at the cell surface. *Nat. Chem.* 3: 582–589. https://doi.org/10.1038/nchem.1090.

6 Watt, F.M. and Huck, W.T. (2013). Role of the extracellular matrix in regulating stem cell fate. *Nat. Rev. Mol. Cell Biol.* 14: 467–473. https://doi.org/10.1038/nrm3620.

7 Stevens, M.M. and George, J.H. (2005). Exploring and engineering the cell surface interface. *Science.* 310: 1135–1138. https://doi.org/10.1126/science.1106587.

8 Dong, H.L., Liu, L., Wang, J.Y. et al. (2020). DNA-based reprogramming strategy of receptor-mediated cellular behaviors: from genetic encoding to nongenetic engineering. *ACS Appl. Bio Mater.* 3: 2796–2804. https://doi.org/10.1021/acsabm.9b01223.

9 Zhu, Y., Jin, K., Mao, X.O. et al. (2003). Vascular endothelial growth factor promotes proliferation of cortical neuron precursors by regulating E2F expression. *FASEB J.* 17: 186–193. https://doi.org/10.1096/fj.02-0515com.

10 Briquez, P.S., Clegg, L.E., Martino, M.M. et al. (2016). Design principles for therapeutic angiogenic materials. *Nat. Rev. Mater.* 1: 15006. https://doi.org/10.1038/natrevmats.2015.6.

11 Hubert, P., Sawma, P., Duneau, J.P. et al. (2010). Single-spanning transmembrane domains in cell growth and cell-cell interactions: more than meets the eye? *Cell Adhes. Migr.* 4: 313–324. https://doi.org/10.4161/cam.4.2.12430.

12 Bleijs, D.A., Binnerts, M.E., van Vliet, S.J. et al. (2000). Low-affinity LFA-1/ICAM-3 interactions augment LFA-1/ICAM-1-mediated T cell adhesion and signaling by redistribution of LFA-1. *J. Cell Sci.* 113: 391–400. https://doi.org/10.1242/jcs.113.3.391.

13 Zou, W. and Chen, L. (2008). Inhibitory B7-family molecules in the tumour microenvironment. *Nat. Rev. Immunol.* 8: 467–477. https://doi.org/10.1038/nri2326.

14 Boussiotis, V.A. (2016). Molecular and biochemical aspects of the PD-1 checkpoint pathway. *N. Engl. J. Med.* 375: 1767–1778. https://doi.org/10.1056/NEJMra1514296.

15 Park, J., Andrade, B., Seo, Y. et al. (2018). Engineering the surface of therapeutic "living" cells. *Chem. Rev.* 118: 1664–1690. https://doi.org/10.1021/acs.chemrev.7b00157.

16 Wu, Z., Xiao, M., Lai, W. et al. (2022). Nucleic acid-based cell surface engineering strategies and their applications. *ACS Appl. Bio Mater.* 5: 1901–1915. https://doi.org/10.1021/acsabm.1c01126.

17 Wang, Q., Cheng, H., Peng, H.S. et al. (2015). Non-genetic engineering of cells for drug delivery and cell-based therapy. *Adv. Drug Delivery Rev.* 91: 125–140. https://doi.org/10.1016/j.addr.2014.12.003.

18 Stephan, M.T. and Irvine, D.J. (2011). Enhancing cell therapies from the outside in: cell surface engineering using synthetic nanomaterials. *Nano Today* 6: 309–325. https://doi.org/10.1016/j.nantod.2011.04.001.

19 Zhang, S.Y., Zhou, Z.R., and Qian, R.C. (2021). Recent progress and perspectives on cell surface modification. *Chem. Asian J.* 16: 3250–3258. https://doi.org/10.1002/asia.202100852.

20 Custodio, C.A. and Mano, J.F. (2016). Cell surface engineering to control cellular interactions. *ChemNanoMat* 2: 376–384. https://doi.org/10.1002/cnma.201600047.

21 Lo, P.K., Karam, P., Aldaye, F.A. et al. (2010). Loading and selective release of cargo in DNA nanotubes with longitudinal variation. *Nat. Chem.* 2: 319–328. https://doi.org/10.1038/nchem.575.

22 Yang, F., Li, Q., Wang, L. et al. (2018). Framework-nucleic-acid-enabled biosensor development. *ACS Sens.* 3: 903–919. https://doi.org/10.1021/acssensors.8b00257.

23 Pinheiro, A.V., Han, D., Shih, W.M. et al. (2011). Challenges and opportunities for structural DNA nanotechnology. *Nat. Nanotechnol.* 6: 763–772. https://doi.org/10.1038/nnano.2011.187.

24 Pan, J., Li, F., Cha, T.G. et al. (2015). Recent progress on DNA based walkers. *Curr. Opin. Biotechnol.* 34: 56–64. https://doi.org/10.1016/j.copbio.2014.11.017.

25 Qu, X., Zhu, D., Yao, G. et al. (2017). An exonuclease III-powered, on-particle stochastic DNA walker. *Angew. Chem. Int. Ed.* 56: 1855–1858. https://doi.org/10.1002/anie.201611777.

26 Guo, W., Hong, F., Liu, N. et al. (2015). Target-specific 3D DNA gatekeepers for biomimetic nanopores. *Adv. Mater.* 27: 2090–2095. https://doi.org/10.1002/adma.201405078.

27 Wei, R., Martin, T.G., Rant, U. et al. (2012). DNA origami gatekeepers for solid-state nanopores. *Angew. Chem. Int. Ed.* 51: 4864–4867. https://doi.org/10.1002/anie.201200688.

28 Zhou, L., Gao, M., Fu, W. et al. (2020). Three-dimensional DNA tweezers serve as modular DNA intelligent machines for detection and regulation of intracellular microRNA. *Sci. Adv.* 6: eabb0695. https://doi.org/10.1126/sciadv.abb0695.

29 Hwang, M.T., Wang, Z., Ping, J. et al. (2018). DNA nanotweezers and graphene transistor enable label-free genotyping. *Adv. Mater.* e1802440. https://doi.org/10.1002/adma.201802440.

30 Fang, X. and Tan, W. (2010). Aptamers generated from cell-SELEX for molecular medicine: a chemical biology approach. *Acc. Chem. Res.* 43: 48–57. https://doi.org/10.1021/ar900101s.

31 Kuai, H., Zhao, Z., Mo, L. et al. (2017). Circular bivalent aptamers enable in vivo stability and recognition. *JACS* 139: 9128–9131. https://doi.org/10.1021/jacs.7b04547.

32 Ji, J.H., Karna, D., and Mao, H.B. (2021). DNA origami nano-mechanics. *Chem. Soc. Rev.* 50: 11966–11978. https://doi.org/10.1039/d1cs00250c.

33 Fan, S.S., Wang, D.F., Kenaan, A. et al. (2019). Create nanoscale patterns with DNA origami. *Small* 15: 1805554. https://doi.org/10.1002/smll.201805554.

34 Pàmies, P. (2013). Lego sets from DNA bricks. *Nat. Mater.* 12: 3–3. https://doi.org/10.1038/nmat3534.

35 Cui, Y., Yan, J., and Wei, B. (2021). Hybrid wireframe DNA nanostructures with scaffolded and scaffold-free modules. *Angew. Chem. Int. Ed.* 60: 9345–9350. https://doi.org/10.1002/anie.202015564.

36 Benson, E., Mohammed, A., Gardell, J. et al. (2015). DNA rendering of polyhedral meshes at the nanoscale. *Nature* 523: 441–U139. https://doi.org/10.1038/nature14586.

37 Benson, E., Mohammed, A., Bosco, A. et al. (2016). Computer-aided production of scaffolded DNA nanostructures from flat sheet meshes. *Angew. Chem. Int. Ed.* 55: 8869–8872. https://doi.org/10.1002/anie.201602446.

38 Gai, L., Zhu, Y., Zhang, C. et al. (2021). Targeting canonical and non-canonical STAT signaling pathways in renal diseases. *Cells* 10: https://doi.org/10.3390/cells10071610.

39 Steinberg, G.R. and Hardie, D.G. (2023). New insights into activation and function of the AMPK. *Nat. Rev. Mol. Cell Biol.* 24: 255–272. https://doi.org/10.1038/s41580-022-00547-x.

40 Hsiao, S.C., Shum, B.J., Onoe, H. et al. (2009). Direct cell surface modification with DNA for the capture of primary cells and the investigation of myotube formation on defined patterns. *Langmuir* 25: 6985–6991. https://doi.org/10.1021/la900150n.

41 You, M., Lyu, Y., Han, D. et al. (2017). DNA probes for monitoring dynamic and transient molecular encounters on live cell membranes. *Nat. Nanotechnol.* 12: 453–459. https://doi.org/10.1038/nnano.2017.23.

42 Vermesh, U., Vermesh, O., Wang, J. et al. (2011). High-density, multiplexed patterning of cells at single-cell resolution for tissue engineering and other applications. *Angew. Chem. Int. Ed.* 50: 7378–7380. https://doi.org/10.1002/anie.201102249.

43 Zhao, W., Schafer, S., Choi, J. et al. (2011). Cell-surface sensors for real-time probing of cellular environments. *Nat. Nanotechnol.* 6: 524–531. https://doi.org/10.1038/nnano.2011.101.

44 Meng, M., Wang, Z., Fathima, A. et al. (2019). Interface-induced magnetic polar metal phase in complex oxides. *Nat. Commun.* 10: 5248. https://doi.org/10.1038/s41467-019-13270-7.

45 Raouane, M., Desmaele, D., Urbinati, G. et al. (2012). Lipid conjugated oligonucleotides: a useful strategy for delivery. *Bioconjugate Chem.* 23: 1091–1104. https://doi.org/10.1021/bc200422w.

46 Liu, H.P., Kwong, B., and Irvine, D.J. (2011). Membrane anchored immunostimulatory oligonucleotides for in vivo cell modification and localized immunotherapy. *Angew. Chem. Int. Ed.* 50: 7052–7055. https://doi.org/10.1002/anie.201101266.

47 Rode, A.B., Endoh, T., Tateishi-Karimata, H. et al. (2013). Real-time monitoring of DNA hybridization kinetics on living cell surfaces. *Chem. Commun.* 49: 8444–8446. https://doi.org/10.1039/c3cc42990c.

48 Dube, D.H. and Bertozzi, C.R. (2003). Metabolic oligosaccharide engineering as a tool for glycobiology. *Curr. Opin. Chem. Biol.* 7: 616–625. https://doi.org/10.1016/j.cbpa.2003.08.006.

49 Mahal, L.K., Yarema, K.J., and Bertozzi, C.R. (1997). Engineering chemical reactivity on cell surfaces through oligosaccharide biosynthesis. *Science.* 276: 1125–1128. https://doi.org/10.1126/science.276.5315.1125.

50 Chandra, R.A., Douglas, E.S., Mathies, R.A. et al. (2006). Programmable cell adhesion encoded by DNA hybridization. *Angew. Chem. Int. Ed.* 45: 896–901. https://doi.org/10.1002/Anie.200502421.

51 Chang, X., Zhang, C., Lv, C. et al. (2019). Construction of a multiple-aptamer-based DNA logic device on live cell membranes via associative toehold activation for accurate cancer cell identification. *JACS* 141: 12738–12743. https://doi.org/10.1021/jacs.9b05470.

52 Selden, N.S., Todhunter, M.E., Jee, N.Y. et al. (2012). Chemically programmed cell adhesion with membrane-anchored oligonucleotides. *JACS* 134: 765–768. https://doi.org/10.1021/ja2080949.

53 Ellington, A.D. and Szostak, J.W. (1990). In vitro selection of RNA molecules that bind specific ligands. *Nature* 346: 818–822. https://doi.org/10.1038/346818a0.

54 Tuerk, C. and Gold, L. (1990). Systematic evolution of ligands by exponential enrichment: RNA ligands to bacteriophage T4 DNA polymerase. *Science.* 249: 505–510. https://doi.org/10.1126/science.2200121.

55 Camorani, S., Crescenzi, E., Gramanzini, M. et al. (2017). Aptamer-mediated impairment of EGFR-integrin αvβ3 complex inhibits vasculogenic mimicry and growth of triple-negative breast cancers. *Sci. Rep.* 7: 46659. https://doi.org/10.1038/srep46659.

56 Berg, K., Lange, T., Mittelberger, F. et al. (2016). Selection and characterization of an α6β4 Integrin blocking DNA Aptamer. *Mol. Ther. Nucleic Acids* 5: e294. https://doi.org/10.1038/mtna.2016.10.

57 Zumrut, H.E., Ara, M.N., Maio, G.E. et al. (2016). Ligand-guided selection of aptamers against T-cell Receptor-cluster of differentiation 3 (TCR-CD3) expressed on Jurkat.E6 cells. *Anal. Biochem.* 512: 1–7. https://doi.org/10.1016/j.ab.2016.08.007.

58 Zhang, D., Zheng, Y., Lin, Z. et al. (2020). Equipping natural killer cells with specific targeting and checkpoint blocking aptamers for enhanced adoptive immunotherapy in solid tumors. *Angew. Chem. Int. Ed.* 59: 12022–12028. https://doi.org/10.1002/anie.202002145.

59 Jiang, Z., Le, N.D.B., Gupta, A. et al. (2015). Cell surface-based sensing with metallic nanoparticles. *Chem. Soc. Rev.* 44: 4264–4274. https://doi.org/10.1039/c4cs00387j.

60 Ke, G., Zhu, Z., Wang, W. et al. (2014). A cell-surface-anchored ratiometric fluorescent probe for extracellular pH sensing. *ACS Appl. Mater. Interfaces* 6: 15329–15334. https://doi.org/10.1021/am503818n.

61 Zeng, S., Liu, D., Li, C. et al. (2018). Cell-surface-anchored ratiometric DNA tweezer for real-time monitoring of extracellular and apoplastic pH. *Anal. Chem.* 90: 13459–13466. https://doi.org/10.1021/acs.analchem.8b03299.

62 Wymann, M.P. and Schneiter, R. (2008). Lipid signalling in disease. *Nat. Rev. Mol. Cell Biol.* 9: 162–176. https://doi.org/10.1038/nrm2335.

63 Di, Z., Zhao, J., Chu, H. et al. (2019). An acidic-microenvironment-driven DNA nanomachine enables specific ATP imaging in the extracellular milieu of tumor. *Adv. Mater.* 31: e1901885. https://doi.org/10.1002/adma.201901885.

64 Zheng, X., Peng, R., Jiang, X. et al. (2017). Fluorescence resonance energy transfer-based DNA nanoprism with a split aptamer for adenosine triphosphate sensing in living cells. *Anal. Chem.* 89: 10941–10947. https://doi.org/10.1021/acs.analchem.7b02763.

65 Tan, K.Y., Li, C.Y., Li, Y.F. et al. (2017). Real-time monitoring ATP in mitochondrion of living cells: a specific fluorescent probe for ATP by dual recognition sites. *Anal. Chem.* 89: 1749–1756. https://doi.org/10.1021/acs.analchem.6b04020.

66 Finger, T.E., Danilova, V., Barrows, J. et al. (2005). ATP signaling is crucial for communication from taste buds to gustatory nerves. *Science.* 310: 1495–1499. https://doi.org/10.1126/science.1118435.

67 Zhao, D., Chang, D.R., Zhang, Q. et al. (2021). Rapid and specific imaging of extracellular signaling molecule adenosine triphosphate with a self-phosphorylating DNAzyme. *JACS* 143: 15084–15090. https://doi.org/10.1021/jacs.1c04925.

68 Qiu, L., Zhang, T., Jiang, J. et al. (2014). Cell membrane-anchored biosensors for real-time monitoring of the cellular microenvironment. *JACS* 136: 13090–13093. https://doi.org/10.1021/ja5047389.

69 Feng, G., Luo, X., Lu, X. et al. (2019). Engineering of nucleic acids and synthetic cofactors as holo sensors for probing signaling molecules in the cellular membrane microenvironment. *Angew. Chem. Int. Ed.* 58: 6590–6594. https://doi.org/10.1002/anie.201901320.

70 Zhao, B., O'Brien, C., Mudiyanselage, A. et al. (2017). Visualizing intercellular tensile forces by DNA-based membrane molecular probes. *JACS* 139: 18182–18185. https://doi.org/10.1021/jacs.7b11176.

71 Tokunaga, T., Namiki, S., Yamada, K. et al. (2012). Cell surface-anchored fluorescent aptamer sensor enables imaging of chemical transmitter dynamics. *JACS* 134: 9561–9564. https://doi.org/10.1021/ja302551p.

72 Zhou, W., Saran, R., and Liu, J. (2017). Metal sensing by DNA. *Chem. Rev.* 117: 8272–8325. https://doi.org/10.1021/acs.chemrev.7b00063.

73 Qu, X., Yang, F., Chen, H. et al. (2017). Bubble-mediated ultrasensitive multiplex detection of metal ions in three-dimensional DNA nanostructure-encoded microchannels. *ACS Appl. Mater. Interfaces* 9: 16026–16034. https://doi.org/10.1021/acsami.7b03645.

74 Qi, L., Xiao, M., Wang, F. et al. (2017). Poly-cytosine-mediated nanotags for SERS detection of Hg(2). *Nanoscale* 9: 14184–14191. https://doi.org/10.1039/c7nr05165d.

75 Xiong, M., Zhu, H., Rong, Q. et al. (2016). A membrane-anchored fluorescent probe for detecting K(+) in the cell microenvironment. *Chem. Commun.* 52: 4679–4682. https://doi.org/10.1039/c6cc00192k.

76 Keshri, P., Zhao, B., Xie, T. et al. (2021). Quantitative and multiplexed fluorescence lifetime imaging of intercellular tensile forces. *Angew. Chem. Int. Ed.* 60: 15548–15555. https://doi.org/10.1002/anie.202103986.

77 Kim, H., Jin, X., Glass, D.S. et al. (2020). Engineering and modeling of multicellular morphologies and patterns. *Curr. Opin. Genet. Dev.* 63: 95–102. https://doi.org/10.1016/j.gde.2020.05.039.

78 Toda, S., Frankel, N.W., and Lim, W.A. (2019). Engineering cell-cell communication networks: programming multicellular behaviors. *Curr. Opin. Chem. Biol.* 52: 31–38. https://doi.org/10.1016/j.cbpa.2019.04.020.

79 Weiss, A. and Littman, D.R. (1994). Signal transduction by lymphocyte antigen receptors. *Cell* 76: 263–274. https://doi.org/10.1016/0092-8674(94)90334-4.

80 Brownlie, R.J. and Zamoyska, R. (2013). T cell receptor signalling networks: branched, diversified and bounded. *Nat. Rev. Immunol.* 13: 257–269. https://doi.org/10.1038/nri3403.

81 Lan, R.Y., Selmi, C., and Gershwin, M.E. (2008). The regulatory, inflammatory, and T cell programming roles of interleukin-2 (IL-2). *J. Autoimmun.* 31: 7–12. https://doi.org/10.1016/j.jaut.2008.03.002.

82 Gartner, Z.J. and Bertozzi, C.R. (2009). Programmed assembly of 3-dimensional microtissues with defined cellular connectivity. *PNAS* 106: 4606–4610. https://doi.org/10.1073/pnas.0900717106.

83 Todhunter, M.E., Jee, N.Y., Hughes, A.J. et al. (2015). Programmed synthesis of three-dimensional tissues. *Nat. Methods* 12: 975–981. https://doi.org/10.1038/nmeth.3553.

84 Qian, R.C., Zhou, Z.R., Guo, W. et al. (2021). Cell surface engineering using DNAzymes: metal ion mediated control of cell-cell interactions. *JACS* 143: 5737–5744. https://doi.org/10.1021/jacs.1c00060.

85 Akbari, E., Mollica, M.Y., Lucas, C.R. et al. (2017). Engineering cell surface function with DNA origami. *Adv. Mater.* 29: https://doi.org/10.1002/adma.201703632.

86 Li, J., Xun, K., Pei, K. et al. (2019). Cell-membrane-anchored DNA nanoplatform for programming cellular interactions. *JACS* 141: 18013–18020. https://doi.org/10.1021/jacs.9b04725.

87 Li, X., Xie, X., Ma, Z. et al. (2018). Programming niche accessibility and in vitro stemness with intercellular DNA reactions. *Adv. Mater.* 30: e1804861. https://doi.org/10.1002/adma.201804861.

88 Qian, R.C., Zhou, Z.R., Wu, Y. et al. (2022). Combination cancer treatment: using engineered DNAzyme molecular machines for dynamic inter- and intracellular regulation. *Angew. Chem. Int. Ed.* 61: e202210935. https://doi.org/10.1002/anie.202210935.

89 Du, Y., Lyu, Y., Lin, J. et al. (2023). Membrane-anchored DNA nanojunctions enable closer antigen-presenting cell-T-cell contact in elevated T-cell receptor triggering. *Nat. Nanotechnol.* https://doi.org/10.1038/s41565-023-01333-2.

90 Xiao, M., Lai, W., Yu, H. et al. (2021). Assembly pathway selection with DNA reaction circuits for programming multiple cell-cell interactions. *JACS* 143: 3448–3454. https://doi.org/10.1021/jacs.0c12358.

91 Shi, P., Zhao, N., Coyne, J. et al. (2019). DNA-templated synthesis of biomimetic cell wall for nanoencapsulation and protection of mammalian cells. *Nat. Commun.* 10: 2223. https://doi.org/10.1038/s41467-019-10231-y.

92 Gao, T., Chen, T., Feng, C. et al. (2019). Design and fabrication of flexible DNA polymer cocoons to encapsulate live cells. *Nat. Commun.* 10: 2946. https://doi.org/10.1038/s41467-019-10845-2.

93 Yu, X.Y., Lin, W., Lai, Q. et al. (2023). ARMOR: auto-assembled resilient biomimetic calcified ornaments for selective cell protection by dual-aptamer-driven hybridization chain reaction. *Angew. Chem. Int. Ed.* https://doi.org/10.1002/anie.202301083.

94 Simon, M.I., Strathmann, M.P., and Gautam, N. (1991). Diversity of G proteins in signal transduction. *Science.* 252: 802–808. https://doi.org/10.1126/science.1902986.

95 Schlessinger, J. (2000). Cell signaling by receptor tyrosine kinases. *Cell* 103: 211–225. https://doi.org/10.1016/s0092-8674(00)00114-8.

96 Chen, H., Xu, W.Y., Shi, H. et al. (2023). DNA-based artificial receptors as transmembrane signal transduction systems for protocellular communication. *Angew. Chem. Int. Ed.* https://doi.org/10.1002/anie.202301559.

97 Chen, S., Xu, Z., Li, S. et al. (2022). Systematic interrogation of cellular signaling in live cells using a membrane-anchored DNA multitasking processor. *Angew. Chem. Int. Ed.* 61: e202113795. https://doi.org/10.1002/anie.202113795.

98 Lv, C., Gu, X.Y., Li, H.W. et al. (2020). Molecular transport through a biomimetic DNA channel on live cell membranes. *ACS Nano* 14: 14616–14626. https://doi.org/10.1021/acsnano.0c03105.

99 Burns, J.R., Seifert, A., Fertig, N. et al. (2016). A biomimetic DNA-based channel for the ligand-controlled transport of charged molecular cargo across a biological membrane. *Nat. Nanotechnol.* 11: 152–156. https://doi.org/10.1038/nnano.2015.279.

100 Chen, H., Zhou, L., Li, C. et al. (2021). Controlled dimerization of artificial membrane receptors for transmembrane signal transduction. *Chem. Sci.* 12: 8224–8230. https://doi.org/10.1039/d1sc00718a.

101 Burns, J.R., Gopfrich, K., Wood, J.W. et al. (2013). Lipid-bilayer-spanning DNA nanopores with a bifunctional porphyrin anchor. *Angew. Chem. Int. Ed.* 52: 12069–12072. https://doi.org/10.1002/anie.201305765.

102 Krishnan, S., Ziegler, D., Arnaut, V. et al. (2016). Molecular transport through large-diameter DNA nanopores. *Nat. Commun.* 7: 12787. https://doi.org/10.1038/ncomms12787.

103 Diederichs, T., Pugh, G., Dorey, A. et al. (2019). Synthetic protein-conductive membrane nanopores built with DNA. *Nat. Commun.* 10: 5018. https://doi.org/10.1038/s41467-019-12639-y.

104 Thomsen, R.P., Malle, M.G., Okholm, A.H. et al. (2019). A large size-selective DNA nanopore with sensing applications. *Nat. Commun.* 10: 5655. https://doi.org/10.1038/s41467-019-13284-1.

105 Teumer, A., Li, Y., Ghasemi, S. et al. (2019). Genome-wide association meta-analyses and fine-mapping elucidate pathways influencing albuminuria. *Nat. Commun.* 10: 4130. https://doi.org/10.1038/s41467-019-11576-0.

106 Rutkowska, A. and Schultz, C. (2012). Protein tango: the toolbox to capture interacting partners. *Angew. Chem. Int. Ed.* 51: 8166–8176. https://doi.org/10.1002/anie.201201717.

107 Putyrski, M. and Schultz, C. (2012). Protein translocation as a tool: the current rapamycin story. *FEBS Lett.* 586: 2097–2105. https://doi.org/10.1016/j.febslet.2012.04.061.

108 Ishida, M., Watanabe, H., Takigawa, K. et al. (2013). Synthetic self-localizing ligands that control the spatial location of proteins in living cells. *JACS* 135: 12684–12689. https://doi.org/10.1021/ja4046907.

109 Umeda, N., Ueno, T., Pohlmeyer, C. et al. (2011). A photocleavable rapamycin conjugate for spatiotemporal control of small GTPase activity. *JACS* 133: 12–14. https://doi.org/10.1021/ja108258d.

110 Bottens, R.A. and Yamada, T. (2022). Cell-penetrating peptides (CPPs) as therapeutic and diagnostic agents for cancer. *Cancers* 14: https://doi.org/10.3390/cancers14225546.

111 Hunt, J.F., Rath, P., Rothschild, K.J. et al. (1997). Spontaneous, pH-dependent membrane insertion of a transbilayer α-helix. *Biochemistry* 36: 15177–15192. https://doi.org/10.1021/bi970147b.

112 Scott, H.L., Nguyen, V.P., Alves, D.S. et al. (2015). The negative charge of the membrane has opposite effects on the membrane entry and exit of pH-low insertion peptide. *Biochemistry* 54: 1709–1712. https://doi.org/10.1021/acs.biochem.5b00069.

113 Moshnikova, A., Moshnikova, V., Andreev, O.A. et al. (2013). Antiproliferative effect of pHLIP-amanitin. *Biochemistry* 52: 1171–1178. https://doi.org/10.1021/bi301647y.

114 Di, Z., Lu, X., Zhao, J. et al. (2022). Mild acidosis-directed signal amplification in tumor microenvironment via spatioselective recruitment of DNA amplifiers. *Angew. Chem. Int. Ed.* 61: e202205436. https://doi.org/10.1002/anie.202205436.

115 Wu, H., Zheng, L., Ling, N. et al. (2023). Chemically synthetic membrane receptors establish cells with artificial sense-and-respond signaling pathways. *JACS* 145: 2315–2321. https://doi.org/10.1021/jacs.2c10903.

116 Barrera, F.N., Fendos, J., and Engelman, D.M. (2012). Membrane physical properties influence transmembrane helix formation. *PNAS* 109: 14422–14427. https://doi.org/10.1073/pnas.1212665109.

117 Desale, K., Kuche, K., and Jain, S. (2021). Cell-penetrating peptides (CPPs): an overview of applications for improving the potential of nanotherapeutics. *Biomater. Sci.* 9: 1153–1188. https://doi.org/10.1039/d0bm01755h.

118 Piffoux, M., Silva, A.K.A., Wilhelm, C. et al. (2018). Modification of extracellular vesicles by fusion with liposomes for the design of personalized biogenic drug delivery systems. *ACS Nano* 12: 6830–6842. https://doi.org/10.1021/acsnano.8b02053.

119 Kunisawa, J., Masuda, T., Katayama, K. et al. (2005). Fusogenic liposome delivers encapsulated nanoparticles for cytosolic controlled gene release. *J. Controlled Release* 105: 344–353. https://doi.org/10.1016/j.jconrel.2005.03.020.

120 Dutta, D., Pulsipher, A., Luo, W. et al. (2011). Engineering cell surfaces via liposome fusion. *Bioconjugate Chem.* 22: 2423–2433. https://doi.org/10.1021/bc200236m.

121 Lin, M., Chen, Y., Zhao, S. et al. (2022). A biomimetic approach for spatially controlled cell membrane engineering using fusogenic spherical nucleic acid. *Angew. Chem. Int. Ed.* 61: e202111647. https://doi.org/10.1002/anie.202111647.

122 Chen, Y., Wu, T., Xie, S. et al. (2023). Orientation-controlled membrane anchoring of bioorthogonal catalysts on live cells via liposome fusion-based transport. *Sci. Adv.* 9: eadg2583. https://doi.org/10.1126/sciadv.adg2583.

123 Hu, S., Zhang, J., Tang, R. et al. (2019). Click-type protein-DNA conjugation for Mn(2+) imaging in living cells. *Anal. Chem.* 91: 10180–10187. https://doi.org/10.1021/acs.analchem.9b02191.

Part II

DNA Nanotechnology for Cell Imaging and Intracellular Sensing

6

Metal-Dependent DNAzymes for Cell Surface Engineering and Intracellular Bioimaging

Ruo-Can Qian[1], Yuting Wu[2], Zhenglin Yang[2], Weijie Guo[3], Ze-Rui Zhou[1], and Yi Lu[2,3]

[1] *Key Laboratory for Advanced Materials & School of Chemistry and Molecular Engineering, East China University of Science and Technology, MeiLong Road No. 130, Shanghai 200237, China*
[2] *Department of Chemistry, The University of Texas at Austin, 105 East 24th Street, Austin, TX 78712, USA*
[3] *Department of Molecular Biosciences, The University of Texas at Austin, 100 East 24th Street, Austin, TX 78712, USA*

6.1 Cellular Surface Engineering and Intracellular Bioimaging Show Great Potential in Biological and Medical Research

Cells provide the basic structural and functional support for biological activity. In multicellular organisms, cells respond to environmental changes via cell signaling pathways, in cooperation with cell surface ligands and intracellular molecules/ions [1, 2]. Such regulated responses are ubiquitous and essential in all types of biological activities such as cell proliferation, cell differentiation, cell aging and death, intracellular biochemistry, and cell metabolism [3–8]. In particular, cellular responses have proven roles in disease progression, including different types of cancer, neurodegenerative diseases, immune disorders, and bacterial infections [9–14]. Therefore, technologies that may visualize and control cell responses are in increasing demand, among which cell surface engineering and intracellular bioimaging are the most impactful directions.

Both intracellular and extracellular microenvironments are highly heterogeneous, in which different types of proteins, glycans, lipid molecules, genetic molecules (DNA/RNA), inorganic molecules, and metal ions are present [15–17]. The diversity of cellular components and their related biological processes has given rise to a range of interdisciplinary strategies aimed at purposeful cell modification or targeted selection of specific cellular molecules. To conjugate specific substances on the cell surface, cell engineering strategies including chemical reactions, metabolic labeling, bioorthogonal reactions, and van der Waals interactions have been developed [18–21]. For example, the negatively charged cell surface enables passive electrostatic interaction between cationic polymers and cell membranes [22–25]. Besides, there are various functional groups expressed on proteins located

DNA Nanotechnology for Cell Research: From Bioanalysis to Biomedicine, First Edition. Edited by Zhou Nie.

on the cell surface or inside the cell, which allow for direct covalent modification [26–28]. For intracellular imaging, *in situ* recognition between sensors and analytes is commonly applied, and a combination of signal amplification strategies is available [29, 30]. These *in situ* signal amplification strategies include both enzymatic and non-enzymatic reactions for high-contrast imaging of intracellular molecules [31–33]. Based on these methods, a series of enzymes and nanomaterials have been applied in detecting intracellular molecules and ions, such as DNA polymerases, CRISPR-Cas systems, nucleases, DNAzymes, DNA-based strand transformations, mesoporous nanostructures, and DNA-modified nanoprobes [34–42].

Recent advances in manipulating the functions of biomolecules on cell surfaces and within cells provide new insights into the significance of cell signaling, intracellular functional molecules, and associated diseases, which could contribute to future medical applications.

6.2 Metal-Specific DNAzymes: A Suitable Choice for Artificial Manipulation of Living Cells

While all previously mentioned strategies achieved great success in cell modification and imaging, DNAzyme-based strategies have shown several specific advantages that are especially attractive. First reported in 1994, a class of DNA molecules possessing enzymatic activities was reported, named deoxryribozymes or DNAzymes [43]. Since then, DNAzymes have become the newest member in the enzyme family with the ability to catalyze a specific chemical reaction, such as RNA substrate cleavage, porphyrin metalation, the Diels–Alder reaction, and amino acid modification [44–51]. Similar to protein enzymes, most DNAzymes exhibit enzymatic activity in the presence of specific metal ions. Unlike other sensor-developing strategies that require multiple rounds of trial-and-error, DNAzymes can be obtained using *in vitro* selection from a large, randomized DNA library with 10^{15} different sequences [52]. After a series of selection steps, DNA sequences with high binding affinity and selectivity for a certain metal ion are enriched, sequenced, and tested to obtain DNAzymes that have high sensitivity and specificity for various metal ions, including Cr^{3+} [53], Ca^{2+} [54], Cd^{2+} [55], Co^{2+} [56], Cu^{2+} [57], Hg^{2+} [58], Mg^{2+} [59], Mn^{2+} [60], Pb^{2+} [43], UO_2^{2+} [61], Zn^{2+} [62–64], Ag^+ [65], Li^+ [66], and Na^+ [67, 68].

Compared with their protein counterparts, DNAzymes have several advantages. First, DNAzymes are more conveniently obtained than protein enzymes. Second, DNAzymes are less susceptible to hydrolysis than protein enzymes, making them more stable. Third, DNAzymes can maintain their enzymatic functions after repeated cycles of denaturing and renaturing. Fourth, unlike most protein enzymes that usually require complex extraction and purification, DNAzymes can be directly chemically synthesized at low cost. In addition to the above advantages, DNAzymes can be easily functionalized by modification of various functional groups, such as fluorescent dyes, quenchers, i-motif sequences, aptamers, and nanoparticles. All these features make DNAzymes particularly attractive for use in biological systems. Therefore, DNAzymes appear to be a good choice for the manipulation and bioimaging of living cells. A typical metal-dependent DNAzyme-based catalytic cleavage reaction is shown in Figure 6.1a [62]. By incorporating fluorescence dyes, DNAzymes can be conveniently applied as imaging indicators for cell–cell interactions (Figure 6.1b) [69] or intracellular analytes (Figure 6.1c) [70].

6.3 Cell Surface Engineering by Programmable DNAzymes

Cell surface engineering can either promote or inhibit cell-environment communications in numerous aspects, including the promotion of cell–cell recognition, regulation of intracellular pathways, protection of therapeutic cells, and sensing of intracellular and extracellular microenvironments. The monitoring of cell metabolism and communication in the natural cell environment is of great significance to molecular cell biology, biochemistry, and biophysics. Especially, the plasma membrane microenvironment is closely related to cell adhesion, signal transduction, proliferation, differentiation, and other cellular physiological processes. Therefore, regulating cell–cell interactions and cell behaviors via cell surface engineering is of significance for biological research. The most widely used approach for cell surface modification is genetic engineering, from which desirable surface receptors can be expressed on the cell membrane [71–73]. Apart from genetic engineering, in recent years, several non-genetic cell-surface engineering methods have been developed for direct cell surface modification, which greatly facilitates the engineering process. For example, programmable DNA structures (such as aptamers and DNAzymes) as recognition elements have been attached onto cell surface for responses to extracellular pH, metal ions, small molecules, and proteins [74–76].

6.3.1 Cell Surface Engineering Using DNAzyme-Based Control Switches

For cell surface engineering using DNAzymes, a general strategy for membrane insertion is to utilize the hydrophobic interactions between lipid anchors and phospholipid bilayers [69, 77–79]. The surfaces of two cell types are inserted with cholesterol tail-modified DNAzyme and its substrate, followed by the assembly of the modified cells via hybridization between DNAzyme and its substrate. In addition, the fluorescent marker modifications enable real-time fluorescence tracking on cell membranes. The susceptibility of DNAzymes to metal-dependent cleavage has been used for dynamic control of cell–cell disassembly on metal ion-responsive substrates.

To further realize more elaborate controls of cell dynamic behavior, the enzyme and substrate strands of Zn^{2+}- and Mg^{2+}-specific DNAzymes have been combined to build DNA double-chain molecular switches enabling 2-factor disassembly control including the AND and OR operations. To perform the logic operation on the target cell surface, a logic gate-responsive DNAzyme nanomachine was

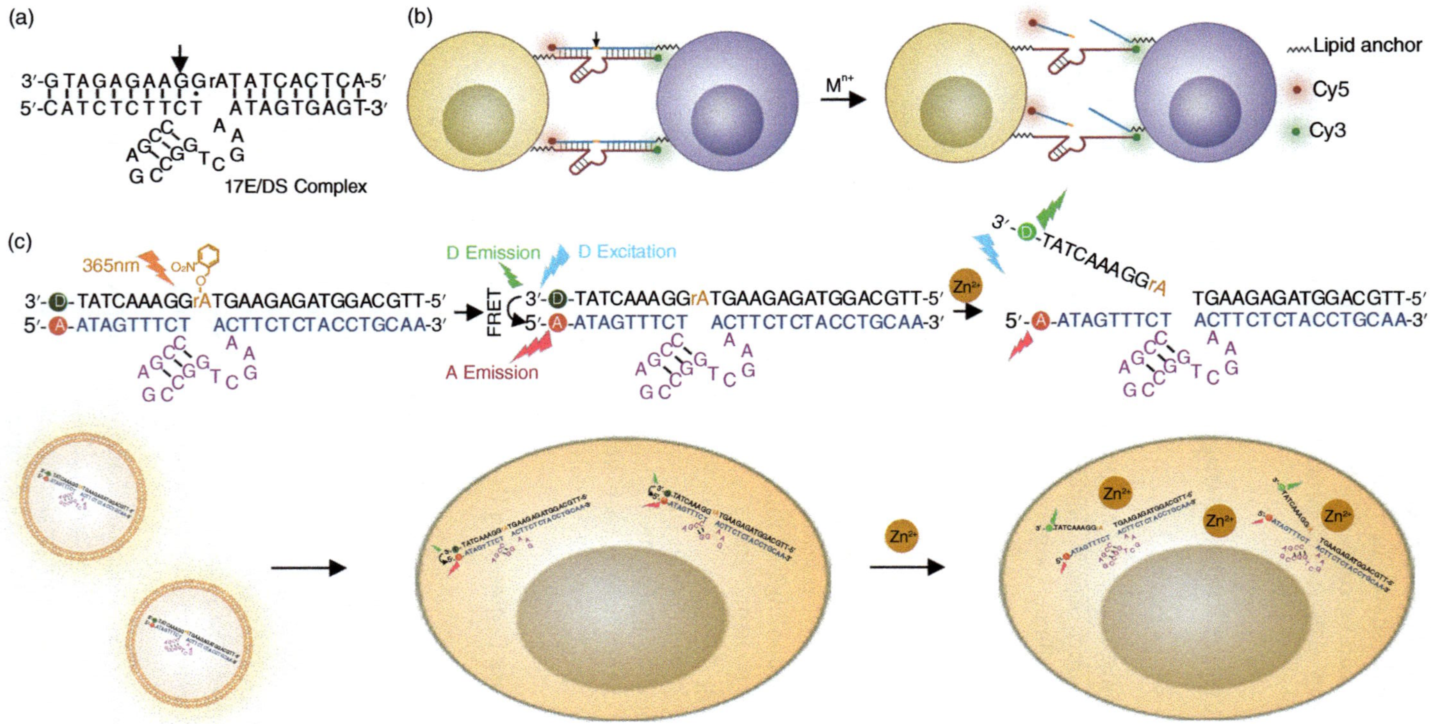

Figure 6.1 Metal-specific DNAzymes. (a) Illustration of a typical DNAzyme-based catalytic cleavage in the presence of Zn^{2+} ions. (b) DNAzyme-based fluorescence imaging of cell–cell interaction. (c) DNAzyme-based FRET imaging of intracellular metal ions. Source: Part (a) is adapted from Li et al. [62], with permission; copyright Oxford University Press, 2000; Part (b) is adapted from Qian et al. [69], with permission; copyright American Chemical Society, 2021; Part (c) is adapted from Hwang et al. [70], with permission; copyright American Chemical Society, 2019.

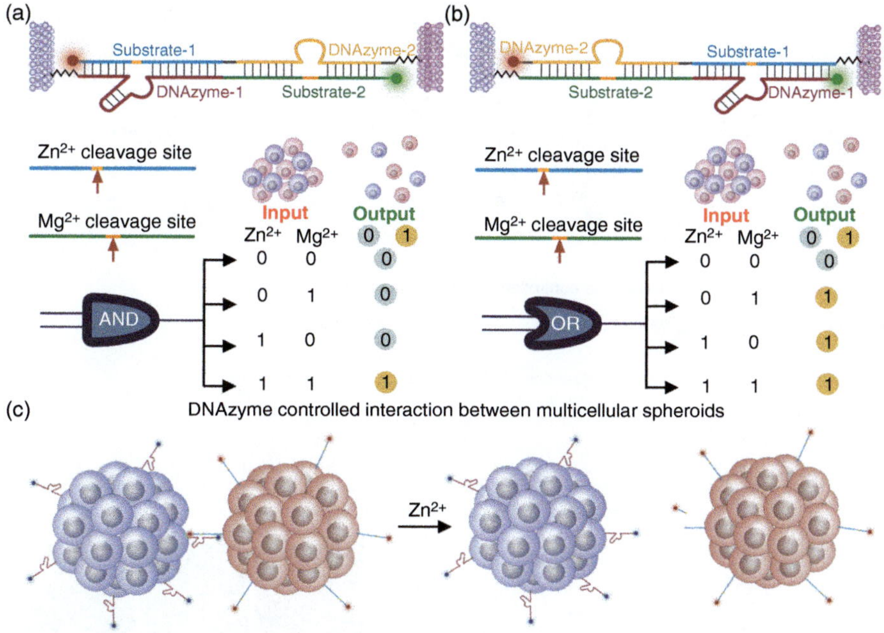

Figure 6.2 DNAzyme-controlled cell–cell interaction. (a) Design of DNAzyme-based control switch for AND operation. (b) Design of DNAzyme-based control switch for OR operation. (c) DNAzyme-controlled interaction between cell spheroids. Source: Parts (a–c) are adapted from Qian et al. [69], with permission; copyright American Chemical Society, 2021.

constructed, consisting of Zn^{2+}-specific DNAzymes, Mg^{2+}-specific DNAzymes, and their substrate strands, which hybridized to form cell assemblies. In the AND operation-responsive DNA double chain control switches, the presence of both input signals, Zn^{2+} and Mg^{2+} ions, caused the double chain to be cut off, releasing T cells from the tumor cell due to structure switching. Correspondingly, in the presence of Zn^{2+} or Mg^{2+} ions, the OR control switch achieved cell disassembly through DNAzyme-controlled substrate cleavage. In other words, with Zn^{2+} and Mg^{2+} serving as inputs 1 and 2, these DNAzyme-based control switches can perform specific recognition and calculation operations on the cell membrane, realizing efficient and accurate imaging analysis of cell–cell interaction. In addition, this DNAzyme nanomachine can also be used for DNAzyme-controlled interaction between cell spheroids. Therefore, this strategy can meet the needs of biomedicine and its application in clinical diagnosis, providing new tools and ideas for accurate diagnosis and treatment of diseases, and biomedical research (Figure 6.2).

6.3.2 Dynamic Inter- and Intra-Cellular Regulation Using Engineered DNAzyme Molecular Machines

Combination cancer therapy has recently emerged as an effective means for promoting cancer apoptosis with an enhanced killing effect [80–82]. Although

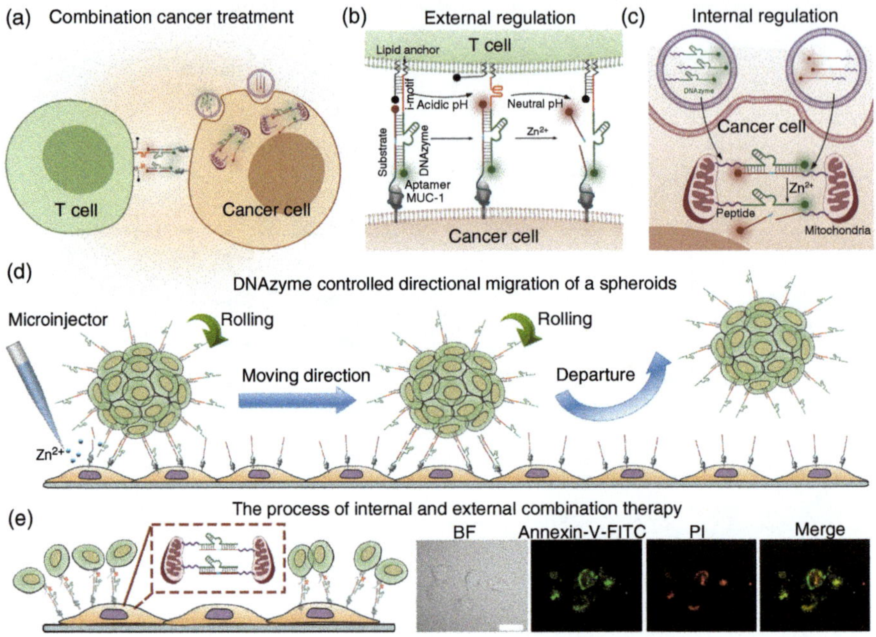

Figure 6.3 Combination cancer treatment by using engineered DNAzyme molecular machines. (a) Both inter- and intracellular regulation are based on engineered DNAzymes. (b) Dynamic intercellular regulations of T-cell/cancer cell interactions. (c) Dynamic intracellular regulations of mitochondria aggregation in cancer cells. (d) The process of a T-cell spheroid migrating by rolling on a cancer cell layer. (e) The combination therapy procedure is based on DNAzyme molecular machines. Source: Adapted from Qian et al. [88], with permission from © John Wiley & Sons, Inc.

some such combination therapeutic methods enable effective treatment, their applications are restricted by severe side effects in normal cells, difficulty of drug delivery, and low immune response [83–85]. In contrast, RNA-cleaving DNAzymes, dependent on metal ions, have been widely used for fluorescence imaging in living cells and are important chemical tools for therapeutic agents for various diseases [70, 86]. In addition, DNAzymes can easily be chemically modified and have been used as mechanical controllers to regulate cell–cell interaction and perform complex dynamic functions [69, 87]. A combination therapeutic strategy based on engineered DNAzyme molecular machines was designed as shown in Figure 6.3a [88]. To achieve external regulation of T cell/cancer cell interactions (Figure 6.3b), this molecular system specifically recognizes cancer cells through a selective aptamer, diverse molecular self-assembly (e.g. assembly of DNAzyme and its substrate), and disassembly pathways under metal ion treatment, and intercellular distance control. Specifically, sequential allosteric activation, in which the conformation of allosteric switches changes from the linear structure (OFF state) to the stem-loop structure (ON state), can be triggered by pH value. To achieve internal regulation of mitochondrial (Figure 6.3c), another DNAzyme-based molecular machine with mitochondria-targeting peptides can be delivered into cancer cells to induce mitochondria aggregation. By modifying cell surfaces and intracellular mitochondria with such engineered DNAzyme molecular machines, both inter- and intra-cellular regulations with programmable selectivity can be achieved.

Moreover, DNAzyme-controlled interactions can also be specifically used for cell spheroids (Figure 6.3d). Through the strong binding affinity and selectivity of a DNAzyme to its substrate, T-cell spheroids can be attached to the surface of the cancer cell layer. With Zn^{2+} ion treatment, the T-cell spheroid is released from the bottom and can subsequently be linked to another location on the cancer cell monolayer. Using this method, DNAzyme molecular machines were applied for point-to-point migration of T-cell spheroids by rolling on the cancer cell layer. In this respect, the interaction between different types of cells based on DNAzyme molecular machines will aid in the design of more potential cancer treatments. This enhanced, synergistic tumor cell-killing effect was thought to be attributed to the cooperation of inter- and intracellular regulations. These results were further confirmed by using Annexin-V-FITC/PI staining as probes to respectively indicate the cancer cell-killing effect under different treatment conditions (Figure 6.3e).

6.3.3 Cell Surface Imaging of Extracellular Signaling Molecule Using DNAzyme Sensors

To accurately measure signaling molecules under high temporal resolution or rapid response kinetics, a self-phosphorylating DNAzyme (denoted SPDz) sensor anchored on the cell surface has been developed for the fluorescence imaging of extracellular ATP transients that occur on a sub-second time scale (Figure 6.4a) [89]. Extracellular nucleotides (mainly ATP) play an important role in many biological processes. ATP, the energy storage molecule in living organisms, is involved in the metabolism of sugars, fats, and proteins. In addition, the abnormal concentration of ATP is associated with a variety of diseases, including hypoxia, ischemia, hypoglycemia, Alzheimer's disease, Parkinson's disease, and some malignant tumors. Thus, accurate and sensitive detection of ATP is imperative for clinical diagnosis and biochemical therapy. To create a cell-surface SPDz sensor, fluorescent SPDz sensors, labeled DK1 (^{F}DK1, internally labeled with FAM or Cy5 on T35), were employed as probes for ATP detection in living cells with high spatiotemporal resolution. The SPDz sensors were modified with cholesterol, which facilitated the sensors to be directly anchored on the cell membrane of MCF-7 cells. The SPDz sensors on the cell surface were activated under various stressful conditions. The SPDz sensors could record robust fluorescence signals at single-cell resolution in response to mechanical stimuli, and the rapid dynamic properties of ATP signaling in astrocytes were captured (Figure 6.4b). Compared with other fluorescence-based ATP sensors, this sensor provides several advantages, including (i) sub-second response kinetics, (ii) extremely high specificity for ATP over other nucleotides, and (iii) micromolar affinities for ATP, traits that are very desirable for the analysis of extracellular ATP as a signaling agent. It is the first report demonstrating a fast-responsive fluorescent DNAzyme sensor anchored on the cell membrane to detect ATP with single, living cell-level spatiotemporal resolution. Subcellular

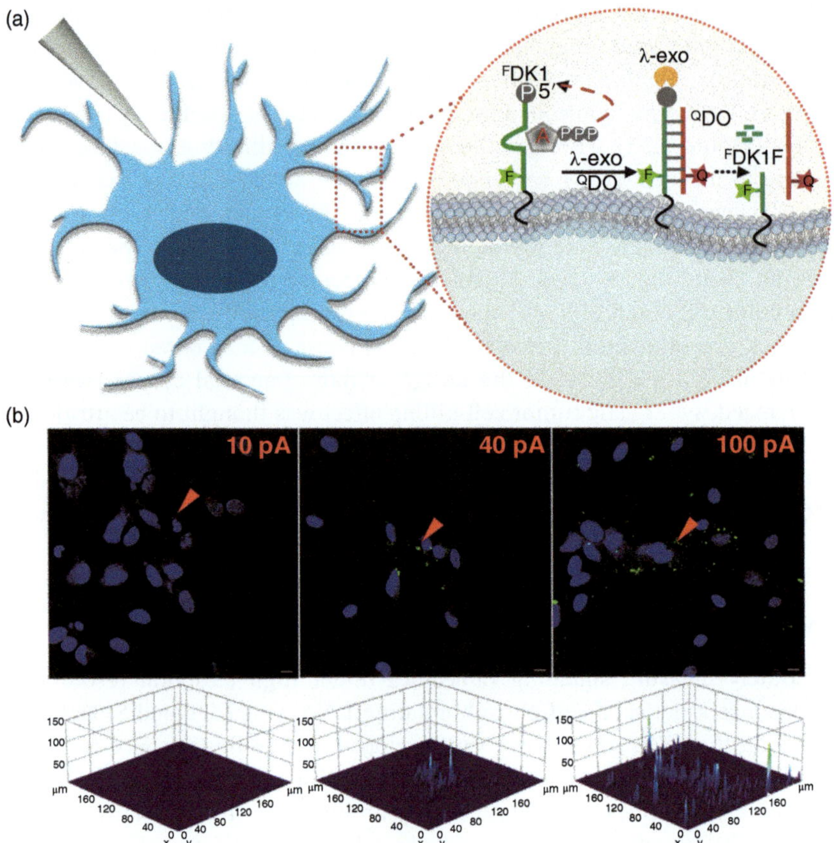

Figure 6.4 SPDz sensor monitoring extracellular ATP. (a) Schematic drawing shows the principle of a cell-surface-anchored SPDz sensor for fluorescence imaging of extracellular ATP. (b) The ATP response in ^{F}DK1-anchored astrocytes under electrical stimulation. Source: Adapted from Zhao et al. [89], with permission from © American Chemical Society.

spatial control could be achieved via designing a DNAzyme sequence to target different subcellular organelles in the future.

6.4 Intracellular Imaging of Metal Ions with DNAzyme-Based Biosensors

6.4.1 DNAzyme-Based Catalytic Beacon for Intracellular Imaging

The DNAzyme was first converted into a catalytic beacon by labeling the DNAzyme with fluorophore and quencher pairs. The fluorophore and quencher pairs can be labeled in three common ways: (i) near the cleavage junction [90], (ii) at the adjacent end of the DNAzyme strand and the substrate strand [91], or (iii) at both ends of the substrate strand (Figure 6.5a–c) [92, 93, 95]. When the DNAzyme sensor

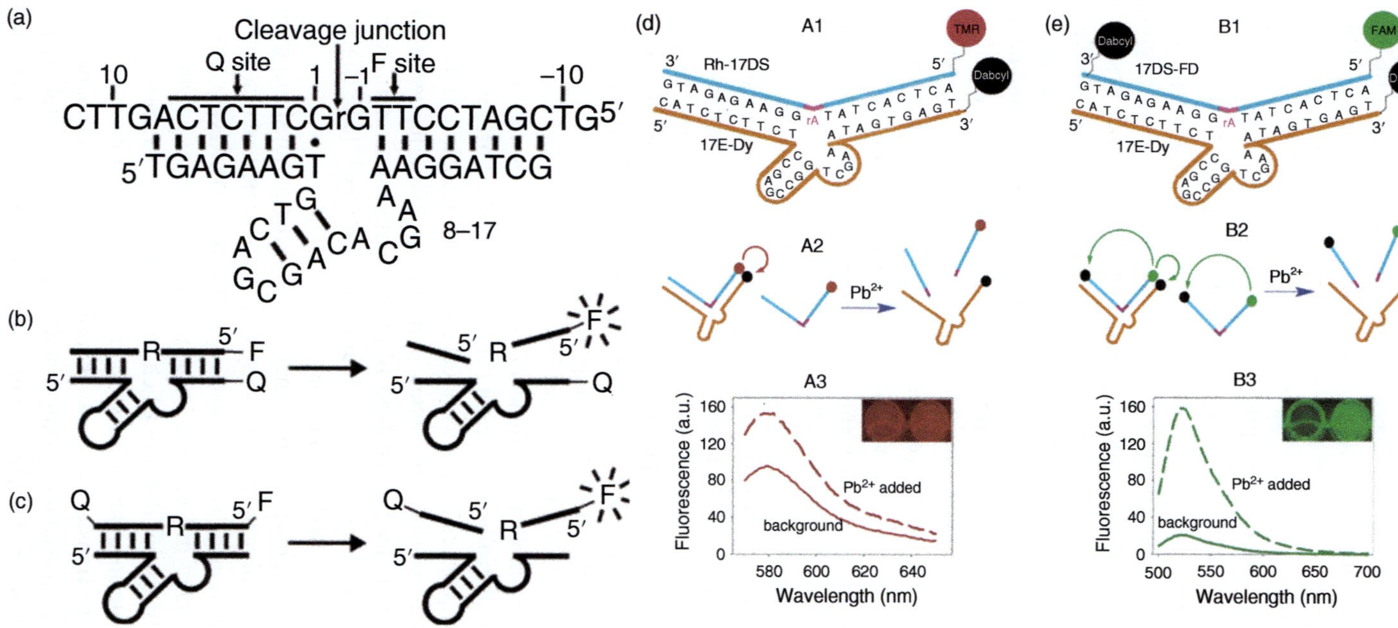

Figure 6.5 Catalytic beacon designs that convert DNAzymes into selective fluorescence sensors for specific metal ions. (a–c) Schematics show different designs of catalytic beacon designs with single fluorophore-quencher pair. (d) Sensors with only one inter-molecular quencher showed high background under room temperature due to partial dissociation of the substrate strand and the enzyme strand. (e) Adding intramolecular quencher lowered the background for metal sensing at elevated temperatures. Source: (a–c) Adapted from Chiuman and Li [93], with permission; copyright Oxford University Press, 2007; (d, e) Adapted from Liu et al. [94], with permission from © American Chemical Society.

is annealed together, the fluorophore is quenched by a nearby quencher and shows a low background. In the presence of the target metal ion, the DNAzyme cleaves the substrate strand, which causes a decrease in the melting temperature between the DNAzyme strand and its cleaved substrate strand. Consequently, the DNAzyme strand and the fragment of the substrate strand dissociate, which increases the distance between the fluorophore and the quencher. This dissociation thus significantly reduces energy transfer between the fluorophore and the quencher, turning on the fluorescence signaling of the catalytic beacon sensor. Due to the selectivity and sensitivity of the DNAzyme, the change in the fluorescence intensity from the sensor is correlated to the concentration of specific metal ions. When labeling the fluorophore and quencher close to the cleavage junction, the sensor has minimal background, but the labeling of the fluorophore and quencher could interfere with the cleavage, making the DNAzyme non-functional. Thus, developing such a catalytic beacon sensor normally requires introducing the fluorophore and quencher into the selection process [93, 94], so that the DNAzyme can be selective and highly active once obtained from the selection. Although introducing the fluorophore and quencher pair into the selection process resolves the issue of potential interactions and inactivation of the DNAzyme sensor, the fluorophore and quencher pair are less flexible and limited to the fluorophore and quencher pairs chosen for the selection. To overcome this limitation, the fluorophore and quencher have also been labeled at the end of the DNA strands and showed less interference with the DNAzyme. This design allowed the modification to be added after the selection and in turn, allowed for the use of different fluorophore and quencher pairs to satisfy the needs of co-staining the DNAzyme sensors with other markers. However, the fluorescence background of this single fluorophore and quencher design increases with elevated sensing temperature, due to the corresponding instability of DNAzyme hybridization (Figure 6.5d) [91, 96]. To improve the signal-to-background noise ratio by lowering the background, an extra intermolecular quencher was added, in addition to the intra-molecular quencher, which quenches the fluorescence signaling from dissociated substrate strand (Figure 6.5e) [97]. Further fine-tuning of the binding arm sequences [98] or conducting *in vitro* selection under certain temperature conditions resulted in better performance with higher detection temperatures. This tunability has enabled multiple applications of the catalytic beacon for sensing different metal ions, including Fe^{2+} [99], Fe^{3+} [99], UO_2^{2+} [61], Cu^{2+} [100], Pb^{2+} [101], Na^+ [67], and Li^+ [66].

In addition to the fluorophore and quencher pair, the energy-transferring designs between different fluorophores were also used for DNAzyme-based sensors. For example, a Förster resonance energy transfer (FRET)-based DNAzyme sensor was developed for sensing Zn^{2+} in Hela cells [70]. Such a FRET sensor provides a reference for calibrating the sensor delivery efficiency and enables ratiometric sensing, which is more accurate in quantification than single fluorophore-based sensors. Other types of energy transfer such as chemiluminescence resonance energy transfer (CRET) were also used for sensing metal ions in environmental samples. In this chapter, we will focus on the sensors with applications in intracellular imaging. More information about sensors for sensing metal ions in environmental samples

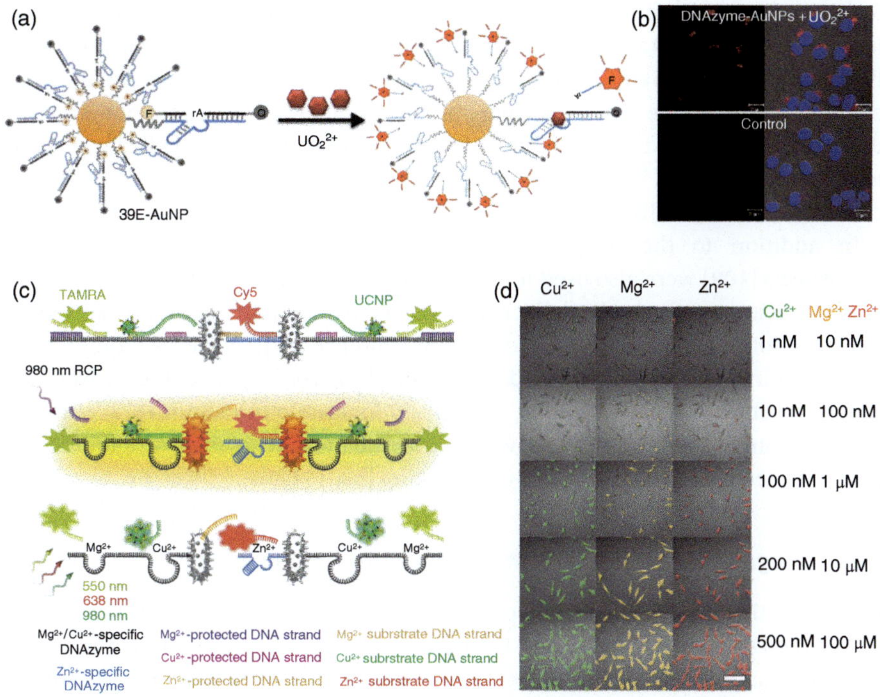

Figure 6.6 Nanoparticle-mediated DNAzyme sensing. (a) Design of DNAzyme-functionalized gold nanoparticles for intracellular sensing of UO_2^{2+}. (b) Imaging of UO_2^{2+} detection in cells. The control group used an inactive DNAzyme sensor or without adding UO_2^{2+} to the active sensor group. (c) Scheme showing the structures of DNA chiral satellite nanoprobes for 980 nm left circularly polarized light-activated intracellular triple-ion detection. (d) Confocal images for sensing different amounts of $MgSO_4$, $ZnSO_4$, and $CuCl_2$ using the nanoprobes in (c). Source: (a, b) Adapted from Wu et al. [106]; (c, d) Reproduced from Gao et al. [107].

can be found in Refs. [102, 103]; information about using DNAzyme to functionalize gold nanoparticles for biosensing can be found in Ref. [104]; and using DNAzymes for personal glucose meter-based point of care detection can be found in Ref. [105].

Intracellular sensing with DNAzyme sensors was first reported using a 13 nm gold nanoparticle (AuNP) to deliver the 39E DNAzyme and achieve the detection of uranyl ions in Hela cells [106]. The DNAzyme contained a thiol at the 3′ end for its conjugation to the AuNP. A Cy3 fluorophore was labeled on the 3′ end of the substrate strand, and a quencher was labeled at the 5′ end of the substrate strand. Without the uranyl ion, the fluorophore was quenched by both the AuNP and the quencher. In the presence of uranyl ion, the DNAzyme cleaved its substrate and caused a decrease in melting temperature between strands. As a result, the fluorophore moved away from the quencher and AuNP, which returned an increase in fluorescence intensity (Figure 6.6a,b). All those above sensors are based on a laser-based activation of the fluorophore, and when the fluorophore is apart from its quencher due to the cleavage reaction of DNAzyme, they show fluorescence signals for detection. In addition to the traditional laser activation of the fluorophores labeled on the DNAzyme sensor, using specific particles can convert special light sources to activate the DNAzyme sensors. For example, the Kuang group conjugated DNAzyme sensors to spiny platinum modified with gold nanorods and up-conversion nanoparticles, which allowed the sensors to be activated by 980 nm, left circular polarized light illumination, and achieved $Zn^{2+}/Mg^{2+}/Cu^{2+}$ sensing in cells (Figure 6.6c,d) [107].

In addition to the gold nanoparticles, gold nanoshells [108] and MnO_2 nanosheets [109] were also used for DNAzyme delivery for cell imaging. DNAzyme catalytic beacons can also be delivered into cells without particles. Multiple delivery reagents including α-helical cationic polypeptides [67], polyethylenimine [99], lipofectamine reagents [110], and TurboFect [66, 108], guided the sensors into different subcellular localizations, such as endoplasmic reticulum [67], lysosome, or cytoplasm [30, 66, 110]. Moreover, DNAzyme nanostructures were found to be able to penetrate into cells for live-cell imaging [53, 111].

6.4.2 Caged DNAzymes for Temporally Controlled Imaging

One limitation of applying the above catalytic beacon sensors in cellular imaging is that the DNAzyme could cleave before entering the cells or reaching the sensing site. For example, if the target metal ion exists in the transfection system, the sensors would be exposed to the metal ion and show signaling before delivery to the desired intracellular location. This limitation not only restricted the application of those fast-reactive catalytic beacons in sensing metal ions in specific subcellular localizations but also limited their applications in tracing metal ions at different time points. To overcome this limitation, different techniques were developed for the reversible caging of the sensors [112]. One of the most widely used caging strategies is to introduce a photolabile protecting group to the DNAzyme sensor to block its activity. 2′-*O*-nitrobenzyl and PC linker are two widely used photolabile groups [112]. Replacing the RNA at the cleavage site with 2′-*O*-nitrobenzyl-modified RNA disturbs the cleavage of DNAzyme (Figure 6.7a), thus inactivating the DNAzyme sensors until needed. To carefully manipulate DNAzyme sensors, the sensor can be activated by 365 nm light, which removes the protection and restores the cleavage site at the desired time point and location. This photocaging method was used on Zn^{2+} [110], Pb^{2+} [110], and Na^+ [67] DNAzymes for temporally controlled sensing in cells and was also used for monitoring metal ions during Aβ aggregation in the cellular microenvironment of brain cells and brain tissues [115]. However, 365 nm light has a very short penetration depth, which is not suitable for deep tissue imaging. To overcome this issue, up-conversion nanoparticles, which can absorb near-infrared (NIR) light and convert it into blue light (365 nm), were used to irradiate the 2′-*O*-nitrobenzyl group. By conjugating the up-conversion nanoparticle with DNAzyme sensors, the Lu group achieved the sensing of Zn^{2+} in living cells and in zebrafish (Figure 6.7b,c) [113]. A PC linker was used to connect a blocking strand to DNAzyme, which disturbs the conformation of active DNAzyme structures. Using up-conversion nanoparticles, Yi et al., were able to

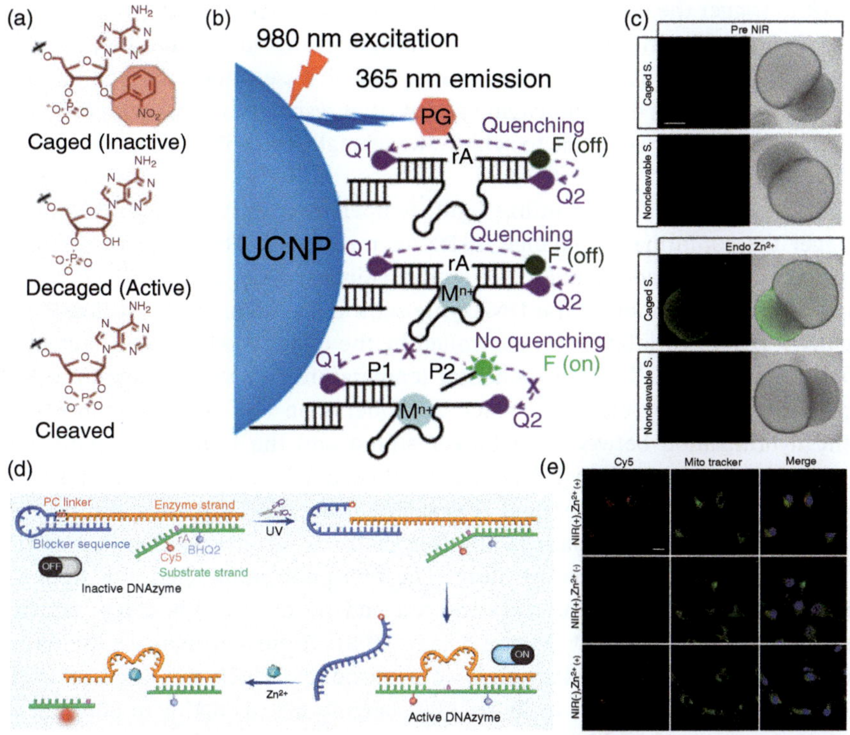

Figure 6.7 Photolabile groups modified DNAzymes for cellular and *in vivo* sensing. (a) Molecular structure of 2′-nitrobenzyl photocage group-based RNA cage and decage. (b) Structure of up-conversion nanoparticle-based NIR DNAzyme probe. Initially, fluorescence of the fluorophore (F) on one end of the substrate strand was quenched by quencher 1 (Q1) and 2 (Q2), and the rA base of the substrate strand was modified and protected by the 2′-nitrobenzyl photocage group. Under NIR light, the 2′-nitrobenzyl photolabile group is removed and restored the rA site for DNAzyme cleavage. In the presence of a metal ion (M^{n+}), the DNAzyme cleaves the rA base and halves the substrate strand. The fluorophore in the substrate fragment dehybridize from the enzyme strand and shows fluorescence turn-on. (c) Images of using the probe in (b) to sense Zn^{2+} in zebrafish embryo. (d) Scheme shows the PC linker-controlled DNAzyme sensor. A short blocker sequence is connected to the enzyme strand with a PC linker. Before light irritation, the blocker sequence disturbs the conformation of the enzyme strand. In the presence of UV light, the PC linker is cleaved, and the short blocker dissociates from the enzyme strand and restores the conformation of the DNAzyme probe. In the presence of Zn^{2+}, the DNAzyme cleaves its substrate to show fluorescence turn-on. (e) Cellular imaging for the PC linker caged sensor. Cy5 signals are from cleaved DNAzyme, which indicates the presence of target metal ions. MitoTracker labels mitochondria. The colocalization of Cy5 and MitoTracker shows the distribution of DNAzyme sensors in the mitochondria. Source: (a–c) Reproduced from Yang et al. [113], with permission. © 2018 American Chemical Society; (d, e) Reproduced from Yi et al. [114], with permission. © 2023 American Chemical Society.

use NIR to trigger the cleavage of PC linker and achieve spatially controlled metal sensing in mitochondria (Figure 6.7d,e). This system was further used for sensing different concentrations of Zn^{2+} in normal and diseased neuronal cell models [114]. Moreover, using different caging groups that respond to lights with different wavelengths, Xu et al. were able to achieve wavelength-controlled selective sensing in cells [116].

Other caging methods also utilize light, temperature, enzyme, and ultrasound to trigger the conformation change in DNAzyme probes. The DNAzyme probes are protected by an additional DNA strand (Linker DNA, Figure 6.8a), which disturbs the conformation of the DNAzyme sensors by DNA–DNA hybridization. This DNA–DNA hybridization is controlled by the difference between their melting temperatures and the environmental temperature. When the environmental temperature is slightly elevated, which is higher than the melting temperature for the hybridization between the Linker strand and the DNAzyme strand but still lower than the melting temperature between DNAzyme and its substrate strand, the additional DNA strand dissociates from the DNAzyme sensor and restores the sensor's conformation. Multiple methods were applied to induce the elevation of environmental temperature. The first publication shed NIR light on gold nanoshells to remove the linker strand and restore the DNAzyme activity for sensing intracellular Zn^{2+} (Figure 6.8a,b) [108]. To further increase the tissue penetration ability, high-intensity focused ultrasound (HIFU) was used to allow non-invasive sensing of Zn^{2+} in living mice (Figure 6.8c,d) [117]. In addition to ultrasound, a magnetic field was also used as a less invasive method for DNAzyme activation. Bakshi et al. conjugated two species of magnetic beads to different components of a multicomponent DNAzyme [118]. The substrate strand was blocked by a hybridizing strand and thus cannot hybridize to the enzyme strand without the help from magnetic field-triggered magnetic beads interactions. With this design, they achieved magnetic field-controlled mRNA sensing in MCF-7 breast cancer cells. The MCF-7 cells, which highly express Twist mRNA, showed a significant increase of fluorescence when applying a magnetic field when compared with HCX cells, cervical epithelial cells, which have low expression of Twist mRNA (Figure 6.8e,f).

Additionally, a protein enzyme was also used to regulate the melting temperature of the three-stranded structure and uncage DNAzyme sensors for temporally controlled sensing in cells [119]. Without the endonuclease I-SceI, the melting temperature between S2 and S3 was 64 °C (Figure 6.9a), which is significantly higher than the temperature for cell culture (normally 37 °C). After I-SceI digestion, the melting temperature of cleaved S3 decreased to 24°, which is lower than the cell culture temperature and allows the dissociation between S3 and S2. Thus, the DNAzyme sensor restored the activity and achieved sensing for Mg^{2+} inside of cells (Figure 6.9b). Similar enzyme-based caging methods were used for sensing metal ions in specific cell types. Yi et al. introduced a short blocking strand, which was conjugated to DNAzyme strand with an AP site to interfere with the hybridization between the substrate strand and enzyme strand of the DNAzyme [120]. When entering cancer cells, an endogenous enzyme, apurinic/apyrimidinic endonuclease 1 (APE1), can

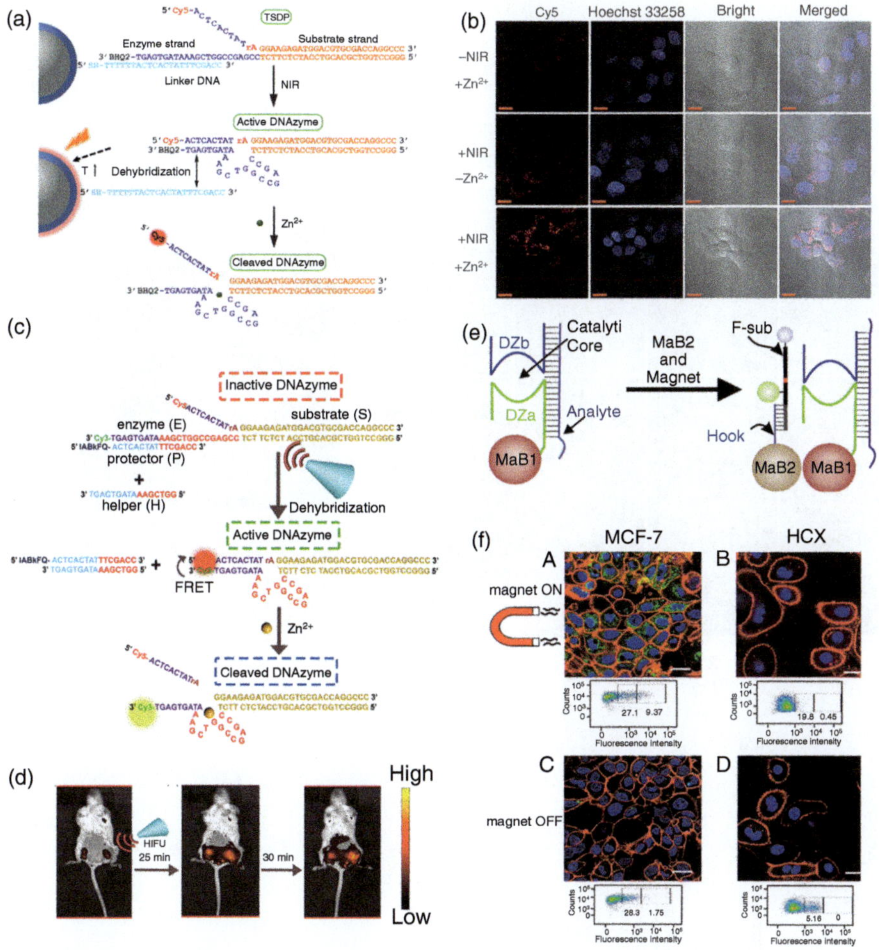

Figure 6.8 DNA–DNA hybridization controls DNAzyme activity for intracellular and *in vivo* imaging. (a) Scheme shows light-controlled sensing with nanoshells. One Linker DNA hybridize interference the conformation of the enzyme strand. When shedding light on the nanoshells, a small heat elevation in the microenvironment of the DNAzyme probe dissociates the Linker DNA from the enzyme strand and restores the conformation of the DNAzyme probe. (b) Cellular images for sensing Zn^{2+} with the DNAzyme probe in (a). (c) Schematic of high-intensity focused ultrasound (HIFU)-controlled Zn^{2+} detection. The protector strand binds to the enzyme strand and disturbs the conformation of the DNAzyme sensor. HIFU restores the activity of the DNAzyme sensor by elevating environmental temperature and dissociating the protector from the DNAzyme probe. (d) Mice imaging using the DNAzyme sensor in (c) for temporally controlled Zn^{2+} sensing in mice. (e) The scheme shows the magnetic controlled sensing for the metal ion. The interaction between DNAzyme and substrate strand is controlled by two different species of magnetic beads under a magnetic field. (f) Magnetic-controlled sensing showed higher Twist mRNA level in MCF-7 cancer cells than HCX, normal epithelial cells. Source: (a, b) Adapted from Wang et al. [108], with permission; copyright Wiley, 2017; (b–d) Adapted from Wang et al. [117], with permission; copyright American Chemical Society, 2022; (e, f) Reproduced from Bakshi et al. [118], with permission. © 2017 American Chemical Society.

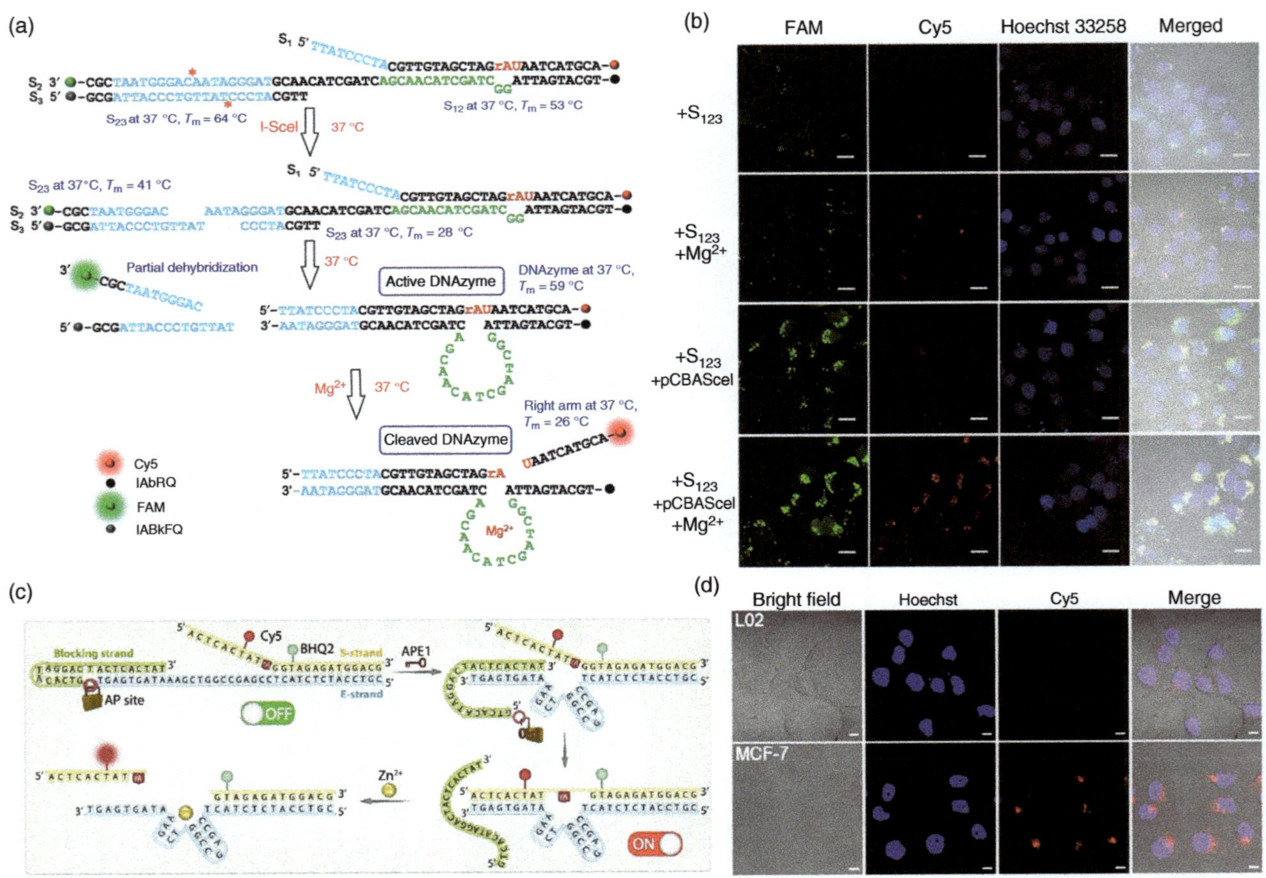

Figure 6.9 Enzyme-controlled metal sensing. (a) Scheme shows the I-SceI-controlled activation of 10–23 DNAzyme for sensing Mg^{2+}. (b) Images of Hela cells using the DNAzyme probe in (a). (c) Scheme shows the APE1-controlled DNAzyme probe. AP-site connects a short-blocking DNA that disturbs DNAzyme conformation. In the presence of APE1, the AP-site is cleaved, and the conformation of DNAzyme sensor restores. (d) The cellular imaging of the DNAzyme probe in (c). In MCF-7 cancer cells, which highly express APE1, the sensor detects Zn^{2+} and shows Cy5 fluorescence signals. In L02, which represents normal cells with low expression of APE1, the DNAzyme probe did not show Cy5 fluorescence. Source: (a, b) Adapted from Lin et al. [119], with permission from © John Wiley & Sons, Inc; (c, d) Adapted from Yi et al. [120], with permission from © John Wiley & Sons, Inc.

cleave the AP site and allow the DNAzyme sensor to form its correct conformation (Figure 6.9c). On the contrary, when entering the normal cells, the expression of APE1 is low, and thus the DNAzyme sensor cannot be activated. With this design, Yi et al. were able to achieve Zn^{2+} sensing in cancer cells but not normal cells in both cell lines and mice (Figure 6.9d).

6.4.3 DNAzyme-Based Sensing with Signal Amplification

One advantage of DNAzyme sensors in metal sensing is the ability to fine-tune their detection range to match the needs for metal sensing. In addition to performing an *in vitro* selection under the desired metal ion concentration, another generalizable way to lower the detection limit is signal amplification. For example, the Lu group has reported a catalytic beacon and molecular beacon design, which utilizes the enzymatic turnover of DNAzymes for signal-amplified sensing (Figure 6.10a) [121]. This design has successfully lowered the detection limit of Pb^{2+} from around 10 nM to 600 pM [91, 101, 121]. A DNA walker is another type of amplification method that utilizes the enzymatic turnover of DNAzymes and has been applied in intracellular sensing. This design uses the same DNAzyme strand to cut multiple substrates and achieve highly sensitive sensing of miRNA in cells [124]. However, the DNAzyme catalytic core is fragile and can be degraded by the intracellular environment during the sensing process. By introducing a circular DNA to the catalytic core, the serum resistance of the DNAzyme walker was increased more than fivefold, which resulted in more than an eightfold enhancement in catalytic activity and sensing performance (Figure 6.10b) [122].

Moreover, catalytic hairpin assembly (CHA), which utilizes a thermodynamic entropy gain process to drive a DNA circuit, can achieve exponential amplification of target sequence and be used to amplify the cleaved product of DNAzymes in cells. The cleaved DNAzyme substrate opens a helper hairpin, which binds to another hairpin strand that is labeled with a fluorophore and a quencher at both ends to show fluorescence turn-on (Figure 6.10c). This CHA-based amplification design lowered the detection limit of sodium DNAzyme sensor (NaA43) from 135 μM [67] to 14 μM and achieved the detection of endogenous Na^{+} inside Hela cells [123].

6.4.4 Genetically Encoded Sensors for Metal Sensing in Living Cells

In addition to the above metal ion sensing methods using DNAzyme sensors conjugated to fluorophores directly, genome-encoded fluorescent proteins are also powerful tools that have expanded the DNAzyme-based toolbox for the visualization of metal ions in cells. Using DNAzymes to cut the mRNA of fluorescence protein and using another fluorescence protein as an internal reference, the Lu group achieved the ratiometric sensing of Zn^{2+} and Mg^{2+} in living cells [125]. Taking advantage of the generalizability of DNAzymes and the well-established nature of fluorescence proteins, we could easily achieve genome-encoded metal sensing with different wavelengths.

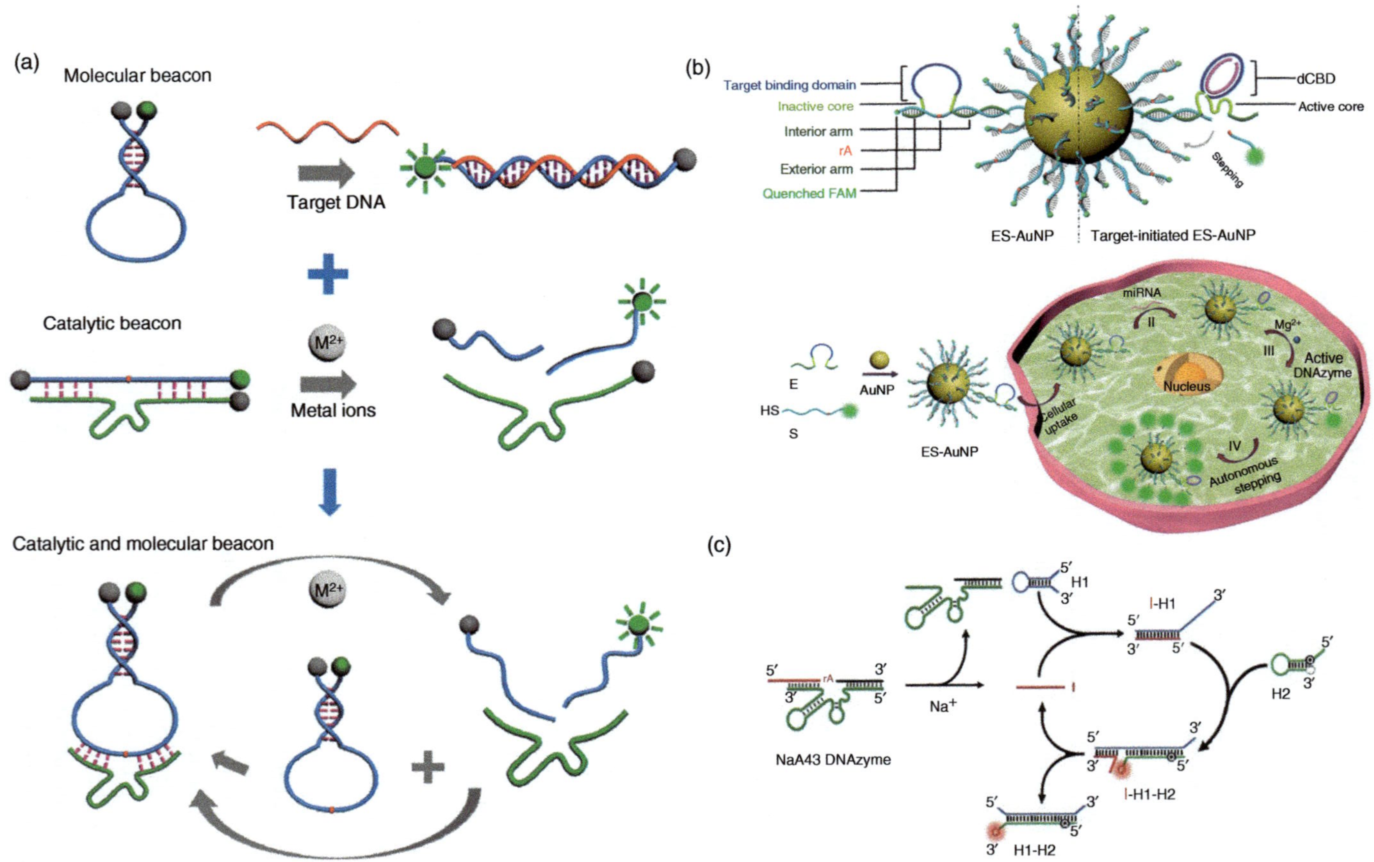

Figure 6.10 Different amplification methods for sensing metal ions in cells. (a) Scheme of the catalytic and molecular beacon for signal amplification [121]. (b) Scheme shows DNA walker-based signal amplification for sensing intracellular miRNA. (c) Scheme shows the catalytic hairpin assembly based signaling amplification for sensing intrinsic Na^+ in cells. Source: (a) Reproduced from Zhang et al. [121], with permission. © 2010 American Chemical Society; (b) Reproduced from Gao et al. [122], with permission. © 2021 American Chemical Society; (c) Reproduced from Wu et al. [123], with permission. © Wiley, 2027.

6.5 Conclusion

DNAzyme sensors have been developed into intracellular metal ion and microRNA sensors in recent years. The catalytic beacon design utilizes energy transfer between DNAzyme strand and substrate strand, as well as the changes in melting temperatures before and after metal-mediated target cleavage. The sensor activity can be controlled by introducing photolabile protection groups, which can be removed by light, or by introducing a blocking strand, which can hybridize with the DNAzyme probe and disturb the conformation. Multiple techniques, including shining light on gold nanoparticles, using HIFU, magnetic fields, and endonucleases, have been used to restore the DNAzyme conformation by elevating environmental temperature or lowering the melting temperature of the blocking strand by cleaving the DNA or linker. Amplification techniques, such as catalytic and molecular beacons or DNA walkers, which utilize the enzymatic turn-over property of DNAzyme, and CHA were used for amplifying the signals and lowering the detection limit of DNAzyme sensors. In addition to using the fluorophore modifications on the DNAzyme sequence, DNAzymes can also target fluorescence proteins and use their fluorescence intensity as a readout, which has expanded the metal sensing toolbox. In summary, DNAzyme sensors are powerful and generalizable tools for metal imaging and sensing inside of cells.

Acknowledgments

The Lu group research described in this chapter has been supported by the US National Institute of Health (GM141931) and the Robert A. Welch Foundation (Grant F-0020). And Qian group research described in this chapter has been supported by the National Natural Science Foundation of China (21977031), the Technology Commission of Shanghai Municipality (2018SHZDZX03), and the Shanghai Science and Technology Committee (22ZR1416800, 23ZR1416100).

References

1 Ghosh, S., Ahmad, R., Zeyaullah, M. et al. (2021). Microbial nano-factories: synthesis and biomedical applications. *Front. Chem.* 9: 626834.

2 Zeng, X., Li, S., and Kohles, S.S. (2011). Multiscale biomechanical modeling of stem cell-extracellular matrix interactions. In: *Advances in Cell Mechanics* (ed. S. Li and B. Sun), 27–53. Heidelberg: Springer.

3 Elmore, S. (2007). Apoptosis: a review of programmed cell death. *Toxicol. Pathol.* 35 (4): 495–516.

4 Gomes, L.R., Menck, C.F.M., and Leandro, G.S. (2017). Autophagy roles in the modulation of DNA repair pathways. *Int. J. Mol. Sci.* 8 (11): 2351.

5 Hotchkiss, R.S. and Nicholson, D.W. (2006). Apoptosis and caspases regulate death and inflammation in sepsis. *Nat. Rev. Immunol.* 6 (11): 813–822.

6 Levoin, N., Jean, M., and Legembre, P. (2020). CD95 structure, aggregation and cell signaling. *Front. Cell Dev. Biol.* 8: 314.

7 Naserian, S., Abdelgawad, M.E., Bakshloo, M.A. et al. (2020). The TNF/TNFR2 signaling pathway is a key regulatory factor in endothelial progenitor cell immunosuppressive effect. *Cell Commun. Signaling* 18 (1): 94.

8 Solano-Gálvez, S.G., Abadi-Chiriti, J., Gutiérrez-Velez, L. et al. (2018). Apoptosis: activation and inhibition in health and disease. *Med. Sci.* 6 (3): 54.

9 Crunkhorn, S. (2016). Ovarian cancer: cyclic peptide inhibits metastases. *Nat. Rev. Drug Discovery* 15 (5): 310.

10 Crunkhorn, S. (2018). Cancer: targeting the ubiquitin pathway. *Nat. Rev. Drug Discovery* 17 (3): 166.

11 Hyer, M.L., Milhollen, M.A., Ciavarri, J. et al. (2018). A small-molecule inhibitor of the ubiquitin activating enzyme for cancer treatment. *Nat. Med.* 24 (2): 186–193.

12 Klein, J.A. and Ackerman, S.L. (2003). Oxidative stress, cell cycle, and neurodegeneration. *J. Clin. Invest.* 111 (6): 785–793.

13 Maibaum, J., Liao, S.M., Vulpetti, A. et al. (2016). Small-molecule factor D inhibitors targeting the alternative complement pathway. *Nat. Chem. Biol.* 12 (12): 1105–1110.

14 McCulloch, T.R., Wells, T.J., and Souza-Fonseca-Guimaraes, F. (2022). Towards efficient immunotherapy for bacterial infection. *Trends Microbiol.* 30 (2): 158–169.

15 Boedtkjer, E. and Pedersen, S.F. (2020). The acidic tumor microenvironment as a driver of cancer. *Annu. Rev. Physiol.* 82 (1): 103–126.

16 Kato, Y., Ozawa, S., Miyamoto, C. et al. (2013). Acidic extracellular microenvironment and cancer. *Cancer Cell Int.* 13 (1): 89.

17 Medema, J.P. and Vermeulen, L. (2011). Microenvironmental regulation of stem cells in intestinal homeostasis and cancer. *Nature* 474 (7351): 318–326.

18 Laughlin, S.T., Agard, N.J., Baskin, J.M. et al. (2006). Metabolic labeling of glycans with azido sugars for visualization and glycoproteomics. *Methods Enzymol.* 415: 230–250.

19 Patra, M., Zarschler, K., and Pietzsch, H.J. (2016). New insights into the pretargeting approach to image and treat tumours. *Chem. Soc. Rev.* 45 (23): 6415–6431.

20 Wang, H., Sobral, M.C., Zhang, D.K.Y. et al. (2020). Metabolic labeling and targeted modulation of dendritic cells. *Nat. Mater.* 19 (11): 1244–1252.

21 Wang, X., Liu, Y., Fan, X. et al. (2019). Copper-triggered bioorthogonal cleavage reactions for reversible protein and cell surface modifications. *JACS* 141 (43): 17133–17141.

22 Erickson, B., DiMaggio, S.C., Mullen, D.G. et al. (2008). Interactions of poly(amidoamine) dendrimers with Survanta lung surfactant: the importance of lipid domains. *Langmuir* 24 (19): 11003–11008.

23 Hong, S., Bielinska, A.U., Mecke, A. et al. (2004). Interaction of poly(amidoamine) dendrimers with supported lipid bilayers and cells: hole formation and the relation to transport. *Bioconjugate Chem.* 15 (4): 774–782.

24 Hong, S., Leroueil, P.R., Janus, E.K. et al. (2006). Interaction of polycationic polymers with supported lipid bilayers and cells: nanoscale hole formation and enhanced membrane permeability. *Bioconjugate Chem.* 17 (3): 728–734.

25 Leroueil, P.R., Berry, S.A., Duthie, K. et al. (2008). Wide varieties of cationic nanoparticles induce defects in supported lipid bilayers. *Nano Lett.* 8 (2): 420–424.

26 Rabuka, D., Forstner, M.B., Groves, J.T. et al. (2008). Noncovalent cell surface engineering: incorporation of bioactive synthetic glycopolymers into cellular membranes. *JACS* 130 (18): 5947–5953.

27 Verma, A. and Stellacci, F. (2009). Effect of surface properties on nanoparticle-cell interactions. *Small* 6 (1): 12–21.

28 Walkey, C.D. and Chan, W.C.W. (2012). Understanding and controlling the interaction of nanomaterials with proteins in a physiological environment. *Chem. Soc. Rev. J.* 41 (7): 2780–2799.

29 Wang, F., Elbaz, J., and Willner, I. (2012). Enzyme-free amplified detection of DNA by an autonomous ligation DNAzyme machinery. *JACS* 134 (12): 5504–5507.

30 Wang, H., He, Y., Wei, J. et al. (2022). Construction of an autocatalytic hybridization assembly circuit for amplified in vivo MicroRNA imaging. *Angew. Chem. Int. Ed.* 61 (19): e202115489.

31 Gong, L., Shan, X., Xu, L. et al. (2022). Intracellular signal amplification for ultrasensitive detection and imaging: progress, challenges, and opportunities. *Anal. Sens.* 2 (5): e202200007.

32 Mudiyanselage, A.P.K.K.K., Yu, Q., Leon-Duque, M.A. et al. (2018). Genetically encoded catalytic hairpin assembly for sensitive RNA imaging in live cells. *JACS* 140 (28): 8739–8745.

33 Xue, C., Zhang, S., Ouyang, C. et al. (2018). Target-induced catalytic assembly of Y-shaped DNA and its application for in situ imaging of MicroRNAs. *Angew. Chem. Int. Ed.* 57 (31): 9739–9743.

34 Chen, J., Zhang, J., Guo, Y. et al. (2011). An ultrasensitive electrochemical biosensor for detection of DNA species related to oral cancer based on nuclease-assisted target recycling and amplification of DNAzyme. *Chem. Commun.* 47 (28): 8004–8006.

35 Gao, F., Lei, J., and Ju, H. (2013). Assistant DNA recycling with nicking endonuclease and molecular beacon for signal amplification using a target-complementary arched structure. *Chem. Commun.* 49 (38): 4006–4008.

36 He, J., Wu, Z., Zhou, H. et al. (2010). Fluorescence aptameric sensor for strand displacement amplification detection of Cocaine. *Anal. Chem.* 82 (4): 1358–1364.

37 Liu, M., Zhang, Q., Chang, D. et al. (2017). A DNAzyme feedback amplification strategy for biosensing. *Angew. Chem. Int. Ed.* 56 (22): 6142–6146.

38 Santiago-Frangos, A., Nemudryi, A., Nemudraia, A. et al. (2022). CRISPR-Cas, Argonaute proteins and the emerging landscape of amplification-free diagnostics. *Methods* 205: 1–10.

39 Shen, W., Deng, H., and Gao, Z. (2012). Gold nanoparticle-enabled real-time ligation chain reaction for ultrasensitive detection of DNA. *JACS* 134 (36): 14678–14681.

40 Wei, J., Wang, H., Wu, Q. et al. (2020). A smart, autocatalytic, DNAzyme biocircuit for in vivo, amplified, microRNA imaging. *Angew. Chem. Int. Ed.* 59 (15): 5965–5971.

41 Zhang, Y., Yang, F., Wei, W. et al. (2022). Self-propelled Janus mesoporous micromotor for enhanced microRNA capture and amplified detection in complex biological samples. *ACS Nano* 16 (4): 5587–5596.

42 Zuo, X., Xia, F., Xiao, Y. et al. (2010). Sensitive and selective amplified fluorescence DNA detection based on Exonuclease III-aided target recycling. *JACS* 132 (6): 1816–1818.

43 Breaker, R.R. and Joyce, G.F. (1994). A DNA enzyme that cleaves RNA. *Cell Chem. Biol.* 1 (4): 223–229.

44 Breaker, R.R. (1997). In vitro selection of catalytic polynucleotides. *Chem. Rev.* 97 (2): 371–390.

45 Cech, T.R. (2002). Ribozymes, the first 20 years. *Biochem. Soc. Trans.* 30 (6): 1162–1166.

46 Hwang, K., Hosseinzadeh, P., and Lu, Y. (2016). Biochemical and biophysical understanding of metal ion selectivity of DNAzymes. *Inorg. Chim. Acta* 452 (1): 12–24.

47 Li, Y. and Breaker, R.R. (1999). Deoxyribozymes: new players in the ancient game of biocatalysis. *Curr. Opin. Struct. Biol.* 9 (3): 315–323.

48 McConnell, E.M., Cozma, I., Mou, Q. et al. (2021). Biosensing with DNAzymes. *Chem. Soc. Rev.* 50 (16): 8954–8994.

49 Morrison, D., Rothenbroker, M., and Li, Y. (2018). DNAzymes: selected for applications. *Small Methods* 2 (3): 1700319.

50 Sai Lau, P. and Li, Y. (2011). Functional nucleic acids as molecular recognition elements for small organic and biological molecules. *Curr. Org. Chem.* 15 (4): 557–575.

51 Zhou, Y. and Zhou, C. (2021). Designing in vivo active DNAzymes. *Nat. Chem.* 13 (4): 299–301.

52 Ihms, H.E. and Lu, Y. (2012). In vitro selection of metal-ion-selective DNAzymes. *Methods Mol. Biol.* 848: 297–316.

53 Zhou, W., Liang, W., Li, D. et al. (2016). Dual-color encoded DNAzyme nanostructures for multiplexed detection of intracellular metal ions in living cells. *Biosens. Bioelectron.* 85: 573–579.

54 Zhou, W., Saran, R., Huang, P.J.J. et al. (2017). An exceptionally selective DNA cooperatively binding two Ca2+ ions. *ChemBioChem* 18 (6): 518–522.

55 Huang, P.J.J. and Liu, J. (2015). Rational evolution of Cd2+-specific DNAzymes with phosphorothioate modified cleavage junction and Cd2+ sensing. *Nucleic Acids Res.* 43 (12): 6125–6133.

56 Bruesehoff, P., Li, J., Augustine, A.J. et al. (2002). Improving metal ion specificity during in vitro selection of catalytic DNA. *Comb. Chem. High Throughput Screen.* 5 (4): 327–335.

57 Carmi, N., Shultz, L.A., and Breaker, R.R. (1996). In vitro selection of self-cleaving DNAs. *Chem. Biol.* 3 (12): 1039–1046.

58 Vannela, R. and Adriaens, P. (2007). In vitro selection of Hg (II) and As (V)-dependent RNA-cleaving DNAzymes. *Environ. Eng. Sci.* 24 (1): 73–84.

59 Santoro, S.W. and Joyce, G.F. (1997). A general purpose RNA-cleaving DNA enzyme. *Proc. Natl. Acad. Sci.* 94 (9): 4262–4266.

60 Fan, H., McGhee, C.E., Lake, R.J. et al. (2023). A highly selective Mn(II)-specific DNAzyme and its application in intracellular sensing. *J. Am. Chem. Soc. Au.* 3 (6): 1615–1622.

61 Liu, J., Brown, A.K., Meng, X. et al. (2007). A catalytic beacon sensor for uranium with parts-per-trillion sensitivity and millionfold selectivity. *Proc. Natl. Acad. Sci.* 104 (7): 2056–2061.

62 Li, J., Zheng, W., Kwon, A.H. et al. (2000). In vitro selection and characterization of a highly efficient Zn(II)-dependent RNA-cleaving deoxyribozyme. *Nucleic Acids Res.* 28 (2): 481–488.

63 Santoro, S.W., Joyce, G.F., Sakthivel, K. et al. (2000). RNA cleavage by a DNA enzyme with extended chemical functionality. *J. Am. Chem. Soc.* 122 (11): 2433–2439.

64 Nelson, K.E., Bruesehoff, P.J., and Lu, Y. (2005). In vitro selection of high temperature Zn^{2+}-dependent DNAzymes. *J. Mol. Evol.* 61 (2): 216–225.

65 Saran, R. and Liu, J. (2016). A silver DNAzyme. *Anal. Chem.* 88 (7): 4014–4020.

66 McGhee, C.E., Yang, Z., Guo, W. et al. (2021). DNAzyme-based lithium-selective imaging reveals higher lithium accumulation in bipolar disorder patient-derived neurons. *ACS Cent. Sci.* 7 (11): 1809–1820.

67 Torabi, S.F., Wu, P., McGhee, C.E. et al. (2015). In vitro selection of a sodium-specific DNAzyme and its application in intracellular sensing. *Proc. Natl. Acad. Sci.* 112 (19): 5903–5908.

68 Torabi, S.F. and Lu, Y. (2015). Identification of the same Na^+-specific DNAzyme motif from two in vitro selections under different conditions. *J. Mol. Evol.* 81 (5): 225–234.

69 Qian, R.C., Zhou, Z.R., Guo, W. et al. (2021). Cell surface engineering using DNAzymes: metal ion mediated control of cell–cell interactions. *J. Am. Chem. Soc.* 143 (15): 5737–5744.

70 Hwang, K., Mou, Q., Lake, R.J. et al. (2019). Metal-dependent DNAzymes for the quantitative detection of metal ions in living cells: recent progress, current challenges, and latest results on FRET ratiometric sensors. *Inorg. Chem.* 58 (20): 13696–13708.

71 Mager, M.D., LaPointe, V., and Stevens, M.M. (2011). Exploring and exploiting chemistry at the cell surface. *Nat. Chem.* 3 (8): 582–529.

72 Custódio, C.A. and Mano, J.F. (2016). Cell surface engineering to control cellular interactions. *ChemNanoMat* 2 (5): 376–384.

73 Li, J., Wang, L., Tian, J. et al. (2020). Nongenetic engineering strategies for regulating receptor oligomerization in living cells. *Chem. Soc. Rev.* 49 (5): 1545–1568.

74 Liu, J.S. and Gartner, Z.J. (2012). Directing the assembly of spatially organized multicomponent tissues from the bottom up. *Trends Cell Biol.* 22 (12): 683–691.

75 Kim, E., Kim, J.C., Min, K. et al. (2018). Rapid and versatile cell aggregate formation using lipid-conjugated heparin. *ACS Appl. Mater. Interfaces* 10 (29): 24431–24439.

76 Mueller, M., Rasoulinejad, S., Garg, S. et al. (2020). The importance of cell–cell interaction dynamics in bottom-up tissue engineering: concepts of colloidal self-assembly in the fabrication of multicellular architectures. *Nano Lett.* 20 (4): 2257–2263.

77 Zhang, Y.M. and Rock, C.O. (2008). Membrane lipid homeostasis in bacteria. *Nat. Rev. Microbiol.* 6 (3): 222–233.

78 Phillips, R., Ursell, T., Wiggins, P. et al. (2009). Emerging roles for lipids in shaping membrane-protein function. *Nature* 459 (7245): 379–385.

79 Simons, K. and Sampaio, J.L. (2011). Membrane organization and lipid rafts. *Cold Spring Harb. Perspect. Biol.* 3 (10): a004697.

80 Hu, C.M.J. and Zhang, L. (2012). Nanoparticle-based combination therapy toward overcoming drug resistance in cancer. *Biochem. Pharmacol.* 83 (8): 1104–1111.

81 Pusuluri, A., Wu, D., and Mitragotri, S. (2019). Immunological consequences of chemotherapy: single drugs, combination therapies and nanoparticle-based treatments. *J. Control. Release* 305: 130–154.

82 Qi, J., Jin, F., Xu, X. et al. (2021). Combination cancer immunotherapy of nanoparticle-based immunogenic cell death inducers and immune checkpoint inhibitors. *Int. J. Nanomedicine* 16: 1435–1456.

83 Blanco, E., Shen, H., and Ferrari, M. (2015). Principles of nanoparticle design for overcoming biological barriers to drug delivery. *Nat. Biotechnol.* 33 (9): 941–951.

84 Zhao, Z., Ukidve, A., Kim, J. et al. (2020). Targeting strategies for tissue-specific drug delivery. *Cell* 181 (1): 151–167.

85 Zhou, Y., Chen, X., Cao, J. et al. (2020). Overcoming the biological barriers in the tumor microenvironment for improving drug delivery and efficacy. *J. Mater. Chem. B* 8 (31): 6765–6781.

86 Liu, J. and Lu, Y. (2002). FRET study of a trifluorophore-labeled DNAzyme. *J. Am. Chem. Soc.* 124 (51): 15208–15216.

87 Chen, L., Luo, S., Ge, Z. et al. (2022). Unbiased enrichment of circulating tumor cells via DNAzyme-catalyzed proximal protein biotinylation. *Nano Lett.* 22 (4): 1618–1625.

88 Qian, R.C., Zhou, Z.R., Wu, Y. et al. (2022). Combination cancer treatment: using engineered DNAzyme molecular machines for dynamic inter- and intracellular regulation. *Angew. Chem.-Int. Edit.* 61 (49): e202210935.

89 Zhao, D., Chang, D., Zhang, Q. et al. (2021). Rapid and specific imaging of extracellular signaling molecule adenosine triphosphate with a self-phosphorylating DNAzyme. *J. Am. Chem. Soc.* 143 (37): 15084–15090.

90 Mei, S.H., Liu, Z., Brennan, J.D. et al. (2003). An efficient RNA-cleaving DNA enzyme that synchronizes catalysis with fluorescence signaling. *J. Am. Chem. Soc.* 125 (2): 412–420.

91 Li, J. and Lu, Y. (2000). A highly sensitive and selective catalytic DNA biosensor for lead ions. *J. Am. Chem. Soc.* 122 (42): 10466–10467.

92 Xiang, Y. and Lu, Y. (2012). Expanding targets of DNAzyme-based sensors through deactivation and activation of DNAzymes by single uracil removal: sensitive fluorescent assay of uracil-DNA glycosylase. *Anal. Chem.* 84 (22): 9981–9987.

93 Chiuman, W. and Li, Y. (2007). Efficient signaling platforms built from a small catalytic DNA and doubly labeled fluorogenic substrates. *Nucleic Acids Res.* 35 (2): 401–405.

94 Liu, Z., Mei, S.H.J., Brennan, J.D., and Li, Y. (2003). Assemblage of signaling DNA enzymes with intriguing metal-ion specificities and pH dependences. *J. Am. Chem. Soc.* 125 (25): 7539–7545.

95 Zhang, X.B., Kong, R.M., and Lu, Y. (2011). Metal ion sensors based on DNAzymes and related DNA molecules. *Annu. Rev. Anal. Chem.* 4 (1): 105–128.

96 Lu, Y., Liu, J., Li, J. et al. (2003). New highly sensitive and selective catalytic DNA biosensors for metal ions. *Biosens. Bioelectron.* 18 (5–6): 529–540.

97 Liu, J. and Lu, Y. (2003). Improving fluorescent DNAzyme biosensors by combining inter- and intramolecular quenchers. *Anal. Chem.* 75 (23): 6666–6672.

98 Nagraj, N., Liu, J., Sterling, S. et al. (2009). DNAzyme catalytic beacon sensors that resist temperature-dependent variations. *Chem. Commun.* 30 (27): 4103–4105.

99 Wu, Y., Torabi, S.F., Lake, R.J. et al. (2023). Simultaneous Fe^{2+}/Fe^{3+} imaging shows Fe^{3+} over Fe^{2+} enrichment in Alzheimer's disease mouse brain. *Sci. Adv.* 9 (16): eade7622.

100 Liu, J. and Lu, Y. (2007). A DNAzyme Catalytic Beacon sensor for paramagnetic Cu^{2+} ions in aqueous solution with high sensitivity and selectivity. *JACS* 129 (32): 9838–9839.

101 Lan, T., Furuya, K., and Lu, Y. (2010). A highly selective lead sensor based on a classic lead DNAzyme. *Chem. Commun.* 46 (22): 3896–3898.

102 Lee, J.H., Wang, Z., and Lu, Y. (2012). DNAzyme-based sensing for metal ions in ocean platform. In: *Molecular Biological Technologies for Ocean Sensing*, 103–116. New Jersey: Humana Press.

103 Nagraj, N. and Lu, Y. (2011). Catalytic nucleic acid biosensors for environmental monitoring. In: *Nucleic Acid Biosensors for Environmental Pollution Monitoring* (ed. I. Palchetti and M. Mascini), 82–98. Cambridge: Royal Society of Chemistry.

104 Xiang, Y., Wu, P., Tan, L.H., and Lu, Y. (2014). DNAzyme-functionalized gold nanoparticles for biosensing. In: *Biosensors Based on Aptamers and Enzymes* (ed. H.-S. Kim and M.B. Gu), 93–120. Heidelberg: Springer.

105 Lan, T., Zhang, J., and Lu, Y. (2016). Transforming the blood glucose meter into a general healthcare meter for in vitro diagnostics in mobile health. *Biotechnol. Adv.* 34 (3): 331–341.

106 Wu, P., Hwang, K., Lan, T., and Lu, Y. (2013). A DNAzyme-gold nanoparticle probe for uranyl ion in living cells. *JACS* 135 (14): 5254–5257.

107 Gao, R., Xu, L., Hao, C. et al. (2019). Circular polarized light activated chiral satellite nanoprobes for the imaging and analysis of multiple metal ions in living cells. *Angew. Chem. Int. Ed.* 58 (12): 3913–3917.

108 Wang, W., Satyavolu, N.S.R., Wu, Z. et al. (2017). Near-infrared photothermally activated DNAzyme-gold nanoshells for imaging metal ions in living cells. *Angew. Chem. Int. Ed.* 56 (24): 6798–6802.

109 Fan, H., Zhao, Z., Yan, G. et al. (2015). A smart DNAzyme–MnO_2 nanosystem for efficient gene silencing. *Angew. Chem. Int. Ed.* 54 (16): 4801–4805.

110 Hwang, K., Wu, P., Kim, T. et al. (2014). Photocaged DNAzymes as a general method for sensing metal ions in living cells. *Angew. Chem. Int. Ed.* 53 (50): 13798–13802.

111 Meng, H.M., Zhang, X., Lv, Y. et al. (2014). DNA dendrimer: an efficient nanocarrier of functional nucleic acids for intracellular molecular sensing. *ACS Nano* 8 (6): 6171–6181.

112 Wu, Y., Yang, Z., and Lu, Y. (2020). Photocaged functional nucleic acids for spatiotemporal imaging in biology. *Curr. Opin. Chem. Biol.* 57: 95–104.

113 Yang, Z., Loh, K.Y., Chu, Y.T. et al. (2018). Optical control of metal ion probes in cells and zebrafish using highly selective DNAzymes conjugated to upconversion nanoparticles. *JACS* 140 (50): 17656–17665.

114 Yi, D., Zhao, H., Zhao, J. et al. (2023). Modular engineering of DNAzyme-based sensors for spatioselective imaging of metal ions in mitochondria. *JACS* 145 (3): 1678–1685.

115 Liu, L., Zhang, M., Rong, Q. et al. (2022). Photoactivatable cell-surface DNAzyme sensor for real-time monitoring of the metal ion-induced Aβ aggregation. *CCS Chem.* 5 (9): 1–13.

116 Xu, X., Xiao, L., Gu, C. et al. (2021). Wavelength-selective activation of photocaged dnazymes for metal ion sensing in live cells. *ACS Omega* 6 (20): 13153–13160.

117 Wang, X., Kim, G., Chu, J.L. et al. (2022). Noninvasive and spatiotemporal control of DNAzyme-based imaging of metal ions in vivo using high-intensity focused ultrasound. *JACS* 144 (13): 5812–5819.

118 Bakshi, S.F., Guz, N., Zakharchenko, A. et al. (2017). Magnetic field-activated sensing of mRNA in living cells. *J. Am. Chem. Soc.* 139 (35): 12117–12120.

119 Lin, Y., Yang, Z., Lake, R.J. et al. (2019). Enzyme-mediated endogenous and bioorthogonal control of a DNAzyme fluorescent sensor for imaging metal ions in living cells. *Angew. Chem. Int. Ed.* 58 (47): 17061–17067.

120 Yi, D., Zhao, J., and Li, L. (2021). An enzyme-activatable engineered DNAzyme sensor for cell-selective imaging of metal ions. *Angew. Chem. Int. Ed.* 60 (12): 6300–6304.

121 Zhang, X.B., Wang, Z.D., Xing, H. et al. (2010). Catalytic and molecular beacons for amplified detection of metal ions and organic molecules with high sensitivity. *Anal. Chem.* 82 (12): 5005–5011.

122 Gao, Y., Zhang, S., Wu, C. et al. (2021). Self-protected DNAzyme walker with a circular bulging DNA shield for amplified imaging of miRNAs in living cells and mice. *ACS Nano* 15 (12): 19211–19224.

123 Wu, Z., Fan, H., Satyavolu, N.S.R. et al. (2017). Imaging endogenous metal ions in living cells using a DNAzyme–catalytic hairpin assembly probe. *Angew. Chem. Int. Ed.* 56 (30): 8721–8725.

124 Xue, C., Zhang, S., Li, C. et al. (2019). Y-shaped backbone-rigidified triangular DNA scaffold-directed stepwise movement of a DNAzyme walker for sensitive microRNA imaging within living cells. *Anal. Chem.* 91 (24): 15678–15685.

125 Xiong, M., Yang, Z., Lake, R.J. et al. (2020). DNAzyme-mediated genetically encoded sensors for ratiometric imaging of metal ions in living cells. *Angew. Chem. Int. Ed.* 59 (5): 1891–1896.

7

DNA Nanomotors for Bioimaging in Living Cells

Hanyong Peng[1,2], Aijiao Yuan[1,2], Hang Xiao[1,2], Zi Ye[1,2], Lejun Liao[3], Shulin Zhao[3], and X. Chris Le[4]

[1] *State Key Laboratory of Environmental Chemistry and Ecotoxicology, Research Center for Eco-Environmental Sciences, Chinese Academy of Sciences, Beijing 100085, China*
[2] *School of Resources and Environmental Sciences, University of Chinese Academy of Sciences, Beijing 101314, China*
[3] *State Key Laboratory for the Chemistry and Molecular Engineering of Medicinal Resources, School of Chemistry and Pharmaceutical Sciences, Guangxi Normal University, Guilin 541004, China*
[4] *Division of Analytical and Environmental Toxicology, Department of Laboratory Medicine and Pathology, Faculty of Medicine and Dentistry, University of Alberta, Edmonton, Alberta T6G 2G3, Canada*

Diverse developments in DNA nanotechnology enable continuing innovations in artificial DNA nanomachines. The operation of DNA nanomachines mimics the function of biological protein motors in nature, such as myosin, dynein, and kinesin. Various nanomachines have been developed, including DNA tweezers [1–3], DNA gears [4, 5], molecular spiders [6], and DNA robots [7]. DNA nanomotors, an emerging class of nanomachines, have the ability to travel along DNA tracks continuously and perform specific functions autonomously [8–11]. DNA nanomotors possess unique features, such as predictability of their structures, integration of multiple components, and transport of functional molecules. The operation of DNA nanomotors can be tuned to release payloads, such as fluorophores, and to generate signal amplification. Therefore, DNA nanomotors have great potential for diverse applications in biosensing and imaging of intracellular molecules [12–14].

A typical DNA nanomotor is built from the assembly of DNA strands, and consists of four components: a track, energy, legs, and cargo [15–18]. A DNA nanomotor can only walk along the track, which defines the walking distance and direction [19–23]. Through chemical or physical reactions, various energy sources have been used to drive the progressive movement of DNA nanomotors. Examples include DNA strand displacement, enzymatic cleavage, light, and conformational changes induced by binding to small molecules [24–28]. To increase the speed of DNA nanomotors along an oriented track, researchers have designed DNA nanomotors with multiple legs or pedals and have achieved a higher turnover rate for each step in a more precise position. Continuous innovations in DNA nanomotors improve their performance in progressivity, speed, and directionality [29–31]. The artificially engineered DNA nanomotors with varying species and DNA sequences have made

DNA Nanotechnology for Cell Research: From Bioanalysis to Biomedicine, First Edition. Edited by Zhou Nie.

extensive contributions to bioanalytical applications for signal amplification in living cells [32–34].

Three projecting single-stranded triple-cross DNA molecules have been used to create a DNA nanomotor, which is designed to move along a one-dimensional (1D) track [35]. The DNA nanomotor glides precisely forward or backward, like an inchworm. Pierce et al. developed DNA nanomotors based on strand displacement reactions [36]. The nanomotor is successfully hybridized on the track when two single-stranded oligonucleotide chains (A1 and A2) are available and complementary to the track strands and the foot of the nanomotor. The walking leg corresponding to the displacement of A1 caused by the addition of the fuel chain D1 is liberated from the track and moves along its 1D track (Figure 7.1A). The low substrate DNA density incurs a speed constraint on the 1D DNA track. Nanomaterials can serve as scaffolds in the fabrication of carbon nanotube (CN) tracks [37]. The DNA nanomotor crawls over the nanotube track at a top speed of 0.1 nm/s and moves more than 3 μm total distance. This is still three orders of magnitude slower than the natural protein motor, though it is significantly faster than previously built DNA nanomotors [27, 35, 36, 38].

The substrate molecules can be located precisely on a two-dimensional (2D) track using DNA origami or DNA-modified planes [31, 39–41]. A "spider" DNA nanomotor with sophisticated 2D track movements like "start," "follow," "reverse," and "stop" was developed by Lund et al. [7]. The DNA spiders can move 50 steps at a rate of 3 nm/min, topographically steering in a 100 nm area. Wickham et al. [42] constructed a directional and processive 100 nm 2D DNA track on which the DNA nanomotor moves 16 steps independently in succession without detaching from the track. The size of each rectangular origami unit is around 90 nm, limiting the walking distance. It is possible to program the assembly of many DNA origami tiles and extend the length of the track [43–45].

The larger surface area on the three-dimensional (3D) track permits the loading of more substrate molecules. Three-dimensional tracks are traditionally constructed on nanomaterials, such as CNs, magnetic nanoparticles, and gold nanoparticles (AuNPs). A binding-induced DNA nanomotor was designed by Zhang et al. [28] to identify nucleic acids and proteins (Figure 7.1B). A high density of DNA substrate was established by hybridizing hundreds of fluorophore-labeled DNA molecules onto a 20 nm AuNP. The activated DNA nanomotor was capable of moving more than 100 steps. Because of the release of multiple fluorophores for signal amplification, detection of streptavidin and PDGF-BB at concentrations as low as 0.5 and 5 pM was achieved using the 3D DNA nanomotor. A 3D DNA nanomotor built on microspheres and magnetic nanoparticles has achieved a detection limit as low as 4.1 fM for biosensing nucleic acids (Figure 7.1C) [9, 46]. In contrast with DNA nanomotors that consume substrate molecules on the track, a regenerative track was developed to detect the Ebola virus [47].

The strand displacement reaction, which is based on the principle of complementary base pairing, tends to a kinetic and thermodynamic equilibrium by forming and breaking hydrogen bonds among oligonucleotide chains [48]. Catalytic hairpin assembly (CHA) is an enzyme-free isothermal technique, which has

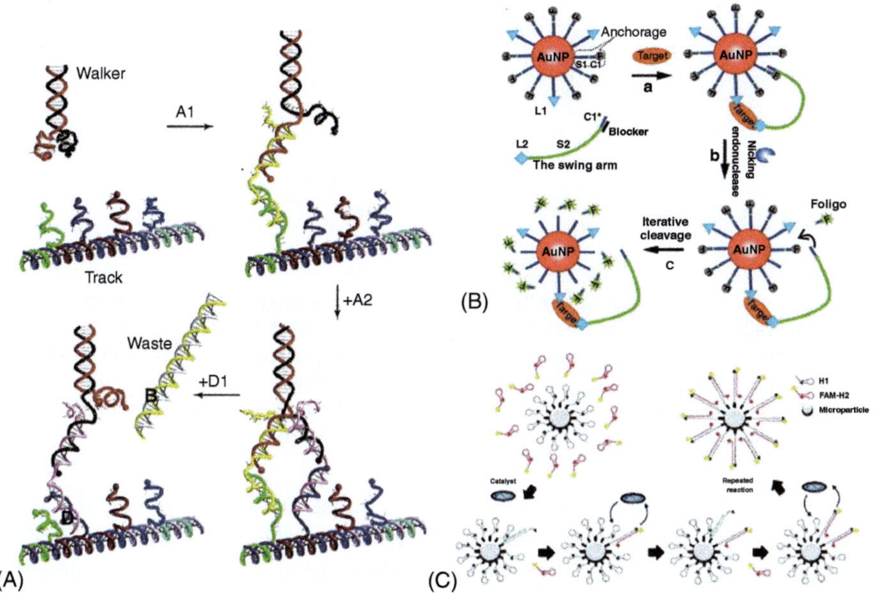

Figure 7.1 (A) A DNA nanomotor travels over a 1D linear track that facilitates molecular transport by toehold-mediated strand displacement processes. Source: Reproduced from Shin and Pierce [36], with permission of 2004 American Chemical Society. (B) A 3D DNA nanomotor that is triggered by specific binding to the target. The affinity binding initiates the on-particle DNA walking and the enzymatic cleavage of the substrate. The iterative cleavage and release of fluorescent probes result in signal amplification. Source: Reproduced from Zhang et al. [28], with permission of 2015 John Wiley & Sons. (C) Upon the addition of a "catalyst" strand, the 3D DNA nanomotor that is self-assembled by catalytic hairpins on a microparticle surface opens selectively (Nanomotor, blue), is successively displaced by another DNA hairpin in solution (in red), and proceeds arbitrarily to the next footing. Source: Reproduced from Jung et al. [9], with permission of 2017 Springer Nature.

been used to progressively drive the DNA nanomotor [49, 50]. Li et al. reported a nonenzymatic 3D DNA nanomotor powered by one-foot CHA reactions [23]. In this CHA-powered system, the hairpin tracks (H1) and an affinity probe are modified on a magnetic nanobead. In the presence of the target gene, the swing arm and affinity probe on the nanobead both bind to the target, thus triggering the oligonucleotide to hybridize with H1 and initiating the CHA. The addition of H2, which hybridizes with the newly exposed sequence of H1, leads to the second branch migration, thus releasing the swing arm and starting the next cycle [23]. CHA has been extensively used as a driving force in DNA walking systems because of its excellent catalytic efficiency and biocompatibility [23, 32, 51–58]. Another enzyme-free strategy, entropy-driven catalysis (EDC), has also been used in DNA walking systems [59]. When combining EDC with DNA nanomotors, the walking strands often act as catalytic strands to enable autonomous walking [21, 60–67]. Yang et al. [21] reported an aptamer/protein proximity binding-triggered molecular walking machine to detect thrombin (Figure 7.2a). The complex consisting of two different aptamers recognizes the target thrombin, which initiates proximity

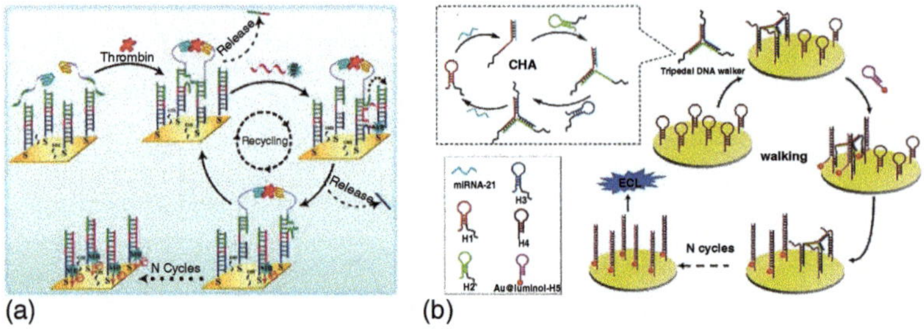

Figure 7.2 DNA nanomotors propelled by various forms of energy. (a) A system of molecular walking is initiated by the displacement of an ssDNA track induced by the proximity binding of an aptamer and a protein. Source: Reproduced from Yang et al. [21], with permission of 2007 American Chemical Society. (b) A CHA-based tripedal DNA nanomotor for electrochemical detection of miRNA-21. Source: Reproduced from Wang et al. [58], with permission of 2020 American Chemical Society.

binding-induced displacement of a single-stranded DNA (ssDNA) of the track modified on the gold electrode. As a consequence, a prelocked domain is exposed and hybridized with the methylene blue (MB)-tagged ssDNA, which serves as a fuel strand. The toehold exchange strand displacement then occurs leading to the release of the walking strand. MB-DNAs are captured by the complex surface to produce an amplified electrochemical signal response to the thrombin. Fuel strands can serve as electrochemical signal molecules for entropic drive and signal output (Figure 7.2b), such as in CHA [58] and hybridization chain reaction (HCR) [68].

Protein enzymes, especially endonucleases, exonucleases, and DNAzymes, are widely used to drive DNA nanomotors. Endonucleases are restriction enzymes that can cleave single- or double-stranded DNA by hydrolyzing phosphodiester bonds in the middle of a polynucleotide chain. Yin et al. [38] devised a unidirectional DNA nanomotor that moves forward autonomously through the alternating action of ligase and restriction endonuclease. Cleavages destroy the previous recognition sites, and ligations create new ones correspondingly, which drives the motor progressively and directionally. Nicking endonucleases have been widely incorporated in enzyme-driven DNA nanomotors [8, 28, 69–72]. Other strategies combine nicking enzymes with AuNPs, which act as fluorescence quenchers. The cleavage catalyzed by the nicking enzyme releases the track fragments from the AuNPs, resulting in fluorescence generation and signal amplification [8, 28, 70, 72, 73].

Oligonucleotide chains can also be digested unidirectionally by exonucleases from the 5′ or 3′ terminus in a single-stranded state, which serves as the driving force of DNA machines. Qu et al. [29] reported a DNA nanomotor that can move unidirectionally through the toehold-mediated strand displacement reaction (Figure 7.3a). The hybridized DNA nanomotors moving on the AuNP-based 3D tracks can be identified by exonuclease III. The exonuclease III digests the track strands from the 3′ terminus on the AuNP.

The DNA nanomotor based on DNA enzymes exhibits greater stability and resilience to a variety of external experimental conditions when compared to its

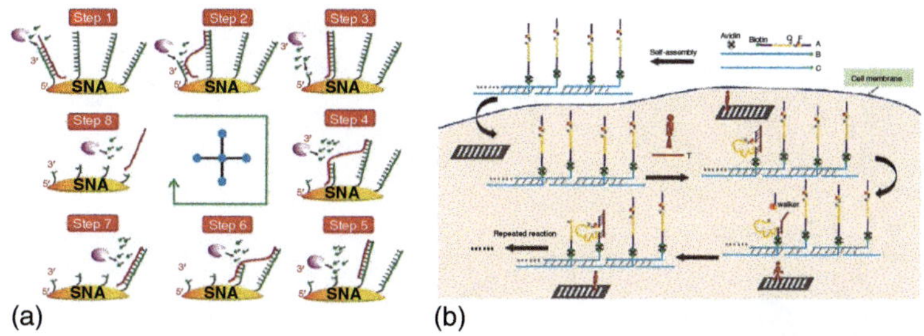

Figure 7.3 (a) A DNA nanomotor propelled by exonuclease III for traversing 3D tracks on AuNPs. Source: Reproduced from Qu et al. [29], with permission of 2017 John Wiley & Sons. (b) An autonomous DNA enzyme nanomotor that creates a single strand (strand A) by combining the enzyme and substrate and interconnecting them *via* rigid DNA nanowires. Source: Reproduced from Yang et al [16], with permission of 2020 American Chemical Society.

protein enzyme-based counterpart. Yang et al. [16] devised an autonomous DNA enzyme nanomotor that creates a single strand (strand A) by combining the enzyme and substrate and interconnecting them *via* rigid DNA nanowires (Figure 7.3b). Strand A hybridizes with an independent walking strand, thus resulting in the formation of a DNAzyme that triggers self-cleavage at the cutting site (rA) on strand A. The fluorescence signal is reinstated following the separation of the fluorescent and quencher groups labeled on either side of the cleavage site. The chain displacement reaction between the DNA nanomotor and the next substrate facilitates the degradation of the enzyme-substrate complex, thereby enabling the autonomous movement of the DNA nanomotor. In contrast, the DNAzyme and walking strand sequences are typically incorporated within an oligonucleotide chain and attached to nanoparticles to facilitate autonomous 3D movement [25, 68, 74, 75].

The energy needed to drive the DNA nanomotor can be generated from conformational changes induced by environmental stimuli [76], such as light, pH, or binding to metal ions or other small molecules. Wang et al. [77] reported a bipedal locomotion device activated by H^+/OH^- and Hg^{2+}/cysteine triggers (Figure 7.4a). The entire system is composed of four interlocking footholds functioning as tracks, a nanomotor equipped with two arms, and a blocker strand containing quenchers. The DNA nanomotor is hybridized with track I and track II, resulting in the formation of a stable thymine (T)-Hg^{2+}-thymidine (T) structure. The binding of Hg^{2+} causes the migration of the DNA nanomotor from track I to track III. The nanomotor can be moved in a bidirectional manner under the precise control of the operator because the i-motif structure and thymine-Hg^{2+}-thymidine structure can be dissociated separately by OH^- and cysteine. Yang et al. [78] modified nucleotide chains with photosensitive molecules and were able to regulate DNA conformation through specific light inputs (Figure 7.4b). The walking strands were functionalized with a desired photosensitizer, pyrene, whereas the 2D DNA tile was assembled through disulfide bond-linked stator strands. Under 350 nm irradiation, the pyrene-modified molecule was excited to induce the cleavage of disulfide bond-linked stator chains *via* electron transfer, thereby exposing new anchoring points on the walking strand to enable

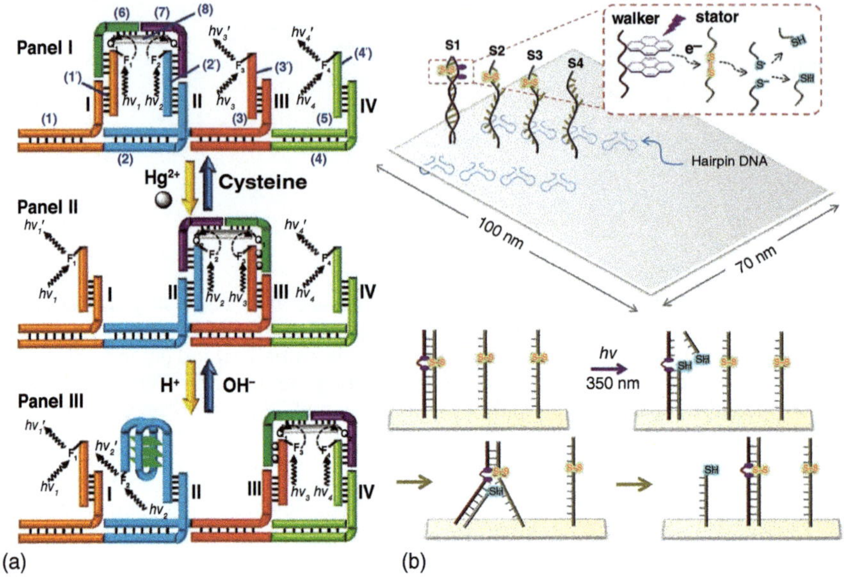

Figure 7.4 (a) A bipedal nanomotor can be activated by H^+/OH^- and Hg^{2+}/cysteine triggers. Source: Reproduced from Wang et al. [77], with permission of 2011 American Chemical Society. (b) A remarkable walking mechanism on a rectangular DNA tile, propelled by light energy and activated by a precise wavelength of light. Source: Reproduced from Yang et al. [78], with permission of 2015 American Chemical Society.

translocation and hybridization with the next stator during branch migration. Thus, the light-driven artificial molecular nanomachine exhibited self-propulsion in the presence of an uninterrupted light source.

DNA nanomotors may be classified as unipedal, bipedal, or multipedal DNA nanomotors on the basis of their mobile legs. Unipedal DNA nanomotors have only one leg to walk on, but they can oscillate in a specific area of their foothold following a set program and walk for a short distance along the track randomly or unidirectionally [51, 54, 79]. Chain replacement reactions and enzymatic reactions are the main drivers of DNA nanomotors. In addition, photoinitiated reactions and chemical stimulation reactions can also be used as drivers of DNA nanomotors. Based on RNA cleavage by DNA polymerase, Cha et al. modified DNA enzyme strands with CdS nanocrystals to walk on linear RNA tracks that adsorb nonspecifically to the surface of CNs (Figure 7.5a) [37]. DNA motors could extract chemical energy from RNA molecules labeled on nanotubes to provide the driving force for subsequent conformational changes. In addition, since the cleavage rate k_{cat} is three orders of magnitude larger than the binding rate $k_{ligation}$, a strand replacement reaction occurs when the free structural domain of DNase is combined with the next substrate, allowing the motor molecule to travel on the nanotube. Observed by visible and near-infrared images, they found that the molecular motor has unidirectional motion of more than 3 μm and a movement speed of 1 nm/min. Wang et al. designed a nicking-assisted monopod DNA nanomotor in combination with DNA origami technology (Figure 7.5b) [70]. In the presence of target DNA and nicking

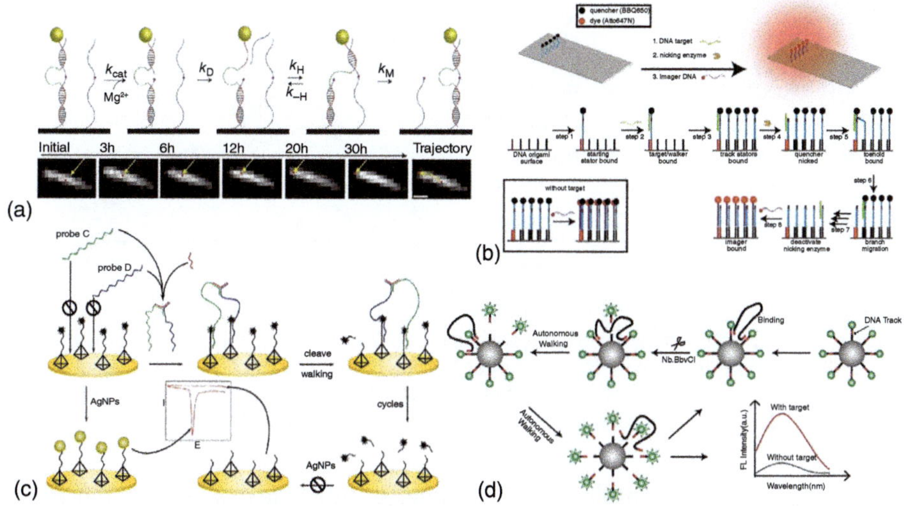

Figure 7.5 The different legs of DNA nanomotors. (a) A DNase-driven monopod DNA nanomotor moving along CN tracks. Source: Reproduced from Cha et al. [37], with permission from Springer Nature. (b) A single-leg nanomotor and its walking on a DNA origami track. Source: Reproduced from Wang et al. [70], with permission of 2017 American Chemical Society. (c) A bipedal DNA nanomotor driven by strand displacement reaction. Source: Reproduced from Chai and Miao [30], with permission of 2019 American Chemical Society. (d) A 3D bipedal DNA nanomotor in combination with EXPAR for the detection of miRNA-21. Source: Reproduced from Yang et al. [80], with permission of 2021 Elsevier.

enzyme, the randomly distributed and moving nanomotor on origami moved one burst molecule per step. After combining the nanomotor with a fluorescently labeled imaging strand, the goal of walking visualization is achieved.

Bipedal DNA nanomotors have two relatively free legs and can walk alternately along the track. Compared to unipedal DNA nanomotors, bipedal DNA nanomotors walk for a longer time with higher stability and a longer travel distance, thereby achieving higher signal amplification. In recent years, bipedal DNA nanomotors have been combined with electrochemical sensors and are widely used in biosensing (Figure 7.5c) [30, 31, 81]. Incorporating exponential amplification reaction (EXPAR) and incision endonuclease, Liu et al. designed a bipedal DNA nanomotor for the detection of intracellular microRNA (miRNA) (Figure 7.5d) [80]. DNA strands with fluorescent labels were assembled on the 3D orbit of polystyrene microspheres. A large number of DNA strands were generated after EXPAR. The presence of nicking endonuclease activated the DNA nanomotor. As a result, it enabled a quenched fluorescent signal as it moved, which could reach the miRNA detection limit of 5.2 fM in 70 min.

Multipedal DNA nanomotors, especially those with three or four legs, are known for their faster walking speed compared to unipedal and bipedal DNA nanomotors [7, 32, 58, 82]. Gu et al. developed a multilegged DNA nanomotor containing three arms and four legs (Figure 7.6a) [83]. The four DNAzyme legs were connected by streptavidin. Each leg could hybridize with and cleave its substrate into two parts,

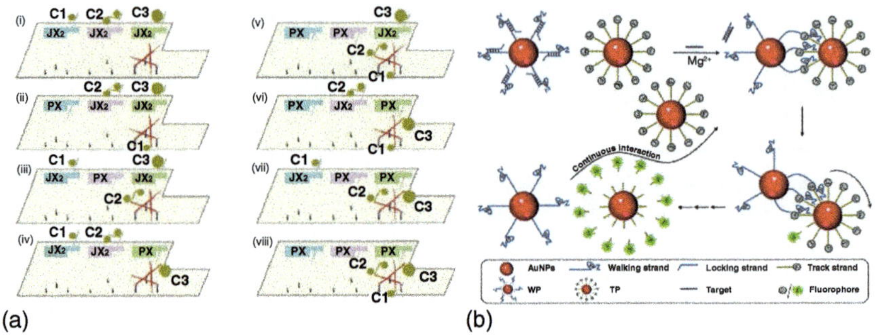

Figure 7.6 (a) A DNA nanomotor that travels on a 2D track. Source: Reproduced from Gu et al. [83], with permission of 2010 Springer Nature. (b) Detection of Zika virus-associated RNA fragments (ZIKV-RNA) using a multistage DNA nanomotor. Source: Reproduced from Zhang et al. [84], with permission of 2020 Royal Society of Chemistry.

which could then be transferred to the next target substrate in a sequential manner. The motion of the multilegged nanomotor was observed using AFM images. Based on the mutual motion of the particles, Zhang et al. [84] conjugated DNA walking strands and track strands containing fluorescence markers on AuNPs to construct spherical DNA nanomotors and 3D DNA tracks. They named these walking particles (WPs) and tracking particles (TPs) (Figure 7.6b), respectively. The spherical DNA nanomotor could recognize and target Zika virus RNA, and then walk along the 3D track, resulting in a large signal accumulation.

The aberrant expression of nucleic acids and proteins is associated with the occurrence of various tumors, cancers, and many other pathological conditions [85–88]. Numerous strategies have been developed for the detection of clinically relevant biomolecules, for example, northern blotting [89], microarray hybridization [90], quantitative reverse transcription-PCR (qRT-PCR) [91] for nucleic acid detection, enzyme-linked immunosorbent assay (ELISA) [92], and capillary electrophoresis [93] for protein detection. However, these in-tube methods make it difficult to meet the detection requirements of intracellular targets due to the minute number of targets, complicated biological matrices, and extraction procedures. Live cell imaging methods can provide spatiotemporal information about targets for long-term tracking and monitoring. Therefore, live cell imaging of these key biomolecules is of great significance for pathological research, early diagnosis, and precise treatment of diseases.

DNA nanomotors, particularly 3D DNA nanomotors based on nanoparticles, have shown excellent walking efficiency, DNA loading capacities, as well as the capability of signal amplification, which are promising for *in situ* imaging of disease-related biomolecules [29, 94, 95]. Although rolling circle amplification (RCA), HCR, and isothermal circular strand displacement polymerization (ICSDP) have been employed for signal amplification *in vivo* [88, 96, 97], these protein enzyme-mediated reactions are susceptible to changes in the surrounding microenvironment and usually have compromised amplification performance [98]. In contrast, DNA nanomotors based on nanoparticles can be transferred into living cells through endocytosis without using transfection reagents. The DNA nanomotors can act as an inherent signal amplifier, which traverses multiple steps autonomously and continuously in the presence of specific stimuli, resulting in cascade reactions to generate an accumulative signal for ultrasensitive detection. The specific stimuli can be pre-designed as endogenous molecules or external light radiation without the need to transport additional fuels into the cells. Furthermore, in the design of DNA nanomotors for live-cell imaging, integrating all the requisite components into a single entity, which can reduce the interference of exogenous substances and avoid the asynchronous behaviors of separate building blocks, improves imaging sensitivity [99]. The analytical performances of diverse DNA nanomotors are listed and compared in Table 7.1.

Table 7.1 Analytical performance of DNA nanomotors.

DNA nanomotors	Linear range	LOD	Target	References
		1D DNA nanomotor		
DNA nanowire	0–1 nM	23 pM	miR-222	[16]
Carbon nanotube	5 fM to 1 nM	1.7 fM	miRNA	[18]
Oligonucleotides	100 fM to 1 nM	58 fM	miRNA	[27]
		2D DNA nanomotor		
Au electrode plane	5 pM to 1 nM	1.7 pM	Thrombin	[21]
Au electrode plane	1.0 fM to 100 fM	0.22 fM	DNA	[30]
Pt nanoparticle-modified glassy carbon electrode	0.50 fM to 10 pM	0.24 fM	miRNA	[31]
Au electrode plane	0–80 nM	0.14 nM	Hg^{2+}	[63]
		3D DNA nanomotor		
AuNPs	1 pM to 1 nM	0.5 pM	mRNA	[19]
Magnetic microparticles	30.77 pM–461.5 pM	6.5 pM	Streptavidin	[32]
AuNPs	0.01–500 pM	90 fM	EBOV gene	[47]
Silica microsphere	0.2–20 nM	60 pM	H5N1 DNA	[53]
AuNPs	10–1000 cells/ml	7 cells/ml	Telomerase	[56]
Spherical microparticles	50 pM to 1 nM	100 pM	HIV	[60]
AuNPs	0.01–0.3 U/ml	0.0067 U/ml	T4 polynucleotide kinase (T4 PNK)	[100]
Carboxylated silica microspheres	10 pg/ml—to 100 ng/ml	1.2 pg/ml[1]	carcinoembryonic antigen (CEA)	[101]
AuNPs	2×10^4 to 2×10^9 particles/ml	8.2×10^3 particles/ml	Tumor-derived exosomes	[102]

Imaging of nucleic acids is important for understanding important cellular processes, such as transcription, translation, transport, degradation, and subcellular localization of RNA, as well as for detecting and studying various genetic diseases [103]. Real-time imaging of dynamic changes in DNA and RNA structures is particularly useful in cancer detection and treatment [104, 105]. miRNAs play a critical role in various physiological processes, such as cell proliferation, differentiation, and apoptosis [106, 107]. Abnormal miRNA expression has been closely related to the occurrence and development of certain diseases, such as malignant tumors [108, 109]. Therefore, accurate monitoring of intracellular miRNA levels is of great significance in basic biological research and disease diagnosis [110].

Although DNA motors have been developed for imaging miRNA [25, 94, 111–114], there are several challenges in using synthetic DNA motors to detect miRNA in living cells. For example, (i) the motor system needs to be efficiently delivered into the target living cells to ensure effective functioning, (ii) the walking of the motor must be autonomous and self-powered without external energy or fuel [19, 115, 116], and (iii) the target intracellular miRNA must initiate the operation of the motors. To address these challenges, Peng et al. reported in 2017 the first 3D DNA nanomotor system that could be operated in living cells (Figure 7.7a) [25]. The motor was constructed on a 20 nm AuNP integrated with DNAzyme and its substrate stands. It can amplify the detection of specific miRNA in individual cancer cells. The entire motor system is composed of hundreds of substrate strands and dozens of DNAzyme nanomotor strands, in which each DNAzyme molecule is silenced by the locking strand that is tuned to respond to a specific intracellular target (miRNA-10b). The locking strand is fluorescently labeled with cyanidin 5 (Cy5), and the substrate strand is labeled with carboxyfluorescein (FAM). The fluorescence from Cy5 and FAM is quenched by an AuNP when the DNAzyme motor is inactive. Once the DNAzyme motor is taken up by the cell, the intracellular miRNA hybridizes with the locked strand through a strand-displacement reaction, releasing the locked strand from the DNAzyme. The unlocked DNAzyme then hybridizes to the substrate on the AuNP. In the presence of cofactor Mn^{2+}, the DNAzyme cleaves the substrate molecule, thus releasing the FAM-labeled fragment from the AuNP surface and restoring its fluorescence. Meanwhile, the DNAzyme hybridizes to the next substrate strand, enabling the motor to walk from one substrate strand to the next. The stepwise walk is repeated autonomously, driving the DNAzyme motor to traverse the AuNP surface. Monitoring the fluorescence of FAM provides real-time imaging of the intracellular operation of the motor in live cells. Since the Peng et al. report in 2017, various DNAzyme nanomotors with specificity for exogenous cofactors (e.g. Mn^{2+}, Zn^{2+}) [74, 117, 118] have been used in the field of live cell imaging.

To improve the stability of the DNAzyme nanomotor against degradation, Gao et al. constructed a self-protecting DNAzyme nanomotor by incorporating a circular bulge DNA protection structure within the DNAzyme catalytic core [98]. They also designed a self-protected ES-AuNP nanoprobe, consisting of an AuNP core and several substrate strands (S) and DNA enzymes (E). The molecular structures of the E/S heterodimeric and functional regions are shown in Figure 7.7b. The

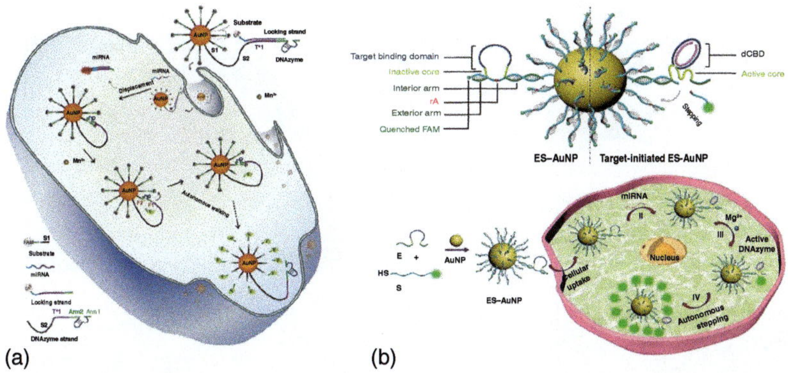

Figure 7.7 Application of DNA nanomotor for imaging nucleic acid in a live cell. (a) Amplified detection and imaging of intracellular microRNA using a 3D DNA motor driven by DNAzyme. Source: Reproduced from Peng et al. [25]/Springer Nature/CC BY 4.0. (b) The assembly of ES-AuNP and its intracellular operation for miRNA imaging. Source: Reproduced from Gao et al. [98], with permission of 2021 American Chemical Society.

catalytically active core of DNAzyme (E) is split in half by the target-binding structural domain, thereby inhibiting the catalytic function. When hybridized to the target miRNA, the two separate parts of the DNAzyme can be brought into the correct position to form the active secondary structure required for the catalytic core. Once activated by the cofactor Mg^{2+}, the DNAzyme cleaves the substrate strand. The FAM-labeled half-fragment (P1) dissociates from the AuNP, leaving the sulfhydrylated half-fragment (P2) on the AuNP surface. The outer arm of the DNAzyme in its dissociated state searches for its next binding site around it on another substrate, while the more stable inner arm remains bound to the surface-modified sulfhydrylated fragment, preventing the nanomotor from falling off the AuNP track. The toehold-mediated strand replacement reaction can then force the inner arm of the DNAzyme to move to the binding site of the next substrate pre-bound by the outer arm to complete the advanced operation. In this way, the nanomotor can walk along the S-track attached to AuNP. Each step forward by the nanomotor contributes to fluorescence enhancement, enabling DNAzyme-catalyzed signal amplification based on multiple transduction reactions. Using ES-AuNP, the authors distinguished the target miRNAs and nontarget species in the cells. The ability to differentiate between diseased and healthy cell lines offers a promising platform for assessing cancer stage and progression, as well as evaluating treatment response and prognosis.

Intracellular protein enzymes, such as nucleic acid exonucleases and endonucleases, are commonly used to construct and power DNA nanomotors by cleaving specific nucleic acid substrates. These motors have the advantages of high specificity and high efficiency [18, 29]. However, the use of protein enzyme-driven DNA nanomotors to detect intracellular biomolecules is limited due to the harsh conditions for protease activity and the difficulty of delivering the protein into the cells. Zhu et al. reported a dual enzyme-driven DNA nanomotor for intracellular miRNA detection and imaging [114]. The target miRNA as the walking strand of the DNA

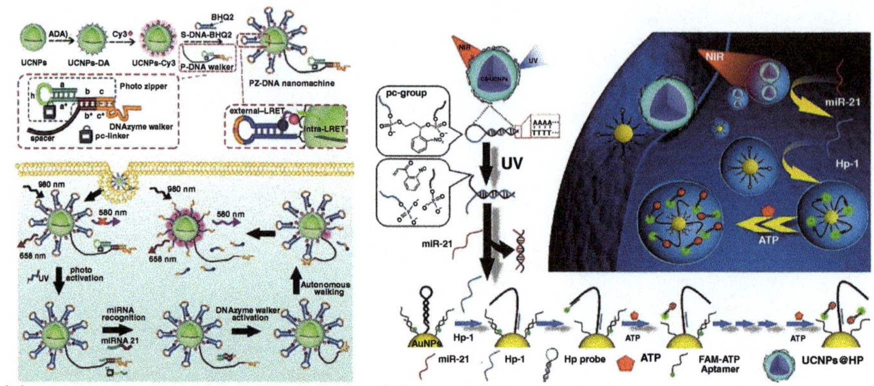

(a) (b)

Figure 7.8 (a) Schematic illustrations of the preparation and activation of the photo-zipper-locked DNA nanomachine, as well as its operation in response to miRNA in living cells. Source: Reproduced from Zhang et al. [121], with permission of 2020 Royal Society of Chemistry. (b) Near-infrared light-controlled, endogenous ATP-driven DNA nanomotor system for microRNA imaging. Source: Reproduced from Ye et al. [122], with permission of 2021 American Chemical Society.

nanomotor walked autonomously under the catalytic effect of two enzymes. This strategy allows imaging of low-expression miRNAs in cells and is expected to be extended to the detection of other functional biological components.

Non-enzymatic walking systems, such as CHA and EDC, can also be delivered into cells using nanomaterials as carriers [26, 119]. To specifically and sensitively detect miRNA-21, Xu et al. developed a DNA nanomotor based on localized catalytic hairpin assembly (LCHA) by assembling two FAM-labeled hairpin DNAs (H1 and H2) on the surface of AuNPs [120]. The DNA nanomotor greatly improves the local concentrations of H1 and H2, thus achieving faster reaction speed and higher efficiency. Zhang et al. designed a photo-zipper-locked miRNA-responsive DNA nanomachine (PZ-DNA nanomachine) based on upconversion nanoparticles (UCNPs) (Figure 7.8a), which used photocleavable DNA strands to block the miRNA recognition region, thereby providing sufficient protection for DNA nanomachines against nonspecific extracellular activation. Following intracellular photoactivation, satisfactory signal amplification for sensitive miRNA imaging is achieved [121]. Multiple emissions from UCNPs have also been used as internal standards to self-calibrate the intracellular miRNA in response to fluorescence signals. The developed PZ-DNA nanomachine enables sensitive imaging of intracellular miRNAs in different cell lines.

The enzyme-free and self-driven nature of endogenous ATP has been harnessed in catalytic assembly-driven DNA nanomotors for real-time *in situ* analysis of functional biomolecules in living cells [123]. Ye et al. developed a near-infrared light-activated DNA nanomotor system driven by endogenous ATP for imaging miRNAs in living cells [122]. To develop a spatiotemporally resolved photocontrollable system, the authors used the commercially available photocleavable o-nitrobenzene group as the photoactivating material and UCNPs as the DNA carrier and photoconversion material (Figure 7.8b). Given the low abundance of intracellular miRNA and the high concentration of ATP, an ATP-driven 3D DNA nanomotor sensor was used as a downstream signal amplification tool for *in situ* detection and imaging of intracellular miRNA with high sensitivity and selectivity.

Integration of protein response units, such as affinity aptamers and enzymes with cleavage or extension activity, into DNA nanomotors, is crucially important for the detection of proteins and enzyme activities. Lv et al. exploited a DNA nanomotor for the detection and activation imaging of APE1, an enzyme associated with the base excision repair (BER) of DNA damage in living cells [120]. This random bipedal DNA nanomotor autonomously traverses AuNP *via* CHA responses. The nanomotor includes a DNA substrate hairpin H0 for obtaining bipedal catalyst DNA, a hairpin probe H1 based on CHA reaction located on the AuNP 3D orbit, and a hairpin probe H2 used for fuel participation in CHA reaction. Under APE1 conditions, the H0 and AP sites release biped catalysts through APEA cleavage and hybridize with two H1, and bipedal DNA nanomotors can move on a single AuNP track through CHA reaction with the help of H2. The movement of the DNA nanomotor makes the fuel probe H2 and the hairpin probe H1 form a cascade assembly, so that Cy5 on H1 is far away from the AuNP surface, thus forming an amplified fluorescence signal for imaging APE1 activity. Compared with the single-legged DNA nanomotor, the bipedal DNA nanomotor has higher sensitivity, as demonstrated by a detection limit of 0.01 U/ml for APE1 activity. Zhou et al. designed a four-armed tetrahedral DNA nanomotor for ultrasensitive detection and imaging of APE1 [124]. The tetrahedral frame and multi-arm structure increase the resistance of nuclease and enhance the biological stability of the DNA nanomotor. Among a number of traditional DNA nanomotors, tetrahedral DNA nanomotors have the fastest walking speed, with an initial speed 4.5 times faster than that of bipedal DNA nanomotors, and the detection limit is as low as 5.5×10^{-3} U/ml for APE1.

Liu et al. constructed a 3D DNAzyme nanomotor for the detection and *in situ* imaging of telomerase activity [125]. All modules are integrated into a single nanoparticle. The walking strands (WS) contain DNA enzyme sequences and telomerase primers, and the tracking strands (TS) contain RNA cleavage sites labeled with FAM. Under the action of telomerase, telomerase primers extend and release DNA enzymes to drive the synthesis of nanomotors to enhance fluorescence signals. Xue et al. developed an integrated synchronous DNA nanodevice that bypassed the freely diffused split-module system to reduce the effective reflection of the increased interaction module concentration. They achieved simultaneous imaging of miRNA and telomerase markers after specific identification of intracellular targets (Figure 7.9a) [99].

The conformation changes of aptamers, induced by protein binding, can initiate a DNA-walking system. Yin et al. developed a 3D DNA nanomotor for the imaging of amyloid β-peptide oligomer (AβO), a promising biomarker for the diagnosis of Alzheimer's disease (AD) (Figure 7.9b) [75]. The Zn^{2+}-dependent DNAzyme walking strands are blocked by the aptamers and the track strands are both anchored on the surface of AuNP. After the integrated nanoprobe is delivered into cells, the aptamer specifically recognizes AβO to form a compact structure, and releases the DNAzyme walking strand. With the assistance of cofactor Zn^{2+}

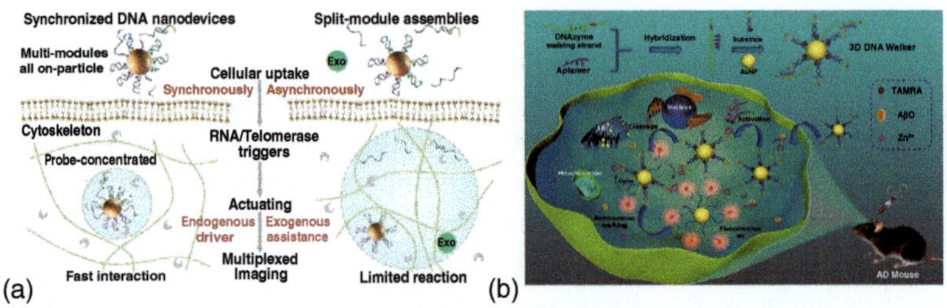

Figure 7.9 Schematic diagrams of DNA nanomotors for protein imaging. (a) All-in-one synchronized DNA nanodevices driven by endogenous enzymes. Source: Reproduced from Xue et al. [99], with permission of 2019 American Chemical Society. (b) A 3D DNA nanomotor nanoprobe for real-time imaging of AβO. Source: Reproduced from Yin et al. [75], with permission of 2020 American Chemical Society.

originally accumulated in the brain of AD patients, the DNAzyme walking strand cleaves the substrate track strand to release the fluorescently labeled DNA fragment. The signal amplification arising from DNAzyme walking and repeated subtracted cleavage results in a detection limit of 22.3 pM. The detection results can be used to distinguish AD mice from wild-type mice; thus, the system holds great potential for the early diagnosis of AD. DNA nanomotors have been applied to the detection of diverse intracellular biomolecules, which are abnormally expressed in tumor cells and related to various diseases (Figure 7.10a) [126–129].

DNA nanomotors have the capability to image small molecules, such as metal ions and protons. Metal ions can serve as cofactors for the initiation of DNAzyme cleavage, which can be exploited to develop DNAzyme nanomotors to image metal ions in living cells [130, 131]. However, many DNAzymes have poor selectivity for metal ions, leading to nonspecific initiation during cell transfection. To address this issue, Cui et al. [130] developed an acid-switchable DNAzyme nanodevice to control the DNAzyme activity in living cells (Figure 7.10b). This device utilized pH-sensitive DNA strands (pH-DSs) and DNAzyme precursors (DPs) to load onto AuNPs. pH-DSs consisted of switchable DNA (SW-DNA) and partially complementary strand 1 (C1). In response to an acidic microenvironment, the conformation of DNA transforms from linear to triplet, resulting in the release of C1. The displacement of the locking strand on DPs by free C1 triggers the conversion of the DNAzyme from "inactive" to "active", which is typically activated by intracellular metal ions (Zn^{2+} or Pb^{2+}). Li et al. [132] developed an APE1 enzyme-activated DNAzyme nanomotor to image intracellular Zn^{2+}. Yang et al. [133] created a DNAzyme-based two-photon AuNP imaging probe (TP-8-17ES-AuNP) capable of detecting Zn^{2+} *in vivo*. The two-photon imaging technique presents a stronger fluorescence signal compared to the single-photon imaging. Light-activated DNA nanomotors have also been used for intracellular imaging of metal ions [134, 135].

Artificially engineered DNA nanomotors can amplify signals through programmable operation, enabling the detection of minute amounts of intracellular molecules. The extraordinary features of directionality, progressivity, and speed of DNA nanomotors make them an effective biosensor capable of directional walking at high speeds. However, careful design is required to fully utilize these advantages

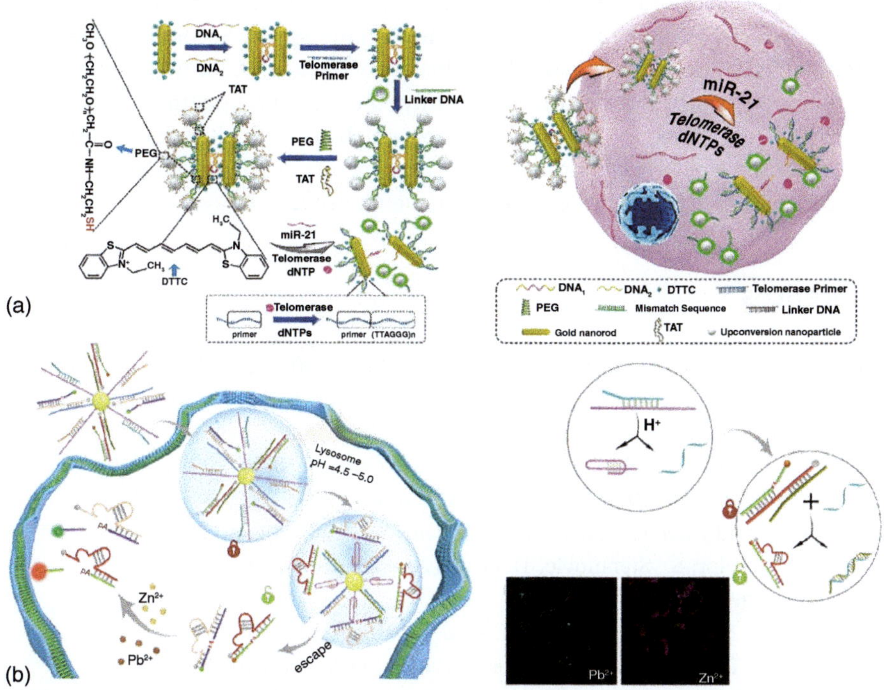

Figure 7.10 (a) A nanomotor system for the detection of dual targets: miRNA and telomerase. Source: Reproduced from Ma et al. [126], with permission of 2017 American Chemical Society. (b) Acid-switchable DNAzyme nanodevice for imaging multiple metal ions in living cells. Source: Reproduced from Cui et al. [130], with permission of 2017 American Chemical Society.

when constructing the DNA motors with motors, legs, and tracks. Various types of motors have been used to drive DNA nanomotors, including DNAzymes, protein enzymes, and self-catalytic assembly systems, such as HCR, CHA, and EDC. To maintain processivity during live cell imaging, simultaneous delivery of all components of DNA nanomotor into the cell is required.

DNA nanomotors can be used to image nucleic acids, proteins, metal ions, and other small intracellular molecules. They offer the advantages of programmable structure, controllable operation, tunable target recognition, and stable storage in the field of live cell imaging. However, bottom-up built DNA nanomotors are still unable to match the performance of natural protein motors that use natural units to build synthetic walking systems. A DNA nanomotor that adopts the cell membrane as a track has been reported, and it can move by floating on the cell surface [13]. Ibusuki et al. used a DNA nanotube as a track to improve the speed of DNA nanomotors 100 times faster than the average [136].

The rapid development of DNA nanomotors opens up new possibilities for live cell imaging. However, to further expand the applicability of DNA nanomotors in live cell imaging, several emerging challenges need to be addressed. Firstly, the functional scaffold of the DNA motor system is difficult to maintain due to varying microenvironments during cellular uptake. The distribution and molar ratio of these DNA elements can affect the walking speed. Secondly, most DNA nanomotors require the supply of a high density of track DNA for motor movement due to the "burnt-bridges" strategy. Thirdly, the walking system usually needs additional "fuel" elements or metal cofactors, which are difficult to deliver into cells synchronously. Lastly, the nanoscale DNA nanomotors are prone to degradation or aggregation during cellular internalization, which needs to be avoided.

References

1 Xu, X., Wang, L., Li, K. et al. (2018). *Anal. Chem.* 90: 3521–3530.
2 Shimron, S., Cecconello, A., Lu, C.H., and Willner, I. (2013). *Nano Lett.* 13: 3791–3795.
3 Yang, W., Shen, Y., Zhang, D. et al. (2019). *Anal. Chem.* 91: 7782–7789.
4 Zhang, P., Lin, Z., Zhuo, Y. et al. (2017). *Anal. Chem.* 89: 1338–1345.
5 Tian, Y. and Mao, C. (2004). *J. Am. Chem. Soc.* 126: 11410–11411.
6 Pei, R., Taylor, S., Stefanovic, D. et al. (2006). *J. Am. Chem. Soc.* 128: 12693–12699.
7 Lund, K., Manzo, A.J., Dabby, N. et al. (2010). *Nature* 465: 206–210.
8 Yang, X., Tang, Y., Mason, S.D. et al. (2016). *ACS Nano* 10: 2324–2330.
9 Jung, C., Allen, P.B., and Ellington, A.D. (2016). *Nat. Nanotechnol.* 11: 157–163.
10 Tomov, T.E., Tsukanov, R., Liber, M. et al. (2013). *J. Am. Chem. Soc.* 135: 11935–11941.
11 You, M.X., Huang, F.J., Chen, Z. et al. (2012). *ACS Nano* 6: 7935–7941.
12 Liu, J., Cui, M., Zhou, H., and Yang, W. (2017). *ACS Sens.* 2: 1847–1853.
13 Li, H., Gao, J., Cao, L. et al. (2021). *Angew. Chem. Int. Ed.* 60: 26087–26095.
14 He, L., Lu, D., Liang, H. et al. (2018). *J. Am. Chem. Soc.* 140: 258–263.
15 Wickham, S.F., Bath, J., Katsuda, Y. et al. (2012). *Nat. Nanotechnol.* 7: 169–173.
16 Yang, Y., He, Y., Deng, Z. et al. (2020). *ACS Appl. Bio. Mater.* 3: 6310–6318.
17 Arredondo, D. and Stefanovic, D. (2020). *Phys. Rev. E* 101: 062101.
18 Xiao, M., Wang, X., Li, L., and Pei, H. (2019). *Anal. Chem.* 91: 11253–11258.
19 Wu, Z., Liu, G.Q., Yang, X.L., and Jiang, J.H. (2015). *J. Am. Chem. Soc.* 137: 6829–6836.
20 Li, B., Ellington, A.D., and Chen, X. (2011). *Nucleic Acids Res.* 39: e110.
21 Yang, J., Dou, B., Yuan, R., and Xiang, Y. (2017). *Anal. Chem.* 89: 5138–5143.
22 Wang, S., Ji, Y., Fu, H. et al. (2019). *Analyst* 144: 691–697.
23 Li, W., Wang, L., and Jiang, W. (2017). *Chem. Commun.* 53: 5527–5530.
24 Hu, H., Zhou, F., Wang, B. et al. (2021). *Nanoscale* 13: 1863–1868.
25 Peng, H., Li, X.F., Zhang, H., and Le, X.C. (2017). *Nat. Commun.* 8: 14378.
26 Yang, X.J., Cui, M.R., Li, X.L. et al. (2020). *Chem. Commun.* 56: 2135–2138.
27 Wang, L., Deng, R., and Li, J. (2015). *Chem. Sci.* 6: 6777–6782.
28 Zhang, H., Lai, M., Zuehlke, A. et al. (2015). *Angew. Chem. Int. Ed.* 54: 14326–14330.
29 Qu, X., Zhu, D., Yao, G. et al. (2017). *Angew. Chem. Int. Ed.* 56: 1855–1858.

30 Chai, H. and Miao, P. (2019). *Anal. Chem.* 91: 4953–4957.

31 Chen, A., Zhuo, Y., Chai, Y., and Yuan, R. (2019). *Chem. Commun.* 55: 13932–13935.

32 Li, N., Du, M., Liu, Y. et al. (2018). *ACS Sens.* 3: 1283–1290.

33 Yehl, K., Mugler, A., Vivek, S. et al. (2016). *Nat. Nanotechnol.* 11: 184–190.

34 Bazrafshan, A., Kyriazi, M.E., Holt, B.A. et al. (2021). *ACS Nano* 15: 8427–8438.

35 Sherman, W.B. and Seeman, N.C. (2004). *Nano Lett.* 4: 1801.

36 Shin, J.S. and Pierce, N.A. (2004). *J. Am. Chem. Soc.* 126: 10834–10835.

37 Cha, T.G., Pan, J., Chen, H. et al. (2014). *Nat. Nanotechnol.* 9: 39–43.

38 Yin, P., Yan, H., Daniell, X.G. et al. (2004). *Angew. Chem. Int. Ed.* 43: 4906–4911.

39 Wang, K., Feng, M., He, M.Q. et al. (2018). *Talanta* 188: 685–690.

40 You, M., Lyu, Y., Han, D. et al. (2017). *Nat. Nanotechnol.* 12: 453–459.

41 Mason, S.D., Tang, Y., Li, Y. et al. (2018). *TrAC Trends Anal. Chem.* 107: 212–221.

42 Wickham, S.F., Endo, M., Katsuda, Y. et al. (2011). *Nat. Nanotechnol.* 6: 166–169.

43 Liber, M., Tomov, T.E., Tsukanov, R. et al. (2015). *Small* 11: 568–575.

44 Tomov, T.E., Tsukanov, R., Glick, Y. et al. (2017). *ACS Nano* 11: 4002–4008.

45 Xing, Y., Liu, B., Chao, J., and Wang, L. (2017). *RSC Adv.* 7: 47425–47434.

46 Li, N., Zheng, J., Li, C. et al. (2017). *Chem. Commun.* 53: 8486–8488.

47 Pang, H., Xu, X., and Jiang, W. (2020). *Sens. Actuators, B* 314: 128053.

48 Simmel, F.C., Yurke, B., and Singh, H.R. (2019). *Chem. Rev.* 119: 6326–6369.

49 Xu, M. and Tang, D. (2021). *Anal. Chim. Acta* 1171: 338523.

50 Peng, H., Newbigging, A.M., Reid, M.S. et al. (2020). *Anal. Chem.* 92: 292–308.

51 Hu, Y. and Chu, X. (2021). *Nanoscale* 13: 1596–1599.

52 Wang, D., Zhou, H., Liu, F., and Li, Y. (2021). *Sens. Actuators, B* 347: 130563.

53 Zhao, X., Zhu, Z., Zou, R. et al. (2021). *Microchem. J.* 170: 106728.

54 Guo, Y., Cao, Q., and Feng, Q. (2020). *Anal. Chim. Acta* 1135: 55–63.

55 Zhang, X.L., Yin, Y., Du, S.M. et al. (2021). *Anal. Chem.* 93: 13952–13959.

56 Shen, Y., Gong, J., Li, S. et al. (2021). *Sens. Actuators, B* 329: 129078.

57 Zhu, D., Huang, J., Lu, B. et al. (2019). *ACS Appl. Mater. Interfaces* 11: 20725–20733.

58 Wang, L., Liu, P., Liu, Z. et al. (2020). *ACS Sens.* 5: 3584–3590.

59 Zhang, D.Y., Turberfield, A.J., Yurke, B., and Winfree, E. (2007). *Science* 318: 1121–1125.

60 Yuan, C., Fang, J., Duan, Q. et al. (2019). *Biosens. Bioelectron.* 133: 243–249.

61 Shen, Y., Li, Z., Wang, G., and Ma, N. (2018). *ACS Sens.* 3: 494–503.

62 Lu, H., Yang, F., Liu, B. et al. (2019). *Nanoscale Horiz.* 4: 472–479.

63 Hu, L., Cui, J., Wang, Y., and Jia, J. (2022). *Anal. Methods* 14: 2504–2510.

64 Zhang, K., Wang, K., Zhu, X., and Xie, M. (2017). *Anal. Chim. Acta* 949: 53–58.

65 He, X., Zeng, T., Li, Z. et al. (2016). *Angew. Chem. Int. Ed.* 55: 3073–3076.

66 Zhang, D.Y. and Winfree, E. (2009). *J. Am. Chem. Soc.* 131: 17303–17314.

67 Jung, C. and Ellington, A.D. (2014). *Acc. Chem. Res.* 47: 1825–1835.

68 Wang, R., Wang, L., Xu, X., and Jiang, W. (2018). *Sens. Actuators, B* 268: 287–292.

69 Bath, J., Green, S.J., and Turberfield, A.J. (2005). *Angew. Chem. Int. Ed.* 44: 4358–4361.

70 Wang, D., Vietz, C., Schroder, T. et al. (2017). *Nano Lett.* 17: 5368–5374.

71 Zhang, Z., Hu, Y., Yuan, W. et al. (2021). *Anal. Chem.* 93: 9939–9948.

72 Cheng, X., Bao, Y., Liang, S. et al. (2021). *Anal. Chem.* 93: 9593–9601.

73 Chen, Y., Xiang, Y., Yuan, R., and Chai, Y. (2015). *Nanoscale* 7: 981–986.

74 Liu, C., Hu, Y., Pan, Q. et al. (2019). *Biosens. Bioelectron.* 136: 31–37.

75 Yin, Y., Chen, G., Gong, L. et al. (2020). *Anal. Chem.* 92: 9247–9256.

76 King, J.J., Irving, K.L., Evans, C.W. et al. (2020). *J. Am. Chem. Soc.* 142: 20600–20604.

77 Wang, Z.G., Elbaz, J., and Willner, I. (2011). *Nano Lett.* 11: 304–309.

78 Yang, Y., Goetzfried, M.A., Hidaka, K. et al. (2015). *Nano Lett.* 15: 6672–6676.

79 Song, L., Zhuge, Y., Zuo, X. et al. (2022). *Adv. Sci.* 9: e2200327.

80 Yang, L., Fang, J., Li, J. et al. (2021). *Anal. Chim. Acta* 1143: 157–165.

81 Liu, C., Hu, Y., Pan, Q. et al. (2020). *Chem. Commun.* 56: 3496–3499.

82 Yao, D., Bhadra, S., Xiong, E. et al. (2020). *ACS Nano* 14: 4007–4013.

83 Gu, H., Chao, J., Xiao, S.J., and Seeman, N.C. (2010). *Nature* 465: 202–205.

84 Zhang, H., Xu, X., and Jiang, W. (2020). *Chem. Sci.* 11: 7415–7423.

85 Huang, M., Yang, J., Wang, T. et al. (2020). *Angew. Chem. Int. Ed.* 59: 4800–4805.

86 Ruan, K., Fang, X., and Ouyang, G. (2009). *Cancer Lett.* 285: 116–126.

87 Yang, Y., Tapias, V., Acosta, D. et al. (2022). *Nat. Commun.* 13: 159.

88 Yang, Z., Zhang, S., Zhao, H. et al. (2018). *Anal. Chem.* 90: 13891–13899.

89 Pall, G.S., Codony-Servat, C., Byrne, J. et al. (2007). *Nucleic Acids Res.* 35: e60.

90 Lee, I., Ajay, S.S., Chen, H. et al. (2008). *Nucleic Acids Res.* 36: e27.

91 Kroh, E.M., Parkin, R.K., Mitchell, P.S., and Tewari, M. (2010). *Methods* 50: 298–301.

92 Bruggink, K.A., Jongbloed, W., Biemans, E.A.L.M. et al. (2013). *Anal. Biochem.* 433: 112–120.

93 Picou, R.A., Kheterpal, I., Wellman, A.D. et al. (2011). *J. Chromatogr. B* 879: 627–632.

94 Liang, C.P., Ma, P.Q., Liu, H. et al. (2017). *Angew. Chem. Int. Ed.* 56: 9077–9081.

95 Xiao, M., Zou, K., Li, L. et al. (2019). *Angew. Chem. Int. Ed.* 58: 15448–15454.

96 Deng, R., Tang, L., Tian, Q. et al. (2014). *Angew. Chem. Int. Ed.* 53: 2389–2393.

97 Ge, J., Zhang, L.L., Liu, S.J. et al. (2014). *Anal. Chem.* 86: 1808–1815.

98 Gao, Y., Zhang, S., Wu, C. et al. (2021). *ACS Nano* 15: 19211–19224.

99 Xue, J., Chen, F., Bai, M. et al. (2019). *Anal. Chem.* 91: 4696–4701.

100 Feng, C., Wang, Z., Chen, T. et al. (2018). *Anal. Chem.* 90: 2810–2815.

101 He, M.Q., Wang, K., Wang, W.J. et al. (2017). *Anal. Chem.* 89: 9292–9298.

102 Yu, Y., Zhang, W.S., Guo, Y. et al. (2020). *Biosens. Bioelectron.* 167: 112482.

103 Zou, X., Shi, Y., Zhu, R. et al. (2019). *Anal. Chem.* 91: 15899–15907.

104 Le, P., Ahmed, N., and Yeo, G.W. (2022). *Nat. Cell Biol.* 24: 815–824.

105 Han, G., Zhao, J., Zhang, R. et al. (2019). *Angew. Chem. Int. Ed.* 58: 7087–7091.

106 Leone, E., Morelli, E., Martino, M.T.D. et al. (2013). *Clin. Cancer Res.* 19: 2096–2106.
107 Flynn, R.A., Pedram, K., Malaker, S.A. et al. (2021). *Cell* 184: 3109–3124.
108 Andorfer, C.A., Necela, B.M., Thompson, E.A., and Perez, E.A. (2011). *Trends Mol. Med.* 17: 313–319.
109 Tang, Y., He, X., Zhou, Z. et al. (2016). *Chem. Commun. (Camb)* 52: 13905–13908.
110 Choi, C.K., Li, J., Wei, K. et al. (2015). *J. Am. Chem. Soc.* 137: 7337–7346.
111 Yang, Y., Huang, J., Yang, X. et al. (2016). *Anal. Chem.* 88: 5981–5987.
112 Wu, Y., Huang, J., Yang, X. et al. (2017). *Anal. Chem.* 89: 8377–8383.
113 Yang, Y., Huang, J., Yang, X. et al. (2017). *Anal. Chem.* 89: 5850–5856.
114 Chen, T., Chen, Y., Chen, H. et al. (2019). *Nano Res.* 12: 1055–1060.
115 Rosi, N.L., Giljohann, D.A., Thaxton, C.S. et al. (2006). *Science* 312: 1027–1030.
116 Seferos, D.S., Giljohann, D.A., Hill, H.D. et al. (2007). *J. Am. Chem. Soc.* 129: 15477–15479.
117 Gao, J.L., Liu, Y.H., Zheng, B. et al. (2021). *ACS Appl. Mater. Interfaces* 13: 31485–31494.
118 Xiong, M., Zhang, M., Liu, Q. et al. (2020). *Chem. Commun.* 56: 2901–2904.
119 Li, J., Wang, J., Liu, S. et al. (2020). *Angew. Chem. Int. Ed.* 59: 20104–20111.
120 Lv, M.M., Liu, J.W., Yu, R.Q., and Jiang, J.H. (2020). *Chem. Sci.* 11: 10361–10366.
121 Zhang, Y., Zhang, Y., Zhang, X. et al. (2020). *Chem. Sci.* 11: 6289–6296.
122 Ye, M., Kong, Y., Zhang, C. et al. (2021). *ACS Nano* 15: 14253–14262.
123 Ma, P.Q., Liang, C.P., Zhang, H.H. et al. (2018). *Chem. Sci.* 9: 3299–3304.
124 Zhou, X.M., Zhuo, Y., Tu, T.T. et al. (2022). *Anal. Chem.* 94: 8732–8739.
125 Liu, X., Zhang, L., Lu, L. et al. (2022). *Analyst* 147: 1968–1975.
126 Ma, W., Fu, P., Sun, M. et al. (2017). *J. Am. Chem. Soc.* 139: 11752–11759.
127 Xu, X., Zhang, P., Zhang, R. et al. (2019). *Chem. Commun.* 55: 6026–6029.
128 Novina, C.D. and Sharp, P.A. (2004). *Nature* 430: 161–164.
129 Wu, L. and Qu, X. (2015). *Chem. Soc. Rev.* 44: 2963–2997.
130 Cui, M.R., Li, X.L., Xu, J.J., and Chen, H.Y. (2020). *ACS Appl. Mater. Interfaces* 12: 13005–13012.
131 Huang, P.J., Lin, J., Cao, J. et al. (2014). *Anal. Chem.* 86: 1816–1821.
132 Lin, Y., Yang, Z., Lake, R.J. et al. (2019). *Angew. Chem. Int. Ed.* 58: 17061–17067.
133 Yang, C., Yin, X., Huan, S.Y. et al. (2018). *Anal. Chem.* 90: 3118–3123.
134 Wang, K.H., Wu, P., Kim, T. et al. (2014). *Angew. Chem. Int. Ed.* 53: 13798–13802.
135 Wang, W., Satyavolu, N.S.R., Wu, Z. et al. (2017). *Angew. Chem. Int. Ed.* 56: 6798–6802.
136 Ibusuki, R., Morishita, T., Furuta, A. et al. (2022). *Science* 375: 1159–1164.

8

Illuminating RNA in Live Cells with Inorganic Nanoparticles-Based DNA Sensor Technology

Fangzhi Yu, Xiangfei Li, Xiulin Yi, and Lele Li

National Center for Nanoscience and Technology, CAS Key Laboratory for Biomedical Effects of Nanomaterials and Nanosafety, Beijing 100190, China

8.1 RNA Detection and Imaging

RNA molecules are not only a decoder that transmits the encoded information from genome into functional proteins but also key regulators that directly participate in diverse fundamental cellular processes [1–3]. RNA metabolism is a complicated, multi-stage process involving transcription, splicing, modification, localization, and degradation. These steps are strictly regulated in both spatial and temporal scales [4, 5]. It has been well documented that dysregulation in RNA metabolism, including abnormal expression and localization, as well as loss of functional modifications, often leads to human diseases [6–8]. Therefore, tracking RNA dynamics is highly desirable to elucidate insightful mechanisms of cell physiology and pathologies.

Benefiting from the Watson-Crick base pairing rule, a variety of technologies that can determine the mean levels of target RNA of specific sequences, including reverse transcriptase-quantitative polymerase chain reaction (RT-qPCR) and other DNA-based approaches, have been widely developed to identify potential functions of RNA of interest. Nevertheless, these detection methods can only be performed in test tubes and need to extract RNA molecules from cell lysate, which sacrifices spatiotemporal precision and cell heterogeneity. Though the fluorescence in situ hybridization (FISH) technique, which utilizes fluorescently labeled DNA probes, enables the visualization of intracellular RNA with a high spatial resolution, it requires cell fixation and cannot provide dynamic information of target RNA in live cells.

Previously, two major challenges heavily limited the applications of DNA-based probes for live-cell RNA imaging. First, negatively charged DNA molecules cannot spontaneously penetrate the cell membrane. Second, DNA-based probes may suffer from nonspecific digestion by endogenous nucleases. Therefore, to efficiently deliver DNA-based probes into cells and to improve their stability during delivery, two pivotal issues that need to be addressed for any strategy aimed at imaging RNA molecules in live cells. During the past two decades, the rapid development of

DNA Nanotechnology for Cell Research: From Bioanalysis to Biomedicine, First Edition. Edited by Zhou Nie.

nanotechnology has provided such desirable carriers for the delivery of DNA-based probes used for RNA imaging. Among these nanocarriers, gold nanoparticles (AuNPs) and upconversion nanoparticles (UCNPs) are two of the most widely used inorganic nanoparticles due to their unique properties. First, both of them have advantages in modification with DNA or other functional ligands via facile operation. Second, their unique optical properties make them very suitable to construct biosensors. For instance, AuNPs can effectively quench the adjacent fluorophores to minimize the background signal. The upconversion luminescence (UCL) of UCNPs can serve as not only a signal output but also a light source to excite neighboring fluorophores and other photo-sensitive moieties. Third, they exhibit satisfactory biocompatibility, which has been tested in a large number of in vivo studies. In this chapter, a comprehensive summary of the recent advances in the design and application of inorganic nanoparticle-based DNA technology for live-cell RNA imaging is provided.

8.2 RNA Imaging Based on Direct Hybridization

UCNPs are one of such nanoparticles widely used to integrate DNA probes for RNA imaging due to their inherent advantages such as unique anti-Stoke shifts, excellent photostability, and low autofluorescence under near-infrared (NIR) light irradiation [9]. Generally, the UCNP-based design for RNA detection is based on the changes in the efficiency of luminescence resonance energy transfer (LRET) between UCNPs and fluorescent acceptors. By preparing carboxylic acid-functionalized UCNPs, Chen et al. constructed a highly sensitive DNA sensor for nucleic acid detection [10]. As shown in Figure 8.1a, the biotin-labeled oligonucleotides could be tethered on the surface of UCNPs due to the interaction between biotin and streptavidin, allowing them to capture target DNA or RNA through hybridization. Importantly, the unpaired region of the target strand could hybridize to TAMRA-labeled DNA, forming a sandwich-shaped hybridization structure. Thus, the distance between UCNPs and TAMRA is close enough for an effective LRET. To improve the LRET efficiency, Ren et al. prepared UCNPs with core–shell structures, in which the emitting ions were confined in the thin shell to shorten their distance to fluorescent acceptors labeled on DNA probes [13]. The core-shell UCNPs were combined with the above sandwich DNA hybridization strategy to develop a LRET-based approach for imaging microRNA (miRNA) in cancer cells and in vivo. Yang et al. utilized similar core–shell UCNPs to load dual-labeled DNA probes complementary to miR-21 via electrostatic adsorption [14]. Hybridization with the target would release the probe from UCNPs. Thus, the LRET between UCNPs and the labeled fluorophores was suppressed, and the UCL was restored. The authors showed that two acceptors quenched UCL and reduced the background more efficiently, leading to an improved detection sensitivity for in situ imaging of miR-21 in MCF-7 cells.

Notably, signal readout dependent on absolute intensity change may lead to inaccurate imaging results caused by various interference factors, including cell auto-fluorescence, complex intracellular environments, and drifts of light sources.

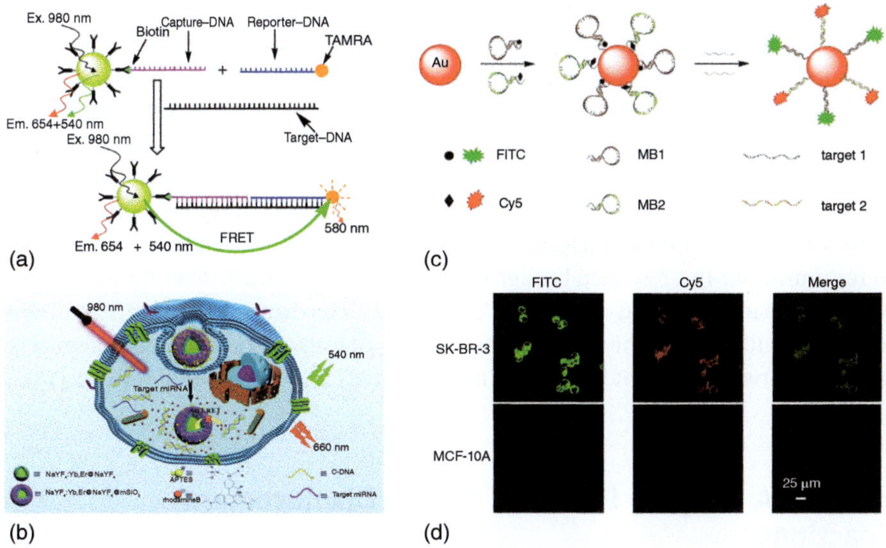

Figure 8.1 (a) Schematic of the formation of sandwich-shape hybridization structure on UCNP's surface for LRET-based sensing of target nucleic acids. Source: Reproduced from Chen et al. [10], with permission of 2008 American Chemical Society. (b) Schematic of ratiometric imaging based on the ratio of the UCL intensity at two different wavelengths. Source: Reproduced from Gong et al. [11], with permission of 2020 Royal Society of Chemistry. (c, d) Bi-color imaging of two tumor-associated mRNAs with an AuNP/MB nanosensor. Source: Adapted from Qiao et al. [12], with permission. © 2011 John Wiley & Sons, Inc.

Ratiometric measurements, which acquire not only signal variations induced by the analyte but also one additional reference signal for normalization, enable more accurate detection with a built-in self-calibration feature [15]. The UCL wavelengths of UCNPs can be precisely modulated through the doping of different lanthanide metal elements, making them particularly suitable for ratiometric imaging. Gong et al. fabricated a versatile UCNP-based ratiometric imaging nanosystem for intracellular miRNA biosensing [11]. The nanosystem was composed of Yb/Er-doped UCNP with a mesoporous silica shell, rhodamine B loaded into the mesopores, and the complementary DNA of target miRNA wrapped around the surface (Figure 8.1b). The loaded rhodamine B completely quenched the green UCL of UCNPs at 540 nm due to the LRET effect but did not affect the red UCL at 660 nm. When the complementary DNA was hybridized to target miRNA, it dissociated from the nanosystem, releasing the encapsulated rhodamine B molecules and recovering the UCL signal at 540 nm. Thus, the ratio of UCL intensity at 540 and 660 nm was employed to afford ratiometric sensing of target miRNA.

Molecular beacons (MBs) have been engineered as a powerful tool for RNA detection since they were first developed by Tyagi and Kramer in 1996 [16]. Hairpin-shaped MB probes are usually dual labeled with a pair of fluorophore and quencher at each end. As such, they would fluoresce after hybridization to the target nucleotide sequence due to the increased distance between the fluorophore and quencher. Notably, this hybridization-induced signal-on mechanism makes MB probes suitable for RNA imaging in live cells through rational design [17]. For example, by assembling two kinds of fluorophore-labeled MB probes on AuNPs, Qiao et al. achieved simultaneous imaging of messenger RNA (mRNA) molecules of two types of tumor-associated genes (survivin and cyclin D1) in breast cancer cells (Figure 8.1c,d) [12]. In this design, AuNPs played three roles: (i) serving as a nanocarrier to deliver MB probes into cells; (ii) providing protection to MB probes from nonspecific nuclease digestion; and (iii) functioning as a quencher to minimize the fluorescence background of MB. Likewise, by loading MB probes on polydopamine-coated AuNPs, Choi et al. constructed Au@PDA–hpDNA nanoprobes to monitor the differentiation status of human mesenchymal stem cells by tracking two important osteogenic marker miRNAs, miR-29b and miR-31 [18].

8.3 RNA Imaging Based on Strand Displacement Reactions

Chemically, enzyme-free strand displacement reactions (SDR) refer to the process through which one single-stranded invader can hybridize to one of the two strands and replace the other pre-hybridized strand of donor duplex [19]. Notably, a single-stranded tail (termly toehold) is required for the complementary strand to initiate the hybridization with the invader strand and to drive SDR. SDR has been well documented and widely utilized for many biological applications [19–22], including molecular imaging. Different from hairpin-shaped MB probes, which are dual-labeled single-stranded DNA, SDR usually adopts double-stranded DNA, the two strands of which are respectively labeled with a fluorophore and a quencher. Importantly, this feature endows SDR with higher programmability and designability, since it can output a displaced strand for initiating downstream reactions, such as cascade amplification [22].

When combined with nanoparticles, SDR has been widely designed to image target RNA molecules in live cells. By modifying DNA probes on AuNPs, Mirkin's group constructed nano-flares to detect mRNA in live cells [23]. As shown in Figure 8.2a, the 3′-thiolated recognition sequence was covalently modified on AuNPs via Au-thiol bond, and the 5′-Cy5-labeled reporter sequence was loaded via hybridization. The Cy5 signal was initially quenched by AuNPs and could be recovered when the reporter sequence is released from AuNPs by the displacement of target mRNA molecules. Later, this group further developed sticky-flares to track the spatial distribution of mRNA targets in live cells [26]. In sticky-flares, the fluorophore was labeled on the complementary sequence, which was pre-hybridized with a strand covalently conjugated on AuNPs. Target recognition and hybridization would make the fluorophore-labeled complementary sequence dissociate from AuNPs. Consequently, the recovered fluorescence signal could quantify the expression level and track the localization of mRNA within cells. By adopting a similar design, Li et al. reported multicolor nano-flares for simultaneous imaging of three tumor-associated

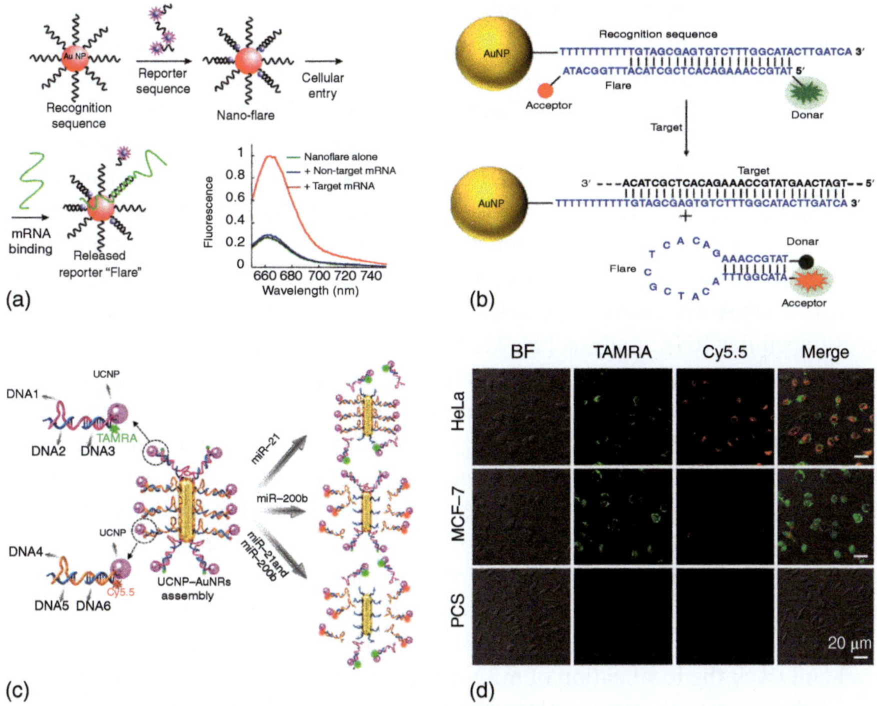

Figure 8.2 (a) Schematic and verification of the SDR-based nano-flare for sensing of mRNA. Source: Reproduced from Seferos et al. [23], with permission of 2007 American Chemical Society. (b) Schematic of FRET nanoflares for intracellular mRNA detection. Source: Reproduced from Yang et al. [24], with permission of 2015 American Chemical Society. (c) Schematic illustrations of the fabrication and working principle of AuNR@UCNP assemblies for the simultaneous imaging of intracellular miR-21 and miR-200b. Source: Qu et al. [25]/National Academy of Science/CC BY 4.0. (d) Application of the AuNR@UCNP assemblies for dual miRNA imaging in different cell lines. Source: Reproduced with permission from Qu et al. [25]. © 2019 National Academy of Sciences/CC BY 4.0.

mRNAs in live cells [27]. Lin et al. constructed a simple and fast-responsive nanoprobe for real-time imaging of miR-630 during cell apoptosis [28].

Yang et al. developed fluorescence resonance energy transfer (FRET)-based nanoflares consisting of AuNP, recognition sequences, and flares to image mRNA molecules in live cells [24]. As illustrated in Figure 8.2b, the recognition sequence was modified on the AuNP surface and was hybridized to the flares, which were designed as a hairpin structure and were labeled with a pair of FRET fluorophores (donor: FAM, acceptor: TAMRA) at two ends. Without target mRNA, the flares were captured by the recognition sequences and exhibited a low FRET efficiency. Target mRNA molecules could displace the flares from the recognition sequences and allow them to switch into a hairpin structure, bringing the donor and acceptor into close proximity and resulting in high FRET efficiency. Notably, compared to the single fluorescence intensity-based nanoflares, the upgraded FRET nanoflares, which rely on a stable DNA hairpin structure, could reduce false positive signals induced by nonspecific responses such as nuclease in complicated intracellular environments.

By using partially complementary DNA strands to link AuNPs and UCNPs, Kuang's group constructed Au-UCNP pyramids that enabled the detection of intracellular miRNA in real time [29]. The UCL of the UCNPs was quenched by AuNPs in the pyramids. The binding of target miRNA with the recognition sequence could result in the disassembly of the pyramid, leading to the recovery of the UCL. In addition, the Au-UCNP pyramids exhibited more sensitive circular dichroism signal for miRNA detection. The authors also developed multifunctional Au-Cu_9S_5-UCNP-Ag_2S nanopyramids for simultaneous detection of miR-203b and miR-21 in live cells through a similar strategic design [30]. Later, this group fabricated DNA-driven AuNR-UCNP satellites and Fe_xCu_ySe@UCNP nanoprobes for intracellular RNA analysis [25, 31, 32]. As shown in Figure 8.2c, AuNR-UCNP was assembled by using six DNA sequences, among which the linking sequences (DNA_2 and DNA_5) were 5′-thiolated and immobilized on the AuNR (gold nanorod) core, the recognition sequences (DNA_1 for miR-21 and DNA_4 for miR-200b) were linked to UCNPs, and the reporter sequences (DNA_3 and DNA_6) were respectively labeled with TAMRA and Cy5.5. By this design, the 980 nm-excited UCL emissions of UCNPs at 540 and 660 nm were largely quenched by the AuNR core, and thereby could not light up the fluorophores. Target recognition could make the fluorophore-labeled complementary sequence dissociate from AuNPs. Consequently, the recovered fluorescence signal could quantify the expression level and track the localization of mRNA within cells. When miR-21 and miR-200b were hybridized to the recognition sequences and replaced the linking sequences, UCNPs would dissociate from the AuNR core, and UCL would be recovered to excite TAMRA and Cy5.5, respectively. This approach exhibited exceptionally high specificity, low limit of detection (LOD), and extended linearity, allowing for quantitative miRNA imaging in live cells with zeptomolar sensitivity (Figure 8.2d). Recently, by assembling UCNPs and Au nanocages with DNA, Meng et al. constructed an all-in-one theranostic nanoplatform for the simultaneous diagnosis and treatment of drug-induced liver injury by imaging the biomarker miR-122 [33].

8.4 Signal-amplified RNA Imaging

In situ imaging of low-abundance, even trace levels of targets in live cells is highly desirable but challenging. A number of signal amplification technologies have been successfully developed and applied to improve the detection sensitivity of nucleic acids over the past decades [34, 35]. In particular, many enzyme-free signal amplification strategies, such as hybridization chain reaction (HCR) and catalyzed hairpin assembly (CHA), are integrated with nanotechnology to achieve amplified RNA imaging in live cells due to their simplicity and universality.

8.4.1 HCR-Based DNA Nanosensors for Amplified RNA Imaging

HCR, developed by Dirks and Pierce in 2004 [36], represents a typical isothermal enzyme-free signal amplification strategy and shows a huge potential in

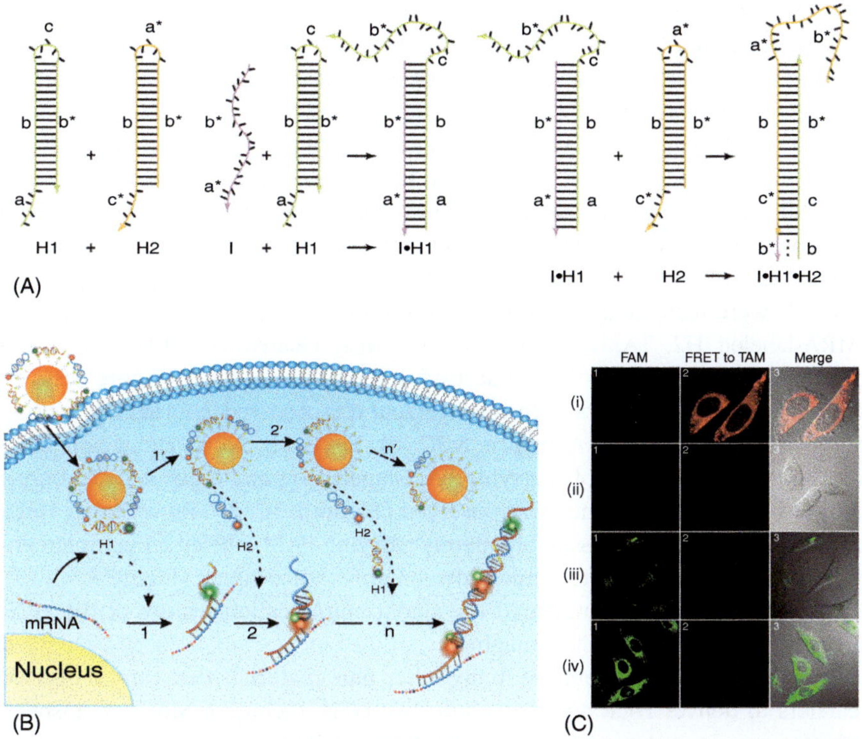

Figure 8.3 (A) Schematic showing the secondary structure of substrates and the processes of HCR. Source: Reproduced from Dirks and Pierce [36], with permission of 2004 National Academy of Science. (B) Illustration of the nanoassembly-based intracellular HCR for in situ imaging of survivin mRNA. Source: Reproduced from Wu et al. [37], with permission of 2015 American Chemical Society. (C) Fluorescence images of (i) HeLa and (ii) C166 cells incubated with nanoassembly carrying FAM-labeled H1 and TAMRA-labeled H2, HeLa cells treated with nanoassembly carrying (iii) FAM-labeled L1 or (iv) FAM-labeled H1 and unlabeled H2. Source: Reproduced with permission from Wu et al. [37]. © 2015 American Chemical Society.

ultrasensitive biosensing. As shown in Figure 8.3A, two metastable DNA hairpins complementary to each other are usually employed to perform HCR. Notably, the hybridization between these two hairpins is thermodynamically favorable but kinetically inhibited due to the secondary structure. Therefore, a single-strand trigger is required to open the first hairpin through a toehold-mediated SDR, initiating a cascade of hybridization events to produce a long double-stranded DNA. HCR has been widely used for ultrasensitive detection of various analytes, especially nucleic acid targets [38]. It is notable that, due to the transient melting of DNA hairpin, HCR may suffer from background leakage in the absence of the trigger. The sequence thermodynamics of DNA hairpins should be carefully assessed to design leak-resistance HCR [39].

The negatively charged DNA hairpins used in HCR are difficult to be directly taken up by cells, thereby requiring an effective delivery carrier for live-cell HCR.

Wu et al. reported the application of HCR for imaging mRNA in live cells [37]. The intracellular HCR strategy relies on the successful delivery of the two hairpin substrates to target cells. To this end, the authors constructed a hybrid nanostructure by using cysteine-terminated cationic peptides to modify AuNPs for electrostatic adsorption of fluorophore-labeled DNA hairpins, achieving intracellular delivery. Survivin mRNA, which is overexpressed in many cancer cells, was selected as the target. After the nanostructure entered into cells, survivin mRNA could initiate the designed HCR, leading to the dissociation of hairpins from the AuNPs and the signal recovery of fluorophores labeled on DNA hairpins (Figure 8.3B). When HeLa cells were incubated with the nanoassembly carrying FAM-labeled H1 and TAMRA-labeled H2, TAMRA could be efficiently excited via FRET under the excitation at 488 nm, suggesting a successful realization of HCR (Figure 8.3C, i). Under identical conditions, no fluorescent signal was observed in C166 cells lacking expression of survivin mRNA (Figure 8.3C, ii). Moreover, when HeLa cells were treated with the nanoassembly carrying FAM-labeled H1 and unlabeled H2, only a remarkable green fluorescence was observed (Figure 8.3C, iv). By contrast, HeLa cells treated with the nanoassembly merely carrying FAM-labeled L1 complementary to target mRNA exhibited much lower fluorescence signal (Figure 8.3C, iii). These results collectively confirmed the signal amplification feature of the HCR strategy for intracellular mRNA imaging.

In addition to AuNPs, there are many other nanomaterials that can be utilized as carriers to deliver HCR substrates into live cells, including MnO_2 nanosheet, graphene oxide, polydopamine-coated ZnO NPs, and carbon nitride nanosheets. Interested readers are referred to Ref. [40] for a systematic summary.

8.4.2 CHA-Based DNA Nanosensors for Amplified RNA Imaging

CHA, developed by Yin et al. in 2008 [41], is another enzyme-free isothermal signal amplification technique that has been widely used in biosensing [42]. Similar to HCR, CHA employs two DNA hairpins that are complementary to each other. Since the complementary segments are caged within intramolecular hairpin stems, spontaneous base pairing between the two substrates is kinetically blocked. With a single-stranded nucleic acid target, the two hairpins are opened successively to form the most thermodynamically favorable DNA duplex through the toehold-mediated SDR (Figure 8.4A). Different from HCR in which one single-stranded trigger induces the formation of one long double-stranded DNA, the target single strand can be released in CHA to initiate the next cycle of duplex assembly, therefore serving as a catalyst to produce many (even >100) DNA duplexes. This strategy can be coupled with various analytical tools for signal-amplificated detection, such as colorimetric and electrochemical technologies [46, 47]. In particular, integrating the CHA strategy with biocompatible nanomaterials is able to image RNA molecules in live cells [48].

By using AuNPs and commercially available lipofectamine-3000 transfection reagent to respectively deliver H1 (labeled with fluorophores) and H2 hairpins, Wang et al. achieved miR-21-catalyzed CHA in several cell lines [49]. By this

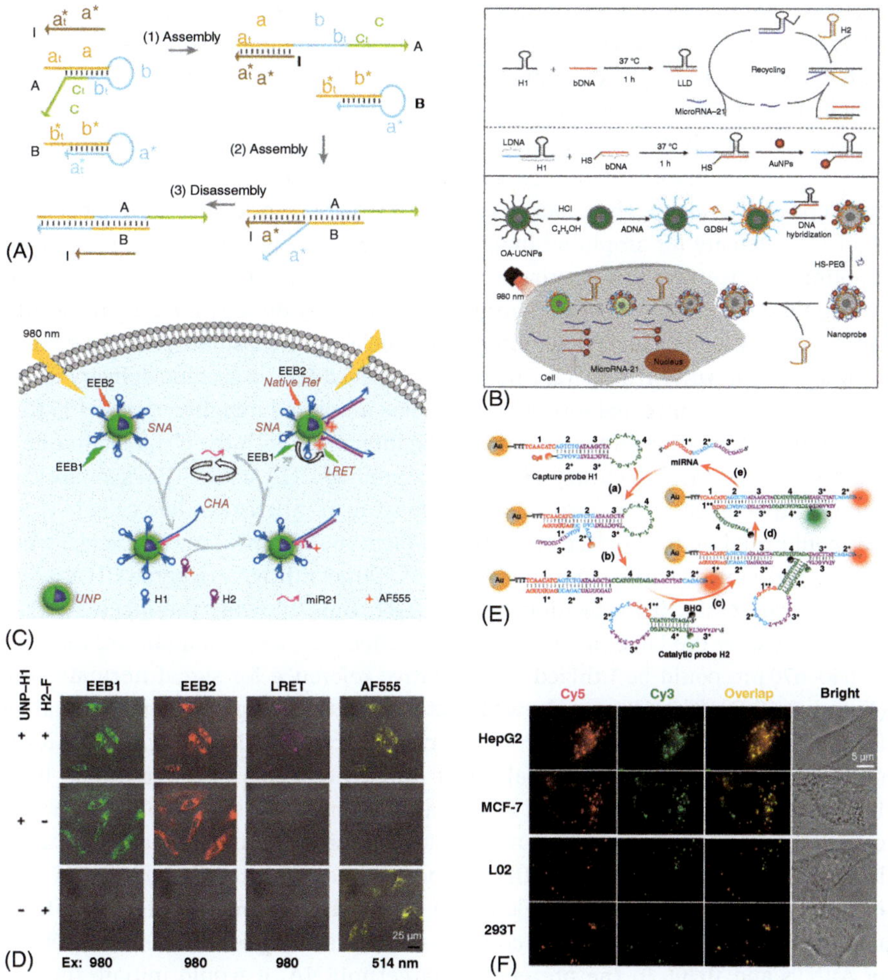

Figure 8.4 (A) Schematic showing the secondary structure of the substrates and the processes of CHA. Source: Reproduced from Yin et al. [41], with permission of 2008 Springer Nature. (B) Schematic illustration of the design of lock-like DNA and the upconversion nanoprobe for monitoring miR-21 in live cells. Source: Reproduced from Zhang et al. [43], with permission of 2018 John Wiley and Sons. (C) Working principle. Source: Reproduced from Huo et al. [44], with permission of 2019 American Chemical Society. (D) Application of the nanoamplicon comparator for enhanced ratiometric imaging of miRNA in live cells. Source: Reproduced from Huo et al. [44], with permission of 2019 American Chemical Society. (E) Working principle and (F) application of the dual-color colocalization CHA nanosystem for imaging miR-21 in live cells. Source: Reproduced with permission from Li et al. [45]. © 2020 American Chemical Society.

strategy, the LOD of miR-21 was ~67 pM, which was about 24.5-fold lower than that of normal nanobeacon without amplification. Compared with AuNPs, the gold nanoflowers possess a horn shape and a high specific surface area, thus exhibiting a higher loading efficiency. Liu et al. designed a flowerlike nanovector loaded with DNA hairpins to perform CHA-based imaging of miR-21 with a LOD of 400 fM [50].

Based on re-engineering of conventional CHA, Zhang et al. designed a lock-like DNA by hybridizing the H1 hairpin with a blocker strand and conjugated it on an AuNP [43]. The conjugated products, together with the H2 hairpins, were further attached to a UCNP for monitoring miR-21 in live cells. In the absence of a target, AuNPs were stably tethered to the UCNP's surface and effectively quenched the UCL signal. Hybridization of miR-21 with the lock-like DNA could initiate CHA and repeatedly release the blocker strand and AuNPs from the UCNP, restoring the UCL intensity for amplified imaging (Figure 8.4B). Benefiting from the signal amplification feature, this approach exhibited a very low LOD (~0.74 fM) *in vitro* and could distinguish cancer cells from normal cells by directly imaging intracellular miR-21. By using cancer cell membrane vesicles to encapsulate FAM-labeled H1 (conjugated on AuNPs) and H2, Chen et al. achieved tumor-specific delivery due to the homing-targeting capability, which enabled amplified imaging of miR-155 [51].

Ratiometric imaging strategies can also be integrated with the CHA technique to improve the detection accuracy. Huo et al. combined CHA with UCNPs to fabricate a nanoamplicon comparator for sensitive imaging of miRNAs in live cells [44]. H1 was conjugated on the surface of UCNPs to form the spherical nucleic acids, while free H2 was labeled with a fluorophore AF555. Target miRNA-triggered CHA would result in successful hybridization of H2 to H1, thus allowing the labeled AF555 to be excited by the UCL of 530–560 nm via effective LRET. In addition, the UCL of 640–670 nm could be utilized as the native reference for signal normalization (Figure 8.4C,D). The authors showed that the relative imaging intensity obtained from the developed nanoamplicon comparator correlated well with the result obtained by RT-qPCR for the analysis of miRNA contents in different cell lines. Li et al. constructed a dual-color CHA nanosystem for in situ amplified imaging of miRNA in live cells [45]. The nanosystem consisted of two components: DNA hairpin H1, which was labeled with Cy5 and modified on AuNPs' surface; and DNA hairpin H2, which was dual labeled with Cy3 and BHQ. Without target miRNA, neither Cy5 nor Cy3 fluorescence could be observed owing to the quenching effects of AuNPs and BHQ. In the presence of target miRNA, it would initiate the programmed CHA and open the two DNA hairpins, inducing a signal increase of both Cy3 and Cy5 (Figure 8.4E). The assembly-induced signals could be counted one by one using a single-molecule imaging apparatus. This dual-color colocalization imaging strategy could avoid false-positive signals to some extent and thereby obtain improved detection accuracy (Figure 8.4F).

8.4.3 Amplified RNA Imaging Based on DNA Nanomachines

Benefiting from unparalleled programmability, DNA has become an attractive material to design and construct smart nanostructures for biological applications. Among them, DNA nanomachines could realize stimuli-responsive nanoscale movement and information transduction, making them widely used in amplified detection of various analytes (e.g. RNA) [52, 53]. According to energy consumption, DNA nanomachines can be classified into two types: fuel-powered and self-powered. Mechanistically, fuel-powered DNA nanomachines are mainly designed based on programmed SDR, while self-powered DNA nanomachines could be fabricated using deoxyribozymes (DNAzymes).

Entropy-driven catalytic reactions, introduced by Zhang et al. in 2007 [54], utilize a series of linear single-stranded DNAs to execute toehold-assisted SDR for catalytic circuits. He et al. combined this entropy-driven strategy with AuNPs and quantum dots (QDs) to develop a catalytic approach for highly sensitive sensing of miRNAs in live cells [55]. As shown in Figure 8.5a, DNA1 molecules were conjugated on AuNPs' surface via gold thiol bond, and QD-labeled DNA2 molecules were tethered through a linker strand (L) with a sandwich hybridization structure. Thus, the fluorescence of QDs was quenched by the nearby AuNP. A single copy of miRNA target could trigger disassembly and fluorescence recovery of multiple QDs through the entropy-driven catalytic reactions as follows: it first hybridized to the linker strand and replaced the QD-labeled DNA2 through SDR mediated by the first toehold, making the second toehold available to the fuel DNA strand for hybridization; hybridization of the fuel DNA to L could release target miRNA to participate in next cycle of reactions,

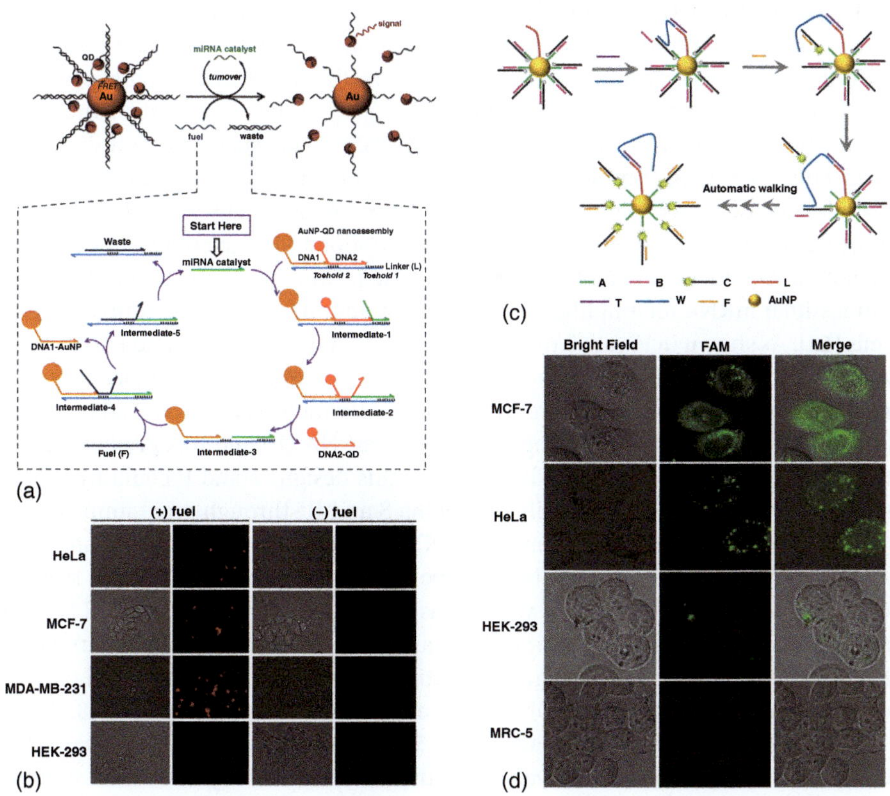

Figure 8.5 (a) Working principle and (b) application of AuNP-QDs nanoassembly for catalytic miRNA imaging in live cells. Source: Adapted from He et al. [55], with permission. © 2016 John Wiley & Sons, Inc. (c) Schematic illustration and (d) application of the designed DNA walker for amplified imaging of miR-21 in live cells. Source: Reproduced with permission from Liang et al. [56]. © 2017 John Wiley & Sons, Inc.

thus yielding amplified imaging signals. With the aid of fuel DNA strands, a 10-fold signal enhancement could be achieved for the imaging of miR-21 in several cell lines (Figure 8.5b). Later, by designing a DNA hairpin dual labeled with FAM and TAMRA as the product of DNA-fueled catalytic nanomachine, Li et al. employed a more accurate FRET signal for sensing miRNA in live cells [57].

Liang et al. employed this entropy-driven strategy to develop a dynamic DNA walker for irreversible catalytic substrate disassembly and applied it to miRNA imaging in live cells [56]. As shown in Figure 8.5c, the nanomachine was constructed by conjugating many three-stranded substrate complexes (A/B/C) and one affinity ligand (L) on an individual AuNP. Substrates A and B co-hybridized to C labeled with FAM, whose fluorescence was initially quenched by the AuNP. In addition, a walking strand (W) and a fuel strand (F) were required to drive the programmed substrate disassembly. In the presence of target (T), it could partly hybridize to L and W, thus tethering W to the AuNP. Importantly, the tethered W could act as a walkable leg to perform entropy-driven catalytic reaction as follows: W interacted with C and displaced B from C, making C available to F for base pairing; hybridization between F and C led to complete disassembly of the A/B/C complex and thereby a recovery of FAM emission; simultaneously, it liberated W to move along the DNA-AuNP track to initiate next cycle of A/B/C disassembly. Thus, the fluorescence signal could be continuously produced for signal-amplified imaging of the target. After functional verification in test tubes, the designed DNA walker was successfully utilized for intracellular imaging of miR-21 in four different cell lines (Figure 8.5d).

Besides co-delivered DNA fuels, high-abundance endogenous molecules can also be utilized as fuels to drive nanomachines for amplified RNA imaging. For example, Li et al. designed a dynamic DNA nanomachine powered by a relatively abundant intracellular mRNA for amplified imaging of low abundance of target miRNA in live cells [58]. As shown in Figure 8.6A, two types of DNA duplexes, R/S and F*/R*, were assembled on an AuNP to fabricate the DNA nanomachine. R was designed to recognize target miRNA (T) and was pre-hybridized to S, which was dual labeled with Cy3 and Cy5. F* was designed to recognize the fuel mRNA (F) and was pre-hybridized to R*, induing a loop in the middle of R*. By this design, T and F could hybridize to R and F*, respectively, gradually displacing S and R* through programmed SDR (Step1). Then, S was released from the AuNP, forming a hairpin structure and generating a FRET signal (Step 2). More importantly, the liberated R* could shift to hybridize with R and displace T, which was reused to repeatedly liberate S for signal amplification (Steps 3–5). Using abundant GAPDH mRNA and low abundance let-7a miRNA as the fuel and target, respectively, the authors demonstrated the signal amplification feature of the DNA nanomachine in MCF-7 cells (Figure 8.6B).

As the energy currency for life, adenosine 5′-triphosphate (ATP) is universally distributed in all cells with a relatively constant level and thereby is an ideal fuel for driving DNA nanomachines. Ma et al. developed a highly integrated DNA nanomachine that could be powered by ATP for amplified imaging of intracellular miRNA without any auxiliary additives [59]. As illustrated in Figure 8.6C, a high density of double-stranded substrates (S/R) were assembled on AuNP's surface as tracks for a swing arm (A) pre-locked with a hairpin structure. The 5′ thiol-modified oligo S was

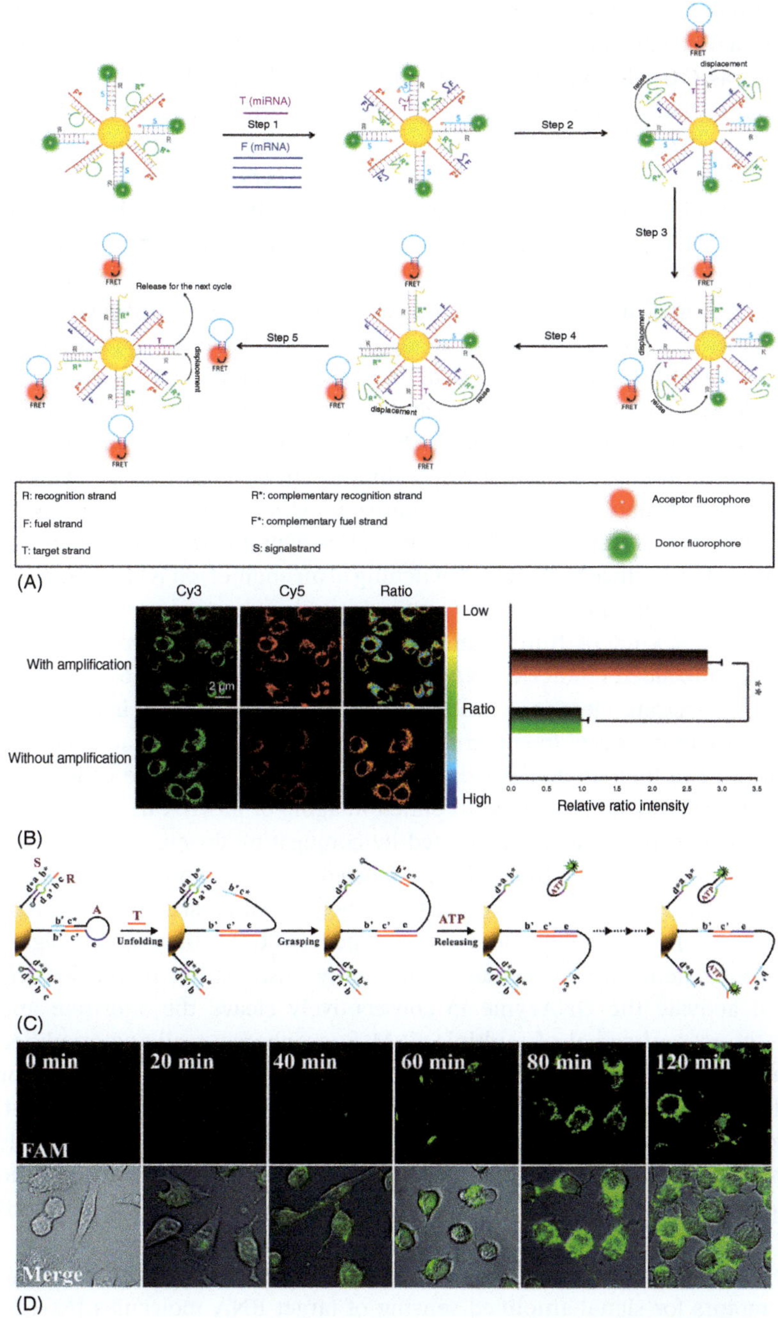

Figure 8.6 (A) Working principle and (B) application of mRNA-powered DNA nanomachine for amplified imaging of intracellular miRNA. Source: Reproduced with permission from Li et al. [58]. © 2020 John Wiley & Sons, Inc. (C) Working principle and (D) operation of ATP-fueled DNA nanomachines for real-time imaging of miR-21 in live cells. Source: Reproduced with permission from Ma et al. [59]. © 2018 Royal Society of Chemistry/CC BY 3.0.

covalently immobilized onto the AuNP surface. The sequence of R was designed to include a toehold (domain c) and a 27-nt ATP aptamer labeled with FAM. The swing arm A was 3′ thiol-modified for tethering and was designed with a stem-loop hairpin structure that could be unfolded by a specific miRNA target (T) through competitive hybridization. Initially, R was stably bound to S, its ATP binding ability was completely inhibited, and the labeled FAM was quenched by the AuNP. The target T could specifically open the hairpin structure of A and liberate the 5′ end as a free-swing arm that could grasp R from S through SDR mediated by the toehold c, exposing more ATP recognition sequence. Thus, intracellular free ATP molecules could induce R to fold into an ATP-aptamer structure and make it dissociate from A, liberating the swing arm A to spread along the S/R tracks and to continuously release FAM-labeled R for signal amplification. Since all components were assembled on a single AuNP, the designed DNA nanomachine operated rapidly inside cells. Real-time monitoring showed that it took about 80 minutes to reach equilibrium with the brightest fluorescence signal responding to intracellular miR-21 of HeLa cells (Figure 8.6D). Notably, different from native protein motors, which are powered by the energy released from ATP hydrolysis, this dynamic DNA nanomachine utilized ATP binding-induced structural switching of an engineered ATP aptamer to drive progressive operation in live cells.

DNAzymes are a kind of functional nucleic acid that can cleave RNA and ribonucleotide-embedded chimeric DNA substrates at a specific site [60, 61]. Due to this unique substrate cleavage capability, DNAzymes have been increasingly integrated with nanoparticles to construct self-powered DNA nanomachines that do not require extra fuels to achieve signal-amplified imaging in live cells. Peng et al. reported a DNAzyme motor for amplified imaging of miRNA in cancer cells [62]. The DNAzyme motor was constructed by conjugating dozens of pre-locked DNAzyme molecules and hundreds of FAM-labeled substrate strands on AuNPs. The target miRNA could hybridize to the Cy5-labeled locking strand and release it from the 8–17 DNAzyme via toehold-mediated SDR, leading to the signal recovery of equivalent Cy5 molecules. Importantly, dissociation of the locking strand could activate the DNAzyme to consecutively cleave the substrate and restore the fluorescent signal of multiple FAM fluorophores, realizing amplified miRNA imaging in live cells. After a 60-minute operation of the DNAzyme motor in MDA-MB-231 cells expressing the target miRNA (miR-10b) at a very low level, the fluorescence intensity of FAM was much higher than that of Cy5, confirming its signal amplification feature. Moreover, little fluorescence could be seen when the motor system was introduced into two control cells (MCF-10a and MCF-7) which are lack of miR-10b, proving its stability and specificity.

This pre-lock and target activation strategy has been widely used to design DNAzyme motors for signal-amplified sensing of target RNA molecules [63–65]. For instance, Yang et al. developed an AuNP-based hairpin-locked DNAzyme probe to sense miRNA in live cells [63]. Without target miRNA, the DNAzyme strand formed a stable hairpin structure through intramolecular pairing, and its RNA cleavage activity was inhibited. Hybridization of target to the DNAzyme strand could open the hairpin and liberate the catalytic core domain to form the active conformation, which then cleaved the self-strand, restoring the quenched fluorescence and recycling the target strand for signal amplification. To improve the loading capacity and the operation efficiency of DNAzyme motor, Kong et al. used the rolling circle amplification technique to extend the tracks conjugated on a magnetic nanobead for the assembly of locked DNAzyme and its substrate [66]. By this design, the DNAzyme motor was able to detect miR-21 with an LOD of 3.57 aM and achieve sensitive imaging in live cells. It is notable that specific metal ions (e.g. Zn^{2+} and Mn^{2+}) are usually necessary as cofactors to activate the catalytic cleavage activity of DNAzymes, while their concentrations inside cells are lower than the required level, limiting the operation efficiency of DNAzyme motors. To address this issue, one routinely used method is to incubate the cells with a specific cation solution. Recently, Hu et al. integrated Zn^{2+}-dependent 8–17 DNAzyme-based DNA walker with zeolitic imidazolate framework-8 (ZIF-8) to construct a self-sufficient DNA motor for imaging miRNA in live cells [67]. After entering cells, ZIF-8 could be efficiently degraded by acidic pH, releasing the DNA motor and the required Zn^{2+}. Thus, it could operate immediately to produce fluorescence signal once the pre-locked DNAzyme was activated by a target miRNA, allowing for real-time imaging inside live cells.

Notably, a DNAzyme can be split into two parts, and its catalytic RNA cleavage activity can be restored once the two components get together, thus permitting the design of DNAzyme-based biosensing with more flexibility [68–70]. Based on this mechanism, Gao et al. designed a self-protected DNAzyme walker (termed ES-AuNP) for imaging intracellular miRNA [71]. A circular bulging DNA fragment (referred to as "target binding domain") was inserted into the catalytic core domain of the DNAzyme to inhibit its catalytic function and to provide an arch-like shield for terminal protection. To construct ES-AuNP, an AuNP was modified with dozens of thiol-modified and FAM-labeled substrates, several of which hybridized with the inactive DNAzymes. Initially, the fluorescence of FAM was quenched by the AuNP core. In the presence of target miRNA, however, it could hybridize to the target binding domain and bring the two separated halves of DNAzyme into the adjacent location to form an active catalytic core, restoring the substrate cleavage capability. Thus, the activated DNAzyme could act as a walking arm to repeatedly cleave the conjugated substrates and release the cleaved products labeled with FAM, leading to an amplified fluorescence signal. The authors showed that ES-AuNP enabled them to sensitively assess the relative expression levels of miRNA in live cells.

8.5 Spatiotemporally Controlled RNA Imaging in Live Cells

Since the expression and translocation of RNA in cells are strictly regulated on both temporal and spatial scales, spatiotemporally selective imaging techniques are highly desirable for tracking these dynamic processes to dissect the critical biological functions of RNA of interest. However, precise imaging of RNA in live cells at a desired time duration and spatial location is a remarkable challenge for conventional DNA-based sensors that are usually designed in an "always-on" manner and respond to target RNA through irreversible mechanisms such as stable hybridization and substrate cleavage. Of note, such biosensors can be immediately turned on once they encounter the analytes during the delivery process and thereby lack spatiotemporal controllability, resulting in significantly compromised imaging resolution. To address this issue, Zhao et al. proposed a conceptual methodology that employs light-activatable biosensors to achieve spatiotemporally controlled bio-imaging and biosensing [72]. Different from "always-active" sensors, the target recognition unit of activatable sensors is initially locked and needs to be triggered by a certain "key" that can be an external stimulus or an endogenous factor.

Due to the advantages of noninvasive modality and ease of operation, external light represents an attractive one of such "keys" to precisely control the biosensors for spatiotemporally selective RNA imaging in live cells. Zhao et al. designed a NIR light-activatable DNA nanosensor that enabled to image miRNA with high spatial and temporal precision [73]. As shown in Figure 8.7a, the DNA nanosensor was constructed by engineering a dual-labeled DNA probe with a photocleavable (PC) linker and integrating it with a UCNP. Without light irradiation, the DNA probe formed a stable hairpin structure containing a PC linker at the loop region and labeled with Cy5 and BHQ2 at the opposite ends. It could not recognize and hybridize to target miRNA and thereby emitted no fluorescence. Ultraviolet (UV) light enabled to cleave the PC linker and switch the intramolecular hybridization of the stem region into an intermolecular one with reduced stability, which allowed the target miRNA to replace the Cy5-labeled fragment and bind to the BHQ2-labeled fragment, resulting in a recovery of Cy5 fluorescence for bio-imaging. Moreover, the UCNP could act as a NIR-to-UV transducer, which absorbed deep-tissue-penetrable NIR light and produced the UV emission to activate the sensing ability of the DNA probe. Using this DNA nanosensor, imaging of miRNA in live cells could be controlled at a chosen time and location through a remote NIR light activation (Figure 8.7b).

This UCL-controlled imaging technology has been widely integrated with other strategies to achieve on-demand RNA imaging in live cells. For instance, Li et al. combined core–shell-structured UCNPs and DNA tetrahedrons carrying a UV-activatable DNA probe for imaging miR-21, allowing them to significantly minimize false positive signals caused by nonspecific responses [76]. Mi et al. constructed UCNP-based photonic nanocircuits that could employ NIR light to control DNA-based molecular computation, realizing correlated molecular imaging of miR-21 and ATP in a Boolean logic manner with high spatial precision [77]. Later, Zhao et al. extended this UCL-controlled strategy from circuit-based multiplexed imaging, which produces one signal output upon multiple inputs, to multicolor imaging, which employs several independent channels to record corresponding targets [74]. As shown in Figure 8.7c, a triangular DNA nano sucker (TDS) was first assembled from three DNA hairpins (PH1, PH2, and PH3, each one contained a PC linker and was labeled with a different fluorophore as indicated) with three DNA adaptors (Q1, Q2, and Q3), respectively, labeled with a quencher and then loaded onto a UCNP to fabricate a UCL-activated DNA nanosensor for multicolor imaging of three miRNA molecules. Upon NIR light irradiation, the UCNP could

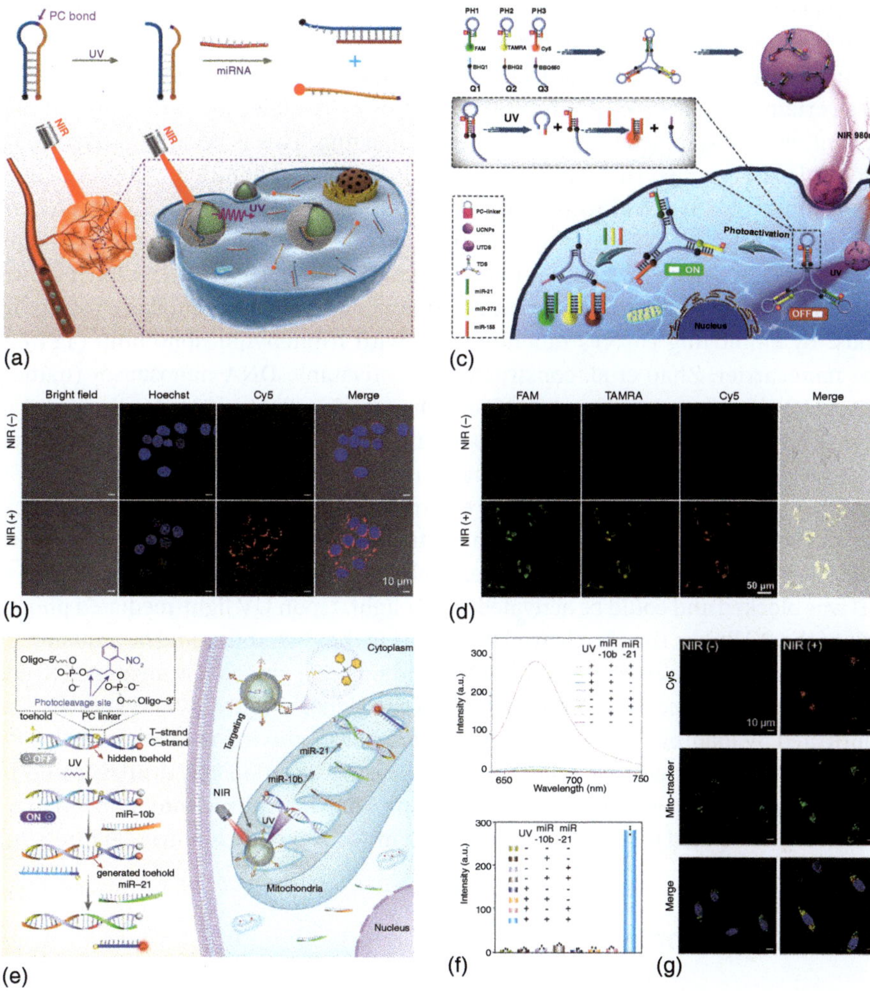

Figure 8.7 (a) Working principle and (b) application of UCNP-based DNA nanosensors for UCL-controlled miRNA imaging. Source: Reproduced with permission from Zhao et al. [73]. © 2019 American Chemical Society. (c) Working principle and (d) application of the UCL-activated DNA nanosystem for multiplexed imaging of three intracellular miRNAs. Source: Reproduced with permission from Zhao et al. [74]. © 2021 American Chemical Society. (e) Working principle and (f and g) demonstration of the photoactivated DNA nanosensor for correlated imaging of miR-10b and miR-21 in mitochondria. Source: Reproduced with permission from Zhao et al. [75]. © 2021 John Wiley & Sons, Inc.

emit UV luminescence to cleave the PC linker inserted in the hairpins, activating their sensing functions for a designated target. It was shown that the designed nanosensor allowed for simultaneous imaging of miRNA-21 (FAM), miRNA-373 (TAMRA), and miRNA-155 (Cy5) in HeLa cells irradiated by 980 nm NIR light (Figure 8.7d). The UCL nanotechnology could be further combined with signal amplification strategies, such as HCR [78], CHA [79], and fuel-powered [80, 81] or self-powered DNA nanomachines [82, 83], to realize more accurate RNA imaging with high sensitivity and spatial selectivity.

Subcellular localization is crucial for RNA molecules to exert their specific roles in a certain cellular process. Precise tracking of RNA molecules with a subcellular resolution is of significance but remains challenging. This is because conventional probes lack an active subcellular localization ability. In addition, "always-active" probes may also respond to the targets encountered during the delivery process. It is notable that UCNPs employed in the above mentioned not only act as a light transducer for sensing activation but also serve as a delivery carrier, which allows for modification with other functional components such as organelle-targeting ligands. By employing UCNPs functionalized with triphenylphosphonium (TPP) as the nanocarrier, Zhao et al. constructed an activatable DNA nanosensor (namely PD-mUC) for controlled imaging of mitochondrial miRNAs [75]. As illustrated in Figure 8.7e, the activatable DNA sensor (denoted as PD) was constructed by annealing a toeholds-bearing strand (T) with a complementary strand (C) containing a PC linker. A stable duplex structure of PD ensured an extremely low fluorescence background because Cy5 labeled at the 3′end of the C-strand was efficiently quenched by BHQ2 labeled at the 5′ end of the T-strand. Notably, the miRNA sensing capability of PD was blocked and could be activated by UV light. Upon UV light-mediated photolysis of the PC linker, the two cleaved fragments of the C-strand would be sequentially replaced by two target miRNAs (miR-10b and miR-21) via toehold-mediated SDRs, leading to a significant increase in fluorescent signal (Figure 8.7f). PD-mUC was fabricated by loading PD onto UCNPs and further modifying the system with TPP, a mitochondria-targeting ligand. Thus, the sensing function of PD-mUC could be selectively activated by remote NIR irradiation after it localized into mitochondria, achieving correlated imaging of miR-10b and miR-21 in mitochondria (Figure 8.7g).

8.6 Conclusion

The integration of DNA programmability and the unique properties of nanomaterials makes it possible to construct advanced DNA nanosensors for sensitive and selective imaging of RNA molecules in live cells. AuNPs and UCNPs are two of the most widely used inorganic nanoparticles for the fabrication of such DNA nanosensors. According to the specific mechanism, RNA imaging can be achieved by direct probe hybridization, toehold-mediated SDR, or DNAzyme-based substrate cleavage. Based on these primary reactions, diverse signal amplification strategies have been proposed to improve detection sensitivity, including HCR, CHA, and entropy-driven catalytic reactions. Furthermore, by rational combination of these strategies with nanoparticles, a variety of DNA nanomachines have been designed for signal-enhanced imaging of low abundance RNA in live cells, such as fuel-powered DNA walkers and self-powered DNAzyme motors. In addition, spatiotemporally selective molecular imaging has emerged as an attractive approach in recent years. By designing activatable DNA sensors and integrating them with functionalized nanoparticles, RNA imaging can be exclusively initiated in selected cells and even in a specific organelle. Together, these DNA nanosensors have been in principle demonstrated to hold tremendous potential in live-cell RNA imaging and are expected to be employed for exploring dynamic and complicated RNA biology.

Acknowledgment

This work was supported by the National Natural Science Foundation of China (22125402, 22204030).

References

1 Rinn, J.L. and Chang, H.Y. (2012). *Annu. Rev. Biochem.* 81: 145–166.
2 Liu, C.X. and Chen, L.L. (2022). *Cell* 185: 2016–2034.
3 Bartel, D.P. (2004). *Cell* 116: 281–297.
4 Buxbaum, A.R., Haimovich, G., and Singer, R.H. (2015). *Nat. Rev. Mol. Cell Biol.* 16: 95–109.
5 Ryder, P.V. and Lerit, D.A. (2018). *Traffic* 19: 496–502.
6 Jobbins, A.M., Haberman, N., Artigas, N. et al. (2022). *Nucleic Acids Res.* 50: 3379–3393.
7 Fernandopulle, M.S., Lippincott-Schwartz, J., and Ward, M.E. (2021). *Nat. Neurosci.* 24: 622–632.
8 Orellana, E.A., Siegal, E., and Gregory, R.I. (2022). *Nat. Rev. Genet.* 23: 651–664.
9 Su, Q., Feng, W., Yang, D., and Li, F. (2017). *Acc. Chem. Res.* 50: 32–40.
10 Chen, Z., Chen, H., Hu, H. et al. (2008). *J. Am. Chem. Soc.* 130: 3023–3029.
11 Gong, L., Liu, S., Song, Y. et al. (2020). *J. Mater. Chem. B* 8: 5952–5961.
12 Qiao, G., Gao, Y., Li, N. et al. (2011). *Chemistry* 17: 11210–11215.
13 Ren, H., Long, Z., Shen, X. et al. (2018). *ACS Appl. Mater. Interfaces* 10: 25621–25628.
14 Yang, L., Zhang, K., Bi, S., and Zhu, J.J. (2019). *ACS Appl. Mater. Interfaces* 11: 38459–38466.
15 Huang, X., Song, J., Yung, B.C. et al. (2018). *Chem. Soc. Rev.* 47: 2873–2920.
16 Tyagi, S. and Kramer, F.R. (1996). *Nat. Biotechnol.* 14: 303–308.
17 Zheng, J., Yang, R., Shi, M. et al. (2015). *Chem. Soc. Rev.* 44: 3036–3055.
18 Choi, C.K.K., Li, J., Wei, K. et al. (2015). *J. Am. Chem. Soc.* 137: 7337–7346.
19 Guo, Y., Wei, B., Xiao, S. et al. (2017). *Quant. Biol.* 5: 25–41.
20 Zhang, D.Y. and Winfree, E. (2009). *J. Am. Chem. Soc.* 131: 17303–17314.
21 Zhang, D.Y. and Seelig, G. (2011). *Nat. Chem.* 3: 103–113.
22 Simmel, F.C., Yurke, B., and Singh, H.R. (2019). *Chem. Rev.* 119: 6326–6369.
23 Seferos, D.S., Giljohann, D.A., Hill, H.D. et al. (2007). *J. Am. Chem. Soc.* 129: 15477–15479.
24 Yang, Y., Huang, J., Yang, X. et al. (2015). *J. Am. Chem. Soc.* 137: 8340–8343.
25 Qu, A., Sun, M., Xu, L. et al. (2019). *Proc. Natl. Acad. Sci. U.S.A.* 116: 3391–3400.

26 Briley, W.E., Bondy, M.H., Randeria, P.S. et al. (2015). *Proc. Natl. Acad. Sci. U.S.A.* 112: 9591–9595.
27 Li, N., Chang, C., Pan, W., and Tang, B. (2012). *Angew. Chem. Int. Ed.* 51: 7426–7430.
28 Lin, H.R., Jiang, Y.J., Cheng, F. et al. (2021). *ACS Appl. Nano Mat.* 4: 13469–13476.
29 Li, S., Xu, L., Ma, W. et al. (2016). *J. Am. Chem. Soc.* 138: 306–312.
30 Li, S., Xu, L., Sun, M. et al. (2017). *Adv. Mater.* 29: e1606086.
31 Jiang, X., Hao, C., Zhang, H. et al. (2021). *ACS Appl. Mater. Interfaces* 13: 41405–41413.
32 Gao, R., Hao, C., Xu, L. et al. (2018). *Anal. Chem.* 90: 5414–5421.
33 Meng, L., Wang, Q., Wang, L. et al. (2021). *Mater. Today Bio* 12: 100157.
34 Zhao, Y., Chen, F., Li, Q. et al. (2015). *Chem. Rev.* 115: 12491–12545.
35 Li, M., Yin, F., Song, L. et al. (2021). *Chem. Rev.* 121: 10469–10558.
36 Dirks, R.M. and Pierce, N.A. (2004). *Proc. Natl. Acad. Sci. U.S.A.* 101: 15275–15278.
37 Wu, Z., Liu, G.Q., Yang, X.L., and Jiang, J.H. (2015). *J. Am. Chem. Soc.* 137: 6829–6836.
38 Bi, S., Yue, S., and Zhang, S. (2017). *Chem. Soc. Rev.* 46: 4281–4298.
39 Li, S., Li, P., Ge, M. et al. (2020). *Nucleic Acids Res.* 48: 2220–2231.
40 Peng, H., Newbigging, A.M., Reid, M.S. et al. (2020). *Anal. Chem.* 92: 292–308.
41 Yin, P., Choi, H.M.T., Calvert, C.R., and Pierce, N.A. (2008). *Nature* 451: 318–322.
42 Liu, J., Zhang, Y., Xie, H. et al. (2019). *Small* 15: e1902989.
43 Zhang, K., Song, S., Huang, S. et al. (2018). *Small* 14: e1802292.
44 Huo, M., Li, S., Zhang, P. et al. (2019). *Anal. Chem.* 91: 3374–3381.
45 Li, B., Liu, Y., Liu, Y. et al. (2020). *ACS Nano* 14: 8116–8125.
46 Li, B., Ellington, A.D., and Chen, X. (2011). *Nucleic Acids Res.* 39: e110.
47 Jiang, Y.S., Li, B., Milligan, J.N. et al. (2013). *J. Am. Chem. Soc.* 135: 7430–7433.
48 Wu, C., Cansiz, S., Zhang, L. et al. (2015). *J. Am. Chem. Soc.* 137: 4900–4903.
49 Wang, J., Huang, J., Quan, K. et al. (2018). *Chem. Commun.* 54: 10336–10339.
50 Liu, J., Du, P., Zhang, J. et al. (2018). *Chem. Commun.* 54: 2550–2553.
51 Chen, Y., Dai, W., Wang, D., and Dong, H. (2022). *Chem. Commun.* 58: 9488–9491.
52 Xu, M. and Tang, D. (2021). *Anal. Chim. Acta* 1171: 338523.
53 Song, L., Zhuge, Y., Zuo, X. et al. (2022). *Adv. Sci.* 9: e2200327.
54 Zhang, D.Y., Turberfield, A.J., Yurke, B., and Winfree, E. (2007). *Science* 318: 1121–1125.
55 He, X., Zeng, T., Li, Z. et al. (2016). *Angew. Chem. Int. Ed.* 55: 3073–3076.
56 Liang, C.P., Ma, P.Q., Liu, H. et al. (2017). *Angew. Chem. Int. Ed.* 56: 9077–9081.
57 Li, D., Zhou, W., Yuan, R., and Xiang, Y. (2017). *Anal. Chem.* 89: 9934–9940.
58 Li, J., Wang, J., Liu, S. et al. (2020). *Angew. Chem. Int. Ed.* 59: 20104–20111.
59 Ma, P.Q., Liang, C.P., Zhang, H.H. et al. (2018). *Chem. Sci.* 9: 3299–3304.
60 Silverman, S.K. (2005). *Nucleic Acids Res.* 33: 6151–6163.
61 Breaker, R.R. and Joyce, G.F. (1994). *Chem. Biol.* 1: 223–229.
62 Peng, H., Li, X.F., Zhang, H., and Le, X.C. (2017). *Nat. Commun.* 8: 14378.

63 Yang, Y., Huang, J., Yang, X. et al. (2017). *Anal. Chem.* 89: 5850–5856.
64 Liu, J., Cui, M., Zhou, H., and Yang, W. (2017). *ACS Sens.* 2: 1847–1853.
65 Liu, C., Hu, Y., Pan, Q. et al. (2019). *Biosens. Bioelectron.* 136: 31–37.
66 Kong, L., Kou, B., Zhang, X. et al. (2021). *Chem. Sci.* 12: 15953–15959.
67 Hu, H., Zhou, F., Wang, B. et al. (2021). *Nanoscale* 13: 1863–1868.
68 Tian, B., Han, Y., Wetterskog, E. et al. (2018). *ACS Sens.* 3: 1884–1891.
69 Qi, L., Zhao, Y., Yuan, H. et al. (2012). *Analyst* 137: 2799–2805.
70 Ali, M.M., Wolfe, M., Tram, K. et al. (2019). *Angew. Chem. Int. Ed.* 58: 9907–9911.
71 Gao, Y., Zhang, S., Wu, C. et al. (2021). *ACS Nano* 15: 19211–19224.
72 Zhao, J., Gao, J., Xue, W. et al. (2018). *J. Am. Chem. Soc.* 140: 578–581.
73 Zhao, J., Chu, H., Zhao, Y. et al. (2019). *J. Am. Chem. Soc.* 141: 7056–7062.
74 Zhao, T., Gao, Y., Wang, J. et al. (2021). *Anal. Chem.* 93: 12329–12336.
75 Zhao, J., Li, Z., Shao, Y. et al. (2021). *Angew. Chem. Int. Ed.* 60: 17937–17941.
76 Li, C.Y., Zheng, B., Kang, Y.F. et al. (2020). *ACS Sens.* 5: 199–207.
77 Mi, Y., Zhao, J., Chu, H. et al. (2021). *Anal. Chem.* 93: 2500–2509.
78 Chu, H., Zhao, J., Mi, Y. et al. (2019). *Angew. Chem. Int. Ed.* 58: 14877–14881.
79 Zhao, X., Zhang, L., Gao, W. et al. (2020). *ACS Appl. Mater. Interfaces* 12: 35958–35966.
80 Ye, M., Kong, Y., Zhang, C. et al. (2021). *ACS Nano* 15: 14253–14262.
81 Shen, Y., Li, Z., Wang, G., and Ma, N. (2018). *ACS Sens.* 3: 494–503.
82 Li, D., Zhao, T., Chen, J. et al. (2022). *Anal. Chem.* 94: 14467–14474.
83 Li, C.Y., Liu, J.X., Yuheng, L. et al. (2022). *Anal. Chem.* 94: 5450–5459.

9

Building DNA Computing System for Smart Biosensing and Clinical Diagnosis

Jiao Yang[1] *and Da Han*[1,2]

[1] *Institute of Molecular Medicine (IMM), Shanghai Key Laboratory for Nucleic Acid Chemistry and Nanomedicine, Renji Hospital, School of Medicine, Shanghai Jiao Tong University, No. 160 Pujian Road, Shanghai 200127, China*
[2] *Zhejiang Cancer Hospital, The Key Laboratory of Zhejiang Province for Aptamers and Theranostics, Hangzhou Institute of Medicine (HIM), Chinese Academy of Sciences, No. 150 Dongfang Street, Hangzhou, Zhejiang 310022, China*

9.1 DNA Computing

DNA nanotechnology was pioneered by Ned Seeman in the early 1980s and has developed rapidly since then [1, 2]. One of the derivative directions of DNA nanotechnology is DNA computing, which borrows the concept from silicon-based computers. Due to the high programmability originating from the precise Watson–Crick principle of base pairing as well as the excellent data storage volume, DNA computing exhibits unique advantages such as vast computing parallelism and good biocompatibility, therefore shedding light on applications ranging from mathematical to biomedical fields. In 1994, Adleman first employed DNA computing to solve the Hamiltonian path problem, which demonstrated the great power of DNA computing for resolving NP-complete problems [3]. Later on, Lipon and coworkers generalized Adleman's work to solve the satisfiability of a Boolean formula [4], further exhibiting the high computing power of DNA [5–7].

DNA computing systems generally consist of three functional modules: inputs, signal processing elements, and outputs. The input and output signals can be binary encoded in DNA. Meanwhile, readable methods such as optical, electrical, and magnetic signals are typically interfaced with DNA or other molecule-based outputs for result delivery [8]. For the signal processing modules, logic gates including YES, NOT, AND, OR, NAND, NOR, XOR, and XNOR (Figure 9.1) are utilized to construct logic circuits for information processing. For instance, in 2003, Ghadiri and coworkers created logic gates including AND, NAND, and INHIBIT based on DNA hybridization and used Förster resonance energy transfer (FRET) to report the results [9]. In 2004, Okamoto and coworkers demonstrated the capability of DNA to accomplish logic functions based on the different charge transport abilities along the strand [10]. In general, building information processing modules with

DNA Nanotechnology for Cell Research: From Bioanalysis to Biomedicine, First Edition. Edited by Zhou Nie.

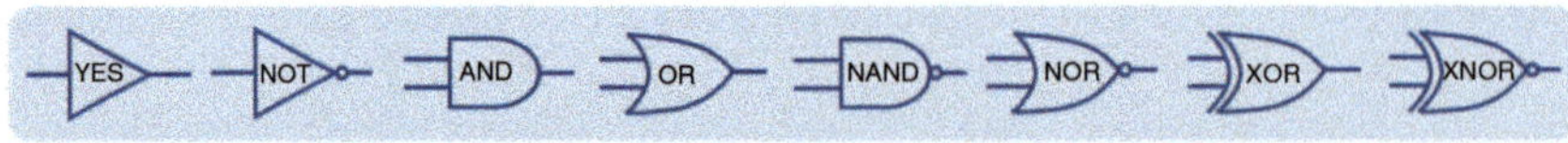

Input		AND	OR	NAND	NOR	XOR	XNOR
0	0	0	0	1	1	0	1
1	0	0	1	1	0	1	0
0	1	0	1	1	0	1	0
1	1	1	1	0	0	0	1

Figure 9.1 Illustrations of basic logic gates and their truth table.

DNA requires functional motifs that can actively make changes in DNA molecules, such as sequences, structures, or hybridization status. We will next introduce some examples of DNA-based logic circuits using different principles, including functional DNA motifs and DNA cascading reactions.

9.1.1 DNA Logic Gates Based on Functional DNA Motifs

DNAzymes are functional nucleic acids with catalytic activities for their substrates. The first DNAzyme that cleaved RNA was reported by Ronald Breaker and Gerald Joyce in 1994 [11]. From then on, diverse DNAzymes have been discovered to catalyze different reactions [12–15]. Generally, the most abundant class of DNAzymes are ribonucleases, which catalyze the cleavage of a ribonucleotide phosphodiester bond through a transesterification reaction, typically with the help of specific metal ions as cofactors [16]. Due to their ability to respond to different inputs and generate controllable outputs, DNAzymes have been utilized to build DNA logic gates. For instance, Stojanovic and coworkers reported a set of deoxyribozyme-based logic gates (Figure 9.2a) [17]. The enzyme strand contained a recognition arm complementary to the substrate of single-stranded DNA (ssDNA) strand and a stem-loop structure for cleavage of the substrate of ssDNA strand. Different combinations of ssDNA sequences as inputs can either regulate the stem-loop catalytic cores or substrate binding, therefore generating logical controls of outputs in response to different inputs. They used this mechanism to build NOT, YES, AND, and XOR logic gates with ssDNAs as inputs. Later on, Stojanovic and coworkers constructed a more complex logic circuit that can perform the calculations of a half-adder, and execution of a molecular automaton called MAYA [21, 22]. It is worth mentioning that those DNAzyme-based logic circuits are based on RNA or chimeric DNA that contain at least one ribonucleotide base serving as substrates, which leads to higher synthesis cost and susceptibility to oligonucleotide degradation. By employing a Cu^{2+}-dependent DNA-cleaving deoxyribozyme, Fan, and coworkers reported robust molecular logic gates, including YES, NOT, and a three-input gate AND(A, NOT(B), NOT(C)), which were composed of DNA only [23]. Willner and

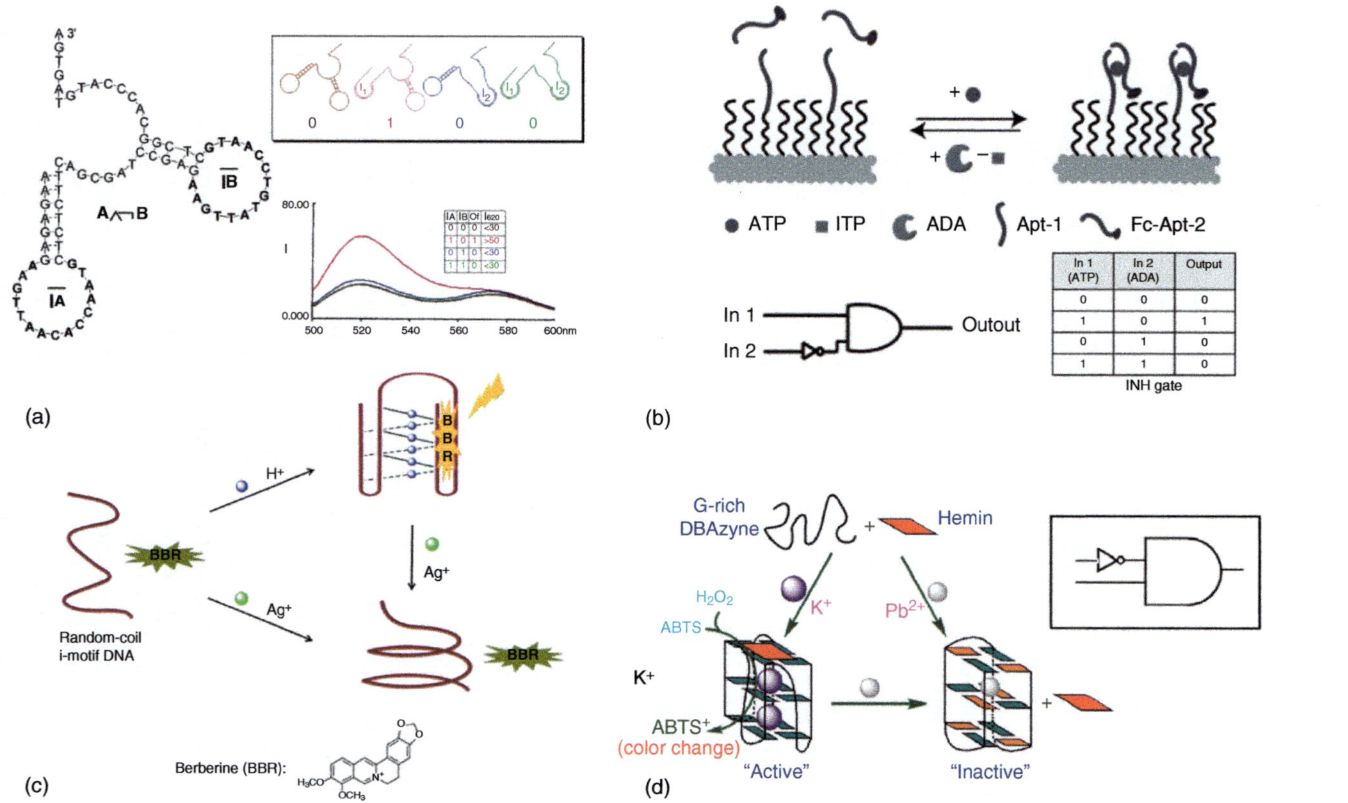

Figure 9.2 DNA logic gates based on functional DNA motifs. (a) Deoxyribozyme-based logic gates with two oligonucleotides I_A and I_B as two inputs and fluorescence intensity as an output. Source: Reproduced from Stojanovic et al. [17], with permission of © 2002 American Chemical Society. (b) Schematic illustration of the ATP split aptamer-based INHIBIT logic gate with an electrochemical output. Source: Reproduced from Feng et al. [18], with permission of John Wiley & Sons. (c) The INHIBIT logic gate based on i-motif structure in response to input H^+ and Ag^+. Source: Reproduced from Xu et al. [19], with permission of the Royal Society of Chemistry. (d) The logic gate based on the G-rich DNAzyme with K^+ and Pb^{2+} as two inputs. Source: Reproduced from Li et al. [20], with permission of © 2009 American Chemical Society.

coworkers constructed DNAzyme-based logic gates with Mg^{2+} as the cofactor [24]. Afterward, they designed artificial pH-programmable DNA logic arrays based on modular libraries of Mg^{2+}- and UO_2^{2+}-dependent DNAzyme subunits and their substrates [25].

Aptamers are oligonucleotides selected by systematic evolution of ligands by exponential enrichment (SELEX) technique with high affinity to their targets such as proteins and small molecules. Similar to antibodies, aptamers can interact with their targets by forming specific 3D structures [26]. Aptamers can be applied in biosensing such as protein detection [27–29], drug delivery [30], and building logical devices [31]. For instance, by combining the RNA aptamers of flavin mononucleotide and theophylline with hammerhead-shaped ribozyme, Breaker and coworkers developed the first AND logic gate using aptamers [32]. Stojanovic and coworkers employed the ssDNA products of deoxyribozyme-based logic gates to control the on- or off-state of aptamers and their targets [33]. Mayer and coworkers explored a split aptamer-based multi-layers of logic gates (INHIBIT and AND gates) that performed a combined XOR logic (Figure 9.2b) [18]. By using thrombin aptamer, Tan and coworkers developed the first threshold-control molecular circuit to manipulate the catalytic activity of thrombin [34]. In addition, the half-adder and half-subtractor based on aptamers were also developed [35].

I-motifs are ssDNA structures that are composed of cytosine-rich sequences. As the N3 in cytosine can be protonated in low pH, the enhanced interaction of $C–C^+$ enables the entire motif to form an intercalated structure via simultaneous Watson–Crick and Hoogsteen base-pairing interactions [36, 37]. It has been proven that several fluorescent dyes, such as crystal violet, $[Ru(phen)_2(dppz)]^{2+}$, and cyanine dye 2,2-diethyl-9-methyl-selenacarbocyanine bromide (DMSB), can bind with i-motif structures, therefore enabling the construction of label-free DNA logic gates in response to specific inputs [19, 38, 39]. Waller and coworkers first reported that Ag^+ could promote the formation of i-motif through $C–Ag^+–C$ base pairing, and the folding was reversible by chelation with cysteine [40]. Based on this study, Luo and coworkers developed a label-free fluorescent DNA INHIBIT logic gate with Ag^+ as an input [41]. Similarly, Pei and coworkers constructed label-free fluorescence DNA logic gates, such as INHIBIT, AND, and OR, based on the specific binding between i-motif and its ligands (Figure 9.2c) [19]. Wei and coworkers developed a triple-output label-free fluorescent DNA logic gate with Ag^+ and cysteine as inputs using a similar mechanism as above [42].

DNA sequences rich in guanine can form a four-stranded structure called G-quadruplex (G4) in the presence of monovalent or divalent cations (K^+, Na^+, Pb^{2+}, Sr^{2+}). In addition, G4 can bind with porphyrin dye through the stacking effect, leading to an enhanced fluorescence intensity [43]. Wang and coworkers constructed a series of unlabeled fluorescent logic gates including YES, NOT, OR, AND, INHIBIT, IMPLICATION, and NAND based on the G4 conformation-induced effect reported by the enhanced fluorescence of protoporphyrin [44]. In addition, G4 can bind with hemin to form G4-hemin complexes that are horseradish peroxidase (HRP) mimicry to catalyze colorimetric reactions, which have been widely applied in the construction of DNA logic gates [45]. For instance, Dong and coworkers constructed an INHIBIT logic gate reported on the formation of G4-hemin complexes, with K^+ and Pb^{2+} as two inputs (Figure 9.2d). With EDTA as another input, the construction of a reversible IMPLICATION gate can be achieved [20]. Combining G4-hemin with logic gate-based operations, Qin and coworkers also developed the OR and INHIBIT logic gates in response to dual inputs [46].

9.1.2 DNA Logic Gates Based on DNA Cascading Reactions

Toehold-mediated strand displacement (TMSD) is developed based on the Gibbs free energy decrease between input strands and output strands during the base displacement process. In specific, a partially hybridized double-stranded DNA (dsDNA) containing a short ssDNA overhang (~5–7 bases, termed toehold) hybridizes with another longer complementary ssDNA (invader strand) via the toehold binding, leading to the release of the short strand from the dsDNA [47]. TMSD exhibits two reaction models, reversible and irreversible (Figure 9.3A), which can either stop or cascade to the next layer of the reaction. Because of the advantages of high programmability and flexibility, TMSD has been utilized to construct various dynamic DNA devices, such as DNA walkers [50, 51], DNA machines [47], and DNA computing systems [48, 49, 52, 53]. In 2006, Winfree and coworkers presented AND, OR, and NOT gates with cascaded TMSD reactions and displayed the experimental implementation of TMSD-based DNA digital logic circuits with multiple cascaded gates [54]. Liu and coworkers constructed XOR and AND logic gates based on TMSD reactions. With those constructed logic gates, a half adder and a full adder were implemented through further cascading reactions [55]. Except for the construction of logic gates, TMSD reactions can be applied to implement complex arithmetic operations. For instance, Winfree and Qian designed a seesaw gate based on well-designed TMSD cascades for operating algorithms including a four-bit square-root calculation [52]. Afterward, Qian and Cherry employed the seesaw DNA gate motif to implement a winner-take-all neural network [56]. In addition, primary mathematical calculations were also realized by TMSD reactions; for instance, Reif and coworkers proposed an architecture for the systematic construction of TMSD-based DNA circuits consisting of addition, subtraction, and multiplication gates for analog computation [57]. Han and coworkers designed DNA-based arithmetical operations such as addition, subtraction, and multiplication for intelligent diagnosis. Specifically, the multiplication operation was realized by hybridizing exact numbers of probes to pre-designed regions of a loop DNA for weight implementation (Figure 9.3B). The summation operation was achieved by using different hybridizing probes for different inputs but triggering the same downstream reactions to generate the summed output (Figure 9.3C). The subtraction was implemented by using two ssDNA that annihilated with each other until only the excess species was left using a cooperative hybridization scheme to report the signal (Figure 9.3D).

In addition to TMSD, hybridization chain reaction (HCR) and catalytic hairpin assembly (CHA) reactions have also been used in the construction of DNA logic gates and molecular circuits. CHA was first reported by Yin et al. [58]. In specific,

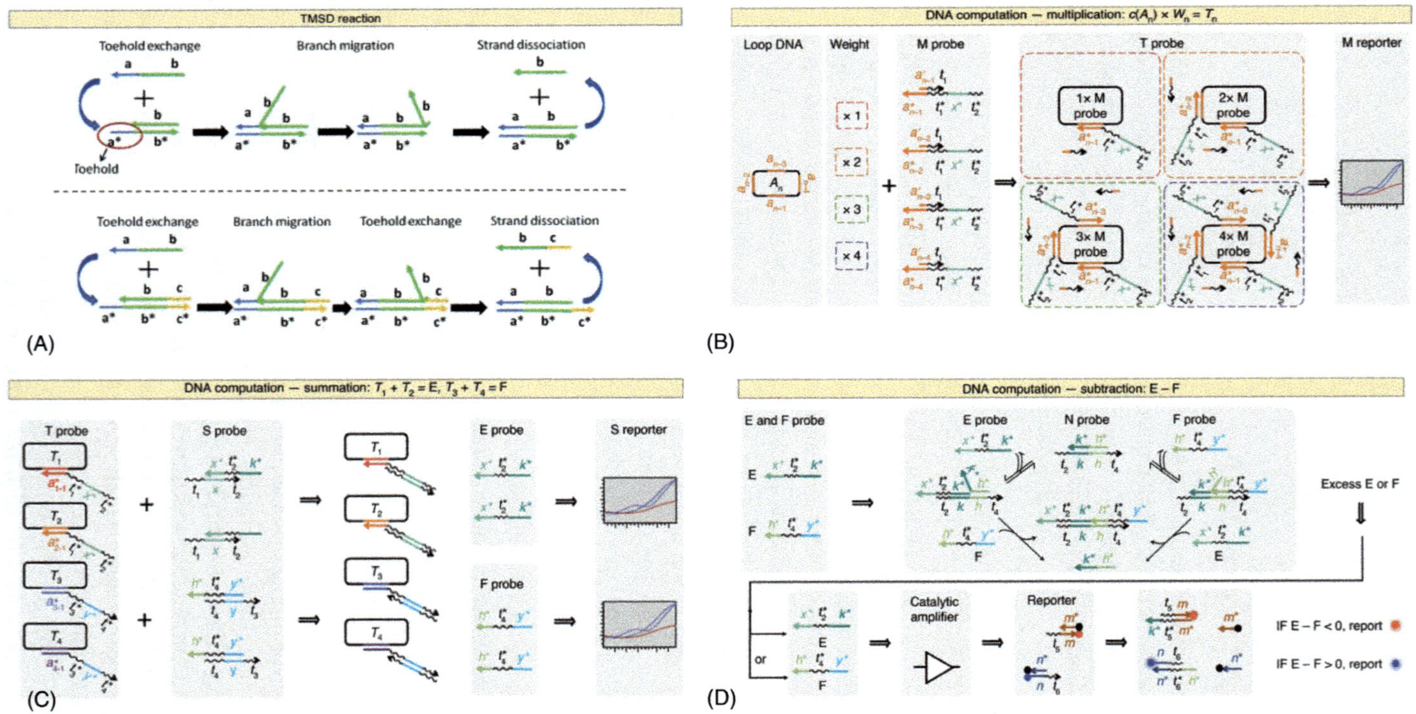

Figure 9.3 DNA computation based on TMSD reactions. (A) Scheme of TMSD reactions. Source: Reproduced from Han et al. [48], with permission of John Wiley & Sons. (B) Scheme for multiplication based on TMSD reactions. (C) Scheme for summation based on TMSD reactions. (D) Scheme for subtraction based on TMSD reactions. Source: Reproduced from Zhang et al. [49], with permission of Springer Nature.

an ssDNA can trigger the hybridization of two partially complementary hairpins. The ssDNA can be recycled post the full hybridization of the hairpins, therefore catalyzing multiple rounds of dsDNA formation by hairpins. HCR is an enzyme-free amplification process involving two complementary hairpins (termed HP1 and HP2) that are unable to hybridize spontaneously. The polymerization of two hairpins can be induced by a short ssDNA as the trigger, and a long dsDNA with multiple strands of hairpins can be formed as the products [59]. Chen and coworkers developed logical DNA devices based on cascaded HCR and CHA to execute typical logic operations such as OR, YES, AND, INHIBIT, NOR, and NAND [60]. Xu and coworkers constructed OR and INHIBIT logic gates based on CHA reactions. Under the cascading INH-OR control, concatenated logic circuit was built with multiple miRNAs as inputs [61].

9.2 DNA-Based Computing Devices for Biosensing

With the capability to convert molecular recognition into detectable signals, DNA-based computing devices exhibit widespread applications in smart biosensing. For instance, thanks to the good parallelism and flexibility of DNA computing, multiple targets can be analyzed simultaneously, and the information of those targets can be output in a programmable manner [62]. According to different application scenarios, we divided the biosensing platforms for biomedical applications into *in vitro* and cellular biosensing.

9.2.1 *In Vitro* Biosensing

Biomarker sensing *in vitro* is a convenient approach and of great significance for disease diagnosis. The inputs of DNA-based computing systems can be DNA, RNA, enzymes, and proteins, which cover diverse species of important biomarkers. By encoding this biomarker-related information (i.e. variety and concentration) into the DNA-computing biosensing platform, the outputs resulting from complex detected information can be reported in a programmable and summarized way by signals such as fluorescent, electrochemical, and colorimetric readout. Consequently, DNA-based computing devices can be widely applied for diagnosis in a rapid, intelligent, and comprehensive manner.

Fluorescent biosensors based on DNA logic gates can be constructed by using fluorescent dyes and labels on DNA strands. To realize simultaneous detection of miRNA 21 and let-7a for precise diagnosis of diseases, Zhou and coworkers designed a DNA AND logic gate based on TMSD and rolling circle amplification (RCA). In the presence of those two miRNAs, the primer of the circular DNA amplification template can be released via TMSD to initiate RCA reactions, and the product of RCA can bind with numerous fluorescent dyes and generate an enhanced fluorescence for dual-target detection [63].

Electrochemical DNA-based logic devices can be developed by modifying the electrode surface with DNA probes labeled with an electrochemical reporter [64, 65].

For instance, Yu and coworkers designed a DNA nanomachine consisting of a DNA walker and tandem strand displacement reactions for DNA-based electrochemical sensing (Figure 9.4A) [66]. In the presence of target DNAs, the DNA walker was freed to trigger the nicking of methylene blue (MB)-labeled DNA probes on the electrode surface, which led to a decreased electrochemical response. The nanomachine was used to construct AND and OR logic gates and exhibited the potential for further applications in clinical diagnosis. In addition, since dsDNA can load various small molecules with electrochemical activity, such as MB [70], $[Ru(NH_3)^6]^{3+}$ [71], doxorubicin hydrochloride [72], and $[(ppy)_2Ir(dppz)]^+PF6^-$, they can be used to build logic gates and report signals by electrochemical readout [73]. For instance, Li and coworkers established the example of an electrochemical DNA logic gate with G4 structures that can load MB to form a G4-MB complex and inhibit the electrochemical diffusion current of MB for generating a sensitive response. Different logic gates including AND, NOR, INHIBIT, IMPLICATION, XOR, and XNOR were established and demonstrated correctly. This study presented the versatility and universality of DNA for constructing an electrochemical platform for biological information processing [74].

Colorimetric logic device-based biosensing platforms were established with DNAzymes that catalyzed chemical and colorimetric reactions. The color change visible to the naked eye makes the colorimetric logic operation have the advantages of simplicity, low cost, and no need for a special instrument. The flexible assembly of DNAzymes with nucleic acid sequences enables the detection of various targets with this design. For instance, Xu and coworkers established a logic gate-based colorimetric biosensor for the detection of DNA by utilizing the HRP-mimicking DNAzyme (G4-hemin) [75]. The dissociative DNAzyme can respond to different concentrations of DNA targets to catalyze the oxidation of 2,2′-azinobis(3-ethylbenzothiazoline)-6-sulfonic acid ($ABTS^{2-}$) and generate a green-colored product that can be differentiated by either a hand-held UV spectrometer or naked eyes, therefore can be applied in point-of-care diagnosis. In addition to colorimetric reactions catalyzed by DNAzymes, spherical nucleic acid (SNA) consists of DNA-modified gold nanoparticles (AuNPs), is often utilized as colorimetric nanomaterials due to their distinct optical properties. Liang and coworkers reported a colorimetric biosensing platform based on DNA circuits involving SNA and TMSD reactions (Figure 9.4B) [67]. The output strand of the upstream TMSD-based DNA circuit served as a trigger to induce the visible SNA aggregate disassembly. By designing two parallel DNA circuits to trigger an SNA disassembly system, an OR gate was constructed for the detection of multiple targets.

By designing the logic gates skillfully, a dual readout biosensing platform can be established. Dong and coworkers constructed the upconversion luminescent (UCL) DNA logic library with dual-modal label-free outputs (both visible lights and luminescence, pale-blue and dark-blue colors were defined as 0 and 1, strong and weak UCL were defined as 0 and 1, respectively) by applying the luminescence of upconversion nanoparticles (UCNPs) and the catalytic capability of G4 DNAzyme toward TMB (3,3′,5,5′-tetramethylbenzidine) oxidation (Figure 9.4C) [68]. When the input DNA strands formed G4, the dark-blue-colored OxTMB (oxidized TMB)

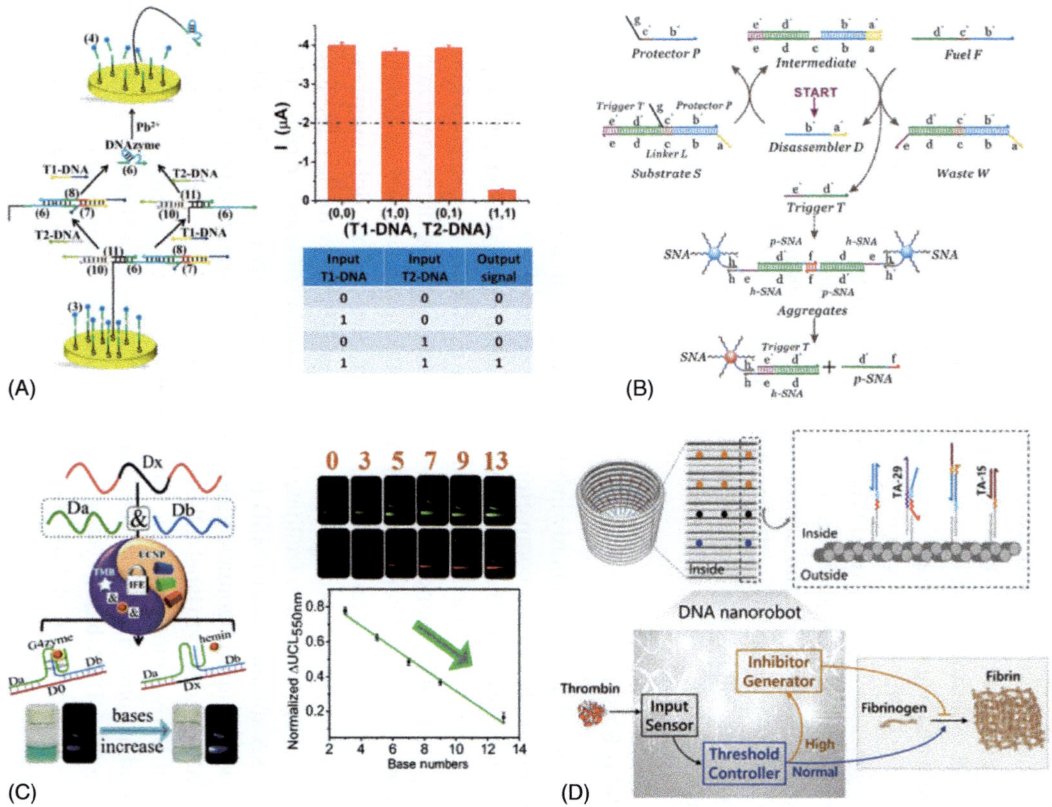

Figure 9.4 DNA computing devices for biosensing implemented *in vitro*. (A) Electrochemical DNA nanomachine based on DNA walker and cascaded TMSD for detection of multiple DNAs. Source: Reproduced from Wang et al. [66], with permission of © 2018 Elsevier. (B) Colorimetric biosensing platform based on visible disassembly of SNA programmed by catalytic DNA circuits. Source: Reproduced from Wei et al. [67], with permission of © 2019 American Chemical Society. (C) Multicolor luminescent DNA logic library and its biosensing application. Source: Reproduced from Fan et al. [68], with permission of John Wiley & Sons. (D) DNA computing device-based nanorobot for intelligently sensing and regulating thrombin. Source: Reprinted from Yang et al. [69], with permission of Wiley.

was produced to quench the UCL emission of UCNPs via inner-filter-effect (IFE). When the input DNA strands did not form G4, the pale-blue TMB cannot quench the UCL emission of UCNPs. Based on the constructed computing devices, they designed a DNA caliper for base-number measurement and obtained a good linear relationship between the UCL peak intensity and the base numbers.

The real-time sensing and regulation of biomarkers are also important for the diagnosis and treatment of disease. Based on a logical circuit with accurate threshold control, Han and coworkers constructed an intelligent DNA nanorobot, which consisted of a barrel-shaped DNA origami (termed DNA barrel) as the framework, and embedded DNA molecular reaction cascades as the computing core to intelligently sense thrombin and regulate thrombin function (Figure 9.4D) [69]. The DNA barrel contained four dsDNAs inside, which served as input sensor, threshold controller, output, and generator, respectively. When the thrombin in human plasma was higher than the threshold value, the downstream cascaded DNA displacement reactions were triggered, and the nanostructure would release inhibitors (TA-15 strand, an aptamer of thrombin) to maintain thrombin at a normal level. This smart platform has great potential to be applied in real-time blood clotting control for clinical applications.

9.2.2 Cellular Biosensing

Biosensing for cellular targets mainly involves the detection of biomarkers on cell membranes and inside cellular compartments. The accurate recognition of cancer cells is significant for cancer diagnosis and intervention. To realize precise targeting of interested cells, multiple aptamer-based logic gates have been applied in the selective recognition of biomarkers on cell membranes to realize high-selective cell labeling and precise drug delivery [34, 76, 77]. In 2014, Tan and coworkers designed a DNA logic robot termed Nano-Claw for logical analysis of multiple membrane markers followed by the autonomous release of drugs [78]. They further developed a general computing platform to facilitate the identification of multiple cell-surface markers via DNA aptamers and TMSD-based logic gates (Figure 9.5a) [79]. Different aptamers with unique oligonucleotide tags can bind to biomarkers on target cells and trigger TMSD reactions to label cells in a predesigned logical manner. Combined with structural DNA nanotechnology, Tan and coworkers engineered a 3D DNA logic gate-based nanomachine for selective labeling of target cells (Figure 9.5b) [80]. Compared to a 2D dsDNA-based system, this 3D DNA nanomachine allowed more volume for the incorporation of different logic units, generating improved accuracy of cell identification.

To enhance the sensitivity of cell isolation, Han and coworkers combined a multiple-aptamer-based AND logic device with HCR amplification to realize selective and sensitive identification for specific cell subtypes (Figure 9.5c). In the simultaneous presence of aptamer-targeting receptors on cell membranes, effective HCR-based amplification can be triggered to amplify the labeling signals [81]. Afterward, this multiple-aptamer-based AND logic device was combined with rolling cycle amplification (RCA) to realize single-step cancer cell isolation with

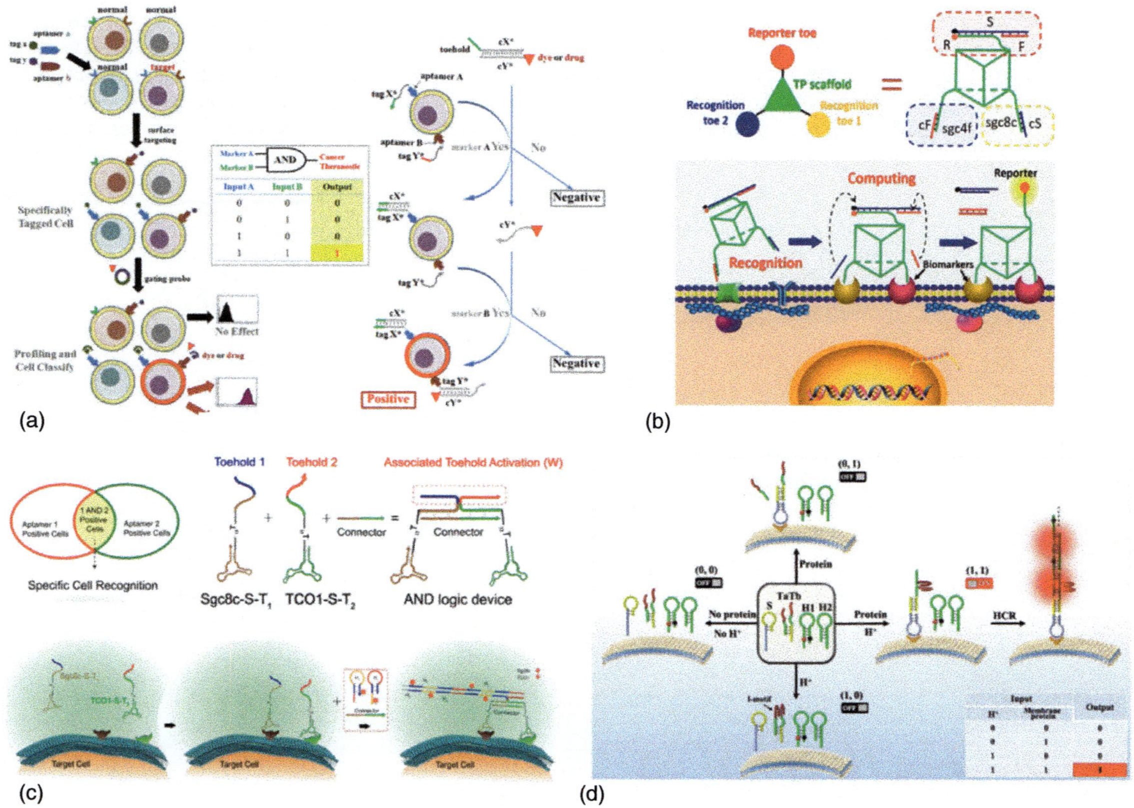

Figure 9.5 DNA computing devices for biosensing implemented on the cell membrane. (a) Multiparameter-based DNA logic platform for selective cell imaging and precise therapeutic application. Source: Reproduced from You et al. [79], with permission of © 2014 American Chemical Society. (b) 3D DNA logic gates nanomachine for selective imaging of target cell. Source: Reproduced from Peng et al. [80], with permission of © 2018 American Chemical Society. (c) Multiple-aptamer-based AND logic device combined with HCR amplification for selective imaging of interested biomarkers on the cell membrane. Source: Reproduced from Chang et al. [81], with permission of © 2019 American Chemical Society. (d) Aptamer and i-motif-based DNA logic gate for accurate and sensitive imaging of cancer cells. Source: Reproduced from Chen et al. [82], with permission of © 2022 American Chemical Society.

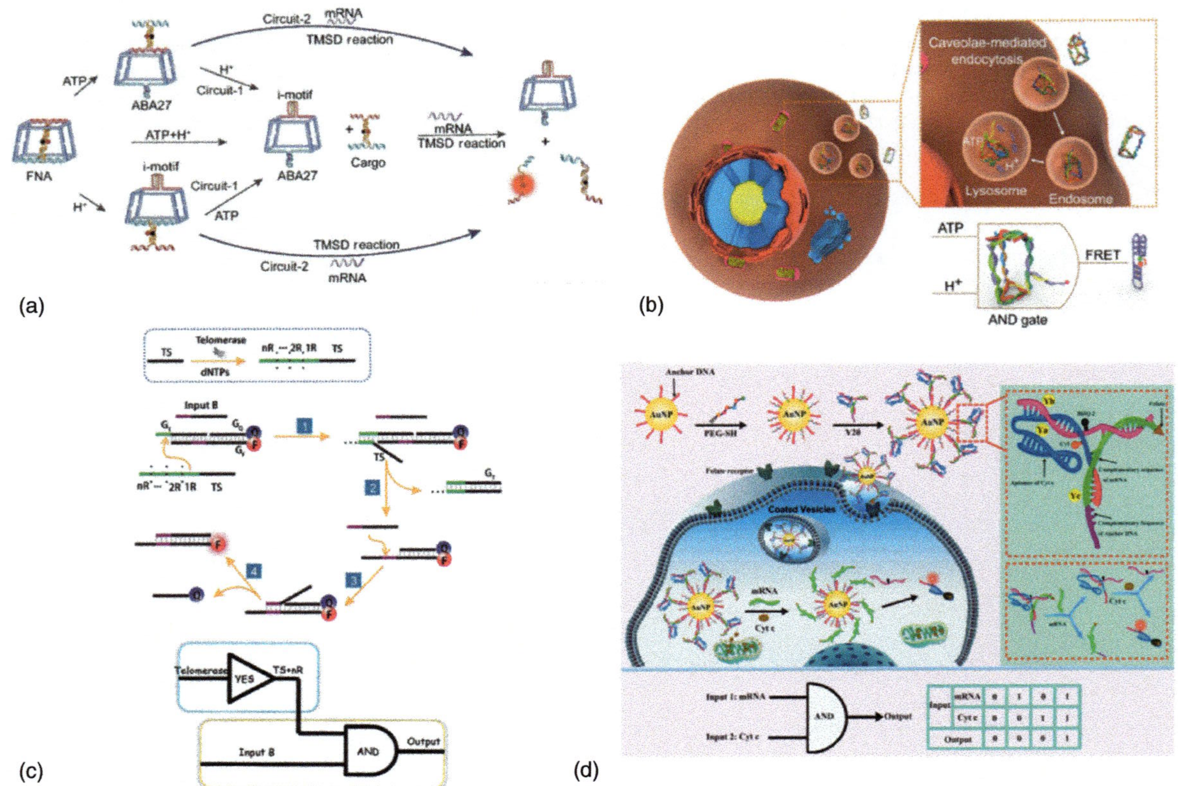

Figure 9.6 DNA computing devices for biosensing implemented within living cells. (a) Imaging of mRNA using i-motif and ATP aptamer-triggered DNA-computation circuit. Source: Reproduced from Wang et al. [85], with permission of John Wiley & Sons. (b) Subcellular imaging using i-motif and ATP-binding aptamer-triggered AND gate. Source: Reproduced from Du et al. [86], with permission of © 2019 American Chemical Society. (c) Imaging of intracellular telomerase activity through cascade DNA logic gates. Source: Wang et al. [87]/Royal Society of Chemistry/CC BY 3.0. (d) Imaging of an apoptotic signaling pathway using a dual-responsive DNA nanodevice. Source: Reproduced from Shi et al. [88], with permission of © 2019 American Chemical Society.

higher sensitivity [83]. Similarly, Wang and coworkers combined HCR amplification with an AND logic device for accurate and sensitive identification of cancer cells by utilizing a split i-motif and a DNA aptamer as recognition elements with high sensitivity and selectivity (Figure 9.5d) [82].

Except for analyzing biomarkers on the surface of the interested cell, DNA logic devices have also been used for the analysis of intracellular biomarkers. Logic gates for imaging endogenous miRNAs in live mammalian cells have been developed since 2013 [84]. Li and coworkers prepared a truncated square pyramid (TSP) cage with an ATP aptamer and an i-motif on its two edges (Figure 9.6a) [85]. Combined with two cascaded gates (OR–AND and AND–AND), the TSP served as logic-controlling units to respond to H^+ and ATP inside cellular compartments and triggered the release of sensing elements for targeted mRNA imaging. In addition to targeting specific types of cells, the DNA logic circuit also enabled the targeting of specific organelles inside the living cells. For instance, Li and coworkers used a framework nucleic acid (FNA) nanodevice to image the lysosomal compartment in cells upon the trigger of ATP and acid PH (Figure 9.6b) [86]. Zhu and coworkers developed a DNA computing approach to image intracellular telomerase activity (Figure 9.6c) [87]. Li and coworkers reported a dual-responsive DNA nanodevice that allowed imaging of an apoptotic signaling pathway in living cells (Figure 9.6d) [88]. The apoptotic signals of manganese superoxide dismutase mRNA and downstream cytochrome c (Cyt c) served as the inputs of the AND logic gate and induced the disassembly of AuNPs for fluorescence reporting. This programmable strategy has the potential to study other signaling pathways and facilitate the advancement of cell biology.

9.3 DNA Computing for Clinical Diagnosis

The combination of DNA computing and bioinformatics methods is used for the screening of biomarkers and the diagnosis of diseases. For instance, DNA computing has applications in the classification of genes, the concentrations and variabilities of which are directly associated with different diseases. In addition, since the detection of a single biomarker is insufficient to give accurate diagnostic information for defined diseases, the molecular computing approach can convert multiple biomarkers into interpretable yes or no answers; therefore, they have great potential for future intelligent and personalized disease diagnosis. The flexible programmability and vast parallelism enable DNA computing to implement more complex algorithms such as addition, subtraction, and multiplication for classification based on multiple markers [49]. Synthetic DNA can also be used to implement neural network algorithms to perform more complex classification, which functions like the brain using neurons to collect information and output results.

Since gene expression changes are associated with a lot of human diseases, Seelig and coworkers reported a molecular multigene classifier for disease diagnostics (Figure 9.7a) [89]. Firstly, the linear computational classifier built using support vector machines (SVMs) was trained by labeled gene expression data *in silico*, and the optimal model with a series of parameters and mathematical functions

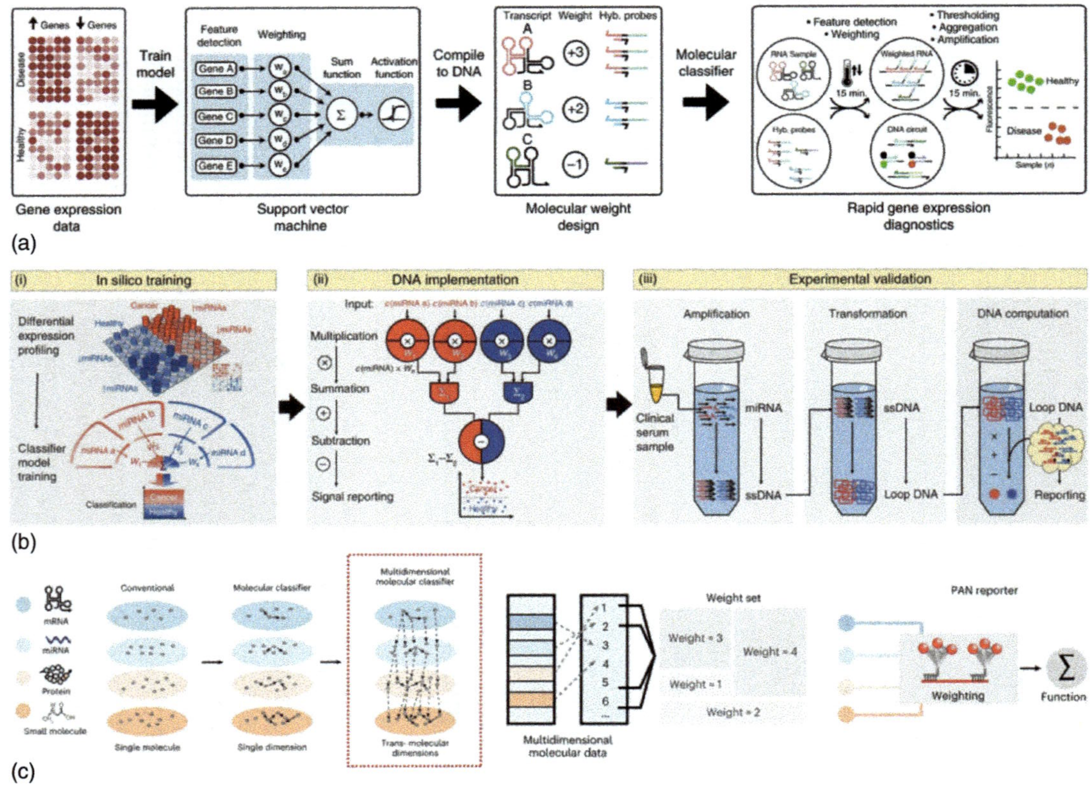

Figure 9.7 DNA computing for intelligent clinical diagnosis. (a) The workflow of the DNA molecular classifier for diagnosing viral and bacterial respiratory infections. Source: Reproduced from Lopez et al. [89], with permission of Springer Nature. (b) The workflow of the DNA molecular classifier analyzing miRNA profiles in clinical serum samples for NSCLC diagnosis. Source: Reproduced from Zhang et al. [49], with permission of Springer Nature. (c) The construction of the DNA-framework-based multidimensional molecular classifiers for prostate cancer diagnosis. Source: Reproduced from Yin [90], with permission of Springer Nature.

was obtained. Then, the classifier was implemented at the molecular level by translating these parameters and functions into a series of molecular interactions between RNA inputs and engineered DNA probes with different weights according to their importance. Finally, the linear classifier was applied in the construction of classifiers for early cancer diagnostics and differentiating viral and bacterial respiratory infections. This molecular computation strategy presented a demonstration of translating *in silico* gene expression classifier into DNA circuitry, with the advantages of no need for an instrument, experimental simplicity, and low cost over classical approaches for high-throughput RNA detection such as qPCR, microarrays, or sequencing. However, due to the low concentration of RNA transcripts in tissue and blood samples, this linear classifier cannot be used for real sample analysis.

To solve this problem, Han and coworkers displayed a classifier to analyze miRNA profiles in clinical serum samples for non-small cell lung cancer (NSCLC) diagnosis (Figure 9.7b) [49]. Similarly, the classifier was trained *in silico* using publicly available miRNA-seq profiles of healthy and NSCLC individuals from The Cancer Genome Atlas with an SVM-based algorithm. Then, a close-to-linear amplification was introduced for the translated molecular interactions to enhance the sensitivity and realize real sample analysis. This close-to-linear amplification was based on linear after the exponential PCR (LATE-PCR), which could maintain the initial ratios of each miRNA input. That means the amplified sample still possesses the miRNA quantity information and will not affect the downstream classification analysis of the original sample. Finally, the classifier achieved rapid and accurate cancer diagnosis using 22 clinical serum samples (8 from healthy people and 14 from people with lung cancer) with an accuracy of 86.4%. This DNA computational platform inspires more clinical applications toward noninvasive disease screening and classification. Afterward, they developed a fully automated DNA computing-based platform for accurate diagnosis of acute respiratory infection (ARI) etiology in four hours with a diagnostic accuracy of 87% (in 80 clinical samples). This automated diagnostic platform achieved a "sample in and result out" flow without the aid of any computer or laboratory specialists [91].

In addition, Han and coworkers presented a DNA computational scheme with switching circuits (SCs) based on TMSD reactions for sensing single-nucleotide-polymorphisms (SNP) variants and detecting SNPs information in clinical samples. Using this programmable DNA molecular computational platform, the automatic and accurate identification of 21 different blood group genotypes from 83 clinical blood samples with 100 % accuracy was achieved in three hours [92]. This strategy can be applied in clinical diagnosis relying on single nucleotide mutations. They further presented a programmable DNA decoding circuit with a signal-transforming approach based on thermostable ligase for genotype-phenotype interpretation. It demonstrated favorable performance in interpreting CYP2C19 genetic profiles into drug responses in 30 human genomic samples with a comparable accuracy of Sanger sequencing [93]. This DNA-powered genetic decoder provides a demonstration to translate complex genetic profiles into actionable healthcare decisions, which can be reconfigured for personalized medication and cancer risk assessment.

Since the existing classifiers coupled with DNA-reaction-based molecular implementation cannot be applied in the analysis of proteins or metabolic small molecules, Zuo and coworkers developed a DNA-encoded molecular classifier that can implement the computational classification of multidimensional molecular clinical data (Figure 9.7c) [90]. The multidimensional molecular classifier for analyzing RNA, proteins, and small molecules was realized by constructing valence-encoded programmable atom-like nanostructure (PAN) reporters with different weights (different numbers of reporting molecules on a PAN). In the presence of the target, the recognition binding event between the target and the probe (ssDNA probe for RNA, antibody for protein, aptamer for small molecule) led to the binding of PAN reporters on the electrochemical chip and triggered a weighted electrochemical signal. The classifier realized precise prostate cancer diagnosis with six biomarkers across three-dimensional datatypes. This study demonstrated the development of multidimensional molecular classifiers to analyze heterogeneous molecular biomarkers for precision diagnosis and therapy.

9.4 Conclusion

DNA molecular computation has been regarded as a powerful tool for biosensing and clinical diagnosis, which can be applied in multiple targets analysis, point-of-care testing (POCT), and biological information processing. Despite multiple advantages, DNA computing for smart biosensing is still facing several challenges. Firstly, the power of DNA-based computing is still in its infancy and hardly competes with the current silicon-based computing scheme. This may limit its further applications in the diagnosis of diseases with complex models. Secondly, signal leakage cannot be avoided inevitably, even in an optimum design, especially for cascading reactions. Thirdly, since a defined DNA computing system is typically only able to be applied to a certain application scenario, a more universal scheme needs to be developed. Although there are some challenges, those bottlenecks will be addressed ultimately with the rapid development of DNA nanotechnology, which will make DNA computation more powerful for human health monitoring.

References

1 Seeman, N.C. and Kallenbach, N.R. (1983). Design of immobile nucleic acid junctions. *Biophys. J.* 44 (2): 201–209.

2 Fu, T.J. and Seeman, N.C. (1993). DNA double-crossover molecules. *Biochemistry* 32 (13): 3211–3220.

3 Adleman, L.M. (1994). Molecular computation of solutions to combinatorial problems. *Science* 266 (5187): 1021–1024.

4 Lipton, R.J. (1995). DNA solution of hard computational problems. *Science* 268 (5210): 542–545.

5 Ouyang, Q., Kaplan, P.D., Liu, S. et al. (1997). DNA solution of the maximal clique problem. *Science* 278 (5337): 446–449.

6 Liu, Q., Wang, L., Frutos, A.G. et al. (2000). DNA computing on surfaces. *Nature* 403 (6766): 175–179.

7 Frutos, A.G., Liu, Q., Thiel, A.J. et al. (1997). Demonstration of a word design strategy for DNA computing on surfaces. *Nucleic Acids Res.* 25 (23): 4748–4757.

8 Fan, D., Wang, J., Wang, E. et al. (2020). Propelling DNA computing with materials' power: recent advancements in innovative DNA logic computing systems and smart bio-applications. *Adv. Sci.* 7 (24): 2001766.

9 Saghatelian, A., Völcker, N.H., Guckian, K.M. et al. (2003). DNA-based photonic logic gates: AND, NAND, and INHIBIT. *J. Am. Chem. Soc.* 125 (2): 346–347.

10 Okamoto, A., Tanaka, K., and Saito, I. (2004). DNA logic gates. *J. Am. Chem. Soc.* 126 (30): 9458–9463.

11 Breaker, R.R. and Joyce, G.F. (1994). A DNA enzyme that cleaves RNA. *Chem. Biol.* 1 (4): 223–229.

12 Chandra, M., Sachdeva, A., and Silverman, S.K. (2009). DNA-catalyzed sequence-specific hydrolysis of DNA. *Nat. Chem. Biol.* 5 (10): 718–720.

13 Flynn-Charlebois, A., Wang, Y., Prior, T.K. et al. (2003). Deoxyribozymes with 2′−5′ RNA ligase activity. *J. Am. Chem. Soc.* 125 (9): 2444–2454.

14 Li, Y. and Breaker, R.R. (1999). Phosphorylating DNA with DNA. *Proc. Natl. Acad. Sci.* 96 (6): 2746–2751.

15 Cuenoud, B. and Szostak, J.W. (1995). A DNA metalloenzyme with DNA ligase activity. *Nature* 375 (6532): 611–614.

16 Ward, W.L., Plakos, K., and DeRose, V.J. (2014). Nucleic acid catalysis: metals, nucleobases, and other cofactors. *Chem. Rev.* 114 (8): 4318–4342.

17 Stojanovic, M.N., Mitchell, T.E., and Stefanovic, D. (2002). Deoxyribozyme-based logic gates. *J. Am. Chem. Soc.* 124 (14): 3555–3561.

18 Feng, L., Lyu, Z., Offenhäusser, A. et al. (2015). Multi-level logic gate operation based on amplified aptasensor performance. *Angew. Chem. Int. Ed.* 54 (26): 7693–7697.

19 Xu, L., Hong, S., Sun, N. et al. (2016). Berberine as a novel light-up i-motif fluorescence ligand and its application in designing molecular logic systems. *Chem. Commun.* 52 (1): 179–182.

20 Li, T., Wang, E., and Dong, S. (2009). Potassium−lead-switched G-quadruplexes: a new class of DNA logic gates. *J. Am. Chem. Soc.* 131 (42): 15082–15083.

21 Stojanović, M.N. and Stefanović, D. (2003). Deoxyribozyme-based half-adder. *J. Am. Chem. Soc.* 125 (22): 6673–6676.

22 Stojanovic, M.N. and Stefanovic, D. (2003). A deoxyribozyme-based molecular automaton. *Nat. Biotechnol.* 21 (9): 1069–1074.

23 Chen, X., Wang, Y.F., Liu, Q. et al. (2006). Construction of molecular logic gates with a DNA-cleaving deoxyribozyme. *Ange. Chem. Int. Ed.* 45 (11): 1759–1762.

24 Elbaz, J., Lioubashevski, O., Wang, F. et al. (2010). DNA computing circuits using libraries of DNAzyme subunits. *Nat. Nanotechnol.* 5 (6): 417–422.

25 Elbaz, J., Wang, F., Remacle, F. et al. (2012). pH-Programmable DNA logic arrays powered by modular DNAzyme libraries. *Nano Lett.* 12 (12): 6049–6054.

26 Sun, H., Zhu, X., Lu, P.Y. et al. (2014). Oligonucleotide aptamers: new tools for targeted cancer therapy. *Mol. Ther. Nucl. Acids* 3: e182.

27 Hamaguchi, N., Ellington, A., and Stanton, M. (2001). Aptamer beacons for the direct detection of proteins. *Anal. Biochem.* 294 (2): 126–131.

28 Douglas, S.M., Bachelet, I., and Church, G.M. (2012). A logic-gated nanorobot for targeted transport of molecular payloads. *Science* 335 (6070): 831–834.

29 Krupenkin, T. and Taylor, J.A. (2011). Reverse electrowetting as a new approach to high-power energy harvesting. *Nat. Commun.* 2 (1): 448.

30 Tian, T., Li, Y., and Lin, Y. (2022). Prospects and challenges of dynamic DNA nanostructures in biomedical applications. *Bone Res.* 10 (1): 40.

31 Amir, Y., Ben-Ishay, E., Levner, D. et al. (2014). Universal computing by DNA origami robots in a living animal. *Nat. Nanotechnol.* 9 (5): 353–357.

32 Jose, A.M., Soukup, G.A., and Breaker, R.R. (2001). Cooperative binding of effectors by an allosteric ribozyme. *Nucleic Acids Res.* 29 (7): 1631–1637.

33 Kolpashchikov, D.M. and Stojanovic, M.N. (2005). Boolean control of aptamer binding states. *J. Am. Chem. Soc.* 127 (32): 11348–11351.

34 Han, D., Zhu, Z., Wu, C. et al. (2012). A logical molecular circuit for programmable and autonomous regulation of protein activity using DNA aptamer–protein interactions. *J. Am. Chem. Soc.* 134 (51): 20797–20804.

35 Andrianova, M. and Kuznetsov, A. (2020). Logic gates based on DNA aptamers. *Pharmaceuticals* 13 (11): 417.

36 Mariottini, D., Idili, A., Vallée-Bélisle, A. et al. (2017). A DNA nanodevice that loads and releases a cargo with hemoglobin-like allosteric control and cooperativity. *Nano Lett.* 17 (5): 3225–3230.

37 Majikes, J.M., Ferraz, L.C.C., and LaBean, T.H. (2017). pH-Driven actuation of DNA origami via parallel I-motif sequences in solution and on surfaces. *Bioconjugate Chem.* 28 (7): 1821–1825.

38 Shi, Y., Sun, H., Xiang, J. et al. (2015). i-Motif-modulated fluorescence detection of silver(I) with an ultrahigh specificity. *Anal. Chim. Acta* 857: 79–84.

39 Huang, D., Yang, C., Yao, Y. et al. (2019). Versatile and homogeneous DNA tetraplex platform for constructing label-free logic devices: from design to application. *Chem. A Eur. J.* 25 (28): 6996–7003.

40 Day, H.A., Huguin, C., and Waller, Z.A.E. (2013). Silver cations fold i-motif at neutral pH. *Chem. Commun.* 49 (70): 7696–7698.

41 Kang, B.H., Gao, Z.F., Li, N. et al. (2016). Thiazole orange as a fluorescent probe: Label-free and selective detection of silver ions based on the structural change of i-motif DNA at neutral pH. *Talanta* 156: 141–146.

42 Lan, Y., Wei, Y., Wei, Y. et al. (2022). Versatile triple-output molecular logic gate for cysteine and silver (I) in foods and the environment based on i-motif DNA modulation. *J. Agric. Food. Chem.* 70 (12): 3608–3617.

43 Guo, Y., Yao, W., Xie, Y. et al. (2016). Logic gates based on G-quadruplexes: principles and sensor applications. *Microchim. Acta* 183 (1): 21–34.

44 Zhu, J., Zhang, L., Li, T. et al. (2013). Enzyme-free unlabeled DNA logic circuits based on toehold-mediated strand displacement and split G-quadruplex enhanced fluorescence. *Adv. Mater.* 25 (17): 2440–2444.

45 Travascio, P., Bennet, A.J., Wang, D.Y. et al. (1999). A ribozyme and a catalytic DNA with peroxidase activity: active sites versus cofactor-binding sites. *Chem. Biol.* 6 (11): 779–787.

46 Liu, S., Ding, J., and Qin, W. (2019). Dual-analyte chronopotentiometric aptasensing platform based on a G-quadruplex/hemin DNAzyme and logic gate operations. *Anal. Chem.* 91 (4): 3170–3176.

47 Yurke, B., Turberfield, A.J., Mills, A.P. et al. (2000). A DNA-fuelled molecular machine made of DNA. *Nature* 406 (6796): 605–608.

48 Han, D., Kang, H., Zhang, T. et al. (2014). Nucleic acid based logical systems. *Chem. A Eur. J.* 20 (20): 5866–5873.

49 Zhang, C., Zhao, Y., Xu, X. et al. (2020). Cancer diagnosis with DNA molecular computation. *Nat. Nanotechnol.* 15 (8): 709–715.

50 Shin, J.-S. and Pierce, N.A. (2004). A synthetic DNA walker for molecular transport. *J. Am. Chem. Soc.* 126 (35): 10834–10835.

51 Pan, J., Li, F., Cha, T.-G. et al. (2015). Recent progress on DNA based walkers. *Curr. Opin. Biotechnol.* 34: 56–64.

52 Qian, L. and Winfree, E. (2011). Scaling up digital circuit computation with DNA strand displacement cascades. *Science* 332 (6034): 1196–1201.

53 Luo, T., Fan, S., Liu, Y. et al. (2021). Information processing based on DNA toehold-mediated strand displacement (TMSD) reaction. *Nanoscale* 13 (4): 2100–2112.

54 Seelig, G., Soloveichik, D., Zhang, D.Y. et al. (2006). Enzyme-free nucleic acid logic circuits. *Science* 314 (5805): 1585–1588.

55 Li, W., Zhang, F., Yan, H. et al. (2016). DNA based arithmetic function: a half adder based on DNA strand displacement. *Nanoscale* 8 (6): 3775–3784.

56 Cherry, K.M. and Qian, L. (2018). Scaling up molecular pattern recognition with DNA-based winner-take-all neural networks. *Nature* 559 (7714): 370–376.

57 Song, T., Garg, S., Mokhtar, R. et al. (2016). Analog computation by DNA strand displacement circuits. *ACS Synth. Biol.* 5 (8): 898–912.

58 Yin, P., Choi, H.M.T., Calvert, C.R. et al. (2008). Programming biomolecular self-assembly pathways. *Nature* 451 (7176): 318–322.

59 Evanko, D. (2004). Hybridization chain reaction. *Nat. Methods* 1 (3): 186–187.

60 Chai, Q., Chen, J., Zeng, S. et al. Closed cyclic DNA machine for sensitive logic operation and APE1 detection. *Small* 1 (3): 2207736.

61 Zhang, S., Li, K.-B., Shi, W. et al. (2019). Resettable and enzyme-free molecular logic devices for the intelligent amplification detection of multiple miRNAs via catalyzed hairpin assembly. *Nanoscale* 11 (11): 5048–5057.

62 Cao, M., Xiong, X., Zhu, Y. et al. (2023). DNA computational device-based smart biosensors. *TrAC, Trends Anal. Chem.* 159: 116911.

63 Chen, Y., Song, Y., Wu, F. et al. (2015). A DNA logic gate based on strand displacement reaction and rolling circle amplification, responding to multiple low-abundance DNA fragment input signals, and its application in detecting miRNAs. *Chem. Commun.* 51 (32): 6980–6983.

64 Yu, S., Wang, Y., Jiang, L.-P. et al. (2018). Cascade amplification-mediated in situ hot-spot assembly for MicroRNA detection and molecular logic gate operations. *Anal. Chem.* 90 (7): 4544–4551.

65 Xia, F., Zuo, X., Yang, R. et al. (2010). Label-free, dual-analyte electrochemical biosensors: a new class of molecular-electronic logic gates. *J. Am. Chem. Soc.* 132 (25): 8557–8559.

66 Wang, K., He, M.-Q., Zhai, F.-H. et al. (2018). Autonomous DNA nanomachine based on cascade amplification of strand displacement and DNA walker for detection of multiple DNAs. *Biosens. Bioelectron.* 105: 159–165.

67 Wei, B., Yao, D., Zheng, B. et al. (2019). Facile strategy for visible disassembly of spherical nucleic acids programmed by catalytic DNA circuits. *ACS Appl. Mater. Interfaces* 11 (22): 19724–19733.

68 Fan, D., Wang, E., and Dong, S. (2019). Upconversion-chameleon-driven DNA computing: the DNA-unlocked inner-filter-effect (DU-IFE) for operating a multicolor upconversion luminescent DNA logic library and Its biosensing application. *Mater. Horiz.* 6 (2): 375–384.

69 Yang, L., Zhao, Y., Xu, X. et al. (2020). An intelligent DNA nanorobot for autonomous anticoagulation. *Ange. Chem. Int. Ed.* 59 (40): 17697–17704.

70 Guo, J., Wang, J., Zhao, J. et al. (2016). Ultrasensitive multiplexed immunoassay for tumor biomarkers based on DNA hybridization chain reaction amplifying signal. *ACS Appl. Mater. Interfaces* 8 (11): 6898–6904.

71 Jia, L.-P., Wang, L.-J., Ma, R.-N. et al. (2018). An electrochemical aptasensor for the highly sensitive detection of 8-hydroxy-2′-deoxyguanosine based on the hybridization chain reaction. *Talanta* 179: 414–419.

72 Song, C., Xie, G., Wang, L. et al. (2014). DNA-based hybridization chain reaction for an ultrasensitive cancer marker EBNA-1 electrochemical immunosensor. *Biosens. Bioelectron.* 58: 68–74.

73 Li, C., Wang, H., Shen, J. et al. (2015). Cyclometalated iridium complex-based label-free photoelectrochemical biosensor for DNA detection by hybridization chain reaction amplification. *Anal. Chem.* 87 (8): 4283–4291.

74 Ge, L., Wang, W., Sun, X. et al. (2016). Versatile and programmable DNA logic gates on universal and label-free homogeneous electrochemical platform. *Anal. Chem.* 88 (19): 9691–9698.

75 Wang, C., Cheng, N., Zhu, L. et al. (2017). Colorimetric biosensor based on a DNAzyme primer and its application in logic gate operations for DNA screening. *Anal. Chim. Acta* 987: 111–117.

76 Zhang, Y., Chen, W., Fang, Y. et al. (2021). Activating a DNA nanomachine via computation across cancer cell membranes for precise therapy of solid tumors. *J. Am. Chem. Soc.* 143 (37): 15233–15242.

77 Han, D., Zhu, G., Wu, C. et al. (2013). Engineering a cell-surface aptamer circuit for targeted and amplified photodynamic cancer therapy. *ACS Nano* 7 (3): 2312–2319.

78 You, M., Peng, L., Shao, N. et al. (2014). DNA "Nano-Claw": logic-based autonomous cancer targeting and therapy. *J. Am. Chem. Soc.* 136 (4): 1256–1259.

79 You, M., Zhu, G., Chen, T. et al. (2015). Programmable and multiparameter DNA-based logic platform for cancer recognition and targeted therapy. *J. Am. Chem. Soc.* 137 (2): 667–674.

80 Peng, R., Zheng, X., Lyu, Y. et al. (2018). Engineering a 3D DNA-logic gate nanomachine for bispecific recognition and computing on target cell surfaces. *J. Am. Chem. Soc.* 140 (31): 9793–9796.

81 Chang, X., Zhang, C., Lv, C. et al. (2019). Construction of a multiple-aptamer-based DNA logic device on live cell membranes via associative toehold activation for accurate cancer cell identification. *J. Am. Chem. Soc.* 141 (32): 12738–12743.

82 Chen, B., Ma, W., Long, X. et al. (2022). Membrane protein and extracellular acid heterogeneity-driven amplified DNA logic gate enables accurate and sensitive identification of cancer cells. *Anal. Chem.* 94 (5): 2502–2509.

83 Gao, Q., Zhao, Y., Xu, K. et al. (2020). Highly specific, single-step cancer cell isolation with multi-aptamer-mediated proximity ligation on live cell membranes. *Angew. Chem. Int. Ed.* 59 (52): 23564–23568.

84 Hemphill, J. and Deiters, A. (2013). DNA computation in mammalian cells: MicroRNA logic operations. *J. Am. Chem. Soc.* 135 (28): 10512–10518.

85 Wang, H., Peng, P., Wang, Q. et al. (2020). Environment-recognizing DNA-computation circuits for the intracellular transport of molecular payloads for mRNA imaging. *Angew. Chem. Int. Ed.* 59 (15): 6099–6107.

86 Du, Y., Peng, P., and Li, T. (2019). DNA logic operations in living cells utilizing lysosome-recognizing framework nucleic acid nanodevices for subcellular imaging. *ACS Nano* 13 (5): 5778–5784.

87 Wang, W., Huang, S., Li, J. et al. (2017). Evaluation of intracellular telomerase activity through cascade DNA logic gates. *Chem. Sci.* 8 (1): 174–180.

88 Shi, H., Wang, Y., Zheng, J. et al. (2019). Dual-responsive DNA nanodevice for the available imaging of an apoptotic signaling pathway in situ. *ACS Nano* 13 (11): 12840–12850.

89 Lopez, R., Wang, R., and Seelig, G. (2018). A molecular multi-gene classifier for disease diagnostics. *Nat. Chem.* 10 (7): 746–754.

90 Yin, F., Zhao, H., Lu, S. et al. (2023). DNA-framework-based multidimensional molecular classifiers for cancer diagnosis. *Nat. Nanotechnol.* 18: 677–686.

91 Ma, Q., Zhang, M., Zhang, C. et al. (2022). An automated DNA computing platform for rapid etiological diagnostics. *Sci. Adv.* 8 (47): eade0453.

92 Zhang, C., Zheng, T., Ma, Q. et al. (2022). Logical analysis of multiple single-nucleotide-polymorphisms with programmable DNA molecular computation for clinical diagnostics. *Angew. Chem. Int. Ed.* 61 (15): e202117658.

93 Liu, J., Zhang, C., Song, J. et al. (2023). Unlocking genetic profiles with a programmable DNA-powered decoding circuit. *Adv. Sci.* 2206343.

10

Intelligent Sense-on-Demand DNA Circuits for Amplified Bioimaging in Living Cells

Yuqiu He[1,2], Zeyue Wang[1,2], Yuqian Jiang[1,2], and Fuan Wang[1,2]

[1] *Wuhan University, Department of Gastroenterology, Zhongnan Hospital of Wuhan University, Wuhan 430072, China*
[2] *Wuhan University, College of Chemistry and Molecular Sciences, Wuhan 430072, China*

10.1 DNA Circuit: The Promising Technique for Bioimaging

As a typical toolkit of biomolecular engineering, DNA nanotechnology aims to fabricate delicate molecular machines with unique functions and to assemble DNA-based molecular circuits capable of performing diverse tasks [1, 2]. In recent years, readily programmable DNA circuits have proceeded quickly due to their increasing importance in analytical and diagnostic applications [3–6]. Similar to the electric current-based circuitry devices that enable signal transmission and conversion, nucleic acid-based circuits execute information processing through molecular hybridization, strand exchange, and catalytic reactions. Currently, DNA circuits have a wide range of applications in biocomputing [7] and cancer therapy [8] and are particularly attractive in bioimaging [9] for the following reasons: (i) Watson–Crick pairing and thermodynamics properties provide the foundation for predicting the folded structures and interactions of nucleic acid molecules, allowing the construction of programmable and customized DNA circuits. (ii) DNA has the advantages of being small, high biocompatibility, and good adaptability in biological microenvironments, which makes it especially suitable for biosensing and bioimaging in the living system. (iii) DNA circuits with excellent amplification properties empower the translation of a single input into multiple outputs, which is especially important for the accurate and sensitive detection of intracellular trace biomarkers. (iv) In addition to interacting with nucleic acid analytes, DNA circuits are easily integrated with functional modules for responding to non-nucleic acid targets, thus making them attractive in the on-demand regulation of signal pathways in living cells.

Concerning these involved components, artificial DNA networks are divided into enzymatic DNA circuits and nonenzymatic DNA circuits. Enzymatic DNA circuits include strand displacement amplification (SDA) [10], loop-mediated

DNA Nanotechnology for Cell Research: From Bioanalysis to Biomedicine, First Edition. Edited by Zhou Nie.

isothermal amplification (LAMP) [11], rolling circle amplification (RCA) [12], and so on [13, 14]. Although the intricate thermocycling processes are eliminated, the implementation of these enzyme-involved circuits needs to be assisted by perishable protein enzymes, which increases the instability and vulnerability of the amplification system [15]. Therefore, it is especially appealing to develop nonenzymatic DNA circuits to cope with the environmental sensitivity and low reproducibility of enzyme-mediated amplification strategies. These nonenzymatic circuits with high anti-interference performance have shown promising applications in bioassays based on their ease of efficient algorithm execution in complex cellular environments. Herein, we will review the molecular reaction mechanisms of classic nonenzymatic DNA circuits and their exquisite integration strategies. The regulation approaches of DNA circuits will also be addressed to help construct intelligent platforms capable of on-demand sensing in living cells and *in vivo*.

10.2 Nonenzymatic DNA Circuits

Nonenzymatic DNA circuits rely on programmable nucleic-acid-alone elements to amplify an initiator-induced signal. According to the different operational mechanisms, nonenzymatic DNA circuits are classified into hybridization chain reaction (HCR) [16], catalytic hairpin assembly (CHA) [17], entropy-driven DNA catalytic reaction (EDR) [18], and DNAzyme-powered catalytic reaction (DZR) [19]. HCR and CHA are driven by the free energy of metastable hairpins hybridization while EDR is fueled by the entropy gain from the liberation of additional nucleic acid strands [20, 21]. These three amplifiers are developed on the basis of toehold-mediated strand displacement reaction. Different from them, DNAzymes are a class of nucleic acids with enzymatic properties that can catalyze various chemical reactions alone [22]. Compared to enzymatic DNA circuits, these enzyme-free strategies are more suitable for implementing bioimaging *in situ* because they do not require the delivery of exotic enzymes for their function.

10.2.1 Hybridization Chain Reaction (HCR)

Dirks and Pierce initially proposed the concept of hybridization chain reaction (HCR) in which an initiator induces the sequential hybridization events of hairpin monomers to generate linear double-stranded DNA copolymers [16]. In the typical HCR system, molecular recognition and signal transduction are coupled to accomplish the target-triggered assembly of DNA nanowires. Figure 10.1A illustrates the principle of the HCR circuit. The simplest system consists of two hairpin monomers H1 and H2. H1 is rationally designed into two domains, where domain a-b is complementary to initiator strand (I) a*-b* and domain c-b* is complementary to sequence c*-b of H2. Note that a sequence a*-b* identical to I strand is initially trapped in hairpin H2. The displacement reaction is initiated when the initiator I hybridizes to the toehold domain a of H1, resulting in the formation of hybrid

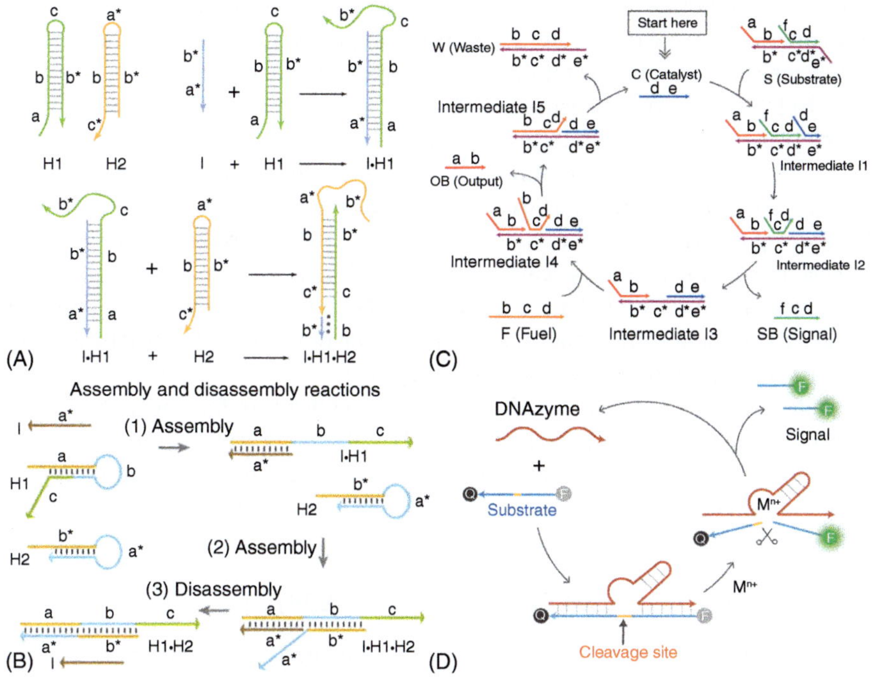

Figure 10.1 (A) Pathway of hybridization chain reaction (HCR). Source: Adapted from Dirks and Pierce [16]. (B) Fundamental of catalytic hairpin assembly (CHA). Source: Reproduced with permission from Yin et al. [17]. © 2008, Nature Publishing Group, with modification. (C) Schematic illustration of entropy-driven DNA catalytic reaction (EDR). Source: Reproduced with permission from Zhang et al. [18]. © 2007, American Association for the Advancement of Science, with modification. (D) Schematic representation of DNAzyme-powered catalytic reaction (DZR). Source: Adapted from Qian et al. [23].

I · H1 and exposure of active terminal sequence c-b*. Subsequently, hairpin H2 is opened by the hybridization of sequence c*-b with sequence c-b*, then releasing the I-analog sequence a*-b* as a new initiator to continue propagating alternating hybridization events between H1 and H2. Ultimately, a nicked double helix polymer is generated to carry out linear amplification of the initiator.

Significant progress has been made in HCR circuit-based sensing platforms to enable *in situ* imaging of RNA in living cells [24, 25]. Furthermore, in order to expand the range of target molecules, aptamer recognition elements are introduced into the HCR circuits. The proposed aptamer trigger mechanism facilitates the visualization of cancer-associated membrane proteins [26] and bioactive species [27], which provides increased insights to explore their biological functions and pathological effects.

10.2.2 Catalytic Hairpin Assembly (CHA)

CHA, a programmable cross-catalytic nucleic acid circuit, was originally designed by Yin et al. in 2008 [17]. The CHA circuit differs slightly from the HCR circuit described above in that the initiator strand is released from the self-assembly DNA duplex after each cycle of the prescribed reaction pathway. The regeneration of the initiator makes CHA able to amplify the signal rapidly and effectively. As shown in Figure 10.1B, two hairpin constituents (H1 and H2) with partially complementary sequences are used to construct the CHA circuit. The spontaneous hybridization of this pair of hairpins is initially hindered until the input of an initiator strand (I). Initiator I contains a domain a* that can hybridize with the domain a of hairpin H1, inducing an H1 conformational transition to expose the sequence b. Then, the sequence b* of hairpin H2 nucleates at sequence b and triggers an assembly reaction to generate an intermediate complex I·H1·H2. Finally, the unstable tripartite complex disassembles into a thermodynamically favorable dsDNA structure, while initiator I is released to catalyze subsequent cycles.

To date, numerous CHA circuits have been conceived for the detection of intracellular nucleic acids [28, 29]. And CHA reactants can be integrated into DNA frameworks (e.g. DNA tetrahedron) to increase their local concentrations and reaction rates [30].

10.2.3 Entropy-Driven DNA Catalytic Reaction (EDR)

In 2007, Zhang and coworkers first reported a catalytic DNA circuit that is driven by the overall entropy of the reaction, with the constant number of base pairs before and after the process [18]. In this strategy, an input strand triggers a series of toehold-assisted branch migrations, eventually releasing the signal strand and output strand while achieving a recycling operation of the input strand. The principle of a canonical EDR is depicted in Figure 10.1C. The reactant species of the EDR system consists of a three-component substrate (S) and a fuel strand (F), where the substrate S is a duplex complex formed by the hybridization of a single-stranded DNA with the signal strand (SB) and output strand (OB). In this reaction, the catalyst strand (C) first binds to the dangling end domain e* on S to form intermediate I1, which is then further transformed to I2 via toehold-mediated strand displacement. Subsequently, the unstable I2 rapidly dissociates into I3 and SB. The metastable structure I3 contains an exposed toehold c*, which can interact with the domain c of F and facilitate the substitution of F for OB and C to form the waste (W) product. The regenerated catalyst strand C initiates a new reaction cycle until any one of the components (F or S) is completely depleted.

Taking advantage of the programmability, modularity, and biocompatibility of the EDR system, various EDR-based biosensors have been established for intracellular analysis [31, 32]. And the substrate-recycling strategy has been constructed for highly sensitive detection of exosomes-enveloped microRNA *in vitro* [33, 34].

10.2.4 DNAzyme-Powered Catalytic Reaction (DZR)

DNAzyme, a man-made DNA enzyme with biocatalytic activity, was first identified by Breaker and Joyce in 1994 through *in vitro* selection methodology [35]. Among them, the DNAzyme with RNA cleavage activity is particularly suitable for fabricating pre-eminent amplifiers with high performance, as it can readily translate a single target binding event into numerous amplicon generation events. Generally, DNAzyme consists of a conserved catalytic domain and two variable substrate recognition arms, which endows it with excellent programmability and predictability. In the presence of a specific cofactor, inactive DNAzyme is converted to an active state to catalyze the transesterification reaction at the cleavage site on the substrate (Figure 10.1D) [23]. As a result, the thermodynamically unstable DNAzyme-cleaved substrate complexes spontaneously dissociate, releasing the catalytic DNAzyme to continuously induce the cleavage reactions of intact substrates.

The multiple turnover rates and facile synthesis properties make DNAzyme a powerful building block for constructing catalytic DNA circuits [36]. To protect DNAzyme from exonuclease degradation, chemically modified nucleotides (e.g. xeno-nucleic acids, inverted nucleotides) were introduced to enhance the biostability of DNAzyme without sacrificing its catalytic activity [37]. The multifunctional nanovectors were also developed to improve the delivery efficiency of DNAzyme and to achieve a self-sufficient cofactor supply [38].

10.3 Intelligent Integrated DNA Circuits for Amplified Bioimaging

Expression level fluctuation of intracellular biomarkers or signaling molecules is closely associated with many physiological and pathological processes. The accurate profiling of disease-associated markers in living cells and *in vivo* can help us monitor disease pathogenesis and progression, which is essential for early diagnosis and prognostic assessment of cancer. However, single-layered DNA circuits have limited signal amplification depth, making it difficult to meet the needs of sensitive detection and efficient bioimaging of low-abundance analytes. Recently, several integrated DNA circuits with polynomial/exponential amplification efficiency have made unprecedented advances in biomolecular imaging applications.

10.3.1 Hybridization-Dependent Cascade DNA Circuits

Headway has been made toward constructing sophisticated nucleic acid reaction networks using individual hybridization-dependent DNA circuits. Considering the programmability and adaptability of DNA circuits, different types of nonenzymatic amplification methods can be ingeniously coupled on demand to construct a cascaded signal amplifier with high sensing performance. For example, a concatenated HCR designed by homogeneously stacking upstream HCR-1 with downstream HCR-2 was developed for visualizing the localization of microRNA in living cells (Figure 10.2A) [39]. The hairpin reactant H2 encodes single-stranded regions at the 5′ and 3′ ends (segments d and e), respectively, independent of the hybridization events of HCR-1. As with a single-layered HCR, the initiator (I) binds to the toehold of H1 and promotes the sequential assembly of H1 and H2, producing a long

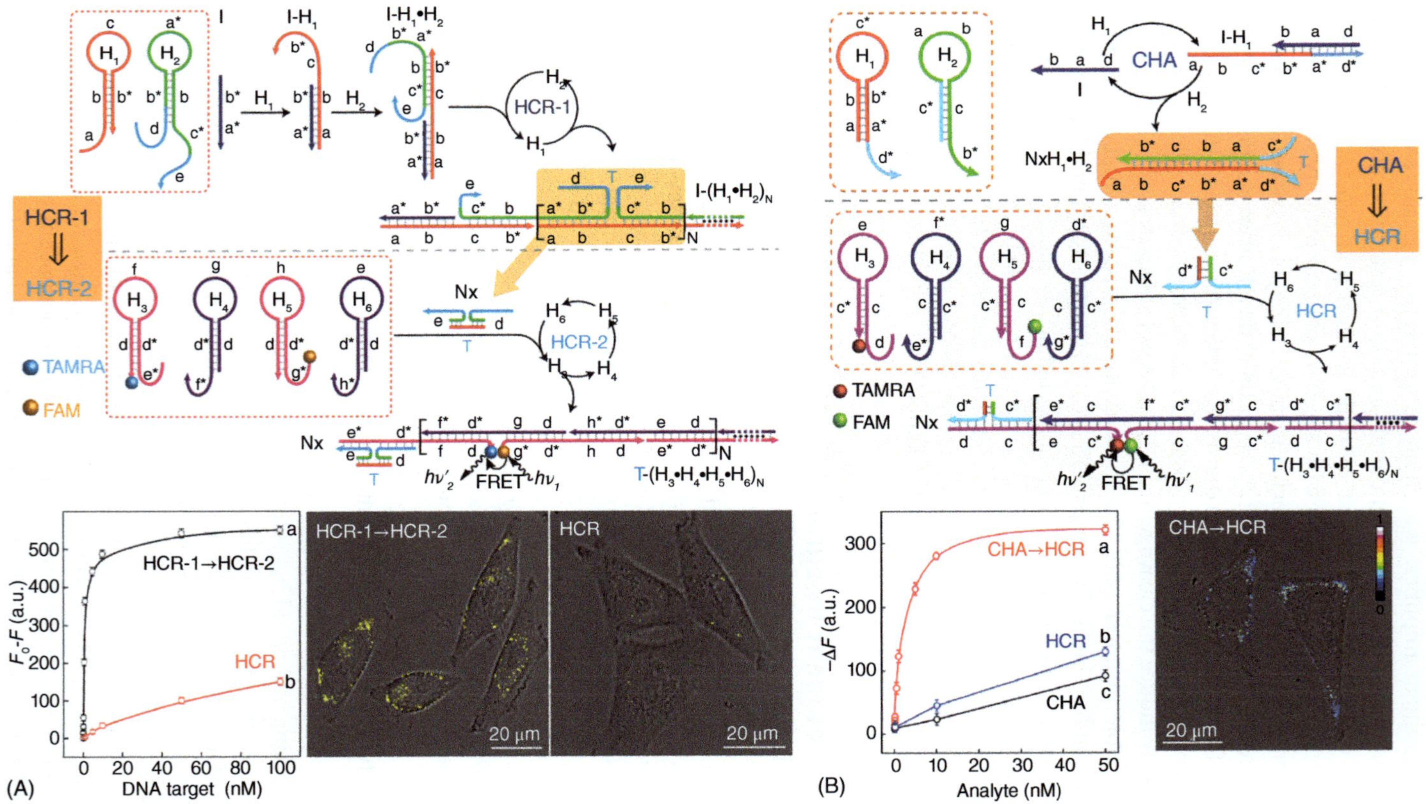

Figure 10.2 (A) Schematic illustration of the concatenated HCR system for microRNA imaging in living cells. Source: Reproduced with permission from Wei et al. [39]. © 2018 Royal Society of Chemistry. (B) Scheme of the cascaded CHA-HCR circuit for amplified microRNA detection. Source: Reproduced with permission from Wang et al. [40]. © 2018 Royal Society of Chemistry.

dsDNA consisting of several repeat units. Meanwhile, the segments d and e of two adjacent H2 in the dsDNA complex are in close proximity, which can act as a reconstitute trigger T to initiate the second layer HCR-2. As a result, an amplified Förster resonance energy transfer (FRET) signal is generated due to the proximity of the fluorophore donor/acceptor pairs in hairpins H3 and H5. By introducing an auxiliary hairpin containing a target recognition sequence and an initiator I sequence, this concatenated HCR is able to accurately identify specific microRNA. Compared with conventional HCR amplifiers, the two-layered HCR significantly improves the amplification efficiency for low-expression microRNAs, offering the possibility of early cancer diagnosis. With a simple design of sensing modules, this concatenated HCR sensing platform is endowed with a variety of functions, such as selective cell discriminations [41], *in situ* monitoring of important small molecules [42], or proteins [43] in living cells.

Another typical example is the cascaded CHA-HCR amplifier. As shown in Figure 10.2B, the trigger sequence T for the downstream HCR is confined in the metastable hairpin substrates of the upstream CHA [40]. Once encountered with the initiator I, the CHA circuit is activated, generating multiple duplexes H1·H2 containing the reassembled sequence T for initiating the subsequent HCR circuit. The cascaded CHA-HCR circuit provides superior detection performance over single-layered CHA or HCR. Moreover, this synergistic amplification platform has excellent imaging capability for potential biomarkers of key diseases [44]. The intracellular delivery efficiency of DNA probes is one of the critical challenges for *in situ* amplification strategies, as negatively charged DNA molecules cannot freely pass through the plasma membrane. To overcome this barrier, Li et al. developed a graphene oxide (GO)-CHA-HCR amplification system that employed GO as a carrier to deliver the components of CHA and HCR into living cells [45]. Afterward, the Xu group proposed a smart DNA/RNA nanosphere-based probe delivery strategy [46]. Compared to traditional inorganic nanomaterials, DNA-based nanocarriers are more biocompatible and easier to deliver on demand.

10.3.2 DNAzyme-Assisted Tandem DNA Circuits

DZR consists of only a DNAzyme sequence and a substrate sequence, making it readily available for integration with motifs of other circuits to construct tandem DNA networks with better amplification capabilities. By grafting the split DNAzyme subunit into the CHA reactants, a two-stage CHA-DNAzyme circuit was proposed by Yang and co-workers (Figure 10.3a) [47]. Specifically, the DNAzyme is initially in an inactivated state because its structure units are trapped in the stem of the hairpin reactants, preventing the formation of a completely catalytically active structure. The target microRNA triggers CHA to assemble copious dumbbell-like amplification products with active DNAzyme at both ends. Then, multiple substrates are cleaved by the intact DNAzyme, which causes the fluorophore to move away from the quencher, thus generating an amplified fluorescent signal. Beyond the combination of the two circuits, a triple-layered CHA-HCR-DNAzyme circuit has been reported for intracellular imaging applications [49].

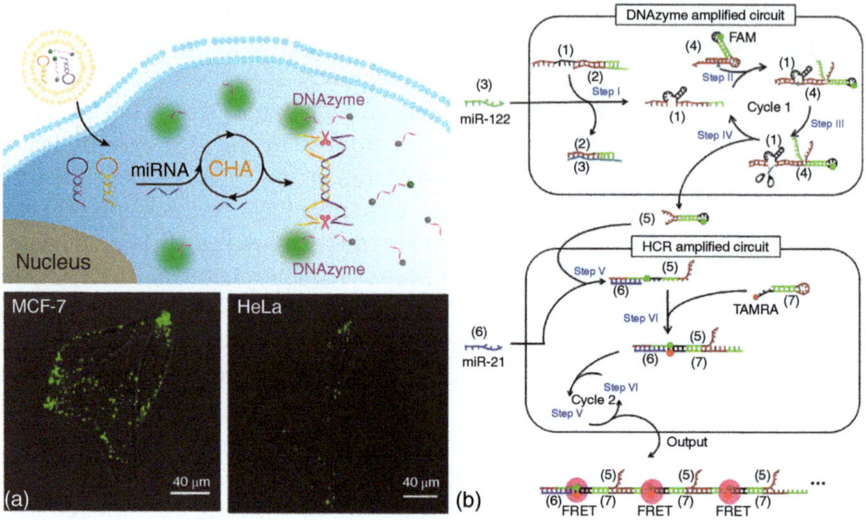

Figure 10.3 (a) Illustration of the integrated CHA-DNAzyme for intracellular microRNA imaging. Source: Reproduced with permission from Yang et al. [47]. © 2018 American Chemical Society. (b) Schematic representation of the DNAzyme-HCR logic DNA circuit controlled by dual-microRNAs. Source: Reproduced with permission from Quan et al. [48]. © 2019 The Royal Society of Chemistry.

However, for the split/recombination strategy, despite the improved amplification depth, the complexity of the system has increased. The split position needs to be meticulously screened; otherwise, the recombinant DNAzyme exhibits unsatisfactory catalytic activity, which is detrimental to the construction of the cascade circuits [50]. To avoid this issue, the Huang group designed a DNAzyme-HCR logic DNA circuit that connected the two amplification modules through a dual stem-loop design [48]. As depicted in Figure 10.3b, DNAzyme (1) is partially hybridized with a blocker strand (2) to inhibit the formation of the active structure. When target miR-122 (3) is present, the duplex (1)/(2) hybridizes with the target via a toehold-mediated strand-displacement reaction to form duplex (2)/(3) while releasing the caged DNAzyme. The hairpin (4) with a dual stem-loop structure contains a substrate strand of DNAzyme and a reactant hairpin (5) of downstream HCR. Then, the active DNAzyme continuously cleaves hairpin (4), yielding hairpin (5) outputs to provide adequate reactants for the HCR reaction. In the presence of another target miR-21, fluorophore FAM-labeled hairpin (5) and fluorophore TAMRA-labeled hairpin (7) are assembled alternately to generate amplified FRET signals. This dual microRNA-controlled concatenated circuit can successfully identify liver cancer cells Huh7 from other cells, which is meaningful to improve the reliability and accuracy of disease diagnosis.

As a typical functional nucleic acid, DNAzyme can be used not only for signal amplification but also for gene repression. The substrate recognition arms of DNAzyme can be flexibly programmed to cleave cancer-specific genes, inhibiting the production of aberrant proteins at the transcriptional or posttranscriptional level. Inspired by the application of DNAzyme in gene regulation, several DNAzyme-based cascaded circuits have also been developed to simultaneously implement *in situ* imaging and cellular gene silencing. For example, Quan et al. explored a multifunctional HCR-DNAzyme nanosystem [51]. The HCR module acts as a signal amplifier to achieve miR-21 imaging, while the therapeutic DNAzyme is selected as an anticancer therapeutic arsenal to inhibit the proliferation and migration of cancer cells by cleaving the EGR-1 mRNA. Gong et al. designed a smart CHA-DNAzyme-based theranostic nanocapsule for microRNA detection and gene therapy [52].

10.3.3 Autocatalysis-Driven Feedback DNA Circuits

In theory, the increase in the number of layers of integrated DNA circuits improves their amplification efficiency. However, the repetitive stacking of DNA circuits causes detrimental effects on the sensing time, thus constraining their application in bioanalysis. Nonspecific hybridization between abundant substrates leads to signal leakage, which also impairs the sensing performance. To address these issues, some autocatalytic DNA circuits consisting of only two modules have emerged to achieve infinite amplification depth. In this reciprocal catalytic strategy, two-layered circuits are directly transformed into multilayered circuits without significantly extending the execution time. For example, a replicating HCR (R-HCR) machinery integrating HCR and DNAzyme was exploited for microRNA imaging (Figure 10.4a) [53]. The hairpin reactants of HCR are functionalized with DNAzyme subunit sequences, which can be assembled into active DNAzyme by trigger T-initiated hybridization events. Different from the two-layered HCR-DNAzyme circuit, the DNAzyme substrates are encoded with feedback elements in the autocatalytic machine rather than as a signal output module. That is, once the substrates are cleaved by the activated DNAzyme, plentiful trigger strands feed back into the HCR process. This cascade DNA circuit can achieve exponential amplification efficiency through synergistic cross-activation procedures, providing a design paradigm for tracing low-expression biomarkers in living cells and *in vivo*.

Despite the conspicuous progress made in the construction of DNAzyme-based autocatalytic machines, there are still urgent challenges that need to be addressed before they can be extensively utilized for *in situ* amplified imaging of biomolecules. The catalytic efficiency of DNAzyme in the intracellular environment is dramatically restricted by the inadequate metal ions supply. Additionally, the inefficient cellular uptake of probes is also a major obstacle to the implementation of these DNA circuits in living systems. Nanomaterial/probe-integrated system provides a possibility of solving these intractable problems. Based on the autocatalytic R-HCR amplifier described above, Wei et al. synthesized a multifunctional honeycomb MnO_2 nanosponge (hMNS) scaffold that works as both a DNA probes carrier and a DNAzyme cofactor (Mn^{2+}) supplier (Figure 10.4b) [38]. In the presence of endogenous glutathione (GSH), hMNS nanocarriers are fragmented to liberate Mn^{2+} and hairpin reactants for the biocircuit. The system has both fluorescence imaging and

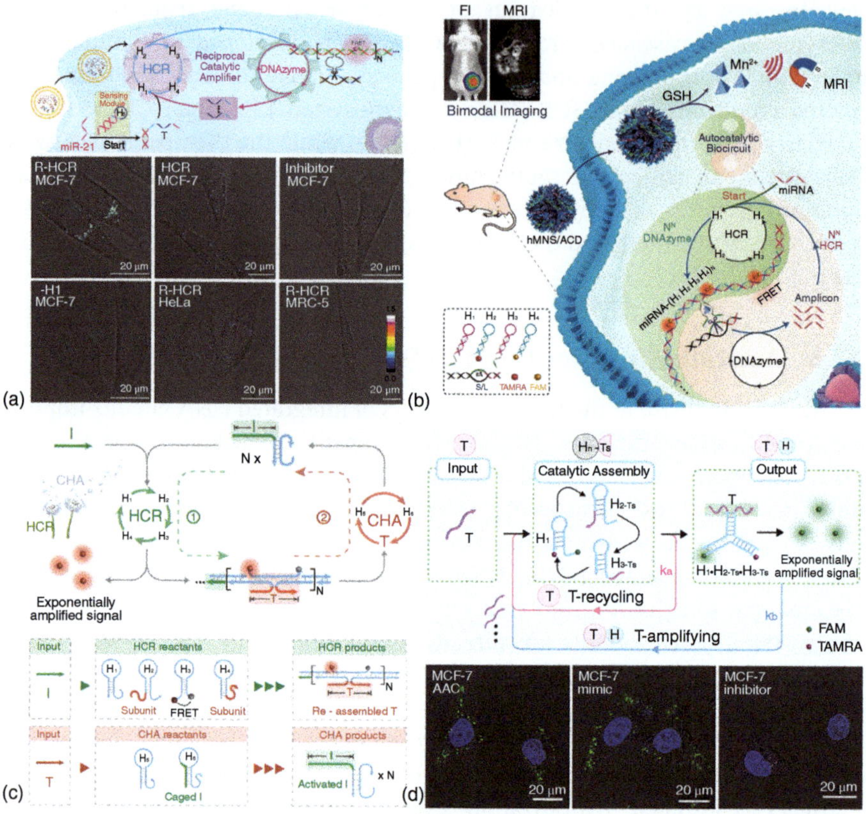

Figure 10.4 (a) Schematic illustration of the microRNA-targeting R-HCR machinery. Source: Reproduced with permission from Wei et al. [53]. © 2020 Oxford University Press. (b) Schematic presentation of the hMNS-sustained autocatalytic R-HCR amplifier for *in vivo* microRNA imaging. Source: Reproduced with permission from Wei et al. [38]. © 2020 John Wiley & Sons, Inc. (c) Reaction procedure of the autocatalytic AHA machine. Source: Reproduced with permission from Ref. [54]. © 2022 Wiley-VCH. (d) Scheme of the AAC system for microRNA visualization in living cells. Source: Reproduced with permission from Li et al. [55]. © 2023 American Chemical Society.

magnetic resonance imaging (MRI) capabilities, as the released paramagnetic Mn^{2+} can also be utilized as an MRI reagent.

Autonomous hybridization-based nucleic acid circuits have also been identified as promising tools for the fabrication of autocatalytic DNA machines, which are less impacted by intracellular metal ion conditions. Wang et al. proposed a dandelion-like autocatalytic hybridization assembly (AHA) circuit composed of a stick-like HCR and a seed-like CHA [54] (Figure 10.4c). With a rational design, initiator I activates the HCR module to produce an incipient nanowire scaffold containing a large number of re-assembled T amplicons. Subsequently, the T amplicons induce the successive assembly of CHA hairpin reactants, generating numerous activated I analogs. Since the reverse feedback feature of I-amplicon, CHA acts as a transducer to continue initiating the HCR pathway. As a result, well-dispersed nanowire products are assembled, accompanied by exponential signal amplification. In addition, Wei et al. constructed an autocatalytic hybridization chain reaction (AHCR) machine by cascading an amplification module HCR-1 and a feedback module HCR-2 [56].

A more attractive strategy is to develop self-stacking autocatalytic DNA circuits with minimal reactant complexity and excellent amplification capabilities (Figure 10.4d) [55]. The autocatalytic assembly circuit (AAC) contains merely three exquisite hairpin probes (H_1, $H_{2\text{-Ts}}$, and $H_{3\text{-Ts}}$) that minimize signal leakage caused by side reactions in complex cascading systems. In the AAC system, H_1 encodes a trigger T recognition sequence, while $H_{2\text{-Ts}}$ and $H_{3\text{-Ts}}$ contain split T fragments, respectively. When trigger T enters the AAC system, the hairpin reactants undergo catalytic DNA assembly reactions to form numerous Y-shaped triplexes carrying reconstituted T sequences. These newly assembled T function as catalysts to accelerate the reaction kinetics, allowing rapid acquisition of exponential signal gain. This strategy breaks through the complexities and signal leakage limitations of traditional integrated DNA circuits and is successfully applied for accurate and robust bioimaging of various microRNAs in living cells and *in vivo*. Similarly, a self-replicate DNA circuit based on HCR has been presented for efficient biosensing [57].

10.4 Stimuli-Responsive DNA Circuits for Reliable Bioimaging

Although the aforementioned autocatalytic circuits with theoretically infinite amplification depth provide unparalleled opportunities to monitor critical biomarkers in biological systems, such ready-to-response DNA circuits still face severe off-site activation because of the wide extracellular/intracellular distribution of these biomolecules. Signal leakage caused by nonspecific interactions of circuitry components in transit may also lead to untrustworthy signal readouts. To further improve the applicability of DNA circuits for bioimaging applications and explore the relationship between biological processes and expression levels of endogenous biomolecules, it is imperative to construct intelligent DNA circuits with dynamic behavior toward specific stimuli. To date, diverse stimuli, including exogenous physical factors and endogenous cues, have been exploited for the modulation of pre-designed DNA circuits. The stimuli-responsive strategy has dramatically facilitated the development of intelligent sense-on-demand DNA circuits.

10.4.1 Photo-Responsive DNA Circuits for Amplified Bioimaging

As noninvasive external stimuli, the photo-regulation strategies offer the possibility of spatiotemporally controllable signal amplification. By applying external light stimuli remotely at a chosen time, DNA circuits can be activated to execute target recognition and bioimaging tasks. For example, the introduction of a photosensitive group into the DZR module prevents the early activation of the DNAzyme sensor

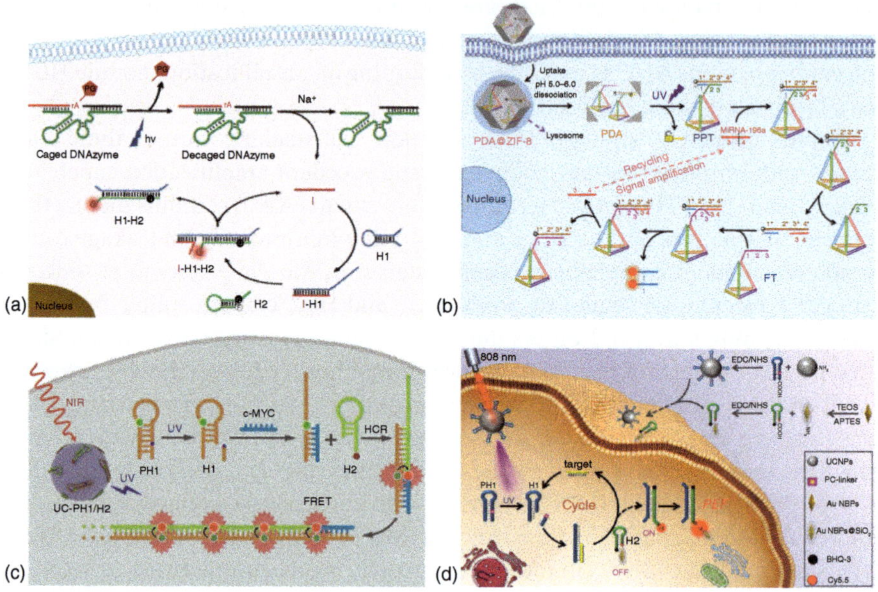

Figure 10.5 (a) Design of the UV light-activated DNAzyme-CHA system for endogenous Na^+ imaging. Source: Reproduced with permission from Wu et al. [58]. © 2017 Wiley-VCH. (b) Scheme of the UV light-responsive EDR system for amplified microRNA imaging. Source: Reproduced with permission from Gao et al. [59]. © 2021 American Chemical Society. (c) Schematic illustration of the 980 nm NIR-controlled HCR circuit for mRNA imaging with spatiotemporal resolution. Source: Reproduced with permission from Chu et al. [60]. © 2019 Wiley-VCH. (d) Schematic presentation of the 808 nm NIR-initiated CHA circuit for spatiotemporal imaging of microRNA. Source: Reproduced with permission from Zhao et al. [61]. © 2022 American Chemical Society.

before reaching the specific site. The Li group proposed a UV light-activated DNAzyme-CHA system for high spatial and temporal resolution imaging of intracellular Na^+ (Figure 10.5a) [58]. 2′-*O*-nitrobenzyl adenosine, a typical photocaged group (PG), is chosen to replace the adenosine ribonucleotide (rA) at the cleavage site, impeding the cleavage of substrate strands by the Na^+-specific DNAzyme. Upon ultraviolet (UV) light irradiation, the 2′-OH group at the scissile site is restored with the removal of PG from the substrate, which allows the transesterification reaction to occur in the presence of Na^+. The released fragment I acts as an initiator to trigger downstream CHA, producing an amplified fluorescent signal. Considering the exact controllability of light, the photocaging strategy enables site-specific metal ions imaging while avoiding undesired activation during the probe delivery process. In addition to chemically modifying the cleavage position of substrate strands, another alternative approach is to modify the conserved nucleotides in the catalytic core of DNAzyme with PGs to control the activity directly by light irradiation [62].

By modifying a photocleavage (PC)-linker into the substrate, the EDR amplifier enables *in situ* light-controlled bioimaging [59]. As shown in Figure 10.5b, the three-component substrate and fuel strand of the EDR system are separately anchored on DNA tetrahedrons to enhance their biostability. All DNA probes are encapsulated in a metal-organic framework (MOF) ZIF-8 and delivered into living cells via endocytosis. Subsequently, ZIF-8 dissociates in the acidic environment of the lysosome, allowing the EDR-sensing elements to be liberated into the cytoplasm. Initially, an additional segment of the locking strand (in yellow) is latched to the EDR substrate through PC linker to impede hybridization of the toehold region with microRNA target. UV irradiation promotes effective photolysis of the PC linker and shedding of the locking strand, which restores the sensing function of the EDR amplifier. Such a UV light-based regulation method enables the artificial control of DNA circuits and enriches the trigger mechanisms. Nevertheless, the high-energy UV light with phototoxic may cause potential cell damage. On the other hand, the shallow tissue penetration depth of UV light also limits its application in bioanalysis.

Upconversion nanoparticles (UCNPs), an emerging class of optical nanomaterials, can selectively convert near-infrared (NIR) light into various shorter-wavelength light by judicious use of lanthanide dopants [63]. Recently, UCNPs have been selected as the NIR-to-UV nanotransducer for photo-responsive DNA circuits. Different from UV light, NIR light as an excitation source can effectively mitigate the photodamage effect and energy loss caused by intense light scattering. Chu et al. designed a NIR-light-initiated HCR for visually fluorescent imaging of native RNA in living cells and tumor tissues (Figure 10.5c) [60]. Lanthanide-ion-doped UCNPs serve as light transducers and probe carriers, which assist the probes to traverse the cell membrane barrier. With NIR irradiation at 980 nm, the characteristic UV emission produced by UCNPs specifically breaks the PC-linker of probe PH1 to form the hairpin H1 with target recognition region. Ultimately, in the presence of the target c-MYC mRNA, HCR is completely activated, enabling amplified bioimaging with ultrahigh spatiotemporal resolution. However, water has a significant absorption of NIR light at 980 nm, accompanied by an overheating effect, which may have adverse effects on the living systems. To tackle this challenge, a NIR (808 nm) light-activated CHA nanodevice was constructed by using Nd^{3+}-sensitized UCNPs (Figure 10.5d) [61]. In particular, the plasmon-enhanced fluorescence (PEF) technique is configured into the CHA nanodevices through the incorporation of gold nanobipyramids@SiO_2, which provides a smart platform for on-demand ultrasensitive bioanalysis.

10.4.2 Enzyme-Activated DNA Circuits for Amplified Bioimaging

Exogenous stimuli-based regulation strategies may perturb exquisite physiological processes and fail to reflect the vital biotransformation information in living organisms. Therefore, there is intense interest in building links between artificial sensing networks and endogenous biological cues. Abnormal homeostasis of diverse endogenous proteases has been recognized to be tightly associated with pathological processes. These enzymes are commonly considered to be disease-related biomarkers, which provide an auxiliary tool for constructing cell-selective imaging pathways. For example, apurinic/apyrimidinic endonuclease 1 (APE1), a type of DNA repair enzyme, plays a biologically crucial role in the base excision repair (BER) pathway [64]. APE1 has been revealed to be overexpressed in the cytoplasm of a variety of tumor cells, whereas in normal cells, it performs repair functions mainly in the nucleus [65]. Motivated by the above properties, APE1 shows fascinating promise in constructing enzyme-activated DNA circuits with cell-type selectivity. Chen et al. proposed an APE1-responsive HCR that enables smart microRNA bioimaging through bioorthogonal regulation (Figure 10.6a) [66]. The hairpin reactant H_2 is caged in a blunt precursor probe $H_{2\text{-P}}$, thus impeding the microRNA-triggered HCR. Upon delivery of the probes into cancer cells, endogenous APE1 in the cytoplasm incises the two AP sites pre-designed in the stem of hairpin $H_{2\text{-P}}$, releasing active H_2 for microRNA imaging. In contrast, HCR amplifiers are consistently inactivated due to insufficient enzymes in the cytoplasm of normal cells. This enzyme-involved engineered DNA circuit can effectively eliminate off-site leakage and detect microRNA in a cell-type-specific manner. Similarly, enzymatically gated CHA was also developed by introducing AP site-modified molecular beacons [68]. Furthermore, to enhance the *in vivo* imaging performance of APE1-activated CHA, ionizable lipid nanoparticles were synthesized for the efficient delivery of DNA probes [69].

Various demethylases, such as FTO [70], the AlkB homolog family, and MGMT [71], have been proven to regulate the activation of DNAzyme-based catalytic circuits by a "repaired and activated" strategy. By mutating the strictly conserved nucleotides in the catalytic core region, the RNA cleaving capacity of DNAzyme is nearly completely abolished, while catalytic activity is restored after the removal of specific chemical modification. Recently, a demethylase (DMase)-driven metal ion sensing platform was constructed by introducing the N^6-methyladenine (m^6A) group to the Zn^{2+}-dependent 8-17E DNAzyme (Figure 10.6b) [67]. The m^6A-caged DNAzyme is initially in the OFF state to eliminate nonspecific signal leakage and off-site activation. In this nucleobase-caging/de-caging manner, the DNAzyme amplifier actuates Zn^{2+}-sensing function in the premise of FTO-mediated methylation repair, thus enabling site-specific bioimaging.

10.4.3 RNA-Stimulated DNA Circuits for Amplified Bioimaging

It is important to establish stimuli-responsive DNA circuits that employ overexpressed nucleic acid molecules in cancer cells as a driving force to foster reliable bioimaging and accurate disease diagnosis. The intracellular nucleic acid molecules regulate hybridization reactions or conformational transitions directly through Watson–Crick interaction, without relying on sophisticated nanomaterials or specific modifications. Wu and coworkers explored a microRNA-unlocked sensing strategy for endogenous Na^+ imaging (Figure 10.7a) [72]. The sensing platform (M-EDz) consists of an upstream microRNA-unlocked EDR amplifier and a downstream Na^+-activated DZR module. When miR-21 and Na^+ are present simultaneously, the M-EDz system is successfully activated to produce amplified output. Specifically, Na^+-specific DNAzyme stably hybridizes with a locked strand (LS) in this design, blocking the normal catalytic activity. In cancer cells, the upregulated miR-21 binds to the toehold region of substrate (MEB) to trigger the

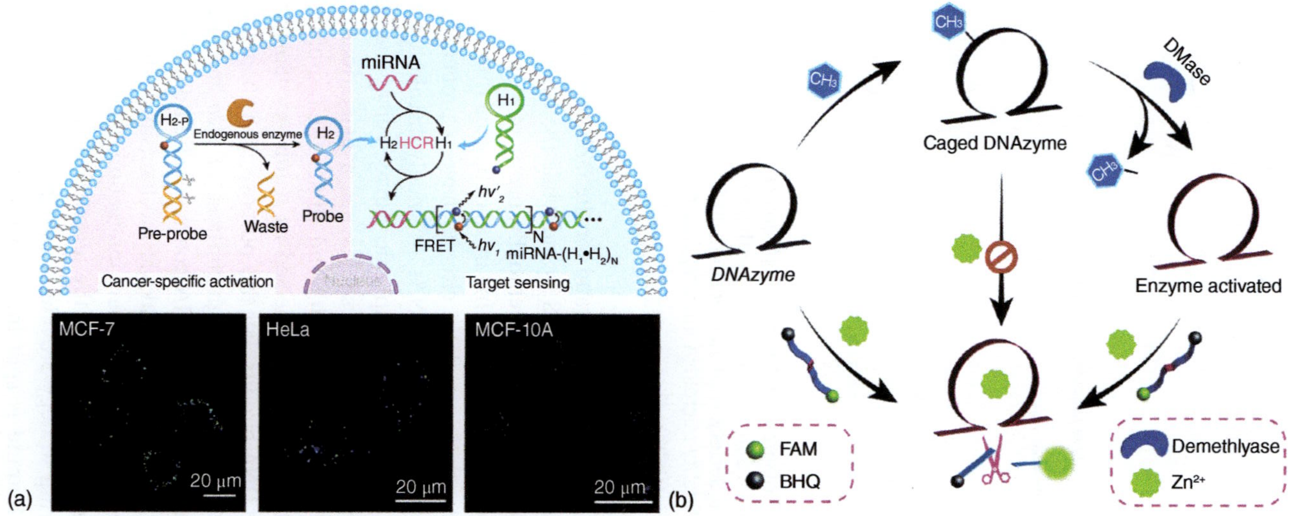

Figure 10.6 (a) Illustration of the on-site activation of the APE1-responsive HCR amplifier for specific microRNA sensing. Source: Reproduced with permission from Chen et al. [66]. © 2021 Royal Society of Chemistry. (b) Schematic presentation of the versatile demethylase-driven DNAzyme for *in situ* imaging of Zn^{2+}. Source: Reproduced with permission from Hong et al. [67]. © 2021 The Royal Society of Chemistry.

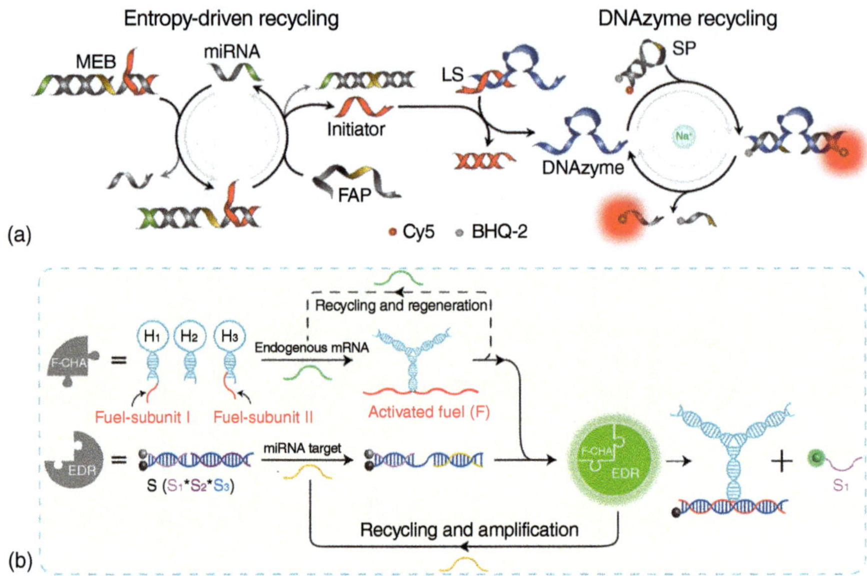

Figure 10.7 (a) Schematic diagram of the microRNA-unlocked M-EDz platform for cell-selective imaging of Na^+. Source: Reproduced with permission from Wu et al. [72]. © 2021 American Chemical Society, with modification. (b) Principle of the mRNA-stimulated CHA-EDR circuit for high-fidelity microRNA imaging. Source: Reproduced with permission from He et al. [73]. © 2022 Wiley-VCH.

EDR circuit and displaces the initiator strand for the DZR module. Then, the free initiator strand forms a duplex with the LS strand while the DNAzyme switches to the active state in the presence of Na^+. Eventually, plenty of molecular beacons SP are cleaved by the active DNAzyme, resulting in an enhanced fluorescent signal.

In fact, microRNA-regulated catalytic circuits have a propensity to turn on before reaching the target site, as microRNAs are widely distributed in organisms, albeit at lower expression levels than specific cancer cells. The generation of background signals above tolerable thresholds has a deleterious effect on the sensitivity and reliability of DNA circuits and limits their usefulness. Clearly, site-specific biomolecules need to be selected as regulatory tools to ensure the on-site execution of the DNA circuits. For this purpose, an orthogonally controlled catalytic circuit was established for high-contrast *in vivo* microRNA imaging by incorporating an mRNA-activated CHA control section and an EDR sensing section (Figure 10.7b) [73]. The main leakage mechanism of the EDR amplifier is the inherent thermodynamically driven interaction between the intact fuel molecules and the ternary substrates in the absence of catalysts. Therefore, the fuel splitting-restructuring strategy provides a valuable solution to the high signal leakage. Here, the fuel strand is split into two subunits (subunit I and subunit II) and grafted into the CHA components H_1 and H_3, respectively, achieving the blockage of EDR. The in-cell-specific mRNA serves as an appealing regulator for initiating CHA-guided fuel recombination and amplification. Subsequently, the regenerated fuel strands are replenished into the downstream EDR system for high-fidelity microRNA imaging. The multiply guaranteed catalytic DNA circuit not only dramatically reduces the cross-talk of circuit elements but also achieves cell-specific activation, thereby ensuring the stability and reliability of the sensing platform. The regulatory paradigm is also applicable to the construction of other promising DNA amplification machines [74].

10.4.4 Other Strategies for Regulating DNA Circuits

Aside from the diverse stimuli mentioned above, multiple pathological parameters, such as glutathione (GSH) and adenosine triphosphate (ATP), have been demonstrated to be available for the engineering of stimuli-responsive DNA circuits. Meng et al. incorporated an aptamer-based recognition element upstream of the conventional HCR amplification element for high-sensitivity imaging of microRNA (Figure 10.8a) [75]. In this work, cancer cell membranes and GSH-responsive MOFs are integrated as probe delivery vehicles to improve the homologous targeting ability and spatiotemporal controllability of this sensing platform. When the nanosystem is internalized into specific cancer cells via membrane-directed targeted delivery, the MOF carrier is rapidly degraded by intracellular GSH and then releases HCR hairpin reactants. The toehold region of the hairpin for microRNA recognition is initially sealed by an ATP aptamer through Watson-Crick base pairing, blocking the sensing function of HCR. Once endogenous ATP binds the aptamer sequence, the HCR units in active modality analyze microRNA and output an amplified fluorescent signal. Benefiting from the ATP-involved direct regulation and GSH-mediated indirect regulation, the nanosystem is capable of spatiotemporally controllable imaging in different cells, which opens a reliable avenue for the investigation of microRNA expression profiles. By altering the sequence of aptamers, we can easily establish the connection between the target and other vital biomolecules, thus providing more possibilities for the extension of DNA circuits.

The discovery of pH-sensitive multistranded DNA motifs has facilitated the development of stimuli-responsive programmable nucleic acid circuits. Different pH values in the tumor microenvironment, cytosol, and organelles are expected to serve as switches to regulate DNA conformation and logical operations. Chao group fabricated a pH-controlled DNAzyme nanomachine with a tunable response window for performing logical sensing in different cellular environments [76]. As shown in Figure 10.8b, four-stranded precursors and several substrates are decorated on AuNP to form the locked pH-DNA nanomachine. The precursor consists of three locking strands and a functional strand that contains both the i-motif sequence and the DNAzyme sequence. When the nanomachine enters the lysosome (pH = 4.5–5.0) via the endocytic pathway, acid-induced i-motif structure formation leads to the release of two locking strands. Then, in the cytoplasm (pH ~ 7.2), the target microRNA binds to locking strand 3, and the DNAzyme sequentially cleaves the substrates for amplified imaging. Such an i-motif structure has also been integrated into the HCR system for pH-driven membrane protein identification [77].

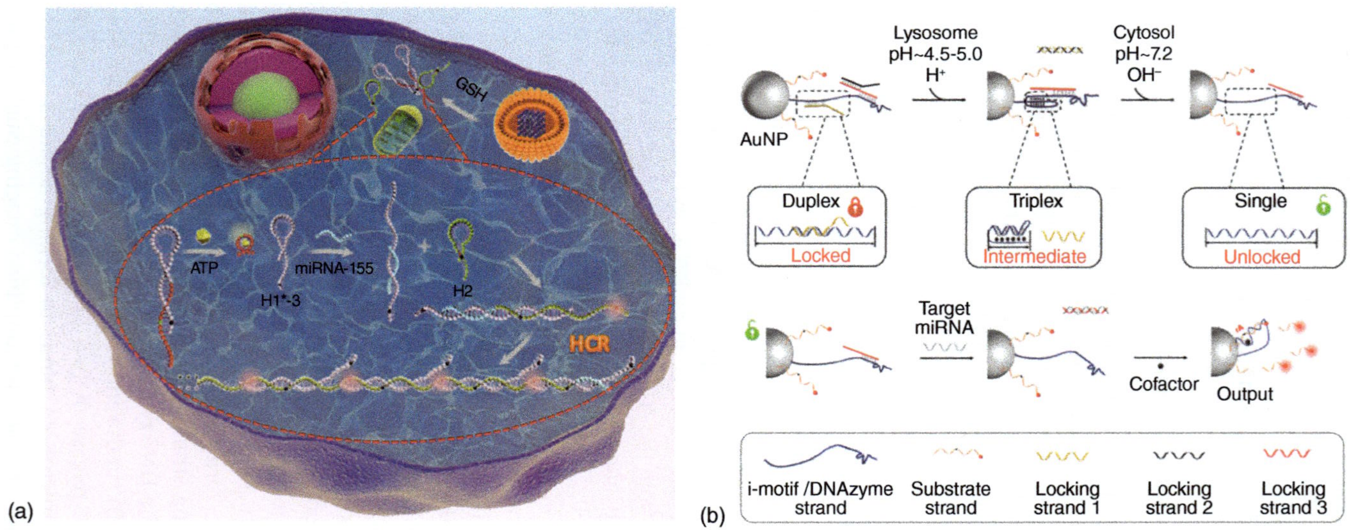

Figure 10.8 (a) Schematic interpretation of the endogenous ATP-controlled HCR circuit for *in vivo* imaging of microRNA. Source: Reproduced with permission from Meng et al. [75]. © 2021 American Chemical Society. (b) Scheme of the pH-stimulated DNAzyme nanomachine for intracellular microRNA sensing. Source: Reproduced with permission from Cui et al. [76]. © 2022 American Chemical Society.

10.5 Conclusion and Perspectives

Readily programmable nonenzymatic DNA circuits with excellent biocompatibility have become powerful biological tools as the scope of bioanalysis expands from *in vitro* detection to *in situ* imaging. Moreover, the integration of intelligent stacking strategies and stimulus-responsive modules makes it possible to fabricate sense-on-demand DNA circuits that play an indispensable role in molecular diagnostics and disease identification. In this chapter, we reviewed the basic mechanisms of four representative enzyme-free DNA circuits, which are the foundation for the applications and renovations of nucleic acid reaction networks. Based on these principles, a series of ingenious cascaded DNA circuits with polynomial or exponential amplification depth have been conceived for ultrasensitive detection and *in situ* imaging of various molecular targets (e.g. cancer-specific nucleic acids, physiologically relevant metal ions, disease-associated proteins). Furthermore, spatiotemporally controlled DNA circuits capable of perceiving environmental stimuli have been developed to eliminate signal leakage from off-site activation and to improve the ability of cell-selective imaging.

Despite remarkable strides made, challenges associated with DNA circuitry for imaging in cellular settings still linger, which probably hinders the advancement of nucleic acid-based diagnostic tools. (i) One of the critical steps in constructing amplification circuits with high sensing performance is the design of nucleic acid probes, as small sequence discrepancies may lead to erroneous signal conversion or severe background signal leakage. To improve the simplicity and usability of nucleic acid circuits, especially for complex cascade systems, it is desirable to develop systematic guidelines for the probe design. (ii) Mainstream approaches for visualizing intracellular molecules rely on fluorophores that are susceptible to the effects of photobleaching, impairing the fluorescence signal and signal-to-noise ratio. The accuracy and reliability of DNA circuits for bioimaging can be strengthened by developing photobleaching-corrected methods, improving fluorescent live-cell imaging techniques, or introducing multimodal analysis techniques. (iii) For *in situ* analysis, delivery technologies need to be introduced to assist the circuit elements to combat biological barriers. Exploring various transfection reagents or nanosystems with high loading efficiency and low biotoxicity will help promote the transition of DNA circuits from basic research to practical applications. (iv) Stimuli-responsive DNA circuits are still in their infancy, with only a finite number of cues identified as regulatory factors, and their amplification efficiency is inadequate for low-abundance analyte detection. Screening for specific biomarkers to engineer control modules and incorporating them into cascaded or autocatalytic circuits could contribute to the on-demand sensing of analytes. (v) Accurate diagnosis and effective therapy are typically built on the identification of multiple biomarkers. Given the achievements of DNA circuits in amplified bioimaging, it is intriguing to develop multiplexed molecular circuits to receive and process diverse biological information and perform complex diagnostic tasks.

Acknowledgments

This work is supported by the National Natural Science Foundation of China (22274121, 22274123, and 22074112) and the Fundamental Research Funds for the Central Universities (2042022kf1175).

References

1 Chen, Y.J., Groves, B., Muscat, R.A. et al. (2015). DNA nanotechnology from the test tube to the cell. *Nat. Nanotechnol.* 10 (9): 748–760.

2 Del Grosso, E., Franco, E., Prins, L.J. et al. (2022). Dissipative DNA nanotechnology. *Nat. Chem.* 14 (6): 600–613.

3 Gines, G., Zadorin, A.S., Galas, J.C. et al. (2017). Microscopic agents programmed by DNA circuits. *Nat. Nanotechnol.* 12 (4): 351–359.

4 Yang, Q., Wang, Y., Liu, T. et al. (2022). Microneedle array encapsulated with programmed DNA hydrogels for rapidly sampling and sensitively sensing of specific microRNA in dermal interstitial fluid. *ACS Nano* 16 (11): 18366–18375.

5 Miao, P. and Tang, Y. (2021). Cascade strand displacement and bipedal walking based DNA logic system for miRNA diagnostics. *ACS Cent. Sci.* 7 (6): 1036–1044.

6 Chen, J., Fu, S., Zhang, C. et al. (2022). DNA logic circuits for cancer theranostics. *Small* 18 (20): 2108008.

7 Wang, F., Lv, H., Li, Q. et al. (2020). Implementing digital computing with DNA-based switching circuits. *Nat. Commun.* 11 (1): 121.

8 Gong, X., Wang, H., Li, R. et al. (2021). A smart multiantenna gene theranostic system based on the programmed assembly of hypoxia-related siRNAs. *Nat. Commun.* 12 (1): 3953.

9 Wu, H., Chen, T.T., Wang, X.N. et al. (2020). RNA imaging in living mice enabled by an *in vivo* hybridization chain reaction circuit with a tripartite DNA probe. *Chem. Sci.* 11 (1): 62–69.

10 Walker, G.T., Fraiser, M.S., Schram, J.L. et al. (1992). Strand displacement amplification-an isothermal, *in vitro* DNA amplification technique. *Nucleic Acids Res.* 20 (7): 1691–1696.

11 Tomita, N., Mori, Y., Kanda, H. et al. (2008). Loop-mediated isothermal amplification (LAMP) of gene sequences and simple visual detection of products. *Nat. Protoc.* 3 (5): 877–882.

12 Lizardi, P.M., Huang, X., Zhu, Z. et al. (1998). Mutation detection and single-molecule counting using isothermal rolling-circle amplification. *Nat. Genet.* 19 (3): 225–232.

13 Compton, J. (1991). Nucleic acid sequence-based amplification. *Nature* 350 (6313): 91–92.

14 Van Ness, J., Van Ness, L.K., and Galas, D.J. (2003). Isothermal reactions for the amplification of oligonucleotides. *Proc. Natl. Acad. Sci. U.S.A.* 100 (8): 4504–4509.

15 Li, B., Ellington, A.D., and Chen, X. (2011). Rational, modular adaptation of enzyme-free DNA circuits to multiple detection methods. *Nucleic Acids Res.* 39 (16): e110.

16 Dirks, R.M. and Pierce, N.A. (2004). Triggered amplification by hybridization chain reaction. *Proc. Natl. Acad. Sci. U.S.A.* 101 (43): 15275–15278.

17 Yin, P., Choi, H.M.T., Calvert, C.R. et al. (2008). Programming biomolecular self-assembly pathways. *Nature* 451 (7176): 318–322.

18 Zhang, D.Y., Turberfield, A.J., Yurke, B. et al. (2007). Engineering entropy-driven reactions and networks catalyzed by DNA. *Science* 318 (5853): 1121–1125.

19 Wang, F., Elbaz, J., Teller, C. et al. (2011). Amplified detection of DNA through an autocatalytic and catabolic DNAzyme-mediated process. *Angew. Chem. Int. Ed.* 50 (1): 295–299.

20 Bai, M., Chen, F., Cao, X. et al. (2020). Intracellular entropy-driven multi-bit DNA computing for tumor progression discrimination. *Angew. Chem. Int. Ed.* 59 (32): 13267–13272.

21 Qing, Z., Xu, J., Hu, J. et al. (2019). *In situ* amplification-based imaging of RNA in living cells. *Angew. Chem. Int. Ed.* 58 (34): 11574–11585.

22 Liu, M., Chang, D., and Li, Y. (2017). Discovery and biosensing applications of diverse RNA-cleaving DNAzymes. *Acc. Chem. Res.* 50 (9): 2273–2283.

23 Qian, R.C., Zhou, Z.R., Guo, W. et al. (2021). Cell surface engineering using DNAzymes: metal ion mediated control of cell–cell interactions. *J. Am. Chem. Soc.* 143 (15): 5737–5744.

24 Wu, Z., Liu, G.Q., Yang, X.L. et al. (2015). Electrostatic nucleic acid nanoassembly enables hybridization chain reaction in living cells for ultrasensitive mRNA imaging. *J. Am. Chem. Soc.* 137 (21): 6829–6836.

25 Li, J., Li, D., Yuan, R. et al. (2017). Biodegradable MnO_2 nanosheet-mediated signal amplification in living cells enables sensitive detection of down-regulated intracellular microRNA. *ACS Appl. Mater. Interfaces* 9 (7): 5717–5724.

26 Tang, J., Lei, Y., He, X. et al. (2020). Recognition-driven remodeling of dual-split aptamer triggering *in situ* hybridization chain reaction for activatable and autonomous identification of cancer cells. *Anal. Chem.* 92 (15): 10839–10846.

27 Zhang, C.H., Wang, H., Liu, J.W. et al. (2018). Amplified split aptamer sensor delivered using block copolymer nanoparticles for small molecule imaging in living cells. *ACS Sens.* 3 (12): 2526–2531.

28 Karunanayake Mudiyanselage, A.P.K.K., Yu, Q., Leon Duque, M.A. et al. (2018). Genetically encoded catalytic hairpin assembly for sensitive RNA imaging in live cells. *J. Am. Chem. Soc.* 140 (28): 8739–8745.

29 Zhao, S.J., Zheng, P., Wu, Z. et al. (2022). DNA-templated bioorthogonal reactions via catalytic hairpin assembly for precise RNA imaging in live cells. *Anal. Chem.* 94 (6): 2693–2698.

30 Wang, Y., Bai, Y., Cao, L.P. et al. (2022). Catalytic hairpin assembled polymeric tetrahedral DNA frameworks for microRNA imaging in live cells. *Biosens. Bioelectron.* 197: 113783.

31 Liang, C.P., Ma, P.Q., Liu, H. et al. (2017). Rational Engineering of a dynamic, entropy-driven DNA nanomachine for intracellular microRNA imaging. *Angew. Chem. Int. Ed.* 56 (31): 9077–9081.

32 He, L., Lu, D., Liang, H. et al. (2018). mRNA-initiated, three-dimensional DNA amplifier able to function inside living cells. *J. Am. Chem. Soc.* 140 (1): 258–263.

33 Zhang, C., Wang, Z., Liu, Y. et al. (2019). Nicking-assisted reactant recycle to implement entropy–driven DNA circuit. *J. Am. Chem. Soc.* 141 (43): 17189–17197.

34 Xia, Y., Huang, Z., Chen, T. et al. (2022). Sensitive fluorescent detection of exosomal microRNA based on enzymes-assisted dual-signal amplification. *Biosens. Bioelectron.* 209: 114259.

35 Breaker, R.R. and Joyce, G.F. (1994). A DNA enzyme that cleaves RNA. *Chem. Biol.* 1 (4): 223–229.

36 Lin, Y., Yang, Z., Lake, R.J. et al. (2019). Enzyme-mediated endogenous and bioorthogonal control of a DNAzyme fluorescent sensor for imaging metal ions in living cells. *Angew. Chem. Int. Ed.* 58 (47): 17061–17067.

37 Yang, K. and Chaput, J.C. (2021). REVEALR: A multicomponent XNAzyme-based nucleic acid detection system for SARS-CoV-2. *J. Am. Chem. Soc.* 143 (24): 8957–8961.

38 Wei, J., Wang, H., Wu, Q. et al. (2020). A smart autocatalytic DNAzyme biocircuit for *in vivo* amplified microRNA imaging. *Angew. Chem. Int. Ed.* 59 (15): 5965–5971.

39 Wei, J., Gong, X., Wang, Q. et al. (2018). Construction of an autonomously concatenated hybridization chain reaction for signal amplification and intracellular imaging. *Chem. Sci.* 9 (1): 52–61.

40 Wang, H., Li, C., Liu, X. et al. (2018). Construction of an enzyme-free concatenated DNA circuit for signal amplification and intracellular imaging. *Chem. Sci.* 9 (26): 5842–5849.

41 Gong, X., Wei, J., Liu, J. et al. (2019). Programmable intracellular DNA biocomputing circuits for reliable cell recognitions. *Chem. Sci.* 10 (10): 2989–2997.

42 Wu, Q., Yang, L., Xie, L. et al. (2022). Modular assembly of a concatenated DNA circuit for *in vivo* amplified aptasensing. *Small* 18 (21): 2200983.

43 Shang, J., Wei, J., Wang, Q. et al. (2020). Adaption of an autonomously cascade DNA circuit for amplified detection and intracellular imaging of polynucleotide kinase with ultralow background. *Biosens. Bioelectron.* 152: 111994.

44 Shang, J., Yu, S., Chen, Y. et al. (2021). Real-time investigation of intracellular polynucleotide kinase using a cascaded amplification circuit. *Anal. Chem.* 93 (46): 15559–15566.

45 Li, H., Zhang, B., He, X. et al. (2021). Intracellular circRNA imaging and signal amplification strategy based on the graphene oxide-DNA system. *Anal. Chim. Acta* 1183: 338966.

46 Song, J., Mou, H.Z., Li, X.Q. et al. (2022). Self-assembled DNA/RNA nanospheres with cascade signal amplification for intracellular microRNA imaging. *Sens. Actuatator, B* 360: 131644.

47 Yang, L., Wu, Q., Chen, Y. et al. (2019). Amplified microRNA detection and intracellular imaging based on an autonomous and catalytic assembly of DNAzyme. *ACS Sens.* 4 (1): 110–117.

48 Quan, K., Li, J., Wang, J. et al. (2019). Dual-microRNA-controlled double-amplified cascaded logic DNA circuits for accurate discrimination of cell subtypes. *Chem. Sci.* 10 (5): 1442–1449.

49 Wang, H., Wang, H., Wu, Q. et al. (2019). A DNAzyme-amplified DNA circuit for highly accurate microRNA detection and intracellular imaging. *Chem. Sci.* 10 (41): 9597–9604.

50 Mokany, E., Bone, S.M., Young, P.E. et al. (2010). MNAzymes, a versatile new class of nucleic acid enzymes that can function as biosensors and molecular switches. *J. Am. Chem. Soc.* 132 (3): 1051–1059.

51 Quan, K., Wang, J., Li, J. et al. (2022). An endogenous miRNA-initiated hybridization chain reaction and subsequent DNAzyme activation for cellular theranostics. *CCS Chem.* 4 (11): 3505–3515.

52 Gong, X., Li, R., Wang, J. et al. (2020). A smart theranostic nanocapsule for spatiotemporally programmable photo-gene therapy. *Angew. Chem. Int. Ed.* 59 (48): 21648–21655.

53 Wei, J., Wang, H., Gong, X. et al. (2020). A proteinase-free DNA replication machinery for *in vitro* and *in vivo* amplified microRNA imaging. *Nucleic Acids Res.* 48 (10): e60.

54 Wang, H., He, Y., Wei, J. et al. (2022). Construction of an autocatalytic hybridization assembly circuit for amplified *in vivo* microRNA imaging. *Angew. Chem. Int. Ed.* 61 (19): e202115489.

55 Li, R., Zhu, Y., Gong, X. et al. (2023). Self-stacking autocatalytic molecular circuit with minimal catalytic DNA assembly. *J. Am. Chem. Soc.* 145 (5): 2999–3007.

56 Wei, J., Shang, J., He, S. et al. (2022). Construction of a homogeneous enzyme-free autocatalytic nucleic acid machinery for high-performance intracellular imaging of microRNA. *CCS Chem.* 4 (11): 3549–3562.

57 Wei, J., Yu, M., Tan, K. et al. (2023). Tailoring a minimal self-replicate DNA circuit for highly efficient intracellular imaging of microRNA. *Small* 19: 2207961.

58 Wu, Z., Fan, H., Satyavolu, N.S.R. et al. (2017). Imaging endogenous metal ions in living cells using a DNAzyme-catalytic hairpin assembly probe. *Angew. Chem. Int. Ed.* 56 (30): 8721–8725.

59 Gao, J.L., Yuheng, L., Liu, J.X. et al. (2021). A photoresponsive and metal-organic framework encapsulated DNA Tetrahedral entropy-driven amplifier for high-performance imaging intracellular microRNA. *Anal. Chem.* 93 (49): 16638–16645.

60 Chu, H., Zhao, J., Mi, Y. et al. (2019). Near-infrared light-initiated hybridization chain reaction for spatially and temporally resolved signal amplification. *Angew. Chem. Int. Ed.* 58 (42): 14877–14881.

61 Zhao, T., Sun, X., Chen, J. et al. (2022). Optically programmable plasmon enhanced fluorescence-catalytic hairpin assembly signal amplification strategy for spatiotemporally precise imaging. *Anal. Chem.* 94 (13): 5399–5405.

62 Young, D.D., Lively, M.O., and Deiters, A. (2010). Activation and deactivation of DNAzyme and antisense function with light for the photochemical regulation of gene expression in mammalian cells. *J. Am. Chem. Soc.* 132 (17): 6183–6193.

63 Chen, G., Qiu, H., Prasad, P.N. et al. (2014). Upconversion nanoparticles: design, nanochemistry, and applications in theranostics. *Chem. Rev.* 114 (10): 5161–5214.

64 Maher, R.L., Wallace, S.S., and Pederson, D.S. (2019). The lyase activity of bifunctional DNA glycosylases and the 3'-diesterase activity of APE1 contribute to the repair of oxidized bases in nucleosomes. *Nucleic Acids Res.* 47 (6): 2922–2931.

65 Zhang, H., Ba, S., Yang, Z. et al. (2020). Graphene quantum dot-based nanocomposites for diagnosing cancer biomarker APE1 in living cells. *ACS Appl. Mater. Interfaces* 12 (12): 13634–13643.

66 Chen, Y., Gong, X., Gao, Y. et al. (2021). Bioorthogonal regulation of DNA circuits for smart intracellular microRNA imaging. *Chem. Sci.* 12 (47): 15710–15718.

67 Hong, C., Wang, Q., Chen, Y. et al. (2021). Intelligent demethylase-driven DNAzyme sensor for highly reliable metal-ion imaging in living cells. *Chem. Sci.* 12 (46): 15339–15346.

68 Shang, Y., Chen, Y., Wang, Q. et al. (2022). Construction of an endogenously activated catalytic DNA circuit for highly robust *in vivo* microRNA imaging. *Nano Today* 45: 101553.

69 Liu, Q., Huang, Y., Li, Z. et al. (2022). An enzymatically gated catalytic hairpin assembly delivered by lipid nanoparticles for the tumor-specific activation of signal amplification in miRNA imaging. *Angew. Chem. Int. Ed.* 61 (51): e202214230.

70 Wang, Q., Tan, K., Wang, H. et al. (2021). Orthogonal demethylase-activated deoxyribozyme for intracellular imaging and gene regulation. *J. Am. Chem. Soc.* 143 (18): 6895–6904.

71 Wang, X., Yi, X., Huang, Z. et al. (2021). "Repaired and Activated" DNAzyme enables the monitoring of DNA alkylation repair in live cells. *Angew. Chem. Int. Ed.* 60 (36): 19889–19896.

72 Wu, Y., Li, Z., Shi, M. et al. (2021). Programmable DNAzyme computing for specific *in vivo* imaging: intracellular stimulus-unlocked target sensing and signal amplification. *Anal. Chem.* 93 (36): 12456–12463.

73 He, S., Yu, S., Li, R. et al. (2022). On-site nonenzymatic orthogonal activation of a catalytic DNA circuit for self-reinforced *in vivo* microRNA imaging. *Angew. Chem. Int. Ed.* 61 (45): e202206529.

74 Yu, S., Shang, J., He, S. et al. (2022). Multiply guaranteed and successively amplified activation of a catalytic DNA machine for highly efficient intracellular imaging of microRNA. *Small* 18 (32): 2203341.

75 Meng, X., Wang, H., Yang, M. et al. (2021). Target-cell-specific bioorthogonal and endogenous ATP control of signal amplification for intracellular microRNA imaging. *Anal. Chem.* 93 (3): 1693–1701.

76 Cui, M.R., Chen, Y., Zhu, D. et al. (2022). Intelligent programmable DNA nanomachines for the spatially controllable imaging of intracellular microRNA. *Anal. Chem.* 94 (30): 10874–10884.

77 Chen, B., Ma, W., Long, X. et al. (2022). Membrane protein and extracellular acid heterogeneity-driven amplified DNA logic gate enables accurate and sensitive identification of cancer cells. *Anal. Chem.* 94 (5): 2502–2509.

11

DNA Nanoscaffolds for Biomacromolecules Organization and Bioimaging Applications

Yuanfang Chen[1,2], Jiayi Li[1,2], and Yuhe R. Yang[1,2]

[1]CAS Key Laboratory of Nanosystem and Hierarchical Fabrication, CAS Center for Excellence in Nanoscience, National Center for Nanoscience and Technology of China, CAS, North Road No. 1, Zhongguancun, Beijing 100190, China
[2]School of Nanoscience and Engineering, University of Chinese Academy of Sciences, Yuquan Road, Beijing 100049, China

11.1 Introduction

Arranging biomacromolecules in a defined order and orientation at the nanoscale is an enduring topic and has demonstrated remarkable success in deciphering the intermolecular interaction of bionetworks and analyzing the mechanisms of biosystems [1]. DNA origami, a technique using hundreds of short DNA (staples) to fold a long DNA (scaffold), can be fabricated into varieties of multidimensional nanostructures with nanometer precision, including planar 2D shapes and patterns [2, 3], 3D polygons [4, 5], multilayer bundles [6, 7], twisted and curved objects [8, 9], dynamic devices [10] and jigsaw-like superstructures [11–13]. Therefore, DNA origami is considered a multifunctional toolbox in biological research, and has been widely utilized as a powerful scaffold for proteins [14–16], fluorophores [17–19], metallic nanoparticles [20, 21], drug molecules [20, 21], lipids [22–24], etc. In this chapter, we provide a concise overview of assembly procedures for DNA-scaffolded biomolecules, including conjugation, assembly, and purification in a step-wise manner. We then explore the applications of DNA nanoscaffolds to orchestrate biomacromolecules, highlighting their significance and potential in studying cellular operation mechanisms. Lastly, we delve into the advancements brought about by DNA nanostructures in various bioimaging techniques, from fluorescence microscopy, and DNA-PAINT to cryogenic electron microscopy.

11.2 Assembly of DNA-Scaffolded Biomacromolecules

As a powerful and universal tool to manipulate biomolecules at the nanoscale, DNA nanoscaffold can meet various desires in biomolecule characterization

Yuanfang Chen and Jiayi Li were contributed equally.

DNA Nanotechnology for Cell Research: From Bioanalysis to Biomedicine, First Edition. Edited by Zhou Nie.

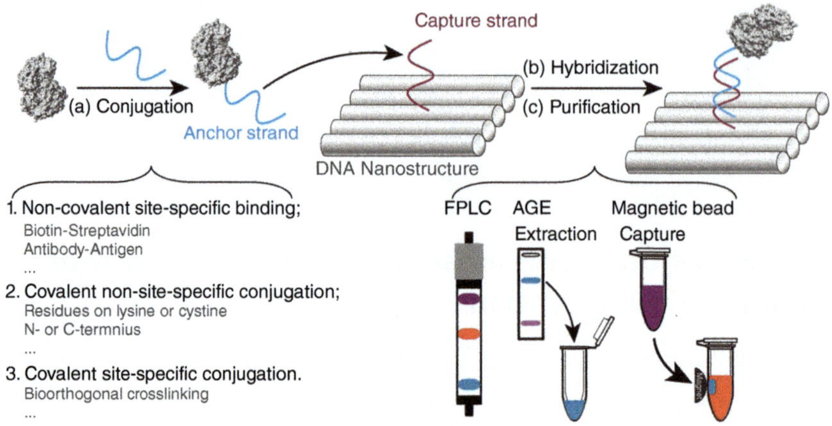

Figure 11.1 Workflow of DNA-protein complexes synthesizing. (a) Conjugation between oligonucleotides and proteins. (b) Hybridization of nanostructures and modified proteins. (c) Purification.

and application. To achieve this, three main steps are required to assemble biomolecule complexes on DNA scaffolds: (i) preparation of DNA-conjugated proteins, (ii) hybridization of DNA-protein conjugates with DNA nanoscaffolds, and (iii) purification (Figure 11.1).

The conjugation between DNA and proteins involves modifications of two parties: (i) The modification strategies for DNA are comparatively versatile due to the autologous stepwise chemical synthesis, and the decoration of diverse functional groups has been commercialized [25], while (ii) protein modification methods are not as standardized [26]. Based on different protein modification strategies, the DNA-protein conjugation methods can be categorized into three classes. The non-covalent site-specific binding methods rely on affinity interactions such as Biotin-Streptavidin [27] or Zinc-finger protein-DNA [28–30]. The covalent non-site-specific conjugation methods exploit chemical reactions between crosslinkers and reactive handles on proteins (e.g. lysine or cystine residues [31], N- or C-terminus). As for covalent site-specific conjugation approaches, one promising strategy is to introduce unnatural amino acids with non-native functional groups that enable bioorthogonal crosslinking [32]. However, introducing unnatural amino acids into proteins requires a series of sophisticated processes, such as editing the target plasmids, selecting suitable orthogonal aminoacyl tRNA synthetase (aaRS)-tRNA pair to ensure their compatibility with the expression system [33].

Assembling a complete nanostructure requires a hybridization of protein-conjugated oligonucleotides with their complementary single strands on the DNA scaffold. The annealing process in the hybridization stage must be precisely controlled to achieve a high assembly yield while retaining the activity of proteins. Here is a common annealing program: 37 °C for five minutes; 36–10 °C with a two minutes per degree decrease; and then store the samples at 4 °C [34]. As a result, capture probes at the length of 10–20 nt with a melting temperature (T_m) below

37 °C are commonly used [35]. Additionally, increasing the number of attachment sites can also elevate the assembly yield [34].

Purification is an essential step that removes the excess proteins after hybridization. In the majority of cases, the protein-functionalized DNA nanostructures are purified by either fast protein liquid chromatography (FPLC) or agarose gel electrophoresis (AGE) depending on the size of the nanostructures [34]. Specific affinity labels can also be harnessed to separate nanostructures from free proteins. For example, by hybridizing magnetic beads-modified oligonucleotides, DNA origami structures can be attracted by a magnetic field, leaving the free proteins available to be excluded [36, 37]. The beads-modified DNA can be removed by a following strand displacement reaction. Besides, Numajiri et al. successfully purified DNA origami structures by exploiting the affinity of Histags to Co^{2+} resin [38].

11.3 Application of DNA Nanoscaffold for Regulation of Enzyme Cascade Reaction

In the cellular environment, the precise spatial organization of the enzyme cascade reaction contributes to the high efficiency of metabolic pathways by facilitating and regulating substrate transport. The self-assembled DNA nanostructures have emerged as a promising platform for functionalizing enzymes due to their well-defined geometry, programmability, addressability at the nanometer scale, and functionalization capabilities [39]. This enables precise control over key parameters, such as position, stoichiometry, and relative distance between biomolecular components.

A combination of enzyme cascades and addressable DNA assembly has proven to be effective in fabricating various stable artificial multi-enzyme complexes with precise control over position, shape, and DNA-based composition [40]. This approach has facilitated efficient and specific control of enzyme distance, substrate diffusion, channeling, and compartmentalization.

The Glucose oxidase (GOx)/Horseradish peroxidase (HRP) pair is the most widely utilized enzyme cascade for investigating the effects of proximity and confinement in DNA nanostructures (Figure 11.2a). In this biochemical enzyme cascade, the primary enzyme GOx catalyzes the oxidation of glucose to gluconic acid, generating hydrogen peroxide (H_2O_2) synchronously. HRP then subsequently consumes the H_2O_2 as substrate, mediating the oxidation of $ABTS^{2-}$ to $ABTS^{-}$. The advantage of this system stems from its accessibility, relative robustness, ease of attaching the enzymes to a DNA scaffold, and ability to monitor its function through H_2O_2 production [46].

11.3.1 Distance Control of Enzyme Cascade

Double-stranded DNA (dsDNA) can serve as a simple DNA nanoscaffold for enzyme assembly and inter-enzyme distance control, thereby facilitating the efficiency of intermediates transfer and enzymatic activity (Figure 11.2b) [47–49]. More intricate

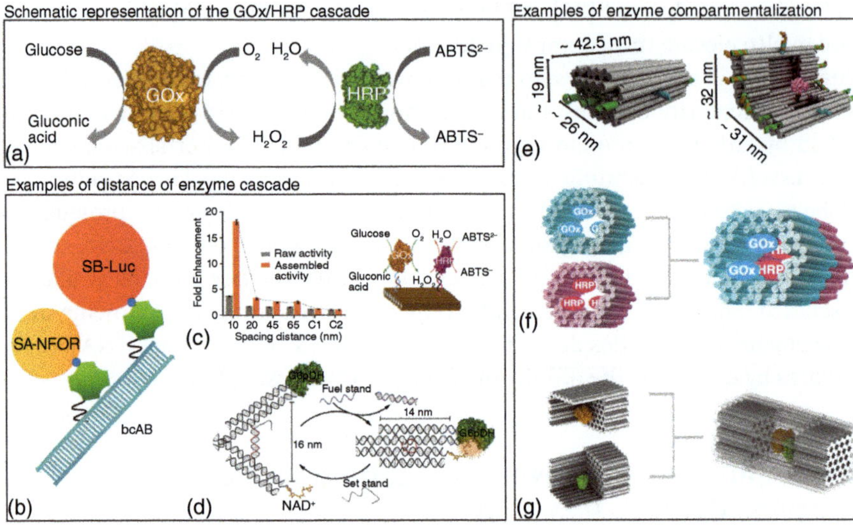

Figure 11.2 Enzyme cascade. (a) Schematic representation of the GOx/HRP cascade system. (b) NAD(P)H: FMN oxidoreductase and luciferase cascade on dsDNA. (c) Organization of a GOx/HRP cascade on a DNA origami with precisely controlled distance. Source: Reproduced with permission from Fu et al. [41]. © 2012 American Chemical Society. (d) Design of a tweezer complex using an enzyme-cofactor system involving glucose-6-phosphate dehydrogenase (G6pDH) and NAD$^+$. Source: Reproduced with permission from Liu et al. [42]/Springer Nature. (e) A DNA-based nano vault that can reversibly open and close to regulate enzyme-substrate accessibility. Source: Grossi et al. [43]/Springer Nature/CC BY 4.0. (f) A modular DNA origami-based enzyme cascade nanoreactor. Source: Reproduced from Linko et al. [44], with permission of the Royal Society of Chemistry. (g) The combination of two nanocages for enclosing enzymes. Source: Zhao et al. [45]/Springer Nature/CC BY 4.0.

geometric patterns of multi-enzyme systems, such as two- and four-hexagon-like [50] or three-point star (3PS) DNA scaffolds [51], were designed for more precise control of the enzyme distances. Yan's group conducted a detailed study on the distance dependency of enzyme cascades by scaffolding the GOx-HRP pair on DNA origami (Figure 11.2c) [41]. Through measuring the activity of different enzyme complexes at different inter-enzyme distances, they found a sharp decrease in cascade activity as the distance increased from 10 to 20 nm, followed by a gradual decrease in activity as the distance further increased to 65 nm. This result can be attributed to the intermediate transfer mechanism, which follows a three-dimensional diffusion model for inter-enzyme distances larger than 20 nm, and dimensionally limited diffusion of the intermediate across connected protein surfaces for closely spaced enzymes. Additionally, a DNA wireframe octahedron was applied to create a library of over 40 arrangements of GOx/HRP, reinforcing the previous findings [52].

In addition to the static nanostructure mentioned above, DNA nanomachines, such as DNA tweezers, offer a new platform to study the enzyme cascade on a dynamic scaffold, capable of reversibly tuning the spatial distance between two functional domains. A typical DNA tweezer contains two DNA double-crossover

(DX) motifs joined by Holliday junctions. It also includes a DNA motor integrated into the middle of the tweezer that drives switching between stem-loop and double helix structures through a strand displacement reaction, allowing it to alternate between open and closed states [53]. This design can magnify the small distance generated by the DNA motor into a substantial spatial separation between the tips of two tweezer arms [54]. For example, Liu et al. attached glucose-6-phosphate dehydrogenase (G6pDH) and NAD^+ to different arms of the DNA tweezer, and the actuation of enzymatic function is achieved by switching the tweezer between open and closed states (Figure 11.2d) [42]. Similar DNA tweezers were also demonstrated to be used to reversibly regulate the GOx/HRP enzyme cascade reaction by controlling their special distance [54].

Subsequently, various modifications were made to the design of DNA tweezers, including changes in dimensions, dynamics of opening and closing, mechanism, and methods of detection. For instance, Alexander and coworkers introduced an ion-activated reconfigurable DNA nanostructure that could undergo large-scale conformational transitions in sub-second timescales, overcoming the limitations of the slow actuation timescales in the strand-displacement mechanism for actuating the tweezer [55]. Additionally, Willner's group represented another DNA tweezer with a distinct structure and mechanism, which could regulate a three-component catalytic cascade that consisted of β-galactosidase (β-Gal), GOx, and the K^+-ion-stabilized hemin-G-quadruplex horseradish peroxidase (HRP)-mimicking DNAzyme [56]. Furthermore, Kou et al. also established a universal electrochemical assay for highly sensitive target DNA detection, using a DNA tweezer that was regenerated and could control the inter-enzyme spacing for regulation [57].

11.3.2 Enzyme Compartmentalization

Metabolic pathways are often organized within membrane-delimited compartments, such as mitochondria, lysosomes, peroxisome, and carboxysomes, or in membrane-free compartments [58]. The physical segregation of chemical reactions within specialized compartments has numerous effects on biochemical reactions, including increasing the local concentration of reactive species, preventing the degradation of delicate compounds, reducing the toxicity of intermediates, protecting encapsulated contents from competing pathways, and enhancing molecular reaction rates [58, 59]. With the advent of DNA nanotechnology, this compartmentalization strategy can be simulated *in vitro* using DNA nanoreactors, in which enzymes can be arranged appropriately and different reaction compounds can be separated from each other.

Taking inspiration from the separation of enzymes and substrates in the natural system to control reaction rate, Andersen and coworkers designed a DNA nanoscale vault (DV) that encapsulates signal enzyme and regulates its catalytic activity by the reversible opening and closing of the DV to control the substrate accessibility to the encapsulated enzymes (Figure 11.2e) [43]. A similar strategy was also used to design a reversible pH-responsive DNA nanocage for enzyme encapsulation, enabling precise control over the accessibility of the enzyme to the substrate and the substrate's dimension to regulate enzyme activity [60]. Based on the volume confinement strategy, Fan and coworkers attached GOx/HRP pairs inside DNA nanotubes, which further reduced the likelihood of H_2O_2 escaping and enhanced the activity of the cascade compared to the surface-confined enzymes [61]. Subsequently, the first modular DNA origami-based volume confinement nanoreactor that can efficiently perform a designed enzyme cascade was achieved by combining different tubular DNA origami units (Figure 11.2f) [44]. Shortly afterward, Zhao et al. developed a general approach to construct DNA nanocage-confined enzymes to enhance their catalytic activity and stability (Figure 11.2g) [45]. Surprisingly, they discovered a fourfold increase when two enzymes were individually encapsulated in separate DNA nanocages. Their hypothesis suggests that the electronegative environment created by the DNA scaffold attracts the hydration layer, stabilizing the active enzyme conformations around the highly ordered water molecules. Furthermore, Xiong et al. used a DNA octahedral scaffold (OS) as a carrier for GOx and discovered that the variation in the local pH environment caused by DNA density could interestingly influence enzymatic activity, experimentally proving the previous suspicion [62].

11.3.3 Directed Substrate Channeling with Swinging Arms

Substrate channeling facilitates the direct transfer of the intermediate from one enzyme to another proximal enzyme, therefore enhancing cascade activity. One such mechanism of substrate channeling is using swinging arms, which involves attaching a covalent prosthetic chemical group to the enzyme complex and allowing it to swing between multiple active sites to enable the transportation of intermediates [63].

To mimic this directed substrate channeling in a non-cellular environment, Yan's group introduced an NAD^+-modified swinging arm between two coupled dehydrogenases, facilitating efficient hydride transfer. They used a DNA double-crossover tile to create a two-enzyme complex, and the activity of the enzyme complex with an NAD^+-modified swinging arm was found to have a 90-fold enhancement compared to the complex without an NAD^+ arm (Figure 11.3a) [63]. By exploiting the programmability of DNA nanostructures, the key parameters including inter-enzyme distance, orientation, and stoichiometry were manipulated for the optimal activity. A complex consisting of four MDH surrounding a single G6pDH, channeled by four NAD^+-modified swinging arms, demonstrated the highest activity, which was approximately threefold higher than those channeled by a single swinging arm positioned between a pair of G6pDH-MDH. More importantly, this channeling strategy greatly improved the selectivity of the catalytic reaction and excluded interference from side reactions by only modifying a swinging arm on the DNA scaffold [63].

On this basis, a further investigation was reported to use the swinging arms to assist the transfer of intermediates (NAD^+) between G6pDH and LDH pairs on a 2D enzyme network (Figure 11.3b) [64]. Additionally, NAD^+-modified DNA swinging arms were also successfully used to directly regulate the two enzyme

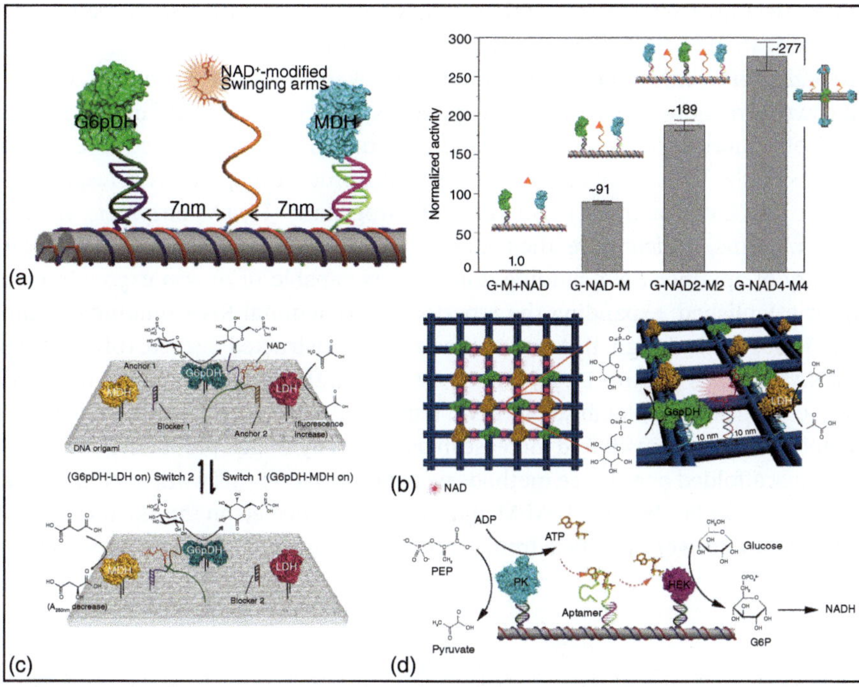

Figure 11.3 Artificial swinging arms facilitate the substrate channeling between multi-enzyme complexes on DNA scaffolds. (a) Design of the NAD$^+$-modified swinging arm provided restricted diffusion of NAD$^+$/NADH between two dehydrogenases. Source: Reproduced with permission from Fu et al. [63]/Springer Nature. (b) Design of a 2D enzyme network. Source: Reproduced with permission from Yang et al. [64]/John Wiley & Sons. (c) Design of an enzyme pathway regulation pathway system on DNA origami. Source: Reproduced with permission from Ke et al. [65]/John Wiley & Sons. (d) Design of an aptamer-modified DNA nanostructure provided the enrichment of ATP molecules between two kinases. Source: Reproduced with permission from Wang et al. [66]. © 2022 American Chemical Society.

pathways (G6PDH-MDH and G6PDH-LDH) through the control of NAD$^+$ substrate channeling and specifically shifting NAD$^+$ between the enzymes (Figure 11.3c) [65]. However, in contrast to the substrate channeling strategy, Fu's group subsequently used aptamers to modify the surface of DNA nanostructure assembled with multi-enzymes for the local enrichment or exclusion of substrate molecules, thereby regulating the activity of an enzyme cascade reaction (Figure 11.3d) [66].

11.3.4 Scaffolded Enzyme Cascade in Living Cells

The *in vivo* assembly of DNA nanoscaffolds and biomacromolecules holds great potential for advancing our understanding of intracellular processes and expanding the applications of DNA nanoscaffolds in cellular systems. However, to date, there has been very limited research on DNA nanostructure-based enzyme cascade in living cells [67]. This is primarily due to the technical challenges associated with achieving isothermal assembly of DNA or RNA scaffolds, as well as the characterization of DNA/RNA assembly *in vivo* [68].

To realize efficient isothermal assembly of RNA scaffolds *in vivo*, in 2011, Camille and coworkers rationally designed a sequence-symmetric RNA building block comprised of dimerization and polymerization domains [69]. For the co-assembly of enzymes on the RNA scaffolds, the PP7 and MS2 protein adaptors were respectively fused to the ferredoxin (Fd) and [FeFe]-hydrogenase (HydA) enzymes, yielding fusion proteins, which were then attached to the RNA aptamer domains [69]. A library of aptamer-RNA binding domain sets capable of *in vivo* expression was further established, expanding the repertoire of functional RNA binding domains [70]. Through this design, they have successfully constructed discrete (0D), 1D, and 2D RNA scaffolds (Figure 11.4a) and co-assembled them with a cascade of Fd and HydA enzymes [69]. Notably, the RNA-templated Fd-HydA cascade 2D network resulted in a significant 48-fold increase in bacteria's hydrogen production compared to the unscaffolded one. Three methods were employed in this work to characterize the RNA assemblies *in vivo*: (i) AFM analysis was performed on the assembled RNA scaffolds, which were purified from lysed cells using a DNA-based precipitation

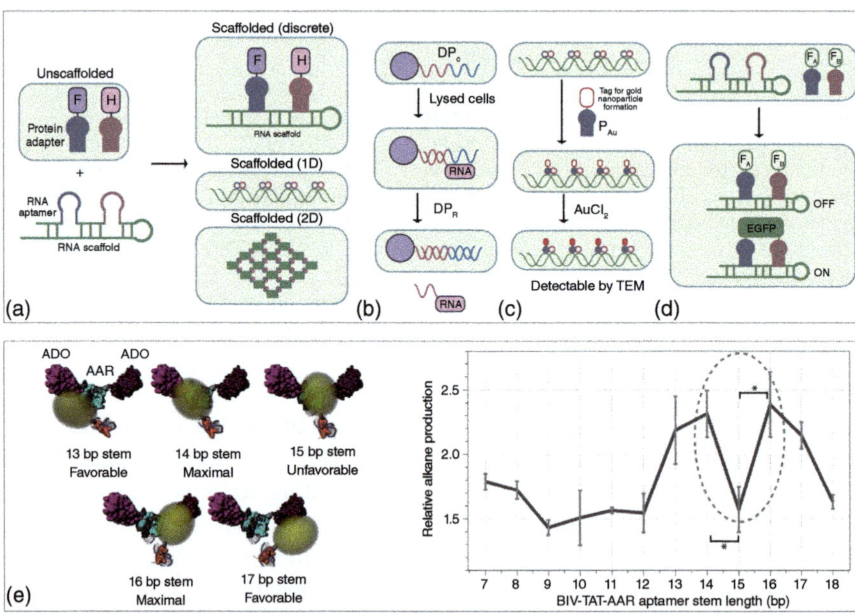

Figure 11.4 Scaffolded enzyme cascade within living cells. (a) Illustration of proteins organized on discrete, 1D, and 2D RNA assemblies, F and H each represents Fd and HydA enzyme. (b) Utilization of DNA-based precipitation method for *in vivo* RNA assembly characterization using a DP_C probe and a release probe DP_R for recovery. (c) TEM analysis providing insights into the structure of 1D and 2D RNA assemblies. (d) Schematic representation of fluorescence protein complementation *in vivo*, FA and FB represent two cleaved GFP fragments, respectively. (e) Length and orientation of aptamers affected metabolic flux increase on scaffolds. Source: Reproduced with permission from Sachdeva et al. [70]/Oxford University Press.

method (Figure 11.4b). (ii) TEM analysis was conducted to visualize 1D and 2D RNA scaffolds *in vivo*, employing gold-binding metallothionin-PP7 fusion proteins (P_{Au}) for tagging (Figure 11.4c). (iii) Fluorescence complementation experiment was carried out to monitor the binding of fusing two cleaved green fluorescent protein (GFP) fragments (FA and FB) when co-expressed with RNA scaffolds (Figure 11.4d). Furthermore, the same group constructed more intricate and flexible RNA scaffolds to organize a two-enzyme (acyl-ACP reductase, AAR, and aldehyde deformylating oxygenase, ADO) pathway [70]. By manipulating the relative orientation through changing aptamer stem length, they discovered that the intracellular cascade efficiency was influenced by the relative orientation of the scaffolded enzymes, with optimal lengths of 14- and 16-base pair aptamers (Figure 11.4e).

Till now, the *in vivo* DNA-scaffolded enzyme reactions have been limited to dsDNA constructs [71–73]. While DNA nanostructures with greater complexity have achieved successful isothermal assembly *in vitro* [74, 75], the realization of such assembly within intracellular environments remains elusive. As an alternative approach, the DNA scaffold can be assembled *in vitro* before cellular internalization, while several key aspects should be addressed: (i) DNA scaffold engineering for lysosomal escape, enabling protein assembly in the cytoplasm or other subcellular locations. (ii) Optimization of DNA-protein assembly methods, such as utilizing DNA aptamers or DNA binding proteins, to achieve high-yield protein immobilization *in vivo*. (iii) Development of advanced imaging techniques to directly visualize DNA scaffolds with assembled enzymes in living cells. Overall, while these few studies are encouraging, further research and development efforts are needed to unlock the full potential of DNA/RNA-scaffolded enzyme cascades in living cells.

11.4 DNA Nanostructures Empowered Bioimaging Technologies

Bioimaging technologies have revolutionized our understanding of cellular processes and structures and continue to be an indispensable tool in biological research, including fluorescence microscopy, super-resolution microscopy, and cryogenic electron microscopy (cryo-EM). As a biocompatible and highly programmable framework, DNA nanostructure has displayed its advantages in promoting bioimaging modalities, such as multiplexed [76] and logical responsive [77] imaging, protein structural analysis [14, 78], etc.

11.4.1 DNA Nanostructures Scaffolded Fluorophore Expansion

DNA nanoscaffolds can arrange fluorophores with high spatial accuracy [79] which can be conveniently applied to fluorescence bioimaging [80–82]. Their diversity and biocompatibility provided access to many advanced features of in-cell or *in vivo* fluorescence imaging. With the introduction of the targeting molecule angiopep-2, Fan's group developed a Dylight 775 carrying DNA probe that can cross the blood–brain barrier and be targeted to the glioma cells for brain tumor imaging [83]

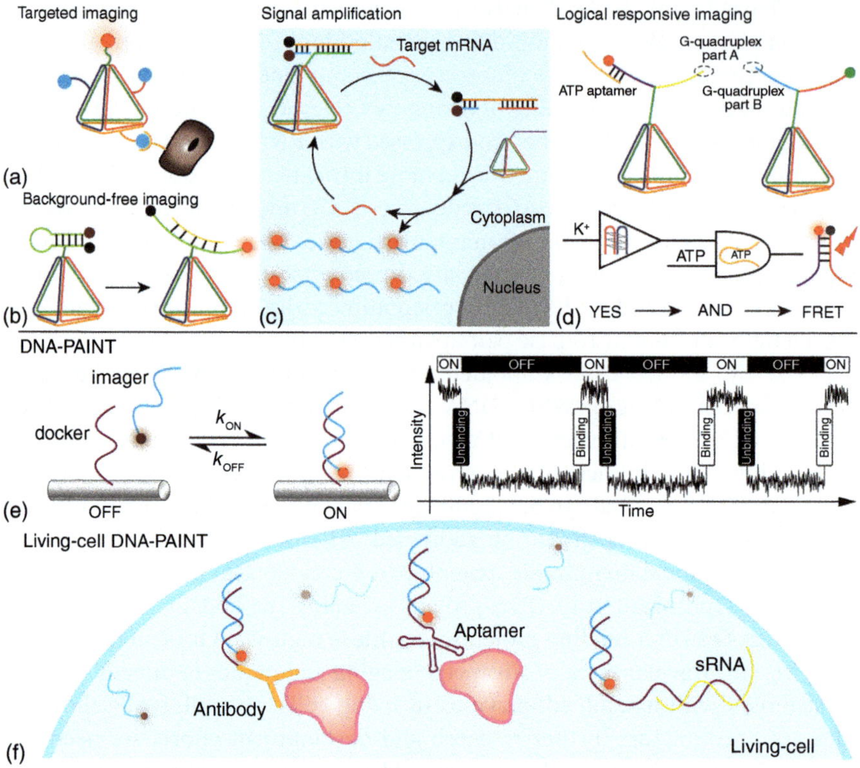

Figure 11.5 DNA nanotechnology expended fluorescence imaging. (a) Brain tumor-targeting imaging probe. (b) Background-free imaging based on FRET. (c) Signal amplifier for ultra-sensitive mRNA detection. (d) Logical responsive DNA probes. (e) The switching of "ON" and "OFF" states in DNA-PAINT strategy (left) and fluorescence intensity sequence diagram of DNA-PAINT (right). Source: Reproduced with permission from Jungmann et al. [84]. © 2010 American Chemical Society. (f) Methods to achieve living-cell DNA-PAINT.

(Figure 11.5a). Fluorescence bioimaging has long been plagued by background noise, while a rationally designed DNA sensor can quench the fluorophore before the target molecules are detected, thus avoiding the interference from background signal [85] (Figure 11.5b). Signal amplification is another advanced application of DNA probes. He et al. designed a DNA amplifier for the ultrasensitive detection of mRNA, which can undergo an autonomous strand displacement circuit initiated by mRNA in living cells (Figure 11.5c) to release enormous fluorescent signals [86]. Moreover, the feature of smart responding bioimaging can be achieved by DNA-based logic gates. Using the reconfigurable tetrahedrons as the framework, taking multiple biomolecules and ions as the inputs with fluorescence as the output, Fan's group has realized the INH, XOR, OR, AND logic operators in living cells [77]. Similarly, Peng's group also designed a smart responding imaging probe controlled by a "YES-AND" logic circuit that can be anchored to the cell membrane and then marks the level of extracellular K^+ and ATP [17] (Figure 11.5d).

11.4.2 DNA-PAINT-Based Super-Resolution Fluorescence Imaging

Due to the extensive exploitation of optical fluorescence microscopy, overcoming the resolution limit caused by light diffraction has drawn increasing interest. As a strategy for super-resolution imaging, the Point Accumulation for Imaging in Nanoscale Topography (PAINT) [87] adopted fluorescence probes that transiently bind to targets to achieve a stochastic on–off fluorescence pattern to circumvent this challenge. DNA-based PAINT (DNA-PAINT, Figure 11.5e) applies fluorophores-modified short oligonucleotides as probes (imager strands) to temporary binding with targets that are labeled with complementary sequences (docking strands) [84, 88]. The programmability of super-resolution imaging has therefore been enabled by the contribution of DNA nanotechnology.

Living-cell DNA-PAINT (Figure 11.5f) can be performed by modifying either the imager or docking strands with special bio-affinity labels. For example, Strauss et al. have developed a protein aptamer with a docking sequence to image the POI in living cells [89]. Additionally, the PAINT strategy can also be utilized to visualize the cellular distribution of small RNAs (sRNAs) by using them as the docking site [90]. Besides, other biomolecule labels such as antibodies, nanobodies, and phalloidin (label to actin fibers) [91] also show promise as the medium for DNA-PAINT.

11.4.3 DNA Nanostructures-Assisted Cryogenic Electron Microscopy Characterization

In the past three decades, cryogenic electron microscopy (cryo-EM) has been applied to determine the structures of biomacromolecules [92]. Single particle cryo-EM single particle technology takes rapidly cryofixed ice layer as the sample for transmission electron microscopy (TEM) imaging, and then computationally produces a 3D density map by averaging a large number of individual 2D particles in random orientations extracted from the obtained electron micrographs. After Mao's group first characterized DNA polyhedron in 2008 [93], various types of DNA nanostructures have been successfully reconstructed by cryo-EM [94–97], providing valuable references for the research of DNA nanoscaffold-assisted cryo-EM analysis of biomacromolecules and cellular structures. Based on the highly addressable nature of DNA nanostructures, a "molecular support" [78] and a "molecular goniometer" [14] scaffold have been designed to precisely control the orientate of individual proteins during cryo-EM imaging.

As a modality of cryo-EM, cryogenic electron tomography (cryo-ET) technology produces a 3D reconstruction from a series of differently tilted microscopic images without averaging, which is favorable for cell investigations. However, identifying small cellular features from a cryo-ET density map is often challenging. In 2021, a tagging strategy for cryo-electron tomography (cryo-ET) presented by Silvester et al. has enabled the demonstration of membrane proteins on the cell surface by using DNA origami "signposts" [98] (Figure 11.6). Here, the DNA frameworks serve as tags that label target proteins on the cellular membrane, thereby enhancing their contrasts in the tomograms. Additionally, the asymmetric shapes provided contextual information that can clearly indicate the location of the targets of interest.

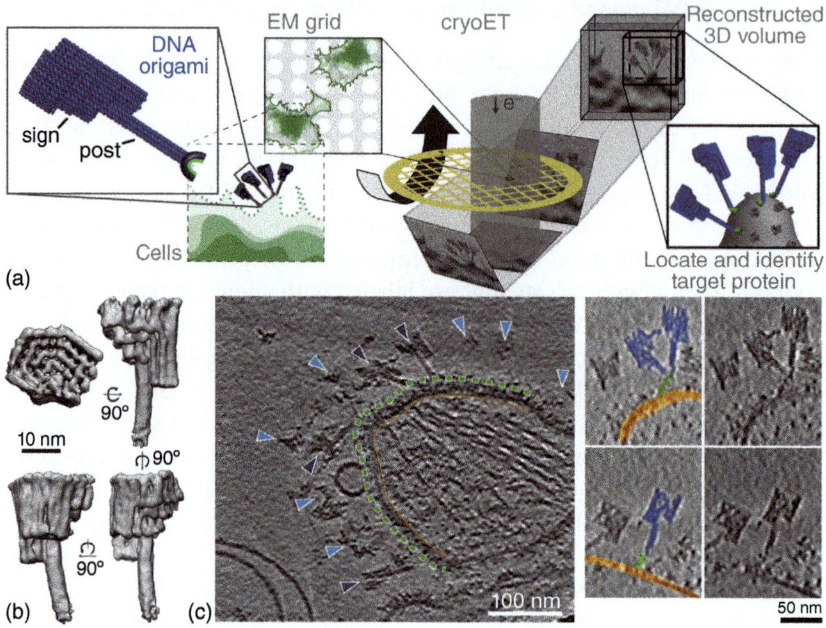

Figure 11.6 DNA origami "signposts" for cellular membrane tagging. Source: Emma Silvester et al. [98]/Reproduced with permission from Elsevier. (a) Schematic diagram of signposts workflow. (b) Structural characterization of signpost origami. (c) Computed tomogram slices.

The biocompatibility and non-cytotoxicity nature of DNA origami further enabled the possibilities of labeling organelles and subcellular regions within the cytoplasm. However, several issues need to be first addressed, including but not limited to cellular uptake, intracellular distribution and intracellular stability of DNA origami.

11.5 Summary and Outlook

The outstanding programmability and addressability of DNA origami make it an ideal nanoscale scaffold. This chapter begins by summarizing the three common steps of constructing DNA origami-scaffolded protein structures. Later, we separately discussed their applications in the study of biomacromolecules and bioimaging technologies through arranging proteins, fluorophores, and other functional groups into unique conformations to achieve special purposes. Here, some of the technical challenges that need to be addressed for better utilization of DNA origami nanoscaffolds are listed below.

Firstly, the stability of the overall complexes needs to be optimized to achieve *in vivo* functionalities. DNA has always been regarded as a biomolecule with excellent cellular biocompatibility and low cytotoxicity [99], but it also faces challenges in maintaining its integrity in biological buffers or cellular environments. Despite that many researchers have investigated and achieved the improved stability of DNA origami [100], the connecting domains between DNA and functional groups have often been neglected, which is in fact the most fragile part of the whole complex. For example, the probes extending from nanoscaffolds that bind the functional groups are possible to be degraded by enzymes. Investigating how factors such as the number, length, and sequence affect the stable duration time of the assembled structures can keep intact can provide a starting point for systematical research.

Secondly, albeit the utilization of DNA scaffold-assisted cryo-EM imaging has been proven feasible, several technological bottlenecks still impede their further applications. (i) It is important to note that the relative motions of proteins on DNA are common and can result in image blurring or even invisibility after averaging during the imaging process. Thus, establishing a more rigid and specific DNA-protein linkage at the sample preparation stage is crucial to constrain the movement. (ii) Although DNA origami does not severely obscure samples like metal tags [101], their delocalized signal originating would still exhibit a shadow overlap with cell signals, interfering with the progress of subsequent 3D reconstruction [14]. Thus, to minimize the image artifacts and maintain a high apparent resolution, a sufficient distance between DNA origami and protein should be ensured. Moreover, the intrinsic flexibility of DNA nanostructures, especially the periphery of structures [95], may also blur the proteins of interest unless the corresponding signal is extracted from the images. Due to the challenges caused by DNA scaffolds in cryo-EM imaging, it is necessary to design more distinguishable DNA scaffolds, where simulation software such as CanDo [102] and oxDNA [103] could be helpful feedback platforms with minimal experimental expenses.

DNA nanotechnology has demonstrated versatile utilizations in biomolecular arrangement, biosensing and bioimaging. Future efforts devoted to addressing the above-mentioned challenges would make DNA nanoscaffold a more efficacious molecular template, allowing it to better serve as a designable molecular platform for intracellular applications.

References

1 Dubey, N.C. and Tripathi, B.P. (2021). Nature inspired multienzyme immobilization: strategies and concepts. *ACS Appl. Bio Mater.* 4 (2): 1077–1114. https://doi.org/10.1021/acsabm.0c01293.

2 Rajendran, A., Endo, M., Katsuda, Y. et al. (2011). Programmed two-dimensional self-assembly of multiple DNA origami jigsaw pieces. *ACS Nano* 5 (1): 665–671. https://doi.org/10.1021/nn1031627.

3 Rothemund, P.W.K. (2006). Folding DNA to create nanoscale shapes and patterns. *Nature* 440 (7082): 297–302. https://doi.org/10.1038/nature04586.

4 Ke, Y., Douglas, S.M., Liu, M. et al. (2009). Multilayer DNA origami packed on a square lattice. *JACS* 131 (43): 15903–15908. https://doi.org/10.1021/ja906381y.

5 Monferrer, A., Kretzmann, J.A., Sigl, C. et al. (2022). Broad-spectrum virus trapping with heparan sulfate-modified DNA origami shells. *ACS Nano*. https://doi.org/10.1021/acsnano.1c11328.

6 Ke, Y., Sharma, J., Liu, M. et al. (2009). Scaffolded DNA origami of a DNA tetrahedron molecular container. *Nano Lett.* 9 (6): 2445–2447. https://doi.org/10.1021/nl901165f.

7 Ke, Y., Voigt, N.V., Gothelf, K.V., and Shih, W.M. (2012). Multilayer DNA origami packed on hexagonal and hybrid lattices. *JACS* 134 (3): 1770–1774. https://doi.org/10.1021/ja209719k.

8 Dietz, H., Douglas, S.M., and Shih, W.M. (2009). Folding DNA into twisted and curved nanoscale shapes. *Forensic Sci.* 325 (5941): 725–730. https://doi.org/10.1126/science.1174251.

9 Han, D., Pal, S., Nangreave, J. et al. (2011). DNA origami with complex curvatures in three-dimensional space. *Forensic Sci.* 332 (6027): 342–346. https://doi.org/10.1126/science.1202998.

10 Ijäs, H., Nummelin, S., Shen, B. et al. (2018). Dynamic DNA origami devices: from strand-displacement reactions to external-stimuli responsive systems. *Int. J. Mol. Sci.* 19 (7).

11 Marras, A.E. (2022). Hierarchical assembly of DNA origami nanostructures. *MRS Commun.* 12 (5): 543–551. https://doi.org/10.1557/s43579-022-00248-8.

12 Sigl, C., Willner, E.M., Engelen, W. et al. (2021). Programmable icosahedral shell system for virus trapping. *Nat. Mater.* 20 (9): 1281–1289. https://doi.org/10.1038/s41563-021-01020-4.

13 Wintersinger, C.M., Minev, D., Ershova, A. et al. (2022). Multi-micron crisscross structures grown from DNA-origami slats. *Nat. Nanotechnol.* https://doi.org/10.1038/s41565-022-01283-1.

14 Aksel, T., Yu, Z., Cheng, Y., and Douglas, S.M. (2021). Molecular goniometers for single-particle cryo-electron microscopy of DNA-binding proteins. *Nat. Biotechnol.* 39 (3): 378–386. https://doi.org/10.1038/s41587-020-0716-8.

15 Seitz, I., Ijas, H., Linko, V., and Kostiainen, M.A. (2022). Optically responsive protein coating of DNA origami for triggered antigen targeting. *ACS Appl. Mater. Interfaces* 14 (34): 38515–38524. https://doi.org/10.1021/acsami.2c10058.

16 Wang, C.-H., Chen, X.-Q., Su, Y.-Y. et al. (2022). Precise regulating T cell activation signaling with spatial controllable positioning of receptors on DNA origami. *Chin. J. Anal. Chem.* 50 (6). https://doi.org/10.1016/j.cjac.2022.100091.

17 Gong, H., Dai, Q., and Peng, P. (2022). Cell-membrane-anchored DNA logic-gated nanoassemblies for in situ extracellular bioimaging. *ACS Appl. Mater. Interfaces*. https://doi.org/10.1021/acsami.2c13735.

18 Li, J., Dai, J., Jiang, S. et al. (2020). Encoding quantized fluorescence states with fractal DNA frameworks. *Nat. Commun.* 11 (1): 2185. https://doi.org/10.1038/s41467-020-16112-z.

19 Woehrstein, J.B., Strauss, M.T., Ong, L.L. et al. (2017). Sub-100-nm metafluorophores with digitally tunable optical properties self-assembled from DNA. *Sci. Adv.* 3 (6): e1602128. https://doi.org/10.1126/sciadv.1602128.

20 Lei, Y.X., Mei, Z.C., Chen, Y.F. et al. (2022). Facile purification and concentration of DNA origami structures by ethanol precipitation. *ChemNanoMat* 8 (7): e202200161. https://doi.org/10.1002/cnma.202200161.

21 Niu, R., Song, C., Gao, F. et al. (2021). DNA origami-based nanoprinting for the assembly of plasmonic nanostructures with single-molecule surface-enhanced Raman scattering. *Angew. Chem. Int. Ed.* 60 (21): 11695–11701. https://doi.org/10.1002/anie.202016014.

22 Iric, K., Subramanian, M., Oertel, J. et al. (2018). DNA-encircled lipid bilayers. *Nanoscale* 10 (39): 18463–18467. https://doi.org/10.1039/c8nr06505e.

23 Piao, J., Zhao, C., and Dong, Y. (2022). DNA nanostructure-assisted nanodiscs provide a toolbox to investigate membrane proteins. *Cell Rep. Phys. Sci.* 3 (6). https://doi.org/10.1016/j.xcrp.2022.100897.

24 Wang, C., Piao, J., Li, Y. et al. (2020). Construction of liposomes mimicking cell membrane structure through frame-guided assembly. *Angew. Chem. Int. Ed. Engl.* 59 (35): 15176–15180. https://doi.org/10.1002/anie.202005334.

25 Integrated DNA Technologies, Inc. (2023). DNA oligos and ultramer(TM) DNA oligos. https://sfvideo.blob.core.windows.net/sitefinity/docs/default-source/flyer/idt-dna-oligos-and-ultramer-dna-oligos-flyer.pdf?sfvrsn=e6221a07_8 (accessed 21 June 2023).

26 Hoyt, E.A., Cal, P.M.S.D., Oliveira, B.L., and Bernardes, G.J.L. (2019). Contemporary approaches to site-selective protein modification. *Nat. Rev. Chem.* 3 (3): 147–171. https://doi.org/10.1038/s41570-019-0079-1.

27 Weber, P.C., Ohlendorf, D.H., Wendoloski, J.J., and Salemme, F.R. (1989). Structural origins of high-affinity biotin binding to streptavidin. *Forensic Sci.* 243 (4887): 85–88. https://doi.org/10.1126/science.2911722.

28 Kurokawa, T., Kiyonaka, S., Nakata, E. et al. (2018). DNA origami scaffolds as templates for functional tetrameric Kir3 K(+) channels. *Angew. Chem. Int. Ed. Engl.* 57 (10): 2586–2591. https://doi.org/10.1002/anie.201709982.

29 Nakata, E., Liew, F.F., Uwatoko, C. et al. (2012). Zinc-finger proteins for site-specific protein positioning on DNA-origami structures. *Angew. Chem. Int. Ed. Engl.* 51 (10): 2421–2424. https://doi.org/10.1002/anie.201108199.

30 Ngo, T.A., Nakata, E., Saimura, M., and Morii, T. (2016). Spatially organized enzymes drive cofactor-coupled cascade reactions. *JACS* 138 (9): 3012–3021. https://doi.org/10.1021/jacs.5b10198.

31 Sakamoto, S. and Hamachi, I. (2019). Recent progress in chemical modification of proteins. *Anal. Sci.* 35 (1): 5–27. https://doi.org/10.2116/analsci.18R003.

32 Marth, G., Hartley, A.M., Reddington, S.C. et al. (2017). Precision templated bottom-up multiprotein nanoassembly through defined click chemistry linkage to DNA. *ACS Nano* 11 (5): 5003–5010. https://doi.org/10.1021/acsnano.7b01711.

33 Zhang, Y., Park, K.Y., Suazo, K.F., and Distefano, M.D. (2018). Recent progress in enzymatic protein labelling techniques and their applications. *Chem. Soc. Rev.* 47 (24): 9106–9136. https://doi.org/10.1039/c8cs00537k.

34 Fu, J., Yang, Y.R., Dhakal, S. et al. (2016). Assembly of multienzyme complexes on DNA nanostructures. *Nat. Protoc.* 11 (11): 2243–2273. https://doi.org/10.1038/nprot.2016.139.

35 Kalendar, R. (2022). A guide to using FASTPCR software for PCR, in silico PCR, and oligonucleotide analysis. In: *PCR Primer Design* (ed. C. Basu), 223–243. Springer US. https://doi.org/10.1007/978-1-0716-1799-1_16.

36 Abe, K., Sugiyama, H., and Endo, M. (2021). Construction of an optically controllable CRISPR-Cas9 system using a DNA origami nanostructure. *Chem. Commun.* 57 (45): 5594–5596. https://doi.org/10.1039/d1cc00876e.

37 Shaw, A., Benson, E., and Hogberg, B. (2015). Purification of functionalized DNA origami nanostructures. *ACS Nano* 9 (5): 4968–4975. https://doi.org/10.1021/nn507035g.

38 Numajiri, K., Yamazaki, T., Kimura, M. et al. (2010). Discrete and active enzyme nanoarrays on DNA origami scaffolds purified by affinity tag separation. *JACS* 132 (29): 9937–9939. https://doi.org/10.1021/ja104702q.

39 Zhang, F., Nangreave, J., Liu, Y., and Yan, H. (2014). Structural DNA nanotechnology: state of the art and future perspective. *JACS* 136 (32): 11198–11211. https://doi.org/10.1021/ja505101a.

40 Jiao, Y., Shang, Y., Li, N., and Ding, B. (2022). DNA-based enzymatic systems and their applications. *iScience* 25 (4): 104018. https://doi.org/10.1016/j.isci.2022.104018.

41 Fu, J., Liu, M., Liu, Y. et al. (2012). Interenzyme substrate diffusion for an enzyme cascade organized on spatially addressable DNA nanostructures. *JACS* 134 (12): 5516–5519. https://doi.org/10.1021/ja300897h.

42 Liu, M., Fu, J., Hejesen, C. et al. (2013). A DNA tweezer-actuated enzyme nanoreactor. *Nat. Commun.* 4: 2127. https://doi.org/10.1038/ncomms3127.

43 Grossi, G., Dalgaard Ebbesen Jepsen, M., Kjems, J., and Andersen, E.S. (2017). Control of enzyme reactions by a reconfigurable DNA nanovault. *Nat. Commun.* 8 (1): 992. https://doi.org/10.1038/s41467-017-01072-8.

44 Linko, V., Eerikainen, M., and Kostiainen, M.A. (2015). A modular DNA origami-based enzyme cascade nanoreactor. *Chem. Commun.* 51 (25): 5351–5354. https://doi.org/10.1039/c4cc08472a.

45 Zhao, Z., Fu, J., Dhakal, S. et al. (2016). Nanocaged enzymes with enhanced catalytic activity and increased stability against protease digestion. *Nat. Commun.* 7: 10619. https://doi.org/10.1038/ncomms10619.

46 Klein, W.P., Thomsen, R.P., Turner, K.B. et al. (2019). Enhanced catalysis from multienzyme cascades assembled on a DNA origami triangle. *ACS Nano* 13 (12): 13677–13689. https://doi.org/10.1021/acsnano.9b05746.

47 Erkelenz, M., Kuo, C.H., and Niemeyer, C.M. (2011). DNA-mediated assembly of cytochrome P450 BM3 subdomains. *JACS* 133 (40): 16111–16118. https://doi.org/10.1021/ja204993s.

48 Niemeyer, C.M., Koehler, J., and Wuerdemann, C. (2002). DNA-directed assembly of bienzymic complexes from in vivo biotinylated NAD(P)H:FMN oxidoreductase and luciferase. *ChemBioChem* 3 (2–3): 242–245. https://doi.org/10.1002/1439-7633(20020301)3:2/3<242::AID-CBIC242>3.0.CO;2-F.

49 Sun, Q., Madan, B., Tsai, S.L. et al. (2014). Creation of artificial cellulosomes on DNA scaffolds by zinc finger protein-guided assembly for efficient cellulose hydrolysis. *Chem. Commun.* 50 (12): 1423–1425. https://doi.org/10.1039/c3cc47215a.

50 Wilner, O.I., Weizmann, Y., Gill, R. et al. (2009). Enzyme cascades activated on topologically programmed DNA scaffolds. *Nat. Nanotechnol.* 4 (4): 249–254. https://doi.org/10.1038/nnano.2009.50.

51 Liu, M., Fu, J., Qi, X. et al. (2016). A three-enzyme pathway with an optimised geometric arrangement to facilitate substrate transfer. *ChemBioChem* 17 (12): 1097–1101. https://doi.org/10.1002/cbic.201600103.

52 Kahn, J.S., Xiong, Y., Huang, J., and Gang, O. (2022). Cascaded enzyme reactions over a three-dimensional, wireframe DNA origami scaffold. *JACS Au* 2 (2): 357–366. https://doi.org/10.1021/jacsau.1c00387.

53 Zhou, C., Yang, Z., and Liu, D. (2012). Reversible regulation of protein binding affinity by a DNA machine. *JACS* 134 (3): 1416–1418. https://doi.org/10.1021/ja209590u.

54 Xin, L., Zhou, C., Yang, Z., and Liu, D. (2013). Regulation of an enzyme cascade reaction by a DNA machine. *Small* 9 (18): 3088–3091. https://doi.org/10.1002/smll.201300019.

55 Marras, A.E., Shi, Z., Lindell, M.G. 3rd, et al. (2018). Cation-activated avidity for rapid reconfiguration of DNA nanodevices. *ACS Nano* 12 (9): 9484–9494. https://doi.org/10.1021/acsnano.8b04817.

56 Hu, Y., Wang, F., Lu, C.H. et al. (2014). Switchable enzyme/DNAzyme cascades by the reconfiguration of DNA nanostructures. *Chem. Eur. J.* 20 (49): 16203–16209. https://doi.org/10.1002/chem.201404122.

57 Kou, B., Chai, Y., Yuan, Y., and Yuan, R. (2018). Dynamical regulation of enzyme cascade amplification by a regenerated DNA nanotweezer for ultrasensitive electrochemical DNA detection. *Anal. Chem.* 90 (18): 10701–10706. https://doi.org/10.1021/acs.analchem.8b00477.

58 Fu, J., Wang, Z., Liang, X.H. et al. (2020). DNA-scaffolded proximity assembly and confinement of multienzyme reactions. *Top. Curr. Chem.* 378 (3): 38. https://doi.org/10.1007/s41061-020-0299-3.

59 Huang, J., Gambietz, S., and Sacca, B. (2022). Self-assembled artificial DNA nanocompartments and their bioapplications. *Small* e2202253. https://doi.org/10.1002/smll.202202253.

60 Kim, S.H., Kim, K.R., Ahn, D.R. et al. (2017). Reversible regulation of enzyme activity by pH-responsive encapsulation in DNA nanocages. *ACS Nano* 11 (9): 9352–9359. https://doi.org/10.1021/acsnano.7b04766.

61 Fu, Y., Zeng, D., Chao, J. et al. (2013). Single-step rapid assembly of DNA origami nanostructures for addressable nanoscale bioreactors. *JACS* 135 (2): 696–702. https://doi.org/10.1021/ja3076692.

62 Xiong, Y., Huang, J., Wang, S.T. et al. (2020). Local environment affects the activity of enzymes on a 3D molecular scaffold. *ACS Nano* 14 (11): 14646–14654. https://doi.org/10.1021/acsnano.0c03962.

63 Fu, J., Yang, Y.R., Johnson-Buck, A. et al. (2014). Multi-enzyme complexes on DNA scaffolds capable of substrate channelling with an artificial swinging arm. *Nat. Nanotechnol.* 9 (7): 531–536. https://doi.org/10.1038/nnano.2014.100.

64 Yang, Y.R., Fu, J., Wootten, S. et al. (2018). 2D enzyme cascade network with efficient substrate channeling by swinging arms. *ChemBioChem* 19 (3): 212–216. https://doi.org/10.1002/cbic.201700613.

65 Ke, G., Liu, M., Jiang, S. et al. (2016). Directional regulation of enzyme pathways through the control of substrate channeling on a DNA origami scaffold. *Angew. Chem. Int. Ed. Engl.* 55 (26): 7483–7486. https://doi.org/10.1002/anie.201603183.

66 Wang, Z., St Iago-Mcrae, E., Ebrahimimojarad, A., et al. (2022). Modulation of enzyme cascade activity by local substrate enrichment and exclusion on DNA nanostructures. *Langmuir*. https://doi.org/10.1021/acs.langmuir.2c02064

67 Lin, C., Rinker, S., Wang, X. et al. (2008). In vivo cloning of artificial DNA nanostructures. *PNAS* 105 (46): 17626–17631. https://doi.org/10.1073/pnas.0805416105.

68 Rajendran, A., Nakata, E., Nakano, S., and Morii, T. (2017). Nucleic-acid-templated enzyme cascades. *ChemBioChem* 18 (8): 696–716. https://doi.org/10.1002/cbic.201600703.

69 Delebecque, C.J., Lindner, A.B., Silver, P.A., and Aldaye, F.A. (2011). Organization of intracellular reactions with rationally designed RNA assemblies. *Forensic Sci.* 333 (6041): 470–474. https://doi.org/10.1126/science.1206938.

70 Sachdeva, G., Garg, A., Godding, D. et al. (2014). In vivo co-localization of enzymes on RNA scaffolds increases metabolic production in a geometrically dependent manner. *Nucleic Acids Res.* 42 (14): 9493–9503. https://doi.org/10.1093/nar/gku617.

71 Baneyx, F. and Mujacic, M. (2004). Recombinant protein folding and misfolding in *Escherichia coli*. *Nat. Biotechnol.* 22 (11): 1399–1408. https://doi.org/10.1038/nbt1029.

72 Conrado, R.J., Wu, G.C., Boock, J.T. et al. (2012). DNA-guided assembly of biosynthetic pathways promotes improved catalytic efficiency. *Nucleic Acids Res.* 40 (4): 1879–1889. https://doi.org/10.1093/nar/gkr888.

73 Ponchon, L. and Dardel, F. (2007). Recombinant RNA technology: the tRNA scaffold. *Nat. Methods* 4 (7): 571–576. https://doi.org/10.1038/nmeth1058.

74 Jungmann, R., Liedl, T., Sobey, T.L. et al. (2008). Isothermal assembly of DNA origami structures using denaturing agents. *JACS* 130 (31): 10062–10063. https://doi.org/10.1021/ja8030196.

75 Zhang, Z., Song, J., Besenbacher, F. et al. (2013). Self-assembly of DNA origami and single-stranded tile structures at room temperature. *Angew. Chem. Int. Ed. Engl.* 52 (35): 9219–9223. https://doi.org/10.1002/anie.201303611.

76 Jungmann, R., Avendano, M.S., Woehrstein, J.B. et al. (2014). Multiplexed 3D cellular super-resolution imaging with DNA-PAINT and Exchange-PAINT. *Nat. Methods* 11 (3): 313–318. https://doi.org/10.1038/nmeth.2835.

77 Pei, H., Liang, L., Yao, G. et al. (2012). Reconfigurable three-dimensional DNA nanostructures for the construction of intracellular logic sensors. *Angew. Chem. Int. Ed. Engl.* 51 (36): 9020–9024. https://doi.org/10.1002/anie.201202356.

78 Martin, T.G., Bharat, T.A., Joerger, A.C. et al. (2016). Design of a molecular support for cryo-EM structure determination. *PNAS* 113 (47): E7456–E7463. https://doi.org/10.1073/pnas.1612720113.

79 Mathur, D., Kim, Y.C., Díaz, S.A. et al. (2020). Can a DNA origami structure constrain the position and orientation of an attached dye molecule? *J. Phys. Chem. C* 125 (2): 1509–1522. https://doi.org/10.1021/acs.jpcc.0c09258.

80 Chao, H., Qiu, L., Zhou, X. et al. (2022). Separation of ofloxacin enantiomers by capillary electrophoresis with fluorescence detection using deoxyribonucleic acid oligonucleotides as chiral selectors. *J. Sep. Sci.* 45 (14): 2699–2707. https://doi.org/10.1002/jssc.202200209.

81 Liu, X., Zhou, X., Xia, X., and Xiang, H. (2020). Catalytic hairpin assembly-based double-end DNAzyme cascade-feedback amplification for sensitive fluorescence detection of HIV-1 DNA. *Anal. Chim. Acta* 1096: 159–165. https://doi.org/10.1016/j.aca.2019.10.051.

82 Pfeiffer, M., Trofymchuk, K., Ranallo, S. et al. (2021). Single antibody detection in a DNA origami nanoantenna. *iScience* 24 (9): 103072. https://doi.org/10.1016/j.isci.2021.103072.

83 Tian, T., Li, J., Xie, C. et al. (2018). Targeted imaging of brain tumors with a framework nucleic acid probe. *ACS Appl. Mater. Interfaces* 10 (4): 3414–3420. https://doi.org/10.1021/acsami.7b17927.

84 Jungmann, R., Steinhauer, C., Scheible, M. et al. (2010). Single-molecule kinetics and super-resolution microscopy by fluorescence imaging of transient binding on DNA origami. *Nano Lett.* 10 (11): 4756–4761. https://doi.org/10.1021/nl103427w.

85 Tay, C.Y., Yuan, L., and Leong, D.T. (2015). Nature-inspired DNA nanosensor for real-time in situ detection of mRNA in living cells. *ACS Nano* 9 (5): 5609–5617. https://doi.org/10.1021/acsnano.5b01954.

86 He, L., Lu, D., Liang, H. et al. (2018). mRNA-initiated, three-dimensional DNA amplifier able to function inside living cells. *JACS* 140 (1): 258–263. https://doi.org/10.1021/jacs.7b09789.

87 van Wee, R., Filius, M., and Joo, C. (2021). Completing the canvas: advances and challenges for DNA-PAINT super-resolution imaging. *Trends Biochem. Sci* 46 (11): 918–930. https://doi.org/10.1016/j.tibs.2021.05.010.

88 Schnitzbauer, J., Strauss, M.T., Schlichthaerle, T. et al. (2017). Super-resolution microscopy with DNA-PAINT. *Nat. Protoc.* 12 (6): 1198–1228. https://doi.org/10.1038/nprot.2017.024.

89 Strauss, S., Nickels, P.C., Strauss, M.T. et al. (2018). Modified aptamers enable quantitative sub-10-nm cellular DNA-PAINT imaging. *Nat. Methods* 15 (9): 685–688. https://doi.org/10.1038/s41592-018-0105-0.

90 Huang, K., Demirci, F., Batish, M. et al. (2020). Quantitative, super-resolution localization of small RNAs with sRNA-PAINT. *Nucleic Acids Res.* 48 (16): e96. https://doi.org/10.1093/nar/gkaa623.

91 Agasti, S.S., Wang, Y., Schueder, F. et al. (2017). DNA-barcoded labeling probes for highly multiplexed Exchange-PAINT imaging. *Chem. Sci.* 8 (4): 3080–3091. https://doi.org/10.1039/c6sc05420j.

92 Bai, X.-c., McMullan, G., and Scheres, S.H.W. (2015). How cryo-EM is revolutionizing structural biology. *Trends Biochem. Sci* 40 (1): 49–57. https://doi.org/10.1016/j.tibs.2014.10.005.

93 He, Y., Ye, T., Su, M. et al. (2008). Hierarchical self-assembly of DNA into symmetric supramolecular polyhedra. *Nature* 452 (7184): 198–201. https://doi.org/10.1038/nature06597.

94 Andersen, E.S., Dong, M., Nielsen, M.M. et al. (2009). Self-assembly of a nanoscale DNA box with a controllable lid. *Nature* 459 (7243): 73–76. https://doi.org/10.1038/nature07971.

95 Bai, X.C., Martin, T.G., Scheres, S.H., and Dietz, H. (2012). Cryo-EM structure of a 3D DNA-origami object. *PNAS* 109 (49): 20012–20017. https://doi.org/10.1073/pnas.1215713109.

96 Kube, M., Kohler, F., Feigl, E. et al. (2020). Revealing the structures of megadalton-scale DNA complexes with nucleotide resolution. *Nat. Commun.* 11 (1): 6229. https://doi.org/10.1038/s41467-020-20020-7.

97 Xu, Y., Yang, Y.R., Shi, Q. et al. (2023). An infectious virus-like particle built on a programmable icosahedral DNA framework. *Angew. Chem. Int. Ed. Engl.* 62 (4): e202214731. https://doi.org/10.1002/anie.202214731.

98 Silvester, E., Vollmer, B., Prazak, V. et al. (2021). DNA origami signposts for identifying proteins on cell membranes by electron cryotomography. *Cell* 184 (4): 1110–1121: e1116. https://doi.org/10.1016/j.cell.2021.01.033.

99 Zhang, Q., Jiang, Q., Li, N. et al. (2014). DNA origami as an in vivo drug delivery vehicle for cancer therapy. *ACS Nano* 8 (7): 6633–6643. https://doi.org/10.1021/nn502058j.

100 Bila, H., Kurisinkal, E.E., and Bastings, M.M.C. (2019). Engineering a stable future for DNA-origami as a biomaterial. *Biomater. Sci. UK* 7 (2): 532–541. https://doi.org/10.1039/C8BM01249K.

101 Fernandez, J.J., Laugks, U., Schaffer, M. et al. (2016). Removing contamination-induced reconstruction artifacts from cryo-electron tomograms. *Biophys. J.* 110 (4): 850–859. https://doi.org/10.1016/j.bpj.2015.10.043.

102 Kim, D.N., Kilchherr, F., Dietz, H., and Bathe, M. (2012). Quantitative prediction of 3D solution shape and flexibility of nucleic acid nanostructures. *Nucleic Acids Res.* 40 (7): 2862–2868. https://doi.org/10.1093/nar/gkr1173.

103 Poppleton, E., Bohlin, J., Matthies, M. et al. (2020). Design, optimization and analysis of large DNA and RNA nanostructures through interactive visualization, editing and molecular simulation. *Nucleic Acids Res.* 48 (12): e72. https://doi.org/10.1093/nar/gkaa417.

Part III

DNA Nanotechnology for Regulation of Cellular Functions

12

Adopting Nucleic Acid Nanotechnology for Genetic Regulation *In Vivo*

Friedrich C. Simmel

Technische Universität München, School of Natural Sciences, Department of Bioscience, Am Coulombwall 4a, 85748 Garching, Germany

12.1 Introduction

Traditionally, DNA nanotechnology [1] was devoted to the realization of intricate nanoscale structures based on the self-assembly of DNA molecules, utilizing their unique sequence-dependent molecular recognition properties. In the well-known DNA origami technique [2–4], a long single-stranded "scaffold" molecule is folded into a specific shape by connecting different domains of the scaffold via DNA staples whose sequence is chosen complementary to the corresponding domain sequences. Origami structures are typically 10–100 nm in size, and their shape can be almost arbitrarily designed.

Next to applications in structural DNA nanotechnology, DNA molecules were also heavily utilized for the realization of dynamical systems – molecular switches, motors, and molecular computing systems. At the core of most of these applications is the controlled switching of nucleic acid assemblies between different states using some external input, which could be physical (light, temperature, electric fields) or chemical (using changes in pH, small molecules, protein, or nucleic acid effectors).

It is often emphasized that the utilization of DNA in this context is "non-biological" as nature does not employ DNA molecules for switching or as a material. Indeed, DNA nanostructures have been used to arrange nanoparticles or fluorophores for nanoelectronic and photonic applications, and DNA nanostructures have been studied for templating materials, or for the proximity-induced enhancement of chemical reactions [5, 6]. DNA molecular devices were employed to develop molecular assembly lines [1] and nanorobotic systems [7–11]. As bare DNA does not have any functional role in non-biological applications, most of these potential applications require chemical modification of the DNA of some sort.

In the past years, nucleic acid nanotechnology has nevertheless increasingly focused on potential biomedical applications, in particular in the context of super-resolution-based bioimaging [12] or the development of smart drugs and drug

DNA Nanotechnology for Cell Research: From Bioanalysis to Biomedicine, First Edition. Edited by Zhou Nie.

delivery systems [7, 10, 13, 14]. Most of these applications still require chemical stabilization and functionalization (e.g. with antibodies) and cannot be realized with bare, non-functionalized nucleic acids.

In this chapter, we will discuss some of the more recent attempts to employ concepts from nucleic acid nanotechnology in the context of genetic engineering, which involves the use of switchable nucleic acid structures for gene regulation and the production of nucleic acid-based devices by gene expression. Importantly, these applications make use of DNA and RNA in their biological context – as genetic templates – and do not necessarily require chemical functionalization.

12.2 Toehold-Mediated Strand Displacement: Switching Nucleic Acids with Nucleic Acids

Switching nucleic acid structures with DNA or RNA effectors is based on the hybridization of the effectors to an appropriate single-stranded input domain with complementary sequence. Notably, such hybridization also leads to a change of mechanical properties, as single-stranded DNA or RNA is much more flexible than their duplex forms. As it is often desired to switch the systems back and forth between two or more distinct states, it is also necessary to be able to remove the effectors again, which is typically achieved using a toehold-mediated strand displacement (TMSD) process (Figure 12.1a) [15, 16]. In TMSD, a DNA strand termed the "invader" strand displaces another strand (the "incumbent") that is initially bound to a target or substrate strand. The invader is designed with a sequence complementary to the target. Initial binding of the invader to the incumbent-substrate complex occurs at a short single-stranded sequence domain called a "toehold." Once the invader strand has bound to the toehold, it can begin to displace the incumbent in a process known as branch migration. Strand displacement continues until the invader strand is fully bound to the target strand and the displaced strand is completely detached.

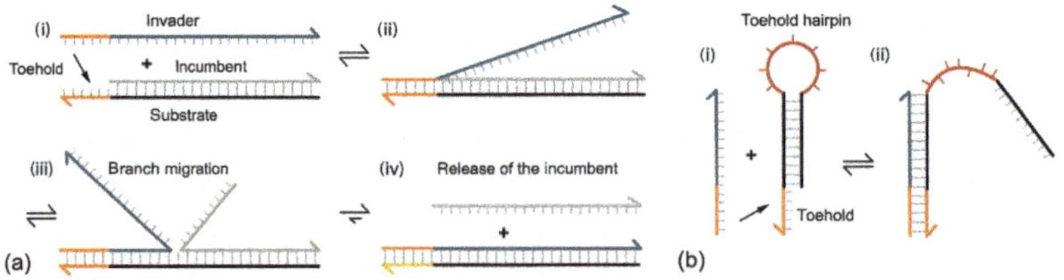

Figure 12.1 Toehold-mediated strand displacement (TMSD). (a) TMSD involves the displacement of an incumbent strand initially hybridized to a substrate strand by an invading strand. The binding of the invader is facilitated by a short single-stranded sequence domain on the substrate called "toehold." The invader then displaces the incumbent in a three-way branch migration process. As the invader can form more base pairs with the substrate than the incumbent, the process is biased in favor of the invader. (b) A similar process can also be used to break secondary structures, e.g. by toehold-mediated invasion of the stem of a hairpin structure. Arrowheads indicate the 3′ ends of the molecules.

The toehold provides the invader with a thermodynamic advantage, which makes the displacement of the incumbent energetically more favorable than the reverse process. Importantly, the toehold sequence simultaneously also acts as a molecular address that can be used to exert TMSD only on specific molecules that are identified through this sequence.

TMSD has turned out to be an extremely versatile method and has been used to create sophisticated dynamical systems with DNA, such as DNA-based molecular switches, molecular motors, and programmable *in vitro* chemical reaction networks, as reviewed, e.g. in Ref. [16].

The TMSD principle can also be applied to rearrange secondary structures (Figure 12.1b) in RNA molecules, which opens up the possibility of creating molecular devices and circuits that can be expressed and operated inside cells. In this context, RNA-based gene regulatory processes are particularly interesting as they can be directly coupled to synthetic RNA-based devices.

RNA-based regulation is found in numerous biological contexts, such as metabolic regulation, developmental processes, and immune responses, and it is found ubiquitously across prokaryotes, eukaryotes, and viruses. Prototypical examples include riboswitches, short RNAs (sRNAs) and CRISPR in bacteria, RNA interference (RNAi), and long noncoding RNAs in eukaryotes that deploy noncoding RNA molecules, along with protein complexes, to regulate gene expression at the transcriptional or translational level.

12.3 Toehold Riboregulators and Related Systems

12.3.1 Riboswitches

An intriguing example of RNA-based gene regulation is riboswitches, which are regulatory RNA elements primarily observed in bacteria (Figure 12.2A). Riboswitches are located in the 5′ untranslated regions (UTRs) of mRNA molecules, where they form folded RNA domains capable of altering their conformation in response to specific metabolites [19, 20]. A typical riboswitch encompasses an aptamer region for metabolite recognition, followed by an expression platform that modulates gene expression. The aptamer structure undergoes a reorganization upon the binding of metabolites such as guanine, adenine, or vitamin B12, triggering a conformational change in the expression platform. This, in turn, influences the expression of the downstream mRNA sequence, which may take place at the transcriptional or translation level. In the case of transcriptional switching, the riboswitch's conformational change results in the formation of a terminator structure in the mRNA, leading to the halting of the RNA polymerase and transcriptional termination. In the case of riboswitches that control gene expression at the translation stage, the binding of a specific metabolite triggers a conformational change that reveals the ribosome binding site (RBS) on the mRNA (see Figure 12.2A). The exposure of the RBS allows the ribosome to bind and initiate the translation process, thereby activating gene expression.

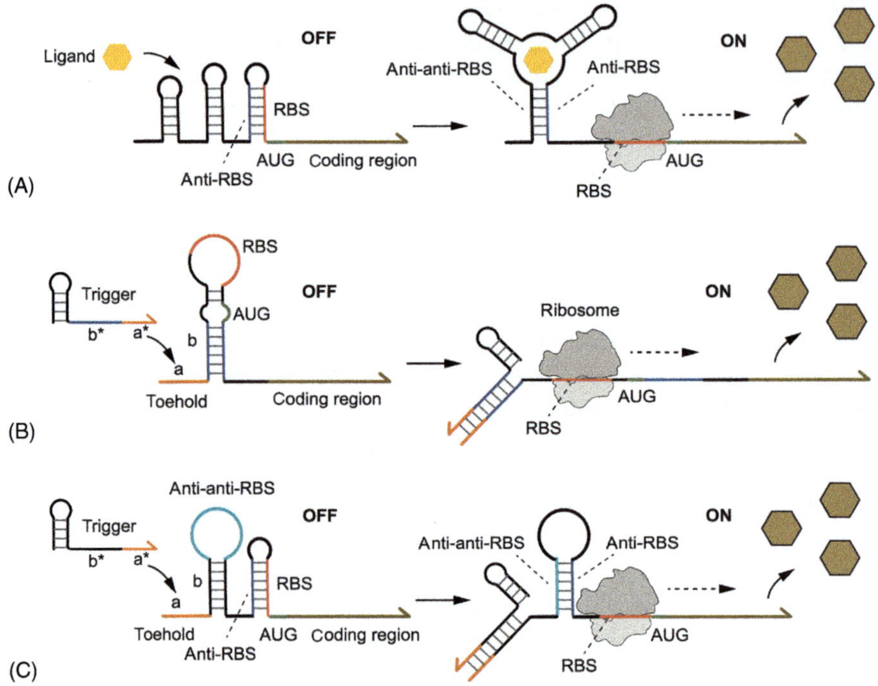

Figure 12.2 Riboswitches and toehold riboregulators. (A) Naturally occurring riboswitches are secondary structures in the 5′ untranslated regions of mRNAs that control transcription or translation processes in response to small-molecule ligands. The figure shows an example of a translational ON switch. In the initial (OFF) state, the RBS on the mRNA is sequestered by an anti-RBS sequence and thus translation is inhibited. In the presence of the ligand, the RNA structure folds into an aptamer that binds the ligand, which also makes the RBS accessible to the ribosome and thus activates the translation of the mRNA. (B) In artificial toehold switches [17], the RBS is sequestered within a hairpin loop, which can be opened by a trigger RNA via TMSD. Hence, translation is activated by an RNA input rather than a small molecule. The hairpin on the trigger stabilizes the RNA against degradation. Source: Adapted from Green et al. [17]. (C) A toehold riboregulator that merges the architecture of a riboswitch (with an anti-RBS-RBS hairpin) and the toehold switch [18]. Activation with an RNA trigger leads to the sequestration of the anti-RBS by an anti-anti-RBS, releasing the RBS and thus activating translation. Source: Adapted from Wang et al. [18].

12.3.2 Translational Switching with Toehold Switches

The regulatory principle of riboswitches has inspired the development of synthetic riboregulators, which, unlike their natural counterparts, are activated by RNA effector molecules rather than small metabolites. An early example of such a synthetic translational regulator involved integrating a custom-designed cis-repressing RNA hairpin motif into the 5′ UTR of an mRNA [21]. This motif sequestered the RBS, thereby obstructing translation. An activating sRNA molecule was introduced that could unfold the repressing RNA hairpin, giving ribosomes access to the RBS and enabling translation of the subsequent sequence, in this case, coding for GFP.

An enhanced version of this synthetic riboregulator with ON/OFF ratios up to several hundredfolds, known as the "toehold switch," was later demonstrated using the TMSD principle for switching (Figure 12.2B) [17]. The translational toehold activator starts in the OFF state, with the RBS initially hidden within a hairpin loop and the translation start codon (AUG) embedded within a bulge in the stem of the hairpin. The hairpin has a 12 nt unstructured sequence at its 5′ end, serving as a toehold for a trigger RNA molecule. Once the toehold switch stem is invaded, the hairpin opens, releasing both the RBS and the start codon. The specific design of the toehold switch bypasses sequence constraints for the trigger RNA, potentially enabling riboregulators to respond to arbitrarily chosen input sequences.

The toehold switch principle can also be implemented for creating translational repressors [22]. Here, without the trigger, the RBS is accessible to the ribosome. The invasion of the toehold by the trigger RNA induces a conformational change, securing the RBS within a stable hairpin, thereby repressing translation. Alternatively, an unstable hairpin containing the RBS in its loop region can be locked by a trigger, turning translation OFF.

A related type of RNA-switchable riboregulator mimics the structure of natural translational riboswitches more closely (Figure 12.2C) [18]. As often found in natural riboswitches, the RBS is hidden within a duplex stem by an anti-RBS sequence. By leveraging the same principle as for the toehold switches, an anti-anti-RBS sequence can be positioned inside the loop of a toehold hairpin preceding the anti-RBS-RBS duplex. The activation by a trigger RNA unfolds the toehold hairpin, enabling the anti-anti-RBS to bind to the anti-RBS, consequently releasing the RBS and initiating translation. This strategy can also be adapted to create translational toehold repressors.

12.3.3 Toehold Switching in Eukaryotes

Creating eukaryotic translational switches requires a strategy distinct from that for prokaryotic systems. In eukaryotes, RNA polymerase II transcribes mRNA, which results in an RNA molecule featuring a 7-methylguanylate cap at its 5′ end and a poly-A tail at the 3′ end. Often, the cap and tail are connected via the poly-A-binding protein (PABP). Although translation initiates in the cap region, there is no specific RBS to employ for building a toehold switch akin to the prokaryotic system. The so-called Kozak sequence found near the start codon of eukaryotic mRNAs contributes to translation initiation, but it is not absolutely required. Attempts to create riboregulators based on Kozak sequences thus resulted in switches with only low performance. An alternative approach for developing eukaryotic toehold switches utilizes cap-independent initiation processes that require specific secondary structural motifs, such as the IRES (internal ribosome entry site). IRES modules, found in viral RNA molecules, facilitate the recruitment of eukaryotic ribosomes for translation. An approach termed eToehold demonstrated successful translational switching using IRES modules from a range of viruses [23]. eToeholds have proven successful in detecting viral RNA or endogenous mRNAs in eukaryotic cells.

12.3.4 Transcriptional Switching

Rather than switching ON or OFF translation processes, similar approaches to toehold switches were also developed for transcriptional regulation. RNA transcriptional activators termed STARs (small transcription activating RNAs) were developed that use antisense RNAs to manipulate an intrinsic transcriptional terminator controlling gene expression [24]. STARs can actually be combined with toehold switches to form RNA regulators that control gene expression both at the transcriptional and translational stages, increasing the overall dynamic range of regulation.

Regulatory mechanisms such as STARs employ trigger RNA molecules that necessarily contain the intrinsic terminator's complementary sequence, which limits the choice of possible RNA inputs. Such sequence limitations can be avoided by sequestering the anti-terminator sequence within a toehold hairpin, similar to the riboswitch-inspired translational toehold switches described above [18]. It is also possible to create synthetic riboregulators based on Rho-dependent transcriptional termination, which can be achieved by controlling access of the termination factor Rho to its binding site on the RNA (the Rho utilization site *rut*). For instance, riboregulators can be designed to sequester an anti-rut sequence within a toehold hairpin in the OFF state, leaving the downstream rut sequence accessible and thus promoting termination. In the presence of an activating trigger RNA, the toehold hairpin is opened via TMSD, releasing the anti-rut sequence, which in turn sequesters the *rut* site. In consequence, Rho factor does not bind to the nascent mRNA, and transcription is not terminated.

12.3.5 Applications as Sensors

Toehold switches have been successfully used in a wide range of biosensing applications, particularly for RNA analytes (Figure 12.3A). Recent developments include electrochemical sensors and bioelectronic sensors for rapid disease diagnostics, including Ebola, Zika, and SARS-CoV-2 [25, 28–30]. A toehold switch variant termed SNIPRs (single-nucleotide-specific programmable riboregulators) was designed to detect single nucleotide variations [31]. While it has been shown feasible to detect RNA species *in vitro*, or RNA molecules transcribed in high abundance from plasmids, detecting *endogenous* RNA molecules *in vivo* remains a significant challenge due to their typically low concentrations and competing degradation processes. A notable exception is the successful detection of endogenous small RNA (sRNA) for RyhB, an sRNA in *Escherichia coli* whose expression depends on iron concentration [17].

Toehold switching can also be applied to create allosteric fluorescent light-up aptamers that respond to the presence of specific input RNA molecules. Rather than sequestering the RBS within a secondary structural motif (as in the case of translational regulators), the sequence of a fluorogenic aptamer such as Spinach, Broccoli, Mango, or Pepper is folded into an inactive conformation, which can be activated by TMSD-induced refolding of the aptamer structure [32–34].

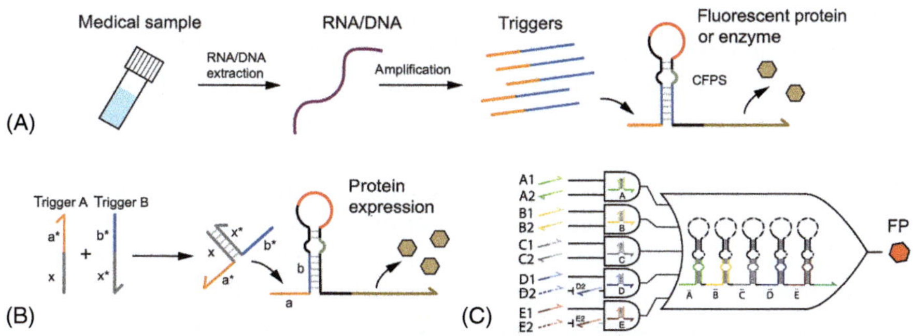

Figure 12.3 Applications for toehold switches. (A) Toehold switches can be used for the amplified detection of nucleic acids as part of a biosensing workflow. Typically, nucleic acids extracted from medical samples have to be amplified due to their extremely low concentrations. The amplicons can then be used as inputs for a toehold switch that further amplifies the signal in a cell-free protein synthesis (CFPS) step. CFPS can be used to produce a fluorescent protein or an enzyme for signal readout. Source: Adapted from Refs [25, 26]. (B) Toehold switches can be equipped with logic gate inputs. For instance, the trigger sequence – composed of a toehold part (sequence a*) and an invading sequence (b*) can be split and distributed onto two separate input molecules. Only when both inputs hybridize and thus form a functional trigger, the toehold switch is activated (AND logic). (C) In order to implement OR logic, multiple toehold switches can be concatenated in series. Source: Adapted from Green et al. [27].

12.3.6 Applications in Biocomputing

RNA-based regulatory mechanisms have great potential for biocomputing applications. As these mechanisms leverage sequence-programmable RNA interactions, they are – at least in principle – more amenable to "rational" design of biocomputing components and circuits than protein-based regulation.

For instance, it has been shown that multiple toehold switches can be serially connected to regulate the simultaneous expression of several genes [17]. In another approach, multi-input OR logic was implemented by controlling the same gene through a series of toehold switches on a single mRNA transcript (Figure 12.3C) [22, 27]. For this serial connection to function, ribosomes binding to upstream RBS sequences must still be able to move through all downstream secondary structures.

AND gate behavior is achievable through composite trigger RNA molecules assembled from several input strands (Figure 12.3B). This method hinges on the requirement of both toehold and incumbent sequence complementarity for TMSD initiation by an invader [27]. Only input molecules containing both elements can initiate TMSD, rendering a complex containing all necessary sequence domains as an active trigger.

Logical negation is possible with surplus anti-trigger RNAs, which are fully complementary to the input triggers. As duplex formation between triggers and anti-triggers is energetically and kinetically more favorable than opening a toehold hairpin, toehold switching is prevented, thereby inhibiting translational activation or repression. Employing a blend of repressors, activators, AND, OR, and NOT logic enables the realization of intricate logical expressions that control gene expression based on a specific input pattern [22, 27, 35].

Another method to achieve AND and OR logic functions uses multi-arm junctions [30] in the input domain of translational riboregulators, which was recently implemented in cell-free diagnostics applications. Contrary to toehold switches, in multi-arm junctions strand, invasion is initiated within a hairpin loop region (termed LIRA for loop-initiated RNA activator).

Translational riboregulators can be coupled with transcriptional switching to realize logic functions. For instance, small transcription activators (STARS) [24] combined with toehold switches can effectively establish AND gates to control gene expression [36]. Similarly, combining Rho-dependent termination/anti-termination with a riboswitch-inspired toehold repressor can execute NOR logic [18].

12.4 Applying Nucleic Acid Nanotechnology to CRISPR and RNA Interference

12.4.1 CRISPR Techniques

Since 2012, the life sciences have been transformed by a range of novel, sequence-programmable molecular tools based on CRISPR-associated processes [37–40]. For instance, CRISPR-Cas9 technology has enabled the precise cutting of double-stranded DNA (dsDNA) molecules at arbitrary sequence locations, which can be freely programmed by the choice of their "guide RNAs" (gRNAs). The CRISPR-associated protein Cas9 binds to gRNA molecules containing a handle sequence (recognized by the protein) and a spacer sequence, which directs the Cas9/gRNA complex to a complementary sequence on a dsDNA target. In addition to sequence complementarity of the protospacer, a short protospacer adjacent motif (PAM) is required on the 3′ side adjacent to the binding sequence (e.g. the sequence 5′-NGG-3′ for Cas9 from Streptococcus pyogenes, also termed SpCas9).

Cas9 can be used to introduce double-strand breaks (DSB) at gRNA-programmed locations in the genome, which can be used for gene editing applications. Repair of the DSB results in the formation of indels (insertions-deletions) or the acquisition of new gene sequences through homology-directed repair (HDR).

Using a catalytically deactivated version of Cas9 (dCas9), dCas9/gRNA complexes can be directed toward a corresponding target sequence without cutting it. This capability is useful for a wide variety of applications, for example, labeling (CRISPR imaging [41, 42]) or gene regulation. For instance, in CRISPR interference, a dCas9/gRNA is used similarly as a transcriptional inhibitor that either blocks the promoter itself or acts as a roadblock for transcriptional elongation [43]. In eukaryotes, transcriptional activation can be achieved by fusing different activator domains, such as VP64 or VP64-p65-Rta (VPR), to dCas9 [44]. Other Cas9 fusions have been developed that can "edit" single bases at specific locations, or precisely introduce new sequences without the need for HDR and without indel formation [45]. Cytosine base editors (CBEs) are fusions of dCas9 with a cytidine deaminase, which results in the conversion of cytosine into uracil, which, upon DNA replication, is converted into a T (thus C → T). Conversely, adenine base editors (ABEs) contain an adenosine deaminase and convert A → G. Prime editors are Cas9 fusions with a reverse transcriptase, which allows the introduction of a short DNA sequence defined by the sequence of a "prime editing guide RNA" (pegRNA) [46].

Next to Cas9, many other CRISPR-associated proteins have become popular for their different PAMs, gRNAs, or other functions. For instance, Cas12a [47] has a similar function as Cas9, but it trims its own guide using an RNase activity that cleaves off any 5′ sequence context [37]. Cas13 [48, 49] and Cas7-11 [50] can be used for RNAi, i.e. they cleave RNA rather than DNA at sequence locations defined by their guide RNA. In contrast to Cas7-11, Cas13 has an RNA-dependent RNase activity that has been used for biosensor applications. Cas1-Cas2 is an integrase complex that has been used for "storage" of dsDNA and RNA sequences in the genome of prokaryotes.

12.4.2 Switchable Guide RNAs

As CRISPR systems center around functional RNA molecules – the gRNAs – they can also be readily modulated by effector RNA molecules to either inhibit or activate their function. A comparatively straightforward approach is to use antisense RNA complementary to the gRNA to inhibit the activity of dCas9/gRNA complexes and recover gene expression of initially repressed genes [51, 52]. In this case, however, the input RNA sequence is necessarily constrained to be at least partly complementary to the guide RNA. Such sequence constraints prevent arbitrary wiring of such switchable gRNAs into larger circuits, and particularly the use of endogenous RNA sequences as inputs.

Various techniques have been developed to reduce the sequence dependence between the gRNA and the effector RNA used as an input. In one approach, the effector RNA was engineered to bind an internal loop in the gRNA rather than the spacer sequence, which has been proven effective even in mammalian cells [53]. Another strategy based on two additional hairpins added to the gRNA was successfully tested *in vitro* in *E. coli* and human cells using synthetic and even endogenous triggers [54].

An alternative approach for the conditional activation of gRNAs is based on RNA strand displacement (Figure 12.4a). For instance, gRNAs for Cas9 were extended at their 5′ end with sequences that fold back onto the spacer [57, 58]. An adjacent toehold allows for the binding of an RNA trigger that binds the extension and releases the target sequence, thereby restoring activity. In this type of conditional gRNA design, however, the trigger needs to displace a domain that is complementary to the spacer, and thus the trigger sequence or spacer sequence is still constrained. Alternatively, Jin et al. extended Cas9 gRNAs at their 3′ end and designed this extension to fold back onto an extension at the 5′ end without pairing the spacer, thereby avoiding fixed sequences on the trigger [55], which was shown to inhibit the activity

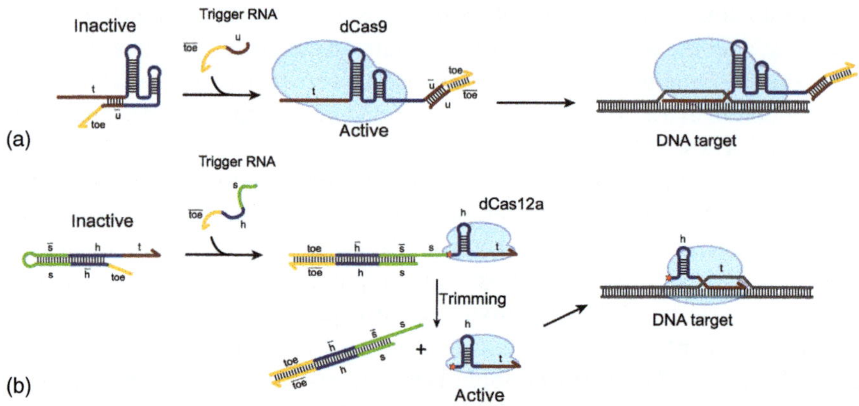

Figure 12.4 Conditional guide RNAs. (a) CRISPR-associated nucleases like Cas9 bind to specific "handle" secondary structures (shown in blue) on their guide RNAs (gRNAs). Binding can be inhibited by preventing the formation of the handle or otherwise sterically blocking it. In the example shown, a gRNA for Cas9 is extended at the 3′ end and folded back onto the target, which renders the gRNA inactive. Opening this "conditional" gRNA with a trigger RNA makes the handle accessible. In the example, catalytically inactive dCas9 is shown, which binds to its cognate DNA target without cleaving it. Source: Adapted from Jin et al. [55]. (b) Conditional guide RNAs for Cas12a make use of the RNA-processing capabilities of the protein. Once bound to the handle, Cas12a cleaves off all sequence context upstream of the position indicated by the red star. In the reaction shown in the figure, the handle h is initially sequestered inside a hairpin. Opening the hairpin with a trigger RNA allows the formation of the handle, which is then trimmed by dCas12a, resulting in a fully functional dCas12a-gRNA complex. Source: Adapted from Oesinghaus and Simmel [56].

of Cas9 *in vitro*. In all these approaches, the activated gRNA constructs contained double-stranded extensions, which likely reduced the binding affinity for Cas9 and thus their activity.

This issue can be completely circumvented by utilizing the RNA processing activity of Cas12a (Figure 12.4b). For instance, Cas12a gRNAs can be extended at their 5′ end and folded back onto the handle sequence to inhibit Cas12a binding. When an RNA trigger restores the handle structure, Cas12a binds and then cleaves off the extension, restoring a wild-type gRNA with full activity. This approach is also amenable to the creation of multi-input AND gates [56], for which several input RNAs have to hybridize together to generate a functional TMSD trigger (similar to that in [27]). The approach can be varied to hybridize "helper" strands to an endogenous RNA that translates the natural RNA sequence into an RNA trigger sequence that can switch the gRNA [54, 56]. An approach to sense natural transcripts in *E. coli* cells without the need for intermediate helper strands using engineered Cas12a gRNAs was demonstrated in Ref. [59].

Rather than using trigger RNAs as inputs and TMSD for activation, various other strategies have been developed to create switchable gRNAs. For instance, gRNAs can be made responsive to small molecules by using aptamer modifications [60, 61] or allosteric ribozymes [62]. In mammalian cells, it is also possible to control CRISPR processes by trimming the gRNAs using miRNA-directed RNA cleavage (Figure 12.5b) [63, 64].

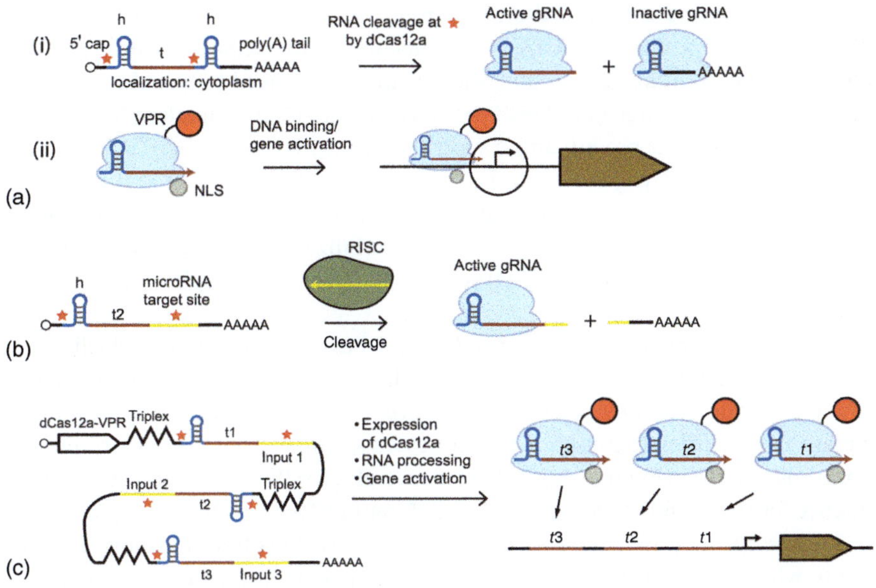

Figure 12.5 Conditional guide RNAs for dCas12a in mammalian (eukaryotic) cells. (a) When the guide RNA is transcribed in eukaryotic cells with RNA polymerase II, the gRNA is capped at the 5'end and augmented with a poly(A) tail, which also promotes the export of the gRNA from the nucleus to the cytoplasm. In the design shown here, the gRNA actually contains two handles, which are both processed by dCas12a. This leads to the formation of an active dCas12a-gRNA complex. In order to exert a gene regulatory function, dCas12a has to be fused to a transcriptional activator domain (VPR) and also equipped with a nuclear localization signal (NLS). The dCas12a-gRNA is then localized in the nucleus, where it activates genes with a corresponding target sequence close to the promoter. (b) Rather than cleaving off the poly(A) tail by dCas12a, this can also be achieved by an RNA-induced silencing complex (RISC) that contains a microRNA. Source: Adapted from Refs [63, 64]. (c) The coding sequence for dCas12a and multiple gRNAs can be put on the same transcript. After the expression of the CRISPR-associated protein, the rest of the transcript can be processed into functional dCas12a-gRNA complexes using, e.g. several microRNA inputs. This can be used for multiplex control of one or several target genes. Source: Adapted from Oesinghaus and Simmel [64].

12.4.3 Implementing More Complex Programs in Mammalian Cells

Switchable gRNAs are potentially promising tools for *in vivo* computing and for exerting spatiotemporal control over CRISPR processes, in particular in mammalian cells. Both for basic research and therapeutic purposes, it is desirable to activate CRISPR only under specific conditions (disease/no disease), only in specific cell types or tissues, or only at specific time points. Conditional gRNAs might provide such capabilities and could be used in conjunction with other control elements such as tissue-specific promoters or the expression of anti-CRISPR proteins.

Utilizing conditional gRNA in mammalian cells comes with a variety of challenges. For instance, RNA molecules are differently localized inside cells depending on the choice of the promoter. Transcription with RNA Pol II leads to capping, poly-adenylation and export of the RNA from the nucleus, whereas transcription with RNA Pol III localizes the RNA in the nucleus (Figure 12.5a). To exert gene regulatory control, both the Cas protein and its guide RNA need to be able to interact and be brought to the nucleus. This is achieved by attaching a nuclear localization signal (NLS) to the Cas protein and transcribing the gRNA with a U6 (RNA Pol III) promoter. Alternatively, gRNAs can be transcribed with an RNA Pol II promoter, followed by cleaving off the poly-A tail, which has been used as a regulatory mechanism in Refs [63, 64]. Removal of the tail destabilizes the RNA, however, which can be counteracted by secondary structural motifs such as triplex-forming RNA sequences [65].

In order to implement more complex, CRISPR-based programs multiple different gRNAs have to be delivered to the cells, which is challenging due to the large size of the eukaryotic promoter regions, which are much larger than the gRNAs they generate. One solution to this issue is to concatenate multiple gRNAs and transcribe them in one stretch (Figure 12.5c). In the case of CRISPR-Cas12a, e.g. the coding sequence for the protein together with 15 different gRNAs was put on a single RNA Pol II transcript [66]. In this case, the mRNA for Cas12a is first translated into the protein, which then further processes the transcripts by binding to the gRNA handles and trimming the gRNAs. Multimodal inputs can be achieved by combining miRNA-mediated processing, ribozyme cleavage, and strand displacement within a single transcript [64].

12.4.4 Combining CRISPR with Origami

Conditional gRNAs did not make use of intricate nucleic acid-based nanostructures so far. However, there have been several attempts to combine DNA or RNA origami with CRISPR techniques. For instance, gene sequences were already folded into origami-like structures, which were also modified with CRISPR-Cas9 target sequences [67]. Cas9-gRNA carrying an NLS could bind to these nanostructures and helped shuttle them into the nucleus, where the nanostructured gene was inserted into the genome by CRISPR-Cas9-mediated HDR. In another approach, single-guide RNA were fused with RNA origami structures (termed sgRNAO), which could be used to exert dCas9-mediated transcriptional control in *Saccharomyces cerevisiae* [68]. Notably, the RNA origami scaffold could be used to localize transcriptional activators close to sgRNAO target site.

12.4.5 MicroRNAs and RNA Interference

We note in short that TMSD was also applied to control RNAi in mammalian cells, where branch migration processes were used to assemble functional siRNAs from precursor duplexes that had been prepared in the test tube and transfected into the cells [69, 70]. As mentioned above, RNAi and CRISPR can also be combined by using the miRNA-directed RISC to process gRNAs containing the corresponding target sequences. miRNAs are particularly interesting as endogenous RNA inputs for nucleic acid devices [71] as different miRNA patterns were previously shown to be specific to certain tissues or types of cancer [72, 73].

12.5 Delivery of Nucleic Acid Devices, *In Vivo* Production, and Challenges for *In Vivo* Operation

12.5.1 Delivery

We wish to briefly (and non-exhaustively) digress on a few other aspects of the *in vivo* operation of nucleic acid devices. Nucleic acid structures may either be generated *in vitro* from synthesized oligonucleotides or from enzymatically generated DNA or RNA molecules, followed by transfection of the structures into cells using similar techniques as those developed for the delivery of gene fragments, i.e. using transfection agents or electroporation.

It is likely that, e.g. compact, origami-like nucleic nanostructures are deformed, disassembled, and degraded inside the cellular milieu owing to the low salt concentrations (compared to standard origami folding conditions) and the presence of nucleases, etc. This might be overcome by internal crosslinking of the structures [74], which, however, makes the encoded information unreadable by enzymes. Other approaches to stabilize DNA nanostructures were based on folding of the structures with polyamines [75], coating with oligolysine [76], or amphiphilic block copolymers [77], but it is not clear what happens to these protective agents once the structures are taken up by the cells. Potentially, deformation or partial disassembly is not even an issue for applications, where only the control of spatial proximity or stoichiometry by the nanostructure is desired. Furthermore, disassembly of nanostructures might even be required for delivering genomic payload and expressing genetic information contained in their sequences [67, 78]. For excellent overviews of challenges associated with the delivery and *in vivo* operation of nucleic acid devices, including immune responses, see, e.g. Refs [79, 80].

12.5.2 *In Vivo* Production

Another approach (which refers to most of the work discussed in this chapter) is to produce nucleic nanostructures, switches, and computational circuits *in situ*, i.e. in cellulo, by transcription of the corresponding genetic templates. This is comparatively straightforward for synthetic riboregulators and conditional gRNAs, as they are comprised of single, rather sRNA molecules. *In vitro* work has recently also demonstrated the co-transcriptional production of RNA strand displacement circuits [81].

DNA origami structures face the problem that they are composed of hundreds of DNA staple strands binding to a long scaffold structure. While large-scale biotechnological production of scaffolds and staples has been demonstrated, isothermal assembly of such structures inside living cells appears extremely challenging. To overcome this problem, single-stranded DNA [82] and RNA origami assemblies [83] have been developed that can fold intramolecularly. In particular, co-transcriptional RNA origami appears promising in this context as it is compatible with gene regulatory strategies used in synthetic biology and due to the comparatively low immunogenicity of RNA compared to DNA [83]. For a review of RNA nanotechnology and its nanomedical applications, see Ref. [84].

12.5.3 Challenges

Multicomponent assemblies as well as the detection of (or binding to) endogenous RNA species face the major challenge that hybridization inside cells is compromised by the low intracellular stability of RNA (in particular in bacteria) and their low concentrations. In the absence of protecting RNA-binding proteins (such as Hfq in bacteria, Cas proteins, or the RISC in eukaryotes), RNA will be degraded too quickly to be able to hybridize to its complementary target sequence. In addition, in eukaryotic cells, RNA species can be localized in different cellular compartments or organelles and might thus be unable to interact in the first place. In order to develop more complex molecular systems composed of multiple RNA components and RNA-based regulatory circuits that integrate endogenous RNA as *in vivo* inputs, a much better understanding and control of the RNA life cycle and its spatial localization in the cell will be required. One approach to overcome this issue in the context of *in vivo* RNA biosensing is to convert the transient appearance of an endogenous RNA signal into a stable, lasting chemical modification of a different molecule. This has been achieved, e.g. by utilizing RNA-guided RNA editing processes by adenosine deaminases [85, 86].

12.6 Conclusion and Outlook

In this chapter, we have discussed how molecular switching principles and nanostructures that were developed in nucleic acid nanotechnology initially in a non-biological context can be "brought back" into the cell and used for applications in genetic engineering and synthetic biology. Most notably, one of the key processes of dynamic nucleic acid nanotechnology – TMSD – has been successfully employed to control gene regulation in bacteria and eukaryotes, adopting strategies that were inspired by naturally occurring riboswitches, CRISPR techniques, or miRNA-based RNAi.

A wide range of biosensing applications have been already developed based on such systems, in particular methods for the detection of disease-associated RNA molecules.

Among the major advantages of nucleic acid-based systems is their comparatively simple sequence-based programmability and potentially quick and easy experimental implementation.

However, the application of "naked" nucleic acid systems inside cells faces manifold challenges that are related to the rapid degradation of DNA and RNA *in vivo*, immune responses, low concentrations and cellular localization, unknown *in vivo* secondary structures, and interactions with DNA or RNA-binding proteins, all of which affect hybridization efficiency within living cells. An improved understanding of these processes will be required to render nucleic acid systems *in vivo* more robust and sensitive to endogenous molecular signals.

Sensing endogenous RNAs could then be used to make synthetic gene regulation, including therapeutic interventions, highly specific to a specific cell or tissue type, dependent on a disease state, or also on other aspects such as cell cycle or metabolic activity.

A promising avenue is the combination of nucleic acid-based switches and computational elements with more complex nucleic acid nanostructures that can be applied *in vivo*. For instance, nucleic acid sequences that encode specific genetic programs and circuits could be folded into DNA or RNA nanostructures whose structure promotes their cellular uptake, improves their stability, or directs them to specific target regions. These structures could be equipped with context-dependent switching mechanisms and loaded with additional molecular payloads, resulting in truly multimodal nanomedical robotic systems.

Acknowledgments

This work was supported by the Deutsche Forschungsgemeinschaft through projects DFG SI 761/5-1 (project no. 453249455) and DFG SI 761/6-1 (project no. 468955252). The author would like to thank Tianhe Wang and Henning Hellmer for the preparation of figures that were modified for this article.

References

1 Seeman, N.C. and Sleiman, H.F. (2017). DNA nanotechnology. *Nat. Rev. Mater.* 3: 17068. https://doi.org/10.1038/natrevmats.2017.68.

2 Rothemund, P.W. (2006). Folding DNA to create nanoscale shapes and patterns. *Nature* 440: 297–302. https://doi.org/10.1038/nature04586.

3 Douglas, S.M., Dietz, H., Liedl, T. et al. (2009). Self-assembly of DNA into nanoscale three-dimensional shapes. *Nature* 459: 414–418. https://doi.org/10.1038/nature08016.

4 Dey, S., Fan, C., Gothelf, K.V. et al. (2021). DNA origami. *Nat. Rev. Methods Primers* 1: 13. https://doi.org/10.1038/s43586-020-00009-8.

5 He, Y. and Liu, D.R. (2010). Autonomous multistep organic synthesis in a single isothermal solution mediated by a DNA walker. *Nat. Nanotechnol.* 5: 778–782. https://doi.org/10.1038/nnano.2010.190.

6 Meng, W., Muscat, R.A., McKee, M.L. et al. (2016). An autonomous molecular assembler for programmable chemical synthesis. *Nat. Chem.* 8: 542–548. https://doi.org/10.1038/nchem.2495.

7 Douglas, S.M., Bachelet, I., and Church, G.M. (2012). A logic-gated nanorobot for targeted transport of molecular payloads. *Science* 335: 831–834. https://doi.org/10.1126/science.1214081.

8 Wickham, S.F., Bath, J., Katsuda, Y. et al. (2012). A DNA-based molecular motor that can navigate a network of tracks. *Nat. Nanotechnol.* 7: 169–173. https://doi.org/10.1038/nnano.2011.253.

9 Thubagere, A.J., Li, W., Johnson, R.F. et al. (2017). A cargo-sorting DNA robot. *Science* 357: eaan6558. https://doi.org/10.1126/science.aan6558.

10 Li, S., Jiang, Q., Liu, S. et al. (2018). A DNA nanorobot functions as a cancer therapeutic in response to a molecular trigger in vivo. *Nat. Biotechnol.* 36: 258–264. https://doi.org/10.1038/nbt.4071.

11 Kopperger, E., List, J., Madhira, S. et al. (2018). A self-assembled nanoscale robotic arm controlled by electric fields. *Science* 359: 296–301. https://doi.org/10.1126/science.aao4284.

12 Schnitzbauer, J., Strauss, M.T., Schlichthaerle, T. et al. (2017). Super-resolution microscopy with DNA-PAINT. *Nat. Protoc.* 12: 1198–1228. https://doi.org/10.1038/nprot.2017.024.

13 Zhao, Y.X., Shaw, A., Zeng, X. et al. (2012). DNA origami delivery system for cancer therapy with tunable release properties. *ACS Nano* 6: 8684–8691. https://doi.org/10.1021/nn3022662.

14 Sigl, C., Willner, E.M., Engelen, W. et al. (2021). Programmable icosahedral shell system for virus trapping. *Nat. Mater.* 20: 1281–1289. https://doi.org/10.1038/s41563-021-01020-4.

15 Zhang, D.Y. and Seelig, G. (2011). Dynamic DNA nanotechnology using strand-displacement reactions. *Nat. Chem.* 3: 103–113. https://doi.org/10.1038/nchem.957.

16 Simmel, F.C., Yurke, B., and Singh, H.R. (2019). Principles and applications of nucleic acid strand displacement reactions. *Chem. Rev.* 119: 6326–6369. https://doi.org/10.1021/acs.chemrev.8b00580.

17 Green, A.A., Silver, P.A., Collins, J.J., and Yin, P. (2014). Toehold switches: de-novo-designed regulators of gene expression. *Cell* 159: 925–939. https://doi.org/10.1016/j.cell.2014.10.002.

18 Wang, T. and Simmel, F.C. (2022). Riboswitch-inspired toehold riboregulators for gene regulation in *Escherichia coli*. *Nucleic Acids Res.* 50: 4784–4798. https://doi.org/10.1093/nar/gkac275.

19 Mandal, M. and Breaker, R.R. (2004). Gene regulation by riboswitches. *Nat. Rev. Mol. Cell Biol.* 5: 451–463. https://doi.org/10.1038/nrm1403.

20 Sherwood, A.V. and Henkin, T.M. (2016). Riboswitch-mediated gene regulation: novel RNA architectures dictate gene expression responses. *Annu. Rev. Microbiol.* 70: 361–374. https://doi.org/10.1146/annurev-micro-091014-104306.

21 Isaacs, F.J., Dwyer, D.J., Ding, C. et al. (2004). Engineered riboregulators enable post-transcriptional control of gene expression. *Nat. Biotechnol.* 22: 841–847. https://doi.org/10.1038/nbt986.

22 Kim, J., Zhou, Y., Carlson, P.D. et al. (2019). De novo-designed translation-repressing riboregulators for multi-input cellular logic. *Nat. Chem. Biol.* 15: 1173–1182. https://doi.org/10.1038/s41589-019-0388-1.

23 Zhao, E.M., Mao, A.S., de Puig, H. et al. (2022). RNA-responsive elements for eukaryotic translational control. *Nat. Biotechnol.* 40: 539–545. https://doi.org/10.1038/s41587-021-01068-2.

24 Chappell, J., Takahashi, M.K., and Lucks, J.B. (2015). Creating small transcription activating RNAs. *Nat. Chem. Biol.* 11: 214–220. https://doi.org/10.1038/nchembio.1737.

25 Pardee, K., Green, A.A., Takahashi, M.K. et al. (2016). Rapid, low-cost detection of Zika virus using programmable biomolecular components. *Cell* 165: 1255–1266. https://doi.org/10.1016/j.cell.2016.04.059.

26 Mousavi, P.S., Smith, S.J., Chen, J.B. et al. (2019). A multiplexed, electrochemical interface for gene-circuit-based sensors. *Nat. Chem.* 12: 48–55. https://doi.org/10.1038/s41557-019-0366-y. PMID - 31767994.

27 Green, A.A., Kim, J., Ma, D. et al. (2017). Complex cellular logic computation using ribocomputing devices. *Nature* 548: 117–121. https://doi.org/10.1038/nature23271.

28 Pardee, K., Green, A.A., Ferrante, T. et al. (2014). Paper-based synthetic gene networks. *Cell* 159: 940–954. https://doi.org/10.1016/j.cell.2014.10.004.

29 Carr, A.R., Dopp, J.L., Wu, K. et al. (2022). Toward mail-in-sensors for SARS-CoV-2 detection: interfacing gel switch resonators with cell-free toehold switches. *ACS Sens* 7: 806–815. https://doi.org/10.1021/acssensors.1c02450.

30 Ma, D., Li, Y., Wu, K. et al. (2022). Multi-arm RNA junctions encoding molecular logic unconstrained by input sequence for versatile cell-free diagnostics. *Nat. Biomed. Eng.* 6: 298–309. https://doi.org/10.1038/s41551-022-00857-7.

31 Hong, F., Ma, D., Wu, K. et al. (2020). Precise and programmable detection of mutations using ultraspecific riboregulators. *Cell* 180: 1018–1032.e16. https://doi.org/10.1016/j.cell.2020.02.011.

32 Wang, T. and Simmel, F.C. (2023). Switchable fluorescent light-up aptamers based on riboswitch architectures. *Angew. Chem. Int. Ed.* e202302858. https://doi.org/10.1002/anie.202302858.

33 Yan, Z., Tang, A.A., Eshed, A. et al. (2023). Rapid and multiplexed nucleic acid detection using programmable aptamer-based RNA switches. *medRxiv*. https://doi.org/10.1101/2023.06.02.23290873.

34 A. Tang, A. Afasizheva, C. Cano, K. Plath, et al. (2023). Optimization of RNA Pepper sensors for the detection of arbitrary RNA targets. https://doi.org/10.1101/2023.06.01.543282.

35 Falgenhauer, E., Muckl, A., Schwarz-Schilling, M., and Simmel, F.C. (2022). Transcriptional interference in toehold switch-based RNA circuits. *ACS Synth. Biol.* 11: 1735–1745. https://doi.org/10.1021/acssynbio.1c00486.

36 Lehr, F.X., Hanst, M., Vogel, M. et al. (2019). Cell-free prototyping of AND-logic gates based on heterogeneous RNA activators. *ACS Synth. Biol.* 8: 2163–2173. https://doi.org/10.1021/acssynbio.9b00238.

37 Jinek, M., Chylinski, K., Fonfara, I. et al. (2012). A programmable dual-RNA-guided DNA endonuclease in adaptive bacterial immunity. *Science* 337: 816–821. https://doi.org/10.1126/science.1225829.

38 Gasiunas, G., Barrangou, R., Horvath, P., and Siksnys, V. (2012). Cas9-crRNA ribonucleoprotein complex mediates specific DNA cleavage for adaptive immunity in bacteria. *Proc. Natl. Acad. Sci. U.S.A.* 109: E2579–E2586. https://doi.org/10.1073/pnas.1208507109.

39 Mali, P., Esvelt, K.M., and Church, G.M. (2013). Cas9 as a versatile tool for engineering biology. *Nat. Methods* 10: 957–963. https://doi.org/10.1038/nmeth.2649.

40 Cong, L., Ran, F.A., Cox, D. et al. (2013). Multiplex genome engineering using CRISPR/Cas systems. *Science* 339: 819–823. https://doi.org/10.1126/science.1231143.

41 Ma, H., Tu, L.C., Naseri, A. et al. (2018). CRISPR-Sirius: RNA scaffolds for signal amplification in genome imaging. *Nat. Methods* 15: 928–931. https://doi.org/10.1038/s41592-018-0174-0.

42 Wang, S., Hao, Y., Zhang, L. et al. (2019). Multiplexed superresolution CRISPR imaging of chromatin in living cells. *CCS Chem.* 1: 278–285. https://doi.org/10.31635/ccschem.019.20180035.

43 Qi, L.S., Larson, M.H., Gilbert, L.A. et al. (2013). Repurposing CRISPR as an RNA-guided platform for sequence-specific control of gene expression. *Cell* 152: 1173–1183. https://doi.org/10.1016/j.cell.2013.02.022. PMID - 23452860.

44 Kiani, S., Chavez, A., Tuttle, M. et al. (2015). Cas9 gRNA engineering for genome editing, activation and repression. *Nat. Methods* 12: 1051–1054. https://doi.org/10.1038/nmeth.3580.

45 Rees, H.A. and Liu, D.R. (2018). Base editing: precision chemistry on the genome and transcriptome of living cells. *Nat. Rev. Genet.* 19: 770–788. https://doi.org/10.1038/s41576-018-0059-1.

46 Anzalone, A.V., Koblan, L.W., and Liu, D.R. (2020). Genome editing with CRISPR-Cas nucleases, base editors, transposases and prime editors. *Nat. Biotechnol.* 38: 824–844. https://doi.org/10.1038/s41587-020-0561-9.

47 Zetsche, B., Heidenreich, M., Mohanraju, P. et al. (2017). Multiplex gene editing by CRISPR-Cpf1 using a single crRNA array. *Nat. Biotechnol.* 35: 31–34. https://doi.org/10.1038/nbt.3737.

48 East-Seletsky, A., O'Connell, M.R., Knight, S.C. et al. (2016). Two distinct RNase activities of CRISPR-C2c2 enable guide-RNA processing and RNA detection. *Nature* 538: 270–273. https://doi.org/10.1038/nature19802.

49 Abudayyeh, O.O., Gootenberg, J.S., Konermann, S. et al. (2016). C2c2 is a single-component programmable RNA-guided RNA-targeting CRISPR effector. *Science* 353: aaf5573. https://doi.org/10.1126/science.aaf5573.

50 Ozcan, A., Krajeski, R., Ioannidi, E. et al. (2021). Programmable RNA targeting with the single-protein CRISPR effector Cas7-11. *Nature* 597: 720–725. https://doi.org/10.1038/s41586-021-03886-5.

51 Lee, Y.J., Hoynes-O'Connor, A., Leong, M.C., and Moon, T.S. (2016). Programmable control of bacterial gene expression with the combined CRISPR and antisense RNA system. *Nucleic Acids Res.* 44: 2462–2473. https://doi.org/10.1093/nar/gkw056.

52 Muckl, A., Schwarz-Schilling, M., Fischer, K., and Simmel, F.C. (2018). Filamentation and restoration of normal growth in *Escherichia coli* using a combined CRISPRi sgRNA/antisense RNA approach. *PLoS One* 13: e0198058. https://doi.org/10.1371/journal.pone.0198058.

53 Hochrein, L.M., Li, H., and Pierce, N.A. (2021). High-performance allosteric conditional guide RNAs for mammalian cell-selective regulation of CRISPR/Cas. *ACS Synth. Biol.* 10: 964–971. https://doi.org/10.1021/acssynbio.1c00037.

54 Lin, J., Liu, Y., Lai, P. et al. (2020). Conditional guide RNA through two intermediate hairpins for programmable CRISPR/Cas9 function: building regulatory connections between endogenous RNA expressions. *Nucleic Acids Res.* 48: 11773–11784. https://doi.org/10.1093/nar/gkaa842.

55 Jin, M., Garreau de Loubresse, N., Kim, Y. et al. (2019). Programmable CRISPR-Cas repression, activation, and computation with sequence-independent targets and triggers. *ACS Synth. Biol.* 8: 1583–1589. https://doi.org/10.1021/acssynbio.9b00141.

56 Oesinghaus, L. and Simmel, F.C. (2019). Switching the activity of Cas12a using guide RNA strand displacement circuits. *Nat. Commun.* 10: 2092. https://doi.org/10.1038/s41467-019-09953-w.

57 Hanewich-Hollatz, M.H., Chen, Z., Hochrein, L.M. et al. (2019). Conditional guide RNAs: programmable conditional regulation of CRISPR/Cas function in bacterial and mammalian cells via dynamic RNA nanotechnology. *ACS Cent. Sci.* 5: 1241–1249. https://doi.org/10.1021/acscentsci.9b00340.

58 Siu, K.H. and Chen, W. (2019). Riboregulated toehold-gated gRNA for programmable CRISPR-Cas9 function. *Nat. Chem. Biol.* 15: 217–220. https://doi.org/10.1038/s41589-018-0186-1.

59 Collins, S.P., Rostain, W., Liao, C., and Beisel, C.L. (2021). Sequence-independent RNA sensing and DNA targeting by a split domain CRISPR-Cas12a gRNA switch. *Nucleic Acids Res.* 49: 2985–2999. https://doi.org/10.1093/nar/gkab100.

60 Kundert, K., Lucas, J.E., Watters, K.E. et al. (2019). Controlling CRISPR-Cas9 with ligand-activated and ligand-deactivated sgRNAs. *Nat. Commun.* 10: 2127. https://doi.org/10.1038/s41467-019-09985-2.

61 Liu, Y., Zhan, Y., Chen, Z. et al. (2016). Directing cellular information flow via CRISPR signal conductors. *Nat. Methods* 13: 938–944. https://doi.org/10.1038/nmeth.3994.

62 Tang, W., Hu, J.H., and Liu, D.R. (2017). Aptazyme-embedded guide RNAs enable ligand-responsive genome editing and transcriptional activation. *Nat. Commun.* 8: 15939. https://doi.org/10.1038/ncomms15939.

63 Wang, X.W., Hu, L.F., Hao, J. et al. (2019). A microRNA-inducible CRISPR-Cas9 platform serves as a microRNA sensor and cell-type-specific genome regulation tool. *Nat. Cell Biol.* 21: 522–530. https://doi.org/10.1038/s41556-019-0292-7.

64 Oesinghaus, L. and Simmel, F.C. (2021). Controlling gene expression in mammalian cells using multiplexed conditional guide RNAs for Cas12a*. *Angew. Chem. Int. Ed.* 60: 23894–23902. https://doi.org/10.1002/anie.202107258.

65 Wilusz, J.E., JnBaptiste, C.K., Lu, L.Y. et al. (2012). A triple helix stabilizes the 3' ends of long noncoding RNAs that lack poly(A) tails. *Genes Dev.* 26: 2392–2407. https://doi.org/10.1101/gad.204438.112.

66 Campa, C.C., Weisbach, N.R., Santinha, A.J. et al. (2019). Multiplexed genome engineering by Cas12a and CRISPR arrays encoded on single transcripts. *Nat. Methods* 16: 887–893. https://doi.org/10.1038/s41592-019-0508-6.

67 Lin-Shiao, E., Pfeifer, W.G., Shy, B.R. et al. (2022). CRISPR-Cas9-mediated nuclear transport and genomic integration of nanostructured genes in human primary cells. *Nucleic Acids Res.* 50: 1256–1268. https://doi.org/10.1093/nar/gkac049.

68 Pothoulakis, G., Nguyen, M.T.A., and Andersen, E.S. (2022). Utilizing RNA origami scaffolds in *Saccharomyces cerevisiae* for dCas9-mediated transcriptional control. *Nucleic Acids Res.* 50: 7176–7187. https://doi.org/10.1093/nar/gkac470.

69 Afonin, K.A., Viard, M., Martins, A.N. et al. (2013). Activation of different split functionalities on re-association of RNA-DNA hybrids. *Nat. Nanotechnol.* 8: 296–304. https://doi.org/10.1038/nnano.2013.44.

70 Groves, B., Chen, Y.J., Zurla, C. et al. (2016). Computing in mammalian cells with nucleic acid strand exchange. *Nat. Nanotechnol.* 11: 287–294. https://doi.org/10.1038/nnano.2015.278.

71 Zhang, C., Zhao, Y., Xu, X. et al. (2020). Cancer diagnosis with DNA molecular computation. *Nat. Nanotechnol.* 15: 709–715. https://doi.org/10.1038/s41565-020-0699-0.

72 Lu, J., Getz, G., Miska, E.A. et al. (2005). MicroRNA expression profiles classify human cancers. *Nature* 435: 834–838. https://doi.org/10.1038/nature03702.

73 Krol, J., Loedige, I., and Filipowicz, W. (2010). The widespread regulation of microRNA biogenesis, function and decay. *Nat. Rev. Genet.* 11: 597–610. https://doi.org/10.1038/nrg2843.

74 Gerling, T., Kube, M., Kick, B., and Dietz, H. (2018). Sequence-programmable covalent bonding of designed DNA assemblies. *Sci. Adv.* 4: eaau1157. https://doi.org/10.1126/sciadv.aau1157.

75 Chopra, A., Krishnan, S., and Simmel, F.C. (2016). Electrotransfection of polyamine folded DNA origami structures. *Nano Lett.* 16: 6683–6690. https://doi.org/10.1021/acs.nanolett.6b03586.

76 Ponnuswamy, N., Bastings, M.M.C., Nathwani, B. et al. (2017). Oligolysine-based coating protects DNA nanostructures from low-salt denaturation and nuclease degradation. *Nat. Commun.* 8: 15654. https://doi.org/10.1038/ncomms15654.

77 Agarwal, N.P., Matthies, M., Gur, F.N. et al. (2017). Block copolymer micellization as a protection strategy for DNA origami. *Angew. Chem. Int. Ed.* 56: 5460–5464. https://doi.org/10.1002/anie.201608873.

78 Kretzmann, J.A., Liedl, A., Monferrer, A. et al. (2023). Gene-encoding DNA origami for mammalian cell expression. *Nat. Commun.* 14: 1017. https://doi.org/10.1038/s41467-023-36601-1.

79 Surana, S., Shenoy, A.R., and Krishnan, Y. (2015). Designing DNA nanodevices for compatibility with the immune system of higher organisms. *Nat. Nanotechnol.* 10: 741–747. https://doi.org/10.1038/nnano.2015.180.

80 Ramakrishnan, S., Ijas, H., Linko, V., and Keller, A. (2018). Structural stability of DNA origami nanostructures under application-specific conditions. *Comput. Struct. Biotechnol. J.* 16: 342–349. https://doi.org/10.1016/j.csbj.2018.09.002.

81 Schaffter, S.W. and Strychalski, E.A. (2022). Cotranscriptionally encoded RNA strand displacement circuits. *Sci. Adv.* 8: eabl4354. https://doi.org/10.1126/sciadv.abl4354.

82 Han, D., Qi, X., Myhrvold, C. et al. (2017). Single-stranded DNA and RNA origami. *Science* 358: eaao2648. https://doi.org/10.1126/science.aao2648.

83 Geary, C., Rothemund, P.W., and Andersen, E.S. (2014). A single-stranded architecture for cotranscriptional folding of RNA nanostructures. *Science* 345: 799–804. https://doi.org/10.1126/science.1253920.

84 Jasinski, D., Haque, F., Binzel, D.W., and Guo, P. (2017). Advancement of the emerging field of RNA nanotechnology. *ACS Nano* 11: 1142–1164. https://doi.org/10.1021/acsnano.6b05737.

85 Jiang, K., Koob, J., Chen, X.D. et al. (2023). Programmable eukaryotic protein synthesis with RNA sensors by harnessing ADAR. *Nat. Biotechnol.* 41: 698–707. https://doi.org/10.1038/s41587-022-01534-5.

86 Kaseniit, K.E., Katz, N., Kolber, N.S. et al. (2023). Modular, programmable RNA sensing using ADAR editing in living cells. *Nat. Biotechnol.* 41: 482–487. https://doi.org/10.1038/s41587-022-01493-x.

13

Cell Membrane Functionalization via Nucleic Acid Tools for Visualization and Regulation of Cellular Receptors

Shan Chen[1], *Jingying Li*[2], *and Huanghao Yang*[3]

[1] *Minjiang University, College of Geography and Oceanography, No.200 Xiyuan Gong Road, Fuzhou 350108, China*
[2] *Fuzhou University, College of Biological Science and Engineering, No.2 Xue Yuan Road, Fuzhou 350108, China*
[3] *Fuzhou University, College of Chemistry, New Cornerstone Science Laboratory, MOE Key Laboratory for Analytical Science of Food Safety and Biology, No.2 Xue Yuan Road, Fuzhou 350108, China*

13.1 Nucleic Acid-Based Functionalization Strategies: From Receptor Information to DNA Probes

Live cells rely on coordinating diverse signaling components to maintain normal cell functions [1]. As the primary mediators, receptors on the cell membrane are of paramount importance in cellular signal transduction. Generally, many types of receptors sense and integrate stimuli from the extracellular environment, which usually initiate complex intracellular signaling cascades and thereby drive appropriate cellular events [2, 3]. However, aberrant receptor signaling is strongly associated with disease progressions, such as cancer and neurodegenerative disorders. For instance, the epidermal growth factor receptor (EGFR), one of the receptor tyrosine kinases (RTKs), is highly influential in cell proliferation performance, while abnormal mutations and elevated expression of EGFR are frequently observed in various cancer types [4]. Another representative example is integrin, which mainly involves cell adhesion behavior. However, cumulative evidence suggests that cancer cells could hijack integrin-mediated signal transduction to support their metastasis [5]. Given the critical and complex roles in cellular activities of receptors, tremendous efforts have been dedicated to deciphering receptor information to elucidate the receptor-mediated signal transduction mechanism, thus contributing to the novel therapeutic design. Many cellular receptors have been identified as reliable biomarkers for the early diagnosis of diseases [6, 7]. Furthermore, receptors provide desirable targets for artificial perturbation of cell signaling owing to their easy accessibility and excellent specificity [8, 9]. As previously reported, over 60% of the approved therapeutic agents target cellular receptors [10]. Of note, increasing current studies indicate that abnormal cell signaling results from changes in the spatial distribution of receptors rather than the obvious alteration of expression

DNA Nanotechnology for Cell Research: From Bioanalysis to Biomedicine, First Edition. Edited by Zhou Nie.

levels [11]. Moreover, post-translational modification (PTM) patterns of the same receptor, such as glycosylation and phosphorylation, differ between the physiological and pathological states, resulting in distinct biological functions [12]. This comprehensive understanding of receptor information has paved the way to new horizons of receptor studies, from the expression levels to spatio-temporal dynamic functions of cellular receptors. To dissect and manipulate elaborate receptor-mediated activities, various molecular probes have been extensively exploited. For precisely decrypting receptor information, specific attachment of probes on the target receptor is an essential prerequisite. Meanwhile, real-time signal feedback is of great necessity for measuring the dynamic receptor functions in live cells. Hence, novel and robust probes with high accuracy and rapid response are in urgent demand.

Nucleic acid has evolved into a promising molecular tool for probing receptors and even manipulating receptor functions. Due to their unique merits of easy programmability, good biocompatibility, and synthetic flexibility, nucleic acid tools have exhibited high potential in exploiting diverse dynamic-assembled DNA probes [13]. The strict Watson–Crick base pairing rule guarantees that the quick DNA assembly process occurred in a predicted manner, which ensures accurate and fast response capability [14]. Moreover, DNA probes could be endowed with receptor targeting capability via easily conjugating receptor-biorecognition elements. The emerging receptor-biorecognition elements allow the accurate installation of DNA probes on a variety of receptors, expanding the diversity of DNA probes applicable to receptor studies [15, 16]. Indeed, the rapid development of DNA technology in receptor studies largely benefits from the advances in nucleic acid-based cell membrane functionalization strategies. As is well known, most hydrophilic DNA sequences are unable to interact with cell membranes directly. Endowing DNA probes with membrane anchoring capability could be achieved mainly via linking DNA to different functional moieties. These moieties generally work through three basic mechanisms, including (i) target recognition, (ii) covalent conjugation of native components on the membrane, and (iii) hydrophobic insertion into the phospholipid bilayer (Figure 13.1a) [17, 19]. Since receptors are important cell membrane components, these nucleic acid-based functionalization strategies, especially target recognition and covalent conjugation, provide great opportunities for establishing connections between DNA probes and receptors. Well-designed DNA sequences or DNA nanostructures could thus be constructed to be further applied in various DNA assembly reactions or DNA-based logic computing for imaging or regulation purposes.

Among nucleic acid-based cell membrane functionalization strategies, target recognition is a common route to attach DNA probes on receptors via modifying DNA with receptor-biorecognition elements [20]. Notably, the aptamer is a typical example of biorecognition elements for direct receptor binding. As a functional nucleic acid, the aptamer can mutually match the spatial conformation of the target receptor by folding into a specific tertiary structure [21]. With the development of systematic evolution of ligands by exponential enrichment (SELEX) techniques, a variety of single-stranded receptor-binding aptamers have been identified with

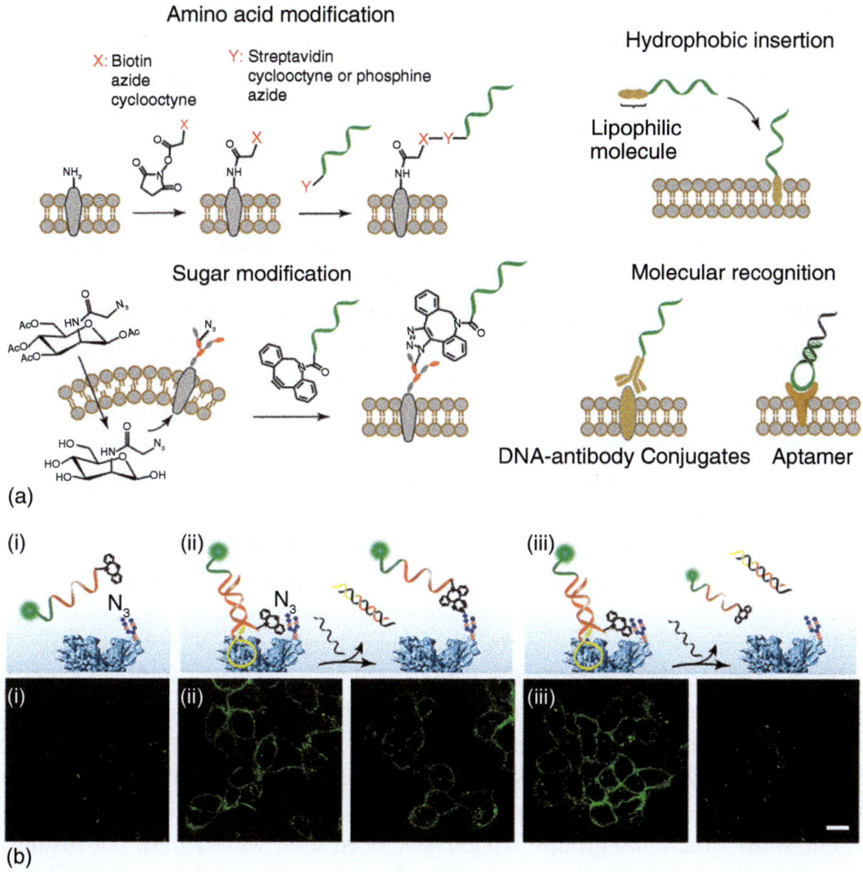

Figure 13.1 Installation of DNA probes on receptors via nucleic acid-based cell membrane functionalization strategies. (a) Nucleic acid-based cell membrane functionalization strategies. Source: Reproduced with permission from Shi and Wang [17]. © 2021 John Wiley & Sons. (b) Aptamer-directed covalent modification of DNA probes on c-Met receptors. Source: Reproduced with permission from Yang et al. [18]. © 2021 American Chemical Society.

high binding affinity and specificity against their targets [22–24]. Moreover, taking advantage of their oligonucleotide nature, aptamers can be easily incorporated into DNA probes by an automated DNA synthesizer [25]. These unique features make the aptamer one of the most popular biorecognition elements for directly installing DNA probes on receptors. In addition to nucleic acid aptamers, some other receptor-biorecognition elements, including antibodies, peptides, and ligands, could be integrated with the DNA probe to support receptor recognition as well. For instance, in an attempt to organize transmembrane tumor necrosis factor (TNF) receptor distributions by nucleic acid tools, short peptides able to recognize TNF receptors were precisely patterned on a DNA origami via click chemistry [26]. Taken together, the conjugation of different biorecognition elements to DNA probes offers diverse, innovative options for selective receptor functionalization.

Covalent conjugation provides another way to tag DNA probes on receptors with high stability. It is noteworthy that the receptors offer various binding sites, such as amino acid residues and glycans, for the covalent attachment of DNA probes [27, 28]. Conventionally, the covalent attachment could be achieved through the reaction between the amino acid residues on the receptors and the small molecules connected to the DNA probes. For instance, thiol- or primary amine-modified DNA probes could be conjugated with succinimidyl 4-(*N*-maleimidomethyl) cyclohexane-1-carboxylate (SMCC) for receptor attachment on living cells, as SMCC is a classic heterobifunctional linker molecule able to react with thiol group and amine group simultaneously [29]. Alternatively, the primary amine groups on receptors could be treated with biotinyl-*N*-hydroxy-succinimide (NHS-biotin), followed by attaching the streptavidin-modified DNA probes through affinity interaction [30]. To simplify the modification procedure, receptors with amine groups are frequently conjugated with DNA probes carrying *N*-hydroxysuccinimide (NHS) ester moieties directly [31, 32]. Nevertheless, amino and thiol groups on receptors are usually involved in the maintenance of the intrinsic structure and function of receptors [33], and thus the conjugation of DNA probes on the amino acid residues may result in the disruption of the receptor activities. In this regard, glycans presented on receptors have been considered as another type of covalent modification site. Glycosylation, the attachment of glycans to proteins, is one of the most common PTM of receptors. The conjugation of glycans with DNA probes may minimize the negative effect on receptor activities [34]. With the assistance of metabolic glycan engineering, various artificial functional groups, such as azides, alkynes, ketones, thiols, alkenes, and diazirines, have been introduced to the glycans of receptors, for the conjugation of DNA probes via click chemistry [35]. Among those functional groups, azides have been used widely [36]. For instance, tetra-acylated *N*-acetylmannosamine ($Ac_4ManNAz$), a synthetic derivative of *N*-acetylmannosamine (ManNAc), can be metabolized by live cells through the sialic acid biosynthesis pathway [37], so that azide groups can be incorporated into sialic acids on receptors for further conjugation. DNA probes containing dibenzocyclooctyne (DBCO), succinimidyl ester DIBO alkyne, or phosphine could covalently connect with the azide groups through click reaction or Staudinger ligation [38, 39].

Despite their high universality, most of the covalent modification strategies described above lack receptor specificity. On that account, receptor-recognition elements should be incorporated into subsequent procedures to realize target receptor analysis. Optionally, receptor-selective covalent modification strategies are desirable [40]. To this end, receptor-biorecognition elements have been used to direct DNA probes to the target receptor. The aptamer is a promising candidate in support of receptor targeting because of its small size and good erasability [41]. In a representative design, the mesenchymal-epithelial transition factor (c-Met)-targeting aptamer and the DBCO-modified DNA strand were assembled into an integrated DNA probe. The c-Met-targeting aptamer directed the DBCO-modified DNA strand close to the pre-introduced azide groups on c-Met receptor. In this way, the DBCO-modified DNA strand enabled site-selective covalent connection between the DNA probe and its target receptor owing to the reliable recognition capability of the aptamer. Upon the conjugation of DNA sequence, aptamers bound to the c-Met receptor could be efficiently removed by hybridization with fully complementary DNA to avoid the occupation of active sites on the receptor. Consequently, specific DNA probe-conjugated receptors were generated for subsequent applications (Figure 13.1b) [18]. Similar aptamer-guided covalent modification strategies have also been exploited to chemically engineer the primary amine groups on selective membrane proteins with a DNA tag in live cells [42]. Given the diversity of receptor-binding aptamers, the covalent chemical conjugation strategies could be easily expanded to other target receptors by using appropriate biorecognition elements. Overall, the nucleic acid-based cell membrane functionalization strategies combined with receptor-biorecognition elements have succeeded in the connection of DNA probes with selective receptors, providing desirable opportunities for cellular receptor visualization and regulation.

13.2 Uncovering Molecular Information of Cellular Receptors

Cellular receptors offer abundant molecular information closely relevant to cell behaviors, mainly involving receptor expression levels, posttranslational modification patterns, and dynamic nanoscale organizations. Profiling receptors' molecular information is highly beneficial for basic cell biology and biomedical applications. Conventional biochemical analytical methods, such as western blotting (WB) and enzyme-linked immunosorbent assay (ELISA), only provide holistic average results and fail to monitor various dynamic processes in live cells [43]. In this regard, nucleic acid tools have inspired a variety of DNA probes compatible with living systems to decrypt the molecular information of cellular receptors [44, 45]. Moreover, diverse DNA-based signal amplification strategies, such as catalytic hairpin assembly (CHA) [46], rolling circle amplification (RCA) [47], and hybridization chain reaction (HCR) [48], have been developed to achieve highly sensitive analysis or detection of the cellular receptor molecular information.

The receptor expression level provides the most intuitive information and holds great importance in the identification of cell types. Aberrant expression of cellular receptors is closely related to abnormal cell functions, thereby making receptors important biomarkers for disease diagnosis [49]. Isolation of cells with specific receptor expression from complex microenvironments has instructive implications in precision medicine, especially when combined with noninvasive analytical procedures like DNA probe-based fluorescence imaging [50, 51]. In the simplest example, the expression level of a target receptor can be qualitatively and/or quantitatively evaluated by the fluorescent signal generated from a fluorescent dye conjugated to the target receptor-binding aptamers [52]. To reduce the background noise and improve signal sensitivity, functional aptamer elements could also be integrated into a DNA self-assembly structure. A free-state DNA self-assembly structure often has a low fluorescent background, while spontaneous conformational reorganization of the DNA structure caused by aptamer-receptor binding events can enhance the fluorescent signal specifically [53, 54]. In some cases, the target receptors can also be detected from normal cells with low abundance, making it challenging to distinguish between normal and aberrant cells like cancer cells [55]. Accordingly, cellular receptor-DNA probe couples providing high identification specificity and accuracy are urgently needed. Benefiting from the programmability and versatility of nucleic acid tools, efforts have been dedicated to constructing DNA probes in response to external cues to improve identification performance [56]. For instance, intercalated-motif (i-motif) DNA probes, i.e. DNA probes with cytosine-rich four-stranded quadruplex structures stabilized by acidic conditions rather than physiological pH, have been proposed for the recognition of target cellular receptors. By implication of i-motif structure with protein tyrosine kinase 7 (PTK-7) receptor-recognition aptamer, the designed DNA probe enables modulating aptamer-PTK-7 affinity based on the pH value of the cellular microenvironment (Figure 13.2a) [57]. Moreover, photo-cleavage units have also been introduced into the DNA probes to impart the aptamer-receptor recognition event in a light-controlled manner [59]. To be mentioned, investigation of a single cellular receptor biomarker may satisfy the need to distinguish between apparently dysregulated cells and normal healthy cells, but it may be far from sufficient to identify different cancer subtype cells owing to the cell heterogeneity in complex microenvironments [60]. Multiple receptors, therefore, should be considered to provide more information.

To profile the expression of multiple cellular receptors, introducing different recognition elements into DNA probes is the primary requirement. DNA assembly reactions or DNA logic operations are often used for the integration of multiple signals and finally produce outputs for computational analysis [61–63]. A typical example is the identification of different leukemia cells. As we know, leukemia contains several subtype cells, such as human acute lymphoblastic leukemia cells CEM, human Burkitt's lymphoma cells Ramos, human acute T leukemia cells Jurkat, and human erythromyeloblastoid leukemia cells K562. These cell types display different receptor expression patterns, making it difficult to be distinguished via a single receptor biomarker. Logic-actuated DNA probes have been applied to address the challenge by constructing logic devices with both recognition and signal readout functions [64]. Multiple DNA aptamers, including Sgc8c, TCO1, and Sgc4f, have been used for the recognition of different cancer biomarkers, with associative DNA assembly processes responsive to signal management through strategies like Boolean logic analysis. Such an approach has succeeded in the differentiation of target cell subtypes from various similar cells (Figure 13.2b) [58].

Current studies have revealed that the PTM of cellular receptors modulates receptor functions [65]. It is worth noting that the PTM pattern of cancer cells usually differs from that of normal cells, while the receptor expression levels may not change significantly [66]. By this token, the research on deciphering receptor molecular information goes further from the expression level to the biological function. Particularly, considerable efforts have shed light on receptor-specific glycosylation, which often varies with receptor status. It has been identified that

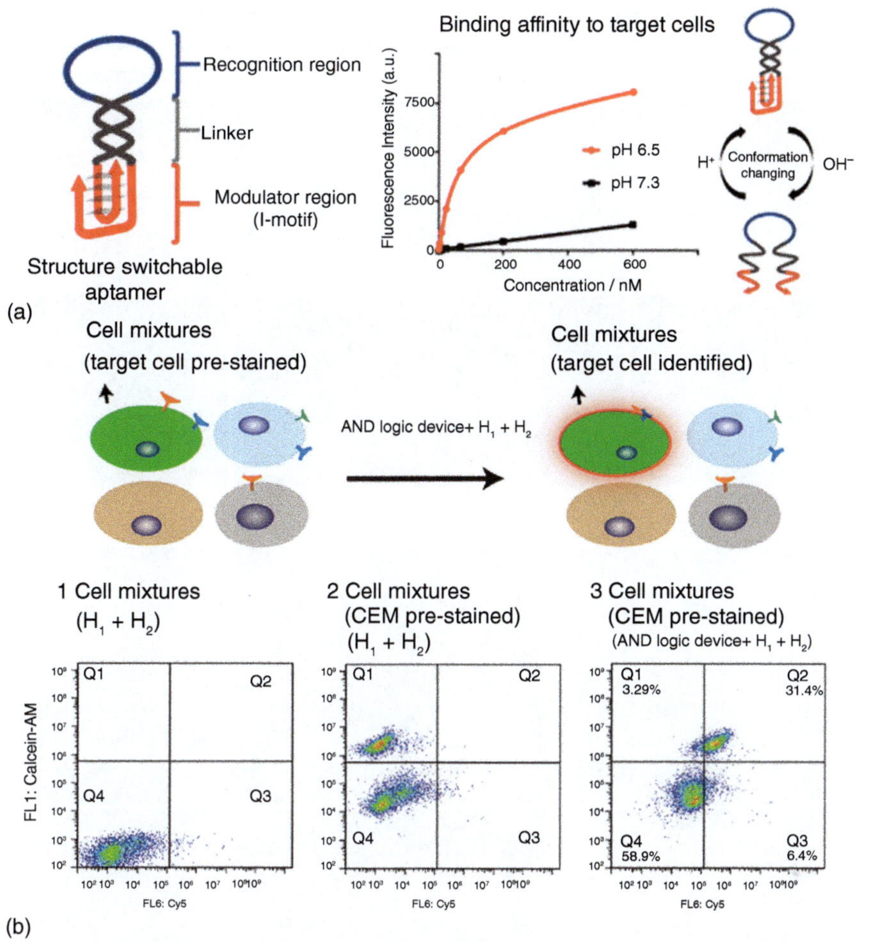

Figure 13.2 Profiling receptor expression levels via nucleic acid tools. (a) A structure-switchable aptamer with i-motif region displayed reconfigurable binding affinity to PTK-7 receptors in response to pH value. Source: Reproduced with permission from Li et al. [57]. © 2018 American Chemical Society. (b) Flow cytometry analysis of dual-aptamer-based AND logic device for CEM cell identification in 1 : 1 : 1 : 1 mixtures of CEM + Ramos + Jurkat + K562 cells. Source: Reproduced with permission from Chang et al. [58]. © 2019 American Chemical Society.

numerous receptors, such as EGFR, CD44, and the notch receptors, bear altered glycan epitopes in breast cancer cells [67, 68]. The key point for visualizing glycosylation in a receptor-specific manner is integrating the information of both target receptors and glycans. In general, live cell imaging strategy relies on the cooperation of two DNA probes, including a glycan-tagging probe and a receptor-recognition probe [69]. The glycan-tagging probe is conjugated with glycan via metabolic oligosaccharide engineering and click chemical reactions, to explore the glycan information using a specific DNA probe. The protein-recognition probe is then employed to attach to the target receptor. Upon sequential binding to the targets,

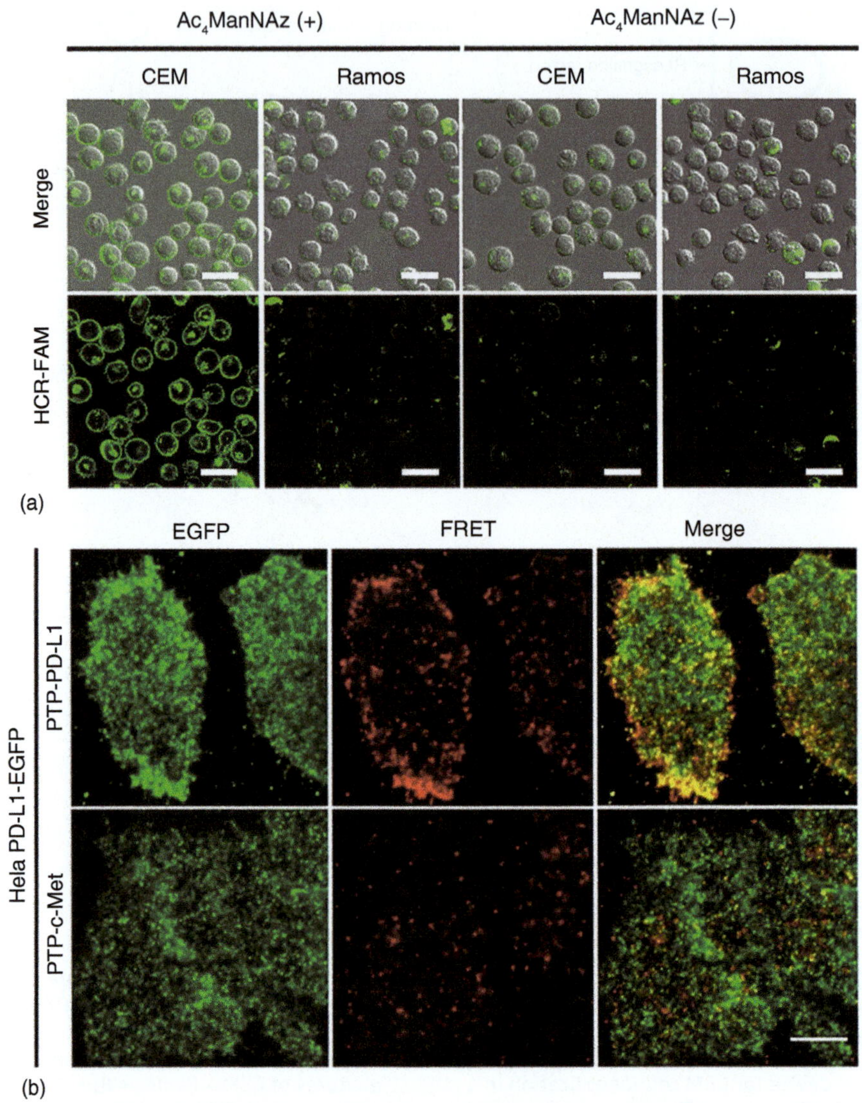

Figure 13.3 Profiling receptor-specific glycosylation patterns via nucleic acid tools. (a) The confocal fluorescence images of PTK-7 glycation on the CEM or Ramos cells using proximity-induced HCR amplification strategy. Source: Reproduced with permission from Li et al. [70]. © 2018 American Chemical Society. (b) The total internal reflection fluorescence images of PD-L1-EGFP and protein-specific glycation on HeLa cells. Source: Reproduced with permission from Shao et al. [71]. © 2022 John Wiley & Sons.

the proximity-induced hybridization between the two DNA probes leads to the structure-switching of elaborately designed DNA probes. It could further trigger HCR assembly to amplify signals like the fluorescence signal coming from sialic acid glycation on PTK-7 (Figure 13.3a) [70]. With a hierarchical DNA dynamic hybridization strategy, distinct monosaccharide patterns of MUC1 receptors on different cells are revealed by visualization of MUC1-specific terminal sialic acid and fucose [72]. An alternative biorthogonal glucose analog has also been developed as a glycation chemical reporter to sidestep complicated metabolic processes and enable *in situ* labeling. In this way, visualization of glucose glycation on PD-L1, integrin, and c-Met receptors has been realized (Figure 13.3b) [71].

Aside from the PTM patterns, much evidence has indicated that receptor nanoscale organizations display important molecular information associated with receptor functions [73, 74]. Dynamic mobile receptors on the cell membrane can interact with different signal components, such as a specific ligand or other membrane proteins, to regulate the receptor clustering pattern and to involve in the signaling activation [75]. Uncovering the spatial distributions of receptors is therefore of great interest in deciphering endogenous receptor activities. Nonetheless, conventional biochemical analyses often lose this dynamic molecular information and are thus unsuitable for live cell applications. To tackle this challenge, nucleic acid tools have been employed to offer diverse innovative approaches to map the dynamic receptor organizations in live cells [76]. Characterization of endogenous membrane protein interactions uncovers the nanoscale spatial organization of membrane proteins, which are intimately linked to intercellular communication and intracellular signaling [77]. For example, a precisely designed aptamer-G-quadruplex/hemin probe was employed for exploring the protein nanoenvironment of the specified receptor. Aptamers acted as receptor-recognition elements to bring G-quadruplex/hemin (Gq/h) to the target c-Met receptor, while Gq/h served as a nucleic acid-based mimic of horseradish peroxidase. Gq/h could catalyze rapidly the conversion of inert phenol-containing substrates into reactive phenoxyl radicals under the support of H_2O_2. The short half-life (<1 ms) and a small labeling radius (<40 nm) ensure the landing of radicals adjacent to the target receptor to achieve proximity labeling. Such a probe, combined with subsequent affinity-based enrichment and proteomic analysis, allows the investigation of the dynamic change of c-Met interaction networks [78]. Notably, receptor dimerization or oligomerization events are often activated by extracellular ligand binding. For instance, the binding of transforming growth factor-β (TGF-β) to its serine and threonine kinase receptor, the TGF-β type II receptor (TβRII), can mediate the dimerization of TβRII on the cell membrane and activate the intracellular downstream signal transduction. Nucleic acid tools have recently offered great opportunities for information liberation of such receptor dimerization process. As preeminent recognition elements, aptamer first guides two designed DNA probes to their corresponding receptor monomers. When receptor dimerization occurs upon TGF-β stimulation, the two aptamer probes are brought into proximity to trigger the DNA assembly behavior. Fluorescent signals are thus generated, indicating the formation of receptor dimers (Figure 13.4a) [79]. On this basis, various DNA nanostructures and signal amplification strategies have been utilized to monitor the ligand-activated receptor dimerization process [80, 81]. In addition to monitoring the lateral spatial distribution of receptors on the cell membrane, the axial spatial distribution of receptors inside cells also attracts attention. For instance, a hairpin DNA probe containing biotin tails was developed to functionalize membrane

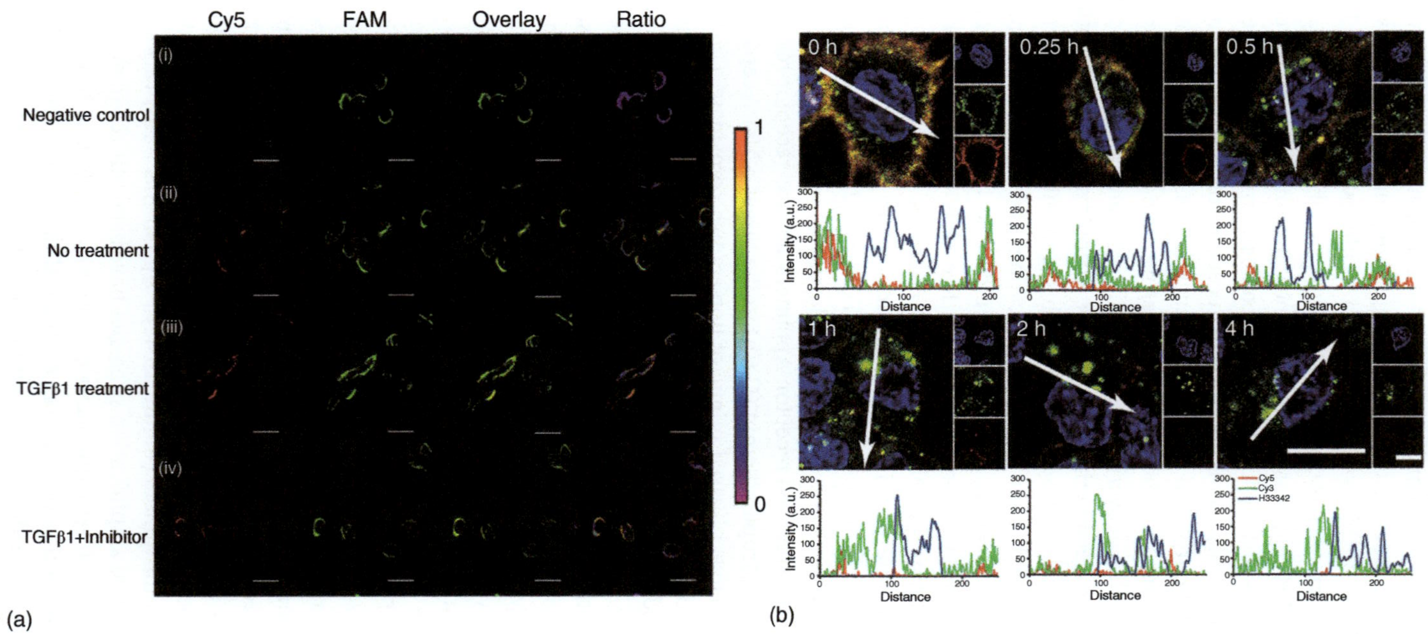

Figure 13.4 Profiling receptor nanoscale organizations via nucleic acid tools. (a) The representative confocal fluorescence images of TβRII dimerization in DU145 cells. Source: Reproduced with permission from Liang et al. [79]. © 2018 American Chemical Society. (b) The axial distribution of c-Met receptors in DU145 cells after HGF stimulation was tracked with confocal fluorescence imaging. Source: Reproduced with permission from Yang et al. [18]. © 2021 American Chemical Society.

proteins via NHS ester-streptavidin linkage. Strong fluorescence emanated from intracellular regions when cells were exposed to shear stress generated from a fluid cell medium, revealing that mechanical stimuli could enhance membrane protein internalization [82]. More specifically, through tagging c-Met receptor glycans with DNA probes covalently, the axial trafficking of c-Met was tracked successfully in living cells. The quantitative results showed that internalization occurred in about 91% of c-Met receptors at four hours after the hepatocyte growth factor (HGF)-induced ligand–receptor interactions (Figure 13.4b) [18]. Collectively, rationally designed DNA probes have paved a new way for noninvasively lighting up receptor molecular information in live cells.

13.3 Governing Cellular Receptors-Mediated Signal Transduction

Given the critical role in physiological function and disease progression, cellular signaling perturbation via regulating receptor activity has attracted widespread attention. Ligand–receptor interactions are the most common activation mechanism involved in receptor-mediated signal transduction, followed by receptor oligomerization or conformational changes [83, 84]. Several receptor-binding aptamers have exhibited great antagonistic potential in interfering with cellular signaling via impeding ligand–receptor interactions. Since aptamer-receptor recognitions depend on matching spatial conformation, there are certain probabilities of blocking the native ligand binding sites. Representatively, the binding of c-Met with aptamer SL1 could interfere with the molecular interaction between HGF and c-Met and thus preclude HGF-induced receptor dimerization and downstream signaling activation [85]. The interference has been further confirmed with in silico calculations (Figure 13.5a) [86]. PD-L1-binding aptamer shared the same molecular mechanism for suppressing PD-L1/PD1 interactions [89]. Of concern, a similar mechanism has been utilized for preventing SARS-CoV-2 infections. The SARS-CoV-2 virus invades host cells via the receptor-binding domain of spike protein (S_{RBD}) on the virus, which binds the human angiotensin-converting enzyme 2 (ACE2) receptor on the host cell membrane. An S_{RBD}-targeting aptamer CoV2-6 was identified by SELEX, which shared the same binding region on S_{RBD} as ACE2 receptors. The presence of aptamer CoV2-6 could disrupt the binding between S_{RBD} and ACE2, so as to inhibit SARS-CoV-2 invasion (Figure 13.5b) [87]. Furthermore, aptamer-mediated regulation could be achieved by sterically altering the active domains of receptors. Integrin $\alpha3\beta1$ is a heterodimeric adhesion receptor for extracellular matrix protein laminin. The integrin-binding aptamer ZAP-1, obtained by artificial-nucleotide-expanded cell-SELEX, displayed the regulation capability toward receptor activity and cellular adhesion behaviors. In this case, the association of ZAP-1 with integrin $\alpha3\beta1$ may stabilize the low-affinity conformers, thereby inhibiting the ligand-binding activity of integrin $\alpha3\beta1$ (Figure 13.5c) [88].

Nonetheless, the regulation capability via a single free aptamer is significantly influenced by the comparative flexibility and small size of aptameric nucleic

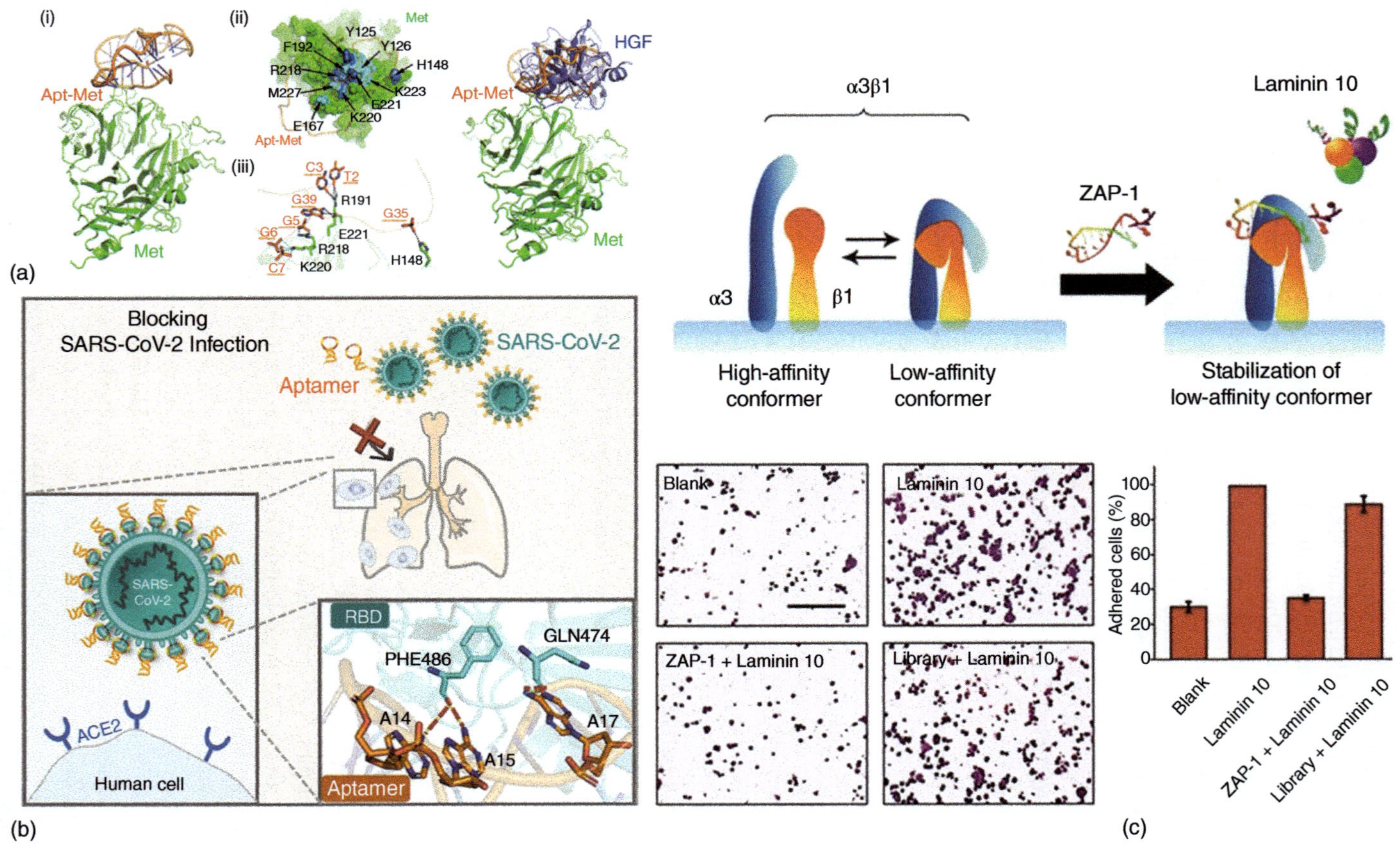

Figure 13.5 Manipulation of receptor-mediated signal transduction via antagonistic nucleic acid aptamers. (a) Structural model of the binding of c-Met with aptamer SL1 predicted by in silico calculations: (i) cartoon representation of the aptamer SL1/c-Met complex; (ii) binding interface and (iii) detailed interaction patterns between aptamer SL1 and c-Met receptor. Source: Reproduced with permission from Wang et al. [86]. © 2019 American Chemical Society. (b) Schematic diagram of S_{RBD}-targeting aptamer CoV2-6 inhibited SARS-CoV-2 infection. Source: Reproduced with permission from Sun et al. [87]. © 2021 John Wiley & Sons. (c) Aptamer ZAP-1 stabilized the low-affinity conformers of integrin α3β1 and inhibited cell adhesion and migration of MDA-MB-231 cells. Source: Reproduced with permission from Tan et al. [88]. © 2019 John Wiley & Sons.

acid [90]. To enhance regulation efficiency and selectivity, a strategy called artificial protein-pairing was proposed [86]. This strategy employed bispecific aptamers to strongly inhibit c-Met-mediated signaling, in which one aptamer binds to c-Met and another attaches to the paired protein. Some cancer biomarkers with high expression levels were chosen as paired proteins, such as transferrin receptor (TfR) and nucleolin. In one case, the simultaneous binding of c-Met and TfR via bispecific aptamer probes brought the two receptors into close proximity. As a result, proximity-induced artificial protein pairing of TfR-c-Met blocked the access to the c-Met receptor of its ligand HGF and thereby suppressed c-Met receptor function. Compared with single aptamer-mediated regulation, the inhibition efficiency of bispecific aptamer probes was improved owing to the increasing steric hindrance and reduced flexibility (Figure 13.6a). Furthermore, recent advances in dynamic DNA assembly have enabled the precise modulation of cellular signaling in response to environmental cues. DNA-based logic computing has also been introduced for reprogramming cells to sense multiple input cues, including light irradiation and DNA sequence, based on the principle of artificial protein-pairing. The modular design and construction of hierarchical DNA logic circuits performed sophisticated computational control of receptor organizations, rewiring c-Met signal transduction to mimic natural complex information processing (Figure 13.6b) [91]. Another attempt at aptamer-induced artificial protein-paring for suppressing ligand–receptor interaction has been applied in modulating neuroprotective signaling. The cellular prion protein (PrP^{C}), a neuroprotective protein, is one of the cell-surface receptors of amyloid β oligomer (AβO) that mediates related neurotoxicity. Previous studies indicated that the dimerization of PrP^{C} could be critical for its trophic activity, while the lifetime of PrP^{C} dimers was significantly reduced in the presence of AβO. Therefore, inducing PrP^{C} dimerization via the artificial protein-paring strategy could impair PrP^{C}–AβO interactions and relieve the neuroinflammation microenvironment by attenuating AβO toxicity (Figure 13.6c) [92]. Moreover, the artificial protein-pairing strategy could advance from bivalent pairing to multiple clustering. With the assistance of HCR and multiple PTK-7-targeting aptamers, clustering of PTK-7 on the cell membrane has been achieved, which disrupted the original interaction between PTK-7 and VEGF-R1 because of the increased steric hindrance (Figure 13.7a) [93].

Additionally, in the case of aptamer-mediated receptor regulation on the cell surface, the cellular endocytosis effect of live cells at physiological temperature has a great impact on regulation efficiency. To alleviate the dilemma, cholesterol-modified duplex DNA stanchion was equipped with a c-Met-binding aptamer to improve regulatory efficacy substantially. In light of the cell surface functionalization of hydrophobic moieties, the duplex DNA stanchion acted as a stable DNA membrane anchor to reduce the flexibility of aptamer binding and prolong its cell surface persistence as well. The modular design of this membrane-anchored DNA probe was further incorporated with a VEGF sensing module to report cellular signaling perturbation in real time. Such an integrated DNA probe with multitasking capability represented a paradigmatic example of the wide versatility of nucleic acid tools (Figure 13.7b) [94].

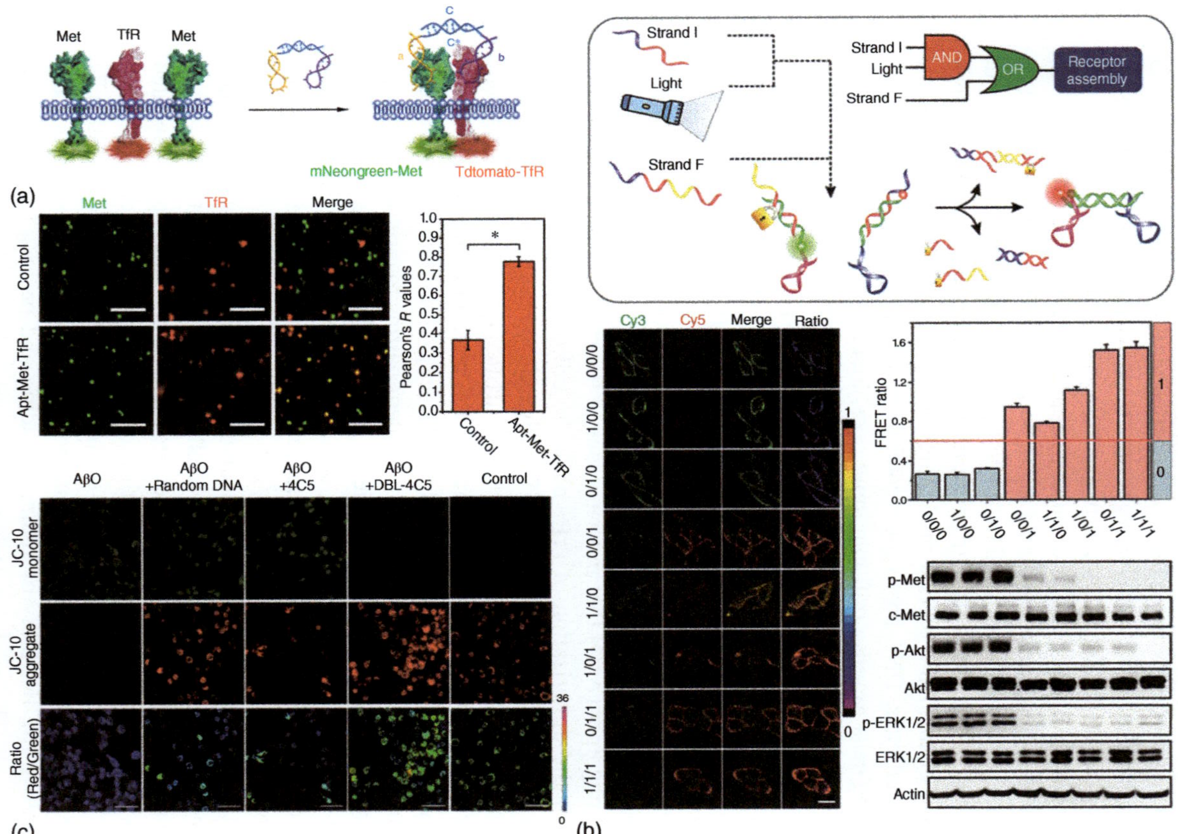

Figure 13.6 Manipulation of receptor-mediated signal transduction via nucleic acid-guided artificial protein-pairing. (a) The total internal reflection fluorescence images of c-Met-TfR colocalization on the cell surface induced by bispecific-aptamer probes. Source: Reproduced with permission from Wang et al. [86]. © 2019 American Chemical Society. (b) The confocal fluorescence images and WB assay of c-Met-mediated signal transduction in DU145 cells after logic gate-actuated receptor assembly. Source: Reproduced with permission from Chen et al. [91]. © 2019 John Wiley & Sons. (c) Suppression of AβO-mediated mitochondrial dysfunction via bispecific aptamer DBL-4C5-induced PrP^C dimerization in BV2 cells was evaluated by measuring mitochondrial membrane potential. Source: Reproduced with permission from Liu et al. [92]. © 2022 American Chemical Society.

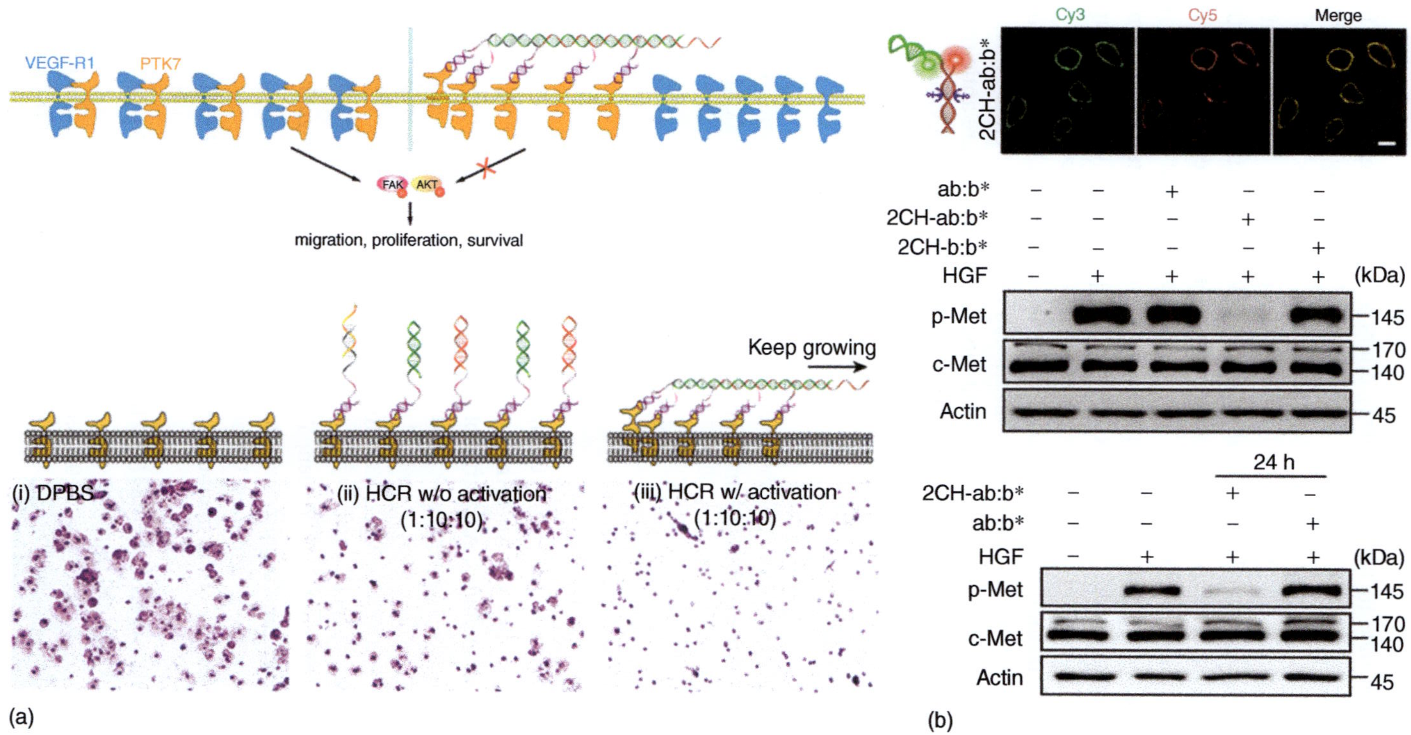

Figure 13.7 (a) Working principle of impeding PTK-7-VEGF-R1 interaction via HCR-driven unidirectional PTK-7 clustering; cell migration under different conditions of H460 cells was measured via transwell assay. Source: Reproduced with permission from Yang et al. [93]. © 2022 American Chemical Society. (b) The confocal fluorescence images of membrane-anchored DNA probe on HeLa cells. WB assays of HGF-induced c-Met phosphorylation after treatments with the DNA probe for 15 minutes or 24 hours. Source: Reproduced with permission from Chen et al. [94]. © 2022 John Wiley & Sons.

13.4 Conclusion

The significant advance in DNA technology offers a powerful chemical tool for decrypting cellular receptor information in live cells. Engineering receptors with elaboratly designed DNA probes via nucleic acid-based cell membrane functionalization technique allows the precise visualization and regulation of specific receptor events. The diverse, dynamic DNA assembly and DNA logic computing expand the utility of nucleic acid tools in receptor studies, from detecting receptor expression levels to monitoring dynamic receptor functions. Further, the exploitation of emerging functional nucleic acids has inspired a variety of DNA antagonistic agents for rewiring receptor-mediated complex cell signaling. On the whole, through visualization and regulation of cellular receptors, versatile nucleic acid tools hold considerable potential for disease diagnosis and precision medicine. To improve the *in vivo* performance of nucleic acid tools under complex dynamic environments, including stability, specificity, and biocompatibility, further studies are anticipated to broaden the scope of such robust tools in practical biomedical applications.

Acknowledgments

This research was supported by the National Key R&D Program of China (No. 2020YFA0210800) and the National Natural Science Foundation of China (Nos. 22207049, 21974022, and 22027805).

References

1 Scott, J.D. and Pawson, T. (2009). Cell signaling in space and time: where proteins come together and when they're apart. *Science* 326 (5957): 1220–1224.
2 Wootten, D., Christopoulos, A., Marti-Solano, M. et al. (2018). Mechanisms of signalling and biased agonism in G protein-coupled receptors. *Nat. Rev. Mol. Cell Biol.* 19: 638–653.
3 Lemmon, M.A. and Schlessinger, J. (2010). Cell signaling by receptor tyrosine kinases. *Cell* 141 (7): 1117–1134.
4 Uribe, M.L., Marrocco, I., and Yarden, Y. (2021). EGFR in cancer: signaling mechanisms, drugs, and acquired resistance. *Cancers* 13 (11): 2748.
5 Kechagia, J.Z., Ivaska, J., and Roca-Cusachs, P. (2019). Integrins as biomechanical sensors of the microenvironment. *Nat. Rev. Mol. Cell Biol.* 20: 457–473.
6 Arteaga, C. (2003). Targeting HER1/EGFR: a molecular approach to cancer therapy. *Semin. Oncol.* 30 (3): 3–14.
7 Wu, X., Liu, H., Han, D. et al. (2019). Elucidation and structural modeling of CD71 as a molecular target for cell-specific aptamer binding. *JACS* 141 (27): 10760–10769.

8 Akiyama, M., Ueki, R., Yanagawa, M. et al. (2019). DNA-based synthetic growth factor surrogates with fine-tuned agonism. *Angew. Chem. Int. Ed.* 60 (42): 22745–22752.

9 Zhang, L., Jing, D., Jiang, N. et al. (2020). Transformable peptide nanoparticles arrest HER2 signalling and cause cancer cell death in vivo. *Nat. Nanotechnol.* 15: 145–153.

10 Christopoulos, A. (2002). Allosteric binding sites on cell-surface receptors: novel targets for drug discovery. *Nat. Rev. Drug Discovery* 1: 198–210.

11 Li, J., Wang, L., Tian, J. et al. (2020). Nongenetic engineering strategies for regulating receptor oligomerization in living cells. *Chem. Soc. Rev.* 49 (5): 1545–1568.

12 Liu, Y.C., Yen, H.Y., Chen, C.Y. et al. (2011). Sialylation and fucosylation of epidermal growth factor receptor suppress its dimerization and activation in lung cancer cells. *Proc. Natl. Acad. Sci.* 108 (28): 11332–11337.

13 Chatterjee, G., Dalchau, N., Muscat, R.A. et al. (2017). A spatially localized architecture for fast and modular DNA computing. *Nat. Nanotechnol.* 12: 920–927.

14 Liu, S., Jiang, Q., Wang, Y. et al. (2019). Biomedical applications of DNA-based molecular devices. *Adv. Healthcare Mater.* 8 (10): 1801658.

15 Shaw, A., Lundin, V., Petrova, E. et al. (2014). Spatial control of membrane receptor function using ligand nanocalipers. *Nat. Methods* 11: 841–846.

16 Wang, M., Yang, D., Lu, Q. et al. (2022). Spatially reprogramed receptor organization to switch cell behavior using a DNA origami-templated aptamer nanoarray. *Nano Lett.* 22 (21): 8445–8454.

17 Shi, P. and Wang, Y. (2021). Synthetic DNA for cell-surface engineering. *Angew. Chem. Int. Ed.* 60 (21): 11580–11591.

18 Yang, W., Nan, H., Xu, Z. et al. (2021). DNA-templated glycan labeling for monitoring receptor spatial distribution in living cells. *Anal. Chem.* 93 (36): 12265–12272.

19 Jones, S.F., Joshi, H., Terry, S.J. et al. (2021). Hydrophobic interactions between DNA duplexes and synthetic and biological membranes. *JACS* 143 (22): 8305–8313.

20 Li, F., Zhang, H., Wang, Z. et al. (2013). Dynamic DNA assemblies mediated by binding-induced DNA strand displacement. *JACS* 135 (7): 2443–2446.

21 Ma, H., Liu, J., Ali, M.M. et al. (2015). Nucleic acid aptamers in cancer research, diagnosis and therapy. *Chem. Soc. Rev.* 44 (5): 1240–1256.

22 Mahlknecht, G., Maron, R., Mancini, M. et al. (2013). Aptamer to ErbB-2/HER2 enhances degradation of the target and inhibits tumorigenic growth. *Proc. Natl. Acad. Sci.* 110 (20): 8170–8175.

23 Wang, J. and Li, G. (2011). Aptamers against cell surface receptors: selection, modification and application. *Curr. Med. Chem.* 18 (27): 4107–4116.

24 Zhang, N., Wang, J., Bing, T. et al. (2022). Transferrin receptor-mediated internalization and intracellular fate of conjugates of a DNA aptamer. *Mol. Ther. Nucleic Acids* 27: 1249–1259.

25 Meng, H.M., Liu, H., Kuai, H. et al. (2016). Aptamer-integrated DNA nanostructures for biosensing, bioimaging and cancer therapy. *Chem. Soc. Rev.* 45 (9): 2583–2602.

26 Wang, Y., Baars, I., Fördös, F. et al. (2021). Clustering of death receptor for apoptosis using nanoscale patterns of peptides. *ACS Nano* 15 (6): 9614–9626.

27 Vogel, K., Glettenberg, M., Schroeder, H. et al. (2013). DNA-modification of eukaryotic cells. *Small* 9 (2): 255–262.

28 Keijzer, J.F. and Albada, B. (2022). DNA-assisted site-selective protein modification. *Biopolymers* 113 (3): e23483.

29 Sacca, B. and Niemeyer, C.M. (2011). Functionalization of DNA nanostructures with proteins. *Chem. Soc. Rev.* 40 (12): 5910–5921.

30 Zhao, W., Schafer, S., Choi, J. et al. (2011). Cell-surface sensors for real-time probing of cellular environments. *Nat. Nanotechnol.* 6: 524–531.

31 Hsiao, S.C., Shum, B.J., Onoe, H. et al. (2009). Direct cell surface modification with DNA for the capture of primary cells and the investigation of myotube formation on defined patterns. *Langmuir* 25 (12): 6985–6991.

32 Yan, X., Zhang, H., Wang, Z. et al. (2018). Quantitative synthesis of protein-DNA conjugates with 1:1 stoichiometry. *Chem. Commun.* 54: 7491–7494.

33 Trads, J.B., Torring, T., and Gothelf, K.V. (2017). Site-selective conjugation of native proteins with DNA. *Acc. Chem. Res.* 50 (6): 1367–1374.

34 Zhang, D., Zheng, Y., Lin, Z. et al. (2020). Equipping natural killer cells with specific targeting and checkpoint blocking aptamers for enhanced adoptive immunotherapy in solid tumors. *Angew. Chem. Int. Ed.* 59 (29): 12022–12028.

35 Whited, J., Zhang, X., Nie, H. et al. (2018). Recent chemical biology approaches for profiling cell surface sialylation status. *ACS Chem. Biol.* 13 (9): 2364–2374.

36 Robinson, P.V., Tsai, C.T., de Groot, A.E. et al. (2016). Glyco-seek: ultrasensitive detection of protein-specific glycosylation by proximity ligation polymerase chain reaction. *JACS* 138 (34): 10722–10725.

37 Chen, Y., Ding, L., Song, W. et al. (2016). Liberation of protein-specific glycosylation information for glycan analysis by exonuclease III-aided recycling hybridization. *Anal. Chem.* 88 (5): 2923–2928.

38 Zhang, X., Li, R., Chen, Y. et al. (2016). Applying DNA rolling circle amplification in fluorescence imaging of cell surface glycans labeled by a metabolic method. *Chem. Sci.* 7: 6182–6189.

39 van Berkel, S.S., van Eldijk, M.B., and van Hest, J.C.M. (2011). Staudinger ligation as a method for bioconjugation. *Angew. Chem. Int. Ed.* 50 (38): 8806–8827.

40 Robinson, P.V., de Almeida-Escobedo, G., de Groot, A.E. et al. (2015). Live-cell labeling of specific protein glycoforms by proximity-enhanced bioorthogonal ligation. *JACS* 137 (33): 10452–10455.

41 Chen, X., Qiu, L., Cai, R. et al. (2020). Aptamer-directed protein-specific multiple modifications of membrane glycoproteins on living cells. *ACS Appl. Mater. Interfaces* 12 (34): 37845–37850.

42 Cui, C., Zhang, H., Wang, R. et al. (2017). Recognition-then-reaction enables site-selective bioconjugation to proteins on live-cell surfaces. *Angew. Chem. Int. Ed.* 56 (39): 11954–11957.

43 Wang, J., Song, J., Zhang, X. et al. (2023). DNA-programed plasmon rulers decrypt single-receptor dimerization on cell membrane. *JACS* 145 (2): 1273–1284.

44 Chen, S., Li, J., Liang, H. et al. (2018). Light-induced activation of c-met signalling by photocontrolled DNA assembly. *Chem. Eur. J.* 24 (60): 15988–15992.

45 Tang, Y., Wang, Z., Yang, X. et al. (2015). Constructing real-time, wash-free, and reiterative sensors for cell surface proteins using binding-induced dynamic DNA assembly. *Chem. Sci.* 6: 5729–5733.

46 Tang, Y., Lin, Y., Yang, X. et al. (2015). Universal strategy to engineer catalytic DNA hairpin assemblies for protein analysis. *Anal. Chem.* 87 (16): 8063–8066.

47 Fan, Y., Li, L., Lu, M. et al. (2019). In situ fluorescent profiling of living cell membrane proteins at a single-molecule level. *Chem. Commun.* 55: 4043–4046.

48 Zhu, G., Zhang, S., Song, E. et al. (2013). Building fluorescent DNA nanodevices on target living cell surfaces. *Angew. Chem. Int. Ed.* 52 (21): 5490–5496.

49 He, X.P. and Tian, H. (2018). Lightening up membrane receptors with fluorescent molecular probes and supramolecular materials. *Chem* 4: 246–268.

50 Jou, A.F., Chou, Y.T., Willner, I. et al. (2021). Imaging of cancer cells and dictated cytotoxicity using aptamer-guided hybridization chain reaction (HCR)-generated G-quadruplex chains. *Angew. Chem. Int. Ed.* 60 (40): 21673–21678.

51 Xie, S., Wang, Z., Fu, T. et al. (2022). Engineering aptamers with selectively enhanced biostability in the tumor microenvironment. *Angew. Chem. Int. Ed.* 61 (31): e202201220.

52 Song, Y., Zhu, Z., An, Y. et al. (2013). Selection of DNA aptamers against epithelial cell adhesion molecule for cancer cell imaging and circulating tumor cell capture. *Anal. Chem.* 85 (8): 4141–4149.

53 Shi, H., He, X., Wang, K. et al. (2011). Activatable aptamer probe for contrast-enhanced in vivo cancer imaging based on cell membrane protein-triggered conformation alteration. *Proc. Natl. Acad. Sci.* 108 (10): 3900–3905.

54 Wang, Y.M., Wu, Z., Liu, S.J. et al. (2015). Structure-switching aptamer triggers hybridization chain reaction on cell surface for activatable theranostics. *Anal. Chem.* 87 (13): 6470–6474.

55 Rudchenko, M., Taylor, S., Pallavi, P. et al. (2013). Autonomous molecular cascades for evaluation of cell surfaces. *Nat. Nanotechnol.* 8: 580–586.

56 Li, L., Tong, R., Chu, H. et al. (2014). Aptamer photoregulation in vivo. *Proc. Natl. Acad. Sci.* 111 (48): 17099–17103.

57 Li, L., Jiang, Y., Cui, C. et al. (2018). Modulating aptamer specificity with pH-responsive DNA bonds. *JACS* 140 (41): 13335–13339.

58 Chang, X., Zhang, C., Lv, C. et al. (2019). Construction of a multiple-aptamer-based DNA logic device on live cell membranes via associative toehold activation for accurate cancer cell identification. *JACS* 141 (32): 12738–12743.

59 Xiao, H., Chen, Y., Yuan, E. et al. (2016). Obtaining more accurate signals: spatiotemporal imaging of cancer sites enabled by a photoactivatable aptamer-based strategy. *ACS Appl. Mater. Interfaces* 8 (36): 23542–23548.

60 Ai, L., Peng, T., Li, Y. et al. (2022). A dual-targeting circular aptamer strategy enables the recognition of different leukemia cells with enhanced binding ability. *Angew. Chem. Int. Ed.* 61 (33): e202109500.

61 You, M., Peng, L., Shao, N. et al. (2014). DNA "nano-claw": logic-based autonomous cancer targeting and therapy. *JACS* 136 (4): 1256–1259.

62 Peng, R., Zheng, X., Lyu, Y. et al. (2018). Engineering a 3D DNA-logic gate nanomachine for bispecific recognition and computing on target cell surfaces. *JACS* 140 (31): 9793–9796.

63 Gao, Q., Zhao, Y., Xu, K. et al. (2020). Highly specific, single-step cancer cell isolation with multi-aptamer-mediated proximity ligation on live cell membranes. *Angew. Chem. Int. Ed.* 59 (52): 23564–23568.

64 Chen, L., Chen, W., Liu, G. et al. (2021). Nucleic acid-based molecular computation heads towards cellular applications. *Chem. Soc. Rev.* 50 (22): 12551–12575.

65 Wagner, K.W., Punnoose, E.A., Januario, T. et al. (2007). Death-receptor O-glycosylation controls tumor-cell sensitivity to the proapoptotic ligand Apo2L/TRAIL. *Nat. Med.* 13: 1070–1077.

66 Takahashi, M., Kizuka, Y., Ohtsubo, K. et al. (2016). Disease-associated glycans on cell surface proteins. *Mol. Aspects Med.* 51: 56–70.

67 Listinsky, J.J., Siegal, G.P., and Listinsky, C.M. (2011). The emerging importance of α-L-fucose in human breast cancer: a review. *Am. J. Transl. Res.* 3 (4): 292–322.

68 Christiansen, M.N., Chik, J., Lee, L. et al. (2014). Cell surface protein glycosylation in cancer. *Proteomics* 14 (4–5): 525–546.

69 Yuan, B., Chen, Y., Sun, Y. et al. (2018). Enhanced imaging of specific cell-surface glycosylation based on multi-FRET. *Anal. Chem.* 90 (10): 6131–6137.

70 Li, J., Liu, S., Sun, L. et al. (2018). Amplified visualization of protein-specific glycosylation in zebrafish via proximity-induced hybridization chain reaction. *JACS* 140 (48): 16589–16595.

71 Shao, Z., Yuan, H., Zhou, Z. et al. (2022). Visualization of protein-specific glycation in living cells via bioorthogonal chemical reporter. *Angew. Chem. Int. Ed.* 61 (41): e202210069.

72 Li, S., Liu, Y., Liu, L. et al. (2018). A hierarchical coding strategy for live cell imaging of protein-specific glycoform. *Angew. Chem. Int. Ed.* 57 (37): 12007–12011.

73 Taylor, M.J., Husain, K., Gartner, Z.J. et al. (2017). A DNA-based T cell receptor reveals a role for receptor clustering in ligand discrimination. *Cell* 169 (1): 108–119.

74 Chu, T.W., Yang, J., Zhang, R. et al. (2014). Cell surface self-assembly of hybrid nanoconjugates via oligonucleotide hybridization induces apoptosis. *ACS Nano* 8 (1): 719–730.

75 Ueki, R., Ueki, A., Kanda, N. et al. (2016). Oligonucleotide-based mimetics of hepatocyte growth factor. *Angew. Chem. Int. Ed.* 55 (2): 579–582.

76 Zhang, K., Gao, H., Deng, R. et al. (2019). Emerging applications of nanotechnology for controlling cell-surface receptor clustering. *Angew. Chem. Int. Ed.* 58 (15): 4790–4799.

77 Zhang, J., Qiu, Z., Fan, J. et al. (2021). Scan and unlock: a programmable DNA molecular automaton for cell-selective activation of ligand-based signaling. *Angew. Chem. Int. Ed.* 60 (12): 6733–6743.

78 Yang, W., Huang, Z., Xu, Z. et al. (2022). Selective and nongenetic peroxidase tag of membrane protein: a nucleic acid tool for proximity labeling. *Anal. Chem.* 94 (2): 1101–1107.

79 Liang, H., Chen, S., Li, P. et al. (2018). Nongenetic approach for imaging protein dimerization by aptamer recognition and proximity-induced DNA assembly. *JACS* 140 (12): 4186–4190.

80 Li, Y., Zhang, X., Pan, W. et al. (2020). A nongenetic proximity-induced fret strategy based on DNA tetrahedron for visualizing the receptor dimerization. *Anal. Chem.* 92 (17): 11921–11926.

81 Yang, T., Xu, L., Liu, S. et al. (2019). Amplified fluorescence imaging of HER2 dimerization on cancer cells by using a co-localization triggered DNA nanoassembly. *Microchim. Acta* 186: 439.

82 He, Z., Zhang, W., Mao, S. et al. (2018). Shear stress-enhanced internalization of cell membrane proteins indicated by a hairpin-type DNA probe. *Anal. Chem.* 90 (9): 5540–5545.

83 Scholler, P., Moreno-Delgado, D., Lecat-Guillet, N. et al. (2017). HTS-compatible FRET-based conformational sensors clarify membrane receptor activation. *Nat. Chem. Biol.* 13: 372–380.

84 Ojosnegros, S., Cutrale, F., Rodriguez, D. et al. (2017). Eph-ephrin signaling modulated by polymerization and condensation of receptors. *Proc. Natl. Acad. Sci.* 114 (50): 13188–13193.

85 Ueki, R. and Sando, S. (2014). A DNA aptamer to c-Met inhibits cancer cell migration. *Chem. Commun.* 50: 13131–13134.

86 Wang, L., Liang, H., Sun, J. et al. (2019). Bispecific aptamer induced artificial protein-pairing: a strategy for selective inhibition of receptor function. *JACS* 141 (32): 12673–12681.

87 Sun, M., Liu, S., Wei, X. et al. (2021). Aptamer blocking strategy inhibits SARS-CoV-2 virus infection. *Angew. Chem. Int. Ed.* 60 (18): 10266–10272.

88 Tan, J., Zhao, M., Wang, J. et al. (2019). Regulation of protein activity and cellular functions mediated by molecularly evolved nucleic acids. *Angew. Chem. Int. Ed.* 58 (6): 1621–1625.

89 Lai, W.Y., Huang, B.T., Wang, J.W. et al. (2016). A novel PD-L1-targeting antagonistic DNA aptamer with antitumor effects. *Mol. Ther. Nucleic Acids* 5: e397.

90 Li, L., Chen, X., Cui, C. et al. (2019). Aptamer displacement reaction from live-cell surfaces and its applications. *JACS* 141 (43): 17174–17179.

91 Chen, S., Xu, Z., Yang, W. et al. (2019). Logic-gate-actuated DNA-controlled receptor assembly for the programmable modulation of cellular signal transduction. *Angew. Chem. Int. Ed.* 58 (50): 18186–18190.

92 Liu, S., Li, S., Lin, J. et al. (2022). Aptamer-induced-dimerization strategy attenuates AβO toxicity through modulating the trophic activity of PrP^C signaling. *JACS* 144 (21): 9264–9270.

93 Yang, X., Yang, L., Yang, D. et al. (2022). In situ DNA self-assembly on the cell surface drives unidirectional clustering of membrane proteins for the modulation of cell behaviors. *Nano Lett.* 22 (8): 3410–3416.

94 Chen, S., Xu, Z., Li, S. et al. (2022). Systematic interrogation of cellular signaling in live cells using a membrane-anchored DNA multitasking processor. *Angew. Chem. Int. Ed.* 61 (11): e202113795.

14

Harnessing DNA Nanotechnology for Nongenetic Manipulation and Functionalization of Cell Surface Receptor

Hexin Nan, Hong-Hui Wang, and Zhou Nie

Hunan University, State Key Laboratory of Chemo/Biosensing and Chemometrics, College of Chemistry and Chemical Engineering, College of Biology, Hunan Provincial Key Laboratory of Biomacromolecular Chemical Biology, Changsha 410082, China

14.1 Introduction

Cells respond to external signals and exhibit corresponding cellular behaviors, such as adhesion, migration, proliferation, differentiation, and apoptosis. These cellular behaviors are generally orchestrated by sophisticated intracellular signal transductions elicited by ligand–receptor interactions on the cell surface [1]. Physiological activation of receptor-mediated signaling pathways is crucial for diverse biological functions such as development, wound healing, and immune responses [2]. Pathological activation of receptors leads to uncontrolled cellular behaviors, which are closely associated with the occurrence and development of cancer and autoimmune diseases [3]. Therefore, precise modulation of cell surface receptors through molecular engineering offers not only an opportunity for a better understanding of cellular processes but also the pathway to the development of cell-based therapeutic strategies against diseases.

Cellular functions are tightly regulated by complex signal transduction pathways mediated by cell surface receptors. It is worth noting that most receptors do not function as monomers but require the formation of nanoclusters to promote downstream intracellular signaling pathways [4, 5]. Upon sensing extracellular stimuli, many types of cell surface receptors transform from freely diffusing monomers into dimers or even higher-order oligomers, including tyrosine kinase receptors (RTKs), integrins, G-protein-coupled receptors (GPCRs), and T cell receptors (TCRs), which can elicit distinct downstream intracellular signaling pathways (Figure 14.1a) [6, 7]. Understanding the molecular mechanisms underlying these processes presents a major challenge for cell biologists. Traditional top-down strategies usually resolve signaling pathways by adjusting the expression levels of one or more genes via knockout or overexpression methods [8]. While these techniques provide a direct way to study the biological components of the target pathway, they have limitations in interrogating signaling dynamics and are often complicated by the crosstalk effects of related pathways [9]. Moreover, synthetic

DNA Nanotechnology for Cell Research: From Bioanalysis to Biomedicine, First Edition. Edited by Zhou Nie.

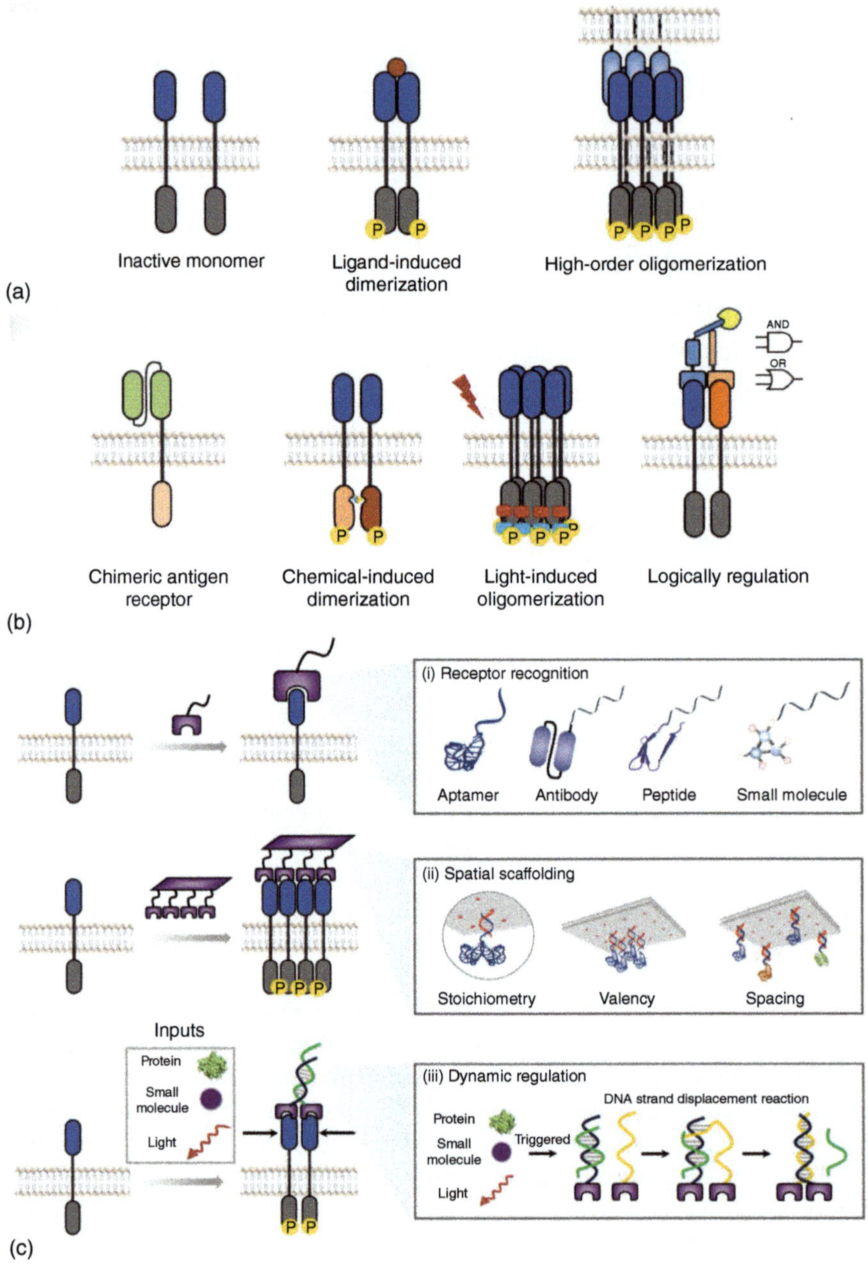

Inactive monomer
Ligand-induced dimerization
High-order oligomerization
(a)
AND
OR
Chimeric antigen receptor
Chemical-induced dimerization
Light-induced oligomerization
Logically regulation
(b)
(i) Receptor recognition
Aptamer
Antibody
Peptide
Small molecule
(ii) Spatial scaffolding
Stoichiometry
Valency
Spacing
Inputs
Protein
Small molecule
Light
(iii) Dynamic regulation
DNA strand displacement reaction
Triggered
(c)

Figure 14.1 Scheme illustrating DNA nanotechnology-driven nongenetic engineering strategies for functionalizing cell surface receptors and modulating cellular signaling. (a) Natural receptor activation mechanism: Upon ligand binding, monomeric receptors undergo dimerization or higher-order oligomerization, regulating the strength of signal activation. (b) The genetic engineering approach involves the construction of synthetic receptors, enabling tailored ligand responses of the CAR, chemical-induced dimerization, light-induced oligomerization, and rewired logic regulation to customize cell signaling. (c) DNA nanotechnology provides customizable modules for nongenetic receptor functionalization to manipulate diverse cell signaling. (i) Receptor recognition modules anchor and manipulate the receptors via receptor-binding aptamers or DNA-conjugated natural or synthetic ligands (antibody, peptide, and small molecule) to the cognate receptors; (ii) Spatial scaffold modules act as a spatial nanoplatform, enabling precise organization of receptor clusters; (iii) Dynamic assembly modules comprise cascaded DNA circuits, allowing for kinetic and intelligent regulation of receptor signaling in response to different external stimuli such as protein, small molecule, and light.

ligands capable of triggering receptor oligomerization have attracted attention of their potential in elucidating cellular mechanisms and advancing clinical therapy. However, their practical biomedical applications remain largely restricted by issues of efficacy and safety [10]. Thus, the exploration of signal activation mediated by receptor self-assembly or co-assembly of multiple receptors in a user-defined way to precisely reprogram signal transduction has emerged as a significant research focus.

Over the last two decades, synthetic biologists have engineered natural receptors into synthetic ones using engineering principles that enable customized sensing and response programs. By introducing engineered components and genetic circuits, advanced synthetic biology strategies can connect extracellular and intracellular cues to user-defined cellular responses [11]. Integrated with information processing tools, these advances further make it possible to program complex and customized cellular functions [12]. Recently, the chimeric receptor engineering approach has been widely employed to synthetically control receptor activation as well as the corresponding downstream signal transduction (Figure 14.1b) [13]. For example, the chimeric antigen receptor (CAR) is designed with three integrated functional domains: an antigen-recognizing extracellular domain, a transmembrane domain, and an intracellular domain. This unique configuration combines antigen-specific recognition with T-cell activation, enabling the CAR-equipped T cells to bind a specific antigen on the surface of the cancer cells and subsequently activate the T-cell signaling for target cell killing [14, 15]. Another example is the synthetic Notch receptor (SynNotch), which is tailored based on the core transmembrane domain of the natural mechanosensitive Notch receptor. This design incorporates an extracellular ligand-binding module and an intracellular transcription module. Upon binding a target ligand, a precise protease cleavage occurs in the SynNotch receptor near the transmembrane region. This cleavage event leads to the translocation of the transcription module into the nucleus, where it activates a target transcription factor for desired gene transcription [16, 17]. Both of above mentioned synthetic biology strategies enable the coupling of user-defined receptor activation to controlled cellular behaviors.

Advances in synthetic biology and cell engineering have emphasized the design of synthetic receptors to serve as molecular tools for researchers aiming to dissect the function of biological systems. The construction of synthetic receptors that respond to flexible, user-controlled inputs, such as small molecules or light, can enhance our understanding of signal networks and facilitate the design of precise cellular responses to specific inputs (Figure 14.1b). To date, a variety of synthetic receptors triggered by small-molecule ligands or light have been developed to probe and reprogram cellular function. For instance, the chemically induced dimerization (CID) strategy is used to control downstream signaling pathways for desired cell behavior, which is achieved by homodimerization or heterodimerization of the synthetic receptors upon small molecule binding [18]. In addition, optogenetic approaches leverage light as a novel physical inducer to modulate protein activity with spatial and temporal precision. Schaffer and colleagues reported a genetic coding tool named clustering indirectly using cryptochrome (CLICR), which involves the fusion of Arabidopsis cryptochrome 2 (Cry2) to the target receptors, allowing for the optical regulation of RTK clustering and subsequent downstream signaling [19].

Moreover, the rational engineering of receptors to integrate intelligently and logically gated protein circuits allows sophisticated and selective manipulation of cell surface receptors, addressing the major challenges in precision medicine (Figure 14.1b). Recently, Baker et al. developed a colocalization-dependent protein switch, Co-LOCKR, which is composed of a cage protein, a lock protein, and an effector protein [20]. The functionality of this system operates with a cage protein sequestering a functional peptide via a latch domain. The peptide is released from sequestration only when the associated cage and lock proteins encounter each other on the cell surface, which causes conformation changes to activate Co-LOCKR and recruit relevant effector proteins. Intriguingly, this approach allows for the execution of AND, OR, and NOT-gate logic operations on the cell surface, giving rise to a single output for cell-specific regulation [20].

Genetically encoded synthetic receptors have opened new avenues for cell-based therapy by empowering researchers to equip cells with novel functionalities. However, synthetic receptors do pose inherent challenges for *in vivo* applications. First, diverse viral vectors, such as adeno-associated viruses (AAVs), which are used for gene delivery, have limitations in their capacity to accommodate the size of recombinant chimeric constructs and infection efficiency [21]. Moreover, it remains challenging to quantitatively control expression levels and ensure the correct conformation of the chimeric proteins [22]. Additionally, genetic modifications can potentially interfere with other endogenous biological processes of the target cells, thereby affecting cellular performance [23]. For *in vivo* applications, there are also significant challenges in achieving targeted delivery to the intended tissues and performing *in situ* genetic modifications of target cells. Lastly, the potential integration of the recombinant DNA into the genome of endogenous cells could lead to uncontrolled risks [24]. Therefore, there is an increasing demand for nongenetic engineering alternatives that would offer a simple design, intelligent response, and strong biocompatibility to enable the reprogramming of receptor-mediated cellular behavior.

In nature, DNA molecules constitute the fundamental genetic material, encoding the crucial genetic information of all living organisms [25]. Exploiting the unparalleled precision and programmability of nucleic acid hybridization, researchers have engineered a wide array of DNA nanostructures with customizable topologies at the nanoscale. These innovative DNA nanostructures have been extensively utilized across a broad spectrum of applications, including biosensing, cell imaging, and drug delivery [26]. Moreover, advancements in dynamic DNA nanotechnology have enabled the construction of diverse stimuli-responsive molecular frameworks, such as DNA logic gates, DNA circuits, DNA nanomachines, and DNA computation [27]. Their unique designs involve the programmed self-assembly of DNA strands, coupled with either DNAzyme catalysis or cascaded strand displacement reactions, establishing molecular devices that can carry out mechanical or dynamic behaviors [28]. Notably, the remarkable progress in DNA nanotechnology in recent years has facilitated the development of DNA nanodevices capable of activating cell surface receptors for artificial regulation of cellular functions.

In this chapter, we mainly focus on the latest advancements in employing DNA nanotechnology for the nongenetic manipulation of cell surface receptors, which allows for precise control over desired cellular behaviors. We will elaborate on the advantages of harnessing the DNA molecule as an actuating machine for receptor manipulation. First of all, DNA has the potential to mimic natural ligands, thereby substituting the role of these ligands in receptor activation. Moreover, the DNA molecule can be intricately tailored into dynamic nanodevices that reprogram natural receptors to perform a range of user-defined functionalities. By integrating DNA circuits and logic gate operations, we can design highly sophisticated DNA nanodevices for intelligent control over receptor function and cellular behaviors with high-order selectivity. Specifically, we also highlight the significant role of DNA-based molecular tools in receptor-mediated therapeutic outcomes *in vivo*. Furthermore, the programmability and spatial addressability in DNA nanostructures, such as DNA origami, underscore the significance of DNA molecules in regulating the nanoscale clustering of receptors for precise cell regulation. Together, the strategies enabled by DNA nanotechnology for receptor manipulation could pave the way for the development of innovative cell-based therapies in the future.

14.2 Principle of DNA-enabled Molecular Engineering for Receptor Regulation

The advent of DNA nanotechnology has revolutionized the role of DNA molecules, transitioning from their inherent function as gene information carriers to being used as the building blocks for molecular engineering. Over recent decades, significant breakthroughs have been made in constructing biocompatible functional nanodevices using DNA molecules [29]. Advancements in functional nucleic acid (FNA) and DNA nanotechnology have established foundational principles for orchestrating essential modules toward rational construction of designer DNA nanodevices aimed at precise receptor regulation. These key modules include: (i) DNA aptamers or DNA-conjugated ligands that serve as specific recognition modules, selectively binding to and manipulating receptors; (ii) DNA nanostructures that can be programmed as spatial scaffold modules to arrange the distribution of receptors at the nanoscale; and (iii) the principle of dynamic DNA assemblies, which could be employed for the kinetic regulation of receptors. By synergizing these DNA-based modules, researchers have developed advanced DNA nanodevices that can not only precisely control receptor activation but also repurpose a receptor non-genetically to perform a novel function.

14.2.1 Recognition Module for Receptor Manipulation

Exploring and manipulating receptors through DNA nanodevices first requires robust, multifunctional, and physiologically compatible recognition modules for binding to cell surface receptors. An ideal DNA-based recognition module is the aptamer, a well-recognized type of FNA that directly interacts with target molecules. Aptamers are single-stranded DNA or RNA molecules, ranging from 15–60 nucleotides, which can self-fold into unique secondary/tertiary structures to specifically recognize a wide array of targets, including small molecules, proteins, cells, and tissues. Aptamers are obtained through a process known as "systematic evolution of ligands by exponential enrichment" (SELEX), which can identify aptamers with high affinity and specificity for their targets [30]. To date, numerous aptamers have been successfully identified and validated that specifically recognize various cell surface receptors, such as RTKs, including mesenchymal epithelial transition receptor (c-Met), fibroblast growth factor receptor (FGFR), vascular endothelial growth factor receptor (VEGFR), insulin receptor (IR), and others [31]. Other well-characterized aptamers include those for integrin [32] and TCR [33]. These receptor-binding aptamers constitute a valuable repertoire of recognition and manipulating modules for DNA-enabled receptor regulation. Besides aptameric recognition modules, both natural and synthetic ligands (antibody, peptide, or small molecule) have been exploited by proper conjugation with DNA strands, thus constituting other types of recognition modules (Figure 14.1c(i)). Notably, these recognition modules open up further opportunities for coupling receptor recognition with other DNA-based regulatory modules, including spatial scaffold modules and dynamic assembly modules.

14.2.2 Spatial Scaffold Module for Receptor Organization

The spatial distribution of receptors plays a crucial role in natural receptor cluster-mediated cell signaling. Therefore, engineering the nanoscale spatial distribution of receptors can significantly facilitate precise receptor regulation [34]. The development of spatial organization modules capable of spatially arranging receptors at the nanoscale would be of great significance for regulating complex and coordinated receptor signaling for desirable cellular behaviors. In 1953, Watson and Crick proposed the principle of complementary base pairing and the double helix structure of DNA, thus paving the way for molecular biology [35].

Accordingly, DNA molecules can be programmed into nanostructures with precise spatial scaffolding and addressability at the nanoscale. In 1982, Seeman pioneered the field by transforming the classical linear assembly mode of DNA into a branched structure, demonstrating that DNA-based molecular engineering could be employed to construct specific, precise, and ordered nanostructures and thereby laid the foundation for DNA nanostructure research [36]. By precisely designing hybridization sequences, diverse DNA nanostructures have been developed, including tiles, tetrahedra, nanotubes, origami, and others [37]. Owing to their predictable, controllable, and addressable features, many different DNA nanostructures have been exploited for biomedical applications, such as drug delivery and biosensing [37, 38]. Significantly, by incorporating the recognition modules into the DNA nanostructures, the accurate stoichiometry, valency, and nanoscale spacing between receptors can be precisely defined, enabling exact spatial regulation over the receptor cluster (Figure 14.1c(ii))[39].

14.2.3 Dynamic Assembly Module for Kinetic Control of Receptor

Ligand-induced receptor activation is a highly orchestrated and dynamic molecular process. Developing reconfigurable modules for dynamical control could facilitate the time-dependent modulation of the oligomeric state of receptors in a user-defined manner. In dynamic DNA nanotechnology, the toehold-mediated strand displacement reaction has become a popular approach for achieving such dynamic control over DNA assembly [40]. In this process, a certain length of single-stranded binding site, known as a toehold, is preserved on a DNA duplex. This toehold can bind to an inputting DNA single strand with a complementary end, thereby initiating a strand displacement reaction and displacing one of the strands from the original DNA duplex (Figure 14.1c(iii)) [41]. Moreover, DNAzymes, which are a type of FNA with enzyme-mimicking catalytic activity, can also serve as dynamic response modules [42]. In the presence of certain cofactors, i.e. ions and amino acids, several DNA/RNA-hydrolyzing DNAzymes can specifically cleave targeted DNA or RNA sequences and release the cleaved strand, which can subsequently initiate a strand displacement reaction. This unique capability of DNAzymes provides another layer of dynamic control in the assembly of DNA nanodevices, which have been exploited for biosensing and therapeutic applications [43, 44]. Furthermore, researchers can construct more complex DNA logic circuits and dynamic chemical reaction networks (CRNs) by integrating multiple strand displacement reactions into cascade reactions [45]. To date, a wide array of DNA strand displacement reactions and their derivatives have been successfully applied to engineer dynamical systems, including DNA circuits, catalytic amplifiers, autonomous molecular motors, and reconfigurable nanomachines. Importantly, these constructed circuits operate as enzyme-free response networks, allowing dynamical control of the reactions by simply adjusting the input and output DNA strands [46]. Together, DNA dynamic assembly modules provide kinetic control over receptor clustering, achieving adaptive or intelligent regulation of receptor function in response [47].

14.3 DNA Nanodevices for Programming Receptor Function

14.3.1 Bivalent Aptamer Mimicking Natural Ligand to Induce Receptor Dimerization

Receptor–ligand interactions are essential events in directing cellular responses to adapt to the extracellular environment. Natural ligands such as growth factors primarily function by binding to and activating their cognate cell surface receptors, subsequently initiating a cascade of intracellular signaling events that regulate cell behavior and fate. Recombinant growth factors, due to their potent biological activity and capacity to promote cell growth and proliferation, have attracted considerable interest in the research fields of wound healing and tissue regeneration, especially in conditions involving organ injury [48]. However, the practical application of recombinant growth factors has shortcomings associated with high production costs, potential risk of contamination, and discrepancies in biological activity across different batches.

In the activation process of most single-pass transmembrane receptors, either natural or recombinant ligands bind and assemble their cognate receptor monomers into dimers or even higher-order oligomers, thereby inducing autophosphorylation to trigger intracellular signaling. In this context, receptor-specific aptamers have emerged as monomeric recognition modules that directly bind to their corresponding receptors. When two copies of the receptor-binding aptamers are conjugated or assembled, they can simulate the activity of natural ligands, thereby inducing receptor dimerization and activation. In 2015, Allen et al. first demonstrated this idea through their work with bivalent aptamers [49]. They engineered a bivalent aptamer for the vascular endothelial growth factor receptor 2 (VEGFR2), separated by an 18-atom PEG spacer. This configuration functioned as a synthetic VEGFR2 agonist, promoting capillary formation in endothelial cells. This study laid the foundation for the development of aptamer-based receptor agonists, providing an innovative strategy for artificially regulating receptor dimerization for signaling activation. Employing similar methodology, Sando et al. developed a DNA-based hepatocyte growth factor (HGF) mimic to activate c-Met, the receptor for HGF (Figure 14.2a) [50]. Specifically, two c-Met aptamer monomers were hybridized through complementary strands at their 5′ termini, thereby creating a bivalent aptamer. This construct could dimerize two c-Met receptors, triggering c-Met signaling in living cells and ultimately promoting the cellular behavior similar to that induced by HGF. Further, they engineered DNA agonist for fibroblast growth factor receptor 1 (FGFR1) (Figure 14.2b) [51]. In this design, two FGFR1-binding DNA aptamers were conjugated to formulate a construct capable of functioning as an analog for basic fibroblast growth factor (bFGF). This engineered DNA agonist was able to maintain the pluripotency and self-renewal activity of human-induced pluripotent stem cells (hiPSCs). In conclusion, these studies collectively illustrate that aptameric recognition modules for a specific receptor can be bivalently conjugated to function as natural ligand mimics, effectively activating the receptor for desired cellular behaviors.

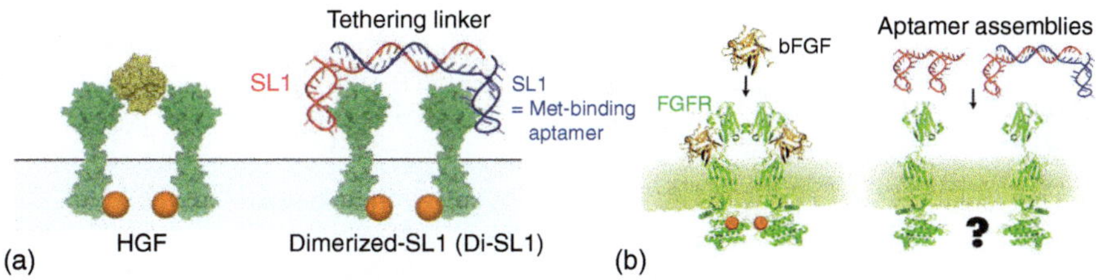

Figure 14.2 Schematic representation of DNA-based bivalent aptamers mimicking natural ligand to induce receptor dimerization. (a) A DNA-based bivalent aptamer mimics hepatocyte growth factor and functions as an agonist to induce c-Met dimerization. Source: Reproduced with permission from Ueki et al. [50]. © 2016 Wiley-VCH. (b) DNA-based bivalent aptamer mimics fibroblast growth factor and acts as an agonist to induce FGFR1 dimerization. Source: Reproduced with permission from Ueki et al. [51]. © 2019 Royal Society of Chemistry.

14.3.2 DNA Nanodevices to Customize Receptor Responsiveness

Engineering receptor responsiveness is indeed a powerful strategy for rewiring receptor signaling to alter cellular functions in response to user-defined cues, such as proteins or small molecules. As a relatively mature technique, CAR-T strategy serves as a prime example of this concept [52]. It involves the construction of a chimeric receptor that is fused with an antibody-based recognition domain, specifically tailored to recognize tumor-associated antigens and trigger T cell activation. This approach illustrates the potential of engineering receptor responsiveness in the development of advanced therapeutics. Alternatively, the programmable assembly of DNA-based recognition modules for specific stimuli and receptors can link cue-specific responsiveness with activation of the target receptor. Exploiting the designability of DNA, Sando et al. developed the DNA aptamer-mediated reprogramming of the interaction partner of RTK (DRIPaR) strategy (Figure 14.3a) [53]. This approach employs a bispecific aptamer that incorporates an aptamer sequence binding to the RTK and another sequence specifically responsive to a user-defined protein as the artificial ligand. Upon addition of external protein molecules, the bispecific aptamers can induce c-Met dimerization and activation by forming a 1 : 2 ternary complex. By introducing different target-recognizing aptamers, the DRIPaR strategy is capable of rewiring RTK signaling to respond to other protein molecules, such as thrombin and platelet-derived growth factor (PDGF). This design offers a powerful approach to customize the receptor signaling pathways to respond to specific external cues as needed.

CID serves as a fundamental tool for cellular engineering, extensively utilized in exploring receptor interactions and assessing subcellular localization of target proteins [56, 57]. This strategy was initially introduced by Crabtree et al. in 1993, where a small molecule was used to trigger the dimerization of proteins fused with FKBP, a protein-binding domain [58]. Despite the potential of CID strategies in receptor regulation, most CID-mediated receptor manipulations still require the use of a genetic engineering approach to construct chimeric protein receptors [57]. Our group developed a DNA-mediated chemically induced

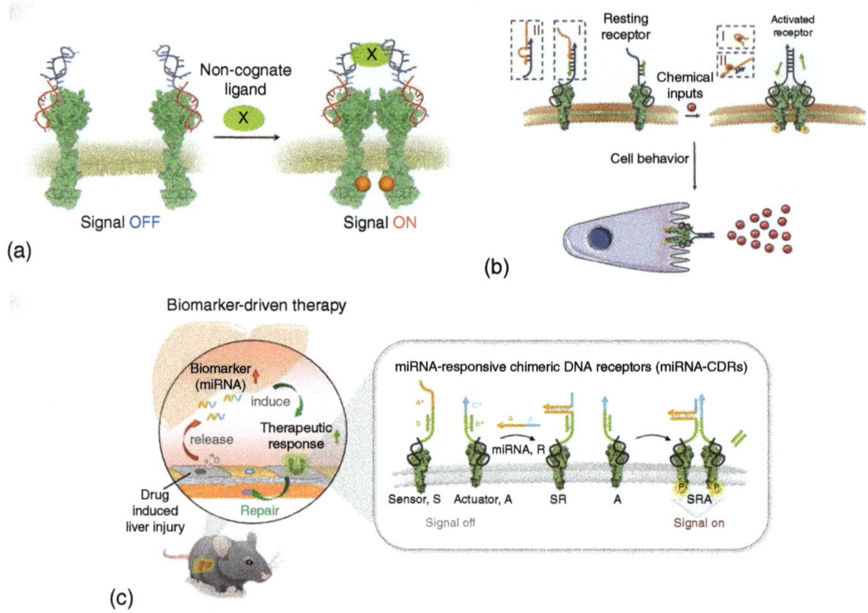

Figure 14.3 Responsive activation of receptor dimerization on the cell membrane by aptamer-based DNA nanodevices. (a) DNA-aptamer-mediated reprogramming of c-Met activation dependent on non-cognate ligands. Source: Reproduced with permission from Ueki et al. [53]. © 2017 American Chemical Society. (b) D-CID nanodevices responsive to user-defined cues for receptor activation and control of cellular behavior. Source: Reproduced with permission from Li et al. [54]. © 2018 Wiley-VCH. (c) Modular and programmable miRNA-responsive chimeric DNA receptor designed for biomarker-driven therapy to treat liver injury. Source: Reproduced with permission from He et al. [55]. © 2023 Wiley-VCH.

dimerization (D-CID) designed for small-molecule-controlled receptor activation and subsequent cell behavior manipulation (Figure 14.3b) [54]. D-CID combines aptameric sensing or DNAzyme-based cleavage with the strand displacement reaction to dimerize the c-Met receptor and trigger downstream signaling to promote cell migration toward small-molecular or ionic cues, including adenosine triphosphate (ATP), histidine, and Zn^{2+}. Furthermore, the multi-input properties of D-CID allow for the orthogonal regulation of chemotaxis in multiple cell populations simultaneously in a targeted and programmable manner. Recently, we designed a modular and programmable miRNA-responsive chimeric DNA receptor (miRNA-CDR), which aimed to facilitate biomarker-driven therapeutic responses (Figure 14.3c) [55]. By incorporating an aptameric receptor recognition module, this miRNA-responsive DNA nanodevice can be integrated with a natural membrane receptor to detect extracellular miRNA levels and subsequently drive receptor activation through dimerization, mediated by a miRNA-triggered complementary strand displacement reaction that induces dynamic DNA assembly. Due to its sequence programmability, the miRNA-CDR platform can be customized to detect and react to specific miRNAs with tunable sensitivity. Importantly, the miRNA-CDR can be tailored to respond orthogonally to designated miRNAs and reprogram specific cellular signaling pathways by targeting distinct RTK receptors, such as c-Met and FGFR1. The functionality of this designer miRNA-CDR was illustrated by redirecting the recognition of elevated miR-122, a biomarker for drug-induced liver injury, to promote c-Met signaling in hepatocytes for biomarker-driven *in situ* liver repair and functional restoration.

Together, the advancement of dynamic DNA nanodevices offers novel approaches to nongenetically convert external chemical cues into receptor signaling, potentially improving the precision of cell-based therapies [59].

14.3.3 Light-Responsive DNA Nanodevices for Spatiotemporal Receptor Regulation

Recent advancements in photochemistry technology and light-responsive nanomaterials have provided exciting opportunities to achieve receptor manipulation with excellent spatiotemporal precision [60]. By integrating light-responsive modules into dynamic DNA nanodevices, it is now possible to modulate cellular behaviors using light as a stimulus. Several innovative DNA-based methodologies have emerged that enable the optical control of receptor oligomerization to elicit cell signaling. Yang et al. developed a light-controlled DNA assembly strategy by introducing an ultraviolet (UV) light photocleavable module (PC-linker, an o-nitrobenzyl moiety), thereby facilitating light-mediated receptor manipulation (Figure 14.4A) [61]. Under UV light exposure, the PC-linker is cleaved, triggering DNA strand replacement and dynamic DNA assembly. This light-responsive process promotes the c-Met dimerization and triggers the downstream intracellular signaling pathway. Similarly, Sando et al. established a photocaged module capable of regulating c-Met dimerization by engineering a DNA aptamer modified with 6-nitropiperonyl-α-methyl (NPM) caged group on the complementary linker sequence at the 3′ terminal of the aptamer sequence (Figure 14.4B) [62]. The caged group temporarily inhibits the complementarity of the linker, and UV light irradiation removes the caged group to restore its hybridization capacity, thereby resulting in c-Met dimerization and activation of downstream signaling pathway. Additionally, Li et al. exploited rolling circle amplification (RCA) method to generate a DNA nanospring scaffold to load multivalent RGD (Arg-Gly-Asp) conjugated DNA, which was anchored on the cell surface as ligands of integrins. Utilizing complementary DNA strands, the extension and contraction of integrin-bound RGD on nanosprings can be tuned via DNA strand displacement reactions, thereby reversibly regulating integrin clustering to modulate cell adhesion [65]. Further, Endo et al. introduced azobenzene group to construct a light-regulated reversible DNA nanosprings polymer system (Figure 14.4C) [63]. When exposed to UV or visible light, the cis-trans conformational change of azobenzene triggers a reversible folding and unfolding of the DNA hairpin structures. This dynamic process induces the contraction and relaxation of DNA nanospring to affect the morphology of mesenchymal stem cells.

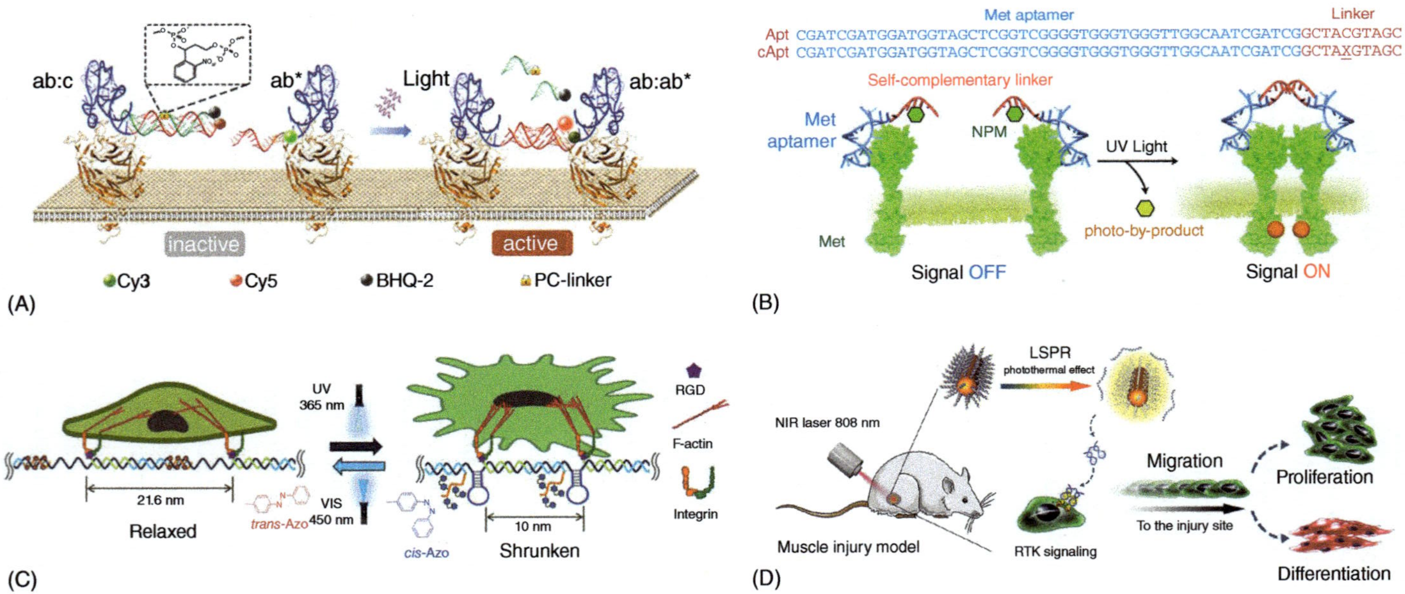

Figure 14.4 Spatiotemporal regulation of receptor using light-controlled assembly of DNA nanodevices. (A) Light-controlled DNA assembly by introducing PC-linker for c-Met activation on the cell membrane. Source: Reproduced with permission from Chen et al. [61]. © 2018 Wiley-VCH. (B) 6-NPM-modified deoxycytidine serves as a caged group to generate photocaged c-Met aptamers that control protein dimerization and activate downstream signaling pathways. Source: Reproduced with permission from Ueki et al. [62]. © 2021 Royal Society of Chemistry. (C) Photoresponsive dynamic DNA polymers modified with azobenzene allow for the reversible adjustment of the distance between RGD ligands on the cell membrane, achieving spatiotemporal regulation of cellular morphology. Source: Reproduced with permission from Sethi et al. [63]. © 2021 Wiley-VCH. (D) Near-infrared light-activated DNA nanodevices are used to induce c-Met dimerization, remotely controlling cell proliferation and differentiation in the treatment of skeletal muscle injury *in vivo*. Source: Reproduced with permission from Wang et al. [64]. © 2019 American Chemical Society.

Despite their effectiveness, the aforementioned strategies suffer from inherent limitations related to the use of UV light, such as low penetration depth and potential phototoxicity. These challenges significantly limit their application *in vivo*. To address these obstacles, our group developed a novel near-infrared light-activated DNA agonist (NIR-DA) system (Figure 14.4D) [64]. Leveraging the photothermal properties of gold nanorods (AuNRs) upon NIR exposure, the NIR-DA system enables NIR light-induced release and activation of the DNA agonist, leading to c-Met dimerization on cell surface and subsequent signal transduction. The NIR-DA system has proven to be effective in regulating cytoskeletal reorganization, cell polarization, and directional migration in live cells. Further, we successfully applied NIR-DA system *in vivo* to mediate c-Met signaling and regulate skeletal muscle satellite cell migration and myogenesis, which are essential processes in skeletal muscle regeneration. The NIR light-responsive strategy enabled deeper tissue penetration, noninvasive manipulation, and minimal photodamage, presenting a promising avenue for the remote control of cell-based therapy.

14.3.4 DNA Nanodevices for Visualization of Receptor Activation

The visualization of receptor dimerization within living cells is essential for exploration of the detailed mechanism involved in intercellular communication and signal transduction. Most commonly, this process is monitored by fusing the receptor with genetically encoded fluorescent proteins. However, the introduction of genetically engineered receptors possibly affects the efficiency of ligand-induced receptor dimerization and interfere with the normal biological processes inside cells. Therefore, there is a growing need to develop nongenetic strategies that allow for real-time monitoring of receptor dimerization. Yang et al. developed an aptamer-based approach for visualization of receptor dimerization using the c-Met monomer-specific aptamer as a recognition probe (Figure 14.5A) [66]. When receptor dimerization occurred in the presence of a ligand, it brought two receptor-anchored aptamer probes into proximity, which then triggered dynamic DNA assembly to detach the quencher group and generate a fluorescent signal. This strategy has been successfully applied to visualize the dimer state of c-Met receptor and transforming growth factor-β type II receptor (TβRII).

To enhance the stability of DNA probes, Tang et al. introduced DNA tetrahedron nanostructures into the probe design and proposed a non-genetic, proximity-induced FRET strategy (Figure 14.5B) [67]. This design allowed for the visualization of c-Met receptor dimers not only on living cells but also within living tumors. Nevertheless, several challenges still remain with DNA fluorescent probe-enabled visualization of receptor activation: (i) the intricate cellular microenvironment can cause a high fluorescence background; (ii) continuous irradiation leads to photobleaching of fluorophores, limiting long-term monitoring; and (iii) the fluorescence signal can be too weak to detect the dimerization of the receptors that are lowly expressed on the cell membrane. To overcome these challenges, Wang et al. recently exploited the surface-enhanced Raman scattering (SERS) principle to visualize the dual homodimerizations of c-Met and TβRII

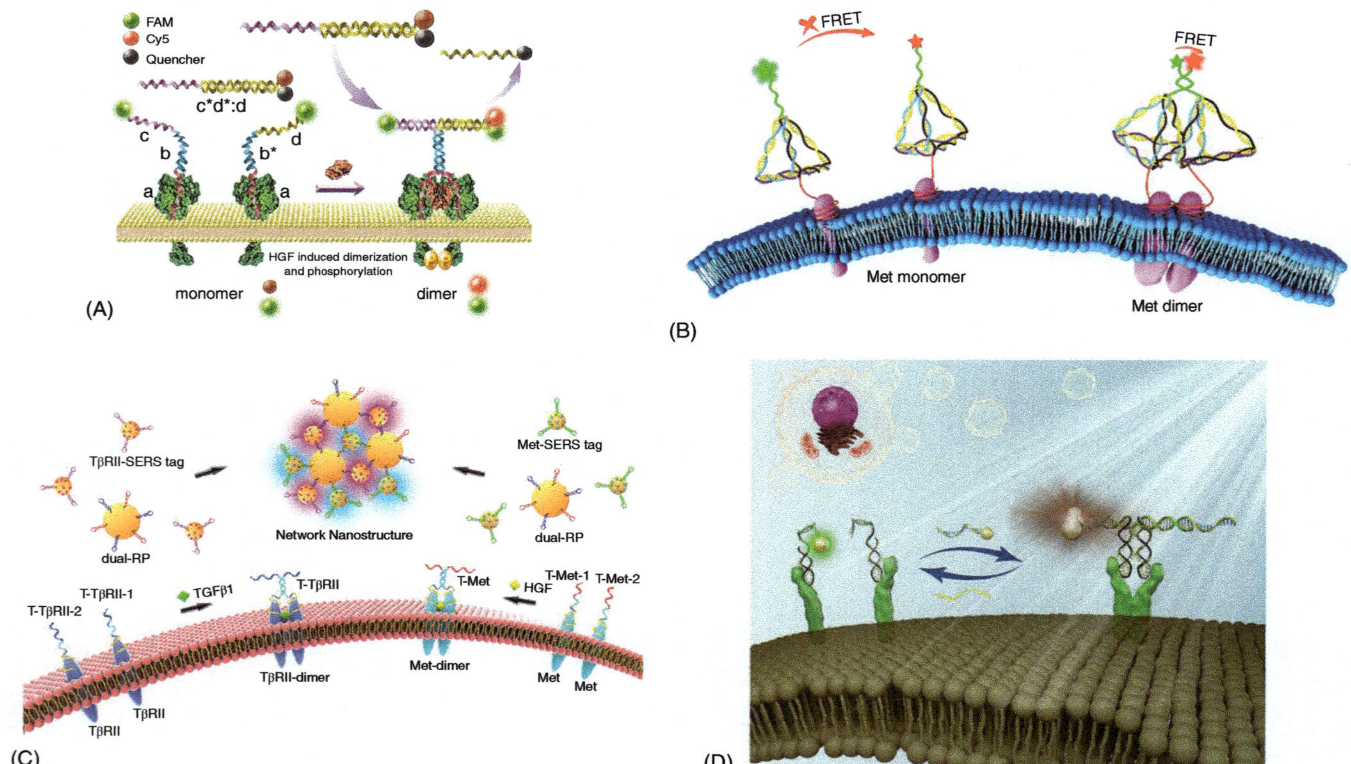

Figure 14.5 DNA nanodevices for visualization of cell surface receptor dimerization. (A) An aptamer recognition and proximity-induced DNA assembly strategy for the visualization of protein dimerization on the cell membrane. Source: Reproduced with permission from Liang et al. [66]. © 2018 American Chemical Society. (B) Observation of protein dimerization by aptamer recognition and proximity-induced FRET imaging with DNA tetrahedron fluorescent probes. Source: Reproduced with permission from Li et al. [67]. © 2020 American Chemical Society. (C) The SERS-based monitoring of protein dimerization by dynamic assembly of AuNPs networks on the cell membrane via proximity ligation-assisted CHA. Source: Reproduced with permission from Dong et al. [68]. © 2022 American Chemical Society. (D) DNA-programmed plasmonic nanoprobes for c-Met dimerization imaging by aptamer recognition, hybridization, and DNA strand displacement reaction. Source: Reproduced with permission from Wang et al. [69]. © 2023 American Chemical Society.

receptors on a single living cell (Figure 14.5C) [68]. Utilizing a double DNA hairpin recognition probe modified on gold nanoparticles (AuNPs), the SERS tags generate signals upon proximity ligation-assisted catalytic hairpin assembly (CHA). This approach is highly sensitive and specific, which allows for accurate identification of single or multiple receptor dimerization under complex cellular environments. To further distinguish and track individual receptors, Xu et al. constructed a plasmon ruler strategy (Figure 14.5D) [69]. In this method, the dimerization of c-Met triggers hybridization between AuNPs-attached single-stranded probes and the receptor-anchored probes, inducing the plasmonic coupling of AuNPs to produce an output signal. Together, the above approaches allow for non-genetic, real-time visualization and tracking of receptor dimerization, opening up opportunities for accurate and long-term monitoring of complex physiological processes.

14.4 Elaborate and Intelligent DNA Nanodevices Reprogramming Receptor Function

Living organisms leverage natural molecular machines to carry out a multitude of essential biological functions. These intricate molecular machines are self-assembled from natural macromolecules such as nucleic acids or proteins. These assemblies rely on weak molecular interactions, such as hydrogen bonds and hydrophobicity. A significant scientific challenge in the realm of molecular engineering lies in the development of artificial molecular machines that can function at the nanoscale for controlled dynamical behavior and intelligent molecular recognition. These artificial machines are designed to emulate their natural counterparts through rational design, accurate synthesis, and regulated assembly of biological or synthetic molecules. Thanks to their unique programmability and controllability, DNA molecules can serve as building blocks to construct automated and intelligent molecular machines, which have recently demonstrated promise in reprogramming the functions of cell surface receptors.

14.4.1 Mechanical Control Over Receptor-Mediated Cellular Behavior

Molecular robots have shown the potential to perform intricate mechanical tasks at the nanoscale, which has been recognized, but their utility in governing cellular signaling and behavior has remained unexplored. In 2021, our group developed a self-propelled DNA molecular robot that operated on the cell membrane to drive cellular migration, making it the first prototype designed to regulate cellular behaviors artificially (Figure 14.6a) [70]. Driven by DNAzyme catalytic processes, this robot moved autonomously across the cell surface, transitioning step-by-step from one receptor-anchored point to another. Every step of the robot led to the dimerization of a pair of c-Met receptors, and the continuous walking accumulated c-Met activation to amplify the cell signaling in an ultrasensitive manner. Interestingly, Monte Carlo simulations indicated that each DNA robot (with an estimated count of about 19,000 per cell) could traverse 95 nm in a 30-minute duration, ultimately propelling a

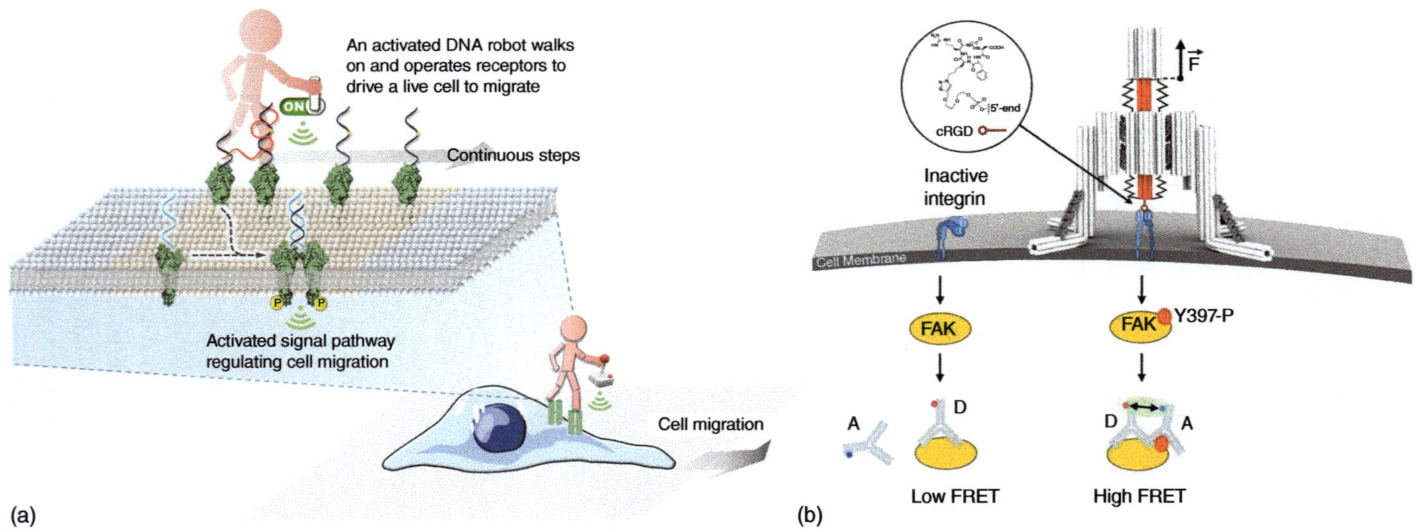

Figure 14.6 Automated DNA nanomachines for mechanical control over receptor-mediated cellular behavior. (a) DNA nanorobot executes desired mechanical molecular operation on the cell surface receptors to activate signal pathways and regulate cell migration. Source: Reproduced with permission from Li et al. [70]. © 2021 Wiley-VCH. (b) Autonomous DNA nano-winch lands and operates on the cell surface to activate integrin signaling. Source: Reproduced with permission from Mills et al. [71]/Springer Nature/CC BY 4.0.

cell to migrate by 87 μm. This process effectively amplified nanometer-scale artificial motions into micrometer-scale cellular movement. We further devised distinct conditional DNA robots to orthogonally control cell migration within mixed-cell populations. This innovative DNA robot serves to drive predetermined cellular behavior, demonstrating vast potential for precisely regulating cellular activities for the development of cell-based therapies.

Besides artificially regulating cellular migration, Bellot et al. recently engineered a DNA origami-based modular nanorobot called Nano-winch, which can autonomously land and operate on the cell surface to manipulate multiple mechanoreceptors (Figure 14.6b) [71]. The cRGD-functionalized Nano-winch could mechanically stimulate integrin receptors on the MCF-7 cell surface between 1 and 15 pN, fine-tuning the level of integrin-mediated downstream phosphorylation of focal adhesion kinase (FAK). Thus, these DNA nanorobots specifically designed to control mechanoreceptors provide promising avenues for investigating a diversity of mechanotransduction circuits in living cells.

14.4.2 Precise Cell Targeting for Selective Receptor Modulation

Protein ligands, such as growth factors and cytokines, play vital roles in regulating cellular signaling and behavior through ligand–receptor interaction on cell surface [72]. The ligand–receptor signaling is crucial in numerous physiological processes, such as tissue regeneration, immune response, and targeted tumor therapy [73]. However, since a given receptor may be widely expressed across a variety of cell types *in vivo*, the corresponding ligand can indiscriminately activate both target and nontarget cells as long as the receptor is presented. This pleiotropy often leads to unintended side effects, such as deactivation, neutralizing effects, and on-target off-cell toxicity, posing significant challenges in clinical application. To address these issues, there is compelling need for the development of artificial ligands capable of precisely recognizing the target cells and selectively activating the desired receptor. In this context, our group designed a programmable and intelligent "scan and unlock" DNA automaton (SUDA) (Figure 14.7) [74]. The SUDA system utilizes embedded DNA-based CRNs on the cell surface to scan and evaluate the molecular profiles of cell-surface proteins using Boolean logic circuits (AND and NIMPLY gates). This approach allows for the programmable and highly selective discrimination of target cell identity (cell-ID). The logic gates integrated into the SUDA system enable high-order discriminative cell targeting and activation of ligand–receptor interactions in a cell-specific manner, which potentially minimizes the on-target off-cell toxicity. By selectively activating ligand–receptor signaling pathways based on cell-surface protein profiles, the SUDA system offers a programmable, scalable, and general approach to improve the precision of cell-specific therapeutics.

The recent advances in intersection of cancer immunology with synthetic biology have given rise to the development of engineering novel macromolecules or cell-based therapies, such as bispecific T-cell engagers (BiTEs) and CAR-T [75]. Over the past decades, these innovative approaches have proven to be effective in cancer treatment, notably hematologic malignancies. However, the complex nature

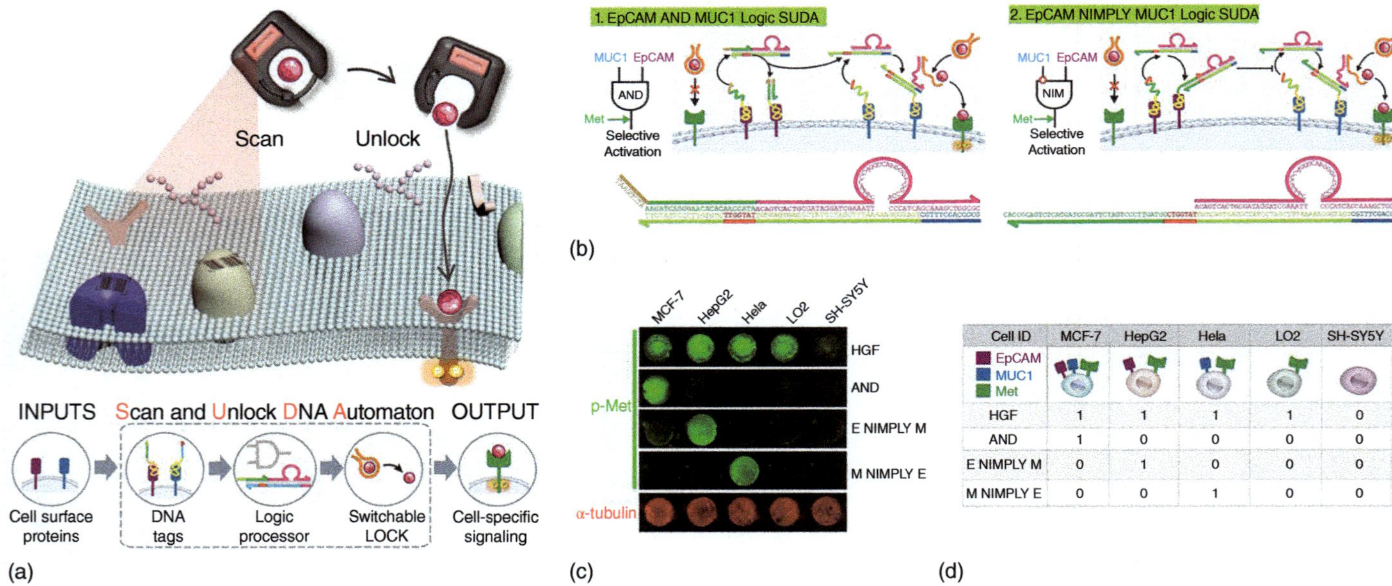

Cell ID	MCF-7	HepG2	Hela	LO2	SH-SY5Y
HGF	1	1	1	1	0
AND	1	0	0	0	0
E NIMPLY M	0	1	0	0	0
M NIMPLY E	0	0	1	0	0

Figure 14.7 Selective activation of receptors based on "scanning" and "unlocking" on the cell surface. (a) Schematic illustration of a programmable DNA molecular automaton for cell-selective activation of ligand-based signaling via an intelligent SUDA system. (b) DNA-based CRNs on the cell surface for high-order cell-specific modulation of ligand–receptor interaction by using Boolean logic circuits (AND and NIMPLY gates). (c) In-cell western blot characterized cell-selective c-Met phosphorylation activation realized by AND or NIMPLY logic-gated DNA molecular automaton. (d) The truth table of AND or NIMPLY logic gates, which processes the logical output for three inputs to EpCAM, MUC1, and c-Met for cell identification and corresponding cell activation. Source: Reproduced with permission from Zhang et al. [74]/Reproduced from Wiley-VCH.

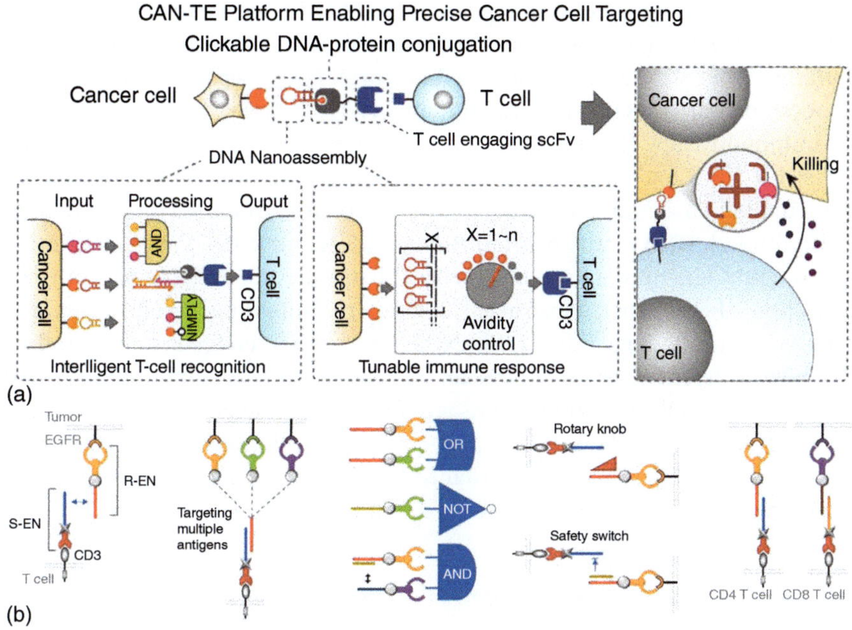

Figure 14.8 DNA-enabled logical recognition to improve T cell killing specificity. (a) A CAN-TE platform enabling intelligent T-cell activation and precise cancer cell targeting. Source: Reproduced with permission from Tang et al. [76]. © 2022 Wiley-VCH. (b) A nanobiohybrid DNA EN-TCR system for the engineered T cells to achieve precisely targeted tumor immunotherapy. Source: Reproduced with permission from Ma et al. [77]. © 2022 American Chemical Society.

of solid tumors still presents challenges, characterized by their heterogeneity, ability to escape antigen targeting, and the off-target toxicity associated with current therapies. Therefore, there is an urgent need to further develop more flexible, intelligent, and adjustable molecular tools to improve the specificity and killing efficacy of targeted cancer immunotherapy. Leveraging the unique advantages of the DNA-based system, our team developed a modular chimeric antibody-nucleic acid T-cell engager (CAN-TE) capable of programmable cancer immunotherapy (Figure 14.8a) [76]. This strategy employed the specific feature of DCV endonuclease to recognize and covalently conjugate with single-stranded DNA with specific sequence. This feature allowed us to construct a modular chimeric molecule, integrating the DCV-αCD3 fusion protein and diverse DNA nanostructures. By incorporating multiple aptamer-antigen recognition systems into dynamic DNA circuits with higher-order logic functions, CAN-TE was able to precisely discriminate cell-ID according to three tumor-associated antigens, which resulted in selective T-cell activation and subsequent cancer eradication. Furthermore, we were able to precisely tune the cancer cell binding avidity by adjusting the valency of the antigen-binding aptamers on the DNA-scaffolds, thus modulating T cell activity for efficient tumor elimination *in vitro* and *in vivo*. Collectively, this modular CAN-TE platform not only allows for intelligent tumor cell recognition through programming DNA circuits with a desirable logic gate but also improves the T-cell killing efficacy with tunable cancer targeting avidity.

Similarly, Ye et al. developed a nanobiohybrid DNA engager-reprogrammed T-cell receptor (EN-TCR) system for the engineered T cells to achieve targeted tumor immunotherapy (Figure 14.8b) [77]. The EN-TCR system consisted of two modules: EN for signaling (S-EN) on T cells and EN for recognition (R-EN) for antigen binding. This unique structure allows precise recognition of tumor antigens by T cells through DNA strand hybridization on S-EN and R-EN, thereby activating T cells to target and eliminate cancer cells. Harnessing the designability of DNA hybridization, they further constructed a series of multiantigen cell recognition circuits with various logic gates (OR, NOT, and AND), endowing independent control over different types of T cells for precise target recognition and gradual adjustment of T cell activation. Together, the DNA-enabled nongenetic T-cell engineering approach enhances targeting precision and effectiveness of T cells, presenting a promising solution to substantially address major challenges in cancer immunotherapy.

14.4.3 Spatial Organization of Nanoscale Receptor Distribution

Receptor clustering in living cells plays a crucial role in cell signaling and physiological responses. The use of DNA nanotechnology has recently emerged as an artificial scaffold to control receptor spatial organization for exploration and manipulation of receptor signaling [78–80]. DNA origami is a class of nanostructures synthesized using bottom-up approach and is characterized by programmable sizes and unique addressability, making it a powerful tool for receptor manipulation with nanoscale precision [81].

First, DNA origami platforms can be utilized to spatially arrange natural ligands to cluster cell surface receptors for interrogation of signaling. Teixeira et al. designed a DNA nanocaliper to precisely control the spatial distribution of Ephrin-A5, thus regulating the distribution of cell surface EphA2 receptor and influencing the invasiveness of breast cancer cells (Figure 14.9a) [82]. Interestingly, they demonstrated that the nanocaliper NC40, with a spacing of 42.9 nm, has a higher EphA2 activation ability compared to the NC100 with the receptor spacing of 101.1 nm. Similarly, Bathe et al. used DNA origami to couple B cell receptor-binding eOD-GT8 antigen via hybridization with the free single-stranded DNA on the origami (Figure 14.9b) [83]. They found that B cell activation was maximized with only 5 antigens and an antigen spacing of 25–30 nm on the surface of a 40 nm viral-like nanoparticle. Furthermore, Teixeira et al. constructed a wireframe DNA origami platform to define the spacing of PD-L1 ligands at the nanoscale, and they revealed that T cell signaling was efficiently inhibited when PD-L1 ligands were spaced approximately 200 nm apart (Figure 14.9c) [84]. Additionally, various DNA origami-based platforms had been utilized to examine the signaling and functionality of both the tumor necrosis factor (TNF) receptor and the TCR [88–91].

Secondly, DNA origami can be used as a playboard to mimic the cell-cell interaction. Vale and colleagues developed a synthetic signaling system to program

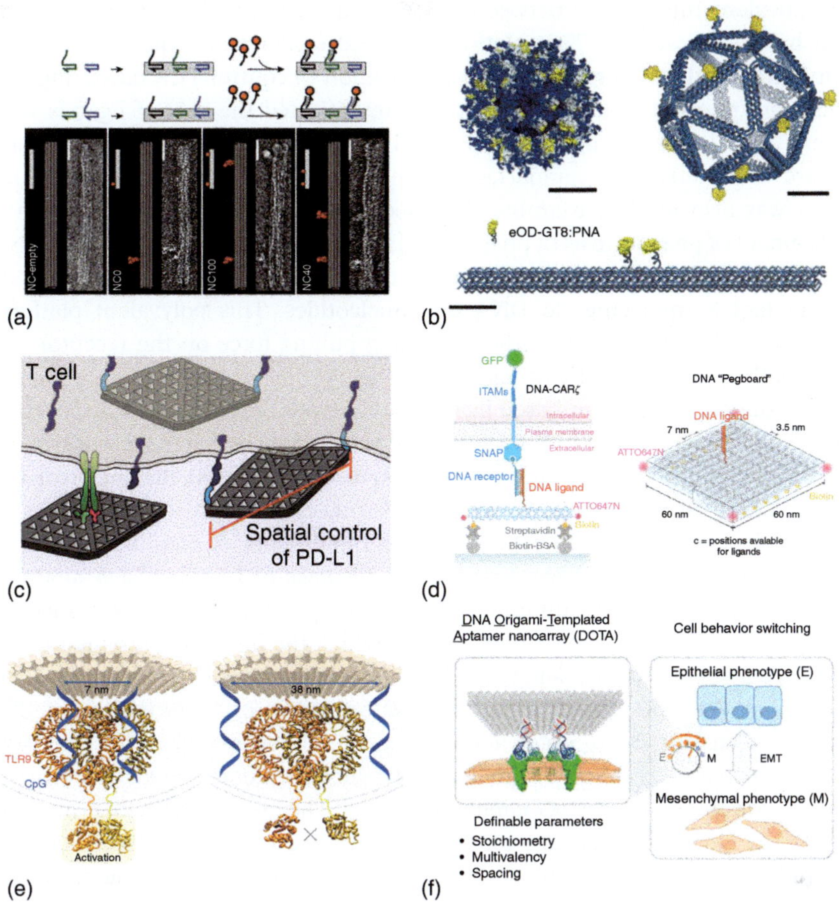

Figure 14.9 DNA origami as a programmable scaffold for the spatial organization of nanoscale receptor distribution. (a) Ephrin ligands modified DNA origami to study the distance-dependent influence of EphA2 receptor activation in human breast cancer cells. Source: Reproduced with permission from Shaw et al. [82]/Reproduced with permission from Springer Nature. (b) Scaffolded DNA origami to control the copy number and spatial distance of HIV immunogens to enhance B-cell activation. Source: Reproduced with permission from Veneziano et al. [83]. © 2020 Springer Nature. (c) DNA origami plate for adjusting nanoscale distance of the PD-L1 ligand to control T cells and regulate cell signaling. Source: Reproduced with permission from Fang et al. [84]/American Chemical Society/CC BY 4.0. (d) DNA origami pegboard for patterned decoration of ligands capable of triggering T cell activation signaling. Source: Reproduced with permission from Dong et al. [85]. © 2021 National Academy of Science. (e) DNA origami presenting CpG motifs for spatially controlled activation of TLR9. Source: Reproduced with permission from Comberlato et al. [86]. © 2022 American Chemical Society. (f) DNA origami-templated aptamer nanoarray that enables spatially reprogrammed receptor organization for cell behavior switching. Source: Reproduced with permission from Wang et al. [87]. © 2022 American Chemical Society.

the multivalent interactions between pMHC and TCRs on the T cells via DNA strand hybridization [92]. This platform allowed facile regulation of the ligand affinity through the strength of DNA hybridization to control TCR clustering and downstream phosphorylation. They further investigated the role of ligands with different affinities in regulating T cell signal transduction by using DNA origami pegboard to modulate the nanoarrangement of ligands (Figure 14.9d) [85]. This strategy was also used to examine the effect of Fcγ receptor clustering on the enhancement of phagocytosis of phagocytes [93]. Hogberg et al. constructed a DNA origami nanopattern to control the nanoscale distribution of Jag1 by covalently coupling Jag1-Fc molecules to DNA oligonucleotides. This polyvalent platform could activate Notch receptor without exerting pulling force on the receptor and subsequently trigger downstream signaling [94].

Apart from natural protein-based ligands, other types of molecules with receptor recognition capabilities can also be employed to organize and activate multivalent receptors, such as RTKs and pattern recognition receptors (PRRs). Bastings et al. utilized DNA origami nanomaterials to precisely array cytosine-phosphate-guanosine (CpG) motifs, matching the spatial crystallographic data of toll-like receptor 9 (TLR9) (Figure 14.9e) [86]. They found that a distance of 7 nm between mounted CpG significantly enhanced immune activation compared with free CpG. Since the flexibility of the linkers affects the interval at which the ligand is presented to the cell surface receptor, various lengths of linkers were introduced between the DNA platform and the CpG ligand to study the effectiveness of TLR9 activation. The results revealed that the receptor activation gradually decreased as the length of the ligand increased. Similarly, Shih et al. introduced a square block DNA origami platform with finely tuned CpG nanospacing [95]. They found that when the CpG spacing was 3.5 nm, the square blocks induced Th1 immune polarization and improved CD4 and CD8 T cell proliferation, activation, and cytokine secretion. These findings indicate that the refined spatial arrangement of CpG may be crucial for enhanced immune activation.

Recently, our group developed an aptameric ligand-presenting DNA nanoplatform for precise organization of RTK, namely DNA origami-templated aptamer nanoarray (DOTA) (Figure 14.9f) [87]. The design principle behind this study involved utilizing a DNA origami structure as a scaffold to spatially orchestrate receptor oligomerization. The DOTA platform facilitated the control of receptor organization by arranging either monomeric aptamer ligands (mALs) or dimeric aptamer ligands (dALs) on the ligand–receptor interface. We proved that the status of c-Met receptor oligomerization, as manipulated through nanoscale spacing and valency, significantly impacted the activation level of c-Met signaling. Furthermore, our studies revealed that receptor oligomerization, as modulated by DOTA, could act as a signaling switch, thus promoting the transition between epithelial and mesenchymal-like cell behaviors. Therefore, this DOTA approach underscores potential advancements in the design of molecular tools for the study of receptor signaling and the creation of innovative strategies for directing cellular functions. More recently, Qi et al. developed a new class of modular and tunable receptors called valency-controlled receptors (VCRs), which harnessed the programmability of DNA origami to present small-molecule moieties [96]. They designed multivalent small-molecule drugs to control the activity of VCRs and incorporated them into CAR architectures to modulate T-cell signal transduction using spatial clustering of drug-regulated receptors that enhanced T-cell sensitivity to low-antigen cancers. Together, DNA origami-based methods hold great promise for controlling receptor clustering and thereby downstream cellular responses, opening new avenues for precise manipulation of cellular signaling.

14.5 Conclusions and Perspectives

Receptor oligomerization plays a critical role in the orchestration of downstream cellular signaling networks and cellular responses. Understanding the effects of ligand spacing and receptor clustering on cell behavior and signaling will provide valuable insights into the fundamental mechanisms of cellular processes and pave the way for the development of novel therapeutic strategies. In this review, we have outlined the latest developments in leveraging DNA nanotechnology for the artificial regulation of receptor-mediated cellular responses. We have elucidated the principles that underpin the design of DNA-based nongenetic engineering strategies for precise control of receptor signaling and have spotlighted the rapidly expanding applications of these strategies in reprogramming receptor-mediated cell behaviors.

Building on the progress made in functional nucleic acid and dynamic DNA nanotechnology, DNA has emerged as a highly promising, intelligent, and logical molecular tool that enables receptor-mediated signaling and its resulting cellular behaviors. The unparalleled programmability and addressability of DNA nanostructures are being utilized to manage receptor activation signaling by precisely arranging the spatial distances of ligands or receptors at the nanoscale. These inventive applications emphasize the significant potential of DNA-based molecular engineering in influencing cellular signaling pathways, thereby establishing the foundation for the development of therapeutics and cell therapies revolving around cell surface receptors. However, the field of DNA-based nongenetic engineering strategies for receptor regulation still faces several challenges. Given the broad array of receptor types on the cell surface, our present collection of aptamers capable of recognizing target receptors is rather limited. Consequently, there is an urgent need to develop aptamers with enhanced affinity and specificity for target receptors. Additionally, thorough mechanistic explorations are required to identify potential fundamental differences in receptor activation mechanisms between natural ligands and artificial DNA nanodevices. Lastly, while DNA nanodevices hold promise for receptor signaling regulation *in vivo*, their specificity and efficiency remain significant challenges, such as limited stability, short circulation duration, and nonspecific protein adsorption, which necessitate further exploration and solutions in future study.

In summary, recent advancements in DNA nanotechnology highlight its potential role in elucidating complex cellular processes and paving an avenue for the nongenetic engineering of receptors with desired sense-and-response program.

Undoubtedly, as DNA nanotechnology continues to evolve, it promises to enrich the realm of chemical biology, offering advanced toolkits for the precise manipulation of receptors to modulate cellular behaviors. By integrating the highly programmable DNA-based modules with the principles of synthetic biology, we anticipate the emergence of more sophisticated nanodevices that will offer smarter solutions for cell-based therapeutics, which will accelerate new developments in regenerative medicine and immunotherapies.

Acknowledgments

This research was supported by the National Key Research and Development Program of China (2020YFA0907500 and 2021YFA0910100 to N.Z.) and the National Natural Science Foundation of China (22034002 and 92253304 to N.Z., 22177030 to W.H.H.).

References

1 Lemmon, M.A. and Schlessinger, J. (2010). Cell signaling by receptor tyrosine kinases. *Cell* 141 (7): 1117–1134.

2 Noriega-Guerra, H. and Freitas, V.M. (2018). Extracellular matrix influencing HGF/c-MET signaling pathway: impact on cancer progression. *Int. J. Mol. Sci.* 19 (11): 3300.

3 Shibuya, M. (2013). Vascular endothelial growth factor and its receptor system: physiological functions in angiogenesis and pathological roles in various diseases. *J. Biochem.* 153 (1): 13–19.

4 Heldin, C.H. (1995). Dimerization of cell surface receptors in signal transduction. *Cell* 80 (2): 213–223.

5 Garcia-Parajo, M.F., Cambi, A., Torreno-Pina, J.A. et al. (2014). Nanoclustering as a dominant feature of plasma membrane organization. *J. Cell Sci.* 127 (23): 4995–5005.

6 Sánchez, M.F. and Tampé, R. (2022). Ligand-independent receptor clustering modulates transmembrane signaling: a new paradigm. *Trends Biochem. Sci* 48 (2): 156–171.

7 Li, J., Wang, L., Tian, J. et al. (2020). Nongenetic engineering strategies for regulating receptor oligomerization in living cells. *Chem. Soc. Rev.* 49 (5): 1545–1568.

8 Ladbury, J.E. and Arold, S.T. (2012). Noise in cellular signaling pathways: causes and effects. *Trends Biochem. Sci* 37 (5): 173–178.

9 Chang, H.J. and Bonnet, J. (2020). Synthetic receptors to understand and control cellular functions. *Methods Enzymol.* 633: 143–167.

10 Kiessling, L.L., Gestwicki, J.E., and Strong, L.E. (2006). Synthetic multivalent ligands as probes of signal transduction. *Angew. Chem. Int. Ed.* 45 (15): 2348–2368.

11 Manhas, J., Edelstein, H.I., Leonard, J.N., and Morsut, L. (2022). The evolution of synthetic receptor systems. *Nat. Chem. Biol.* 18 (3): 244–255.

12 Lim, W.A. (2022). The emerging era of cell engineering: Harnessing the modularity of cells to program complex biological function. *Science* 378 (6622): 848–852.

13 Arber, C., Young, M., and Barth, P. (2017). Reprogramming cellular functions with engineered membrane proteins. *Curr. Opin. Biotechnol.* 47: 92–101.

14 Hong, M., Clubb, J.D., and Chen, Y.Y. (2020). Engineering CAR-T cells for next-generation cancer therapy. *Cancer Cell* 38 (4): 473–488.

15 June, C.H. and Sadelain, M. (2018). Chimeric antigen receptor therapy. *N. Engl. J. Med.* 379 (1): 64–73.

16 Gordon, W.R., Zimmerman, B., He, L. et al. (2015). Mechanical allostery: evidence for a force requirement in the proteolytic activation of Notch. *Dev. Cell* 33 (6): 729–736.

17 Morsut, L., Roybal, K.T., Xiong, X. et al. (2016). Engineering customized cell sensing and response behaviors using synthetic notch receptors. *Cell* 164 (4): 780–791.

18 Voß, S., Klewer, L., and Wu, Y.W. (2015). Chemically induced dimerization: reversible and spatiotemporal control of protein function in cells. *Curr. Opin. Chem. Biol.* 28: 194–201.

19 Bugaj, L.J., Spelke, D.P., Mesuda, C.K. et al. (2015). Regulation of endogenous transmembrane receptors through optogenetic Cry2 clustering. *Nat. Commun.* 6 (1): 6898.

20 Lajoie, M.J., Boyken, S.E., Salter, A.I. et al. (2020). Designed protein logic to target cells with precise combinations of surface antigens. *Science* 369 (6511): 1637–1643.

21 Wang, D., Tai, P.W., and Gao, G. (2019). Adeno-associated virus vector as a platform for gene therapy delivery. *Nat. Rev. Drug Discovery* 18 (5): 358–378.

22 Silva, D.A., Yu, S., Ulge, U.Y. et al. (2019). De novo design of potent and selective mimics of IL-2 and IL-15. *Nature* 565 (7738): 186–191.

23 Phillips, M.I. and Tang, Y.L. (2008). Genetic modification of stem cells for transplantation. *Adv. Drug Delivery Rev.* 60 (2): 160–172.

24 Aguilera, A. and García-Muse, T. (2012). R loops: from transcription byproducts to threats to genome stability. *Mol. Cell* 46 (2): 115–124.

25 Travers, A. and Muskhelishvili, G. (2015). DNA structure and function. *FEBS J.* 282 (12): 2279–2295.

26 Wang, M., Li, X., He, F. et al. (2022). Advances in designer DNA nanorobots enabling programmable functions. *ChemBioChem* 23 (18): e202200119.

27 Chen, Y.J., Groves, B., Muscat, R.A., and Seelig, G. (2015). DNA nanotechnology from the test tube to the cell. *Nat. Nanotechnol.* 10 (9): 748–760.

28 Kishi, J.Y., Schaus, T.E., Gopalkrishnan, N. et al. (2018). Programmable autonomous synthesis of single-stranded DNA. *Nat. Chem.* 10 (2): 155–164.

29 Seeman, N.C. and Sleiman, H.F. (2017). DNA nanotechnology. *Nat. Rev. Mater.* 3 (1): 1–23.

30 Li, L., Xu, S., Yan, H. et al. (2021). Nucleic acid aptamers for molecular diagnostics and therapeutics: advances and perspectives. *Angew. Chem. Int. Ed.* 60 (5): 2221–2231.

31 Camorani, S., Crescenzi, E., Fedele, M., and Cerchia, L. (2018). Oligonucleotide aptamers against tyrosine kinase receptors: Prospect for anticancer applications. *Biochim. Biophys. Acta* 1869 (2): 263–277.

32 Berg, K., Lange, T., Mittelberger, F. et al. (2016). Selection and characterization of an α6β4 integrin blocking DNA Aptamer. *Mol. Ther. Nucl. Acids* 5: e294.

33 Zumrut, H.E., Ara, M.N., Maio, G.E. et al. (2016). Ligand-guided selection of aptamers against T-cell receptor-cluster of differentiation 3 (TCR-CD3) expressed on Jurkat. E6 cells. *Anal. Biochem.* 512: 1–7.

34 Hu, Y.S., Cang, H., and Lillemeier, B.F. (2016). Superresolution imaging reveals nanometer-and micrometer-scale spatial distributions of T-cell receptors in lymph nodes. *Proc. Natl. Acad. Sci.* 113 (26): 7201–7206.

35 Watson, J.D. and Crick, F.H. (1953). Molecular structure of nucleic acids: a structure for deoxyribose nucleic acid. *Nature* 171 (4356): 737–738.

36 Seeman, N.C. (1982). Nucleic acid junctions and lattices. *J. Theor. Biol.* 99 (2): 237–247.

37 Li, J., Fan, C., Pei, H. et al. (2013). Smart drug delivery nanocarriers with self-assembled DNA nanostructures. *Adv. Mater.* 25 (32): 4386–4396.

38 Bujold, K.E., Lacroix, A., and Sleiman, H.F. (2018). DNA nanostructures at the interface with biology. *Chemistry* 4 (3): 495–521.

39 Sun, L., Shen, F., Qu, Y., and Liu, Z. (2023). Functional DNA as a molecular tool in regulating immunoreceptor-ligand interactions. *JACS Au* 3 (7): 1820–1834.

40 Seelig, G., Soloveichik, D., Zhang, D.Y., and Winfree, E. (2006). Enzyme-free nucleic acid logic circuits. *Science* 314 (5805): 1585–1588.

41 Simmel, F.C., Yurke, B., and Singh, H.R. (2019). Principles and applications of nucleic acid strand displacement reactions. *Chem. Rev.* 119 (10): 6326–6369.

42 Yang, J., Wu, R., Li, Y. et al. (2018). Entropy-driven DNA logic circuits regulated by DNAzyme. *Nucleic Acids Res.* 46 (16): 8532–8541.

43 McConnell, E.M., Cozma, I., Mou, Q. et al. (2021). Biosensing with DNAzymes. *Chem. Soc. Rev.* 50 (16): 8954–8994.

44 Yan, J., Ran, M., Shen, X., and Zhang, H. (2023). Therapeutic DNAzymes: from structure design to clinical applications. *Adv. Mater.* 2300374.

45 Li, J., Green, A.A., Yan, H., and Fan, C. (2017). Engineering nucleic acid structures for programmable molecular circuitry and intracellular biocomputation. *Nat. Chem.* 9 (11): 1056–1067.

46 Chen, L., Chen, W., Liu, G. et al. (2021). Nucleic acid-based molecular computation heads towards cellular applications. *Chem. Soc. Rev.* 50 (22): 12551–12575.

47 Fan, J., Wang, H.H., Xie, S. et al. (2020). Engineering cell-surface receptors with DNA nanotechnology for cell manipulation. *ChemBioChem* 21 (3): 282–293.

48 Barrientos, S., Stojadinovic, O., Golinko, M.S. et al. (2008). Growth factors and cytokines in wound healing. *Wound Repair Regen.* 16 (5): 585–601.

49 Ramaswamy, V., Monsalve, A., Sautina, L. et al. (2015). DNA aptamer assembly as a vascular endothelial growth factor receptor agonist. *Nucleic Acid Ther.* 25 (5): 227–234.

50 Ueki, R., Ueki, A., Kanda, N., and Sando, S. (2016). Oligonucleotide-based mimetics of hepatocyte growth factor. *Angew. Chem. Int. Ed.* 55 (2): 579–582.

51 Ueki, R., Atsuta, S., Ueki, A. et al. (2019). DNA aptamer assemblies as fibroblast growth factor mimics and their application in stem cell culture. *Chem. Commun.* 55 (18): 2672–2675.

52 Srivastava, S. and Riddell, S.R. (2015). Engineering CAR-T cells: design concepts. *Trends Immunol.* 36 (8): 494–502.

53 Ueki, R., Atsuta, S., Ueki, A., and Sando, S. (2017). Nongenetic reprogramming of the ligand specificity of growth factor receptors by bispecific DNA aptamers. *J. Am. Chem. Soc.* 139 (19): 6554–6557.

54 Li, H., Wang, M., Shi, T. et al. (2018). A DNA-mediated chemically induced dimerization (D-CID) nanodevice for nongenetic receptor engineering to control cell behavior. *Angew. Chem. Int. Ed.* 130 (32): 10383–10387.

55 He, F., Wang, M., Wang, J. et al. (2023). An extracellular miRNA-responsive artificial receptor via dynamic DNA nano-assembly for biomarker-driven therapy. *Angew. Chem. Int. Ed.* e202305227.

56 Gordley, R.M., Bugaj, L.J., and Lim, W.A. (2016). Modular engineering of cellular signaling proteins and networks. *Curr. Opin. Struct. Biol.* 39: 106–114.

57 DeRose, R., Miyamoto, T., and Inoue, T. (2013). Manipulating signaling at will: chemically-inducible dimerization (CID) techniques resolve problems in cell biology. *Eur. J. Physiol.* 465: 409–417.

58 Spencer, D.M., Wandless, T.J., Schreiber, S.L., and Crabtree, G.R. (1993). Controlling signal transduction with synthetic ligands. *Science* 262 (5136): 1019–1024.

59 Dong, H., Liu, L., Wang, J. et al. (2020). DNA-based reprogramming strategy of receptor-mediated cellular behaviors: from genetic encoding to nongenetic engineering. *ACS Appl. Bio Mater.* 3 (5): 2796–2804.

60 Wang, Z., Lu, X., and Xing, B. (2019). Precise cell behaviors manipulation through light-responsive nano-regulators: recent advance and perspective. *Theranostics* 9 (11): 3308.

61 Chen, S., Li, J., Liang, H. et al. (2018). Light-induced activation of c-Met signalling by photocontrolled DNA assembly. *Chem. A Eur. J.* 24 (60): 15988–15992.

62 Ueki, R., Hayashi, S., Tsunoda, M. et al. (2021). Nongenetic control of receptor signaling dynamics using a DNA-based optochemical tool. *Chem. Commun.* 57 (48): 5969–5972.

63 Sethi, S., Hidaka, K., Sugiyama, H., and Endo, M. (2021). Non-invasive regulation of cellular morphology using a photoswitchable mechanical DNA polymer. *Angew. Chem. Int. Ed.* 60 (37): 20342–20349.

64 Wang, M., He, F., Li, H. et al. (2019). Near-infrared light-activated DNA-agonist nanodevice for nongenetically and remotely controlled cellular signaling and behaviors in live animals. *Nano Lett.* 19 (4): 2603–2613.

65 Zhang, K., Deng, R., Sun, Y. et al. (2017). Reversible control of cell membrane receptor function using DNA nano-spring multivalent ligands. *Chem. Sci.* 8 (10): 7098–7105.

66 Liang, H., Chen, S., Li, P. et al. (2018). Nongenetic approach for imaging protein dimerization by aptamer recognition and proximity-induced DNA assembly. *J. Am. Chem. Soc.* 140 (12): 4186–4190.

67 Li, Y., Zhang, X., Pan, W. et al. (2020). A nongenetic proximity-induced FRET strategy based on DNA tetrahedron for visualizing the receptor dimerization. *Anal. Chem.* 92 (17): 11921–11926.

68 Dong, C., Fang, X., Xiong, J. et al. (2022). Simultaneous visualization of dual intercellular signal transductions via SERS imaging of membrane proteins dimerization on single cells. *ACS Nano* 16 (9): 14055–14065.

69 Wang, J., Song, J., Zhang, X. et al. (2023). DNA-programed plasmon rulers decrypt single-receptor dimerization on cell membrane. *J. Am. Chem. Soc.* 145 (2): 1273–1284.

70 Li, H., Gao, J., Cao, L. et al. (2021). A DNA molecular robot that autonomously walks on the cell membrane to drive cell motility. *Angew. Chem. Int. Ed.* 133 (50): 26291–26299.

71 Mills, A., Aissaoui, N., Maurel, D. et al. (2022). A modular spring-loaded actuator for mechanical activation of membrane proteins. *Nat. Commun.* 13 (1): 3182.

72 Krippner-Heidenreich, A., Tubing, F., Bryde, S. et al. (2002). Control of receptor-induced signaling complex formation by the kinetics of ligand/receptor interaction. *J. Biol. Chem.* 277 (46): 44155–44163.

73 Jones, S.A. and Jenkins, B.J. (2018). Recent insights into targeting the IL-6 cytokine family in inflammatory diseases and cancer. *Nat. Rev. Immunol.* 18 (12): 773–789.

74 Zhang, J., Qiu, Z., Fan, J. et al. (2021). Scan and unlock: a programmable DNA molecular automaton for cell-selective activation of ligand-based signaling. *Angew. Chem. Int. Ed.* 133 (12): 6807–6817.

75 Edeline, J., Houot, R., Marabelle, A., and Alcantara, M. (2021). CAR-T cells and BiTEs in solid tumors: challenges and perspectives. *J. Hematol. Oncol.* 14: 1–12.

76 Tang, R., Fu, Y.H., Gong, B. et al. (2022). A chimeric conjugate of antibody and programmable DNA nanoassembly smartly activates T cells for precise cancer cell targeting. *Angew. Chem. Int. Ed.* 61 (36): e202205902.

77 Ma, P.Q., Liu, T.X., Li, H.D. et al. (2022). Nano-biohybrid DNA engager that reprograms the T-cell receptor. *J. Am. Chem. Soc.* 144 (49): 22458–22469.

78 Zhang, K., Gao, H., Deng, R., and Li, J. (2019). Emerging applications of nanotechnology for controlling cell-surface receptor clustering. *Angew. Chem. Int. Ed.* 58 (15): 4790–4799.

79 Zhang, K., Ma, Y., Wang, D. et al. (2022). In vivo activation of T-cell proliferation by regulating cell surface receptor clustering using a ph-driven interlocked DNA nano-spring. *Nano Lett.* 22 (5): 1937–1945.

80 Bi, S., Chen, W., Fang, Y. et al. (2023). Cancer cell-selective membrane receptor clustering driven by VEGF secretion for in vivo therapy. *J. Am. Chem. Soc.* 145 (9): 5041–5052.

81 Dey, S., Fan, C., Gothelf, K.V. et al. (2021). DNA origami. *Nat. Rev. Methods Primers* 1 (1): 13.

82 Shaw, A., Lundin, V., Petrova, E. et al. (2014). Spatial control of membrane receptor function using ligand nanocalipers. *Nat. Methods* 11 (8): 841–846.

83 Veneziano, R., Moyer, T.J., Stone, M.B. et al. (2020). Role of nanoscale antigen organization on B-cell activation probed using DNA origami. *Nat. Nanotechnol.* 15 (8): 716–723.

84 Fang, T., Alvelid, J., Spratt, J. et al. (2021). Spatial regulation of T-cell signaling by programmed death-ligand 1 on wireframe DNA origami flat sheets. *ACS Nano* 15 (2): 3441–3452.

85 Dong, R., Aksel, T., Chan, W. et al. (2021). DNA origami patterning of synthetic T cell receptors reveals spatial control of the sensitivity and kinetics of signal activation. *Proc. Natl. Acad. Sci.* 118 (40): e2109057118.

86 Comberlato, A., Koga, M.M., Nüssing, S. et al. (2022). Spatially controlled activation of toll-like receptor 9 with DNA-based nanomaterials. *Nano Lett.* 22 (6): 2506–2513.

87 Wang, M., Yang, D., Lu, Q. et al. (2022). Spatially reprogramed receptor organization to switch cell behavior using a DNA origami-templated aptamer nanoarray. *Nano Lett.* 22 (21): 8445–8454.

88 Wang, Y., Baars, I., Fordos, F., and Hogberg, B. (2021). Clustering of death receptor for apoptosis using nanoscale patterns of peptides. *ACS Nano* 15 (6): 9614–9626.

89 Berger, R.M., Weck, J.M., Kempe, S.M. et al. (2021). Nanoscale FasL organization on DNA origami to decipher apoptosis signal activation in cells. *Small* 17 (26): 2101678.

90 Hellmeier, J., Platzer, R., Eklund, A.S. et al. (2021). DNA origami demonstrate the unique stimulatory power of single pMHCs as T cell antigens. *Proc. Natl. Acad. Sci.* 118 (4): e2016857118.

91 Rushdi, M.N., Pan, V., Li, K. et al. (2022). Cooperative binding of T cell receptor and CD4 to peptide-MHC enhances antigen sensitivity. *Nat. Commun.* 13 (1): 7055.

92 Taylor, M.J., Husain, K., Gartner, Z.J. et al. (2017). A DNA-based T cell receptor reveals a role for receptor clustering in ligand discrimination. *Cell* 169 (1): 108–119.

93 Kern, N., Dong, R., Douglas, S.M. et al. (2021). Tight nanoscale clustering of Fcγ receptors using DNA origami promotes phagocytosis. *Elife* 10: e68311.

94 Smyrlaki, I., Fördös, F., Rocamonde Lago, I. et al. (2022). Notch engagement by Jag1 nanoscale clusters indicates a force-independent mode of activation. *bioRxiv*, 2022-11.

95 Zeng, Y.C., Young, O.J., Wintersinger, C.M. et al. (2023). Fine tuning of CpG spatial distribution with DNA origami for improved therapeutic cancer vaccination. *bioRxiv* 2023-07.

96 Finn, P., Chavez, M., Chen, X. et al. (2023). Programmable drug control of receptor valency modulates the potency of cell therapeutics. *bioRxiv*, 2023-01.

15

DNA-Based Cell Surface Engineering for Programming Multiple Cell–Cell Interactions

Mingshu Xiao, Yueyang Sun, Li Li, and Hao Pei

East China Normal University, Shanghai Key Laboratory of Green Chemistry and Chemical Processes, School of Chemistry and Molecular Engineering, Shanghai Frontiers Science Center of Genome Editing and Cell Therapy, 500 Dongchuan Road, Shanghai 200241, China

15.1 DNA Nanotechnology: The Tool of Choice for Programming Cell–Cell Interactions

Cell–cell interactions refer to the direct interactions between cell surfaces and play an important role in various cell activities (e.g. cell communication, cell adhesion, and cell junction) [1, 2]. Proper cell–cell interactions are also the first prerequisite for the progression of many physiological processes in multicellular organisms (e.g. tissue formation, neurotransmission, and immune response) [3, 4]. Although cell–cell interactions show high dynamicity in their duration and complexity in their composition, the surface and intracellular biomolecules (e.g. adhesion proteins and surface receptor–ligand pairs) that participate in these intercellular interaction events ensure cells properly communicate with each other and guide cellular spatiotemporal organization [3]. Conversely, disruptions and/or alterations to cellular interactions could induce severe pathophysiological effects. A key example is that the interactions between cancer, endothelial, stromal, and immune cells in the tumor microenvironment dictate tumor progression [5, 6]. Hence, the ability to regulate cell–cell interactions across different cell types could facilitate untangling the intricacy of these interactions and the signaling pathways they modulate, thus providing insight into the biological processes that ensue upon cell–cell engagement and leading to the development of cell-based therapeutics for tissue engineering and immunotherapy [7–9].

More simple-yet-effective tools are required to regulate the spatial arrangements of cells, with our understanding of the importance of complexity in cellular interactions increasing. Inspired by cell-surface molecules-modulated cell–cell interactions in natural biosystems [10], different cell surface engineering strategies have emerged with the goal of artificial manipulation of intercellular interactions [11–13]. By deconstructing the complexity of cell–cell interactions, these approaches not only offer a powerful means to investigate cell–cell interaction biology but also open up great opportunities for therapeutic development. In the

DNA Nanotechnology for Cell Research: From Bioanalysis to Biomedicine, First Edition. Edited by Zhou Nie.

past several decades, genetic engineering has been an important approach to enhancing cell–cell interactions through genetic manipulation for modulating the expression of desired cell surface receptors (e.g. CAR T cells) [14–17]. Despite its robustness, the applications of this approach are restricted by several drawbacks [18, 19], such as laborious protocols, inefficient gene transfer, and safety concerns. To evade the genetic modification-related limitations, tremendous efforts have been devoted to developing non-genetic cell surface engineering strategies for directing cell–cell interactions [20–22], in which many functional materials (e.g. proteins, nucleic acids, nanoparticles, and polymers) are employed as recognition/targeting elements to artificially engineer cell surface. One of them is the utility of DNA in cell surface engineering. Compared to genetic engineering approach, these non-genetic engineering strategies are more transient and/or reversible and applicable to different cell types. Moreover, non-genetic strategies are developed around lipid-, glycan-, and protein-based modifications, while the genetic engineering approach focuses primarily on protein expression.

As a generic building block, DNA has numerous distinct advantages over other materials for cell surface engineering [23–26]. First, DNA oligonucleotides are facile to synthesize with high purity and low prices using standard phosphoramidite chemistry [27]. Second, owing to DNA hybridization following the Watson–Crick base pairing rules, DNA materials possess predictable physical and/or chemical properties [28, 29]. Third, DNA materials also provide the desired rigidity and flexibility by choosing single- or double-stranded DNA or their combination. Fourth, DNA sequences with designability offer much flexibility in designing and building desired functional materials [30]. Fifth, some specific DNA molecules (i.e. DNA aptamers and DNAzymes) with independent structures show high binding specificity and affinity to target molecules and enable the execution of specific nongenetic functions [31–33]. Sixth, additional functions can be conferred on DNA molecules through chemical modification with functional groups either inside or at the ends [34, 35]. Collectively, these prominent properties make DNA molecules an appealing functional material for cell surface engineering [36, 37]. Based on the functionalization with DNA molecules, the DNA-engineered cells could be allowed to interact with other cells through the acquired functions of molecular recognition (i.e. DNA hybridization and aptamer-receptor binding). Accordingly, the applications of DNA-based cell surface engineering in precise regulation of cell–cell interactions have been fertilized. This has led to the development of cell-based immunotherapy and tissue engineering.

15.2 Modifying Cell Surface with DNA

To achieve precise regulation of cell–cell interactions via DNA-based cell surface engineering, the prerequisite is to modify cell surface with DNA for conferring recognition or targeting capability on these host cells. Currently, several commonly used approaches to attaching DNA to cell membranes have been developed for cell surface engineering, including chemical modification (Figure 15.1a) [38, 39],

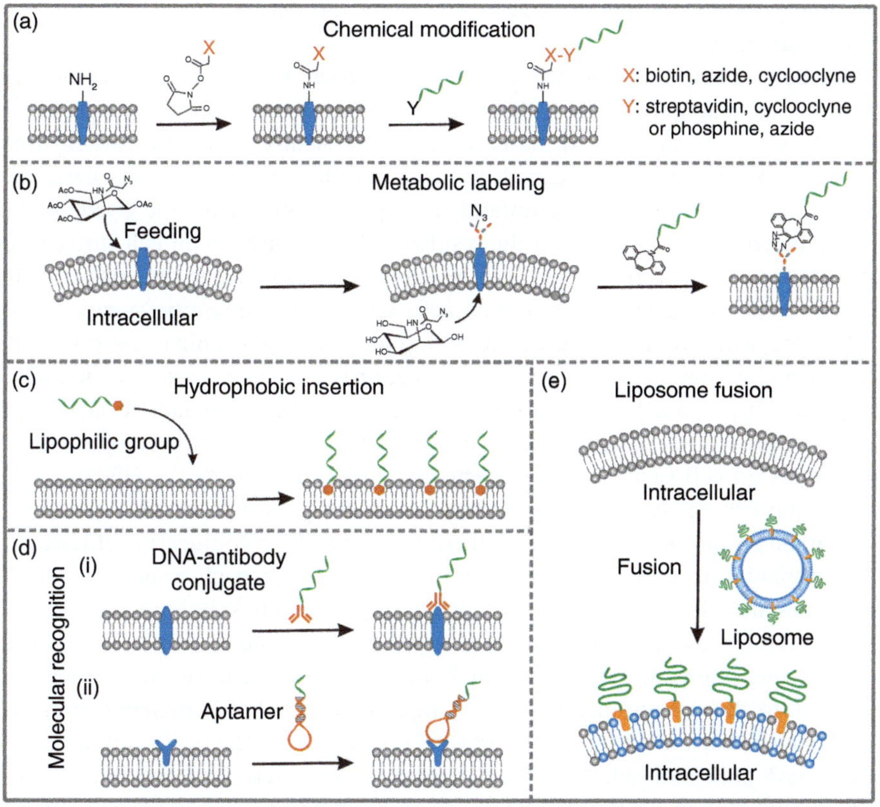

Figure 15.1 Approaches to modifying cell surface with synthetic DNA, including (a) chemical modification, (b) metabolic labeling, (c) hydrophobic insertion, (d) molecular recognition ((i) antibody-DNA conjugates or (ii) aptamers binding to receptors), and (e) liposome fusion.

metabolic labeling (Figure 15.1b) [40–43], hydrophobic insertion (Figure 15.1c) [44, 45], molecular recognition (Figure 15.1d) [23, 26], and liposome fusion (Figure 15.1e) [46].

The membrane component (e.g. proteins) contains specific functional groups (e.g. amines, thiols, and diols) that can be attractive targets for the covalent conjugation of ligands. Hence, a straightforward approach to cell surface engineering is direct chemical conjugation of DNA molecules to these functional groups on membrane (Figure 15.1a). Functionalization of lysine residues (or other surface amines) with *N*-hydroxy-succinimide (NHS) ester-modified DNA is a typical example of chemical modification [40, 47], where a bifunctional crosslinker is often required to convert the amino group into a binding or intermediate reactive group. Another example of modification is covalent conjugation of maleimide-modified DNA [48, 49], which is synthesized through the functionalization of thiol or amine group-modified DNA with succinimidyl 4-(*N*-maleimidomethyl) cyclohexane-1-carboxylate (SMCC) or its derivatives to the cysteine residue on the cell surface. Despite minimal influence on cell phenotype or homing ability, this approach lacks selectivity and may compromise membrane protein functions [50].

Metabolic labeling is another appealing approach to cell surface engineering through the reaction of DNA oligonucleotides with metabolically incorporated sugar moieties (Figure 15.1b). Based on the promiscuity of natural carbohydrate biosynthesis pathways, the sugar substrates containing functional groups (e.g. azide, ketone, and thiol) are facile to covalently incorporate into membrane glycoproteins during metabolism [51, 52], and the resultant glycoconjugates on cell surface can be further used for cell labeling. For example, the azido sugar moieties that are displayed on cell surface after feeding cells with azide-containing sugars can be conjugated with DNA molecules containing alkyne or phosphine group through Staudinger ligation (azide-phosphine conjugation) or click chemistry (azide-alkyne conjugation) [41, 53, 54]. Owing to its simple operability, metabolic labeling is facile to tether DNA to cell surface. Moreover, the reaction with glycoproteins or glycolipids on cell surface shows specificity, which minimizes the effects on the membrane protein functions. However, the diversity in glycosylation pattern of membrane glycoproteins results in difficulty in controlling conjugation efficiency.

Hydrophobic insertion is a favorable noncovalent approach to tethering DNA molecules to cell membrane (Figure 15.1c) [55]. As phospholipid bilayer with cholesterol localizing between the phospholipid molecules makes interior of plasma membrane lipophilic, the hydrophobic component is thus allowed to embed into the membrane. Therefore, different hydrophobic moiety-conjugated DNA molecules, including cholesterol, tocopherol, C18 chain, diperfluorodecyl, and diacyl phospholipid, have been developed to insert into the lipid bilayer through hydrophobic interactions [56, 57], while leaving hydrophilic DNA strands extending from cell surface. Compared to covalent conjugation, this approach possesses several advantages [58], including high anchoring efficiency, cell type independence, and negligible effects on membrane integrity and cellular functions. Nonetheless, the stability of modified DNA conjugates on cell surface should be thoroughly considered given their passive dissociation and cellular internalization [59].

Another noncovalent approach to anchoring DNA on cell surface is molecular recognition (Figure 15.1d), relying on the recognition of membrane receptors by antibodies or aptamers for ligand–receptor binding. Based on antibody–receptor interactions, one can use DNA–antibody conjugates to recognize and bind the cell receptors (e.g. epidermal growth factor receptor, epithelial cell adhesion molecule, and human epidermal growth factor receptor 2) for cell surface engineering (Figure 15.1d(i)) [60, 61]. Apart from DNA–antibody conjugates, DNA aptamers can be directly employed to modify cell surface (Figure 15.1d(ii)) due to their capability of specifically binding membrane proteins by folding into specific secondary and tertiary structures [62, 63]. Benefiting from the designability of DNA sequences, aptamers can be designed with an extended DNA tail for other uses yet without the requirement of additional conjugation. Moreover, the anchored aptamers can even be triggered to dissociate from the cell surface through hybridization with their complementary sequence [64]. Therefore, these features make it flexible and reversible to tune the procedure of molecular recognition-based cell surface engineering. Despite DNA molecules showing good membrane-anchoring stability, this approach has the issue of cell endocytosis.

Fusion of lipid membranes is a natural process in normal cell biology. Inspired by this process, liposome fusion has been readily adapted for applications in cell surface engineering [65–67]. Typically, the prepared liposomes are modified with DNA before engineering cell surface. After incubation with cells, DNA-modified liposomes would spontaneously fuse with the cell membrane so that the corresponding DNA molecules could be displayed on cell surface (Figure 15.1e). Based on liposome fusion, the Xing group used liposomal spherical nucleic acid with DNA strands packed on the surface to engineer external and internal cell surfaces [68], realizing spatially controlled cell surface engineering. Overall, this approach is convenient, efficient, and cell type-independent, but parameters (e.g. membrane components of liposomes and surface coverage of DNA) should be optimized for efficient fusion [69, 70].

15.3 Programming Cell–Cell Interactions by DNA Nanotechnology

DNA nanotechnology has long been employed for the assembly of complex structures of nanoparticles [71–74], and these DNA-based self-assembled constructs can inform design rules for cell assembly in spite of their size being significantly smaller than cells. With the development of DNA-based cell surface engineering strategies, DNA nanotechnology has been also applied to the complex arrangements of cells. In this field, ligand–receptor binding (Figure 15.2a) and DNA hybridization (Figure 15.2b) are two useful methods for artificial manipulation of cell–cell interactions. In addition, host cells can be equipped with synthetic DNA circuits and work as living computers that can be programmed to interact with targeted cells based on multiple external signals (e.g. ions and antigens) (Figure 15.2c). Under external stimulus, such engineered cells can execute sophisticated sensing and recognize multifaceted features of targeted cells [75], thus promoting intercellular interactions. Therefore, DNA circuit-based regulation has emerged as a high-order means to regulate specific cell–cell interactions. On these bases, DNA nanotechnology has offered us a toolkit for programming cell–cell interactions, which improves our understanding of biological cell–cell interactions and provides powerful tools to test hypotheses about intercellular interactions.

15.3.1 Ligand–Receptor Binding-Based Cell–Cell Interactions

One straightforward method to regulate cell–cell interactions is ligand–receptor binding, which relies on the specific interaction between ligand–receptor pairs. By installing the ligands (e.g. aptamers and antibodies) on cell surface, the ligand-engineered host cells enable specific binding to relevant receptor-expressing targeted cells but do not react to off-target cells. On this basis, this method can be

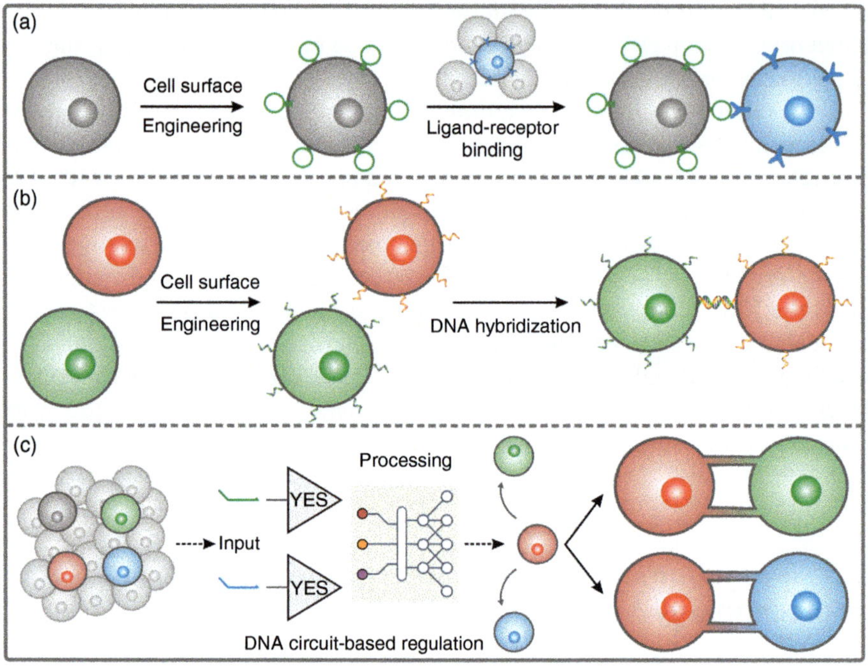

Figure 15.2 DNA nanotechnology-enabled cell–cell interactions, including (a) ligand–receptor binding-based cell–cell interactions, (b) DNA hybridization-based cell–cell interactions, and (c) DNA circuit-regulated cell–cell interactions.

applied to enhanced cell targeting and/or adoptive immunotherapy [76–79]. For example, the Tan group used diacyl lipid–DNA aptamer conjugates (denoted as lipo-aptamer probes) to modify immune cells for specific cell targeting (Figure 15.3a) [80]. In this work, the lipo-aptamer probes are comprised of three distinct segments, including a synthetic diacyllipid tail that serves as the membrane anchor, a PEG linker that allows DNA to extend out from the cell surface, and an aptamer that recognizes and binds to specific cancer cells. Benefiting from their delicate design, the lipo-aptamer probe-modified natural killer (NK) cells showed enhanced interactions with and cytotoxicity to target cancer cells based on the aptamer-receptor binding, holding great potential in adoptive cell therapy. In a similar work, CD30-specific aptamer-engineered NK cells were reported for cell-specific adaptive immunotherapy [84]. To enhance adoptive immunotherapy in solid tumors, Zhang et al. equipped NK cells with dual-aptamers through metabolic labeling, including specific targeting and checkpoint-blocking aptamers (i.e. TLS11a aptamer and PDL1-specific aptamer) (Figure 15.3b) [81]. Here, TLS11a aptamer serves as a targeting ligand for hepatocellular carcinoma cancer cells, and PDL1-specific aptamer is allowed for regulation of PD1/PDL1 signaling. Due to their synergistic effect, the dual aptamer-equipped NK cells exhibited high specificity and induced remarkable upregulation of PDL1 expression in tumor cells, leading to higher cytokine secretion and enhanced checkpoint blockade for improved therapeutic efficacy in solid tumors.

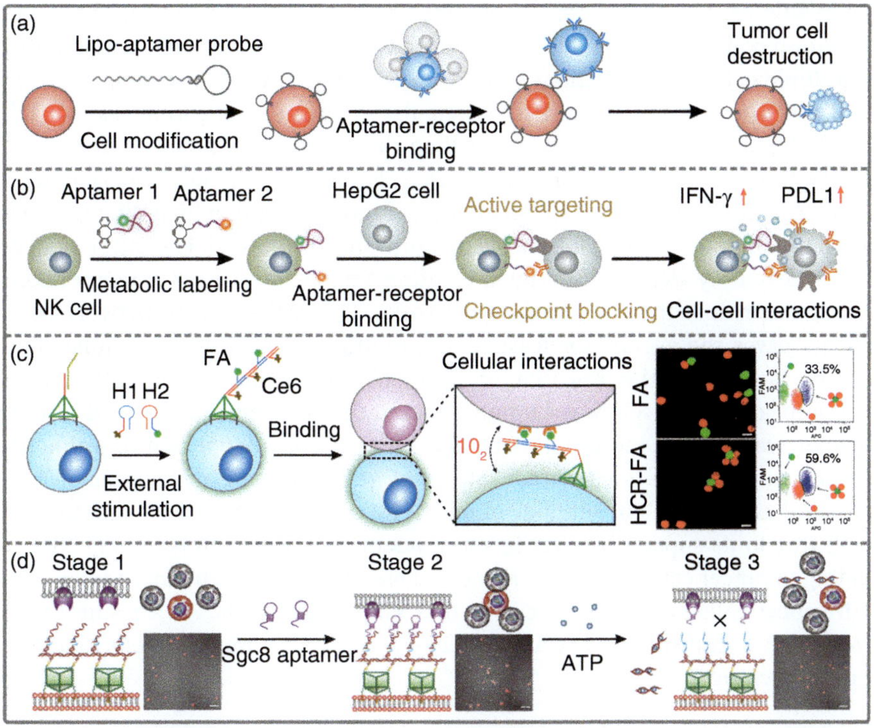

Figure 15.3 Ligand–receptor binding-based cell–cell interactions. (a) Targeting cancer cells (blue) with aptamer-modified immune cells (red). Source: From Xiong et al. [80]. Reprinted with the permission of Wiley-VCH. (b) Dual-aptamer-engineered NK cells for enhanced adoptive cell immunotherapy. Source: From Zhang et al. [81]. Reprinted with the permission of Wiley-VCH. (c) Mediating cell–cell interactions using DNA-based dynamic mimicry of membrane proteins. Source: From Li et al. [82]. Reprinted with the permission of American Chemical Society. (d) Manipulation of multiple cell–cell interactions by tunable DNA scaffold networks. Source: From Guo et al. [83]. Reprinted with the permission of Wiley-VCH.

Despite its simplicity, this method primarily relying on monovalent binding of aptamer to receptor might result in relatively weak cellular interactions. To ensure robust cellular interactions, scaffolds are often used to assemble ligands for multivalent binding [85, 86]. One typical example is DNA nanostructure scaffold. By functionalizing cell surfaces with DNA tetrahedrons based on hydrophobic insertion, Tan and colleagues constructed cell surface-anchored DNA nanoarchitectures (Figure 15.3c) [82]. In addition to their high anchoring stability, the membrane-anchored DNA nanoarchitectures could realize the initiation of DNA assembly (i.e. hybridization chain reaction [HCR]) under external stimulation, in which recognition module (i.e. folic acid [FA]) is assembled onto cell surface to equip the cell with cell-type-specific binding. Accordingly, this membrane-anchored DNA nanostructure enables the manipulation of cellular interaction based on the multiple binding in response to environmental changes, mimicking the dynamic behavior of membrane proteins. This work provides a new idea to engineer therapeutic cells with customized sensing/response pathways. In another work, this group constructed tunable DNA scaffold networks for flexible manipulation of multiple cell–cell interactions (Figure 15.3d) [83]. The DNA scaffold networks consist of a three-dimensional DNA triangular prism and a one-dimensional DNA branched polymer. By precisely controlling these two DNA modules, they could achieve flexible and tunable live cell encapsulation that promotes the stability of DNA structural networks on the membrane. Moreover, both a high-efficiency and specific manipulation of cell–cell interactions and the reversible control of intercellular assembly were realized by integrating aptamers (i.e. Sgc8 aptamer and/or ATP aptamer) into the DNA scaffold networks. This DNA scaffold network-based strategy paves the avenue for feasible and multiple control of the cellular interactions, facilitating the research on cell behavior manipulation, intercellular communication, and cell-based immunotherapy. Besides, *in situ* synthesis of DNA polymer scaffolds on the cell surface also offers a useful approach to promoting cell–cell interactions. For instance, Shi et al. [87] synthesized a supramolecular aptamer-based polyvalent antibody mimic (PAM) on the surface of NK cells through HCR. Due to its superiority over its monovalent counterpart in powering NKs for binding to cancer cells, the PAM-engineered NK cells showed enhanced interaction with cancer cells and more efficiency in killing cancer cells. As a result, this *in situ* synthesis strategy could be applicable to functionalize immune cells for cancer immunotherapy.

15.3.2 DNA Hybridization-Based Cell–Cell Interactions

DNA hybridization is also one universal method to govern cell–cell interactions because of the inherent interaction between complementary DNA sequences [88, 89]. Typically, one type of cell is required to be modified with the synthetic DNA oligonucleotides and then mixed with another type of cell that is labeled with the complementary DNA sequence. Based on DNA hybridization, these engineered cells could be precisely assembled in the pools of cells. Moreover, benefiting from its specificity of cellular interactions, the DNA hybridization technique enables control over assembly of multicellular tissues or organoids at single cell level [90], and this method has also shown its utility in a wide range of cell types (e.g. mammalian cells, bacteria, and yeast) [91]. For example, the Gartner group proposed a DNA-programmed assembly of cells (DPAC) strategy for the development of a mammary tissue model [92–94]. Using this DPAC strategy, they could build multicellular tissues with a defined size, shape, composition, and spatial heterogeneity, opening up the opportunities to construct *in vitro* organ and disease models for the study of many diseases.

The cell–cell interactions that rely on direct hybridization of cell surface-anchored oligonucleotides might incur too much steric hindrance to DNA hybridization, resulting in relatively weak cell–cell interactions. To address this issue, DNA nanostructure-based probes are often attached to the cell surface to assist DNA hybridization between cells, thus achieving robust cell–cell interactions. The Wang group developed polyvalent DNA polymer structures on the surface of live cells

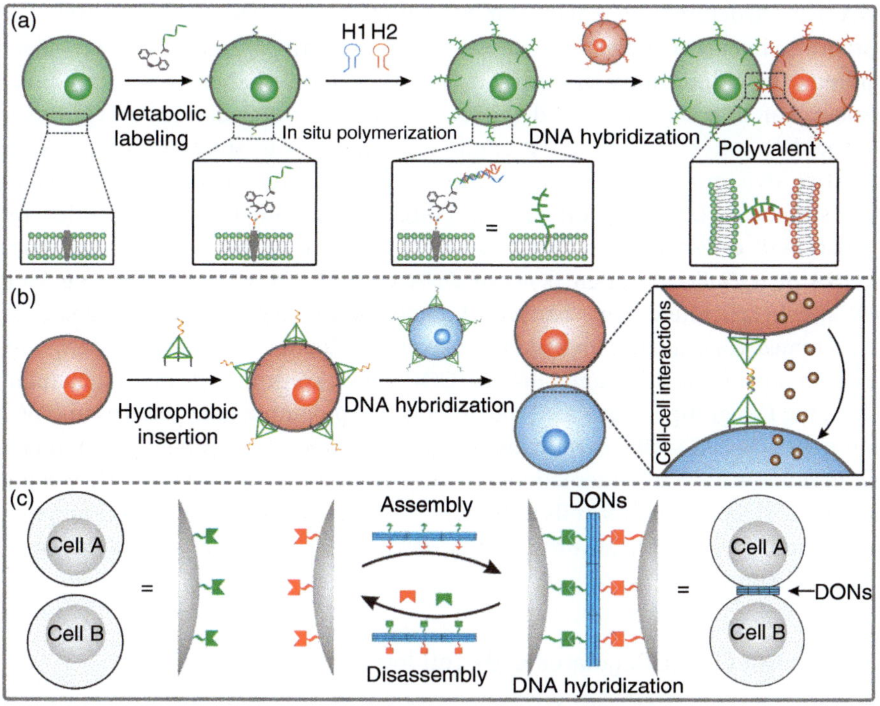

Figure 15.4 DNA hybridization-based cell–cell interactions. (a) *In situ*-formed DNA polymers on the cell surface promote cell–cell interactions. Source: From Shi et al. [95]. Reprinted with the permission of Wiley-VCH. (b) Cell-membrane-anchored tetrahedral DNA probes for programming cell–cell interactions. Source: From Li et al. [96]. Reprinted with the permission of American Chemical Society. (c) Programming cell–cell communications with engineered cell origami clusters. Source: From Ge et al. [97]. Reprinted with the permission of American Chemical Society.

through *in situ* polymerization (Figure 15.4a) [95]. In this work, two different DNA polymers that have the same backbone but display complementary branches could form polyvalent hybridization to direct cell assembly, and the DNA polymers that protrude from membrane to the environment minimize nonspecific cell–cell interactions and reduce potential steric hindrance to DNA hybridization [98]. As a result, strong and effective cell–cell interactions were realized using this polyvalent engineering. In another work, Tan and colleagues used 3D tetrahedral DNA nanostructures to engineer the cell surface for programming cellular interactions (Figure 15.4b) [96]. Compared to conventional linear DNA constructs, the tetrahedral DNA probes show higher membrane-anchoring stability (~100-fold) and higher target accessibility (~2.5-fold). Based on the hybridization of pendant DNA probes, this cell-membrane-anchored DNA nanoplatform enables specific, effective, and tunable connections between cells. This work presents a powerful and effective strategy to study multicellular communication networks.

Another commonly used strategy for decreasing steric hindrance to DNA hybridization is to exploit DNA structures (e.g. double-stranded DNA and

DNA nanostructures) as linkers for regulation of cell assembly [99–101]. For instance, Schoenit et al. [99] designed a DNA-E-cadherin hybrid linker by covalent binding of complementary DNA strands to a truncated E-cadherin with a modified extracellular domain. They used this functional DNA-E-Cadherin hybrid linker to modulate cell–cell adhesion strength and applied it to study its influence on collective dynamics in migrating epithelial monolayers. In addition to duplex DNA structures, DNA origami nanostructures (DONs) are also a great option for regulation of cell–cell interactions [97, 102]. Because of the inherent properties of DNA, it is facile to design pendant DNA on both sides of DONs. Using such DONs, cell–cell interactions between homotypic and heterotypic cells could be programmed via sequence-specific DNA hybridization. For instance, by designing DONs to organize cell origami clusters (COCs) with controlled geometric configuration, the Fan group investigated the influence of programmable patterns of COCs on the types of cell–cell communications (e.g. gap junctions, tunneling nanotubes, and immune/tumor cell interactions) (Figure 15.4c) [97]. On this basis, they could organize T cells and cancer cells with prescribed ratios and geometry and program *in vitro* immune responses, proposing a new idea to engineer and study cancer immunotherapy.

15.3.3 DNA Circuit-Regulated Cell–Cell Interactions

DNA can serve as building blocks to construct DNA circuits, which could reprogram the cells to sense external stimuli (e.g. ssDNA, proteins, and metal ions) and perform specific functions [103–106]. Generally speaking, these DNA circuits integrate the target-triggered DNA switches that contain response elements (e.g. aptamer and DNAzyme). Due to their responsiveness, DNA switches could change their conformation through binding to their targets under external stimuli and subsequently trigger computing of DNA circuits for dynamic DNA self-assembly or disassembly [107–110]. The DNA logic circuits here function as a computer processor to handle inputs and program these engineered cells for precise recognition of targeted cells. Therefore, rational design and fabrication of dynamic DNA circuits might provide a higher-order controlled approach to regulating cell–cell interactions with precision and predictability. For example, Xiao et al. reported DNA probabilistic circuits to programmably regulate receptor clustering for enhanced NK cell recognition and promoted NK cell–cancer cell interactions (Figure 15.5a) [111]. They constructed the cell surface-anchored tetrahedral DNA nanostructure-based artificial receptors (ARs) that consist of three functional modules, including a recognition module, anchoring module, and binding module, and introduced DNA probabilistic circuit into the system to produce its output for control over AR clustering. Moreover, AR clustering could be modulated by altering different inputs. They demonstrated that the resulting AR clusters show improved membrane-anchoring stability and higher binding affinity to target cancer cells compared with monomers. Due to enhanced recognition capability of NK cells through use of DNA probabilistic circuits to program AR clustering, the NK cell-cancer cell interactions could be promoted for efficient cancer cell killing. This study paves new avenues for precise control over

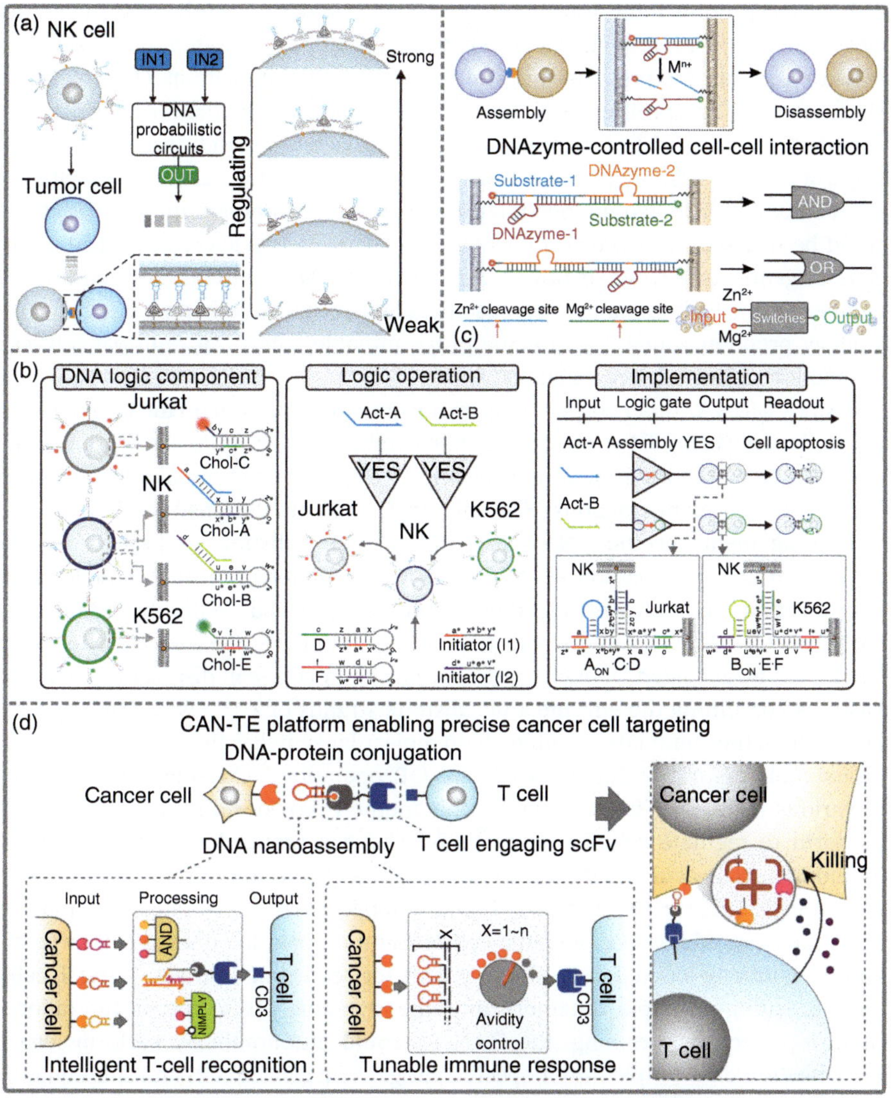

Figure 15.5 DNA circuit-regulated cell–cell interactions. (a) Programming receptor clustering with DNA probabilistic circuits for enhanced NK cell–cancer cell interactions. Source: From Xiao et al. [111]. Reprinted with the permission of Wiley-VCH. (b) Assembly pathway selection with DNA reaction circuits for programming multiple cell–cell interactions. Source: From Xiao et al. [112]. Reprinted with the permission of American Chemical Society. (c) Control of cell–cell interactions by DNAzyme-based switches. Source: From Qian et al. [113]. Reprinted with the permission of American Chemical Society. (d) A chimeric antibody-nucleic acid T-cell engager for precise cancer cell targeting. Source: From Tang et al. [114]. Reprinted with the permission of Wiley-VCH.

cellular recognition and the development of cell-based immunotherapy. In another work, the Pei group designed stem-loop-integrated DNA hairpin motif-based DNA reaction circuits for programming multiple cell–cell interactions (Figure 15.5b) [112]. By adding specific ssDNA for allosteric activation, their designed DNA circuits were allowed to program molecular self-assembly and disassembly pathways, showing assembly pathway selection. After modifying cell surfaces with the DNA circuits, controlled multicellular self-assembly with programmable selectivity could be realized. As a proof of concept, they showed the selective control of NK cells targeting two types of tumor cells using the DNA circuits for cell surface engineering, which enhances cell-specific adaptive immunotherapy efficacy. This strategy proposes a new idea for the programmable control of multiple cell–cell interactions in complex reaction pathways.

Besides ssDNA, other targets (e.g. metal ions and proteins) can also serve as inputs for triggering DNA circuit-regulated cellular interactions. For instance, Lu and colleagues reported metal ion-specific DNAzyme-based circuits for elaborate control of cell–cell interactions (Figure 15.5c) [113]. They designed two DNA chains containing metal-specific DNAzymes and/or their substrate as building blocks for cell surface modification via hydrophobic interactions. Through double-chain hybridization, the cell–cell assembly could be achieved. With the addition of specific cofactor metal ion, the substrate at the ribonucleotide cleavage site would be cleaved by activated DNAzyme, thus resulting in cell–cell disassembly. On this basis, by combining the enzyme and substrate strands of Zn^{2+}- with Mg^{2+}-specific DNAzymes, they built DNA double-chain molecular switches that enable two-factor disassembly control including AND and OR operations. Considering availability of various metal-specific DNAzymes, this method provides a smart and versatile platform to regulate cellular interactions. The Nie group developed a chimeric antibody-nucleic acid T-cell engager (CAN-TE) platform to redirect T cells against cancer cells (Figure 15.5d) [114]. The developed CAN-TE conjugate contains a DNA nanoassembly-based versatile cell recognition arm for Precise cell targeting and an antibody-based T-cell engaging arm for recruiting and activating T cells. By integrating multiple aptamers into the dynamic DNA circuit for tailored multi-input logic processing, the CAN-TE could autonomously perform logical interrogation on multiple tumor-specific antigens on cancer cells and subsequently trigger selective T-cell activation. In addition, the valence of the antigen-binding aptamers could be precisely defined on DNA nanostructure to tune the targeting avidity of T cells for "on demand" immune response. Taken together, this study provides a versatile and programmable strategy to modulate the cell–cell recognition for T cell immunotherapy, paving the way for advancing precise cell-based immunotherapy.

15.4 Conclusion

DNA-based cell surface engineering is a powerful means for programming cell–cell interactions. A variety of approaches to functionalizing cell surface with synthetic

DNA have been developed for this purpose, including chemical modification, metabolic labeling, hydrophobic insertion, molecular recognition, and liposome fusion. Based on the functionalization of DNA molecules, the modified cells can acquire new functions of molecular recognition that do not necessarily exist in nature. Therefore, they can recognize other cells independent of their types. These functions of promoting cell recognition afforded by DNA nanotechnology provide much flexibility to enable the specific and controlled study of cellular interactions between different cell types, and thus significant advances in the fields of tissue engineering and cell-based immunotherapy have been made. Given the versatility and flexibility of DNA nanotechnology-enabled cell–cell interactions, there is no doubt that many future applications in the study of human health and disease will be realized using this approach.

Acknowledgments

This work was supported by the National Science Foundation of China (grant nos. 22174046, 22104038) and the National Key Research and Development Program of China for International Science & Innovation Cooperation Major Project between Governments (2018YFE0113200).

References

1 Chao, G., Wannier, T.M., Gutierrez, C. et al. (2022). helixCAM: a platform for programmable cellular assembly in bacteria and human cells. *Cell* 185 (19): 3551–3567.

2 Bechtel, T.J., Reyes-Robles, T., Fadeyi, O.O. et al. (2021). Strategies for monitoring cell–cell interactions. *Nat. Chem. Biol.* 17 (6): 641–652.

3 Belardi, B., Son, S., Felce, J.H. et al. (2020). Cell–cell interfaces as specialized compartments directing cell function. *Nat. Rev. Mol. Cell Biol.* 21 (12): 750–764.

4 Oslund, R.C., Reyes-Robles, T., White, C.H. et al. (2022). Detection of cell–cell interactions via photocatalytic cell tagging. *Nat. Chem. Biol.* 18 (8): 850–858.

5 Hanahan, D. and Weinberg, R.A. (2011). Hallmarks of cancer: the next generation. *Cell* 144 (5): 646–674.

6 Waldman, A.D., Fritz, J.M., and Lenardo, M.J. (2020). A guide to cancer immunotherapy: from T cell basic science to clinical practice. *Nat. Rev. Immunol.* 20 (11): 651–668.

7 Rafiq, S., Hackett, C.S., and Brentjens, R.J. (2020). Engineering strategies to overcome the current roadblocks in CAR T cell therapy. *Nat. Rev. Clin. Oncol.* 17 (3): 147–167.

8 Darvin, P., Toor, S.M., Sasidharan Nair, V. et al. (2018). Immune checkpoint inhibitors: recent progress and potential biomarkers. *Exp. Mol. Med.* 50 (12): 1–11.

9 Madrigal, J.L., Schoepp, N.G., Xu, L. et al. (2022). Characterizing cell interactions at scale with made-to-order droplet ensembles (MODEs). *Proc. Natl. Acad. Sci.* 119 (5): e2110867119.

10 Stevens, M.M. and George, J.H. (2005). Exploring and engineering the cell surface interface. *Science* 310 (5751): 1135–1138.

11 Csizmar, C.M., Petersburg, J.R., and Wagner, C.R. (2018). Programming cell-cell interactions through non-genetic membrane engineering. *Cell Chem. Biol.* 25 (8): 931–940.

12 Ma, J.-Y., Wang, Y.-X., Huang, Y. et al. (2022). Chemical–biological approaches for the direct regulation of cell–cell aggregation. *Aggregate* 3 (2): e166.

13 Wu, Z., Xiao, M., Lai, W. et al. (2022). Nucleic acid-based cell surface engineering strategies and their applications. *ACS Appl. Bio Mater.* 5 (5): 1901–1915.

14 Toda, S., Blauch, L.R., Tang, S.K. et al. (2018). Programming self-organizing multicellular structures with synthetic cell-cell signaling. *Science* 361 (6398): 156–162.

15 Morsut, L., Roybal, K.T., Xiong, X. et al. (2016). Engineering customized cell sensing and response behaviors using synthetic notch receptors. *Cell* 164 (4): 780–791.

16 Sadelain, M., Rivière, I., and Riddell, S. (2017). Therapeutic T cell engineering. *Nature* 545 (7655): 423–431.

17 Shah, N.N. and Fry, T.J. (2019). Mechanisms of resistance to CAR T cell therapy. *Nat. Rev. Clin. Oncol.* 16 (6): 372–385.

18 Kohn, D.B., Sadelain, M., and Glorioso, J.C. (2003). Occurrence of leukaemia following gene therapy of X-linked SCID. *Nat. Rev. Cancer* 3 (7): 477–488.

19 Bonifant, C.L., Jackson, H.J., Brentjens, R.J. et al. (2016). Toxicity and management in CAR T-cell therapy. *Mol. Ther. Oncol.* 3: 16011.

20 Custódio, C.A. and Mano, J.F. (2016). Cell surface engineering to control cellular interactions. *ChemNanoMat* 2 (5): 376–384.

21 Lee, J.C., Donahue, N.D., Mao, A.S. et al. (2020). Exploring maleimide-based nanoparticle surface engineering to control cellular interactions. *ACS Appl. Nano Mater.* 3 (3): 2421–2429.

22 Stephan, M.T. and Irvine, D.J. (2011). Enhancing cell therapies from the outside in: Cell surface engineering using synthetic nanomaterials. *Nano Today* 6 (3): 309–325.

23 Shi, P. and Wang, Y. (2021). Synthetic DNA for cell-surface engineering. *Angew. Chem. Int. Ed.* 60 (21): 11580–11591.

24 Fan, J., Wang, H.-H., Xie, S. et al. (2020). Engineering cell-surface receptors with DNA nanotechnology for cell manipulation. *ChemBioChem* 21 (3): 282–293.

25 Xiong, M., Liu, Q., Tang, D. et al. (2020). "Apollo Program" in nanoscale: landing and exploring cell-surface with DNA nanotechnology. *ACS Appl. Bio Mater.* 3 (5): 2723–2742.

26 Shi, S., Chen, J., Wang, X. et al. (2022). Biointerface engineering with nucleic acid materials for biosensing applications. *Adv. Funct. Mater.* 32 (37): 2201069.

27 Beaucage, S.L. and Iyer, R.P. (1992). Advances in the synthesis of oligonucleotides by the phosphoramidite approach. *Tetrahedron* 48 (12): 2223–2311.

28 Hong, F., Zhang, F., Liu, Y. et al. (2017). DNA origami: scaffolds for creating higher order structures. *Chem. Rev.* 117 (20): 12584–12640.

29 Xiao, M., Lai, W., Man, T. et al. (2019). Rationally engineered nucleic acid architectures for biosensing applications. *Chem. Rev.* 119 (22): 11631–11717.

30 Dong, Y., Yao, C., Zhu, Y. et al. (2020). DNA functional materials assembled from branched DNA: design, synthesis, and applications. *Chem. Rev.* 120 (17): 9420–9481.

31 Keefe, A.D., Pai, S., and Ellington, A. (2010). Aptamers as therapeutics. *Nat. Rev. Drug Discovery* 9 (7): 537–550.

32 Fang, X. and Tan, W. (2010). Aptamers generated from cell-SELEX for molecular medicine: a chemical biology approach. *Acc. Chem. Res.* 43 (1): 48–57.

33 Xu, W., He, W., Du, Z. et al. (2021). Functional nucleic acid nanomaterials: development, properties, and applications. *Angew. Chem. Int. Ed.* 60 (13): 6890–6918.

34 El-Sagheer, A.H. and Brown, T. (2010). Click chemistry with DNA. *Chem. Soc. Rev.* 39 (4): 1388–1405.

35 Chen, Y., Hong, T., Wang, S. et al. (2017). Epigenetic modification of nucleic acids: from basic studies to medical applications. *Chem. Soc. Rev.* 46 (10): 2844–2872.

36 Li, L., Liu, S., Zhang, C. et al. (2022). Recent advances in DNA-based cell surface engineering for biological applications. *Chem. Eur. J.* 28 (69): e202202070.

37 Dong, H., Liu, L., Wang, J. et al. (2020). DNA-based reprogramming strategy of receptor-mediated cellular behaviors: from genetic encoding to nongenetic engineering. *ACS Appl. Bio Mater.* 3 (5): 2796–2804.

38 Koyfman, A.Y., Braun, G.B., and Reich, N.O. (2009). Cell-targeted self-assembled DNA nanostructures. *J. Am. Chem. Soc.* 131 (40): 14237–14239.

39 Zhao, W., Loh, W., Droujinine, I.A. et al. (2011). Mimicking the inflammatory cell adhesion cascade by nucleic acid aptamer programmed cell-cell interactions. *FASEB J.* 25 (9): 3045–3056.

40 Hsiao, S.C., Shum, B.J., Onoe, H. et al. (2009). Direct cell surface modification with DNA for the capture of primary cells and the investigation of myotube formation on defined patterns. *Langmuir* 25 (12): 6985–6991.

41 Chandra, R.A., Douglas, E.S., Mathies, R.A. et al. (2006). Programmable cell adhesion encoded by DNA hybridization. *Angew. Chem. Int. Ed.* 45 (6): 896–901.

42 Saxon, E. and Bertozzi, C.R. (2000). Cell surface engineering by a modified Staudinger reaction. *Science* 287 (5460): 2007–2010.

43 Selden, N.S., Todhunter, M.E., Jee, N.Y. et al. (2012). Chemically programmed cell adhesion with membrane-anchored oligonucleotides. *J. Am. Chem. Soc.* 134 (2): 765–768.

44 Chidchob, P., Offenbartl-Stiegert, D., McCarthy, D. et al. (2019). Spatial presentation of cholesterol units on a DNA cube as a determinant of membrane protein-mimicking functions. *J. Am. Chem. Soc.* 141 (2): 1100–1108.

45 Xiao, M., Lai, W., Wang, X. et al. (2018). DNA mediated self-assembly of multicellular microtissues. *Microphysiol. Syst.* 2 (1): 1–17.

46 Dutta, D., Pulsipher, A., Luo, W. et al. (2011). Synthetic chemoselective rewiring of cell surfaces: generation of three-dimensional tissue structures. *J. Am. Chem. Soc.* 133 (22): 8704–8713.

47 Sarkar, D., Vemula, P.K., Teo, G.S.L. et al. (2008). Chemical engineering of mesenchymal stem cells to induce a cell rolling response. *Bioconjugate Chem.* 19 (11): 2105–2109.

48 Saccà, B. and Niemeyer, C.M. (2011). Functionalization of DNA nanostructures with proteins. *Chem. Soc. Rev.* 40 (12): 5910–5921.

49 Stephan, M.T., Moon, J.J., Um, S.H. et al. (2010). Therapeutic cell engineering with surface-conjugated synthetic nanoparticles. *Nat. Med.* 16 (9): 1035–1041.

50 Vogel, K., Glettenberg, M., Schroeder, H. et al. (2013). DNA-modification of eukaryotic cells. *Small* 9 (2): 255–262.

51 Mahal, L.K., Yarema, K.J., and Bertozzi, C.R. (1997). Engineering chemical reactivity on cell surfaces through oligosaccharide biosynthesis. *Science* 276 (5315): 1125–1128.

52 Rillahan, C.D., Antonopoulos, A., Lefort, C.T. et al. (2012). Global metabolic inhibitors of sialyl- and fucosyltransferases remodel the glycome. *Nat. Chem. Biol.* 8 (7): 661–668.

53 Baskin, J.M., Prescher, J.A., Laughlin, S.T. et al. (2007). Copper-free click chemistry for dynamic in vivo imaging. *Proc. Natl. Acad. Sci.* 104 (43): 16793–16797.

54 Chang, P.V., Prescher, J.A., Hangauer, M.J. et al. (2007). Imaging cell surface glycans with bioorthogonal chemical reporters. *J. Am. Chem. Soc.* 129 (27): 8400–8401.

55 Loew, M., Springer, R., Scolari, S. et al. (2010). Lipid domain specific recruitment of lipophilic nucleic acids: a key for switchable functionalization of membranes. *J. Am. Chem. Soc.* 132 (45): 16066–16072.

56 You, M., Lyu, Y., Han, D. et al. (2017). DNA probes for monitoring dynamic and transient molecular encounters on live cell membranes. *Nat. Nanotechnol.* 12 (5): 453–459.

57 Liu, H., Kwong, B., and Irvine, D.J. (2011). Membrane anchored immunostimulatory oligonucleotides for in vivo cell modification and localized immunotherapy. *Angew. Chem. Int. Ed.* 50 (31): 7052–7055.

58 Jin, C., He, J., Zou, J. et al. (2019). Phosphorylated lipid-conjugated oligonucleotide selectively anchors on cell membranes with high alkaline phosphatase expression. *Nat. Commun.* 10 (1): 2704.

59 Borisenko, G.G., Zaitseva, M.A., Chuvilin, A.N. et al. (2009). DNA modification of live cell surface. *Nucleic Acids Res.* 37 (4): e28.

60 Agasti, S.S., Liong, M., Peterson, V.M. et al. (2012). Photocleavable DNA barcode–antibody conjugates allow sensitive and multiplexed protein analysis in single cells. *J. Am. Chem. Soc.* 134 (45): 18499–18502.

61 Bailey, R.C., Kwong, G.A., Radu, C.G. et al. (2007). DNA-encoded antibody libraries: a unified platform for multiplexed cell sorting and detection of genes and proteins. *J. Am. Chem. Soc.* 129 (7): 1959–1967.

62 Ren, K., Liu, Y., Wu, J. et al. (2016). A DNA dual lock-and-key strategy for cell-subtype-specific siRNA delivery. *Nat. Commun.* 7 (1): 13580.

63 Zhu, G., Zhang, S., Song, E. et al. (2013). Building fluorescent DNA nanodevices on target living cell surfaces. *Angew. Chem. Int. Ed.* 52 (21): 5490–5496.

64 Li, L., Chen, X., Cui, C. et al. (2019). Aptamer displacement reaction from live-cell surfaces and its applications. *J. Am. Chem. Soc.* 141 (43): 17174–17179.

65 O'Brien, P.J., Elahipanah, S., Rogozhnikov, D. et al. (2017). Bio-orthogonal mediated nucleic acid transfection of cells via cell surface engineering. *ACS Cent. Sci.* 3 (5): 489–500.

66 Dutta, D., Pulsipher, A., Luo, W. et al. (2011). Engineering cell surfaces via liposome fusion. *Bioconjugate Chem.* 22 (12): 2423–2433.

67 Sun, L., Gao, Y., Wang, Y. et al. (2018). Guiding protein delivery into live cells using DNA-programmed membrane fusion. *Chem. Sci.* 9 (27): 5967–5975.

68 Lin, M., Chen, Y., Zhao, S. et al. (2022). A biomimetic approach for spatially controlled cell membrane engineering using fusogenic spherical nucleic acid. *Angew. Chem. Int. Ed.* 61 (1): e202111647.

69 Sato, Y.T., Umezaki, K., Sawada, S. et al. (2016). Engineering hybrid exosomes by membrane fusion with liposomes. *Sci. Rep.* 6 (1): 21933.

70 Löffler, P.M.G., Ries, O., Rabe, A. et al. (2017). A DNA-programmed liposome fusion cascade. *Angew. Chem. Int. Ed.* 56 (43): 13228–13231.

71 Alivisatos, A.P., Johnsson, K.P., Peng, X. et al. (1996). Organization of 'nanocrystal molecules' using DNA. *Nature* 382 (6592): 609–611.

72 Park, S.Y., Lytton-Jean, A.K.R., Lee, B. et al. (2008). DNA-programmable nanoparticle crystallization. *Nature* 451 (7178): 553–556.

73 Jones, M.R., Seeman, N.C., and Mirkin, C.A. (2015). Programmable materials and the nature of the DNA bond. *Science* 347 (6224): 1260901.

74 Seeman, N.C. and Sleiman, H.F. (2017). DNA nanotechnology. *Nat. Rev. Mater.* 3 (1): 17068.

75 Allen, G.M. and Lim, W.A. (2022). Rethinking cancer targeting strategies in the era of smart cell therapeutics. *Nat. Rev. Cancer* 22 (12): 693–702.

76 Guillerey, C., Huntington, N.D., and Smyth, M.J. (2016). Targeting natural killer cells in cancer immunotherapy. *Nat. Immunol.* 17 (9): 1025–1036.

77 Liu, C.-G., Wang, Y., Liu, P. et al. (2020). Aptamer-T cell targeted therapy for tumor treatment using sugar metabolism and click chemistry. *ACS Chem. Biol.* 15 (6): 1554–1565.

78 Gong, L., Li, Y., Cui, K. et al. (2021). Nanobody-engineered natural killer cell conjugates for solid tumor adoptive immunotherapy. *Small* 17 (45): 2103463.

79 Zhang, X.-J., Zhao, Z., Wang, X. et al. (2023). A versatile strategy for convenient circular bivalent functional nucleic acids construction. *Nat. Sci. Rev.* 10 (2): nwac107.

80 Xiong, X., Liu, H., Zhao, Z. et al. (2013). DNA aptamer-mediated cell targeting. *Angew. Chem. Int. Ed.* 52 (5): 1472–1476.

81 Zhang, D., Zheng, Y., Lin, Z. et al. (2020). Equipping natural killer cells with specific targeting and checkpoint blocking aptamers for enhanced adoptive immunotherapy in solid tumors. *Angew. Chem. Int. Ed.* 59 (29): 12022–12028.

82 Li, J., Xun, K., Zheng, L. et al. (2021). DNA-based dynamic mimicry of membrane proteins for programming adaptive cellular interactions. *J. Am. Chem. Soc.* 143 (12): 4585–4592.

83 Guo, Z., Zhang, L., Yang, Q. et al. (2022). Manipulation of multiple cell–cell interactions by tunable DNA scaffold networks. *Angew. Chem. Int. Ed.* 61 (7): e202111151.

84 Yang, S., Wen, J., Li, H. et al. (2019). Aptamer-engineered natural killer cells for cell-specific adaptive immunotherapy. *Small* 15 (22): 1900903.

85 Liu, X., Yan, H., Liu, Y. et al. (2011). Targeted cell–cell interactions by DNA nanoscaffold-templated multivalent bispecific aptamers. *Small* 7 (12): 1673–1682.

86 Zhang, K., Deng, R., Sun, Y. et al. (2017). Reversible control of cell membrane receptor function using DNA nano-spring multivalent ligands. *Chem. Sci.* 8 (10): 7098–7105.

87 Shi, P., Wang, X., Davis, B. et al. (2020). In situ lsynthesis of an aptamer-based polyvalent antibody mimic on the cell surface for enhanced interactions between immune and cancer cells. *Angew. Chem. Int. Ed.* 59 (29): 11892–11897.

88 Furst, A.L., Klass, S.H., and Francis, M.B. (2019). DNA hybridization to control cellular interactions. *Trends Biochem. Sci* 44 (4): 342–350.

89 Teramura, Y., Chen, H., Kawamoto, T. et al. (2010). Control of cell attachment through polyDNA hybridization. *Biomaterials* 31 (8): 2229–2235.

90 Liu, J.S. and Gartner, Z.J. (2012). Directing the assembly of spatially organized multicomponent tissues from the bottom up. *Trends Cell Biol.* 22 (12): 683–691.

91 Furst, A.L., Smith, M.J., and Francis, M.B. (2018). New techniques for the generation and analysis of tailored microbial systems on surfaces. *Biochemistry* 57 (21): 3017–3026.

92 Todhunter, M.E., Jee, N.Y., Hughes, A.J. et al. (2015). Programmed synthesis of three-dimensional tissues. *Nat. Methods* 12 (10): 975–981.

93 Gartner, Z.J. and Bertozzi, C.R. (2009). Programmed assembly of 3-dimensional microtissues with defined cellular connectivity. *Proc. Natl. Acad. Sci.* 106 (12): 4606–4610.

94 Liu, J.S., Farlow, J.T., Paulson, A.K. et al. (2012). Programmed cell-to-cell variability in Ras activity triggers emergent behaviors during mammary epithelial morphogenesis. *Cell Rep.* 2 (5): 1461–1470.

95 Shi, P., Zhao, N., Lai, J. et al. (2018). Polyvalent display of biomolecules on live cells. *Angew. Chem. Int. Ed.* 57 (23): 6800–6804.

96 Li, J., Xun, K., Pei, K. et al. (2019). Cell-membrane-anchored DNA nanoplatform for programming cellular interactions. *J. Am. Chem. Soc.* 141 (45): 18013–18020.

97 Ge, Z., Liu, J., Guo, L. et al. (2020). Programming cell–cell communications with Eengineered cell origami clusters. *J. Am. Chem. Soc.* 142 (19): 8800–8808.

98 Richards, E., Li, S., Chen, N. et al. (2014). Polymerization of affinity ligands on a surface for enhanced ligand display and cell binding. *Biomacromolecules* 15 (12): 4561–4569.

99 Schoenit, A., Lo Giudice, C., Hahnen, N. et al. (2022). Tuning epithelial cell–cell adhesion and collective dynamics with functional DNA-E-cadherin hybrid linkers. *Nano Lett.* 22 (1): 302–310.

100 Liu, J., Li, M., Li, F. et al. (2021). Reconstructing soma–soma synapse-like vesicular exocytosis with DNA origami. *ACS Cent. Sci.* 7 (8): 1400–1407.

101 Gao, F., Yang, D., Xu, F. et al. (2022). Promoting cell fusion by polyvalent DNA ligands. *Nano Lett.* 22 (7): 3018–3025.

102 Akbari, E., Mollica, M.Y., Lucas, C.R. et al. (2017). Engineering cell surface function with DNA origami. *Adv. Mater.* 29 (46): 1703632.

103 Qu, X., Wang, S., Ge, Z. et al. (2017). Programming cell adhesion for on-chip sequential Boolean logic functions. *J. Am. Chem. Soc.* 139 (30): 10176–10179.

104 Qian, R.-C., Zhou, Z.-R., Wu, Y. et al. Combination cancer treatment: using engineered DNAzyme molecular machines for dynamic inter- and intracellular regulation. *Angew. Chem. Int. Ed.* https://doi.org/10.1002/ange.202210935.

105 Zhang, J., Qiu, Z., Fan, J. et al. (2021). Scan and unlock: a programmable DNA molecular automaton for cell-selective activation of ligand-based signaling. *Angew. Chem. Int. Ed.* 60 (12): 6733–6743.

106 Wang, D., Li, S., Zhao, Z. et al. (2021). Engineering a second-order DNA logic-gated nanorobot to sense and release on live cell membranes for multiplexed diagnosis and synergistic therapy. *Angew. Chem. Int. Ed.* 60 (29): 15816–15820.

107 Tang, Q., Lai, W., Wang, P. et al. (2021). Multi-mode reconfigurable DNA-based chemical reaction circuits for soft matter computing and control. *Angew. Chem. Int. Ed.* 60 (27): 15013–15019.

108 Cangialosi, A., Yoon, C., Liu, J. et al. (2017). DNA sequence-directed shape change of photopatterned hydrogels via high-degree swelling. *Science* 357 (6356): 1126–1130.

109 Xiao, M., Lai, W., Wang, F. et al. (2019). Programming drug delivery kinetics for active burst release with DNA toehold switches. *J. Am. Chem. Soc.* 141 (51): 20354–20364.

110 Xiong, X., Xiao, M., Lai, W. et al. (2021). Optochemical control of DNA-switching circuits for logic and probabilistic computation. *Angew. Chem. Int. Ed.* 60 (7): 3397–3401.

111 Xiao, M., Lai, W., Yao, X. et al. (2022). Programming receptor clustering with DNA probabilistic circuits for enhanced natural killer cell recognition. *Angew. Chem. Int. Ed.* 61 (28): e202203800.

112 Xiao, M., Lai, W., Yu, H. et al. (2021). Assembly pathway selection with DNA reaction circuits for programming multiple cell–cell interactions. *J. Am. Chem. Soc.* 143 (9): 3448–3454.

113 Qian, R.-C., Zhou, Z.-R., Guo, W. et al. (2021). Cell surface engineering using DNAzymes: metal ion mediated control of cell–cell interactions. *J. Am. Chem. Soc.* 143 (15): 5737–5744.

114 Tang, R., Fu, Y.-H., Gong, B. et al. (2022). A chimeric conjugate of antibody and programmable DNA nanoassembly smartly activates T cells for precise cancer cell targeting. *Angew. Chem. Int. Ed.* 61 (36): e202205902.

16

Designer DNA Nanostructures and Their Cellular Uptake Behaviors

Jing Ye, Donglei Yang, Chenzhi Shi, Fei Zhou, and Pengfei Wang

Institute of Molecular Medicine, Department of Laboratory Medicine, Shanghai Key Laboratory for Nucleic Acid Chemistry and Nanomedicine, Center for DNA Information Storage, Renji Hospital, School of Medicine, Shanghai Jiao Tong University, Shanghai 200127, China

16.1 Introduction

Cell membranes are widely recognized for their ability to effectively block the entry of harmful substances and impede the absorption of useful substances, such as drugs, by cells. Currently, drugs can be classified into two distinct categories – chemical molecules [1–3] and biomolecular drugs (for example, siRNA [4, 5] and proteins [6, 7], etc.). Chemical molecules are often hampered by poor solubility and are susceptible to toxic reactions and negative side effects. Biomolecular drugs, on the other hand, suffer from reduced stability and are vulnerable to degradation by various biological enzymes, thereby reducing their capacity to effectively target their intended site of action. These drawbacks hinder the drugs from achieving peak effectiveness. Additionally, traditional delivery systems primarily consist of inorganic nanomaterials [8–11], polymeric materials [12–14], and bionanomaterials [15–17], and some of them have been approved for use by the FDA. Although inorganic nanomaterials demonstrate noticeable clinical effectiveness, they present significant imperfections. Specifically, they cannot undergo degradation in living organisms and have a tendency to accumulate in liver and kidney tissues, causing toxic damage to health [18, 19]. While certain polymeric materials can degrade, the exact rate of degradation remains uncertain, and the possible toxicity of the byproducts formed from the degradation process remains a matter of debate. Viral vectors have limitations with regard to the types of cargo they can transport, primarily only allowing for the delivery of short DNA strands, and there is a possibility of such vectors causing mutations, carcinogenesis, and toxic effects through random insertions [20]. Hence, there is a pressing need for the creation of carrier materials that possess exceptional cell permeability, high loading capacity, and exceptional biocompatibility.

DNA, a substance found naturally in most organisms, serves as the carrier of genetic information. One unique characteristic of DNA is its ability to form a double helix structure, typically referred to as a B-type structure, based on the

DNA Nanotechnology for Cell Research: From Bioanalysis to Biomedicine, First Edition. Edited by Zhou Nie.

principle of base complementary pairing (G-C, A-T). This structure has a diameter of around 2 nm, has 3.4 nm (or 10.5 bp) of helix loop per turn, and can persist up to approximately 50 nm or 150 bp, making DNA a relatively rigid polymer at the nanoscale [21]. As well as Watson-Crick base pairing, DNA possesses numerous other properties, including rapid and straightforward synthesis, predictable intermolecular interactions, versatile multisite modification strategies, locally stiff polymers, and a high density of functional groups [22–25]. The inherent properties of DNA mentioned above enable its self-assembly into structures of different complexity, size, and shape, including one-, two-, or three-dimensional forms. DNA nanotechnology has found numerous applications in various fields, including engineering, information storage, biomedicine, and others [26–32]. The unique physicochemical properties of nanocarriers make them show great potential in the field of biomedicine. Structural DNA is mainly used in biosensing [33–35], bioimaging [34, 36, 37], disease diagnosis [38, 39], and drug delivery [21, 40, 41]. The aim of this review is to investigate the impact of DNA nanostructure geometry as a drug delivery platform on cytokinesis. Various design methods of DNA nanostructures, such as DNA origami, single-stranded DNA tiles, and dynamic DNA structures, will be discussed, along with an overview of the types and mechanisms of endocytosis, including phagocytosis, macropinocytosis, and protein-mediated internalization pathways. Additionally, the mechanisms of how different morphologies of DNA nanostructures enter the cell and their resulting cellular effects will be summarized.

16.2 DNA Nanotechnology

16.2.1 The Beginning of DNA Nanotechnology

Two primary methods of nanostructure fabrication are top-down and bottom-up approaches. The former involves sculpting from a solid block, while the latter assembles structures with nanounits. Nadrian C. Seeman, the founder of structural DNA nanotechnology, discovered that branched junctions could combine through single-stranded overhangs, or "sticky ends," to create 3D crystalline materials [42] (Figure 16.1a). Seeman's original goal was to create a three-dimensional DNA lattice that could help orient protein molecules, which can be characterized using XRD. Over time, more robust nanostructures have been developed using various units, such as double-crossover tile [43] (Figure 16.1b), cross-shaped tile [44], three-point [45], and six-point star motifs [46] (Figure 16.1c), that increase larger structure stability and enable precise topology control.

Numerous three-dimensional structures have been developed to expand the applications of DNA nanotechnology. Many polyhedral DNA nanostructures, including tetrahedra [47], dodecahedra [48], and cubes [49] (Figure 16.1d), have been created using just a few strands. In 2009, Seeman's research team achieved a significant

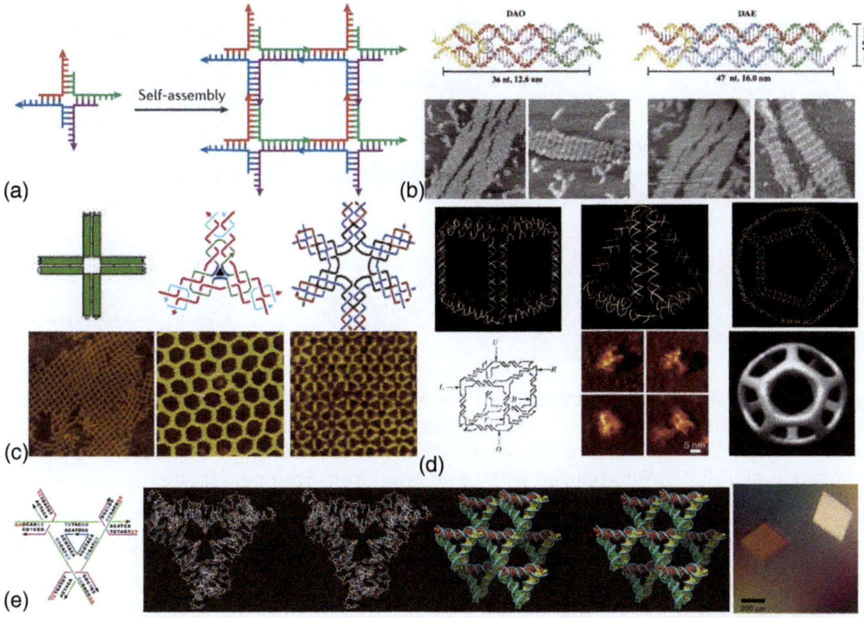

Figure 16.1 Designs of DNA nanostructures are constructed by using a fixed number of DNA strands. (a) The original proposal of DNA lattice assembled by DNA junctions. Source: Adapted from Seeman [42]. (b) DNA lattice assembled by two different double-crossover motifs. Source: Winfree et al. [43]/Reproduced with permission from Springer Nature. (c) Periodic DNA 2D array constructed by different motifs. Source: Adapted from Refs [44–46]. (d) Polyhedral DNA nanostructure, from left to right: cube, tetrahedron, dodecahedron. Source: He et al. [47–49]/Reproduced with permission from Springer Nature. (e) Tensegrity triangle DNA tiles are used to construct macroscopic crystals. Source: Zheng et al. [50]/Reproduced with permission from Springer Nature.

milestone in "bottom-up" nanostructure fabrication by assembling a macroscopic crystal (~50 μm) with DNA strands [50] (Figure 16.1e). This achievement serves as a bridge between the nano-world and the macro-world and represents a breakthrough in the field.

16.2.2 DNA Origami

In 2006, Rothemund unveiled a groundbreaking technique in the realm of structural DNA nanotechnology, known as "DNA origami" [51], which greatly expanded the potential and feasibility of constructing intricate DNA nanostructures. Using a multitude of short single-stranded DNAs (ssDNAs), this technique employs the folding of a lengthy M13 phage DNA, a native single-stranded DNA, into specific shapes (Figure 16.2a). This method is highly efficient, easy to design, and boasts high yields without requiring staple strand purification, leading to rapid developments in the field.

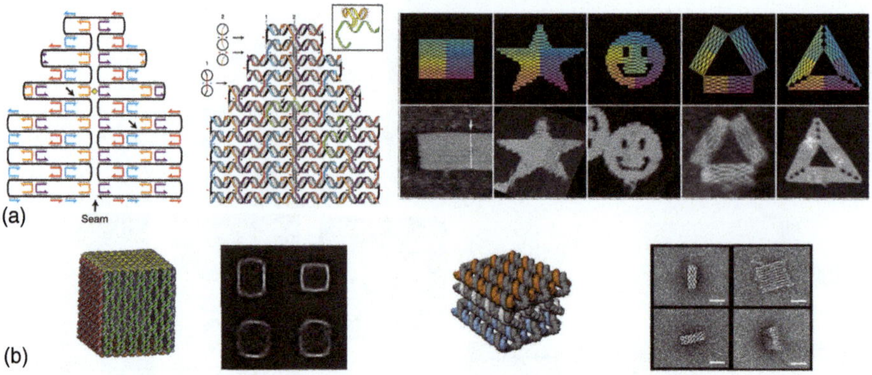

Figure 16.2 DNA origami structures. (a) DNA origami proposed by Rothemund and several 2D structures constructed by this method. Source: Rothemund et al. [51]/Reproduced with permission from Springer Nature. (b) Two examples of DNA 3D origami structures. Source: Andersen et al. [52, 53]/Reproduced with permission from Springer Nature.

Additionally, various 3D DNA nanostructures have been produced using DNA origami techniques, including tubes [52] and honeycomb lattices [53] (Figure 16.2b). Through manipulation of the number of nucleotides per turn, Yan and coworkers [54] and Shih and coworkers [55] were able to create curved and twisted DNA origami structures. More intricate DNA origami structures have been designed using CaDNAno [56] – a computer software specifically for DNA origami design. Yan's research team successfully constructed the first complex wireframe DNA origami nanostructures, with subsequent groups expanding their range. Högberg's team improved origami stability in biological assays by utilizing single helices instead of close-packed helices as construction elements [57].

The examples provided demonstrate the exceptional adaptability of DNA origami. However, there are still some hindrances that need to be addressed. A significant limitation is the length restriction imposed on the scaffold strand, which sets a ceiling on the size of DNA origami. Despite several efforts made to overcome this obstacle – such as hierarchical assembly [58], multi-scaffold approaches [59], and artificially lengthened single-stranded DNAs [60] – they also present fresh challenges.

16.2.3 Single-Stranded DNA Tiles

To enhance the flexibility of DNA nanotechnology and surmount the limitations of DNA origami, Yin's team proposed the single-stranded DNA tiles (SST) approach [61], which employs single strands as the basic units for building DNA nanostructures. Instead of using double-stranded DNA, the SST approach uses single strands as the fundamental building blocks to create DNA nanostructures (Figure 16.3a). By amalgamating hundreds of SSTs, they created a molecular canvas where different pieces can be chosen and self-assembled into diverse structures. They subsequently employed SSTs to construct intricate 3D structures [62], surpassing the size of those assembled by the DNA origami method [63] (Figure 16.3b).

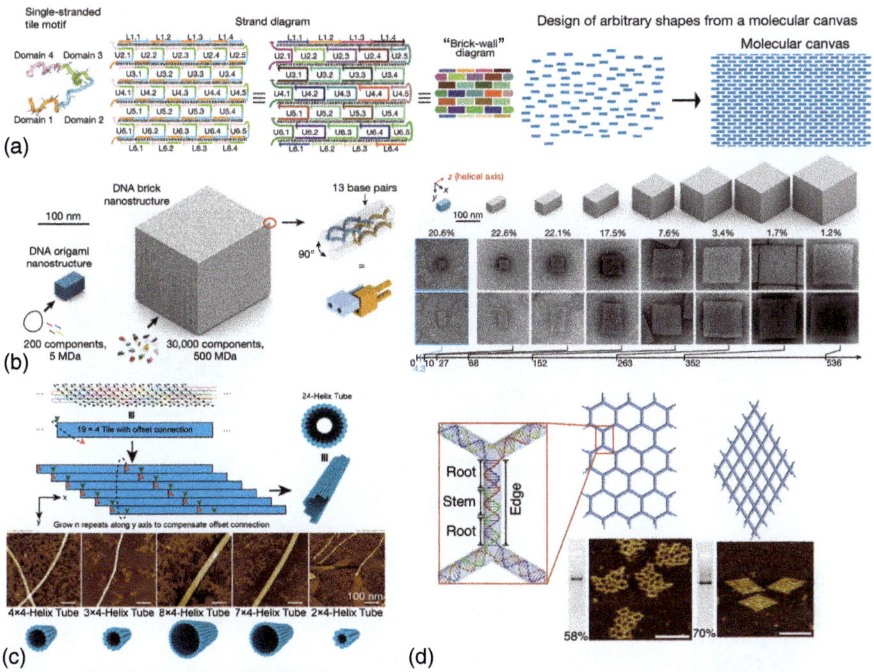

Figure 16.3 DNA nanostructures constructed by single-stranded DNA tiles (SST). (a) SST method proposed by Yin's group. Source: Rothemund et al. [61]/Reproduced with permission from Springer Nature. (b) SST method was extended to assemble varieties of DNA 3D structure. Source: Ong et al. [62, 63]/Reproduced with permission from Springer Nature. (c) DNA 3D structure breaking the size limitation of DNA origami. Source: Zhang et al. [64]/Reproduced with permission from American Chemical Society. (d) Complex wireframe DNA nanostructures from simple building blocks. Source: Wang et al. [65]/ Reproduced with permission from Springer Nature.

In a similar vein, various research teams have been successful in creating increasingly intricate DNA structures. One team led by Ke achieved tube-shaped DNA [64] with varying diameters by utilizing an offset connection (Figure 16.3c), while Wei's group overcame size limitations related to DNA origami by assembling multiple wireframe structures [65] (Figure 16.3d).

16.2.4 Dynamic DNA Structures

The utilization of dynamic DNA structures has significantly increased in the biomedical field, as well as in fundamental research, due to their various applications. To manipulate DNA structures, several fundamental driving forces have been employed, which include molecular binding, strand displacement, and blunt-end effect.

Douglas and his team showcased the potential of nano-transporters by presenting a concept demonstration of a nanorobot [66] that could efficiently transport molecular payloads. In addition, Dietz's group utilized the blunt-end effect [67] and shape configuration to create a nano-motor powered by ATP [68], leading to the development of innovative DNA structures.

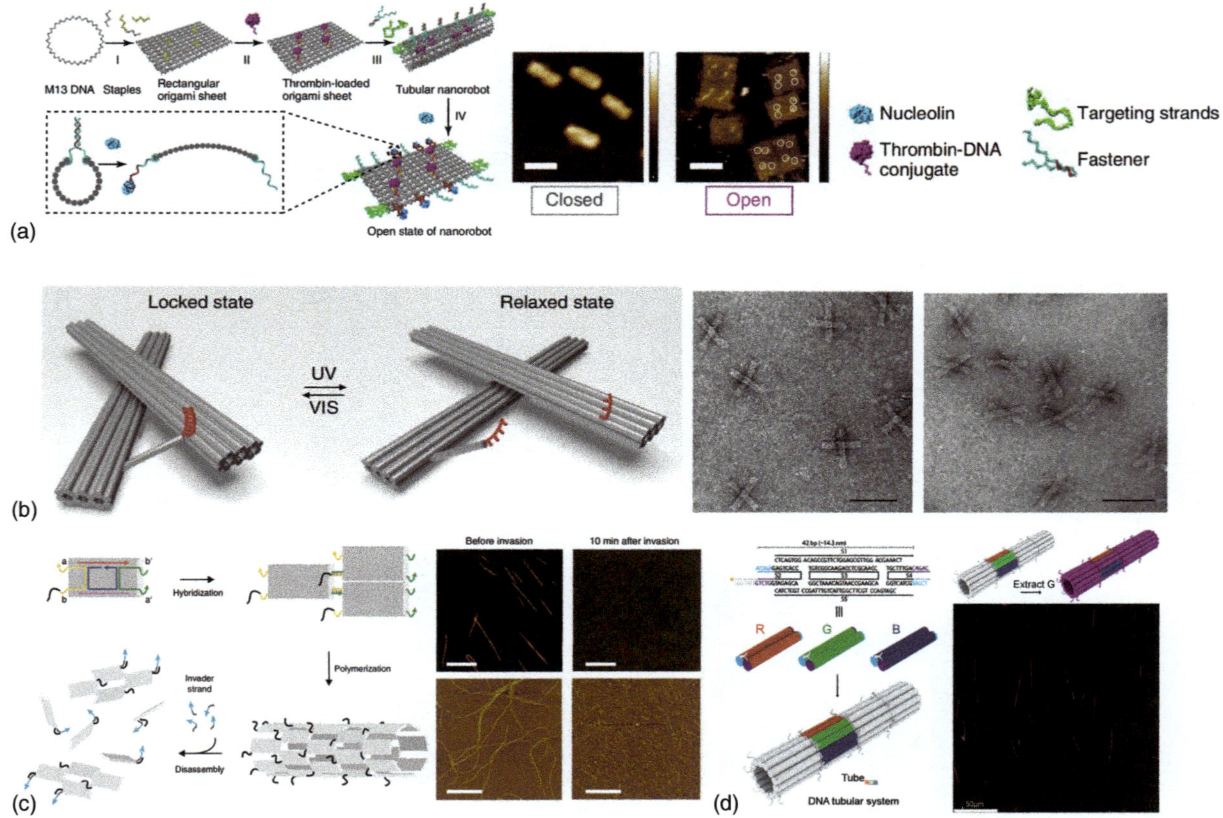

Figure 16.4 Dynamic DNA nanostructures. (a) DNA origami structures used as drug delivery vehicles. (b) Dynamic DNA nanostructures responsive to different wavelengths of light. (c) Dynamic control of DNA nanostructure self-assembly through strand displacement. (d) DNA-Based Molecular System controlled by strand displacement and motif transformation. Source: Green et al. [69]/Reproduced with permission from Springer Nature.

In addition, various research teams have demonstrated the great potential of responsive architectures that can effectively respond to various signals, including nucleic acids and light, for applications in the fields of biology and medicine. [69–72] (Figure 16.4).

16.3 Pathways of Cell Endocytosis

There are two primary mechanisms by which external substances can enter cells: direct diffusion or endocytosis. However, in the present delivery system, only viruses and exosomes [73] are able to enter cells by directly fusing with the cell membrane in order to exert their functions. The process of Herpes simplex viruses (HSV) entering a host cell involves several steps, culminating in membrane fusion between the virus and the cell [74]. Simultaneously, modifications have been implemented in the nanoparticle (NP) system to attain a comparable route of entry into the cell [75–79]. Although there have been advancements in delivery systems, the majority still rely on endocytosis for cellular uptake. This pathway initiates with the creation of a small, sphere-shaped vesicle that buds off the cell membrane. Upon merging with the cell membrane, molecules are absorbed into the cell and transported to their designated location or organelle within it. The creation of the vesicle is facilitated by the folding inward of the cell membrane and is powered by ATP.

Having a comprehensive understanding of endocytosis, a complex process with multiple mechanisms, can greatly aid in improving delivery systems and maximizing delivery efficiency. Currently, there is a growing consensus on the five major types of endocytosis, which include: (1) Clathrin-mediated endocytosis (CME), (2) clathrin-independent endocytosis (CIE), (3) phagocytosis, (4) macropinocytosis, and (5) caveolae-mediated endocytosis [80] (Figure 16.5).

16.3.1 Clathrin-Mediated Endocytosis

The clathrin-mediated endocytic pathway has been extensively researched, making it the most comprehensively understood endocytic pathway to date. This process is utilized by all mammalian cells to internalize a diverse range of substances such as proteins, lipids, carbohydrates, and macromolecules. Clathrin, a protein that forms a honeycomb lattice at the neck of an invaginated pit on the plasma membrane, drives the formation of a vesicle that encapsulates the internalized material within the cell [81]. The start of clathrin assembly on the plasma membrane is prompted by different adaptors and accessory proteins that identify particular cargo molecules and attract clathrin to the suitable site. The process of clathrin-mediated endocytosis (CME) is relatively uncomplicated and can be categorized into three partially overlapping and sequential phases: (i) constructing the clathrin coat; (ii) bending and severing the membrane; and (iii) removing the vesicle coating [82]. The initiation of clathrin coat assembly begins with the clustering of endocytic coat proteins on the inner leaflet of the plasma membrane, followed by the recruitment of other coat proteins from the cytosol. This promotes membrane bending and alters membrane

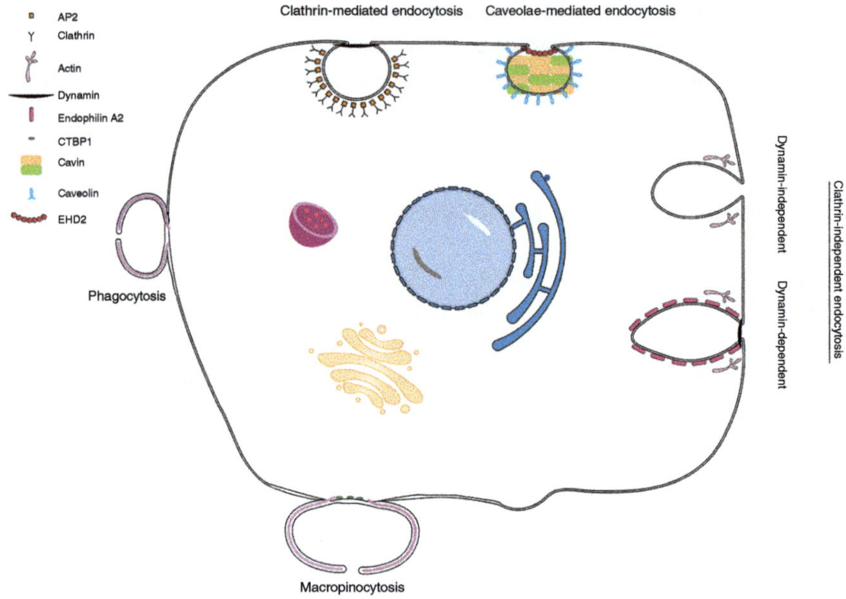

Figure 16.5 Overview of cell endocytosis pathways. There are six major pathways of endocytosis and important related proteins in cells, including clathrin-mediated endocytosis, caveolae-mediated endocytosis, clathrin-independent/dynamin-dependent endocytosis, clathrin-independent/dynamin-independent endocytosis, macropinocytosis, and phagocytosis.

stress, ultimately resulting in the formation of a clathrin-coated pit from a flat cell membrane. Actin polymerization works in cooperation with coat and scission proteins to cut the neck of the membrane invagination, further shaping the membrane. Finally, the endocytic protein machinery is disassembled through uncoating, which releases the cargo-filled vesicle. It is noteworthy that clathrin-coated vesicles are approximately 100 nm in diameter [83], but recent studies have demonstrated that not only bacteria but also structures exceeding 1 um in diameter can utilize this mechanism to enter the cell and undergo further trafficking [84]. The reason for this may be the cooperation between the latter and clathrin, resulting in the creation of actin-rich pedestals that promote uptake. Additional research has demonstrated that internalization occurs through the elongation of the clathrin-coated pit [85].

CME plays a vital part in multiple physiological functions like receptor signaling and internalization, nutrient uptake, and synaptic vesicle recycling. Any malfunctioning in CME can lead to the development of numerous diseases, including neurodegenerative ailments and viral infections [86–88]. To gain deeper insights into these processes and formulate potential therapeutic approaches, it is imperative to comprehend the molecular mechanisms that govern CME and its regulation.

16.3.2 Clathrin-Independent Endocytosis

Although CME is the most recognized form of endocytosis, there exist various clathrin-independent endocytic pathways that also participate in this process. Clathrin-independent endocytic mechanisms can be classified into dynamin-dependent and dynamin-independent pathways, according to whether or not they require the presence of the protein dynamin for vesicle scission. Dynamin is a large GTPase that plays a key role in the endocytic process by facilitating the constriction of vesicles from the plasma membrane [89].

Research has demonstrated that the use of dynamin-dependent CIE pathways plays a vital role in lipid trafficking and regulation of signaling pathways. These pathways are responsible for facilitating the uptake of IL-2 via the RhoA-dependent pathway and the endocytosis of the EGFR (epidermal growth factor receptor), particularly when exposed to high concentrations of EGF (epidermal growth factor) [90, 91]. The extensively studied pathway is known as FEME, which is a dynamin-dependent and clathrin-independent mechanism of endocytosis. This pathway is capable of swiftly and effectively internalizing selected plasma membrane proteins, including receptors and ion channels, as well as some pathogens like viruses and toxins [92, 93]. FEME relies on the recruitment of endophilin, a protein with a BAR domain that binds to and deforms the plasma membrane, leading to the formation of inward membrane protrusions or invaginations. These endophilin-deformed pits trap cargo molecules and then pinch off from the membrane in a dynamin-dependent manner to form endocytic vesicles [92]. Unlike CME, which depends on the assembly of a clathrin coat and the recruitment of several adaptor proteins, FEME requires only endophilin and dynamin for its activation and progression. The molecular mechanisms underlying FEME are still not fully understood, but studies have suggested that several factors may contribute to its regulation. For example, the level of phosphatidylinositol 4,5-bisphosphate (PIP2), a phospholipid present in the plasma membrane, has been shown to affect FEME activity, possibly through the recruitment of endophilin. In addition, some studies have suggested that other accessory proteins, such as amphiphysin or synaptojanin, may also play a role in FEME by modulating endophilin function or dephosphorylating PIP2 [92].

The physiological and pathological significance of FEME also remains to be fully elucidated, but it has been implicated in various cellular processes, such as receptor trafficking, nutrient uptake, and membrane repair. Additionally, FEME has been shown to play a role in the entry of certain pathogens, such as Chikungunya virus [94] and Shiga toxin [95], leading to the speculation that this pathway may represent a viable drug target for viral infection. However, further research is needed to determine how FEME fits into the complex landscape of endocytic pathways and how it contributes to normal cell biology and disease states.

On the other hand, dynamin-independent CIE pathways, such as CLIC/GEEC endocytosis, do not require dynamin for vesicle formation. CLIC/GEEC endocytosis is characterized by the uptake of extracellular fluid and membrane proteins in flat, non-specialized regions of the plasma membrane. This pathway involves the polymerization of actin filaments, which induce membrane curvature and promote vesicle formation. The molecular mechanism of CLIC/GEEC endocytosis involves the formation of tubular invaginations of the plasma membrane, which subsequently pinch off to form cargo-containing vesicles [80]. This process is thought to be initiated by proteins such as GRAF1, which interact with and deform the plasma membrane [96]. After invagination, the vesicles are transported rapidly to the early endosomal compartment, where they undergo fusion and subsequent maturation. Other examples of dynamin-independent CIE mechanisms include flotillin-dependent endocytosis and Arf6-dependent endocytosis. Flotillin-dependent endocytosis occurs via the formation of lipid microdomains in the plasma membrane [97, 98], whereas Arf6-dependent endocytosis utilizes the small GTPase Arf6 to control the formation of endocytic vesicles and regulate membrane dynamics [99].

Comprehending the underlying molecular mechanisms of clathrin-independent endocytosis is critical in our comprehension of various physiological processes, such as nutrient absorption, intracellular signaling, and immune responses [100, 101]. Moreover, CIE pathways have been linked to multiple disorders, including cancer and viral infections, indicating a promising avenue for treatment intervention. However, it is worth noting that my knowledge base is only updated until April 10th, 2023, and new research may have been conducted since then.

16.3.3 Phagocytosis

Phagocytosis is a special kind of endocytosis, which is a process by which certain cells in the body, such as macrophages and neutrophils [102, 103], engulf and digest harmful particles like bacteria, viruses, and cellular debris. What's more, it is a conservative evolutionary process [104, 105].

The process begins with the recognition and binding of the target particle by receptors on the phagocyte cell surface, followed by the formation of a phagosome around the particle [103, 106]. Phagocytosis acts as a means of obtaining nutrients in Protozoa. During this process, external nutrients are engulfed by phagosomes through phagocytosis and then broken down into smaller molecules by lysosomes to support the organism's life activities [107]. During Metazoa, numerous phagocytosis events serve crucial roles, such as defending against pathogens and maintaining tissue balance. Moreover, phagocytes possess the capacity to identify, engulf, and handle targets exhibiting diverse composition and morphology [108, 109].

Phagocytosis is frequently observed in neutrophils and macrophages. Studies have indicated that macrophages are capable of phagocytosing tissue-damaging materials, such as myelin debris in cases of multiple sclerosis (MS), within the central nervous system (CNS) of the brain [107, 110]. Outer segment phagocytosis (OSP) is a tightly controlled biological mechanism in which the retinal pigment epithelium (RPE) cells adjacent to the photoreceptor outer segments (OS) devour and dispose of their tips in a cyclic pattern. Properly regulated OSP is critical for maintaining healthy photoreceptors and preserving vision. However, alterations in OSP have been linked to age-related macular degeneration (AMD) and inherited retinal diseases (IRD) [111, 112].

16.3.4 Macropinocytosis

Macropinocytosis is also a process by which cells engulf large amounts of extracellular fluid and particles. It involves the formation of membrane protrusions that extend and fuse, creating large vesicles called macropinosomes [113, 114]. Macropinocytosis, which has been found in numerous cell types such as dendritic cells and macrophages, as well as cancer cells [115], is an important branch of endocytosis, and it has developed rapidly in many fields of research.

The role of macropinocytosis differs depending on the type of cell, with its involvement spanning nutrient absorption, antigen display, and removal of pathogens. A majority of our understanding of this process is derived from research done on amoebae, intracellular parasites, immune cells, and multiple forms of cancer [105]. First, it was described by Warren Lewis in the 1930s as a nutrient uptake mechanism in mammalian macrophages. Compared with receptor-mediated forms of endocytosis, macropinocytosis differs in the following ways: types of extracellular material for uptake, signaling cascade, and niche uses between cell types. It has also been previously shown that macropinocytosis has a regulatory function on MHC-II class-restricted antigen presentation in dendritic cells, which is important for DC function [116]. Numerous research studies have revealed that endocytosis impacts the transmission of signals and the transportation of molecular proteins during the onset and progression of diseases. For instance, the molecule UBAP2, discovered by Xiong et al., was found to stimulate macropinocytosis and regulate mutated activated KRAS (GTP-bound RAS) in pancreatic cancer cells [117]. Hence, there is a need for further in-depth research on the mechanism of endocytosis. Macropinocytosis is linked to various aspects of tumorigenesis, such as cell proliferation, growth, survival, and metastasis in cancer, and therefore warrants further investigation [118]. An instance is when pancreatic cancer cells that have been transformed by Ras can enhance their macropinocytosis function to take in and break down albumin, which serves as a crucial source of glutamine. Glutamine is one of the scarcer nutrients in the tumor microenvironment (TME) [119].

16.3.5 Caveolin-Mediated Endocytosis

Caveolae-mediated endocytosis is a process by which cells take up extracellular material via the invagination of specialized membrane regions called caveolae [120]. Caveolae are unique domains of the plasma membrane that are characterized by invaginations and consist of caveolin proteins, GPI-anchored proteins, cholesterol, and glycosphingolipids. There are three varieties of caveolin proteins, namely caveolin 1 (CAV-1), caveolin 2 (CAV-2), and caveolin 3 (CAV-3), all of which share a similar amino acid composition and structure. As critical controllers present in a diverse range of cells, they exhibit universal expression and govern a wide range of physiological processes, including but not limited to endocytosis, transcellular transport, cell proliferation, and signal transduction.

Caveolae-mediated endocytosis generally results from the interaction between ligands and receptors on the cell surface, which triggers the creation and uptake of caveolae [121–123].

Besides CME, cells also employ caveolin-mediated endocytosis to eliminate aging, damaged, or infected cells and pathogens, thereby sustaining cell health and homeostasis, as well as contributing to biological activities like immune responses.

16.4 Analysis of DNA Nanostructures' Cellular Uptake Behaviors

DNA nanostructures have been found to affect cellular uptake, and this is largely due to their size, shape, and surface properties, and this process is also influenced by cell type.

16.4.1 Effect of Size and Shape on Cellular Uptake

Research has explored the impact of DNA nanostructures on cellular absorption, and various elements have been pinpointed to influence their uptake rate. The size of DNA particles is one of the critical factors, as studies have shown that smaller particles are more readily taken up by both cancer and non-cancer cells. This is largely due to the higher surface area-to-volume ratio of small particles, which allows for a greater number of interactions with the cell membrane. Additionally, the shape of DNA NPs is another vital factor influencing cellular uptake. Research has determined that rod-shaped particles are more effectively absorbed by cells than spherical particles, likely because their slender shape can more easily penetrate the cell membrane and enter the cell's cytoplasm.

Wang et al. [124] (Figure 16.6a) construct four DNA-origami nanostructures (DONs) with programmed sizes and shapes, including small tetrahedron(ST), 4 * 2 * 11 per edge; small rod(SR), 4 * 4 * 32; large tripod tetrahedron (LT), 7.2 * 12 * 47 per arm; and large rod (LR), 8 * 8 * 127. They investigate the internalization efficiency of these origami structures with different sizes and shapes across various cell lines. Interestingly, they discovered that the internalization pathways differed depending on the cell type, primarily through scavenger receptor-mediated and caveolin-mediated pathways. They further found that larger rod-shaped structures exhibited higher efficiency of cellular uptake due to their greater interaction with membrane receptors. In their study, Rajwar et al. [125] created various 3D DNA nanostructures with different sizes and geometries to investigate the interactions between cell membranes and differing topologies of DNA. The team constructed tetrahedrons (TD, 14.7 nm), cubes (24.4 nm), icosahedrons (ID, 32.7 nm), and buckyballs (BB, 53.8 nm) and found that DNA structures entered cells via a clathrin-mediated process. Additionally, the TD structure demonstrated a higher efficiency of cellular entry compared to the other geometries designed. A group of researchers led by Zhang et al. [126] (Figure 16.6b) has developed three types of DNA origami structures (triangle, tube, and square) as vehicles for drug delivery and investigated their potential for targeted treatment of tumors. According to their findings, the triangle origami exhibited the best drug delivery efficiency and tumor-targeting performance. Zhao et al. [127] (Figure 16.6c) designed DNA nanostructures exhibiting varying degrees of global distortions to seek optimal drug delivery efficiency. They found that by designing structures with varying degrees of twists, drug release can be adjusted and controlled in a rational manner. Peng et al. [128] conducted a study on the design of frame nucleic acids (FNAs) which involved the creation of three varying shapes, namely tetrahedron, triangular prism,

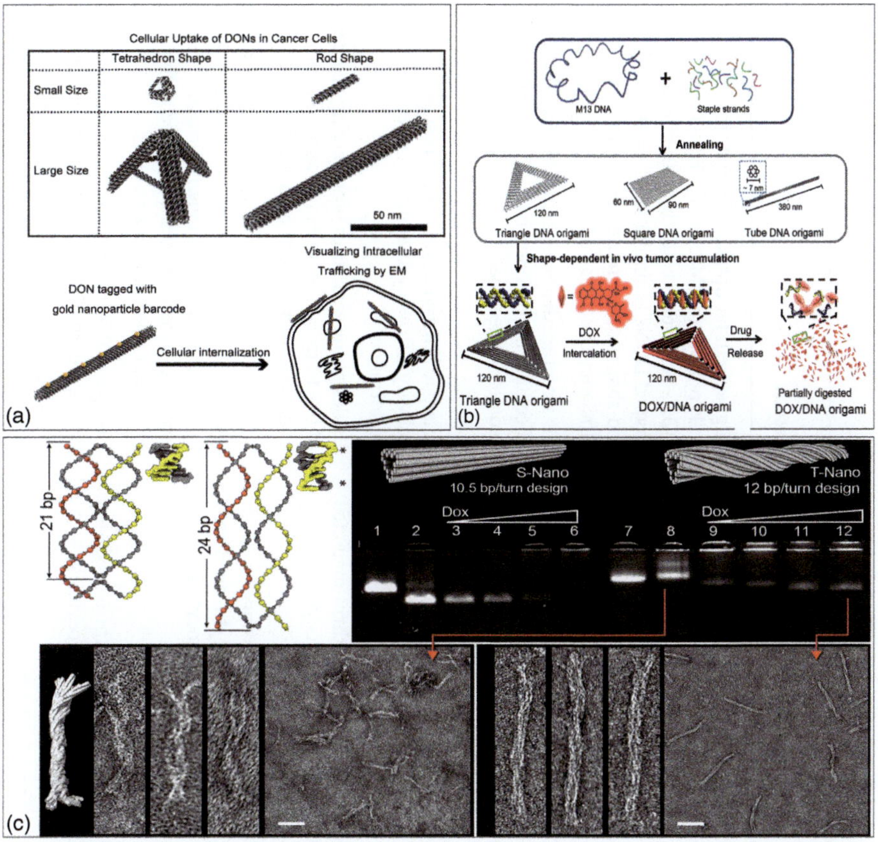

Figure 16.6 Studies on DNA nanostructures with specific sizes and shapes and cellular uptake behaviors. (a) Internalization efficiency of two shapes of DNA nanostructures with different sizes. Source: Reproduced with permission from Martin et al. [122]. © American Chemical Society. (b) Internalization efficiency and tumor accumulation of three different shapes of DNA nanostructures. Source: Reproduced with permission from Wang et al. [124]. © American Chemical Society. (c) Drug encapsulation and internalization efficiency of DNA nanostructures with different levels of twisting. Source: Rajwar et al. [125]/Reproduced with permission from American Chemical Society.

and cube. The researchers discovered that the primary means of internalizing FNAs is through the scavenger receptor-mediated endocytosis pathway and that the efficiency of internalization is primarily influenced by the corner angles of the FNAs. Specifically, the tetrahedron shape exhibited a greater frequency of cell entry as compared to the triangular prism and cube shapes, owing to its narrow corner angle. A research team led by Bastings et al. [129] created a set of 11 unique DNA origami structures with dimensions ranging from 50–400 nm. These structures were labeled with cy5 to investigate their internalization pathway and efficiency in three different cell lines. The team observed that compact DNA nanostructures were more readily internalized than elongated structures with high aspect ratios. The detailed information is summarized in Table 16.1.

Table 16.1 A summary of examples of different DNA nanostructures with different sizes and shapes entering cells by different pathways.

Size and shape				
DNs	**Size, shape**	**Cell type**	**Endocytosis pathway**	**Results**
DNA-origami nanostructures (DON) [124]	Small tetrahedron, small rod, large tripod tetrahedron, large rod	H1299 and DMS53 lung cancer cell lines	Clathrin-mediated endocytosis	Higher cellular uptake efficiency in H1299 cells relative to DMS53 cells. Rod-shaped DONs performed better than tetrahedron-shaped DONs of similar sizes. Larger, and rod-shaped structures exhibited higher efficiency of cellular uptake.
3D DNA nanostructures [125]	Tetrahedrons (TD, 14.7 nm), cubes (24.4 nm), icosahedrons (ID, 32.7 nm), and buckyballs (BB, 53.8 nm)	HeLa, MDA-MB-231, Rpe1, HEK293T, KB3, SUM-159-A and SHSY5Y	Clathrin-mediated endocytosis	TD structure demonstrated a higher efficiency of cellular entry compared to the other geometries designed.
DNA origami structures [126]	Triangle, tube, and square	MDA-MB-231	Clathrin-mediated endocytosis	Triangle origami exhibited the best drug delivery efficiency and tumor-targeting performance.
DNA nanostructures [127]	varying degrees of global distortions	Breast cancer cell lines MCF-7, MDA-MB-231, and MDA-MB-468	Clathrin-mediated endocytosis	By designing structures with varying degrees of twists, drug release can be adjusted and controlled in a rational manner.
Frame Nucleic Acids (FNAs) [128]	Tetrahedron, triangular prism, and cube	HeLa, A549, HL7702, and RAW264.7	Scavenger receptor-mediated endocytosis	The efficiency of internalization is primarily influenced by the corner angles of the FNAs. Tetrahedron shape exhibited a greater frequency of cell entry as compared to the triangular prism and cube shapes.
DNA origami structures [129]	Dimensions ranging from 50–400nm	HUVE cells, BMDCs, and HEK293 cells	Clathrin-mediated endocytosis	Compact DNA nanostructures were more readily internalized than elongated structures with high aspect ratios.

In general, the efficiency of cell entry is greatly influenced by the shape and size of DNA nanostructures. It is widely accepted that DNA nanostructures with a smaller size and more compact structure show higher efficiency in entering cells. This knowledge can guide the design of DNA nanostructures as delivery systems to enhance their delivery efficiency.

16.4.2 Effect of Surface Modifications on Cellular Uptake

DNA surface modifications can have a significant effect on internalization efficiency in different ways, depending on the type of modification. Here are some common modifications:

Charge: DNA molecules typically have a negative charge due to the phosphate backbone, while cell membranes have a negatively charged surface due to sialic acid and heparin sulfate residues. Modification of the DNA surface to be neutral or positively charged can improve the interaction with cell membranes and enhance internalization. For example, positively charged polymers or peptides have been used to coat DNA particles, increasing their uptake efficiency. This is likely due to electrostatic interactions between the positive charges of the coating and the negatively charged cell membrane. A representative method is that the addition of cationic lipids to the surface of DNA molecules improves their uptake by cells compared to unmodified DNA. The positively charged lipids interacted with the negatively charged cell surface, promoting DNA internalization. Cationic lipid–DNA complexes have many applications. For example, it was concentrated on gene therapy purposes to solve drawbacks of viral vectors, such as safety problems. The use of cationic lipid–DNA complexes for gene therapy purposes has been widely and deeply optimized. In addition, adjusting the surface charge of DNA using carboxymethyl chitosan (CMC) significantly improved its internalization in breast cancer cells. The researchers found that the negatively charged DNA bound strongly to the positively charged CMC and formed a stable nanocomplex, which enhanced cellular uptake. Shi et al. [130] provided an effective method to create DNA nanostructures below 100 nm with defined charge patterns and explore whether the density and location of charges affect the cellular uptake efficiency of NPs.

Hydrophobicity: DNA molecules are hydrophilic while the cell membranes are hydrophobic; thus, modification of the DNA surface to be more hydrophobic can increase the interaction with the membrane and therefore enhance internalization. For example, many cationic polymers such as polyethyleneimine (PEI) contain hydrophobic groups such as alkyl chains that can promote the uptake of DNA into cells. This occurs through the formation of complexes between the cationic polymer and the negatively charged DNA, which can then interact with the cell membrane. The ability of PEI and PEI-Arg to promote the uptake of DNA and enhance gene expression within COS-7 cells was evaluated by measuring the percentage of cells transfected and the level of genetic activity per cell [131]. Moreover, cell-penetrating peptides (CPPs) are small, hydrophobic peptides that can complex with DNA to aid in cell uptake. They can enter cells by endocytosis, where they are typically trapped in the endosomes. However, their hydrophobic nature enables them to promote endosomal escape and facilitate DNA internalization. Ye et al. [132] reported that the cooperative effects of CPP-Tat and fusion peptide HA2 enabled successful uptake by cells, endosomal escape, and targeted nuclear entry, resulting in effective delivery of nucleic acids. Furthermore, conjugating DNA with hydrophobic lipids such as cholesterol, alkyl chains, or fatty acids can increase the hydrophobicity of the DNA molecule, enhancing its interaction with the lipid-rich cell membrane. This can improve its internalization efficiency and enable it to bypass endosomal entrapment. Santos et al. [133] developed a novel series of gene delivery vectors that combines a medium-sized generation 5 PAMAM dendrimer (with amine endings) at the core with random peripheral links to hydrophobic chains of varying lengths (12–16 carbon alkyl chains) and quantities (with an average of 4.2–9.7) to capitalize on dendrimers' cationic nature and utilize lipids' affinity for biological membranes. Brglez et al. [134] reported that designed acridine derivatives bearing esterified fatty acids or oligo-ethyleneglycol side chains can be used to alter the surface properties of DNA origami nanostructures.

Specificity: Surface modifications can lead to the binding of DNA molecules to certain receptors on the cell membrane. This specificity can affect the internalization efficiency because if the DNA molecule binds to a receptor that triggers internalization, it can lead to a higher uptake of DNA into the cell. Some DNA sequences feature target motifs that bind to specific receptors or aptamers on the cell surface, leading to their internalization through receptor-mediated endocytosis. Schaffert et al. [135] showed a considerable enhancement in cellular absorption within an established cancer cell line by altering the design of a flat DNA origami structure with transferrin (Tf), which is an iron transport protein. Zhang et al. [136] synthesized fluorophore-conjugated DNA aptamers that target the human Tf receptor (named HG1-9) were used to visualize the internalization, intracellular transportation, and pH environments of aptameric conjugates at the nanoscale. In addition, DNA-binding peptides can be conjugated to DNA molecules, allowing the DNA to penetrate cell membranes and be taken up by cells. In addition, certain DNA strands designed for photochemical internalization (PCI) can be used to treat cancer. PCI makes use of photosensitizing agents (PS) that are lysed by light inside the tumor cell, facilitating DNA release. Nishiyama et al. [137] proposed a three-part structure consisting of DNA, positively charged peptides, and negatively charged phthalocyanine dendrimers with light-reactive properties. Xu et al. [138] developed a helical polypeptide with embedded light-reactive photosensitizer designed for efficient gene delivery, utilizing multivalency and enhanced intracellular uptake while minimizing toxicity through light-strengthened endosomal escape. Udomprasert et al. [139] reported that MUC1 aptamer modification in our nanostructures increased cellular uptake in MUC1-high MCF-7. The more detailed information about DNA nanostructures with surface modifications affecting cell uptake can be seen in Table 16.2.

Overall, the effect of DNA surface modifications on internalization efficiency is determined by a combination of factors, including the type of cell, the type of modification, and the mechanism of internalization. Therefore, it is essential to consider the specific application when choosing a DNA surface modification. Zhai found that

Table 16.2 A summary of examples of different DNA nanostructures with different surface modifications entering cells by different pathways.

	Surface modifications			
DNs	**Surface modification**	**Cell type**	**Endocytosis pathway**	**Results**
DNA complexed with polyethylenimine (PEI) and PEI-polyarginine PTD [131]	Polyethylenimine (PEI) and PEI-polyarginine PTD	COS-7 cells	Clathrin-mediated process	PEI-Arg maintained a high cellular DNA uptake efficiency and facilitated as much as 2-fold more DNA internalization compared to PEI alone.
DNA nanostructure below 100 nM [130]	Photo-responsive monomer	HeLa cells	Clathrin-mediated process	The positive charges located on the edges of nanostructures facilitate more efficient cellular uptake in comparison to the central counterparts.
DNA origami [134]	Acridine derivatized with side chains containing esterified fatty acids or oligo(ethylene glycol) residues	L929 fibroblasts and HeLa cells	Clathrin-mediated process	Intercalator binding does not affect the structural intactness but leads to altered surface properties of the highly negatively charged nanostructures.
Three similar DNA origami nanostructures, disc, donut, and sphere [139]	MUC1 aptamer	MCF-7 and MDA-MB-231 cells.	Clathrin-mediated process	MUC1 aptamer modification in our nanostructures increased cellular uptake in MUC1-high MCF-7.

guanidinylated chitosan (GCS) material created stronger bonds with DNA at normal pH levels than chitosan. Using the COS-7 cell line, the efficacy of transfection for GCS was assessed, and GCS polyplexes demonstrated higher efficacy and less toxicity than chitosan.

The surface properties of DNA NPs can also affect their cellular uptake. For example, positively charged polymers or peptides have been used to coat DNA particles, increasing their uptake efficiency. This is likely due to electrostatic interactions between the positive charges of the coating and the negatively charged cell membrane. Additionally, modifications to the DNA backbone, such as the inclusion of hydrophobic moieties or thiol groups, have also been shown to enhance cellular uptake.

16.4.3 Effect of Other Aspects on Cellular Uptake

Many other factors affect the efficiency of DNA nanostructures entry into cells, including cholesterol trimming, choice of loading cargo, etc. The following is a demonstration of the relevant research.

Abbas et al. [140] constructed deoxyribonucleic acid-nanothread (CPT-DNA-NT) with a diameter of 50–150 nm and a length of 300–600 nm to carry cisplatin into Hela cells, and this type of DNs enter cell through the CME pathway because DNA-NT as polyanionic ligand activates SRs on the cell surface. Arulkumaran et al. [141] designed DNA nanobarrels composed of six DNA duplexes with a size of 9 * 5 * 5 nm containing 3,1, or 0 cholesterol anchors to selectively bind to immune cells. The DNs binding 400-fold selectively via membrane anchors to white blood cells were used as vaccines to play a role in medicine. The white blood cells have been shown to induce micropinocytosis in response to external substances. Jones et al. [142] also explored whether equipping DNA with different hydrophobic anchors affected the targeting ability of DNs to cell membrane. They found that different surface modifications are subject to different conditions, which wound also guide the design of membrane-targeted DNA nanostructures. Jorge et al. [143] constructed DNA tetrahedron (Td) incorporating 5-fluoro-2′-deoxyuridine (FdUn) oligomers to enhance the ability to trigger apoptosis in colorectal cancer cells with cholesterol as an internalization helper that enters cells through caveolin-mediated endocytosis.

Liang et al. [144] used four 55-base ssDNA strands to construct a tetrahedral DNA nanostructure (TDN) with each vertex labeled with cyanine-3 (cy3), and they explored the mechanism of entry of the DNs into cells. With the help of confocal imaging, they found that TDNs were internalized by a caveolin-dependent pathway. Jorge et al. [143] constructed rectangle DNA origami that were used as the carrier of 5-fluoro-2′-deoxyuridine (FdUn) oligomers and they found that these DNs show enhanced cytotoxicity and a higher ability to trigger apoptosis in cells. This DNA origami enters the cells usually via the caveolin-mediated endocytosis. Zhao et al. [145] designed a novel protein drug delivery platform which used the rectangular DNA origami as the carrier of cytotoxic protein ribonuclease (RNase) to deliver RNase into the cytoplasm and kill the tumor cells. This DNA origami–protein complex with the size of 100 * 60 nm is mainly through CME into cells to ship treatment efficacy. Liu et al. [146] constructed a DNA origami nanobox with a size of 36 * 36 * 42 nm and functionalized it with AS1411 aptamers and doxorubicin to understand the effects of surface density and distribution of ligands on the cell surface to achieve optimal cellular regulation. Leung et al. [147] constructed a DNA-based nanostructure called ChloropHore, which was composed of a 61-base pair DNA duplex containing a Cl-reporter domain – Clensor and a pH reporter domain (I-switch). The structure is internalized by the scavenger receptor-mediated endocytic pathway and can be used as a monitor for lysosomal diseases thanks to the special domain. For more detailed information on these examples, please refer to Table 16.3.

Cellular uptake of DNA nanostructures is mainly achieved through the process of endocytosis mediated by the CME, yet there is evidence indicating that other endocytic pathways may also play important roles (Figure 16.7). The internalization of exogenous materials by cells is the result of several endocytic pathways working together, and this is a complex process that requires further research to uncover its underlying principles.

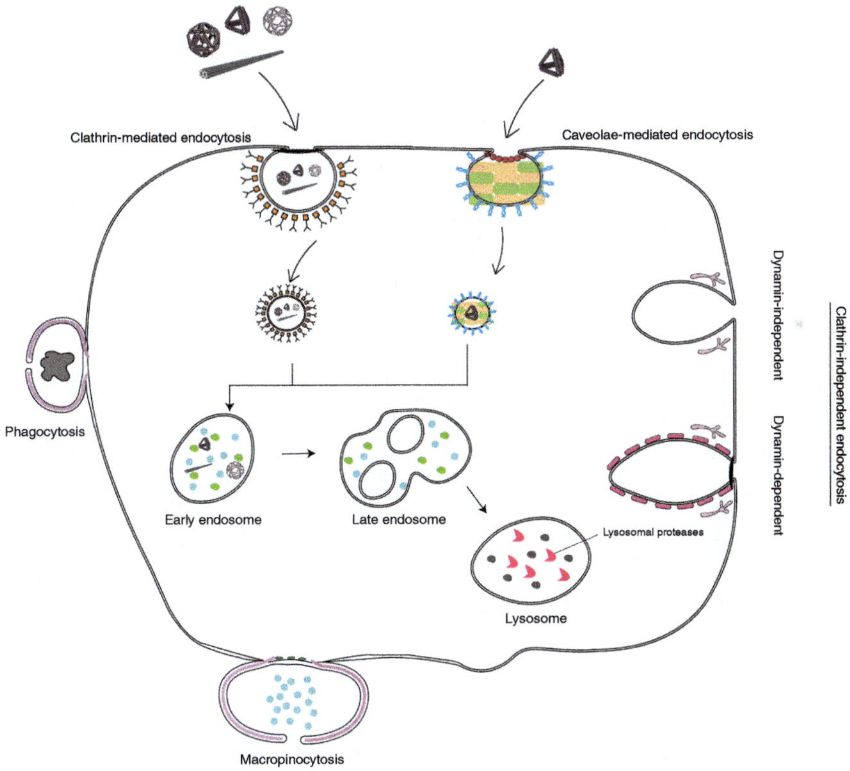

Figure 16.7 A summary on the pathways through which DNA nanostructures enter cells. Most DNA nanostructures are taken up by cells through clathrin-mediated endocytosis, yet there is also evidence suggesting that the cellular uptake of DNA nanostructures may be mediated by multiple endocytic pathways, which is the result of the combined action of several different pathways.

Table 16.3 A summary of examples of different DNA nanostructures with other aspects entering cells by different pathways.

	Other aspects			
DNs	**Other aspects**	**Cell type**	**Endocytosis pathway**	**Results**
DNA nanobarrels [141]	Six DNA duplexes with a size of 9 * 5 * 5 nm containing 3,1, or 0 cholesterol anchors	White Blood Cell	Clathrin-mediated process	The DNs bind with a 400-fold selectively via membrane anchors to white blood cells.
Tetrahedral DNA nanostructure (TDN) [144]	Four 55-base ssDNA strands	HeLa cells and COS-7 cells	Caveolin-dependent pathway	Explore the mechanism of entry of the DNs into cells.
Rectangle DN origami [143]	The carrier of 5-fluoro-2′-deoxyuridine (FdUn) oligomers	HTB-38 and HCC2998 cells	Caveolin-mediated endocytosis	Enhanced cytotoxicity and higher ability to trigger apoptosis in cells.
DNA duplex with modifications [142]	Different hydrophobic anchors (e.g. cholesterol terminal, ethane or hexane terminal and center)	HeLa cells	Clathrin-mediated process	Different surface modifications are subject to different conditions, which will also guide the design of membrane-targeted DNA nanostructures.
DNA origami–protein complex [145]	100 * 60 nm	MCF-7	Clathrin-mediated endocytosis	Deliver RNase into the cytoplasm and kill the tumor cells.
DNA origami nanobox [146]	36 * 36 * 42 nm and functionalize it with AS1411 aptamers and doxorubicin.	MCF-7	Clathrin-mediated endocytosis	Understand the effects of surface density and distribution of ligands on the cell surface to achieve optimal cellular regulation.

In summary, DNA nanostructures can have a significant impact on cellular uptake, and their size, shape, and surface properties all play important roles in this process. Further studies are needed to fully understand the complex mechanisms underlying the uptake of DNA nanomaterials, which could pave the way for more efficient DNA-based delivery systems for biomedical applications.

References

1 Liu, M., Ma, W., Li, Q. et al. (2019). *Cell Proliferation* 52: e12511.
2 Xu, R., Li, Y., Zhu, C. et al. (2023). *Adv. NanoBiomed. Res.* 3: 2200119.
3 Yang, H., Duan, Z., Liu, F. et al. (2023). *ACS Macro Lett.* 12: 295–301.
4 Zhang, H., Zhang, H., Demirer, G.S. et al. (2020). *Nat. Protoc.* 15: 3064–3087.
5 Liu, Q., Wu, D., He, B. et al. (2023). *Attenuating endothelial leakiness with self-assembled DNA nanostructures for pulmonary arterial hypertension. Nanoscale Horizons* 8: 270–278. https://doi.org/10.1039/d2nh00348a.
6 Saccà, B., Meyer, R., Erkelenz, M. et al. (2010). *Angew. Chem. Int. Ed.* 49: 9378–9383.
7 Alvisi, N. and de Vries, R. (2023). *Mater. Today Bio* 19: 100580.
8 Liu, Q., Zhan, C., and Kohane, D.S. (2017). *Bioconjugate Chem.* 28: 98–104.
9 Hirschbiegel, C.-M., Zhang, X., Huang, R. et al. (2023). *Adv. Drug Delivery Rev.* 195: 114730.
10 Zhao, X., Xu, S., Jiang, Y. et al. (2023). *Chem. Eng. J.* 454: 140440.
11 Kumar, V.B., Ozguney, B., Vlachou, A. et al. (2023). *J. Phys. Chem. B* 127: 1857–1871.
12 Rokaya, D., Skallevold, H.E., Srimaneepong, V. et al. (2023). *Shape memory polymeric materials for biomedical applications: an update. J. Compos. Sci.* 7. https://doi.org/10.3390/jcs7010024.
13 Kotsuchibashi, Y. (2020). *Polym. J.* 52: 681–689.
14 Huang, S., Kong, X., Xiong, Y. et al. (2020). *Eur. Polym. J.* 141: 110094.
15 Chang, C., Guo, W., Yu, X. et al. (2023). *Mater. Today Bio* 20: 100612.
16 Harisa, G.I. and Alanazi, F.K. (2014). *Saudi Pharm. J.* 22: 504–515.
17 Liu, Y., Wang, Y., Zhuang, D. et al. (2012). *J. Colloid Interface Sci.* 377: 197–206.
18 Magrez, A., Kasas, S., Salicio, V. et al. (2006). *Nano Lett.* 6: 1121–1125.
19 De Jong, W.H., Hagens, W.I., Krystek, P. et al. (2008). *Biomaterials* 29: 1912–1919.
20 Keles, E., Song, Y., Du, D. et al. (2016). *Biomater. Sci.* 4: 1291–1309.
21 Hu, Q., Li, H., Wang, L. et al. (2019). *Chem. Rev.* 119: 6459–6506.
22 Acuna, G.P., Möller, F.M., Holzmeister, P. et al. (2012). *Science* 338: 506–510.
23 Angelin, A., Weigel, S., Garrecht, R. et al. (2015). *Angew. Chem. Int. Ed.* 54: 15813–15817.
24 Elbaz, J., Lioubashevski, O., Wang, F. et al. (2010). *Nat. Nanotechnol.* 5: 417–422.

25 Marras, A.E., Zhou, L., Su, H.J., and Castro, C.E. (2015). *Proc. Natl. Acad. Sci. U. S. A.* 112: 713–718.

26 Heuer-Jungemann, A. and Linko, V. (2021). *ACS Cent. Sci.* 7: 1969–1979.

27 Dawson, K.A. and Yan, Y. (2021). *Nat. Nanotechnol.* 16: 229–242.

28 Hao, Y., Li, Q., Fan, C., and Wang, F. (2021). *Small Struct.* 2: 2000046.

29 Ren, Y., Zhang, Y., Liu, Y. et al. (2022). *Small Methods* 6: 2101335.

30 Wang, J., Li, Y., and Nie, G. (2021). *Nat. Rev. Mater.* 6: 766–783.

31 Mitchell, M.J., Billingsley, M.M., Haley, R.M. et al. (2021). *Nat. Rev. Drug Discovery* 20: 101–124.

32 Pei, H., Zuo, X., Zhu, D. et al. (2014). *Acc. Chem. Res.* 47: 550–559.

33 Pei, H., Zuo, X., Pan, D. et al. (2013). *NPG Asia Mater.* 5: e51–e51.

34 Huang, R., He, N., and Li, Z. (2018). *Biosens. Bioelectron.* 109: 27–34.

35 Raveendran, M., Lee, A.J., Sharma, R. et al. (2020). *Nat. Commun.* 11: 4384.

36 Yang, Q., Chang, X., Lee, J.Y. et al. (2022). *ACS Appl. Bio Mater.* 5: 4652–4667.

37 Zeng, Z., Zhou, R., Sun, R. et al. (2021). *Biosens. Bioelectron.* 173: 112814.

38 Chen, T., Ren, L., Liu, X. et al. (2018). DNA nanotechnology for cancer diagnosis and therapy. *Int. J. Mol. Sci.* 19. https://doi.org/10.3390/ijms19061671.

39 Zhang, F., Nangreave, J., Liu, Y., and Yan, H. (2014). *J. Am. Chem. Soc.* 136: 11198–11211.

40 Linko, V., Ora, A., and Kostiainen, M.A. (2015). *Trends Biotechnol.* 33: 586–594.

41 Li, J., Fan, C., Pei, H. et al. (2013). *Adv. Mater.* 25: 4386–4396.

42 Seeman, N.C. (1982). *J. Theor. Biol.* 99: 237–247.

43 Winfree, E., Liu, F., Wenzler, L.A., and Seeman, N.C. (1998). *Nature* 394: 539–544.

44 Yan, H., Park, S.H., Finkelstein, G. et al. (2003). *Science* 301: 1882–1884.

45 He, Y., Chen, Y., Liu, H. et al. (2005). *J. Am. Chem. Soc.* 127: 12202–12203.

46 He, Y., Tian, Y., Ribbe, A.E., and Mao, C. (2006). *J. Am. Chem. Soc.* 128: 15978–15979.

47 Goodman, R.P., Schaap, I.A.T., Tardin, C.F. et al. (2005). *Science* 310: 1661–1665.

48 He, Y., Ye, T., Su, M. et al. (2008). *Nature* 452: 198–201.

49 Chen, J. and Seeman, N.C. (1991). *Nature* 350: 631–633.

50 Zheng, J., Birktoft, J.J., Chen, Y. et al. (2009). *Nature* 461: 74–77.

51 Rothemund, P.W. (2006). *Nature* 440: 297–302.

52 Andersen, E.S., Dong, M., Nielsen, M.M. et al. (2009). *Nature* 459: 73–76.

53 Douglas, S.M., Dietz, H., Liedl, T. et al. (2009). *Nature* 459: 414–418.

54 Han, D., Pal, S., Nangreave, J. et al. (2011). *Science* 332: 342–346.

55 Dietz, H., Douglas, S.M., and Shih, W.M. (2009). *Science* 325: 725–730.

56 Douglas, S.M., Marblestone, A.H., Teerapittayanon, S. et al. (2009). *Nucleic Acids Res.* 37: 5001–5006.

57 Benson, E., Mohammed, A., Gardell, J. et al. (2015). *Nature* 523: 441–444.

58 Yao, G., Zhang, F., Wang, F. et al. (2020). *Nat. Chem.* 12: 1067–1075.

59 Zhang, F., Jiang, S., Wu, S. et al. (2015). *Nat. Nanotechnol.* 10: 779–784.

60 Veneziano, R., Ratanalert, S., Zhang, K. et al. (2016). *Science* 352: 1534.

61 Wei, B., Dai, M., and Yin, P. (2012). *Nature* 485: 623–626.

62 Ke, Y., Ong, L.L., Shih, W.M., and Yin, P. (2012). *Science* 338: 1177–1183.

63 Ong, L.L., Hanikel, N., Yaghi, O.K. et al. (2017). *Nature* 552: 72–77.

64 Zhang, Y., Chen, X., Kang, G. et al. (2019). *J. Am. Chem. Soc.* 141: 19529–19532.

65 Wang, W., Chen, S., An, B. et al. (2019). *Nat. Commun.* 10: 1067.

66 Douglas, S.M., Bachelet, I., and Church, G.M. (2012). *Science* 335: 831–834.

67 Gerling, T., Wagenbauer, K.F., Neuner, A.M., and Dietz, H. (2015). *Science* 347: 1446–1452.

68 Ketterer, P., Willner, E.M., and Dietz, H. (2016). *Sci. Adv.* 2: e1501209.

69 Green, L.N., Subramanian, H.K.K., Mardanlou, V. et al. (2019). *Nat. Chem.* 11: 510–520.

70 Yang, D. and Wang, P. (2021). *ACS Appl. Mater. Interfaces* 13: 41004–41011.

71 Li, S., Jiang, Q., Liu, S. et al. (2018). *Nat. Biotechnol.* 36: 258–264.

72 Kuzyk, A., Yang, Y., Duan, X. et al. (2016). *Nat. Commun.* 7: 10591.

73 Yang, Y., Hong, Y., Nam, G.-H. et al. (2017). *Adv. Mater.* 29: 1605604.

74 Campadelli-Fiume, G., Menotti, L., Avitabile, E., and Gianni, T. (2012). *Curr. Opin. Virol.* 2: 28–36.

75 Verma, A. and Stellacci, F. (2010). *Small* 6: 12–21.

76 Donahue, N.D., Acar, H., and Wilhelm, S. (2019). *Adv. Drug Delivery Rev.* 143: 68–96.

77 Haley, B. and Frenkel, E. (2008) Nanoparticles for drug delivery in cancer treatment, https://doi.org/10.1016/j.urolonc.2007.03.015 *Urol. Oncol. Semin. Original Invest.* 26, 57-64.

78 Zhao, J. and Stenzel, M.H. (2018). *Polym. Chem.* 9: 259–272.

79 Gessner, I. and Neundorf, I. (2020). Nanoparticles modified with cell-penetrating peptides: conjugation mechanisms, physicochemical properties, and application in cancer diagnosis and therapy. *Int. J. Mol. Sci.* 21. https://doi.org/10.3390/ijms21072536.

80 Rennick, J.J., Johnston, A.P.R., and Parton, R.G. (2021). *Nat. Nanotechnol.* 16: 266–276.

81 Kaksonen, M. and Roux, A. (2018). *Nat. Rev. Mol. Cell Biol.* 19: 313–326.

82 Wideman, J.G., Leung, K.F., Field, M.C., and Dacks, J.B. (2014). *Cold Spring Harbor Perspect. Biol.* 6: a016998.

83 Ehrlich, M., Boll, W., Van Oijen, A. et al. (2004). *Cell* 118: 591–605.

84 Veiga, E., Guttman, J.A., Bonazzi, M. et al. (2007). *Cell Host Microbe* 2: 340–351.

85 Li, Z., Sun, L., Zhang, Y. et al. (2016). *ACS Macro Lett.* 5: 1059–1064.

86 Dijkstra, A.A., Ingrassia, A., de Menezes, R.X. et al. (2015). *PLoS One* 10: e0128651.

87 Kett, L.R. and Dauer, W.T. (2016). *Mov. Disord.* 31: 1433–1443.

88 Schubert, K.O., Föcking, M., Prehn, J.H.M., and Cotter, D.R. (2012). *Mol. Psychiatry* 17: 669–681.

89 Mayor, S. and Pagano, R.E. (2007). *Nat. Rev. Mol. Cell Biol.* 8: 603–612.

90 Sandvig, K., Kavaliauskiene, S., and Skotland, T. (2018). *Histochem. Cell Biol.* 150: 107–118.

91 Lamaze, C., Dujeancourt, A., Baba, T. et al. (2001). *Mol. Cell.* 7: 661–671.

92 Boucrot, E., Ferreira, A.P.A., Almeida-Souza, L. et al. (2015). *Nature* 517: 460–465.

93 Hohendahl, A., Talledge, N., Galli, V. et al. (2017). *Elife* 6: e26856.

94 Bernard, E., Solignat, M., Gay, B. et al. (2010). *PLoS One* 5: e11479.

95 Sandvig, K., Bergan, J., Kavaliauskiene, S., and Skotland, T. (2014). *Prog. Lipid Res.* 54: 1–13.

96 Sathe, M., Muthukrishnan, G., Rae, J. et al. (2018). *Nat. Commun.* 9: 1835.

97 Asp, N., Pust, S., and Sandvig, K. (2014). *Biochim. Biophys. Acta* 1843: 1987–1996.

98 Meister, M. and Tikkanen, R. (2014). *Membranes (Basel)* 4: 356–371.

99 Grant, B.D. and Donaldson, J.G. (2009). *Nat. Rev. Mol. Cell Biol.* 10: 597–608.

100 Posor, Y., Eichhorn-Grünig, M., and Haucke, V. (2015). *Biochim. Biophys. Acta, Mol. Cell Biol. Lipids* 1851: 794–804.

101 Howes, M.T., Mayor, S., and Parton, R.G. (2010). *Curr. Opin. Cell Biol.* 22: 519–527.

102 Niedergang, F. and Grinstein, S. (2018). *Curr. Opin. Cell Biol.* 50: 57–63.

103 Lim, J.J., Grinstein, S., and Roth, Z. (2017). *Front. Cell Infect. Microbiol.* 7: 191.

104 Flannagan, R.S., Jaumouillé, V., and Grinstein, S. (2012). *Ann. Rev. Pathol. Mech. Dis.* 7: 61–98.

105 Mulcahy, L.A., Pink, R.C., and Carter, D.R. (2014). *J. Extracell. Vesicles* 3.

106 Richards, D.M. and Endres, R.G. (2014). *Biophys. J.* 107: 1542–1553.

107 Lancaster, C.E., Ho, C.Y., Hipolito, V.E.B. et al. (2019). *Biochem. Cell Biol.* 97: 21–29.

108 Wolf, M.T., Wu, X.R., and Huang, C.L. (2013). *Kidney Int.* 84: 130–137.

109 Fang, G., Cao, W., Chen, L. et al. (2021). *J. Cell. Physiol.* 236: 5725–5741.

110 Zhou, T., Li, Y., Li, X. et al. (2022). *Nat. Commun.* 13: 6233.

111 Moran, A.L., Fehilly, J.D., Floss Jones, D. et al. (2022). *FASEB J.* 36: e22556.

112 Kevany, B.M. and Palczewski, K. (2010). *Physiology* 25: 8–15.

113 King, J.S. and Kay, R.R. (2018). *Phil. Trans. R. Soc. B* 374: 20180158.

114 Condon, N.D., Heddleston, J.M., Chew, T.L. et al. (2018). *J. Cell Biol.* 217: 3873–3885.

115 Kerr, M.C. and Teasdale, R.D. (2009). *Traffic* 10: 364–371.

116 Liu, Z. and Roche, P.A. (2015). *Front. Physiol.* 6.

117 Means, N., Elechalawar, C.K., Chen, W.R. et al. (2022). *Mol. Aspects Med.* 83: 100993.

118 Mack, N.A., Whalley, H.J., Castillo-Lluva, S., and Malliri, A. (2011). *Cell Cycle* 10: 1571–1581.

119 Bohdanowicz, M., Schlam, D., Hermansson, M. et al. (2013). *Mol. Biol. Cell* 24 (1700–1712): S1701–S1707.

120 Sui, Z.H., Xu, H., Wang, H. et al. (2017). *Front. Cell Infect. Microbiol.* 7: 400.

121 Razzak, M. (2014). *Nat. Rev. Nephrol.* 10: 242.

122 Martin, W.F., Tielens, A.G.M., Mentel, M. et al. (2017). *Microbiol. Mol. Biol. Rev.* 81.

123 Copp, W., Pontarelli, A., and Wilds, C.J. (2021). *ChemBioChem* 22: 2237–2246.

124 Wang, P., Rahman, M.A., Zhao, Z. et al. (2018). *J. Am. Chem. Soc.* 140: 2478–2484.
125 Rajwar, A., Shetty, S.R., Vaswani, P. et al. (2022). *ACS Nano* 16: 10496–10508.
126 Zhang, Q., Jiang, Q., Li, N. et al. (2014). *ACS Nano* 8: 6633–6643.
127 Zhao, Y.-X., Shaw, A., Zeng, X. et al. (2012). *ACS Nano* 6: 8684–8691.
128 Peng, X., Fang, S., Ji, B. et al. (2021). *Nano Lett.* 21: 6946–6951.
129 Bastings, M.M.C., Anastassacos, F.M., Ponnuswamy, N. et al. (2018). *Nano Lett.* 18: 3557–3564.
130 Shi, Y., Xu, X., Yu, H. et al. (2022). *Nano Lett.* 22: 5330–5338.
131 Doyle, S.R. and Chan, C.K. (2007). *Genet. Vaccines Ther.* 5: 1–12.
132 Ye, S.-f., Tian, M.-m., Wang, T.-x. et al. (2012). *Nanomed. Nanotechnol. Biol. Med.* 8: 833–841.
133 Santos, J.L., Oliveira, H., Pandita, D. et al. (2010). *J. Controlled Release* 144: 55–64.
134 Brglez, J., Nikolov, P., Angelin, A., and Niemeyer, C.M. (2015). *Chem. Eur. J.* 21: 9440–9446.
135 Schaffert, D.H., Okholm, A.H., Sørensen, R.S. et al. (2016). *Small* 12: 2634–2640.
136 Zhang, N., Wang, J., Bing, T. et al. (2022). *Mol. Ther. Nucleic Acids* 27: 1249–1259.
137 Nishiyama, N., Iriyama, A., Jang, W.-D. et al. (2005). *Nat. Mater.* 4: 934–941.
138 Xu, X., Li, Y., Liang, Q. et al. (2018). *ACS Appl. Mater. Interfaces* 10: 256–266.
139 Udomprasert, A., Wootthichairangsan, C., Duangrat, R. et al. (2022). *ACS Appl. Bio Mater.* 5: 2262–2272.
140 Abbas, M., Baig, M., Zhang, Y. et al. (2021). *J. Pharm. Anal.* 11: 330–339.
141 Arulkumaran, N., Lanphere, C., Gaupp, C. et al. (2021). *ACS Nano* 15: 4394–4404.
142 Jones, S.F., Joshi, H., Terry, S.J. et al. (2021). *J. Am. Chem. Soc.* 143: 8305–8313.
143 Jorge, A.F., Aviñó, A., Pais, A.A.C.C. et al. (2018). *Nanoscale* 10: 7238–7249.
144 Liang, L., Li, J., Li, Q. et al. (2014). *Angew. Chem. Int. Ed.* 53: 7745–7750.
145 Zhao, S., Duan, F., Liu, S. et al. (2019). *ACS Appl. Mater. Interfaces* 11: 11112–11118.
146 Liu, K., Xu, C., and Liu, J. (2020). *J. Mater. Chem. B* 8: 6802–6809.
147 Leung, K., Chakraborty, K., Saminathan, A., and Krishnan, Y. (2019). *Nat. Nanotechnol.* 14: 176–183.

Part IV

DNA Nanotechnology for Cell-Targeted Medical Applications

17

Toward Production of Nucleic Acid Nanostructures in Life Cells and Their Biomedical Applications

Mengxi Zheng, Victoria E. Paluzzi, Cuizheng Zhang, and Chengde Mao

Purdue University, Department of Chemistry, West Lafayette, IN 47907, USA

17.1 DNA Nanostructures

Nucleic acids, including DNA and RNA molecules, have been widely used as materials for building artificial structures in recent years. In 1982, Ned C. Seeman first proposed the construction of DNA arrays using immobilized Holliday junctions with sticky-end cohesion [1]. Since then, DNA nanotechnology has rapidly developed, and DNA nanostructures have been applied in many fields, including clinical medicine, biomaterials, cancer diagnosis, and treatment [2].

DNA nanotechnology relies on the base-pairing between complementary sequences. These two complementary sequences will hybridize to form a DNA duplex following Watson–Crick rules, where adenine associates with thymine by two hydrogen bonds, while cytosine and guanine form three hydrogen bonds. When they hybridize, they will form a right-handed double helix, where each base pair rotates about 34.3° with a rise of 3.32 Å along the axis. The resulting double helix has 10.5 base pairs in each turn, and the length of one turn is approximately 3.4 nm. These properties make DNA a unique material for building nanostructures with high predictability and unparalleled programmability.

17.1.1 Strategies of DNA Nanostructures Construction

Over the past 20 years, many approaches for building DNA nanostructures have been demonstrated, including the DNA tile-based strategy, DNA origami strategy, single-stranded tile strategy, and single-stranded DNA (ssDNA) folding strategy (Figure 17.1).

All these strategies have their own advantages and disadvantages. The tile-based DNA self-assembly uses fewer DNA strands to first form a single unit, called a motif. Thousands of motifs are associated with each other by sticky-end cohesion to form uniform patterns. This approach is cost-effective but can only lead to high-symmetric, repeating patterns (Figure 17.1a) [3]. The DNA origami approach generates well-defined DNA nanostructures with arbitrary shapes, but it is costly

DNA Nanotechnology for Cell Research: From Bioanalysis to Biomedicine, First Edition. Edited by Zhou Nie.

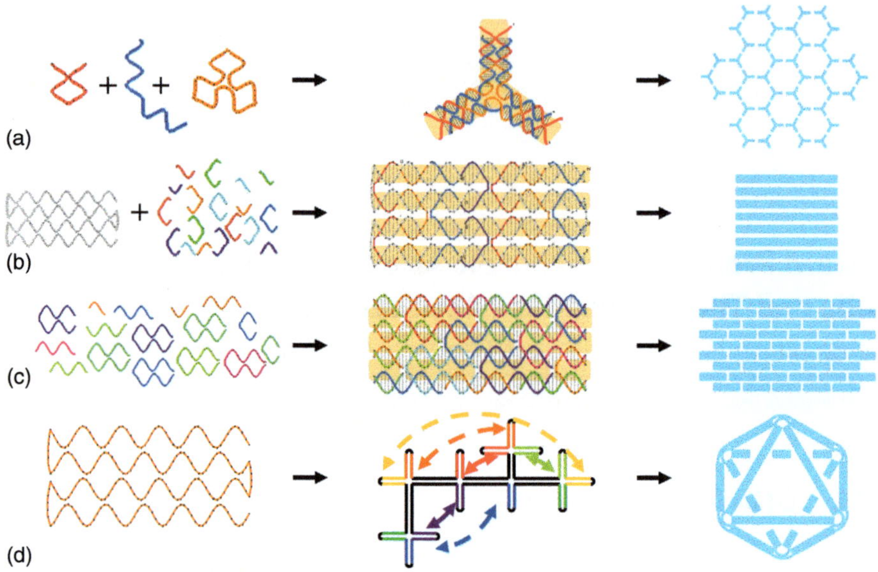

Figure 17.1 Strategies to construct DNA nanostructures. (a) Tile-based self-assembly. (b) DNA origami. (c) DNA single-stranded tile self-assembly. (d) Single-stranded DNA folding.

and has low productivity (Figure 17.1b) [4]. Similarly, DNA single-stranded tile self-assembly can also be used to synthesize arbitrary-shaped 2D and 3D DNA nanostructures, but it is not cost-effective (Figure 17.1c) [5]. The single-stranded DNA folding strategy uses one long single DNA strand folded into designed structures. With the advantages of easy amplification, fewer structural defects, and simple preparation, single-stranded DNA nanostructures still cannot reach the same level of complexity as origami structures (Figure 17.1d) [6].

17.1.2 Production of ssDNA Nanostructures in Living Cells

As mentioned, one of the most remarkable advantages of single-component DNA nanostructures compared to tile-based DNA nanostructures is that they are easy to amplify. In 2004, Joyce et al. constructed a DNA octahedron structure via ssDNA folding [6]. The production of the 1669-nt DNA single strand was achieved by constructing a recombinant plasmid containing the 1669-nt DNA sequences, followed by cloning in *Escherichia coli* (*E. coli*) (Figure 17.1d). The amplified plasmids in double-strand formation were then digested by blunt-end endonuclease *EcoR*V and nicking endonuclease *N.BbvC*I. To further purify the digested product, alkaline agarose gel electrophoresis was used to separate the target DNA fragments and their complementary parts.

Although the yield of duplicating plasmids is high, the subsequent step of converting double-stranded DNA into single strands is tedious and expensive. To overcome this issue, Yan et al. demonstrated an approach in 2008 to replicate ssDNA using a helper phage, as demonstrated in Figure 17.2a [8]. In this strategy, DNA structure

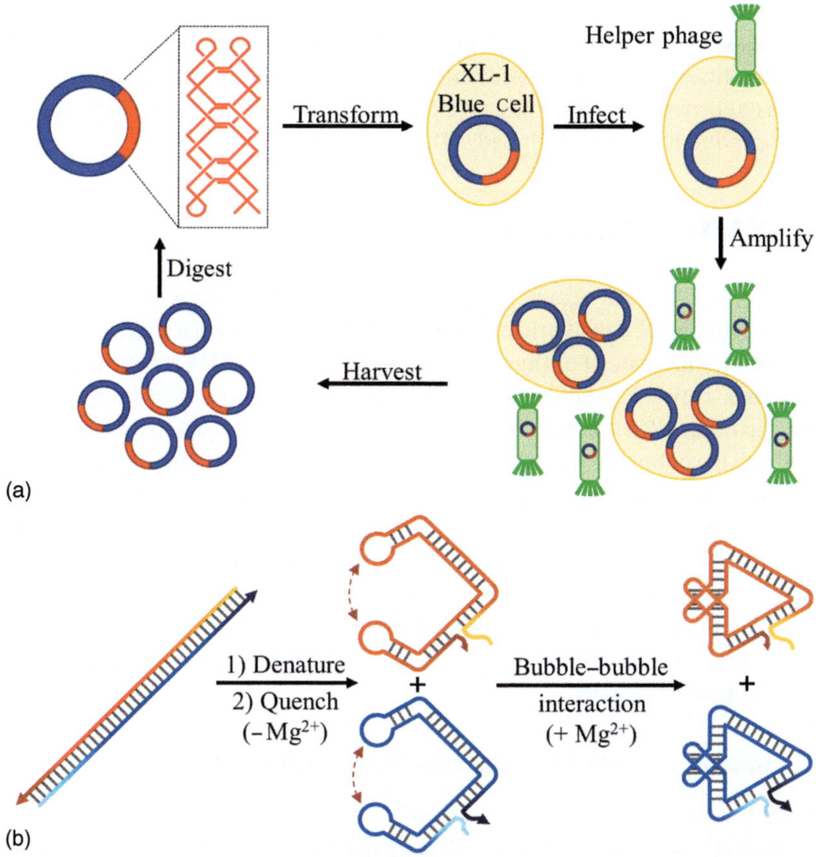

Figure 17.2 Single-stranded DNA nanostructures and the amplification approaches. (a) Schematic drawing showing the *in vivo* replication of a DNA nanostructure using helper phage. (b) Heating-quench method generates identical structures with both DNA sequences from one duplex. Source: Reproduced with permission from Zheng et al. [7]. © 2012 American Chemical Society.

fragments and M13 fragments were first inserted into a double-stranded plasmid and transferred into *E. coli*. Subsequently, the cells were infected with a helper phage, M13KO7. Since M13KO7 preferentially packages single-stranded phagemids containing the M13 origin, the infected cells were then amplified in the medium. Finally, the phages were harvested, and the ssDNA was extracted and digested. Although mutations might occur during amplification in M13 phages, this method has been successfully applied to construct various DNA nanostructures.

In addition to the solution of replicating ssDNA via a helper phage, Mao et al. proposed a method that utilizes both strands in a double-stranded DNA duplex to construct two individual DNA nanostructures (Figure 17.2b) [7]. According to the design principles, the structures are dominated by kinetically stable hairpin structures. The loops of the hairpins further interact with each other and result in the formation of well-defined nanostructures. Since the designed structures are hairpin-rich and kinetically stable, by thermally denaturing the duplex and rapidly decreasing the temperature, both strands in one duplex can separately form identical structures instead of hybridizing into a thermodynamically stable double strand. This approach is compatible with DNA cloning methods and thus enables large-scale production of the designed ssDNA nanostructures.

17.2 RNA Nanostructures

Similar to DNA, RNA can also be programmed to construct RNA nanostructures. RNA, just as with DNA, can hybridize based on Watson–Crick base-pairing, with the exception being that instead of thymine binding to adenine, RNA has uridine binding to adenine. However, there are more key differences between RNA and DNA. In terms of structure, RNA structures are not only based on the base-pairing between Watson–Crick bases but also on wobble base pairs that exist in nature to form a large number of RNA tertiary structures or even more complex nanostructures. Moreover, RNA duplexes are A-form instead of B-form for DNA duplexes; each base pair rotates about 32.7° with a rise of 2.56 Å along the axis. The resulting double helix has 11 base pairs in each turn; the length of one turn is approximately 2.8 nm, and the diameter is 0.3 nm wider than DNA helices, which is about 2.3 nm. In addition, RNA can carry out versatile biological functions, similar to proteins, which enables the engineering of functional nanostructures.

17.2.1 Strategies of RNA Nanostructures Construction

The assembly of RNA nanostructures usually involves the utilization of naturally occurring tertiary structures of RNA. Designed RNA molecules typically contain hairpin structures, and the loops of these hairpins can further form inter- or intramolecular tertiary structures, leading to the bottom-up assembly of multiple RNA molecules. In 2013, Mao et al. constructed a series of nanocages by reengineering a native biological RNA motif: the packaging RNA of the phi29 bacteriophage [9]. In their design in Figure 17.3a, each RNA tile first associates into a trimer or tetramer through natural RNA interactions, then the polymers further dimerize via sticky-end cohesion to form the desired RNA nanocages.

In 2017, Yin et al. put forward the concept of single-stranded origami (ssOrigami) [10]. Both ssDNA origami and ssRNA origami structures were demonstrated, with lengths between 1868-nt and 6337-nt (Figure 17.3b). The approach involves first forming a partially paired double-stranded intermediate, which then folds into the desired shape. This universal approach pushes the limits of ssDNA and RNA techniques, allowing for the construction of large ssRNA structures longer than 2000-nt. As well, this technique utilized the inherent identities of sense and anti-sense oligos that were transcribed from the same dsDNA to then fold into the same ssOrigami.

In 2020, Weizmann et al. created an artificial RNA motif called the branched kissing loop (bKL) by combining the native packaging RNA motif of phi29 bacteriophage with an RNA duplex (Figure 17.3c) [11]. With the help of the bKL motif,

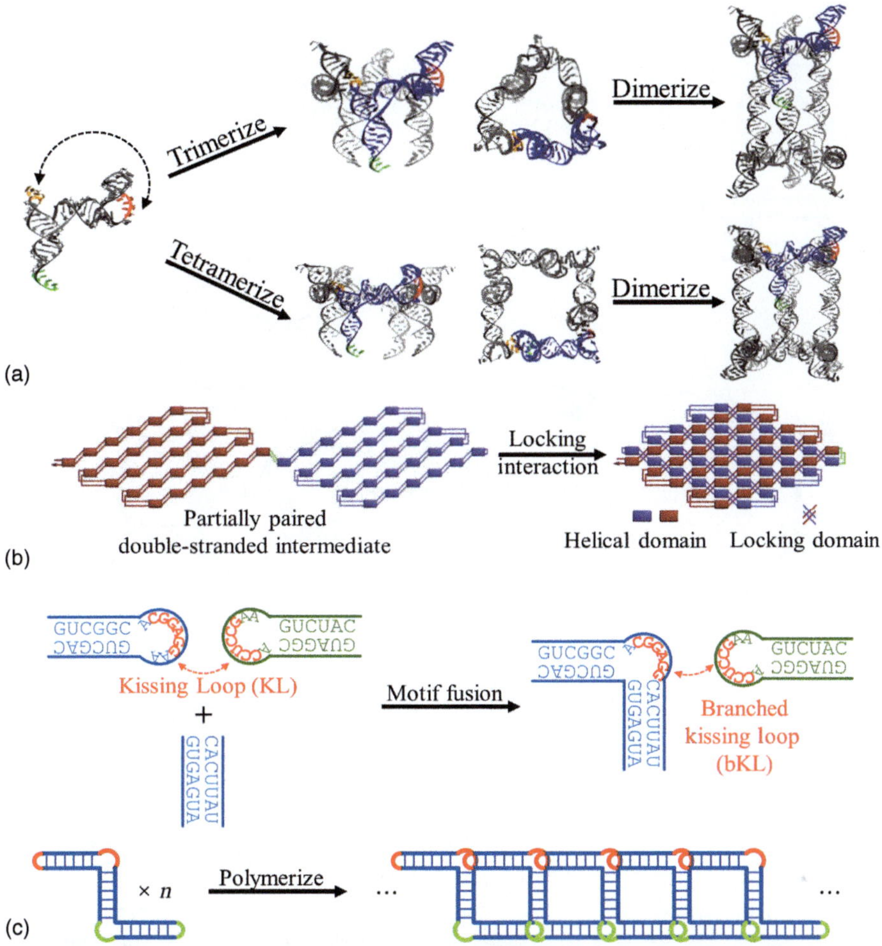

Figure 17.3 The self-assembly of RNA nanostructures. (a) Self-assembly of RNA nanoprisms by re-engineering pRNA. Source: Reproduced with permission from Hao et al. [9]. © 2014 Springer Nature. (b) The scheme of an RNA origami formed by a long single-stranded RNA. Source: Reproduced with permission from Han et al. [10], American Association for the Advancement of Science. (c) The design of a 1D RNA ladder formed by a one-stranded Z-tile containing branched kissing loop interaction.

designed RNA tiles could homo-oligomerize into various 2D and 3D nanostructures, such as tetrameric RNA nanocages, ladders, and honeycomb arrays. The success of the bKL in constructing these complex RNA nanostructures opens up new avenues for creating even more intricate nanostructures using both naturally occurring and derived RNA motifs.

17.2.2 Production of ssRNA Nanostructures in Living Cells

Like DNA architecture, RNA nanostructures can also be achieved by folding a long single-stranded RNA (ssRNA). However, unlike DNA, which typically forms

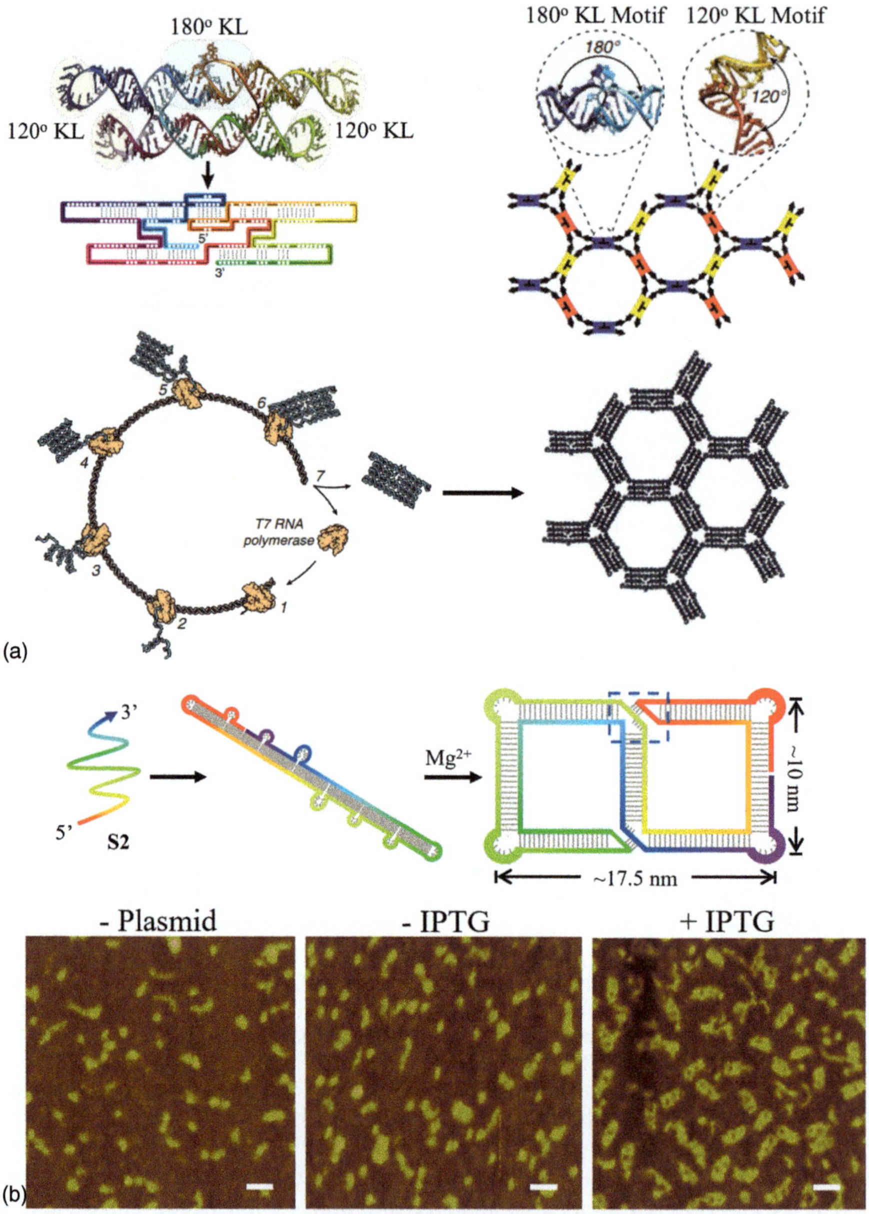

Figure 17.4 The *in vivo* production of RNA nanostructures. (a) The scheme of cotranscriptional folding pathway for the RNA tiles. Source: Reproduced with permission from Geary et al. [12], American Association for the Advancement of Science. (b) The folding, cloning, and *in vivo* expression of RNA nanostructures (Scale bar: 20 nm). Source: Li et al. [13]/Reproduced with permission from Springer Nature/CC BY 4.0.

duplexes in nature, RNA molecules exist as single-stranded formations after being transcribed from a DNA template. Therefore, ssRNA nanostructures have unique advantages over ssDNA, with one predominant advantage being that they are easier to produce than ssDNA. Additionally, it is possible to self-assemble RNA-based nanostructures inside living organisms without the need for extra steps such as enzyme treatment, physical denaturation, and purification that are required for generating ssDNA. As a result, the productivity of ssRNA is also relatively higher than ssDNA.

In 2014, Geary et al. designed a series of ssRNA tiles ranging from 156-nt to 660-nt, containing a pair of intermolecular RNA interactions and a pair of intramolecular cohesions, also known as kissing loops [12]. In Figure 17.4a, the assembled RNA tiles form hexagonal lattices or rectilinear lattices via intermolecular interactions. The arrays could be obtained not only by traditional annealing processes but also cotranscriptional methods. This means that the RNA tiles can spontaneously associate into 2D patterns after transcription, revealing the possibilities of *in vivo* self-assembly of RNA nanodevices.

In 2018, Mao et al. developed a ssRNA folding strategy that enables the formation of RNA nanostructures without getting kinetically trapped [13]. Among several RNA structures that were demonstrated, a double-squared structure was successfully cloned and expressed inside cells, similar to the generation of recombinant proteins (Figure 17.4b). Upon lysing the cells and examining the native solution, a large number of well-shaped nanostructures were observed by AFM imaging. This approach shows the possibility of large-scale *in vivo* production of RNA nanostructures. Furthermore, it facilitates the study of *in vivo* biomedical applications of RNA nanotechnology, such as RNA machines that can be expressed to regulate cellular processes in a living organism.

17.3 Applications

The field of nucleic acid-based architecture has been well-established for decades, with thousands of DNA/RNA nanostructures ranging from nanometers to microns being applied in various fields. However, the study of *in vivo* production of DNA/RNA nanostructures is still limited. Despite the challenges, this technique is still highly attractive. Producing and assembling RNA nanodevices inside cells eliminates the need for the delivery of exogenous DNA/RNA nanostructures, which can potentially have negative effects on cells. Furthermore, the efficiency of delivery and the stability of the structure during delivery could also affect the experimental results.

In 2011, Aldaye et al. demonstrated a strategy to control the biosynthetic pathway for hydrogen production using RNA assemblies [14]. In this design, the expressed RNA tiles could self-assemble into 1D or 2D arrays, each containing two RNA aptamer regions that could individually bind with a hydrogenase protein and a ferredoxin protein. The [FeFe]-hydrogenase and ferredoxin proteins worked as a circuit, catalyzing the reduction of protons to hydrogen through electron transfer.

Due to the interaction between the RNA aptamers and proteins, the proteins localized on the RNA structures, and the concentration of the reaction product increased nearby. Thus, the circuits facilitated hydrogen production, and output increased 11 ± 2.8- and 48 ± 1.5-fold with the expression of 1D and 2D scaffolds, respectively (Figure 17.5a). By controlling the spatial distance of proteins, the fluorescence of split Green fluorescent protein (GFP) could also be controlled, indicating the versatility of interfering with biological processes using *in vivo*-expressed RNA nanostructures.

In 2017, Yin et al. demonstrated the implementation of complex cellular logic computation using *in vivo*-assembled RNA nanodevices (Figure 17.5b) [15]. The key component of this method is an RNA-based device called gate RNA. Once expressed in the cell, this RNA can take a specific RNA fragment as an input signal and generate proteins as an output signal. The gate RNA achieves this function by encoding the ribosome binding site (RBS) and starting the codon in a hairpin structure. In the "off" state, the RBS is hindered and cannot bind with the ribosome, resulting in no protein output. When the gate RNA meets the input signal, an RNA fragment with a specific sequence, the gate RNA switches to the "on" state. The hairpin structure opens, exposing the RBS and starting the codon. Subsequently, the ribosome can bind with RBS and activate the translation process. This strategy can perform not only if/else computations but also multi-input AND, OR, and A AND (NOT B) circuits by changing the gate RNA construction. The efficiency of these logic gates was confirmed by the expression level of GFP proteins, demonstrating the potential applications of smart RNA nanodevices in other biological systems, such as prokaryotic hosts and mammalian cells.

In vivo-expressed RNA is also a promising tool for detecting small molecules in living cells. In 2012, Jaffery et al. described RNA sensors that can bind with both target molecules and the fluorophore 3,5-difluoro4-hydroxybenzylidene imidazolinone (DFBHI) to exhibit a green, fluorescent signal, as seen in Figure 17.5c [16]. Prior to this work, Jaffery and his coworkers discovered an RNA aptamer named Spinach, which binds to DFBHI and releases green fluorescence similar to GFP [17]. Thus, by fusing Spinach with another RNA aptamer targeting specific molecules, the re-engineered RNA structure can identify desired targets and give out green fluorescence. This fluorescence imaging technique was then used to study metabolite dynamics in live cells. The intracellular levels of two small molecules, *S*-adenosylmethionine (SAM) and 5′-diphosphate (ADP), were monitored using this strategy. By exploring other small-molecule aptamers, this approach can be extended to other biological studies and track a wide range of targets in native cells, including small molecules, proteins, and even other RNA molecules. In addition, different RNA sequences were discovered to interact with other fluorophores and emit fluorescent signals with various colors, from wavelengths of 400–600 nm [18]. The combination of target molecules and fluorophores allows people to monitor multiple targets simultaneously.

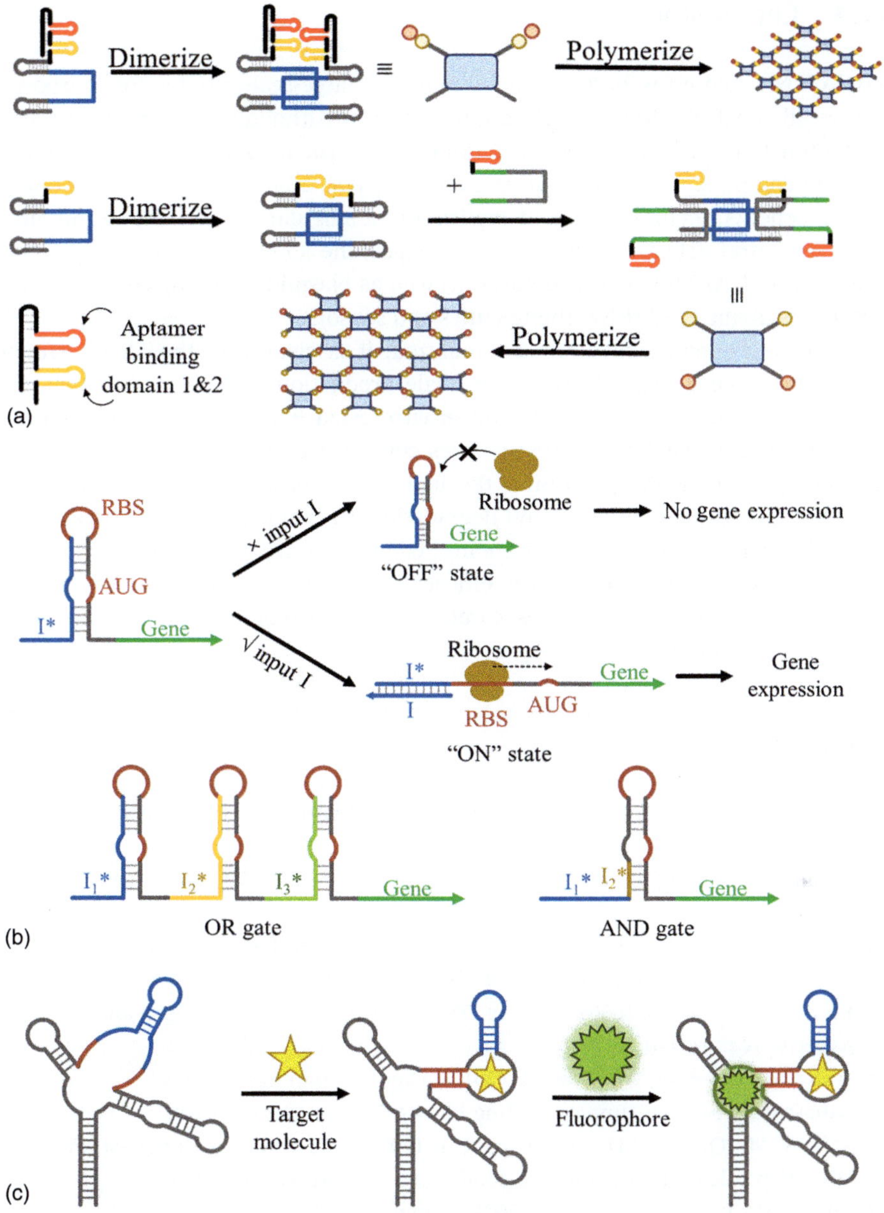

Figure 17.5 The application of *in vivo*-expressed RNA nanostructures. (a) *In vivo*-expressed RNA arrays for protein organization. (b) *In vivo* computation using synthetic RNA nanodevices. (c) Imaging of target molecules by *in vivo*-produced RNA molecules.

17.4 Conclusion

ssDNA/RNA folding is particularly suitable for large-scale production purposes among all the DNA/RNA construction methods. Although the complexity and programmability of ssDNA/RNA folding nanostructures are not comparable to multi-component nanostructures such as DNA origami, this strategy offers unique benefits, including ease of replication, a singular component strand, and fewer structural defects. With these advantages, the ability to amplify large- and small-scale DNA/RNA nanostructures, as well as plasmids in living cells, is a step closer to realizing DNA/RNA therapeutics and sensors.

When ssRNA nanostructures are analyzed, it is clear why they have gained attention in recent years. *In vivo* large-scale production of RNA nanostructures, *in vivo* expressed RNA structures, and *in vivo* tested RNA logic gates provided a proof-of-concept that RNA nanostructures could serve as functional nanodevices and sensors to regulate or monitor the biological process inside the native cell. However, due to the regulation and degradation mechanisms of living organisms, the applications of RNA structures are limited by the downstream nuclease activity that could potentially disrupt their efficacy and function. Therefore, to develop more complex *in vivo* RNA devices, careful research is required to ensure accurate expression of RNA structures.

References

1 Seeman, N.C. (1982). Nucleic acid junctions and lattices. *J. Theor. Biol.* 99: 237–247.

2 Seeman, N.C. and Sleiman, H.F. (2017). DNA nanotechnology. *Nat. Rev. Mater.* 3: 1–23.

3 He, Y., Chen, Y., Liu, H. et al. (2005). Self-assembly of hexagonal DNA two-dimensional (2D) arrays. *J. Am. Chem. Soc.* 127: 12202–12203.

4 Rothemund, P.W.K. (2006). Folding DNA to create nanoscale shapes and patterns. *Nature* 440: 297–302.

5 Wang, W., Chen, S., An, B. et al. (2012). Complex wireframe DNA nanostructures from simple building blocks. *Nature* 485: 623–626.

6 Shih, W.M., Quispe, J.D., and Joyce, G.F. (2004). A 1.7-kilobase single-stranded DNA that folds into a nanoscale octahedron. *Nature* 427: 618–621.

7 Zheng, M., Li, Z., Liu, L. et al. (2021). Kinetic DNA self-assembly: simultaneously co-folding complementary DNA strands into identical nanostructures. *J. Am. Chem. Soc.* 143 (48): 20363–20367.

8 Lin, C., Rinker, S., Wang, X. et al. (2008). In vivo cloning of artificial DNA nanostructures. *Proc. Natl. Acad. Sci. U. S. A.* 105: 17626–17631.

9 Hao, C., Li, X., Tian, C. et al. (2014). Construction of RNA nanocages by re-engineering the packaging RNA of Phi29 bacteriophage. *Nat. Commun.* 5: 3890.

10 Han, D., Qi, X., Myhrvold, C. et al. (2017). Single-stranded DNA and RNA origami. *Science* 358: eaao2648.

11 Liu, D., Geary, C.W., Chen, G. et al. (2020). Branched kissing loops for the construction of diverse RNA homooligomeric nanostructures. *Nat. Chem.* 12: 249–259.

12 Geary, C., Rothemund, P.W.K., and Andersen, E.S. (2014). A single-stranded architecture for cotranscriptional folding of RNA nanostructures. *Science* 345: 799–804.

13 Li, M., Zheng, M., Wu, S. et al. (2018). In vivo production of RNA nanostructures via programmed folding of single-stranded RNAs. *Nat. Commun.* 9: 2196.

14 Delebecque, C.J., Lindner, A.B., Silver, P.A., and Aldaye, F.A. (2011). Organization of intracellular reactions with rationally designed RNA assemblies. *Science* 333: 470–474.

15 Green, A.A., Kim, J., Ma, D. et al. (2017). Complex cellular logic computation using ribocomputing devices. *Nature* 548: 117–121.

16 Paige, J.S., Nguyen-Duc, T., Song, W., and Jaffrey, S.R. (2012). Fluorescence imaging of cellular metabolites with RNA. *Science* 335: 1194–1194.

17 Paige, J.S., Wu, K.Y., and Jaffrey, S.R. (2011). RNA mimics of green fluorescent protein. *Science* 333: 642–646.

18 Strack, R.L., Song, W., and Jaffrey, S.R. (2014). Using spinach-based sensors for fluorescence imaging of intracellular metabolites and proteins in living bacteria. *Nat. Protoc.* 9: 146–155.

18

Engineering Nucleic Acid Structures for Programmable Intracellular Biocomputation

Na Wu[1], *Pengyan Hao*[1], *Chunhai Fan*[2], *and Yongxi Zhao*[1,3]

[1] *Xi'an Jiaotong University, Institute of Analytical Chemistry and Instrument for Life Science, The Key Laboratory of Biomedical Information Engineering of Ministry of Education, School of Life Science and Technology, Xianning West Road, Xi'an, Shaanxi 710049, China*
[2] *Shanghai Jiao Tong University, School of Chemistry and Chemical Engineering, Frontiers Science Center for Transformative Molecules and National Center for Translational Medicine, Shanghai 200240, China*
[3] *Xi'an Jiaotong University, Frontier Institute of Science and Technology, 99 Yanxiang Road, Xi'an, Shaanxi 710049, China*

Intracellular biomolecules, including nucleic acids, proteins, and so on, are spatiotemporally arranged to form interacting networks for performing biofunctions via logic operations [1]. Inspired by that, nucleic acids, enzymes, and other biomolecules have been applied by chemists or biologists to build logic circuits for potential applications in extracellular and intracellular biocomputing. In 1994, Adleman proposed a DNA molecular computation strategy for highly complex NP-complete problems [2]. A directed Hamiltonian problem for a seven-vertex graph was solved using parallel hybridization of DNA in solution. For all the possible path configurations produced in solution, gel electrophoresis and DNA sequencing analysis were performed to fish out the correct solution for the Hamiltonian path problem. This started the beginning of DNA computation and exhibited its great potential. Subsequently, DNA-based computing was used to solve a variety of difficult problems in computer science. In view of its advantage with huge parallelism computing, DNA computation could yield faster computational speed over electronic-based computers in these large NP problems [3–5]. Nucleic acid polymer is a promising material for developing computational elements due to its precise base complementary pairing principle and programmability. Rich active sites for chemical reactions on DNA make it easy to couple with chemical functional groups, organic fluorescent dyes, quantum dots, and so on, which facilitate the reading of signals. Environmental stimuli-responsive logic gates are designed by exploiting functional nucleic acids, including aptamers, DNAzymes, and RNAzymes. Along with the rapid development of DNA nanotechnology, addressable DNA nanostructures provide platforms for precise spatiotemporal arrangement of circuit elements analogous to the circuit boards in silica-based computers. On the one hand, greatly improved speed and efficiency for logic computing benefit from the spatially restricted environment on DNA breadboard, which increases the orthogonality of

DNA Nanotechnology for Cell Research: From Bioanalysis to Biomedicine, First Edition. Edited by Zhou Nie.

DNA reactions (DNA strand displacement reactions (SDRs), DNAzyme cutting reactions, and hybridization chain reactions (HCRs)); on the other hand, the pattern on DNA breadboard formed after experiencing logic operations could be observed under atomic force microscopy, which provides a new way for signal reading.

Logic gates as the core of digital computing are exploited to transform a set of inputs into outputs via logical operation. In a binary system, "1" represents "YES" and "0" represents "NO." The basic logic gates include YES, OR, AND, NOT, NOR, XOR, NAND, and XNOR gates. NOT gate inverts all the signal values; on the contrary, YES gate keeps the original signal values. AND gate means that only when all the inputs are "1," the output can be "1." OR gate represents that as long as one input has "1" in two/multiple inputs, the output will be "1." Other gates could be built via the combination of the above gates (Figure 18.1). These logic gates are applied to design logic circuits for implementing arithmetic operations, such as half-adder and half-subtractor.

Constructing DNA logic gates is an essential prerequisite for DNA molecular computer design. DNA logic operation mainly relies on specific reactions, including SDR [6], DNAzyme cutting reactions [7], HCR [8], and so on. Toehold-mediated SDR is the fundamental basis of DNA computing, which bridges inputs and outputs [9]. In a classic toehold-mediated SDR, a single-stranded DNA (ssDNA) S1 serves as invading strand, and a double-stranded DNA (dsDNA) S2 includes an exposed toehold domain 1. S1 hybridizes with S2 at domain 1 and then triggers the branch migration to release output 2* (Figure 18.2a). This reaction can be used as a building block in a circuit, with input S1 from the upstream reaction as well as output 2* triggering downstream responses. In order to broaden the applications of DNA circuits, inputs have been expended to physical and chemical elements, including optical

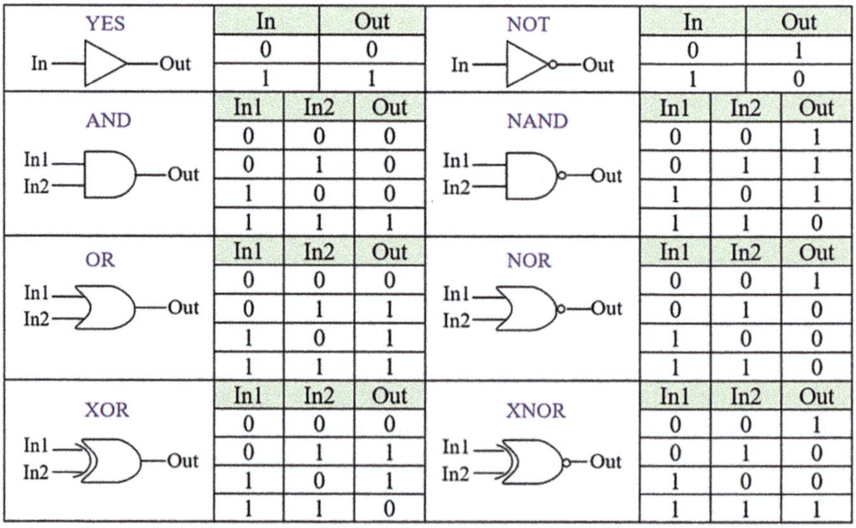

YES In	Out	NOT In	Out
0	0	0	1
1	1	1	0

In1	In2	AND Out	NAND Out	OR Out	NOR Out	XOR Out	XNOR Out
0	0	0	1	0	1	0	1
0	1	0	1	1	0	1	0
1	0	0	1	1	0	1	0
1	1	1	0	1	0	0	1

Figure 18.1 Elementary logic gates and several composite logic gates, including YES, NOT, AND, OR, NAND (NOT AND), NOR (NOT OR), XOR (Exclusive OR), and XNOR (NOT XOR).

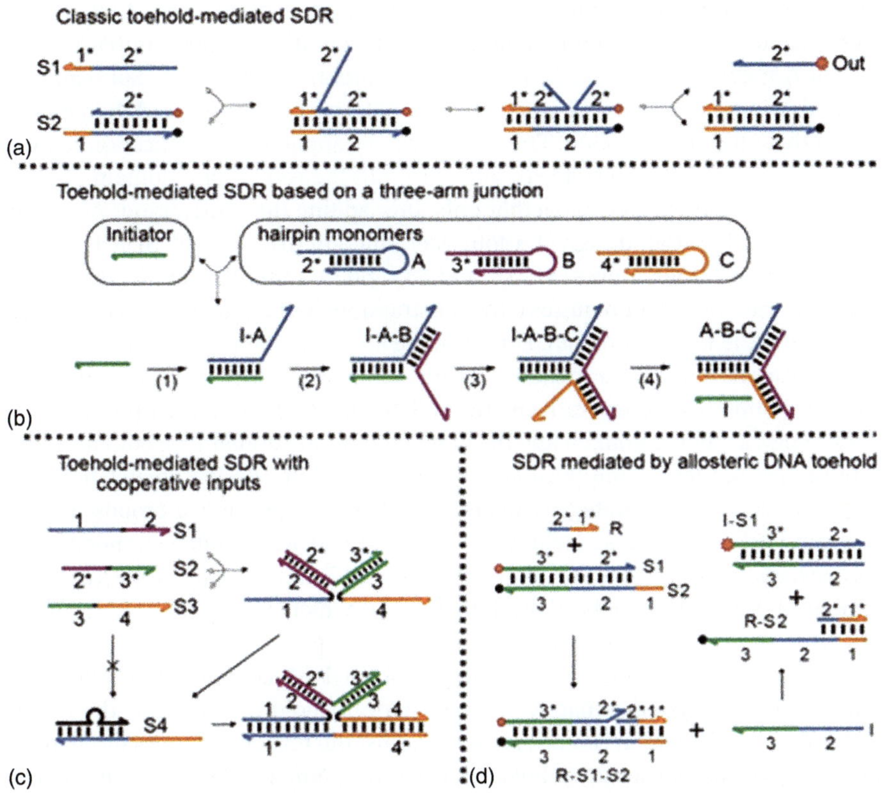

Figure 18.2 Representative toehold-mediated strand displacement reactions (SDRs) and controllable activation and regulation of toeholds: (a) Classic toehold-mediated SDR. (b) Toehold-mediated SDR with toeholds hidden in stem-loop structures. (c) Toehold-mediated SDR with separated recognition domain. (d) SDR mediated by allosteric DNA toehold. Source: Reproduced from Lv et al. [9] with permission; © John Wiley & Sons, 2021.

radiation, temperature, pH, metal ions, small molecules, proteins, nucleotide chains, et al. The appearance of inputs will trigger the conformational change of the recognition motifs, making initially inactive toeholds active. For example, though designing the length of hybridized nucleotide chains and the GC content, the appropriate temperature can be chosen to trigger the separation of two DNA strands. UV irradiation can open dsDNA due to introduced azobenzene modifications on bases. Recognition motifs, including i-motif or CG-C^+, are pH-responsive, and aptamers are small molecule- or protein-responsive.

Controllable activation and regulation of a toehold are important. Before the appearance of inputs, recognition motifs are required to be stable in nonbinding conformations (inactive toeholds). Many strategies were developed to ensure the beginning equilibrium is stable at nonbinding states. A stem-loop structure consisting of a ssDNA-loop domain and a dsDNA-stem domain is often used to deactivate the recognition motif. Yin et al. exploited this strategy to design programmable DNA self-assembly pathways [10]. Toehold-mediated SDR is based on a three-arm junction consisting of an initiator (I) and three hairpin monomers (A, B, and C). Each hairpin includes a toehold and a hidden recognition domain. Initiator opens hairpin A mediated by toehold to release recognition domain on A. Subsequently, released recognition domain triggers the next SDR with hairpin B and then hairpin C until the initiator is released. This system can be applied in amplification requirements (Figure 18.2b). A second strategy is to split the recognition domain into two DNA sequences that can get together only after adding the third strand. According to this idea, Zhu et al. designed a four-way junction-driven DNA SDR [11]. This system includes three inputs (S1, S2, and S3) and an output strand hybridized on S4. The release of output requires a trigger strand including the chain displacement domain and the toehold domain. However, the above two domains exist in inputs S1 and S3, respectively. The formation of a three-strand complex (S1–S2–S3) allows the above two domains to cooperatively trigger SDR with S4 to release output strand. In this strategy, a toehold domain and the branch migration domain are both unrelated to the target sequence (the third strand), which endows design greater flexibility (Figure 18.2c). A third strategy is an allosteric toehold mechanism proposed by the Li group [12]. This system includes a regular strand R, a double-stranded S1–S2 complex, and an input strand I. The regular strand R includes an allosteric toehold domain (1^*) and a short binding domain (2^*), which binds to S1–S2 complex, triggering the exposure of a new toehold on S1. Exposed new toehold allows the invasion of input strand to replace S2 and form I–S1 complex, thus the signal is released. This system separates the allosteric toehold domain and the branch migration domain, which can be applied to adjust the reaction kinetics by changing the length of the allosteric toehold or invading motif. Furthermore, this system is suitable to combine with other controllable toehold-activation strategies for building complexed DNA logic gates or circuits (Figure 18.2d).

Controlling the kinetics of DNA reactions is crucial to DNA circuits and computing networks. The introduction of catalyst can push the equilibrium of a DNA reaction to outputs. In 2007, Zhang et al. presented an entropy-driven DNA catalytic gate strategy for an amplifying circuit (Figure 18.3a) [13]. In this circuit, a ssDNA served as fuel, a dsDNA including output signal and toehold as substrate, and another ssDNA as catalyst input. Catalyst input hybridized to the substrate, triggering signal and new toehold released, then fuel strand hybridized to substrate was initiated via new toehold displacing catalyst input and output. The released catalyst input went back to the system to catalyze the next substrate. This reaction is propelled by the entropic gain of the released molecule avoiding introducing special secondary structures that expand the flexibility in the design of complex circuits. Qian and colleagues reported a "seesaw" gate mode that is suitable for large-scale circuits incorporating thousands of gates compared with previous tens. In this strategy, thresholding and catalysis were applied in every logic operation to control the kinetics as well as stop signal degradation [15]. Toehold-based reversible SDR was employed to implement fast and reliable function in large logic circuits (Figure 18.3b). Every gate comprises input (I), gate-output (G–O), threshold (Th), and fuel (catalyst). Input strand bond to the toehold of G–O to initiate the displacement reaction and the subsequent release of output and generation of

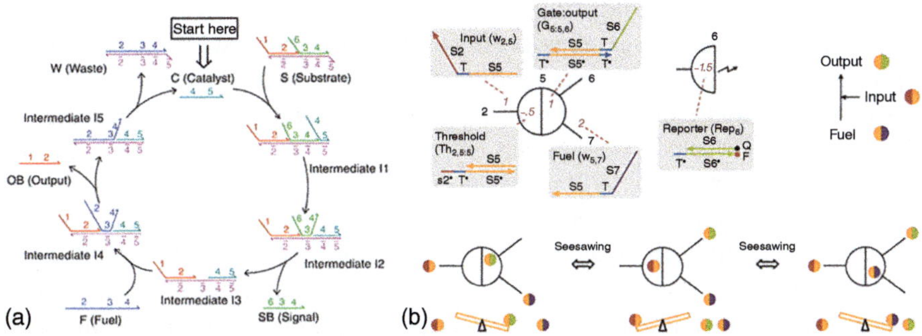

Figure 18.3 Strategies for controlling the kinetics of DNA circuits. (a) Entropy-driven DNA catalytic gate strategy with an input as catalyst for the release of output and then back to next catalytic reaction. Source: Reproduced from Zhang et al. [13] with permission from 2007, American Association for the Advancement of Science. (b) Design strategy of seesaw gates. Threshold and catalysis are employed in this system to control the reaction kinetic as well as prevent signal from degradation. Source: Reproduced from Qian and Winfree [14] with permission from 2011, American Association for the Advancement of Science.

gate-input. This seesaw reaction was reversible. Input strands were consumed by Th strands until a threshold. Thresholding is much faster than seesaw reaction due to the longer toehold, so this competition determines that seesaw reaction can occur only after the input exceeds the threshold. Fuel reacted with gate-input to release input and then released input back to cycle until exhausting G-O and Th. Signal intensity rested on the concentration difference between threshold and gate-output, which is independent in each seesaw unit, making signal intensity avoid attenuation with the increase in the layer of cascade circuits. Seesaw gates can run AND or OR operations simply by regulating the concentration of Th; therefore, all the Boolean logic operations can be implemented via the combination of the built dual-rail logic gates. This makes seesaw strategy more powerful in the construction of large circuits than previous strand displacement mode. Based on this, Qian et al. implemented a four-bit square-root circuit comprising 130 DNA strands [14]. Furthermore, neural networks were constructed using seesaw-based arbitrary linear threshold circuits that successfully implemented a Hopfield associative memory [16]. This work demonstrated that DNA strand displacement cascade circuits had the potential for autonomous recognition, response, and decision-making. However, both linear-threshold circuits and Hopfield networks only recognized a small number of molecular patterns. Subsequently, Qian group developed winner-take-all neural networks on the basis of seesaw DNA gate motif. A winner was screened out by annihilator in competitive reactions, demonstrating the decision of a neural network. Each pattern included 20 distinct DNA strands to recognize one of the handwritten numbers "1" to "9." This network showed the increased recognition ability up to 9 patterns as well as the powerful classification capacity in sophisticated samples via "memory" function [17].

Digital computing is realized through arithmetic operations exploiting logic functions. Adder and half-adder are the basic logic circuits in digital computing. Constructing circuits capable of implementing half-adder operation is critical for digital computing, including DNA computing. Stojanovic designed deoxyribozyme-based logic gates as a half-adder to perform complex logic functions [18]. Benenson et al. demonstrated a DNA-based molecular automaton, in which DNA restriction and ligation enzymes acted as hardware and DNA molecules as software. DNA molecules were coded to control the transition cycles to realize molecular computing [19]. Willner group reported a DNA computing platform using a library of DNAzyme subunits based on the modular coupling of logic gates and a half-adder/half-subtractor system. They implemented multilayered gate cascades, fan-out gate, and parallel logic gate computing with the unique merits of modularity and scalability [20]. Large-scale circuits appeared in fulfilling sophisticated logic operations often consisting of a mass of DNA strands, which easily cause the leakage reactions and slow down the signal restoration. In response to this issue, Song et al. reported a kind of DNA logic circuit with single-stranded logic gates using DNA polymerase [21]. Only single DNA strands were employed simplified the logic system and increased the operation speed. The authors built large-scale logic circuits via cascading strategies and implemented the square-root function of four-bit input numbers with only 37 DNA strands in 25 minutes. Catalytic circuits require reactant recycling for sustained reactions. Zhang and colleagues created entropy-driven reactant recycling DNA circuits with the aid of enzymatic reactions [22]. Active reactant F1:F2:S occurred SDR with catalyst I and fuel F to release product F:S. Product F:S reverted back to the active state (active reactant F1:F2:S) after being nicked by nicked enzyme Nt.BbvCl, then back to the circuit to take part in the next reaction. The single-layered catalytic and two-layered circuits were both implemented using this strategy, demonstrating its simplicity and versatility in the design of entropy-driven DNA circuits.

Analog computation based on DNA circuits was reported and applied to perform complex signal processing [23]. To simulate the molecular networks in biological organisms, Chen et al. reported a DNA-only chemical control system to implement complex signal processing [24]. The reaction elements (short DNA strands) were derived from plasmid DNA, with DNA sequences as the inputs and outputs. This system realized the expected stoichiometry and mass action kinetics of chemical reactions with the DNA SDR as the language of chemistry. They implemented several building blocks for distributed control algorithm and then combined them to achieve consensus between multiple agents. Srinivas et al. built an enzyme-free nucleic acid dynamical system consisting of a chemical reaction network of a "rock-paper-scissors" oscillator [25]. These systems could perform complex information processing and control functions via DNA strand displacement oscillator, which had great potential for the applications in synthetic organelles or artificial cells. Compared with DNA machines navigating along defined tracks, DNA walkers running with unprogrammed tracks are uniquely scalable. Jung and coworkers created a stochastic DNA walker using catalytic hairpin assembly (CHA) reactions on DNA-coated microparticle surfaces [26]. Single catalyst, double catalyst, and tetra catalyst were tested or simulated. Furthermore, the walker behaviors on an unprogrammed, inhomogeneous surface were observed by fluorescence microscopy showing the properties like amorphous chemical reaction network computations.

Predator–prey molecular ecosystems simulation was performed by programming DNA interactions under enzymatic catalysis [27].

The operation order of DNA computer in solution is completely dependent on the complicated sequence design, which adds complexity of logic circuits and limits the universal application of DNA computer. Inspired by silicon-based computer circuit board and precisely arranging circuit modules to implement correct computation, DNA nanostructures represented by DNA origami served as breadboard for DNA molecular computation. Unique addressability and precise location function allow DNA breadboard to program the operation path. Computing speed and reliability are both improved due to kinetically optimized reaction rate as well as spatially blocked false steps under confined space. Liu et al. employed two distinct DNA strands as two input numbers, implemented multiply operations, and then displayed the results on DNA origami computer [28]. Different from logic computing, this system obtained the results via direct screening from a library containing results. Chatterjee et al. demonstrated the superiority of spatial constraints based on a DNA origami scaffold in fast logic computing [29]. This strategy was named DNA domino architecture, and DNA hairpins were located on DNA substrate along the designed route. Free input strand opened the input hairpin to initiate the HCRs among intermediate hairpins along preset trajectory with the addition of fuel strands until the output hairpin was opened. The activated output hairpin reacted with the reporter complex to generate fluorescence signal. Identical DNA hairpins could be reused in this system due to reactions between adjacent hairpins. The computing speed was improved from hours to minutes in time compared to the circuits operated in solution. Wang et al. built DNA switching circuits exploiting SDR reactions to perform digital computing [30]. Arbitrary logic functions were realized with combinative DNA switches, presenting high computing speed. Full-adder and square-rooting functions were further implemented, requiring relatively few computing elements compared with other logic strategies. Exploiting the high programmability and biocompatibility of DNA, DNA robots are expected to perform complex nanomechanical tasks, especially in future biomedicines and intelligent therapeutics. So far, developed DNA robots generally could implement a single function represented by walking along a single direction with appending tasks, including picking up loads or selecting a path. The critical challenges for exploiting DNA robots mainly focus on modularity and algorithm simplicity, analogous to those that have been improved for expanding computing in DNA circuits. The simple module facilitates complicated functions, whereas modularity allows designs for new functions. Qian group developed a cargo-sorting DNA robot to implement cargo sorting by exploiting a simple algorithm and three modular building blocks (Figure 18.4a) [31]. Multiple cargos in two types from beginning in arbitrary locations on DNA origami templates were delivered to destinations with an arm and hand domain for cargo pickup and a leg and foot domain for walking. Several features existed in this system: first, sorting was realized by algorithm with no recognition function required by robot, so this system could be applied to multiple cargos; second, the walking of robots on origami with no need for energy support made it free from specific routes, allowing cargos initially located at arbitrary points and multiple cargos sorted; third, scaling up the system due to DNA origami

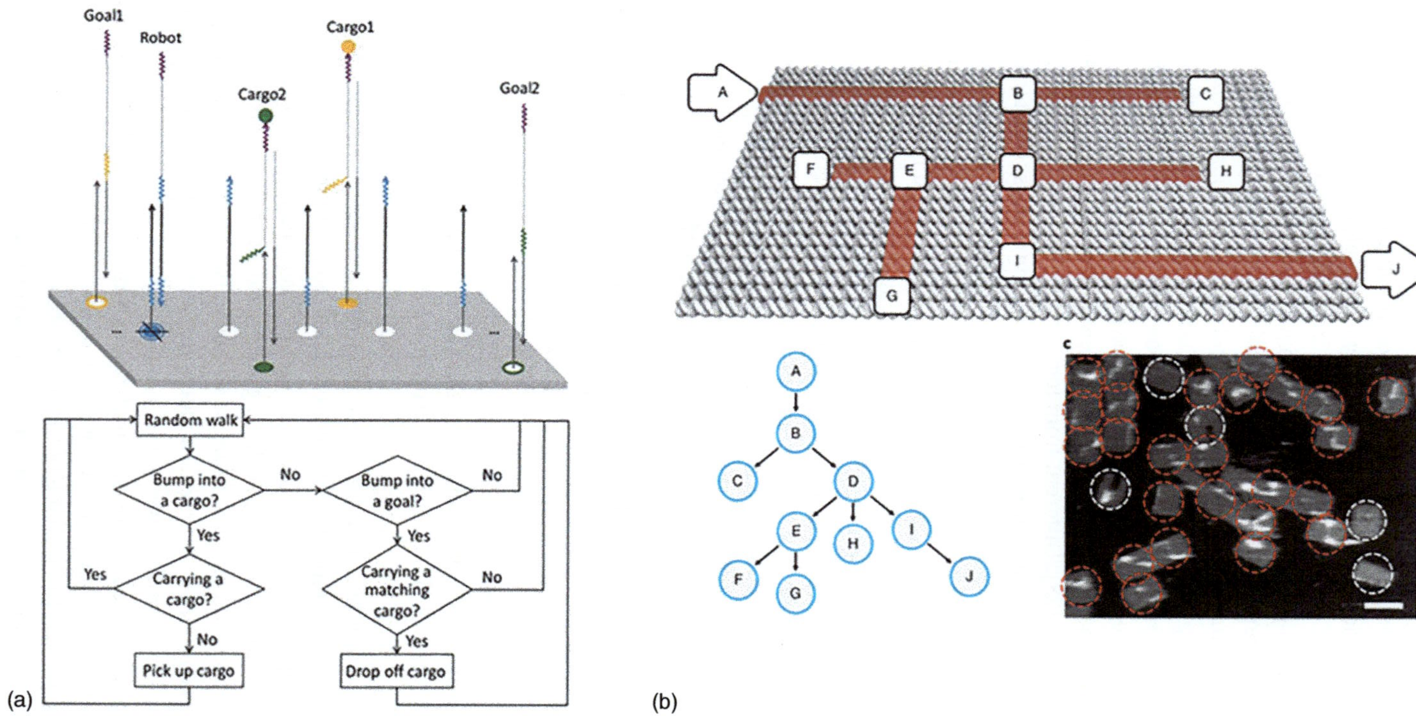

Figure 18.4 Schemes of logic algorithm-assisted DNA robots on DNA origami platforms. (a) DNA robot implementing cargo-sorting task with one arm for picking up and unloading cargos while one leg for random walking. A cargo-sorting algorithm supports the robot sorting multiple cargos simultaneously or multiple robots collectively working on the same task. Source: Reproduced from Thubagere et al. [31] with permission from 2017, AAAS. (b) DNA navigator exploring the correct path based on HCR reactions through a tree predesigned on an origami platform. Reproduced from Chao et al. [32] with permission from 2019, Nature Publishing Group.

platform allows multiple cargos delivered simultaneously as well as parallelism of multiple tasks. Chao et al. presented a DNA navigator system via the design of a simple connected maze as a ten-vertex rooted tree (Figure 18.4b) [32]. Benefiting from the spatial addressability of DNA origami, a parallel depth-first search was realized in this system consisting of a single entry (A), a single exit (J), and four dead-ends (C, F, G, and H). Each navigator probed the possible paths on origami platform to find the solution via a proximal strand exchange cascade method.

In situ studies in living cells can provide more detailed and accurate biological information, and nucleic acid-based molecular computing machines show the potential for in situ studies in living cells. However, considering the complexity of metabolic networks in living systems, nucleic acid logic devices that work well in test-tube studies may be problematic in living cells [33]. Next, we will retrospect the latest advances in the application of nucleic acid logic machines in living cells, which are mainly used to measure intracellular microenvironment and monitor and control intracellular biological processes.

The abundance, localization, and activity of intracellular ions or small molecules, nucleic acids, and protein complexes are closely related to cell proliferation, metabolism, signal transduction, and other activities [34]. Sensing and monitoring changes in the intracellular microenvironment is of great significance for basic biological research as well as for early screening and treatment of diseases. However, previous probe molecules were always in the ON state and diffused throughout the cell leading to a high background, so the specific measurements of the intracellular microenvironment are still challenging [35]. Nucleic acid-based molecular logic devices have advantages such as programmability and versatile responsiveness and have been developed to carry out logic programs in complex cellular environments. In this section, we sum up the outstanding progress of nucleic acid-based molecular computing devices in the field of intracellular microenvironment measurement.

The measurement of a single biomarker within the cell, ignoring the expression heterogeneity of intracellular markers, may lead to the untrustworthiness of the research results [36, 37]. Based on this, scientists have developed a series of nucleic acid-based intelligent probes that logically respond to multiple targets. Pei et al. integrated tetrahedral DNA nanostructures (TDNs) into DNA computation to establish DNA logic gates that can respond to different intracellular targets, including different ions, small molecule components, and short DNA complementary strands [38]. As shown in Figure 18.5a, dynamic sequences (i-motif, ATP-binding aptamer, T-rich mercury-specific oligonucleotide (MSO), and hairpin structures) were assembled to the edges of TDN, and the device underwent different conformational changes in response to different combinations of target signals, which in turn translated into changes in fluorescence intensity, corresponding to logical operations as INH, XOR, OR, and AND logic gates, respectively.

Nucleic acid nanodevices are taken up by cells into the endosome and transported to the lysosome, during which the pH of the endosome undergoes characteristic changes from 6.2 (early endosomes) to 5.5 (late endosomes) and 5 (lysosomes) [41]. Therefore, nucleic acid logic devices can also enable multiple molecular monitoring within subcellular compartments. In 2019, Du et al. constructed a lysosomal

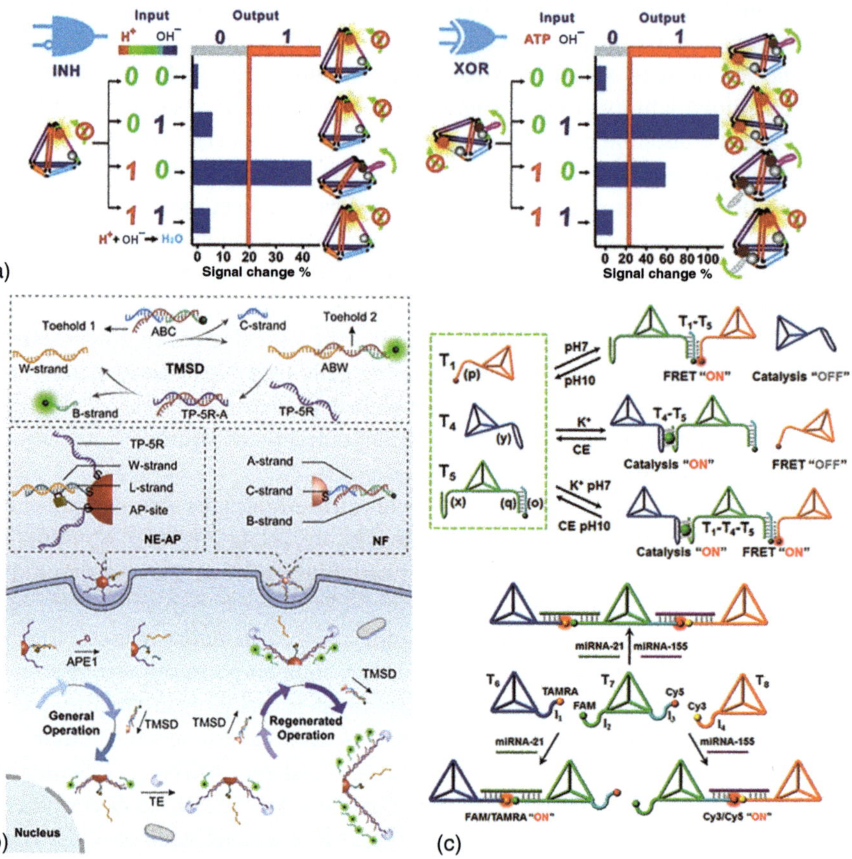

Figure 18.5 Nucleic acid-based logic devices used to measure the intracellular microenvironment. (a) Develop logic devices based on DNA tetrahedron that can respond to different intracellular targets. Source: Reproduced from Pei et al. [38] with permission from 2012, Wiley-VCH. (b) Construction of DNA nanodevices for diagnostic imaging of APE1 in living cells. Source: Reproduced from Zhang et al. [39] with permission from 2023, Wiley-VCH. (c) Tetrahedral mixtures were used for the sensing and imaging of multiple miRNAs in cells. Source: Reproduced from Zhou et al. [40] with permission from 2021, Wiley-VCH.

recognition framework nucleic acid (FNA) logic device in order to simultaneously detect pH and ATP in the lysosome [42]. By introducing pH-responsive i-motif and ATP-binding aptamer as logic control units into the DNA triangular prism (DTP) nanoscaffold, which responds to the "AND" gate input of endogenous protons and ATP in the lysosomal subcellular compartment, triggering the conformational change of the logic device to output the fluorescence signal. Their group also used nucleic acid logic devices to carry out bionic research at the subcellular level. The bimolecular i-motif and split ATP aptamer were assembled on the apex of the two TDN branches, the TDN assembly triggered by the "AND" logic gate was achieved in the acidic environment of the lysosome compartment. Large-scale frame structures were formed in response to the subcellular microenvironment, simulating the formation of spider silk fibers in nature [43].

In addition to ions and small molecules, proteins are also important components of the cellular microenvironment and the material basis of life activities. Monitoring the abundance and activity of proteins is important for basic chemical biology research and diagnostic therapy [44]. Wang et al. reported a method to detect intracellular telomerase activity using cascade logic gates [45]. The logic device was delivered into the cell by the transfection reagents, and the telomerase substrate (TS) probe was activated by telomerase and extended as the output of the "YES" gate (TS + nR), which together with the input B-strand performed the "AND" gate to initiate the toe-mediated SDR. This cascade logic operation separated the fluorescent group from the quenched group-modified DNA duplex and output fluorescent signals that enabled imaging of intracellular telomerase. Chen et al. used MnO_2 nanoflakes to promote the uptake of DNA logical probes by cells [46], the "AND" gate signal input of DNAzyme and intracellular telomerase led to the toe exchange assembly of the nucleic acid chain. The correct conformational folding made DNAzyme have higher catalytic cracking ability, and the fluorescence signal released by the catalyst was used to monitor telomerase activity. Shen et al. have developed a dual 3D DNA walker based on catalytic hairpin assembly of DNAzyme to detect intracellular telomerase activity [47]. Hairpins were assembled on the surface of gold nanoparticles (AuNPs), and when incubated with telomerase, the telomerase primer was extended to form a slender repetitive sequence (TR), which randomly moved along the designed track on the surface of AuNPs for chain replacement, and finally assembled to form a complete DNAzyme, which was the first layer of 3D DNA walker triggered by telomerase. Then, the assembled DNAzyme autonomously moved along the surface of another AuNP to cut the DNA hairpin as the second layer of 3D DNA walker triggered by DNAzyme. Based on the synergistic interaction between the above two layers of DNA walkers, FAM-labeled DNA fragments were finally released after cutting the hairpin, and the telomerase activity was analyzed by detecting the fluorescence signal. Recently, Zhang et al. constructed an ingenious DNA nanodevice, which was used for the diagnosis and imaging of intracellular apurinic/apyrimidinic endonuclease 1 (APE1) [39]. As shown in Figure 18.5b, DNA regenerative nanodevices consist of nanoeffectors (NE-AP) and nanofuelers (NFs), which initiated chain replacement reactions in response to intracellular APE1. Telomerase extended the TP-5R strand as an extension orbital for regenerative nanodevices, and the DNA walker performed target trigger movements along the track, generating a cumulative amplified fluorescence signal through which intracellular APE1 activity was detected in situ. This study demonstrated the regeneration of biomarker-assisted nanomachines in living cells for the first time, providing a new research basis for the development of more suitable personalized DNA nanomachines.

As a biological macromolecule, nucleic acids (such as miRNA and mRNA) are the main vectors for storing and transmitting genetic information. The occurrence of some diseases is closely related to the disorder of RNA [48], so it is necessary to understand the expression and regulation of RNA. To explore RNA expression profile variations, scientists gate nucleic acid-based computations that allow multiple RNA inputs to perform Boolean logic gate operations, activating computational cascades to produce controlled outputs. For example, the nucleic acid logic device designed by Hemphill et al. can respond to the "AND" gate activation of intracellular miRNA-21 and miRNA-122, replace the double-stranded modified with fluorescent groups and quenching groups, and generate fluorescence output to detect the expression pattern of intracellular miRNA [49]. Due to the low abundance of RNA in living cells, researchers have combined nucleic acid logic gates with amplification strategies to improve the sensitivity of RNA detection. For example, Gong et al. developed an isothermal, non-enzymatically reconfigurable dual HCR logic computing system that can sensitively analyze multiple intracellular miRNAs simultaneously [50]. Only by inputting miRNAs simultaneously, both HCR1 and HCR2 in the computing circuit can be activated and output magnified FRET fluorescence signals. They implemented logical calculations for the initiation of multiple endogenous miRNAs, including binary logic gates as well as more advanced cascaded logic circuits. The introduction of HCR in the logical computational circuit ensures continuous recognition of the target and co-amplification of the signal, improving the feasibility and reliability of in situ detection in complex physiological environments. Quan et al. designed a dual amplification cascade logic circuit consisting of DNAzyme amplification module and HCR amplification module for the simultaneous detection of endogenous miRNA-122 and miRNA-21. The "AND" gate of double miRNA started the amplification module in the logic device, and the cell subtypes were identified by analyzing the difference of output fluorescence signals [51]. Recently, Li et al. created a simpler and more efficient biocomputing platform for intracellular nucleic acid imaging [52]. What is reported is a non-enzymatic autocatalytic logic circuit for DNA assembly triggered by a self-stacking target miRNA, where both the endogenous miRNA and the newly assembled Y-shaped structure act as catalysts throughout the reaction, resulting in a positive feedback reaction network with exponential signal gain.

Studies have shown that 3D DNA nanostructures have excellent cellular internalization and stability [53], so researchers have used DNA tetrahedra for logic sensing of intracellular multiple miRNAs. For example, Lin and co-workers constructed a DNA tetrahedron-based logic device for intracellular double miRNA detection [54]. They designed the TDN as a scaffold by integrating recognition and computation modules into the three vertices of the TDN, and only simultaneous input of dual miRNAs could initiate consecutive toe-mediated strand substitution to release fluorescently labeled nucleic acid strands. Considering that the high background signal of a single TDN logic device may interfere with the sensitive inspection of intracellular miRNAs, Zhou et al. utilized TDNs as functional units in response to logic gate signal inputs reconfigurable into tetrahedral dimeric and trimeric nanostructures [40]. As shown in Figure 18.5c, they applied this platform to the sensing and imaging of multiple miRNAs in cancer cells, which can address the above-mentioned limitation of high background signals.

Nucleic acid-based logic devices are also capable of detecting mRNAs in the intracellular microenvironment. Groves et al. constructed "AND" and "OR" logic gates based on four-way strand exchange reactions, and when interacting with multiple natural mRNAs, the designed probes amplify the fluorescent signal to image the strand replacement reaction in the cytoplasm for in situ detection of mRNAs [55]. Of course, some researchers have also introduced DNA nanostructures into logic devices to improve the sensitivity and accuracy of multiple mRNA detection [56–58].

Numerous studies have shown that nucleic acid-based computational devices are capable of performing logical operations in complex intracellular environments to monitor biological processes at the molecular level. For example, Ren et al. reported a genetic hybridization cascade amplification technique that mapped the subcellular distribution of target miRNA with extremely high sensitivity [59]. They divided the RNA fluorescence report molecule Broccoli into two split aptamer fragments, which were connected to the ends of the two hairpin fuel chains (H1 and H2), respectively, and they could not spontaneously hybridize to activate Broccoli fluorescence. H1 and H2 were generated by transcription using the cell's genetic coding system. HCR amplification of H1 and H2 occurred under the activation of the target miRNA, resulting in the reorganization of numerous Broccoli activation chains, thus activating the amplified fluorescence signal. The induced fluorescence signal is associated with the target miRNA, which can be used to track the target in living cells. In addition, the logic device based on nucleic acid can also realize dynamic detection of horizontal correlation between multiple types of biomarkers. For example, Zhan et al. reported a near-infrared light-mediated DNA logic nanomachine that can perform bio-orthogonal cascade-gated imaging of intracellular miRNA and metal ions [60]. This logic device integrates ternary DNAzyme and UV light-cleavable DNA scaffolds onto the surface of nanoparticles, and can trigger a hybrid chain reaction when intracellular Zn^{2+} and external UV light coexist or endogenous miRNA targets exist, and then output enhanced FRET fluorescence signals. Through the construction of combinatorial logic circuit (AND/OR), the dynamic monitoring of the correlation between intracellular miRNA and Zn^{2+} expression level was realized. It is known that the interconnection of different biomarkers controls a variety of biological processes of disease development and progress [61, 62], which provides a new perspective to study related biological processes. Shi et al. assembled Y-shaped DNA onto the surface of AuNPs and constructed a double-responsive DNA logic device [63]. As shown in Figure 18.6A, only when the upstream mRNA and downstream cytochrome c in the apoptosis pathway are input at the same time, the nanomachine can execute the "AND" logic gate operation, and then the Y-shaped DNA on the surface of AuNP is broken down by chain displacement, and the output fluorescence signal can visualize the apoptosis signal pathway in the cell.

It is reported that the RNA expression level and enzyme activity in living cells are in dynamic change and may change significantly under different cell types and conditions [65]. Therefore, monitoring the expression of these biological molecules can help us distinguish cell types or states. For instance, Bai et al. developed a multivalent DNA logic circuit driven by intracellular entropy to realize the simultaneous calculation and analysis of intracellular telomerase and two miRNAs [64]. As shown in Figure 18.6B, the DNA logical probe consists of three parts, including three amplification probes (AMPs), a preassembled multi-arm substrate probe (SP),

(A)

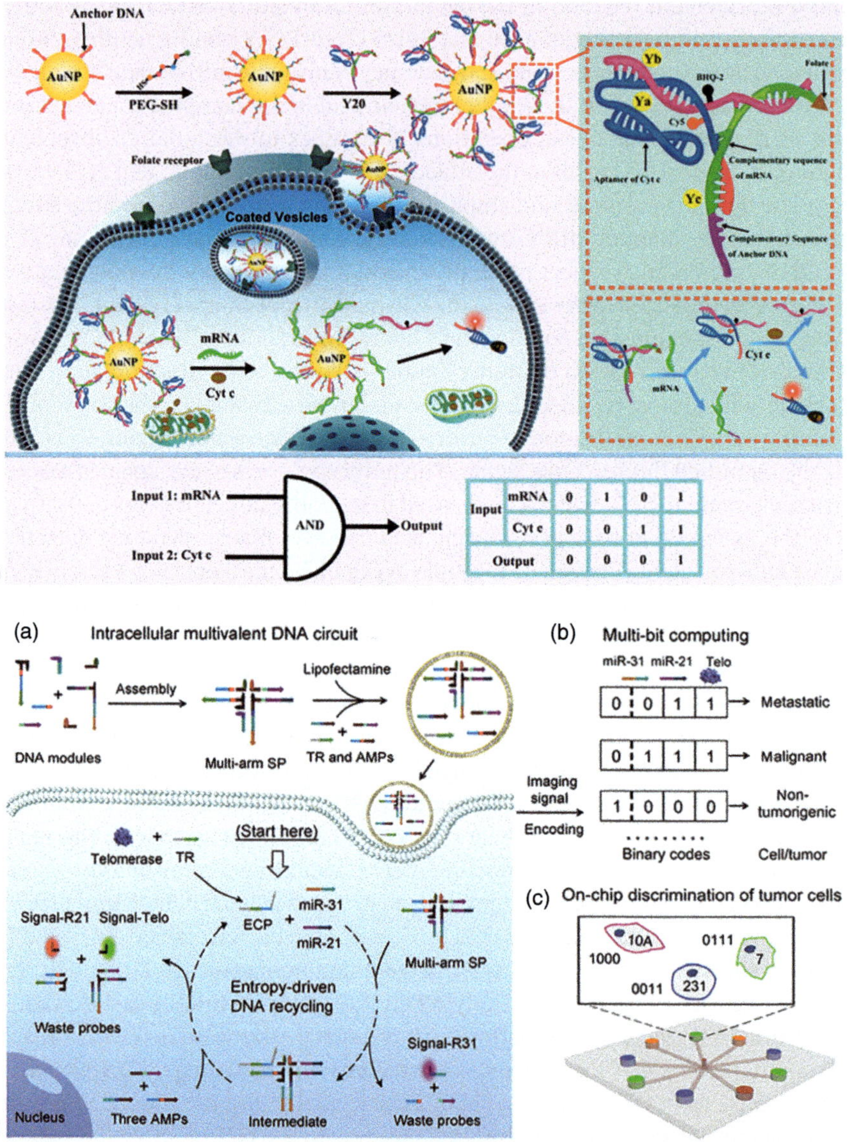

(B)

Figure 18.6 Nucleic acid-based logic devices used to monitor intracellular biological processes. (A) DNA logic devices were constructed to visualize apoptosis signaling pathways in cells. Source: Reproduced from Shi et al. [63] with permission from 2019, American Chemical Society. (B) An intracellular entropy-driven multivalent DNA logic circuit was designed to distinguish non-tumorigenic, malignant, and metastatic breast cells and to monitor tumor progression. Source: Reproduced from Bai et al. [64] with permission from 2020, Wiley-VCH.

and a telomerase recognition (TR) primer. After the DNA probe was delivered to the living cell through the lipofectamine, the TR primer can be extended by telomerase to produce enzyme catalytic product (ECP), and then ECP, miRNA-21, and miRNA-31 triggered their respective chain displacement reaction to produce amplified fluorescence signals. The fluorescence signal is encoded into multi-bit binary code to distinguish non-oncogenic, malignant, and metastatic breast cells and monitor tumor progression.

In addition, Yang et al. integrated three recognition modules and three reporting modules onto the hexahedral DNA framework structure to form three "AND" logic gates for cell subtype identification based on miRNA bispecific recognition [66]. The recognition module consists of strands J and K, and J chains can be released after recognizing intracellular-specific miRNA. The reporting module consists of strands F, T, and Q. The dual-specific miRNA input performed an "AND" Boolean logic operation, and the recognition module released both types of J-strands, which turned on the specific reporting module to output the fluorescent signal through the chain replacement reaction. This design can be extended to the differentiation of various cell types by regulating the type of miRNA, and the process of cell differentiation and development can be tracked at the molecular level.

A key goal of developing nucleic acid-based logic machines is to transform them into in vivo applications. Some researchers have integrated the engineering logic system of multi-input information processing and precise regulation and braking in living cells to detect the cell state and make precise in situ regulation in complex physiological environment. Although there is still a long way to go to realize the mature application of logic system in vivo, many excellent research results have been published. For example, Xie et al. constructed an RNAi-based multi-input logic circuit for identifying and selectively killing specific cancer cells [67]. Multiple intracellular cancer-related miRNAs served as input to initiate the AND-like logic function. Only when the expression level of endogenous miRNAs matched the expression profile of miRNAs in HeLa cells, can the expression of hBax protein of pro-apoptotic gene (hBax) be regulated, thus inducing HeLa cell apoptosis. Similarly, Matsuura et al. used RNA-binding proteins (RBP) to create a miRNA-based logic device to recognize cancer cells and initiate apoptosis by regulating protein expression of target genes [68]. L7Ae in this work is an RBP capable of interacting with the kink-turn (Kt) motif of the 5′-UTR of the associated gene, thus effectively inhibiting the translation of the output gene. Therefore, the successful synthesis of RBP is controlled by using multiple miRNAs as inputs, and then the translation of associated genes into output only in target cells is artificially regulated. As shown in Figure 18.7A, in the ON state, L7Ae protein translation was inhibited by multiple input miRNAs, which converted the fluorescent reporter gene (EGFP) or proapoptotic gene (hBax) into an output protein. Studies have shown that the simultaneous input of multiple cancer-related miRNAs in living cells can activate the "AND" logic gate to translate corresponding proteins for cell identification or selective elimination of cancer cells. Compared with previous single miRNA input systems, this "AND" logic gate multi-input system can more accurately identify target cells and regulate cell fate. This logic circuit provides a powerful tool for

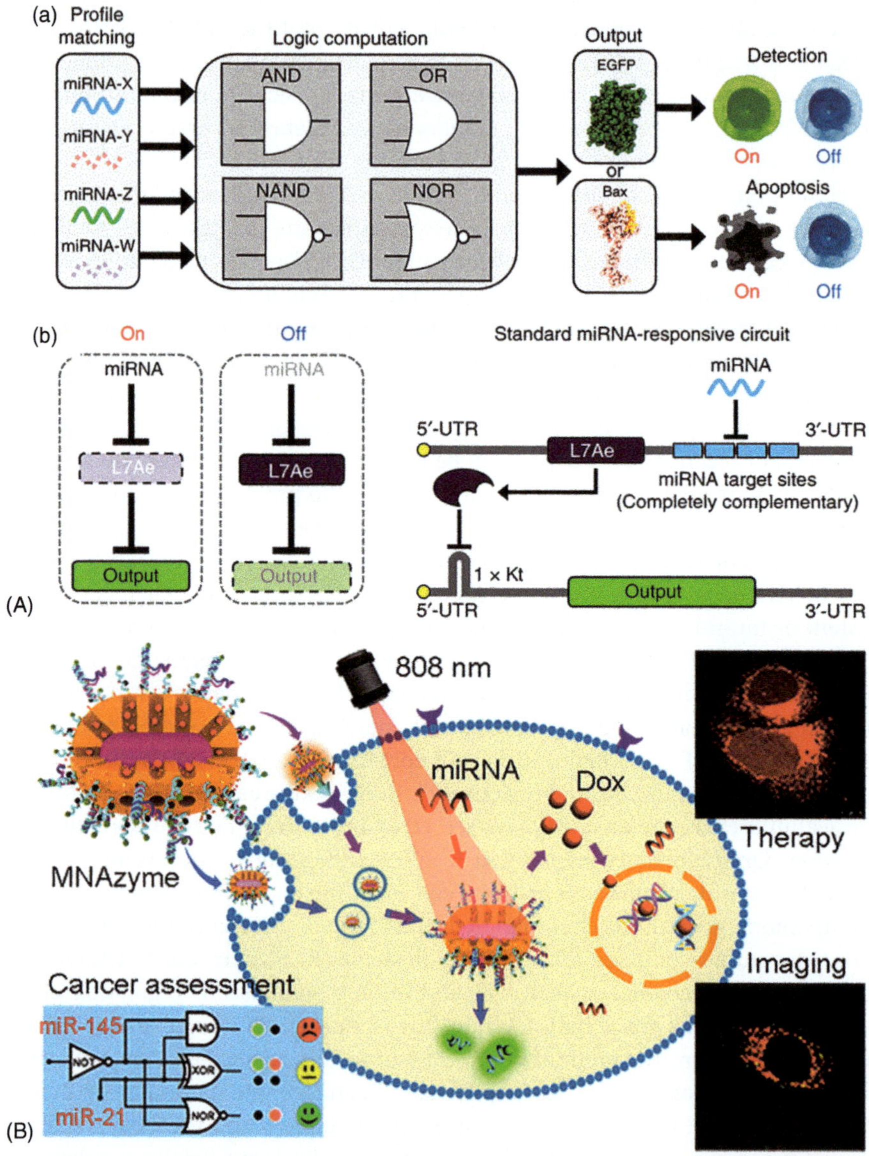

Figure 18.7 Nucleic acid-based logic devices used to regulate intracellular biological processes. (A) A miRNA-based logic apparatus was created using RNA-binding protein (RBP) to recognize cancer cells and initiate apoptosis by regulating protein expression of target genes. Source: Matsuura et al. [68]/ Springer Nature / CC BY 4.0. (B) Multicomponent nucleases (MNAzymes) were assembled on the surface of mesoporous silica-coated gold nanorods (MSGRs) to construct programmable photothermal nanodevices for multiple intracellular detections and control of drug release. Source: Reproduced from Zhang et al. [69] with permission from 2015, American Chemical Society.

future gene therapy applications. The CRISPR-Cas9 system is a very powerful tool for gene regulation and gene editing [70]. A lot of research has been done to achieve response activation of the CRISPR-Cas9 system. For example, Liu and colleagues developed an "AND" gate logic circuit based on CRISPR-Cas9 [71]. Only when both cancer-specific promoters were active in test cells can they drive the transcription of Cas9 mRNA and sgRNA, allowing Cas9/sgRNA complexes to bind to LacI genes via Watson–Crick base pairing and cut DNA sequences, resulting in the output of cellular functional genes, including hBax, p21, and E-cadherin. This logic circuit controls the output of cell phenotypes by regulating the corresponding genes, specifically inducing the apoptosis of bladder cancer cells, inhibiting the growth of bladder cancer cells, or reducing cell movement. Some researchers use logical operation to trigger the release of controlled drugs in specific cells to achieve the purpose of artificially regulating cell fate. As shown in Figure 18.7B, Zhang et al. assembled multicomponent nucleases (MNAzymes) on the surface of mesoporous silica-coated gold nanorods (MSGRs) to create programmable photothermal nanomachines for multiple intracellular detections and control of drug release [69]. Two probes responding to miR-21 and miR-145 were designed for logical manipulation through sequence adjustment for cancer risk assessment. The MNAzyme motif, on the other hand, acted as a goalie, locking adriamycin into the hole. When nanodevices were recognized and internalized into cancer cells guided by aptamers, intracellular miRNAs activated MNAzymes to catalyze substrate cleavage and convert near-infrared light to heat to optimize enzymatic efficiency, inducing gradual dissociation of the MNAzyme motif from the pore, thus achieving controlled drug release.

In conclusion, we summarize the nucleic acid-based computations and their applications in living cells. Scientists take functional nucleic acids that respond to biochemical stimuli as DNA logic gates and construct the computational core of synthetic circuit through reasonable assembly of several logic gates [72]. If combined with a chain exchange reaction, the input signal can be converted into the release of nucleic acid chains. Given the self-assembly and programmable nature of nucleic acids, logic machines designed based on nucleic acids have excellent characteristics such as flexible operation, easy regulation, and predictability, providing potential for in-situ research in living cells. The researchers use the device to sense microenvironmental information in cells to detect cell status and make precise in-situ controls in complex physiological environments.

Despite the extensive developments, there are still challenges to be overcome. Firstly, the intracellular environment is complex, and the structural stability and functional integrity of nucleic acid-based logic machines are a problem when they perform calculations in the intracellular environment. Nucleic acid nanomachines are degraded into small pieces in non-neutral pH or special environments where nucleases are present (e.g. lysosomes). In addition to not functioning properly in the cell, these pieces may interact with intracellular miRNAs or other factors to regulate cell function, raising questions about biosafety. Secondly, unlike the homogeneous solution in test tubes, the intracellular environment is highly heterogeneous, and the movement and distribution of foreign substances are actively regulated by cells in time and space. As a result, these nucleic acid devices are difficult to locate in the right place as expected. These nucleic acid devices are coupled to specific peptide sequences to improve the internalization efficiency of nucleic acid components and to achieve desired intracellular localization. Thirdly, the computational speed of nucleic acid molecular logic machines is confined by chemical reaction kinetics, the spurious interaction and leakage reaction between nucleic acid sequences also slow down the computational speed. In addition to the speed of calculation, the accuracy of calculation also needs to be considered, the low abundance of intracellular molecules leads to the output of false positive signals. In order to solve the problem of slow calculation speed and low accuracy, researchers need to optimize the design strategy. The field is in a phase of rapid development, and we believe that understanding and addressing these challenges will lead to the development of more sophisticated nucleic acid-based computing devices to meet the growing demand for biochemical technologies, diagnostics, and therapeutics.

References

1 Benenson, Y. (2012). Biomolecular computing systems: principles, progress and potential. *Nat. Rev. Genet.* 13 (7): 455–468.

2 Adleman, L.M. (1994). Molecular computation of solutions to combinatorial problems. *Science* 266 (5187): 1021–1024.

3 Ogihara, M. and Ray, A. (2000). DNA computing on a chip. *Nature* 403 (6766): 143–144.

4 Lipton, R.J. (1995). DNA solution of hard computational problems. *Science* 268 (5210): 542–545.

5 Braich, R.S., Chelyapov, N., Johnson, C. et al. (2002). Solution of a 20-Variable 3-SAT problem on a DNA computer. *Science* 296 (5567): 499–502.

6 Zhang, D.Y. and Winfree, E. (2009). Control of DNA strand displacement kinetics using toehold exchange. *J. Am. Chem. Soc.* 131 (47): 17303–17314.

7 Wu, Y., Li, Z., Shi, M. et al. (2021). Programmable DNAzyme computing for specific in vivo imaging: intracellular stimulus-unlocked target sensing and signal amplification. *Anal. Chem.* 93 (36): 12456–12463.

8 Figg, C.A., Winegar, P.H., Hayes, O.G. et al. (2020). Controlling the DNA hybridization chain reaction. *J. Am. Chem. Soc.* 142 (19): 8596–8601.

9 Lv, H., Li, Q., Shi, J. et al. (2021). Biocomputing based on DNA strand displacement reactions. *ChemPhysChem* 22 (12): 1151–1166.

10 Yin, P., Choi, H., Calvert, C.R. et al. (2008). Programming biomolecular self-assembly pathways. *Nature* 451: 318–322.

11 Zhu, J., Zhang, L., Dong, S. et al. (2013). Four-way junction-driven DNA strand displacement and its application in building majority logic circuit. *ACS Nano* 7 (11): 10211–10217.

12 Yang, X., Tang, Y., Traynor, S.M. et al. (2016). Regulation of DNA strand displacement using an allosteric DNA toehold. *J. Am. Chem. Soc.* 138 (42): 14076–14082.

13 Zhang, D.Y., Turberfield, A.J., Yurke, B. et al. (2007). Engineering entropy-driven reactions and networks catalyzed by DNA. *Science* 318 (5853): 1121–1125.

14 Qian, L. and Winfree, E. (2011). Scaling up digital circuit computation with DNA strand displacement cascades. *Science* 332 (6034): 1196–1201.

15 Qian, L. and Winfree, E. (2011). A simple DNA gate motif for synthesizing large-scale circuits. *J. R. Soc. Interface* 8 (62): 1281–1297.

16 Qian, L., Winfree, E., and Bruck, J. (2011). Neural network computation with DNA strand displacement cascades. *Nature* 475 (7356): 368–372.

17 Cherry, K.M. and Qian, L. (2018). Scaling up molecular pattern recognition with DNA-based winner-take-all neural networks. *Nature* 559 (7714): 370–376.

18 Stojanović, M.N. and Stefanović, D. (2003). Deoxyribozyme-based half-adder. *J. Am. Chem. Soc.* 125 (22): 6673–6676.

19 Benenson, Y., Adar, R., Paz-Elizur, T. et al. (2003). DNA molecule provides a computing machine with both data and fuel. *Proc. Natl. Acad. Sci. U.S.A.* 100 (5): 2191–2196.

20 Elbaz, J., Lioubashevski, O., Wang, F. et al. (2010). DNA computing circuits using libraries of DNAzyme subunits. *Nat. Nanotechnol.* 5 (6): 417–422.

21 Song, T., Eshra, A., Shah, S. et al. (2019). Fast and compact DNA logic circuits based on single-stranded gates using strand-displacing polymerase. *Nat. Nanotechnol.* 14 (11): 1075–1081.

22 Zhang, C., Wang, Z., Liu, Y. et al. (2019). Nicking-assisted reactant recycle to implement entropy-driven DNA circuit. *J. Am. Chem. Soc.* 141 (43): 17189–17197.

23 Song, T., Garg, S., Mokhtar, R. et al. (2016). Analog computation by DNA strand displacement circuits. *ACS Synth. Biol.* 5 (8): 898–912.

24 Chen, Y.J., Dalchau, N., Srinivas, N. et al. (2013). Programmable chemical controllers made from DNA. *Nat. Nanotechnol.* 8 (10): 755–762.

25 Srinivas, N., Parkin, J., Seelig, G. et al. (2017). Enzyme-free nucleic acid dynamical systems. *Science* 358 (6369): eaal2052.

26 Jung, C., Allen, P.B., and Ellington, A.D. (2016). A stochastic DNA walker that traverses a microparticle surface. *Nat. Nanotechnol.* 11 (2): 157–163.

27 Fujii, T. and Rondelez, Y. (2013). Predator–prey molecular ecosystems. *ACS Nano* 7 (1): 27–34.

28 Liu, H., Wang, J., Song, S. et al. (2015). A DNA-based system for selecting and displaying the combined result of two input variables. *Nat. Commun.* 6 (1): 10089.

29 Chatterjee, G., Dalchau, N., Muscat, R.A. et al. (2017). A spatially localized architecture for fast and modular DNA computing. *Nat. Nanotechnol.* 12 (9): 920–927.

30 Wang, F., Lv, H., Li, Q. et al. (2020). Implementing digital computing with DNA-based switching circuits. *Nat. Commun.* 11 (1): 121.

31 Thubagere, A.J., Li, W., Johnson, R.F. et al. (2017). A cargo-sorting DNA robot. *Science* 357 (6356): eaan6558.

32 Chao, J., Wang, J., Wang, F. et al. (2019). Solving mazes with single-molecule DNA navigators. *Nat. Mater.* 18 (3): 273–279.

33 Church, G.M., Elowitz, M.B., Smolke, C.D. et al. (2014). Realizing the potential of synthetic biology. *Nat. Rev. Mol. Cell Biol.* 15 (4): 289–294.

34 Grecco, H.E., Schmick, M., and Bastiaens, P.I.H. (2011). Signaling from the living plasma membrane. *Cell* 144 (6): 897–909.

35 Lee, M.H., Han, J.H., Kwon, P.S. et al. (2012). Hepatocyte-targeting single galactose-appended naphthalimide: a tool for intracellular thiol imaging in vivo. *J. Am. Chem. Soc.* 134 (2): 1316–1322.

36 Leung, K., Chakraborty, K., Saminathan, A. et al. (2019). A DNA nanomachine chemically resolves lysosomes in live cells. *Nat. Nanotechnol.* 14 (2): 176–183.

37 Li, F., Li, J., Dong, B.J. et al. (2021). DNA nanotechnology-empowered nanoscopic imaging of biomolecules. *Chem. Soc. Rev.* 50 (9): 5650–5667.

38 Pei, H., Liang, L., Yao, G.B. et al. (2012). Reconfigurable three-dimensional DNA nanostructures for the construction of intracellular logic sensors. *Angew. Chem. Int. Ed.* 51 (36): 9020–9024.

39 Zhang, Q.Y., Wang, Y.H., Wang, W.J. et al. (2023). A telomerase-assisted strategy for regeneration of DNA nanomachines in living cells. *Angew. Chem. Int. Ed.* 62 (6): e202213884.

40 Zhou, Z.X., Fan, D.Q., Wang, J.B. et al. (2021). Triggered dimerization and trimerization of DNA tetrahedra for multiplexed miRNA detection and imaging of cancer cells. *Small* 17 (6): 2007355.

41 Wang, C.S., Zhao, T., Li, Y. et al. (2017). Investigation of endosome and lysosome biology by ultra pH-sensitive nanoprobes. *Adv. Drug Delivery Rev.* 113: 87–96.

42 Du, Y., Peng, P., and Li, T. (2019). DNA logic operations in living cells utilizing lysosome-recognizing framework nucleic acid nanodevices for subcellular imaging. *ACS Nano* 13 (5): 5778–5784.

43 Peng, P., Du, Y., Zheng, J. et al. (2019). Reconfigurable bioinspired framework nucleic acid nanoplatform dynamically manipulated in living cells for subcellular imaging. *Angew. Chem. Int. Ed.* 58 (6): 1648–1653.

44 Wu, Y.Q. and Wang, Y.X. (2019). Protein circuits reprogram cells. *Nat. Chem. Biol.* 15 (2): 96–97.

45 Wang, W.J., Huang, S., Li, J.J. et al. (2017). Evaluation of intracellular telomerase activity through cascade DNA logic gates. *Chem. Sci.* 8 (1): 174–180.

46 Chen, F., Bai, M., Cao, K. et al. (2017). Programming enzyme-initiated autonomous DNAzyme nanodevices in living cells. *ACS Nano* 11 (12): 11908–11914.

47 Shen, Y., Gong, J.M., Li, S.Q. et al. (2021). Enzyme-free dual-DNA walker based on catalytic hairpin assembled DNAzyme for sensing telomerase activity. *Sens. Actuators, B* 329: 129078.

48 Wolin, S.L. and Maquat, L.E. (2019). Cellular RNA surveillance in health and disease. *Science* 366 (6467): 822–826.

49 Hemphill, J. and Deiters, A. (2013). DNA computation in mammalian cells: microRNA logic operations. *J. Am. Chem. Soc.* 135 (28): 10512–10518.

50 Gong, X., Wei, J., Liu, J. et al. (2019). Programmable intracellular DNA biocomputing circuits for reliable cell recognitions. *Chem. Sci.* 10 (10): 2989–2997.

51 Quan, K., Li, J., Wang, J.L. et al. (2019). Dual-microRNA-controlled double-amplified cascaded logic DNA circuits for accurate discrimination of cell subtypes. *Chem. Sci.* 10 (5): 1442–1449.

52 Li, R., Zhu, Y., Gong, X. et al. (2023). Self-stacking autocatalytic molecular circuit with minimal catalytic DNA assembly. *J. Am. Chem. Soc.* 145 (5): 2999–3007.

53 Ding, H.M., Li, J., Chen, N. et al. (2018). DNA nanostructure-programmed like-charge attraction at the cell-membrane interface. *ACS Cent. Sci.* 4 (10): 1344–1351.

54 Lin, Q., Wang, A.M., Liu, S.Y. et al. (2020). A DNA tetrahedron-based molecular computation device for the logic sensing of dual microRNAs in living cells. *Chem. Commun.* 56 (39): 5303–5306.

55 Groves, B., Chen, Y.J., Zurla, C. et al. (2016). Computing in mammalian cells with nucleic acid strand exchange. *Nat. Nanotechnol.* 11 (3): 287–294.

56 Wang, S., Xia, M.C., Liu, J. et al. (2017). Simultaneous imaging of three tumor-related mRNAs in living cells with a DNA tetrahedron-based multicolor nanoprobe. *ACS Sens.* 2 (6): 735–739.

57 Song, L., Xiao, M.S., Lai, W. et al. (2021). Intracellular logic computation with framework nucleic acid-based circuits for mRNA imaging. *Chin. J. Chem.* 39 (4): 947–953.

58 Zhong, L., Cai, S.X., Huang, Y.Q. et al. (2018). DNA octahedron-based fluorescence nanoprobe for dual tumor-related mRNAs detection and imaging. *Anal. Chem.* 90 (20): 12059–12066.

59 Ren, K.W., Wu, R., Mudiyanselage, A. et al. (2020). In situ genetically cascaded amplification for imaging RNA subcellular locations. *J. Am. Chem. Soc.* 142 (6): 2968–2974.

60 Zhan, J.Y., Liu, Z., Liu, R. et al. (2022). Near-infrared-light-mediated DNA-logic nanomachine for bioorthogonal cascade imaging of endogenous interconnected microRNAs and metal ions. *Anal. Chem.* 94 (48): 16622–16631.

61 Barrenas, F., Chavali, S., Alves, A.C. et al. (2012). Highly interconnected genes in disease-specific networks are enriched for disease-associated polymorphisms. *Genome Biol.* 13 (6): R46.

62 Gebert, L.F.R. and MacRae, I.J. (2019). Regulation of microRNA function in animals. *Nat. Rev. Mol. Cell Biol.* 20 (1): 21–37.

63 Shi, H., Wang, Y.X., Zheng, J. et al. (2019). Dual-responsive DNA nanodevice for the available imaging of an apoptotic signaling pathway in situ. *ACS Nano* 13 (11): 12840–12850.

64 Bai, M., Chen, F., Cao, X.W. et al. (2020). Intracellular entropy-driven multi-bit DNA computing for tumor progression discrimination. *Angew. Chem. Int. Ed.* 59 (32): 13267–13272.

65 Shay, J.W. and Wright, W.E. (2019). Telomeres and telomerase: three decades of progress. *Nat. Rev. Genet.* 20 (5): 299–309.

66 Yang, Q.Q., Yang, F., Dai, W.H. et al. (2021). DNA logic circuits for multiple tumor cells identification using intracellular microRNA molecular bispecific recognition. *Adv. Healthcare Mater.* 10 (21): 2101130.

67 Xie, Z., Wroblewska, L., Prochazka, L. et al. (2011). Multi-input RNAi-based logic circuit for identification of specific cancer cells. *Science* 333 (6047): 1307–1311.

68 Matsuura, S., Ono, H., Kawasaki, S. et al. (2018). Synthetic RNA-based logic computation in mammalian cells. *Nat. Commun.* 9: 4847.

69 Zhang, P.H., He, Z.M., Wang, C. et al. (2015). In situ amplification of intracellular microRNA with MNAzyme nanodevices for multiplexed imaging, logic operation, and controlled drug release. *ACS Nano* 9 (1): 789–798.

70 Miller, J.B., Zhang, S.Y., Kos, P. et al. (2017). Non-viral CRISPR/Cas gene editing in vitro and in vivo enabled by synthetic nanoparticle co-delivery of Cas9 mRNA and sgRNA. *Angew. Chem. Int. Ed.* 56 (4): 1059–1063.

71 Liu, Y.C., Zeng, Y.Y., Liu, L. et al. (2014). Synthesizing and gate genetic circuits based on CRISPR-Cas9 for identification of bladder cancer cells. *Nat. Commun.* 5: 5393.

72 Li, J., Green, A.A., Yan, H. et al. (2017). Engineering nucleic acid structures for programmable molecular circuitry and intracellular biocomputation. *Nat. Chem.* 9 (11): 1056–1067.

19

DNA Supramolecular Hydrogels for Biomedical Applications

Ziwei Shi[1,2], *Yuanchen Dong*[1,2], *and Dongsheng Liu*[3]

[1] *Chinese Academy of Sciences, Institute of Chemistry, CAS Key Laboratory of Colloid, Interface and Chemical Thermodynamics, Beijing 100190, China*
[2] *University of Chinese Academy of Sciences, Beijing 100049, China*
[3] *Tsinghua University, Department of Chemistry, Key Laboratory of Organic Optoelectronics & Molecular Engineering of the Ministry of Education, Beijing 100084, China*

19.1 Introduction

Hydrogels are high-water-retainable materials with three-dimensional (3D) polymeric networks [1]. Due to its unique structure and mechanical properties, hydrogels have become the most popular artificial tools for biomedical applications [2], such as biosensing, drug delivery, immunotherapy, cell culture, and tissue engineering. The hydrophilic polymer networks of the hydrogel can be made of a variety of building blocks, including natural polymers and synthetic polymers. Natural polymers such as proteins [3], hyaluronic acid [4], and polysaccharides [5] are well known for their low cost as well as good biocompatibility. However, the risk of batch-to-batch differences, limited mechanical properties, and immune/inflammatory responses have been a concern for the potential application. On the other hand, synthetic polymers offer greater batch control and can be designed to achieve desired mechanical and other customized properties. However, for now, only a few synthetic polymers, including PEG [6], PVA [7], and PNIPAM [8], can be used as the backbone due to their biocompatibility and biodegradability [9].

DNA supramolecular hydrogels are cross-linked by DNA assemblies, which endow them with advantages including programmable constructions, easy modification, enzyme responsiveness, and injectability. Since DNA chains were grafted onto polyacrylamide backbones to form the hydrogel network in 1996, the accessibility of DNA as a crosslinking unit has been widely demonstrated [10]. In 2009, Liu et al. constructed the first pH-responsive supramolecular DNA hydrogel based on noncovalent interactions [11] and then constructed a DNA supramolecular hydrogel based on base pairing in their follow-up work [12]. The supramolecular DNA hydrogel performed excellent characteristics including thixotropy, shear shining, self-healing, excellent molecular permeability, programmability, biocompatibility,

DNA Nanotechnology for Cell Research: From Bioanalysis to Biomedicine, First Edition. Edited by Zhou Nie.

and biodegradability. These properties promote the broad applications of DNA supramolecular hydrogels in various fields.

In this chapter, we will summarize the classification and formation strategy of DNA supramolecular hydrogels and then focus on their recent progress in biomedical applications, including biosensing, drug delivery, immunotherapy, cell culture, and tissue engineering.

19.2 Classification and Preparation of DNA Supramolecular Hydrogels

According to the composition, DNA supramolecular hydrogels can be basically divided into two categories: (i) pure DNA supramolecular hydrogels, which are made entirely of DNA; (ii) hybrid DNA hydrogels, which are cross-linked by DNA strands in polymers or nanomaterials. In this section, we will outline the formation strategies for DNA supramolecular hydrogels based on their composition.

19.2.1 Pure DNA Supramolecular Hydrogels

In 2009, Liu's group reported the first pure supramolecular DNA hydrogel through a one-pot strategy (Figure 19.1A) [11]. In their study, three DNA strands were partially complementary to each other and could assemble into a Y-shaped DNA nanostructure with three-half i-motif sequences dangling out at all ends as interlocking domains. When the pH is lowered to 5, intermolecular i-motif structures form between scaffolds to form the hydrogel. The formation of i-motif structures is reversible and rapid; therefore, the hydrogel could be switched between gel and solution states to controllably trap and release cargo within 1 min by adjusting the environmental pH value. Although the gelation condition limits its application in physiological conditions, this work innovatively develops a novel pure supramolecular DNA hydrogel with excellent dynamic characteristics and reversible acidic responsiveness, and one-pot assembly provides a simpler preparation strategy.

To broaden the application of DNA supramolecular hydrogels, in 2011, Liu's group reported a pure DNA hydrogel formed by two types of building blocks, including Y-shaped nanostructures and linear linkers (Figure 19.1B) [12]. By simply mixing two components at a proper ratio under physiological conditions, the hydrogel could form within seconds based on the hybridization of sticky ends. The introduction of restriction sites could bring specific enzymatic responsiveness of the hydrogel. In addition to the Y-shaped structures [15, 16], X- [17, 18], T- [19], Takumi-shaped [20], and linear double-stranded DNA (dsDNA) [21–23] structures have also been applied to prepare pure DNA hydrogel scaffolds, enabling multifunctional modification of different cargos or stimuli. Therefore, this common strategy not only inherits advantageous properties, including second-order gel formation and dynamic crosslinking, but also demonstrates the potential to regulate mechanical strength and nuclease responsiveness through adjusting sequence. Furthermore, it exhibits excellent gel

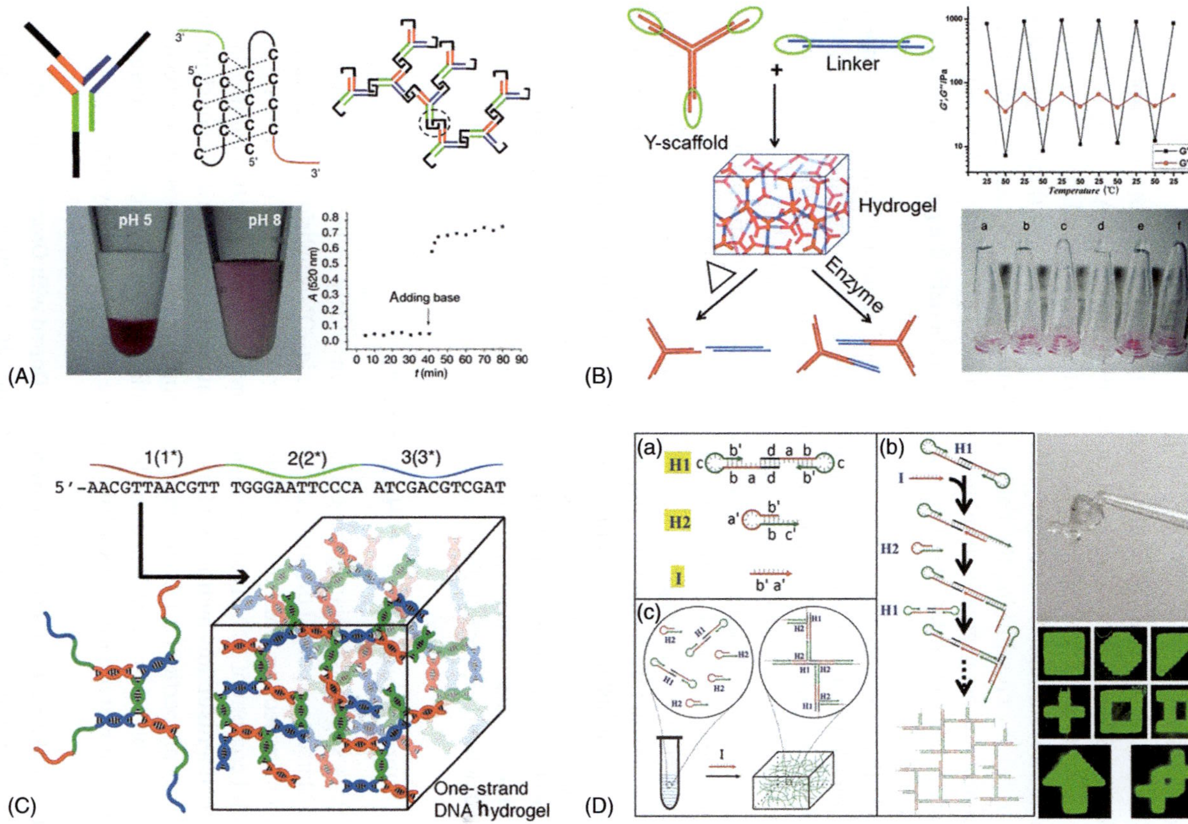

Figure 19.1 Formation of pure DNA supramolecular hydrogels. (A) pH-triggered DNA hydrogels for controlled gold nanoparticle release. Source: Cheng et al. [11]/Reproduced with permission from John Wiley & Sons, Inc. (B) DNA supramolecular hydrogels with designable thermal and enzymatic responsiveness. Source: Xing et al. [12]/Reproduced with permission from John Wiley & Sons, Inc. (C) One-strand multidomain DNA hydrogel with self-complementary palindromic sequence. Source: Reproduced with permission from Jiang et al. [13]. © 2016, Wiley-VCH. (D) DNA initiator triggered cHCR for the self-assembly of patterned DNA hydrogels. Source: Wang et al. [14]/Reproduced with permission from John Wiley & Sons, Inc.

formation ability under physiological conditions, which lays a solid foundation for its future biomedical applications both *in vitro* and *in vivo*.

Supramolecular DNA hydrogels are usually designed with multi-strands, which are expensive and labor intensive. In 2016, Ke et al. prepared DNA supramolecular hydrogels using only one DNA strand (Figure 19.1C) [13]. In their design, the DNA strand was divided into three or more domains, and the sequence of each domain was palindromic. The hydrogel could be formed by crosslinking of the complementary domains. Later, Liu et al. also reported a pH-triggered DNA hydrogel formed by one DNA strand with similar design [24]. The single-strand DNA could form DNA nanowires through two self-complementary sequences and then self-assemble into hydrogels by crosslinking the half i-motif structures under acidic conditions. The one-strand strategy greatly simplifies the synthesis procedure and improves the homogeneity between different batches, which may benefit the preparation of DNA hydrogel.

To bring the preparation process with spatial and temporal control, a clamped hybridization chain reactions (cHCR) strategy has also been reported recently [14]. After adding an initiator strand to solution containing two hairpin strands, as illustrated in Figure 19.1D, hairpin structures could be opened and initiate cascade reactions of DNA hybridization, which could lead to the formation of hydrogel with great spatial and temporal control. Based on similar strategy, Liu et al. developed a pH-triggered DNA hydrogel based on HCR strategy [25]. They introduced a cytosine-rich structure to hybridize with the initiator in a neutral solution. Once the solution is adjusted to pH 5.0, the complementary structure could form C-G·C+ triplex structure, resulting in the release of the initiator I and the formation of a 3D hydrogel network. This study proposes a gelation strategy that enables precise control and pH responsiveness, providing a promising alternative carrier for the controlled delivery of various loadings.

19.2.2 Hybrid Supramolecular DNA Hydrogels

Hybrid DNA supramolecular hydrogels have been widely investigated since Nagahara and Matsuda reported the first polyacrylamide–DNA hybrid hydrogel. In such hybrid supramolecular hydrogels, DNA is generally used as the crosslinking unit, so the solid content of DNA is lower. In addition, different dopants combined with DNA components endow hydrogels with different properties, such as responsiveness and memory properties [26], broadening their application scenarios. In this section, we will focus on DNA supramolecular hydrogels doped with organic polymers or inorganic nanomaterials and DNA-based double network (DN) hydrogels.

19.2.2.1 Hybrid Supramolecular DNA Hydrogels Doped with Organic Polymers

The most commonly used backbone materials are organic polymers, such as polyacrylamide, protein, polypeptide, etc. In 2015, Willner et al. prepared a pH-stimulated shape-memory DNA hydrogel (Figure 19.2a) [27]. In their design, two kinds of DNA strands, a self-complementary sequence and a half i-motif sequence, were grafted on polyacrylamide backbones. Accordingly, at pH 5.0, the

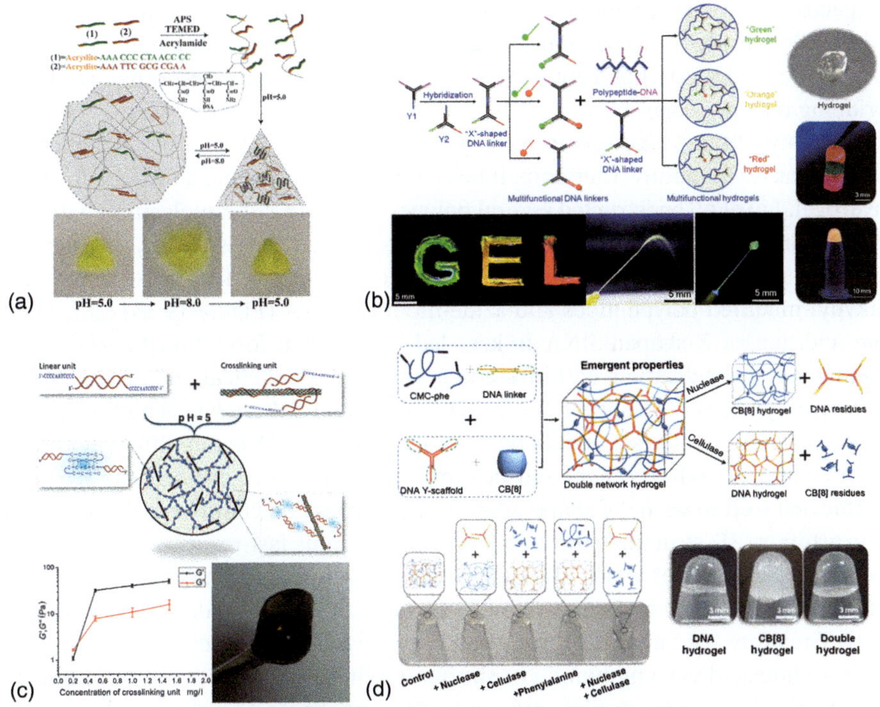

Figure 19.2 Formation of hybrid DNA supramolecular hydrogels: (a) pH-stimulated DNA-polyacrylamide hydrogels exhibit shape-memory properties. Source: Guo et al. [27]/Reproduced with permission from John Wiley & Sons, Inc. (b) Injectable, self-heling polypeptide–DNA hybrid hydrogels with rationally designed multimodification sites. Source: Li et al. [28]/Reproduced with permission from John Wiley & Sons, Inc. (c) Formation of DNA-SWNT hybrid hydrogels. Source: Cheng et al. [29]/Reproduced with permission from Royal Society of Chemistry. (d) Responsive double network hydrogels of interpenetrating DNA and CB [8] host–guest supramolecular systems. Source: Li et al. [30]/Reproduced with permission from John Wiley & Sons, Inc.

hydrogel could be formed through cooperative cross-linking of two points in a triangle-shaped mold. When the pH increased to 8.0, the i-motif structures were dissociated, and hydrogel changed to a shapeless, quasi-liquid state. However, the weak duplex bridges could maintain the entanglement of the polymer chains, preserve the half i-motif sequence in appropriate spatial orientations, and provide the memory code to recover the original shape; thus, the shape of the hydrogel could be recovered in acidic conditions. Furthermore, Willner et al. designed another K^+-triggered shape-memory polyacrylamide–DNA hydrogel [31]. The triangle-shaped hydrogel could be stabilized by DNA duplex and K^+-stabilized G-quadruplex structure. Crown-ether (CE) was also introduced to competitively capture K^+, thus turning the hydrogels into a soft and shapeless state. In another work, by introducing a cytosine-cytosine mismatch to complementary DNA sequences, Ag^+ responsiveness of shape-memory hydrogels could be achieved [32]. The addition of the Ag^+/cysteamine led to the formation/collapse of the network, respectively. This shape-memory hydrogel provided a universal strategy [33], and two kinds of bridge DNA sequences were incorporated into the polymer backbones: one bridged DNA tether acting as responsive stimuli, and the other acting as bridging memory code.

The polypeptide is known to have good biocompatibility and biodegradability, and are easier to modify; therefore, it has emerged as promising backbone for DNA hydrogel. Liu et al. constructed several polypeptide-DNA supramolecular hydrogels [28, 34, 35] and systematically explored their properties. For example, they synthesized the polypeptide-DNA conjugate via Cu^+-catalyzed "click chemistry" between alkynyl-modified polypeptides and azide-modified DNA (Figure 19.2b) [28]. Then, the addition of X-shaped DNA linkers led to hydrogel formation through complementary hybridization within seconds under physiological conditions. When different hydrogel blocks were contacted, they would adhere on each other within minutes, and the interfaces could be eliminated upon DNA chain diffusion after 24 hours, showing extraordinary self-healing properties. Furthermore, the hydrogel performed well in terms of shape-adaptive properties and could be directly written into arbitrary 3D structures through a syringe due to its fast-responding thixotropic properties. In another work, Liu et al. demonstrated that the mechanical strength of DNA–polypeptide hydrogels is positively related to the stability of DNA linkers [34]. Therefore, G' values of these hydrogels may vary from several tens of Pa to several thousand via lengthening or shortening the DNA linker length or inserting mismatch site. These studies systematically investigate the properties of DNA supramolecular hydrogels, including self-healing, thixotropy, shape adaptability, and their corresponding mechanisms, as well as factors that affect their mechanical properties. These findings establish a robust theoretical foundation for the future applications of DNA supramolecular hydrogels.

19.2.2.2 Hybrid Supramolecular DNA Hydrogels Doped with Inorganic Nanomaterials

In addition to organic polymers, other novel nanomaterials like carbon materials (carbon nanotubes, graphene oxide [GO], and carbon dots), magnetic nanoparticles (MNPs) [36], metal nanoparticles [37], and quantum dots [38] have been doped into hydrogels to bring versatile features. For example, Shi et al. reported a strategy for a 3D self-assembled hybrid graphene oxide sheets (GO)/DNA hydrogel [39]. The fish sperm ssDNA connected adjacent GO sheets via strong noncovalent interaction during heating process, and the hydrogel showed high mechanical strength, environmental stability, dye-loading capacity, and self-healing properties. Liu et al. further constructed a DNA/single-walled nanotube (SWNT) hydrogel (Figure 19.2c) [29]. After SWNTs were dispersed in water by DNA wrapping through π–π interactions [40], the network was crosslinked due to the formation of intermolecular i-motif structures when the pH was lower than 6.3. Carbon materials bring hydrogels with excellent conductivity, broadening its application in the electronic field. Later, Liu et al. first introduced MNPs to the DNA supramolecular hydrogel [36]. The imparted magnetic responsiveness enabled hydrogels to be remotely controlled by the magnetic field, showing the potential of DNA hydrogels as smart materials.

19.2.2.3 DNA-Based Double Network Hydrogels

Another typical hybrid supramolecular hydrogel is DN hydrogel, which shows high mechanical strength and toughness. To achieve the contrasting properties of DN hydrogels, the first network is often rigid and brittle, while the second network is soft and ductile [41], so the stress transfer between the two interpenetrating networks leads to dissipate energy to enhance mechanical properties. For example, Liu et al. first developed an interpenetrating DN hydrogel based on entirely supramolecular interactions (Figure 19.2d) [30]. The first network was formed by DNA hybridization between Y structures and linkers, and the second network was constructed through host–guest interaction between cucurbit[8]uril (CB[8]) and phenyl alanine-functionalized carboxymethyl cellulose (CMC-phe). This hydrogel-maintained shear-thinning and thixotropic properties, increased mechanical properties, thermal stability, and excellent stretchability. In addition, networks could be selectively or completely destroyed by use of nuclease and/or cellulase. DNA shows great potential as the building blocks for the construction of DN hydrogel network because physical crosslinking of DNA imparts hydrogel stimuli-responsiveness and self-healing/recovery properties, which are generally ineffective in classical covalent DN hydrogels due to slow polymer chain diffusion. Therefore, DNA-based DN hydrogels have great development prospects.

19.3 Biomedical Application of DNA Supramolecular Hydrogels

By employing rational design of DNA sequences, DNA supramolecular hydrogels can be endowed with multiple stimulus responsiveness and adjustable mechanical properties. The abundant functional groups in DNA have enabled the preparation of customized DNA supramolecular hydrogels, which are promising for various applications. In this section, we will discuss biomedical applications of DNA supramolecular hydrogels, including bioanalysis, drug delivery, immunotherapy, cell culture, and tissue engineering.

19.3.1 DNA Supramolecular Hydrogels for Bio-sensing

Qualitative and quantitative detection of various analytes *in vitro* and *in vivo* is essential in many biomedical fields, from environmental monitoring to food safety to clinical diagnosis [42]. A typical design strategy for DNA supramolecular hydrogel biosensors employs synthetic polymers as scaffolds, with functional DNA sequences like aptamers, DNAzyme, i-motif or antisense DNA as responsive entities to extend the molecular recognition capability. Upon addition of target molecules, the responsive moiety would dissociate and release the encapsulated cargo, producing a readout signal or inducing changes in hydrogel properties, thus presenting the existence of target molecules, including ions, biological small molecules, and nucleic acid.

Colorimetric method is the preferred detection method for most DNA supramolecular hydrogel biosensors. In 2008, Tan et al. demonstrated a general method for fast and easy engineering of target-responsive hydrogels based on aptamers [43]. Two DNA single chains, strands A and B, were grafted onto the polyacrylamide chains, respectively. Upon addition of a linker termed as LinkerAdap, it would hybridize with strands A and B, inducing the crosslinking of polyacrylamide chain to form the hydrogel. Notably, the LinkerAdap contained an adenosine aptamer sequence that included the toehold and six complementary bases with strand B. After the adenosine was added, the aptamer would bind to adenosine and release the strand B, resulting in the dissolution of the hydrogel. Gold nanoparticles could be incorporated into the hydrogel to visualize this transition process.

Later, in 2010, similar to the previous preparation method, Tan et al. established a colorimetric platform for visual detection of cocaine with an amylose–iodine complex as a visual indicator (Figure 19.3a) [44]. In their study, amylase was first mixed with polymer strands A and B (PS-A and PS-B) and encapsulated in the hydrogel upon the addition of Linker-Apt. When cocaine was added, the hydrogel was destroyed, and the amylase was released into the upper solution containing the amylose–iodine complex, and its color could change from dark to yellow in the presence of amylase. Therefore, the color change exhibited a positive correlation with the amount of cocaine. Further, this strategy has been extended to the detection of toxins like microcystin-LR [48], ochratoxin A [49], aflatoxin B1 [50], and T-2 toxin [51] for food safety, molecules like adenosine [52], miRNA [53], cocaine, and ATP [54, 55] for diagnosis, and heavy metal ions like Pb^{2+} [56, 57], Cu^{2+} [58], and UO^{2+} [59] for environmental detection.

Colorimetry often provides semiquantitative analysis and relatively low sensitivity, and to overcome such limitations, different signal amplification strategies based on electrochemistry [45], surface-enhanced Raman scattering (SERS) [46, 60, 61], and the capillary behavior [62] have been developed. For example, Liu et al. reported an electrochemical biosensor for the detection of microRNA that is specific to lung cancer (Figure 19.3b) [45]. Ferrocene-tagged DNAs, which served as recognition probes, could crosslink DNA-grafted polyacrylamide to form hybrid hydrogel. The hybrid hydrogel was immobilized on an indium tin oxide/polyethylene terephthalate (ITO/PET) electrode, and upon hybridization of the recognition probes with the target miRNA, the hydrogel dissociated, resulting in the loss of the ferrocene tag and thus a decrease in measurable current. The sensor could detect miR-21 at a concentration of 5 nM (1 pmol) with a linear readout range of 10 nM to 50 μM. In addition, the recognition probe can be applied to a variety of other targets due to its flexible sequence design.

As an excellent signal amplification strategy, SERS has also been combined with DNA supramolecular hydrogels for the detection of different miRNAs for cancer screening [60, 61]. For example, Wang et al. constructed a biosensor platform combined with previously targeted responsive DNA hydrogel for sensitive detection of alpha-fetoprotein (AFP), which is an important biomarker of hepatocellular carcinoma (Figure 19.3c) [46]. The aptamer could specifically recognize AFP and accurately control the release of immunoglobulin G (IgG) in the hydrogel. Thereafter, the released IgG was captured by SERS probes and biofunctional magnetic beads, resulting in the change of Raman signal in the supernatant after magnetic separation. This

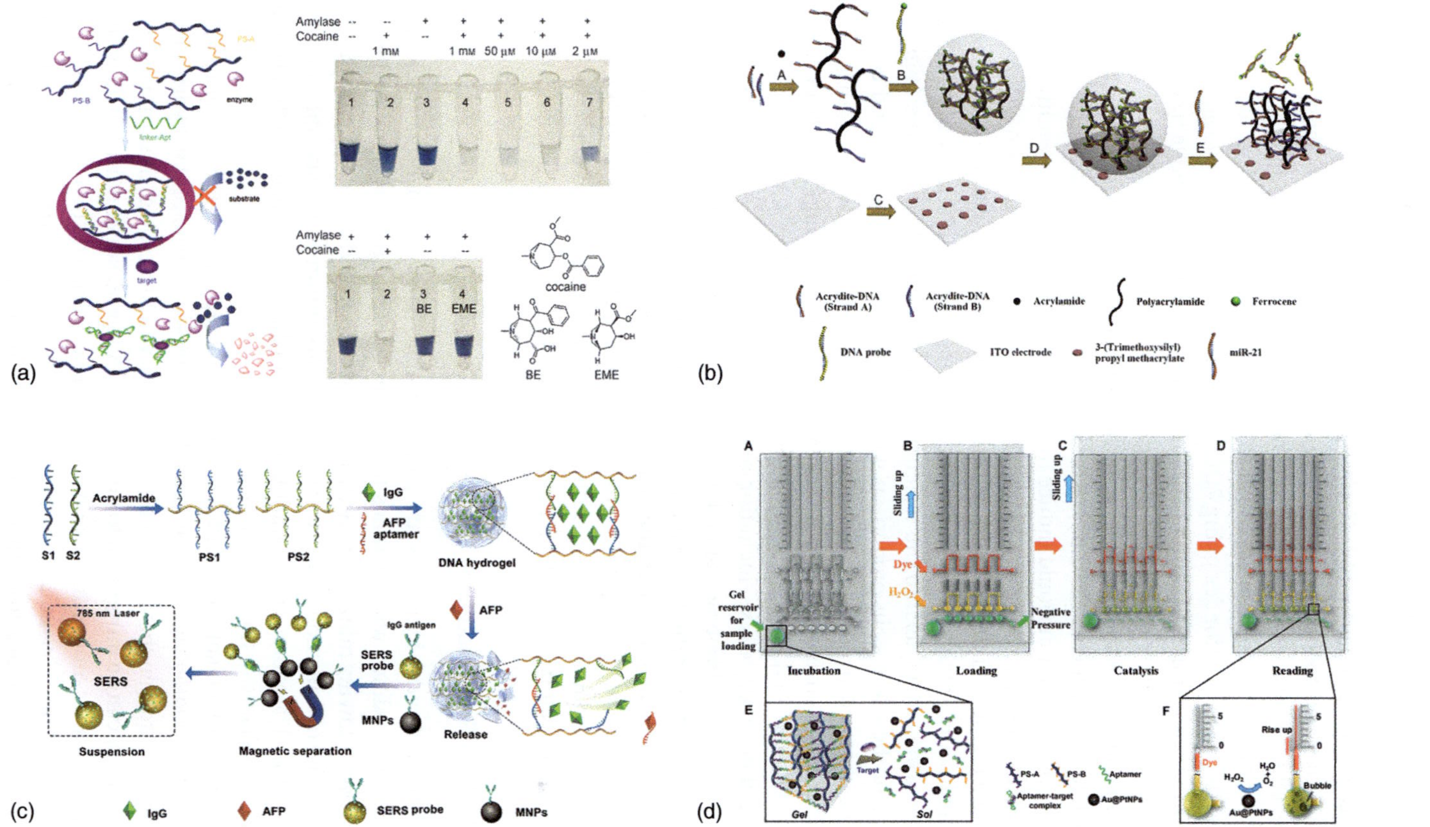

Figure 19.3 DNA supramolecular hydrogels for biosensing: (a) the enzyme-caged hydrogel combined with aptamer achieving gel–sol transition for visual detection of cocaine. Source: Reproduced with permission from Zhu et al. [44]. © 2010 Wiley-VCH. (b) Electrochemical DNA hydrogel biosensor to detect target miR-21. Source: Reproduced with permission from Liu et al. [45]. © 2018 Elsevier. (c) Aptamer-responsive DNA functionalized hydrogels using SERS method for sensitive detection of AFP. Source: Reproduced with permission from Wang et al. [46]. © 2020 American Chemical Society. (d) The working principle of the Au@Pt nanoparticle-encapsulated target-responsive hydrogel with a volumetric bar-chart chip readout for visual quantitative detection. Source: Reproduced with permission from Zhu et al. [47]. © 2014 Wiley-VCH.

method could be used for quantitative detection of AFP, with a detection limit of 50 pg/ml.

In recent years, considerable efforts have been made to develop point-of-care testing (POCT) devices that meet the requirements of stability, portability, ease of storage, and low cost. In particular, volumetric or phase change properties of DNA supramolecular hydrogels in response to stimulus are considered as ideal signal transduction strategies for designing POCT devices. For example, Zhu et al. designed a device termed volumetric bar-chart chip (V-chip), where quantitative results were displayed as bar charts directly on the device [63]. They integrated an Au core/Pt shell nanoparticle (Au@PtNP), which was entrapped in DNA hydrogel, with a volumetric bar-chart chip (V-Chip) as a visual quantitative readout (Figure 19.3D) [47]. Target introduction would induce the hydrogel to break down, resulting in the release of the entrapped Au@PtNPs into the supernatant solution. Then by applying a negative pressure and sliding up, the supernatant was brought into contact with H_2O_2 and could catalyze H_2O_2 decomposition to generate O_2. Subsequently, red ink was pushed upwards into the top channel by the O_2 bubbles that the migrated distance of the ink was in proportion with the target concentration, and cocaine could be detected at a concentration lower than 1 μM. The V-Chip device has allowed for the convenient detection of other targets, including ochratoxin A [49] and Aflatoxin B1 [64], with a detection limit of 1.27 and 1.77 nM, respectively.

In addition, other devices, including personal glucose meters (PGMs) [53, 55, 65] and microfluidic paper-based analytic device (μPAD) [66, 67], have also been combined with DNA supramolecular hydrogels to expand the application scenarios of POCT. The rationale is that binding of the target causes the hydrogel network to collapse, releasing the encapsulant and producing a quantitative result by the naked eye. Therefore, POCT offers a fast, simple, and low-cost detection platform in the absence of laboratory infrastructure and expertise.

19.3.2 DNA Supramolecular Hydrogels for Drug Delivery

DNA supramolecular hydrogels are biocompatible, biodegradable, injectable and can be easily modified with modules for specifically identifying and responding to targets, thus allowing minimally invasive administration at the target sites with controlled release of drugs and high drug concentrations, minimal toxicity and immunogenicity [68]. Therefore, DNA supramolecular hydrogels have been employed as carriers for delivery of functional molecules.

Small molecules, especially anticancer chemotherapy drugs, are widely used in drug therapy. Some drugs can be loaded into DNA supramolecular hydrogels to treat cancer through chemical binding with DNA duplexes, e.g. doxorubicin. For instance, DOX can be introduced to polymer-DNA hybrid supramolecular network. After DNA strands are modified on backbones like carboxymethyl chitosan (CMC) chains [69] or polyacrylamide [70], the addition of aptamers initiates the hybridization of oligonucleotides to form the hydrogels. Introduction of target molecules results in breakdown of hydrogel network and unloading of DOX for tumor therapy. Recently, synergistic therapy has gained extensive attention due to enhanced efficacy against cancer. Guo et al. combined DNA modified polyacrylamide with MXene nanosheets to form a DOX-loaded MX-DNA supramolecular hydrogel [71]. Under near-infrared (NIR) light irradiation, MXene nanosheets converted the NIR energy into heat, and temperature rise triggered the gel-to-solution transition of hydrogels and led to the localized release of DOX, achieving photothermal-chemo synergistic cancer therapy.

Some hydrophobic small molecule drugs are difficult to disperse in the hydrogel network, thus affecting the loading efficiency and *in vivo* therapeutic effect. To solve this problem, covalently conjugation of small molecules onto the DNA strands has been developed [72, 73]. For example, DNA and RNA strands were integrated with floxuridine either by solid-phase synthesis or via enzyme-mediated transcription due to structural similarity between floxuridine and thymine. Then nucleic acid strands assembled into Y-shaped structures and further formed nanogels via hybridization, which could be efficiently uptake by cancer cells and then released the therapeutic agents, exhibiting enhanced anticancer effect compared to floxuridine-containing ssDNA/ssRNA *in vitro*. This strategy can also be applied for other nucleoside analogue drugs. Later, the same group reported another general strategy for the drug-modified DNA supramolecular hydrogel (Figure 19.4a) [72]. They grafted multiple camptothecin (CPT), a hydrophobic chemotherapeutic drug, onto the backbone of the phosphorothioate (PS) oligonucleotide with a disulfide linkage, then CPT-grafted DNAs were assembled into Y-shaped motif and assembled into hydrogels. After injection, hydrogel would disassemble into nanogels by enzymatic degradation, thus penetrating tumor tissues, being internalized by cancer cells, and releasing CPT rapidly in response to GSH, which enhanced anti-tumor efficacy and prevented cancer recurrence *in vivo*.

Another strategy for drug delivery through DNA supramolecular hydrogels is to load small molecules into aggregates like liposomes or microspheres. When cargos are similar to/larger than mesh sizes, cargos will be trapped, and be released only when surrounding hydrogels collapse, thus achieving sustained release of drugs. Based on this strategy, Zhu et al. fabricated a DNA/poly(lactic-*co*-glycolicacid) (PLGA) hybrid hydrogel for water-insoluble ophthalmic therapeutic delivery (Figure 19.4b) [74]. After encapsulating dexamethasone (DEX), which can attenuate inflammatory symptom in allergic conjunctivitis (AC), into PLGA nanoparticles, then cargos were loaded into the hydrogel. This hydrogel system prolonged DEX retention time to 24h due to gradual release in ophthalmic cells and tissues, and the efficacy of once daily treatment was much better than that of multiple treatments with commercial DEX drops.

In addition to small molecules, large molecules like functional protein drugs can also be stabilized and delivered for therapy. DNA hydrogels can impede penetration of proteins like proteases, protecting protein bioactivity from degradation by enzymes that diffuse inward. Based on previous work proposed by Liu's group [35], Gacanin et al. prepared a human serum albumin (HSA) derived protein-DNA hybrid supramolecular hydrogel functionalized with recombinant C3 toxin for local treatment of bone diseases [76]. The hydrogel maintained original bioactivity of the protein, whose further controlled release due to DNase degradation of hydrogels

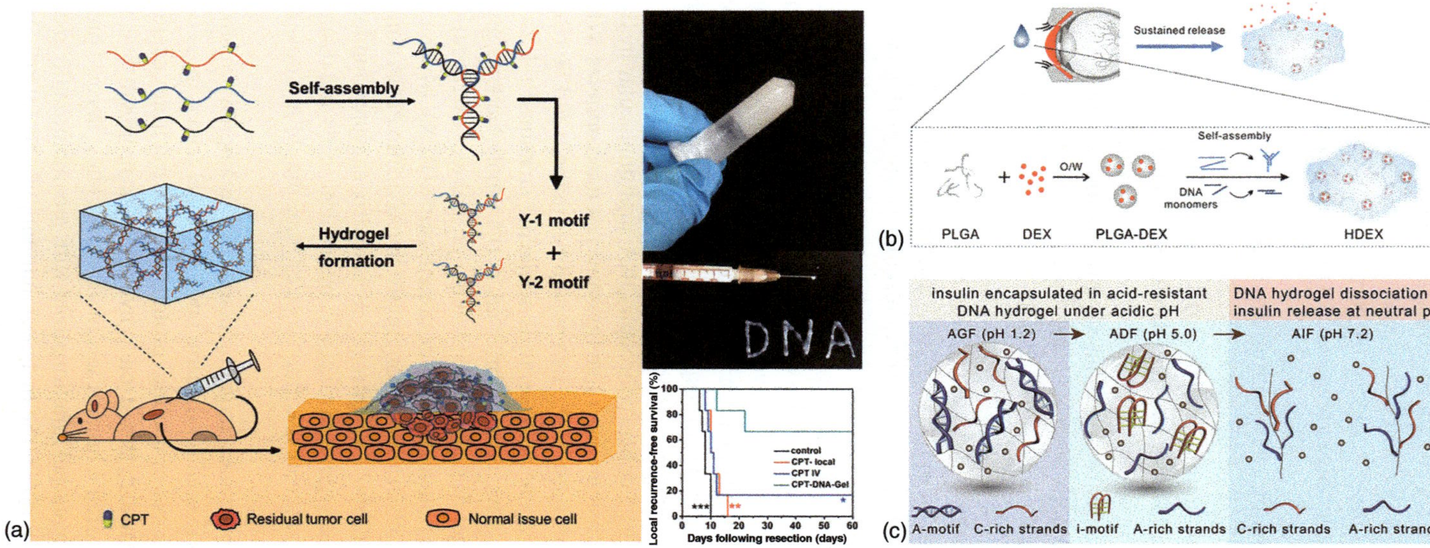

Figure 19.4 DNA supramolecular hydrogels for drug delivery. (a) Injectable CPT-conjugated DNA hydrogel to prevent tumor recurrence. Source: Zhang et al. [72]/Reproduced with permission from American Chemical Society. (b) Preparation of PLGA-DEX DNA hydrogels for sustained DEX release in ophthalmic tissues. Source: Reproduced with permission from Ren et al. [74]. © 2019 American Chemical Society. (c) Acid-resistant and physiological pH-responsive DNA hydrogel for oral insulin delivery. Source: Reproduced with permission from Hu et al. [75]. © 2022 American Chemical Society.

promoted skeletal repair of bone diseases and local improvement of bone quality in osteoporosis bones. Ying et al. recently have reported a novel acid-resistant DNA hydrogel for oral insulin delivery (Figure 19.4c) [75]. Adenine (A)-rich and cytosine (C)-rich oligonucleotides were conjugated onto polyacrylamide chains, then formed the hydrogel and encapsulated insulin in acidic pH (1.2–6.0), cross-linked by parallel A-motif duplex configurations (pH 1.2–3.0) and i-motif structures (pH 4.0–6.0). Under physiological pH, the separation of A-motif and i-motif would induce collapse of DNA hydrogel. Therefore, oral drug delivery was achieved in diabetic rats, because the hydrogel maintained intact in the stomach (pH 1.2), duodenum (pH 5.0), and broke down in small intestine (pH 7.2), where insulin would be released and absorbed.

Functional nucleic acids, such as antisense RNA/DNA, and small interfering RNA (siRNA), have become promising clinical reagents in gene therapy. DNA supramolecular hydrogels can load DNA or RNA via hybridization or direct integrating into structural units, protecting naked nucleic acids from rapid degradation *in vivo*. In 2015, Tan et al. reported a DNA nanohydrogels through hybridization for targeted and GSH-responsive gene therapy [77]. Aptamers, disulfide linkages, and therapeutic genes (antisense DNA, DNAzyme) were incorporated into building units. After actively targeting cells and being cleaved by the intracellular reductive environment, nanohydrogels could release therapeutic genes to silence mRNA expression and inhibit proliferation, metastasis of cancer cells. Zhang et al. incorporated siRNA to crosslink DNA-grafted polycaprolactone (DNA-gPCL) brushes to form spherical nanogels [78]. With great resistance to RNase degradation, enhanced cell uptake, and targeted delivery to tumor sites, this nanogel showed excellent anticancer effects both *in vitro* and *in vivo*. Later, based on similar design, they loaded Cas9 and single guide RNA (Cas9/sgRNA) complex on DNA-g-PCL via the hybridization of DNA and the tail of gRNA. Then, the remaining DNA-grafted PCL chains were crosslinked by DNA linkers, which led to the nanogel formation. Cas9/sgRNA nanogels realized the efficient target gene editing *in vitro* and successfully knocked down the EGFP expression in HeLa cells. However, synthetic materials may induce potential toxicity. Therefore, recently, this group has constructed a novel pure nucleic acid nanogel based on the hybridization of sticky ends on the DNA-tetrahedra and siRNA for gene therapy [79]. Without extra synthetic materials, this nanogel effectively knocked down the target gene expression with excellent biocompatibility.

19.3.3 DNA Supramolecular Hydrogels for Immunotherapy

Immunotherapy is a method of treating diseases by managing body's native immune system [80], which can not only eliminate both local and distant metastatic tumors but can also prevent relapse of tumors through long-term immune memory effects [81]. In recent years, DNA supramolecular hydrogels have shown great potential in immunotherapy, which can target organs like lymph nodes (LNs), lymphoid residing cells and their intracellular compartments, function as surrogate lymphoid-like tissues, and deliver multiple biomolecules that can modulate the humoral and cellular immune response [82].

DNA hydrogels can be immunologically active when they contain immunostimulatory sequences, e.g. CpG motifs, which are common adjuvants in immunotherapy of various diseases [83, 84]. In 2011, Nishikawa et al. first proved better immunostimulatory effects of CpG DNA hydrogels compared to CpG DNA [85]. Further, the same group developed a DNA supramolecular hydrogel formed by CpG motif-containing polypodna with ovalbumin (OVA) as a model antigen (Figure 19.5a) [86]. The self-gelling and injectable CpG DNA hydrogel showed a high immunological potency and low toxicity as a vaccine adjuvant with few adverse events. To prolong the time of continuous antigen release and increase the specific immune response, the author optimized CpG DNA hydrogel system and proposed different strategies including cationizing the antigen to enhance the electrostatic interaction with DNA [89], modifying DNA with cholesterol to enhance the interaction between the antigen and the scaffold through hydrophobic interaction [90], and incorporating chitosan to make the hydrogel stronger and more stable with longer degradation time [91, 92].

CpG DNA supramolecular hydrogels can be used as adjuvants to activate the immune system and as carriers to deliver and protect antigens, thus they have great potential in cancer immunotherapy [93]. Umeki et al. designed a CpG DNA hexapodna hydrogel which combined tumor antigens with antigen presenting

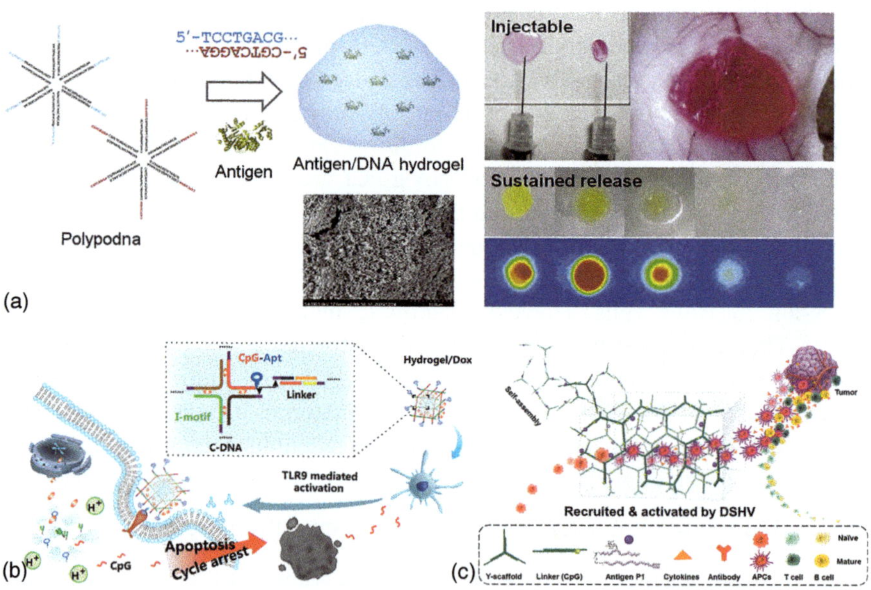

Figure 19.5 DNA supramolecular hydrogels for immunotherapy. (a) Injectable, self-gelling, biodegradable, and immunomodulatory CpG DNA hydrogel for antigen delivery. Source: Nishikawa et al. [86]/Reproduced with permission from Elsevier. (b) CpG-MUC1-hydrogel/Dox for combined antitumor therapy. Source: Reproduced with permission from Shao et al. [87]. © 2018 American Chemical Society. (c) Recruitment and activation of host APCs for robust immune response and antitumor effects *in vivo* induced by the DSHV system. Source: Reproduced with permission from Wei et al. [88]. © 2019 American Chemical Society.

cells (APCs) [94]. The results showed that DNA supramolecular hydrogels injected into tumor could effectively trigger tumor antigen-specific cellular immunity, thus producing strong anti-tumor effect. Yata et al. further introduced this system with gold nanoparticles to realize photothermal immunotherapy of tumors [95]. Sun et al. developed a DNA nanogel vaccine combine with CpG ODN and OVA [96], they proved that the vaccination of the DNA nanogel could induce efficient LN accumulation and inhibit murine melanoma tumor growth.

Recently, Shao et al. have developed an injectable DNA supramolecular hydrogel vaccine (DSHV) (Figure 19.5b) [87]. The MUC1 glycopeptide molecule was connected into a single chain with the T cell epitope of tetanus toxoid and modified the lysine residue, which was loaded into the CpG-containing DNA supramolecular hydrogel network based on its electrostatic interaction. After DSHV was injected into tumor-bearing mice, it could mimic the function of LN to recruit and activate APCs, which distinguished the previous system, thus generating strong immune response and anti-tumor effect. Duan et al. also integrated pH-responsive i-motif and targeted MUC1 adaptor-immune adjuvant CpG into size-controlled DNA nanogels. DOX was inserted into the base pairs of CpG-MUC1 to yield a CpG-MUC1/DOX hydrogel (Figure 19.5c) [88]. After internalization, the hydrogel was dissociated under acidic conditions to induce tumor cell apoptosis, and then CpGs were released to trigger vast cytokine secretion from immune cells, thus further inducing an obvious anti-tumor immune response against malignancy.

In addition to CpG ODNs, guanosine- and uridine-rich ssRNA (GU-rich RNA) can recognize Toll-like receptor 7 (TLR7) and Toll-like receptor 8 (TLR8) and can also induce strong immune responses [97]. However, the instability of RNA limits further applications. To avoid rapid elimination of RNA, Komura et al. developed a GU-rich hexapod-based DNA/RNA hydrogel (RDgel) consisting of 20-mer PS-stable GU-rich RNA [98]. The results showed that it could continuously release GU-rich RNA and induce an effective immune response against colonic tumor cells.

19.3.4 DNA Supramolecular Hydrogels for 3D Cell Culture

Due to a similar network topology and water content to the extracellular matrix (ECM), hydrogels have emerged as the most popular artificial 3D ECM providing biochemical and biophysical cues to regulate cell activities and tissue/organ functions. However, the small meshes in the conventional chemical hydrogels may be unfavorable to cell proliferation and migration. In contrast, the DNA molecule is known for its long persistence length, which can exclude the small meshes in the hydrogel and therefore bring good permeability to nutrition for 3D cell culture. Thus, Liu et al. trapped single cells in microwells and then sealed them with the DNA supramolecular hydrogel as a cover for single-cell analysis (Figure 19.6a) [99]. After cells were trapped in the PDMS microwells, the DNA supramolecular hydrogel cover was formed by adding a Y-scaffold and a linker on top of the chip. To confirm the permeability of hydrogel cover, they used sodium fluorescein molecules as probes and found that they could diffuse through the cover in minutes, and

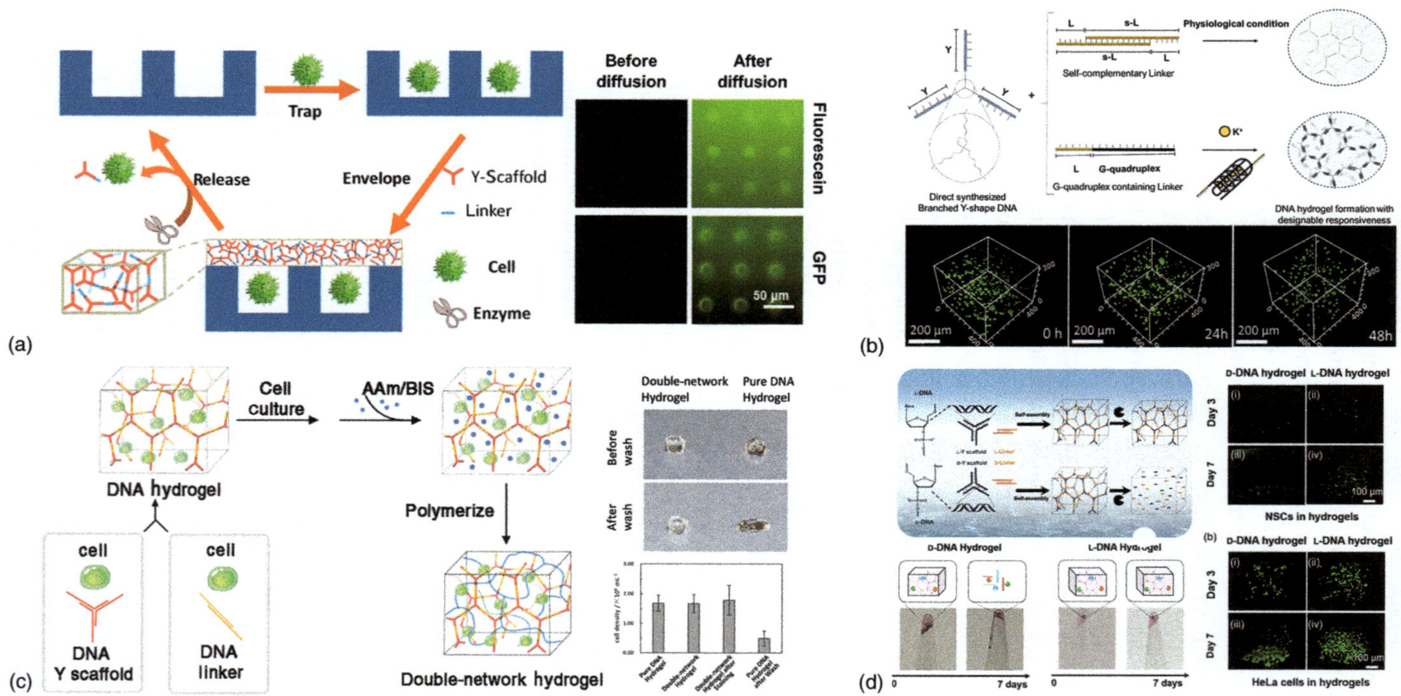

Figure 19.6 DNA supramolecular hydrogels for 3D cell culture. (a) A triggered DNA hydrogel cover to envelop, release single cells and diffuse sodium fluorescein or GFP protein. Source: Jin et al. [99]/Reproduced with permission from John Wiley & Sons, Inc. (b) Dendritic DNA self-assembled hydrogel for 3D cell culture. Source: Yang et al. [100]/Reproduced with permission from American Chemical Society. (c) Formation of a DNA–PAAm double-network hydrogel and its application for 3D cell imaging. Source: Cao et al. [101]/Reproduced with permission from American Chemical Society. (d) Formation and 3D cell culture of D-DNA and biostable L-DNA hydrogels. Source: Yang et al. [102]/John Wiley & Sons, Inc.

larger molecules such as green fluorescent protein (GFP) with 22 kDa could also go through the network. Furthermore, the DNA supramolecular hydrogel overlay could be specifically opened by cleaving restriction enzymes, thus releasing cells on demand. After 24 hours culture, $98 \pm 1\%$ of NIH 3T3 cells (fibroblast cell line) remained alive. Therefore, the DNA supramolecular hydrogel is permeable for nutrients and keeps cells viable.

In addition, DNA supramolecular hydrogels are excellent 3D support for cells to adhere to and grow on. For example, Yang et al. have recently constructed a supramolecular DNA hydrogel based on chemically branched DNA (Figure 19.6b) [100]. Based on previous work, they replaced Y-shaped pure DNA structures with commercially available branched Y-shaped DNA whose three arms were conjugated with oligoethylene glycol chain. By adding linkers, the supramolecular hydrogel could be formed, and the K^+ responsiveness could be realized by introducing intermolecular G-quadruplex sequence to the linker. Compared to multistrand system, this strategy can simplify the hydrogel composition and preparation processes, and it is feasible for 3D cell culture because noncovalent supramolecular interactions provide structural dynamics such as self-healing, shear-thinning, and thixotropy for cell growth and migration.

Weak mechanical properties of DNA supramolecular hydrogels usually limit their application in 3D cell culture. To overcome this challenge, Liu's group has recently demonstrated a mechanical reinforcement strategy by replacing small amounts of linkers with polymeric multiple-unit linker (PMUL), which covalently incorporate multiple supramolecular units into a polymeric backbone to crosslink mechanically reinforced DNA supramolecular (MRDS) hydrogel [103]. Under nearly the same solid content, MRDS hydrogels showed higher mechanical properties (up to nearly three times) with dynamic behaviors, compared to previous Y-L DNA supramolecular hydrogels, because PMUL could trap the DNA crosslinkers in a kinetically interlocked state through the synergetic effect of multiple-unit interaction without changing single interaction, thus reinforcing MRDS hydrogels. They confirmed the excellent cellular biocompatibility of MRDS hydrogel, where HeLa cells could survive and proliferate well; therefore, tunable mechanical properties made it possible to explore the influence of hydrogel strength on cell culture and differentiation.

With the fast development of DNA supramolecular hydrogels in 3D cell culture, more concerns have arisen on the fast hydrolysis of DNA under physiological conditions. To solve these problems, Cao et al. developed a double hydrogel network for cell culture and observation (Figure 19.6c) [101]. After cells were encapsulated in Y scaffold-based DNA hydrogel network, the second polyacrylamide network was formed by introducing AAm/BIS monomers for in situ covalent polymerization. Compared to single-component hydrogels, the DN hydrogel had better mechanical properties and tolerated high temperatures up to 70 °C. Specially, under repeated washing, DN hydrogel could maintain the shape and the 3D location of cells, while pure DNA hydrogel would collapse and lead to cell sedimentation; thus, the interpenetrating network showed potential in 3D immunostaining and imaging. Yang et al. developed a novel L-DNA hydrogel (Figure 19.6d) [102]. L-DNA hydrogels performed identical sequences to previously reported D-DNA hydrogels, which endowed them with similar physical properties. However, their deoxynucleotide was mirror symmetric, and only D-deoxyribose would work in biological processes. Therefore, the L-DNA hydrogel could obviously alleviate inflammation responses and was more stable against nuclease degradation, both *in vitro* and *in vivo*. After cells were cultured in hydrogel for seven days, cells descended at the bottom in the D-DNA hydrogel due to degradation while retaining their 3D distributions in the L-DNA hydrogel; hence, the L-DNA hydrogel is a better cell culture platform for long-term applications.

19.3.5 DNA Supramolecular Hydrogels for Tissue Engineering

Tissue engineering refers to the development of biological substitutes for the functional repair or regeneration of tissues or organs [104], thus overcoming current limitations in organ transplantation [105] and reducing risk of immune rejection [106]. While artificial ECM can imitate the mechanical strength of natural soft tissues, DNA supramolecular hydrogels can be used as self-support scaffolds for cell adhesion and proliferation, vehicles to deliver cells to a designated location in the body, fillers to fill defects and promote healing while preventing infection, thus showing great potential in tissue engineering.

19.3.5.1 Construction of Artificial Tissues *In Vitro*

Applying the layer-by-layer deposition of biomaterials onto a computer-controlled build platform, 3D bioprinting can construct highly complex artificial tissue-like materials with precise structure and functional versatility *in vitro* [107]. Liu and Shu et al. realized in situ multilayer 3D living cell bioprinting with the DNA-polypeptide hybrid supramolecular hydrogel (Figure 19.7a) [108]. In this study, two bio-inks, bio-ink A composed of polypeptide–DNA conjugates mixed with living cells and bio-ink B composed of the complementary DNA, were co-printed through the dual-nozzle printer. Printed hydrogels were uniform constructs with no obvious gaps between layers and maintained their initial shape without collapse due to self-healing and non-shrinking/swelling properties (Figure 19.7b) and high mechanical strength (Figure 19.7c). Specially, the cells in the printed hydrogel had high viability up to 98.8% (Figure 19.7d). Therefore, DNA supramolecular hydrogels can be used as an outstanding ink material for 3D printing with extraordinary biocompatibility and biodegradability.

Later, Liu et al. developed a new "brick-to-wall" strategy to construct tissue-like structures based on pure DNA hydrogel [109]. The interface between hydrogel bricks could be eliminated by DNA chain exchange reactions, so 3D target structures could be constructed by simply attaching different cell-containing bricks together, and cells could migrate between different connected hydrogel bricks. Meanwhile, the damaged cells in bricks could be excluded from the manufacturing process, preventing defects in the final products. This strategy provides a novel way to avoid cell damage during the manufacturing process and also has good prospects in artificial tissue fabrication.

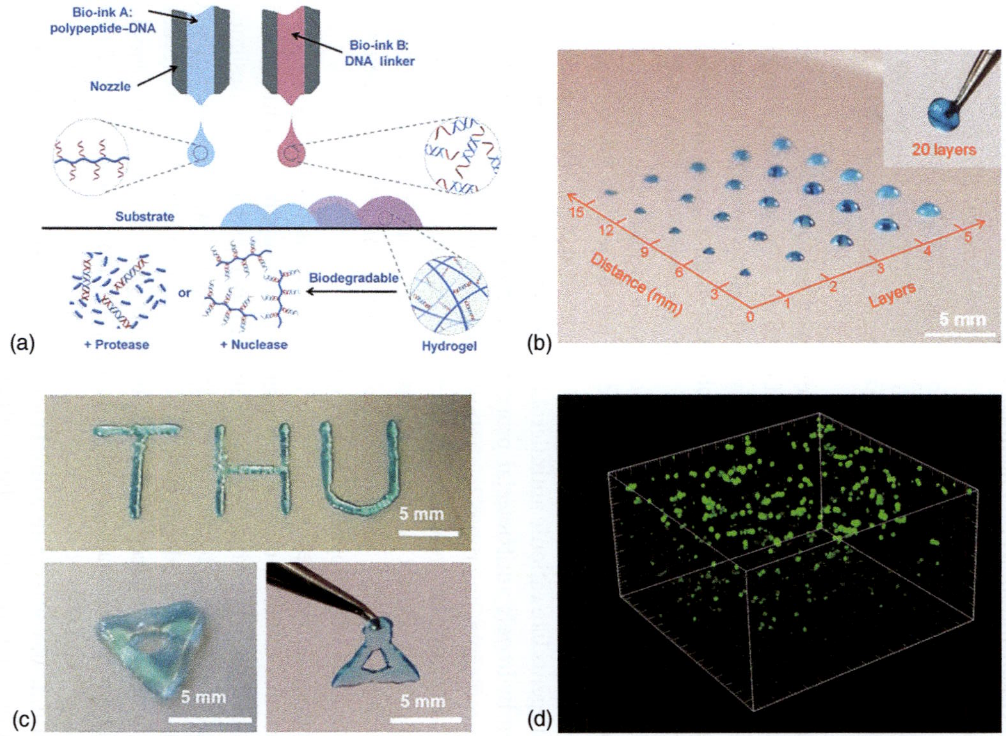

Figure 19.7 DNA supramolecular hydrogels for construction of artificial tissues *in vitro*: (a) 3D bioprinting of the polypeptide–DNA hydrogel to fabricate arbitrarily designed 3D structures; (b) array of multiple layer dots formed via repeating printing layer-by-layer; (c) the letters "THU" printed with five layers and a triangle with ten layers, which is strong enough to be picked up with tweezers; (d) 3D stacks of AtT-20 cells printed in a hydrogel with FDA staining in green. Source: Li et al. [108]/Reproduced with permission from John Wiley & Sons, Inc.

19.3.5.2 Tissue Engineering

DNA supramolecular hydrogels have also been applied for tissue regeneration *in vivo*. Firstly, the injectability and self-healing properties of DNA supramolecular hydrogels allow them to fill and cover defects with irregular shapes and depths. Secondly, the mechanical strength can be easily adjusted for as appropriate mechanical support. Thirdly, biodegradability allows hydrogels to be transplanted to the lesion without additional removal, thus avoiding secondary injury. In addition, large meshes of pure DNA supramolecular hydrogel network allow nutrients and growth factors to permeate and quickly diffuse, promoting cell differentiation, proliferation, and migration, leading to tissue regeneration. Good biocompatibility also avoids concerns about potential biological toxicity. Hence, DNA supramolecular hydrogels can work as scaffolds for in-situ regeneration of tissues in the body, providing a versatile platform.

DNA supramolecular hydrogels have been demonstrated to be effective in wound dressings for skin healing. In 2015, Lee et al. fabricated a tannic acid (TNA) DNA hydrogel with good hemostatic capability (Figure 19.8a) [110]. Tannic acid (TA) could work as a molecular glue whose polyphenols could link to the phosphate backbone of DNA through hydrogel bonds and further form DNA hydrogel. TNA hydrogel was degradable, extensible upon pulling, and adhesive to tissues, therefore showed great hemostatic ability after being applied on the wound (53s for hemostasis) compared to the group (133s for hemostasis) without treatment. Introduction of agents like cationic antimicrobial peptides [114], silver nanoclusters [115] can bring hydrogels antibacterial properties. For example, Yang et al. used another small-molecule microbicide, tetrakis (hydroxymethyl) phosphonium sulfate (THPS), to crosslink DNA via hydrogen bond and electrostatic interaction, thus forming an antibacterial wound dressing and accelerating wound healing [116]. In the above work [114, 115], both used salmon sperm DNA, thus enabling large-scale production of hydrogel with lower cost.

Chronic inflammation caused by diabetes may hinder tissue healing. In order to reduce inflammation and promote tissue regeneration, hydrogels with antioxidant, immunomodulatory, and anti-inflammatory properties have been designed. Wang et al. enveloped the cytokine interleukin-33 (IL-33), which contributes to wound healing by triggering adaptive immune responses, in DNA supramolecular hydrogels assembled by Y-shaped scaffolds and linkers [117]. DNA supramolecular hydrogels could retain the bioactivity of IL-33, and the sustained release IL-33 for at least seven days could resolve local inflammation in the diabetic wound, improve tissue regeneration, and accelerate wound closure *in vivo*. They also found that the DNA hydrogel itself has the antioxidant property, which could scavenge excessive ROS. Li et al. encapsulated cytokine interleukin-10 (IL-10) into an osteogenic DNA supramolecular hydrogel (ILGel) based on Y-shaped scaffolds and linkers [118]. Under diabetic conditions, ILGel could promote M2 phenotype polarization, attenuate periodontal inflammation, and trigger osteogenesis of mesenchymal stem cells (MSCs), thus accelerating alveolar bone formation and defect healing. Jing et al. designed the 8-arm PEG/DNA hybrid hydrogel through base pairing to deliver stem cells from apical papilla-derived exosomes (SCAP-Exo) [119]. This system could

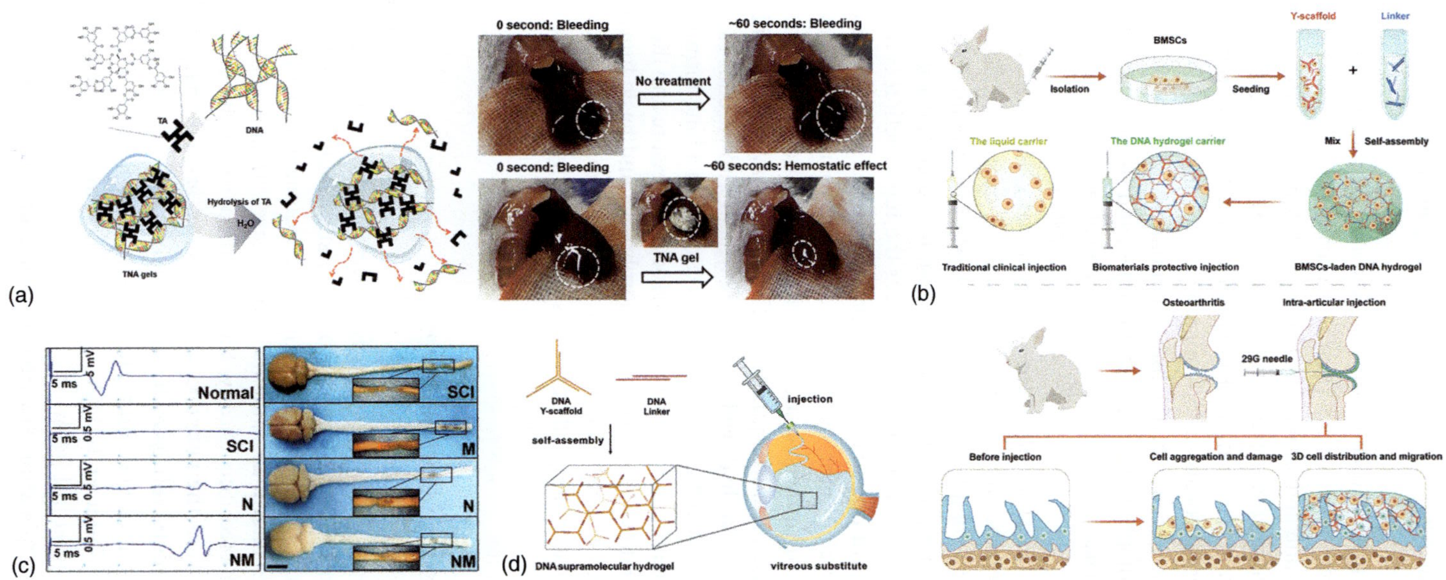

Figure 19.8 DNA supramolecular hydrogels for tissue engineering *in vivo*. (a) TNA hydrogel formation, degradation, and excellent hemostatic ability. Source: Shin et al. [110]/Reproduced with permission from John Wiley & Sons, Inc. (b) Preparation of DNA supramolecular hydrogels containing bone marrow mesenchymal stem cells (BMSCs). Source: Reproduced with permission from Yuan et al. [111]. © 2021 Wiley-VCH. (c) Postoperative motor evoked potential (MEP) trace after highly permeable DNA hydrogel injection and the central nervous system specimens of all experimental groups after 8 weeks. Source: Yuan et al. [112]/Reproduced with permission from John Wiley & Sons, Inc. (d) DNA supramolecular hydrogel and the injection process into vitreous cavity. Source: Reproduced with permission from Jin et al. [113]. © 2021 Wiley-VCH.

response to the elevated pathological cue (MMP-9) in diabetic microenvironment; thus, control release of SCAP-Exo could enhance angiogenesis and osteogenesis in diabetic bone defects.

DNA supramolecular hydrogels can also work as scaffold for in-situ regeneration of other tissues in the body. For example, Yuan et al. used the pure DNA supramolecular hydrogel to support and deliver homologous neural stem cells (NSCs) and realized repair of a 2 mm spinal cord gap in rats (Figure 19.8b) [111]. Eight weeks after delivery, the rats restored basic hindlimb function with detectable motor-evoked potentials. High permeability enabled both migration of implanted NSCs to the host tissue and migration of host cells into the implanted materials, promoting communication, proliferation, differentiation of both implanted and endogenous stem cells, thus helping to form renascent neural networks. Yan et al. applied the same hydrogel system for the injection of MSCs for the therapy of osteoarthritis (OA) in rabbit models (Figure 19.8c) [112]. DNA supramolecular hydrogels have good lubrication, and permeable properties and can protect MSCs from cell damage caused by shear force during injection and long-term joint frictions. After treatment, BMSC-laden DNA supramolecular hydrogels could promote formation of quality cartilage, reduce osteophyte, and normalize subchondral bone under the high friction condition of OA with more effective anti-inflammatory and prochondrogenic effects, thus obviously enhancing MSCs therapy efficacy. Recently, Jin et al. reported the accessibility of DNA supramolecular hydrogels as artificial vitreous substitutes (Figure 19.8d) [113]. Due to strong mechanical strength, shear-shinning behavior and rapid recoverability, DNA supramolecular hydrogels could be injected into vitreous cavity and provide adequate support. Two months of injection confirmed a long-term tamponade effect of hydrogels, maintenance of normal intraocular pressure and the original location of the retina, and the recovery of the morphology and basic functions of the operated eyes without obvious inflammation. Furthermore, physical properties including density, refractive index, and light transmittance of hydrogel were similar to those of vitreous.

19.4 Conclusions and Perspectives

In conclusion, compared with the covalent hydrogels, physical interactions bring DNA supramolecular hydrogels dynamic characteristics including self-healing, shear shinning, and rapid thixotropic properties. The sequence of DNA is tunable and can be easily functionalized; thus, the mechanical strength and responsiveness of hydrogels can be adjusted. Specifically, the rigid property of the double-helix DNA brings large meshes in the network, which ensures extraordinary molecule permeability and guarantees cell growth, proliferation, and migration in tissue engineering. Hence, combined with intrinsically biocompatible and biodegradable properties, the all-in-one character of DNA supramolecular hydrogels has promoted a versatile platform in biomedical fields.

While the promising application potential of DNA supramolecular hydrogels has been demonstrated, there are still some challenges that need to be solved.

Firstly, how to further improve the mechanical strength while maintaining high permeability is challenging, which may be necessary for the high-modulus tissues/organs. Therefore, new gelation strategies are still needed, such as introducing a more rigid DNA skeleton, doping dynamic organic/inorganic materials, or constructing DNA-based double-network hydrogels. Secondly, the accurate development of responsive hydrogels for different application scenarios is important for controlled drug release, cell culture, and tissue regeneration, and therefore, development and selection of biocompatible triggers are critical. Third, high cost of synthetic DNA on a large scale is still the main obstacle to the practical application, so more effective technologies with simple preparation process and lower cost are still in urgent need.

References

1 Gačanin, J., Synatschke, C.V., and Weil, T. (2020). Biomedical applications of DNA-based hydrogels. *Adv. Funct. Mater.* 30 (4): 1906253.

2 Chen, J., Zhu, Y., Liu, H. et al. (2020). Tailoring DNA self-assembly to build hydrogels. *Top. Curr. Chem.* 378 (2): 32.

3 Jonker, A.M., Löwik, D.W.P.M., and van Hest, J.C.M. (2012). Peptide- and protein-based hydrogels. *Chem. Mater.* 24 (5): 759–773.

4 Highley, C.B., Prestwich, G.D., and Burdick, J.A. (2016). Recent advances in hyaluronic acid hydrogels for biomedical applications. *Curr. Opin. Biotechnol.* 40: 35–40.

5 Zhu, T., Mao, J., Cheng, Y. et al. (2019). Recent progress of polysaccharide-based hydrogel interfaces for wound healing and tissue engineering. *Adv. Mater. Interfaces* 6 (17): 1900761.

6 Zhu, J. (2010). Bioactive modification of poly(ethylene glycol) hydrogels for tissue engineering. *Biomaterials* 31 (17): 4639–4656.

7 Adelnia, H., Ensandoost, R., Shebbrin Moonshi, S. et al. (2022). Freeze/thawed polyvinyl alcohol hydrogels: present, past and future. *Eur. Polym. J.* 164: 110974.

8 Tang, L., Wang, L., Yang, X. et al. (2021). Poly(*N*-isopropylacrylamide)-based smart hydrogels: design, properties and applications. *Prog. Mater Sci.* 115: 100702.

9 Seliktar, D. (2012). Designing cell-compatible hydrogels for biomedical applications. *Science* 336 (6085): 1124–1128.

10 Nagahara, S. and Matsuda, T. (1996). Hydrogel formation via hybridization of oligonucleotides derivatized in water-soluble vinyl polymers. *Polym. Gels Networks* 4 (2): 111–127.

11 Cheng, E., Xing, Y., Chen, P. et al. (2009). A pH-triggered, fast-responding DNA hydrogel. *Angew. Chem. Int. Ed.* 48 (41): 7660–7663.

12 Xing, Y., Cheng, E., Yang, Y. et al. (2011). Self-assembled DNA hydrogels with designable thermal and enzymatic responsiveness. *Adv. Mater.* 23 (9): 1117–1121.

13 Jiang, H., Pan, V., Vivek, S. et al. (2016). Programmable DNA hydrogels assembled from multidomain DNA strands. *ChemBioChem* 17 (12): 1156–1162.
14 Wang, J., Chao, J., Liu, H. et al. (2017). Clamped hybridization chain reactions for the self-assembly of patterned DNA hydrogels. *Angew. Chem. Int. Ed.* 56 (8): 2171–2175.
15 Hartman, M.R., Yang, D., Tran, T.N.N. et al. (2013). Thermostable branched DNA nanostructures as modular primers for polymerase chain reaction. *Angew. Chem. Int. Ed.* 52 (33): 8699–8702.
16 Stoev, I.D., Cao, T., Caciagli, A. et al. (2020). On the role of flexibility in linker-mediated DNA hydrogels. *Soft Matter* 16 (4): 990–1001.
17 Xiang, B., He, K., Zhu, R. et al. (2016). Self-assembled DNA hydrogel based on enzymatically polymerized DNA for protein encapsulation and enzyme/DNAzyme hybrid cascade reaction. *ACS Appl. Mater. Interfaces* 8 (35): 22801–22807.
18 Kandatsu, D., Cervantes-Salguero, K., Kawamata, I. et al. (2016). Reversible gel–sol transition of a photo-responsive DNA gel. *ChemBioChem* 17 (12): 1118–1121.
19 Um, S.H., Lee, J.B., Park, N. et al. (2006). Enzyme-catalysed assembly of DNA hydrogel. *Nat. Mater.* 5 (10): 797–801.
20 Nishida, Y., Ohtsuki, S., Araie, Y. et al. (2016). Self-assembling DNA hydrogel-based delivery of immunoinhibitory nucleic acids to immune cells. *Nanomed. Nanotechnol. Biol. Med.* 12 (1): 123–130.
21 Nöll, T., Schönherr, H., Wesner, D. et al. (2014). Construction of three-dimensional DNA hydrogels from linear building blocks. *Angew. Chem. Int. Ed.* 53 (32): 8328–8332.
22 Jia, H., Shi, J., Shao, Y. et al. (2017). DNA hydrogels formed of bended DNA scaffolds and properties study. *Chin. J. Polym. Sci.* 35 (10): 1307–1314.
23 Shi, J., Jia, H., Chen, H. et al. (2019). Concentration insensitive supramolecular polymerization enabled by kinetically interlocking multiple-units strategy. *CCS Chem.* 1 (3): 296–303.
24 Shi, J., Jia, H., and Liu, D. (2017). pH-Responsive supramolecular hydrogel based on one short strand DNA. *Acta Polym. Sin.* 1: 135–142.
25 Li, Y., Chen, J., Dong, Y. et al. (2020). Construction of pH-triggered DNA hydrogels based on hybridization chain reactions. *Chem. Res. Chin. Univ.* 36 (2): 243–246.
26 Shi, J., Shi, Z., Dong, Y. et al. (2020). Responsive DNA-based supramolecular hydrogels. *ACS Appl. Bio Mater.* 3 (5): 2827–2837.
27 Guo, W., Lu, C.-H., Orbach, R. et al. (2015). pH-Stimulated DNA hydrogels exhibiting shape-memory properties. *Adv. Mater.* 27 (1): 73–78.
28 Li, C., Chen, P., Shao, Y. et al. (2015). A writable polypeptide-DNA hydrogel with rationally designed multi-modification sites. *Small* 11 (9–10): 1138–1143.
29 Cheng, E., Li, Y., Yang, Z. et al. (2011). DNA-SWNT hybrid hydrogel. *Chem. Commun.* 47 (19): 5545–5547.

30 Li, C., Rowland, M.J., Shao, Y. et al. (2015). Responsive double network hydrogels of interpenetrating DNA and CB[8] host–guest supramolecular systems. *Adv. Mater.* 27 (21): 3298–3304.

31 Lu, C.-H., Guo, W., Hu, Y. et al. (2015). Multitriggered shape-memory acrylamide–DNA hydrogels. *J. Am. Chem. Soc.* 137 (50): 15723–15731.

32 Guo, W., Qi, X.-J., Orbach, R. et al. (2014). Reversible Ag+-crosslinked DNA hydrogels. *Chem. Commun.* 50 (31): 4065–4068.

33 Guo, W., Lu, C.-H., Qi, X.-J. et al. (2014). Switchable bifunctional stimuli-triggered poly-*N*-isopropylacrylamide/DNA hydrogels. *Angew. Chem. Int. Ed.* 53 (38): 10134–10138.

34 Li, C., Zhou, X., Shao, Y. et al. (2017). A supramolecular hydrogel with identical cross-linking point density but distinctive rheological properties. *Mater. Chem. Front.* 1 (4): 654–659.

35 Wu, Y., Li, C., Boldt, F. et al. (2014). Programmable protein–DNA hybrid hydrogels for the immobilization and release of functional proteins. *Chem. Commun.* 50 (93): 14620–14622.

36 Ma, X., Yang, Z., Wang, Y. et al. (2017). Remote controlling DNA hydrogel by magnetic field. *ACS Appl. Mater. Interfaces* 9 (3): 1995–2000.

37 Eguchi, Y., Kato, T., Tanaka, T. et al. (2017). A DNA–gold nanoparticle hybrid hydrogel network prepared by enzymatic reaction. *Chem. Commun.* 53 (43): 5802–5805.

38 Zhang, L., Jean, S.R., Ahmed, S. et al. (2017). Multifunctional quantum dot DNA hydrogels. *Nat. Commun.* 8 (1): 381.

39 Xu, Y., Wu, Q., Sun, Y. et al. (2010). Three-dimensional self-assembly of graphene oxide and DNA into multifunctional hydrogels. *ACS Nano* 4 (12): 7358–7362.

40 Zheng, M., Jagota, A., Semke, E.D. et al. (2003). DNA-assisted dispersion and separation of carbon nanotubes. *Nat. Mater.* 2 (5): 338–342.

41 Chen, Q., Chen, H., Zhu, L. et al. (2016). Engineering of tough double network hydrogels. *Macromol. Chem. Phys.* 217 (9): 1022–1036.

42 Khajouei, S., Ravan, H., and Ebrahimi, A. (2020). DNA hydrogel-empowered biosensing. *Adv. Colloid Interface Sci.* 275: 102060.

43 Yang, H., Liu, H., Kang, H. et al. (2008). Engineering target-responsive hydrogels based on aptamer–target interactions. *J. Am. Chem. Soc.* 130 (20): 6320–6321.

44 Zhu, Z., Wu, C., Liu, H. et al. (2010). An aptamer cross-linked hydrogel as a colorimetric platform for visual detection. *Angew. Chem. Int. Ed.* 49 (6): 1052–1056.

45 Liu, S., Su, W., Li, Y. et al. (2018). Manufacturing of an electrochemical biosensing platform based on hybrid DNA hydrogel: taking lung cancer-specific miR-21 as an example. *Biosens. Bioelectron.* 103: 1–5.

46 Wang, Q., Hu, Y., Jiang, N. et al. (2020). Preparation of aptamer responsive DNA functionalized hydrogels for the sensitive detection of α-fetoprotein using SERS method. *Bioconjugate Chem.* 31 (3): 813–820.

47 Zhu, Z., Guan, Z., Jia, S. et al. (2014). Au@Pt nanoparticle encapsulated target-responsive hydrogel with volumetric bar-chart chip readout for quantitative point-of-care testing. *Angew. Chem. Int. Ed.* 53 (46): 12503–12507.

48 Wu, P., Li, S., Ye, X. et al. (2020). Cu/Au/Pt trimetallic nanoparticles coated with DNA hydrogel as target-responsive and signal-amplification material for sensitive detection of microcystin-LR. *Anal. Chim. Acta* 1134: 96–105.

49 Liu, R., Huang, Y., Ma, Y. et al. (2015). Design and synthesis of target-responsive aptamer-cross-linked hydrogel for visual quantitative detection of ochratoxin A. *ACS Appl. Mater. Interfaces* 7 (12): 6982–6990.

50 Zhao, M., Wang, P., Guo, Y. et al. (2018). Detection of aflatoxin B1 in food samples based on target-responsive aptamer-cross-linked hydrogel using a handheld pH meter as readout. *Talanta* 176: 34–39.

51 Sun, Y., Li, S., Chen, R. et al. (2020). Ultrasensitive and rapid detection of T-2 toxin using a target-responsive DNA hydrogel. *Sens. Actuators, B* 311: 127912.

52 Lin, Y., Wang, X., Sun, Y. et al. (2019). A chemiluminescent biosensor for ultrasensitive detection of adenosine based on target-responsive DNA hydrogel with Au@HKUST-1 encapsulation. *Sens. Actuators, B* 289: 56–64.

53 Si, Y., Li, L., Wang, N. et al. (2019). Oligonucleotide cross-linked hydrogel for recognition and quantitation of MicroRNAs based on a portable glucometer readout. *ACS Appl. Mater. Interfaces* 11 (8): 7792–7799.

54 Yin, B.-C., Ye, B.-C., Wang, H. et al. (2012). Colorimetric logic gates based on aptamer-crosslinked hydrogels. *Chem. Commun.* 48 (9): 1248–1250.

55 Yan, L., Zhu, Z., Zou, Y. et al. (2013). Target-responsive "sweet" hydrogel with glucometer readout for portable and quantitative detection of non-glucose targets. *J. Am. Chem. Soc.* 135 (10): 3748–3751.

56 Mao, Y., Li, J., Yan, J. et al. (2017). A portable visual detection method based on a target-responsive DNA hydrogel and color change of gold nanorods. *Chem. Commun.* 53 (47): 6375–6378.

57 Chu, J., Chen, C., Li, X. et al. (2021). A responsive pure DNA hydrogel for label-free detection of lead ion. *Anal. Chim. Acta* 1157: 338400.

58 Lin, H., Zou, Y., Huang, Y. et al. (2011). DNAzyme crosslinked hydrogel: a new platform for visual detection of metal ions. *Chem. Commun.* 47 (33): 9312–9314.

59 He, X., Zhou, X., Liu, W. et al. (2020). Flexible DNA hydrogel SERS active biofilms for conformal ultrasensitive detection of uranyl ions from aquatic products. *Langmuir* 36 (11): 2930–2936.

60 Si, Y., Xu, L., Wang, N. et al. (2020). Target MicroRNA-responsive DNA hydrogel-based surface-enhanced raman scattering sensor arrays for MicroRNA-marked cancer screening. *Anal. Chem.* 92 (3): 2649–2655.

61 He, Y., Yang, X., Yuan, R. et al. (2017). Switchable target-responsive 3D DNA hydrogels as a signal amplification strategy combining with SERS technique for ultrasensitive detection of miRNA 155. *Anal. Chem.* 89 (16): 8538–8544.

62 Li, Y., Ma, Y., Jiao, X. et al. (2019). Control of capillary behavior through target-responsive hydrogel permeability alteration for sensitive visual quantitative detection. *Nat. Commun.* 10 (1): 1036.

63 Song, Y., Zhang, Y., Bernard, P.E. et al. (2012). Multiplexed volumetric bar-chart chip for point-of-care diagnostics. *Nat. Commun.* 3 (1): 1283.

64 Ma, Y., Mao, Y., Huang, D. et al. (2016). Portable visual quantitative detection of aflatoxin B1 using a target-responsive hydrogel and a distance-readout microfluidic chip. *Lab Chip* 16 (16): 3097–3104.

65 Gao, X., Li, X., Sun, X. et al. (2020). DNA tetrahedra-cross-linked hydrogel functionalized paper for onsite analysis of DNA methyltransferase activity using a personal glucose meter. *Anal. Chem.* 92 (6): 4592–4599.

66 Wei, X., Tian, T., Jia, S. et al. (2016). Microfluidic distance readout sweet hydrogel integrated paper-based analytical device (μDiSH-PAD) for visual quantitative point-of-care testing. *Anal. Chem.* 88 (4): 2345–2352.

67 Wei, X., Tian, T., Jia, S. et al. (2015). Target-responsive DNA hydrogel mediated "stop-flow" microfluidic paper-based analytic device for rapid, portable and visual detection of multiple targets. *Anal. Chem.* 87 (8): 4275–4282.

68 Zhang, R., Billingsley, M.M., and Mitchell, M.J. (2018). Biomaterials for vaccine-based cancer immunotherapy. *J. Controlled Release* 292: 256–276.

69 Zhao, M., Zhang, Y., Yuan, S. et al. (2019). ATP responsive DNA nanogels grown on biocompatible branches for anticancer drug delivery. *Soft Matter* 15 (18): 3655–3658.

70 Wang, Z., Xia, J., Cai, F. et al. (2015). Aptamer-functionalized hydrogel as effective anti-cancer drugs delivery agents. *Colloids Surf., B* 134: 40–46.

71 He, P., Du, X., Cheng, Y. et al. (2022). Thermal-responsive MXene-DNA hydrogel for near-infrared light triggered localized photothermal-chemo synergistic cancer therapy. *Small* 18 (40): 2200263.

72 Zhang, J., Guo, Y., Pan, G. et al. (2020). Injectable drug-conjugated DNA hydrogel for local chemotherapy to prevent tumor recurrence. *ACS Appl. Mater. Interfaces* 12 (19): 21441–21449.

73 Ma, Y., Liu, H., Mou, Q. et al. (2018). Floxuridine-containing nucleic acid nanogels for anticancer drug delivery. *Nanoscale* 10 (18): 8367–8371.

74 Ren, N., Sun, R., Xia, K. et al. (2019). DNA-based hybrid hydrogels sustain water-insoluble ophthalmic therapeutic delivery against allergic conjunctivitis. *ACS Appl. Mater. Interfaces* 11 (30): 26704–26710.

75 Hu, Y., Gao, S., Lu, H. et al. (2022). Acid-resistant and physiological pH-responsive DNA hydrogel composed of A-motif and i-motif toward oral insulin delivery. *J. Am. Chem. Soc.* 144 (12): 5461–5470.

76 Gačanin, J., Kovtun, A., Fischer, S. et al. (2017). Spatiotemporally controlled release of rho-inhibiting C3 toxin from a protein–DNA hybrid hydrogel for targeted inhibition of osteoclast formation and activity. *Adv. Healthcare Mater.* 6 (21): 1700392.

77 Li, J., Zheng, C., Cansiz, S. et al. (2015). Self-assembly of DNA nanohydrogels with controllable size and stimuli-responsive property for targeted gene regulation therapy. *J. Am. Chem. Soc.* 137 (4): 1412–1415.

78 Ding, F., Mou, Q., Ma, Y. et al. (2018). A crosslinked nucleic acid nanogel for effective siRNA delivery and antitumor therapy. *Angew. Chem. Int. Ed.* 57 (12): 3064–3068.

79 Xue, H., Ding, F., Zhang, J. et al. (2019). DNA tetrahedron-based nanogels for siRNA delivery and gene silencing. *Chem. Commun.* 55 (29): 4222–4225.

80 Martin-Liberal, J., Ochoa de Olza, M., Hierro, C. et al. (2017). The expanding role of immunotherapy. *Cancer Treat. Rev.* 54: 74–86.

81 Xie, Z., Shen, J., Sun, H. et al. (2021). Polymer-based hydrogels with local drug release for cancer immunotherapy. *Biomed. Pharmacother.* 137: 111333.

82 Singh, A. and Peppas, N.A. (2014). Hydrogels and scaffolds for immunomodulation. *Adv. Mater.* 26 (38): 6530–6541.

83 Krieg, A.M., Yi, A.-K., Matson, S. et al. (1995). CpG motifs in bacterial DNA trigger direct B-cell activation. *Nature* 374 (6522): 546–549.

84 Kumagai, Y., Takeuchi, O., and Akira, S. (2008). TLR9 as a key receptor for the recognition of DNA. *Adv. Drug Delivery Rev.* 60 (7): 795–804.

85 Nishikawa, M., Mizuno, Y., Mohri, K. et al. (2011). Biodegradable CpG DNA hydrogels for sustained delivery of doxorubicin and immunostimulatory signals in tumor-bearing mice. *Biomaterials* 32 (2): 488–494.

86 Nishikawa, M., Ogawa, K., Umeki, Y. et al. (2014). Injectable, self-gelling, biodegradable, and immunomodulatory DNA hydrogel for antigen delivery. *J. Controlled Release* 180: 25–32.

87 Shao, Y., Sun, Z.-Y., Wang, Y. et al. (2018). Designable immune therapeutical vaccine system based on DNA supramolecular hydrogels. *ACS Appl. Mater. Interfaces* 10 (11): 9310–9314.

88 Wei, H., Zhao, Z., Wang, Y. et al. (2019). One-step self-assembly of multifunctional DNA nanohydrogels: an enhanced and harmless strategy for guiding combined antitumor therapy. *ACS Appl. Mater. Interfaces* 11 (50): 46479–46489.

89 Umeki, Y., Mohri, K., Kawasaki, Y. et al. (2015). Induction of potent antitumor immunity by sustained release of cationic antigen from a DNA-based hydrogel with adjuvant activity. *Adv. Funct. Mater.* 25 (36): 5758–5767.

90 Umeki, Y., Saito, M., Takahashi, Y. et al. (2017). Retardation of antigen release from DNA hydrogel using cholesterol-modified DNA for increased antigen-specific immune response. *Adv. Healthcare Mater.* 6 (20): 1700355.

91 Nomura, D., Saito, M., Takahashi, Y. et al. (2018). Development of orally-deliverable DNA hydrogel by microemulsification and chitosan coating. *Int. J. Pharm.* 547 (1): 556–562.

92 Ishii-Mizuno, Y., Umeki, Y., Onuki, Y. et al. (2017). Improved sustained release of antigen from immunostimulatory DNA hydrogel by electrostatic interaction with chitosan. *Int. J. Pharm.* 516 (1): 392–400.

93 Mellman, I., Coukos, G., and Dranoff, G. (2011). Cancer immunotherapy comes of age. *Nature* 480 (7378): 480–489.

94 Umeki, Y., Saito, M., Kusamori, K. et al. (2018). Combined encapsulation of a tumor antigen and immune cells using a self-assembling immunostimulatory DNA hydrogel to enhance antigen-specific tumor immunity. *J. Controlled Release* 288: 189–198.

95 Yata, T., Takahashi, Y., Tan, M. et al. (2017). DNA nanotechnology-based composite-type gold nanoparticle-immunostimulatory DNA hydrogel for tumor photothermal immunotherapy. *Biomaterials* 146: 136–145.

96 Sun, J., Pan, M., Liu, F. et al. (2020). Immunostimulatory DNA nanogel enables effective lymphatic drainage and high vaccine efficacy. *ACS Mater. Lett.* 2 (12): 1606–1614.

97 Heil, F., Hemmi, H., Hochrein, H. et al. (2004). Species-specific recognition of single-stranded RNA via toll-like receptor 7 and 8. *Science* 303 (5663): 1526–1529.

98 Komura, F., Okuzumi, K., Takahashi, Y. et al. (2020). Development of RNA/DNA hydrogel targeting toll-like receptor 7/8 for sustained RNA release and potent immune activation. *Molecules* 25 (3): 728.

99 Jin, J., Xing, Y., Xi, Y. et al. (2013). A triggered DNA hydrogel cover to envelop and release single cells. *Adv. Mater.* 25 (34): 4714–4717.

100 Yang, B., Zhao, Z., Pan, Y. et al. (2021). Shear-thinning and designable responsive supramolecular DNA hydrogels based on chemically branched DNA. *ACS Appl. Mater. Interfaces* 13 (41): 48414–48422.

101 Cao, T., Jia, H., Dong, Y. et al. (2020). In situ formation of covalent second network in a DNA supramolecular hydrogel and its application for 3D cell imaging. *ACS Appl. Mater. Interfaces* 12 (4): 4185–4192.

102 Yang, B., Zhou, B., Li, C. et al. (2022). A biostable L-DNA hydrogel with improved stability for biomedical applications. *Angew. Chem.* 134 (30).

103 Li, Y., Ding, Y., Yang, B. et al. (2022). Reinforcing DNA supramolecular hydrogel with polymeric multiple-unit linker. *CCS Chem.* 0 (0): 1–11.

104 Black, C.K., Termanini, K.M., Aguirre, O. et al. (2018). Solid organ transplantation in the 21 st century. *Ann. Transl. Med.* 6 (20): 409–409.

105 Lee, K.Y. and Mooney, D.J. (2001). Hydrogels for tissue engineering. *Chem. Rev.* 101 (7): 1869–1880.

106 Xu, F., Dawson, C., Lamb, M. et al. (2022). Hydrogels for tissue engineering: addressing key design needs toward clinical translation. *Front. Bioeng. Biotechnol.* 10: 849831.

107 Pedde, R.D., Mirani, B., Navaei, A. et al. (2017). Emerging biofabrication strategies for engineering complex tissue constructs. *Adv. Mater.* 29 (19): 1606061.

108 Li, C., Faulkner-Jones, A., Dun, A.R. et al. (2015). Rapid formation of a supramolecular polypeptide–DNA hydrogel for in situ three-dimensional multilayer bioprinting. *Angew. Chem. Int. Ed.* 54 (13): 3957–3961.

109 Wang, Y., Shao, Y., Ma, X. et al. (2017). Constructing tissuelike complex structures using cell-laden DNA hydrogel bricks. *ACS Appl. Mater. Interfaces* 9 (14): 12311–12315.

110 Shin, M., Ryu, J.H., Park, J.P. et al. (2015). DNA/tannic acid hybrid gel exhibiting biodegradability, extensibility, tissue adhesiveness, and hemostatic ability. *Adv. Funct. Mater.* 25 (8): 1270–1278.

111 Yuan, T., Shao, Y., Zhou, X. et al. (2021). Highly permeable DNA supramolecular hydrogel promotes neurogenesis and functional recovery after completely transected spinal cord injury. *Adv. Mater.* 33 (35): 2102428.

112 Yan, X., Yang, B., Chen, Y. et al. (2021). Anti-friction MSCs delivery system improves the therapy for severe osteoarthritis. *Adv. Mater.* 33 (52): 2104758.

113 Jin, Y., Li, Y., Song, S. et al. (2022). DNA supramolecular hydrogel as a biocompatible artificial vitreous substitute. *Adv. Mater. Interfaces* 9 (5): 2101321.

114 Obuobi, S., Tay, H.K.-L., Tram, N.D.T. et al. (2019). Facile and efficient encapsulation of antimicrobial peptides via crosslinked DNA nanostructures and their application in wound therapy. *J. Controlled Release* 313: 120–130.

115 Geng, J., Yao, C., Kou, X. et al. (2018). A fluorescent biofunctional DNA hydrogel prepared by enzymatic polymerization. *Adv. Healthcare Mater.* 7 (5): 1700998.

116 Jiang, X., Li, M., Guo, X. et al. (2019). Self-assembled DNA-THPS hydrogel as a topical antibacterial agent for wound healing. *ACS Appl. Bio Mater.* 2 (3): 1262–1269.

117 Wang, Z., Li, W., Gou, L. et al. (2022). Biodegradable and antioxidant DNA hydrogel as a cytokine delivery system for diabetic wound healing. *Adv. Healthcare Mater.* 11 (21): 2200782.

118 Li, W., Wang, C., Wang, Z. et al. (2022). Physically cross-linked DNA hydrogel-based sustained cytokine delivery for in situ diabetic alveolar bone rebuilding. *ACS Appl. Mater. Interfaces* 14 (22): 25173–25182.

119 Jing, X., Wang, S., Tang, H. et al. (2022). Dynamically bioresponsive DNA hydrogel incorporated with dual-functional stem cells from apical papilla-derived exosomes promotes diabetic bone regeneration. *ACS Appl. Mater. Interfaces* 14 (14): 16082–16099.

20

Rolling Circle Amplification-Based DNA Nanotechnology for Cell Research

Nachuan Song, Yiwen Chu, Xun You, and Dayong Yang

Tianjin University, School of Chemical Engineering and Technology, Frontiers Science Center for Synthetic Biology, Key Laboratory of Systems Bioengineering (MOE), 135 Yaguan Road, Jinnan District, Tianjin, 300350, China

20.1 Introduction

Deoxyribonucleic acid (DNA) is a macromolecular polymer consisting of deoxynucleotides. Each deoxynucleotide monomer includes one of four nitrogen-containing nucleobases: adenine (A), thymine (T), cytosine (C), or guanine (G), a deoxyribose, and a phosphate group [1]. As an important component of genetic molecules, DNA stores genetic information and encodes the amino acid sequences that constitute proteins to realize the biological functions of cells. In addition to its original biological functions, DNA has also been explored as a building block to construct artificial nanostructures in the past decades [2]. Since Nadrian C. Seeman proposed "DNA nanotechnology" *via* complementary base pairing in the 1980s [3–5], the characteristics of DNA are widely used to construct nanostructures with controllable size and precise structure. On the one hand, the self-assembly of DNA forms DNA hydrogel, DNA polyhedron, and DNA crystal through complementary base pairing [6, 7]. On the other hand, the assembly of DNA and other molecules constructs amphiphilic DNA [8], metal-DNA cage [9–11], and DNA-peptide network through functional moieties and chemical bonds [12]. In particular, DNA possesses the properties of sequence programmability [13–15], molecule recognition ability [16–18], and biocompatibility [19–22], which endows DNA with more aspects of applications, such as drug and gene delivery [23–25], therapeutics [18], diagnosis [26–31], protein engineering [32–34], and cell engineering [35–39].

Rolling circle amplification (RCA) is an isothermal enzymatic reaction that uses circle DNA as a template to produce a large number of repetitive single-stranded DNA (ssDNA) [40, 41]. Through precise molecular design, functional nucleic acid sequences are integrated on circle DNA templates, including DNAzymes [42], aptamers [43], antisense oligonucleotides (ASOs) [44], i-motif [45], and G-quadruplex structures, to endow DNA with more biological functions [18]. Based on ultra-long ssDNA produced by RCA, through precisely controlling the concentration of reactants and reaction time in the RCA process, different

DNA Nanotechnology for Cell Research: From Bioanalysis to Biomedicine, First Edition. Edited by Zhou Nie.

products are generated, such as DNA hydrogel and DNA nanoparticles [46]. In this chapter, we focus on the design, synthesis, and applications of cell research using RCA-based DNA nanotechnology in recent years. We first summarize the principle and synthetic methods of RCA and the characteristics of two typical RCA products (DNA hydrogel and DNA nanocomplex). The applications of DNA nanotechnology in the field of cell research are elaborated, including cell separation and nucleic acid drug delivery. At the end of this chapter, the challenges, opportunities, and future development of RCA-based DNA nanotechnology are prospected.

20.2 Principle and Synthetic Methods of RCA

20.2.1 Principle

RCA was an enzyme-catalyzed method for synthesizing ssDNA in three steps: cyclization, ligation, and amplification. In general, a typical RCA reaction requires four ingredients: (i) a circular DNA template; (ii) a short-stranded DNA primer complementary to the template; (iii) deoxynucleotide triphosphates (dNTPs), the raw material for synthesizing ssDNA; and (iv) enzymes and homologous buffer liquids required for the reaction [47, 48].

At the beginning of the reaction, the 5′-phosphorylated linear template and the primer were first mixed in an equimolar ratio. The two ends of the linear template were complementary to the primer, respectively. The mixture of liner template and primer was annealed at high temperature to form a circularized structure with gaps, called Circ-DNA precursor. Circ-DNA precursor formed a phosphodiester bond between the 5′ phosphate end and the 3′ hydroxyl end catalyzed by T4 ligase, connecting the gap and forming Circ-DNA. Subsequently, the ssDNA with extra-large molecular weight was synthesized with dNTPs as raw material and phi 29 DNA polymerase (Figure 20.1).

To ensure the smooth operation of RCA reactions, the length of DNA template was generally 70–110 bases, as an excessively long template reduced the efficiency of DNA ligase reaction. The length of the template complementary to the primer should preferably be about 1/3 of the entire template. In addition to the design of functional sequences such as aptamers, the template sequence should avoid the formation of a hairpin structure. Last but not least, the Gibbs free energy change (ΔG) when the primer is bound to the template should be lower than that when the hairpin structure is formed.

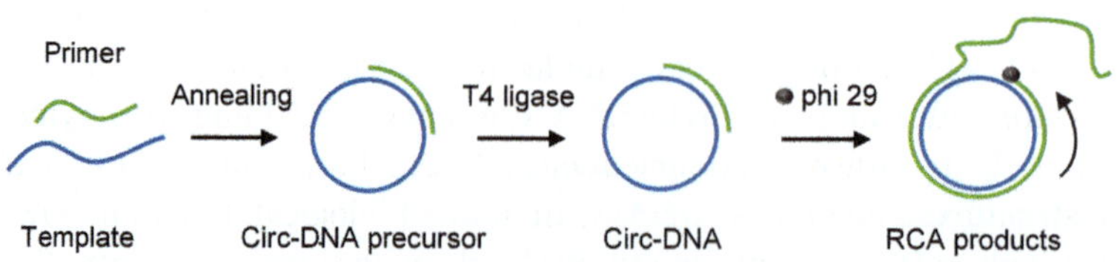

Figure 20.1 Schematic illustration of the process of rolling circle amplification.

20.2.2 DNA Hydrogel

As a typical representative of DNA functional materials, DNA hydrogels not only had high water content, adjustable mechanical strength, and shape plasticity but also integrated the biological functions of DNA, which played an important role in cell capture, release, and delivery [49]. RCA reaction was a common and efficient strategy for synthesizing DNA hydrogels. The preparation of RCA-based DNA hydrogels generally consisted of three essential stages: (i) preparation of precursor solution containing circ-DNA as templates for RCA; (ii) preparation of ultra-long ssDNA as a precursor for DNA hydrogels *via* RCA reaction; and (iii) preparation of the hydrogel *via* cross-linking. In order to adapt to different application requirements, a variety of cross-linking strategies have been developed, such as the entanglement of DNA chains, multi-primed chain amplification, and hybridization with functional moieties [48].

20.2.3 DNA Nanoparticles

In the presence of metal ions, ssDNA could be compressed into a DNA nanocomplex containing multi-functional sequences. Another DNA nanoparticle, DNA nanoflower was formed directly in the process of RCA. DNA nanoflowers were formed by co-crystallization using DNA-mediated inorganic nucleation and growth processes, with porous flower-like surface morphology. The long ssDNA that gradually extended in the RCA reaction was often used to intervene in the nucleation and crystallization process of pyrophosphate by-products to synthetic DNA nanoflowers. Metal cofactors of DNA polymerase, such as magnesium ions (Mg^{2+}), could bind to the RCA reaction by-product pyrophosphate (PPi^{4-}) to form water-insoluble magnesium pyrophosphate (Mg_2PPi). As the reaction proceeded, Mg_2PPi gradually accumulated. Mg_2PPi precipitates acted as primary crystal nuclei to nucleate and grow *in situ* on long-stranded DNA and eventually formed porous, flower-like nanostructures [50]. Different metal cofactors, reaction concentrations, and reaction times could affect the rate of crystallization reaction and the apparent morphology of DNA nanoflowers.

20.3 RCA-Based DNA Nanotechnology for Cell Separation

Cell separation technology was a prerequisite for the application of some functional cells, such as T lymphocytes (T-cells) and stem cells. At present, the widely used cell separation techniques include density gradient centrifugation [51, 52], fluorescence-activated cell sorting (FACS), and magnetic-activated cell sorting (MACS) [53, 54]. However, the mechanical forces generated and the labels introduced in the application of these technologies could cause damage or contamination to the cells [55]. In addition, these cell isolation techniques needed to rely on the help of other systems and adjuvants to achieve the subsequent incubation and application of separated cells [56]. To overcome these disadvantages while improving the specificity of cell separation, DNA nanotechnology has been used for cell recognition and separation. Due to the unique advantages of DNA molecules, such as biocompatibility, nanoscale controllability, programmability, and molecular recognition ability [2], DNA molecules could be used as a matrix material to construct cell capture networks for specific cell separation and *in situ* culture, and further for various subsequent applications. DNA networks, a novel material, were extensively hydrated and provided an elastic, semi-wet, and three-dimensional matrix for cell culture. The porous nature of the DNA network not only facilitated cell entry, but also provided good nutrient transport and waste removal for coated cells [57]. In addition, the synthesis conditions of the DNA network were mild and able to avoid cell damage by the reaction to the greatest extent [58, 59]. These DNA nanotechnology-based cell isolation methods not only provided a mild biocompatible environment for cells but also took advantage of the specificity of nucleic acid aptamers to cells to achieve cell separation [60–62]. To better capture cells specifically, DNA aptamers have been involved and played an important role [63]. DNA aptamers were a class of highly specific DNA sequences that could selectively bind to specific targets [64]. For example, DNA aptamers were used to target proteins overexpressed on the membrane surface of cancer cells, resulting in efficient capture and sensitive detection of circulating tumor cells (CTCs), enabling specific separation of cancer cells [65, 66]. In addition to the separation of tumor cells, DNA aptamers were also used to separate lymphocyte subpopulations from peripheral blood and successfully achieve efficient and traceless isolation of lymphocytes [60].

The DNA field has advanced rapidly, mostly due to the precise control of ordered DNA nanostructures. The utilization of nonordered, bulk-scale DNA materials (e.g. DNA hydrogels) provided a tangible material to interface with cells, bodies, and the environment, facilitating real-world applications. DNA materials for cell separation were usually presented in the form of hydrogels. Hydrogels are hydrophilic polymers formed by physical or chemical cross-linking [67]. As a special soft and wet material, the 3D network structure of hydrogel was similar to that of extracellular matrix, which could be used as an ideal biomaterial for cell separation. DNA hydrogel was a three-dimensional network structure formed by highly cross-linked DNA strands in aqueous solution, which could realize the unified fusion of the skeleton function of hydrogel and the biological function of DNA. According to the method of hydrogel synthesis, DNA hydrogels could be classified into chemical and physical hydrogels. Among them, physical hydrogels were hydrogels formed under the non-covalent bond between DNA molecules. The non-covalent bond mainly included the complementary base pairing between DNA double-strands and the winding between ultra-long DNA strands. Using circular DNA as a template, ultra-long single-stranded DNA (ssDNA) with periodic repeats was obtained by introducing phi29 DNA polymerase to displace the DNA in the template. After that, a secondary primer was added to initiate a multi-primer chain amplification reaction, and the combination of these two amplification reactions could prepare a hydrogel with special properties. Furthermore, through double RCA, two different ssDNA were mixed to form DNA hydrogel *via* complementary base pairing and winding. Because the DNA hydrogel prepared by RCA technology not only had good biocompatibility but also could specifically recognize and capture cells by introducing aptamers, it performed good application potential for cell separation, downstream analysis, and regenerative medicine.

As a biofunctional polymer, DNA has attracted much attention due to its molecular recognition ability, sequence programmability, and biocompatibility. In recent years, DNA has been applied in cell separation, 3D cell culture, and tissue engineering, which has opened up a new way for applied research in the biomedical field. Focusing on the preparation, performance, and biological application of double RCA DNA networks, the efficient and specific capture and controlled release of bone marrow mesenchymal stem cells (BMSCs) by double RCA DNA network in response to nuclease was explored [57]. The complementary sequence of Apt19S was designed on template 1, and DNA strand 1 containing multiple repeats of Apt19s was obtained by RCA to ensure the specific anchoring of Apt19s to BMSCs (Figure 20.2a). Template 2 was designed with sequences partially complementary to template 1, which were also obtained by RCA. The hybridization of DNA strand 1 and DNA strand 2 transformed the solution into a 3D network. Among the DNA sequences used to capture BMSC, DNA strand 1 containing the aptamer Apt19S sequence had a high affinity for ALPL protein on the BMSC membrane, thus achieving specific anchoring of BMSC. Through the double RCA process, BMSCs were specifically captured from bone marrow, and the separation of BMSCs was achieved. The high water content and porous and soft matrix of the DNA network provided a good 3D microenvironment for cell culture. At the same time, the DNA network was digested by nucleases in a controlled manner, thereby deconstructing the DNA network into fragments and releasing the captured BMSCs. In addition, factors such as capture time, cell number, and DNA network volume had important effects on the efficiency of the DNA network to capture BMSCs and achieve separation. With the increase of incubation time, the capture rate and the number of captured cells increased, and reached a plateau at 20 minutes, indicating that the capture process was completed in a relatively short time, which reflected the high capture efficiency of the system (Figure 20.2b). The incubation time was extended to 60 minutes, and the capture rate remained at 87.5%. Through the study of the relationship between the capture rate of BMSCs/the number of captured cells and the number of cells/network capacity, it could be known that when the network volume and incubation time were the same, with the increased number of cells, the number of captured cells gradually increased, but the capture rate decreased (Figure 20.2c). The reason for this phenomenon was that the DNA network was saturated and the network capacity was maximized for the enrichment of BMSCs. As the volume of the DNA network increased, the efficiency of cell separation and the number of captured cells gradually increased (Figure 20.2d). The increment of captured BMSCs depended on the volume of the DNA network, suggesting that the majority of BMSCs were encapsulated in the network rather than adhering to the surface. Based on the above results, it could be concluded that these double RCA DNA networks could specifically and efficiently capture BMSCs. In addition, this study attempted to separate BMSCs from

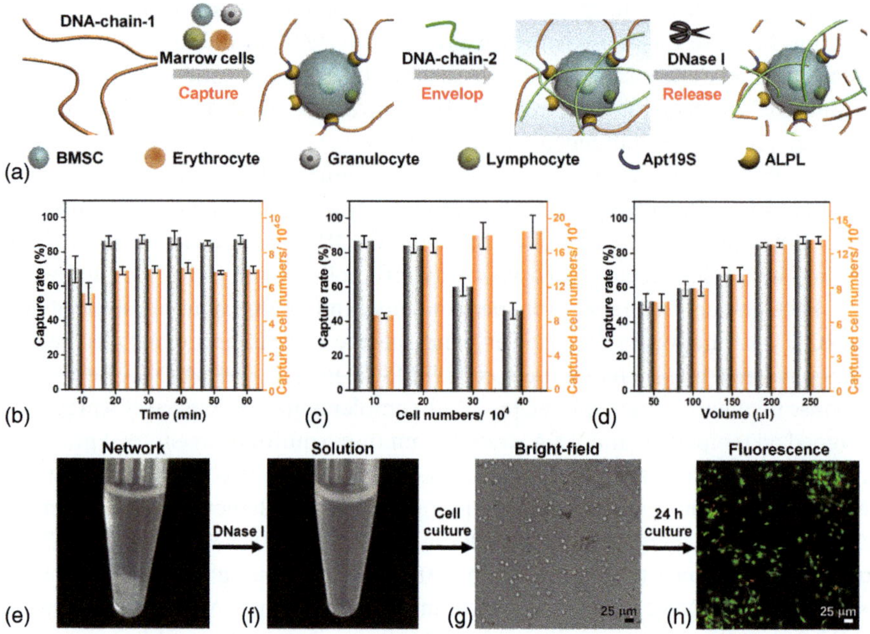

Figure 20.2 Physically cross-linked DNA network for stem cell fishing. (a) The process of capture, envelop, and release: (1) capture, DNA-chain-1 was incubated with BMSCs for cell capture via anchoring with aptamer Apt19S; (2) envelop, DNA-chain-2 was introduced in the cell-containing DNA-chain-1 solution to trigger the formation of DNA network; and (3) release, DNA network can be digested by nuclease DNase I to release BMSCs. (b–d) Capture rate and captured cell numbers of BMSCs in the DNA network with the increase of time (b), cell numbers (c), and volume of network (d). Data represented as mean ± SD, $n = 3$. (e) BMSCs-enveloped DNA network in PBS buffer. (f) DNase I digested the DNA network to solution. (g) Microscopy image of released BMSCs. (h) Fluorescence microscopy image of BMSCs after culturing for 24 hours. Living and dead cells were stained with Calcein-AM (green) and propidium iodide (PI, red), respectively. Source: Chi Yao et al. [57]/Reproduced from American Chemical Society.

the mixed cell system, and the results showed that the DNA network could achieve efficient separation of BMSCs in the mixed cell system. Although it was important to achieve cell separation, release after separation was also an important evaluation factor for future applications of DNA materials. The DNA network was gently degraded by the nuclease DNase I. As the DNA network was degraded, the captured BMSC could be released in a controlled way. The DNA network of encapsulated BMSC was transferred to 1×PBS buffer, which performed a white transparent gel (Figure 20.2e). After adding DNase I, the gel was degraded, and the cells were also dispersed in the buffer (Figure 20.2f). Then the BMSCs were transferred to the confocal dish, and the BMSCs were evenly distributed and showed intact morphology under inverted fluorescence microscopy (Figure 20.2g). After being cultured at 37 °C in a 5% CO_2 incubator for 24 hours, the released BMSCs adhered to the wall and proliferated, showing a spindle shape. After live-dead cell staining, live cells were stained green by Calcein AM, and dead cells were stained red by PI. Images of inverted fluorescence microscopy showed that most of the field of view was live cells (Figure 20.2h). The above experimental data demonstrated that the DNA network not only had good biocompatibility but was also free of cell damage during the whole process of capture, separation, and release. This study took advantage of the designability of DNA to generate multiple reproducible aptamers by RCA, which could target and recognize BMSCs and improve the efficiency of BMSCs capture and separation.

SELEX technology enabled to select and amplify aptamers from a random aptamer library *in vitro*. The nucleic acid aptamers produced by SELEX showed binding affinity and specificity to a variety of target molecules. DNA aptamers had the advantages of low cost, small molecular weight, and strong thermal stability. The emergence of DNA aptamers endowed DNA materials with the molecular recognition ability and also provided a favorable guarantee for the application of DNA materials in the field of cell separation. A double RCA-amplified DNA physical cross-linking network was explored for the specific capture, separation, and release of T lymphocytes in a responsive manner in immunotherapy (Figure 20.3a) [56]. The DNA networks could be precisely degraded in a tumoral inflammatory environment, where the proteolytic enzyme could restore the activity to recognize and cleavage the complementary base pairing in DNA networks. To visualize the capture of T lymphocytes by DNA networks, T lymphocytes were stained with CM-DiI and DNA networks were stained with SYBR Green II, respectively. The inverted fluorescence microscopy was used for imaging in the green channel and the red channel, respectively. It was clearly observed that T lymphocytes were successfully captured and attached to the DNA network (Figure 20.3b). Changes in the volume of the solution were particularly important for the effect of cell capture and separation yield. With the increase of volume from 40 to 200 μl, the capture rate of T-cells increased from 10.4% to 91.8%, and the corresponding isolation yield ranged from 4.1% to 86.2% (Figure 20.3c). To further explore the potential of a DNA network-based T lymphocyte capture system in future clinical applications, a mouse model of melanoma *in situ* was established. The tumor size of mice was measured every two days after drug injection. The tumor volume in the DNA-material-injected group grew slower than in the PBS group, demonstrating the remarkable immunotherapy effect (Figure 20.3d). Two weeks later, the mice were sacrificed and the tumors were weighed (Figure 20.3e). The group containing both therapeutic sequences and the captured tumor-infiltrating T-cells (TIT cells) showed the best inhibition of tumor growth, suggesting a synergistic therapeutic effect of DNA material and T lymphocytes. Based on the efficient isolation of commercial T lymphocyte lines from artificial mixed cells, TIT cells were further separated from tumor tissue digests with high efficiency. Animal experiments were conducted to explore the application value of this system in the biomedical field, which provided a new strategy for cell separation and tumor treatment.

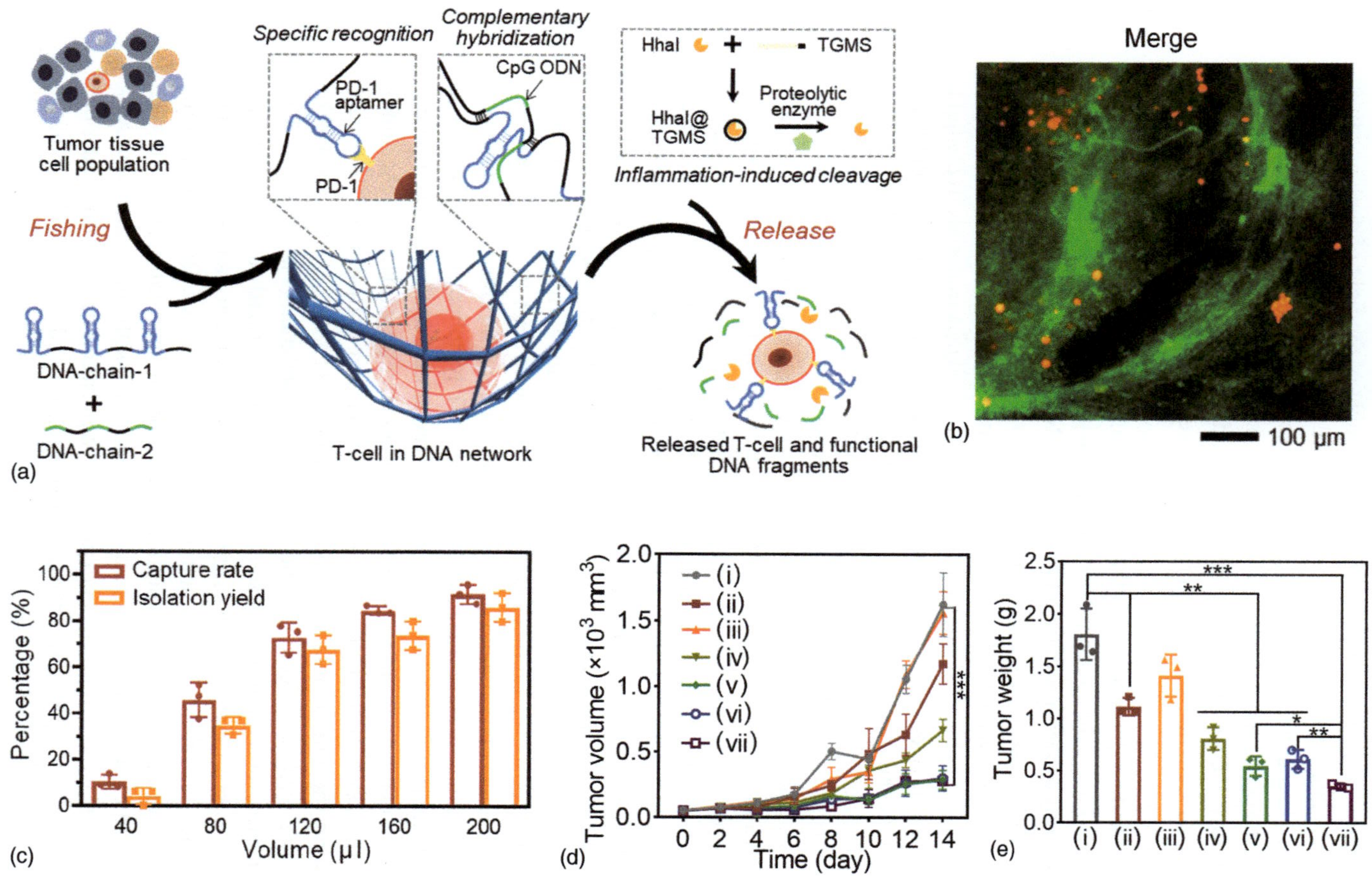

Figure 20.3 T lymphocyte-captured DNA network for localized immunotherapy (a) Construction of DNA network and cell capture of tumor-infiltrating T-Cell. (b) Fluorescence images of captured T-cells (stained with CM-DiI) in DNA network (stained with SYBR Green II). Data represented as mean ± SD, $n = 3$. (c) Evaluation of the capture rate and isolation yield with the increase of the volume of long-strand DNA solution. The cell suspension was 100 μl with a concentration of 10^5 cells/ml. Data represented as mean ± SD, $n = 3$. (d) Tumor weight of mice was measured every other day after the first injection. Data represented as mean ± SD, $n = 3$, $^{***}P < 0.001$. (e) Mice were euthanized 14 days after tumor implantation. Tumors of mice were isolated and weighed. Data represented as mean ± SD, $n = 3$, $^{*}P < 0.05$, $^{**}P < 0.01$, $^{***}P < 0.001$. (i) PBS, (ii) DNA-chain-1, (iii) DNA-chain-2, (iv) DNA network, (v) TIT, (vi) DNA-chain-1 and TIT, and (vii) DNA network and TIT. Source: Chi Yao et al. [56]/Reproduced with permission from American Chemical Society.

20.4 RCA-Based DNA Nanotechnology for Nucleic Acid Drug Delivery

With the elucidation of human gene expression regulation and cancer pathogenesis, nucleic acid pharmacotherapy has emerged as a promising therapeutic strategy for the treatment of various diseases, such as genetic disorders and cancer. Unlike traditional small molecules or monoclonal antibodies, nucleic acid drugs consisted of sequence-specific nucleotides that reprogrammed cells at the genetic level and provided an effective strategy to control protein synthesis [68–71]. Nucleic acid-based drugs mainly included ASOs, small interfering RNA (siRNA), single-guided RNA (sgRNA), microRNA (miRNA), and immunostimulatory nucleic acids [72, 73]. However, the biological instability of nucleic acid drugs has seriously hindered their clinical application and development. Rapid and effective *in vivo* delivery into target cells was critical for the clinical application of nucleic acid drugs. Various non-viral nanocarriers have been developed for nucleic acid drug delivery, such as liposomal and polymeric carriers [74, 75]. Among the many non-viral nanocarriers, DNA nanocarriers, due to their inherent biocompatibility, molecular recognition ability, unique responsiveness, and sequence programmability, showed advantages in nucleic acid drug delivery [16–18]. The DNA nanocarriers that were synthesized by RCA could integrate functional sequences such as aptamers and nucleic acid drugs with a large number of repetitions. RCA-based DNA nanotechnology for nucleic acid drug delivery has been extensively explored in biological and biomedical applications.

The clustered regularly interspaced short palindromic repeats (CRISPRs) technology has been shown to enable efficient gene editing *in vivo* for the treatment of a wide range of diseases. However, the delivery of CRISPR/Cas9 to the target was still a challenge. A RCA-based proton-activatable DNA nanosystem that enabled the co-delivery of Cas9/sgRNA RNP and DNAzyme for combined gene therapy was reported [76]. Through the RCA reaction, an ultra-long ssDNA with repeated sgRNA recognition sequences, DNAzyme, and HhaI cleavage site was synthesized (Figure 20.4a). The sgRNA recognition sequence efficiently combined Cas9/sgRNA ribonucleoprotein (Cas9/sgRNA RNP) *via* complementary base pairing, thereby specifically recognizing and cleaving specific gene fragments to achieve down-regulation of the associated protein expression (PLK1). DNAzyme was a catalytically active ssDNA sequence that effectively cleaved mRNA in cancer cells and has been explored as a promising therapeutic agent. The designed DNAzyme was expected to down-regulate the expression of EGR-1 protein. Both gene editing in nucleus and gene silencing in cytoplasm achieved the combined gene therapy. In the nanosystem, the introduction of manganese ions (Mn^{2+}) not only acted as a cofactor of DNAzyme but also served to compress long ssDNA, forming nanocomplex (DNC). The acid-degradable polymer (polyglycerol dimethacrylate)-coated HhaI enzyme (GHhaI) was eventually absorbed to the DNC *via* electrostatic interactions, forming a proton-activated nanocomplex (H-DNC). The effect of Mn^{2+} on Cas9 protein was further studied, and the secondary structure of Cas9 was characterized by circular dichroism (CD) under different Mn^{2+} conditions. The results

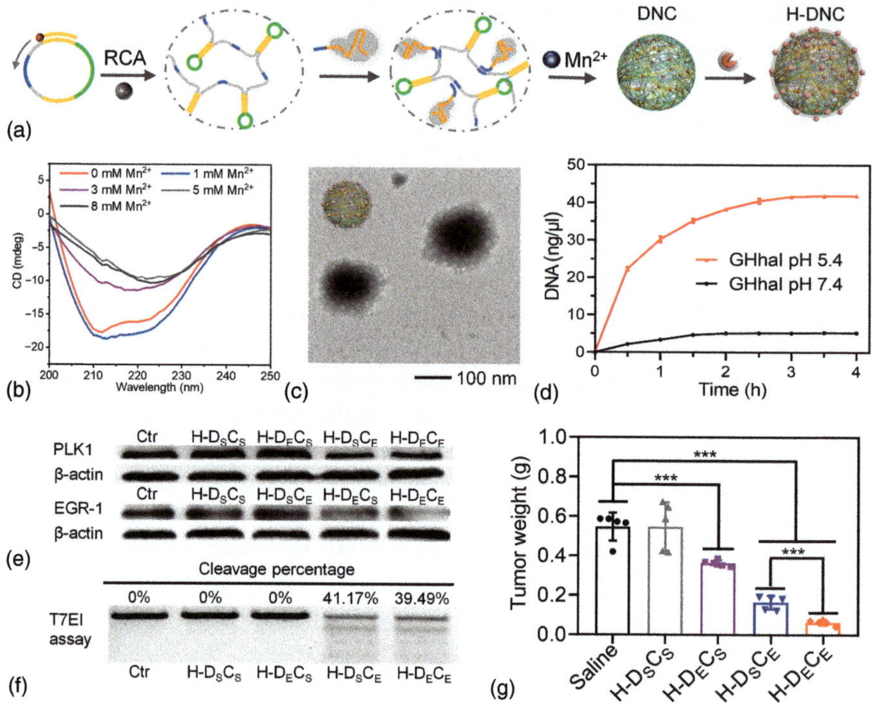

Figure 20.4 RCA-based proton-activatable DNA nanocomplex for the co-delivery of Cas9/sgRNA RNP and DNAzyme. (a) Construction of RCA-based proton-activatable DNA nanocomplex. (b) Circular dichroism (CD) spectra characterize the secondary structure of Cas9 at different concentrations of Mn^{2+}. (c) TEM images of H-DNC. (d) DNA release profiles of H-DNC at different pH. Data represented as mean ± SD, $n = 3$. (e) Western blot (WB) analysis with PLK1 and EGR-1 proteins in MCF-7 cells after different treatments. (f) T7EI assay of PLK1 in MCF-7 cells after treatment with H-DNC. (g) The average tumor weight in groups treated with different drug formulations. Data represented as mean ± SD, $n = 5$, $^{***}P < 0.001$. Source: Prof. Feng Li et al. [76]/Reproduced with permission from John Wiley & Sons, Inc.

showed that the secondary structure of Cas9 protein was well maintained under 1 mM Mn^{2+} (Figure 20.4b). TEM images show that H-DNC had a typical spherical structure with a diameter of about 150 nm (Figure 20.4c). The proton-activating properties of H-DNC were tested by incubating it in pH 7.4 and pH 5.4 mediums, respectively. After three hours of incubation, DNA was gradually released and reached a stable level in the medium at pH 5.4, while only a small fraction of DNA was released in the medium at pH 7.4, indicating that proton triggered the release of HhaI for its subsequent activity of cleavage (Figure 20.4d). After the internalization of H-DNC into cancer cells and entry into the lysosome, the polymer coating on HhaI was degraded in an acidic environment. HhaI recognized and cleaved the complementary sites in H-DNC for the release of Cas9/sgRNA RNP and DNAzyme. Western blot (WB) results showed that the combination treatment strategy resulted in a significant decrease in the expression of PLK1 and EGR-1 proteins (Figure 20.4e), and the T7EI assay demonstrated the gene cleavage efficacy of Cas9/sgRNA RNP (Figure 20.4f). The cytotoxicity results likewise revealed the powerful combined therapeutic effect of Cas9/sgRNA RNP and DNAzyme. As shown by the size of the tumors excised from the mice after 18 days of treatment, H-$D_E C_E$ group (DNAzyme was designed for EGR-1 silencing, and sgRNA was designed for PLK1 gene recognition) showed the most effective tumor suppression *in vivo* (Figure 20.4g). This RCA-based nanocomplex was expected to be used to deliver other gene editing tools and nucleic acid drugs.

In response to the demand for precision medicine, the precise initiation and sequential control of chemical events in the therapeutic process was a key issue in the optimization of drug efficacy and a challenge for the field of nanobiotechnology and biomedicine. DNAzyme, a ssDNA with catalytic function, was a promising therapeutic agent. DNAzyme could catalyze the precise cleavage of intracellular DNA or RNA and down-regulate disease-associated proteins to achieve therapeutic effects. Recently, a DNA nannoflower (DNC-ZMF) containing cascade DNAzymes and promoter-like Zn–Mn–Ferrite (ZMF) for combined gene/chemokinetic therapy of cancer has been reported [77]. DNA nanoflower with multiple functional aptamers was synthesized by RCA, and ZMF was adsorbed on DNA nanoflower *via* electrostatic interactions to form DNC-ZMF (Figure 20.5a). The introduction of AS1411 aptamer enhanced DNC-ZMF uptake in tumor cells. Two types of DNAzymes were designed for sequence cascade enzyme reactions. DNAzyme-1 catalyzed the self-cleavage of ultra-long DNA strands with Zn^{2+} as a cofactor to produce DNAzyme-2-containing fragments; DNAzyme-2 catalyzed the cleavage of mRNA with Mn^{2+} as a cofactor, leading to the down-regulation of early growth response protein 1 (EGR-1), thus inhibiting tumor cell proliferation and promoting tumor cell apoptosis (Figure 20.5b). The SEM and high-resolution TEM images showed that the average size of DNC-ZMF was about 200 nm (Figure 20.5c) and performed a typical crystal structure (Figure 20.5d). After cellular uptake, the role of DNAzyme in DNC-ZMF was investigated. DNC-ZMF released Zn^{2+}, Mn^{2+}, and Fe^{2+} under the action of an acidic tumor microenvironment and glutathione, which triggered the cascade cleavage of DNAzymes. To assess the down-regulation expression of EGR-1, another three different groups were selected, including DP-ND-ZMF (without DNAzyme-1), DP-NR-ZMF (without DNAzyme-2), and DP-nApt-ZMF (without aptamer). Western blot (WB) analysis showed that the DNC-ZMF group had the lowest expression of EGR-1 (Figure 20.5e). Annexin V-FITC/PI apoptosis analysis by flow cytometry demonstrated that the DNC-ZMF group had an excellent apoptosis effect (Figure 20.5f). Furthermore, *in vivo* therapeutic efficacy of DNC-ZMF-mediated combination therapy was evaluated in MCF-7 tumor-bearing mice. After 21 days of treatment with different drug preparations, respectively, the mice were euthanized, and the tumors were excised and weighed. The results showed that the antitumor activity of DNC-ZMF was higher than other groups, and the tumor mass was reduced by 70.4% compared to the PBS group (Figure 20.5g). For assessment of the gene silencing effect of DNC-ZMF *in vivo*, the relative expression of EGR-1 mRNA in tumor tissues was further detected by qRT-PCR analysis. EGR-1 mRNA levels were significantly decreased in DNC-ZMF-treated mice compared to other treatments, reaching a 60.0% downregulation (Figure 20.5h). This DNA-based

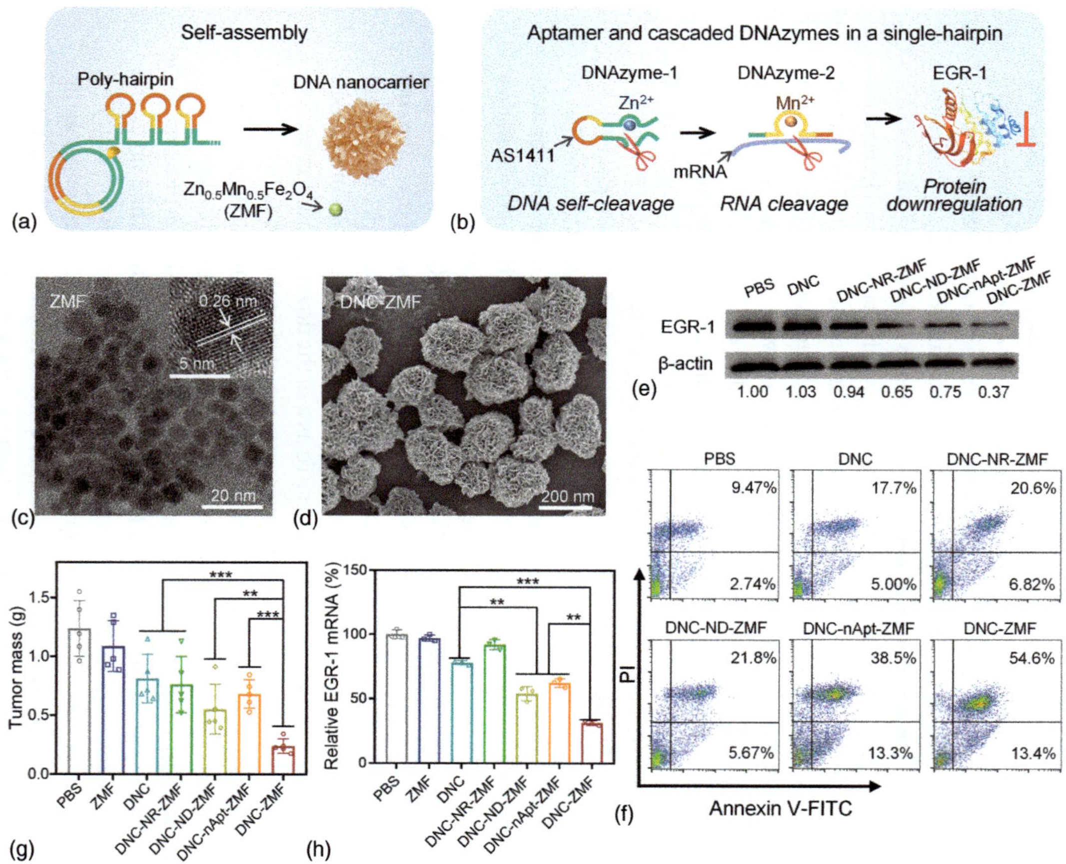

Figure 20.5 RCA-based nanocomplexes containing cascaded DNAzymes and promoter-Like Zn–Mn–Ferrite for Combined Gene/Chemo-dynamic Therapy. (a) Molecular design and preparation of DNA nanocomplex (DNC-ZMF). (b) Cascade catalytic cleavage in DNAzymes. (c) TEM images of ZMF. (d) SEM images of DNC-ZMF. (e) Western blot (WB) analysis of EGR-1 proteins in cells after different treatments. (f) Apoptosis analysis of cells treated with different drug formulations using flow cytometry with Annexin V-FITC/PI assay (g) The tumor weight in groups treated with different drug formulations. Data represented as mean ± SD, $n = 5$, $^{**}P < 0.01$, $^{***}P < 0.001$. (h) Detection of relative EGR-1 mRNA expression levels in MCF-7 tumors treated with different treatments by qRT-PCR. Data represented as mean ± SD, $n = 5$, $^{**}P < 0.01$, $^{***}P < 0.001$. Source: Yao et al. [77]/Reproduced with permission from John Wiley & Sons, Inc.

nanoflower provides a promising nanoplatform for clinical applications of nucleic acid-based materials to advance precision medicine.

The catalytic activity of DNAzyme was highly dependent on the cofactors. The production of sufficient metal ion cofactors in cancer cells was crucial for the clinical application of DNAzyme. On the other hand, the catalytic activity of DNAzyme could be enhanced with the increased temperature. Therefore, an appropriate elevation of the local temperature would improve the catalytic performance of DNAzyme *in vivo*. Based on this, a synergistic DNA-polydop-amine-MnO_2 nanoflower (DP-PM) for NIR-light-powered gene therapy was developed [78]. DNA particles (DPs) were synthesized by RCA reaction with ultra-long DNA strands containing a large number of repeated DNAzyme units (targeting Egr-1). PM was prepared *via* a redox reaction of $KMnO_4$ and dopamine, which assembled on the surface of DPs by π-π stacking interactions of polydopamine with DNA bases to form DP-PM (Figure 20.6a). The SEM image showed that the average size of DP-PM was about 50 nm (Figure 20.6b). The component elements of DP-PM were analyzed by X-ray photoelectron spectra (XPS), which further proved the successful preparation of DP-PM (Figure 20.6c). The photothermal conversion performance of DP-PM was evaluated by 808 nm radiation. The results showed that the solution temperature increased with increasing concentration of DP-PM solution upon 808 nm laser radiation (Figure 20.6d). Photostability was an essential indicator of photothermal conversion materials. The photothermal stability of DP-PM was assessed by the exposure to five consecutive "on"/"off" cycles of laser irradiation. The results showed DP-PM had superior reproducibility over the five cycles, demonstrating considerable stability to laser irradiation (Figure 20.6e). DP-PM accumulated at the tumor site, and under NIR light radiation, polydopamine on the photothermal conversion material DP-PM converted NIR light radiation to heat energy, inducing a temperature rise at the tumor site and enhancing the catalytic activity of DNAzyme. Meanwhile, the high concentration of glutathione (GSH) in MCF-7 cells decomposed PM, released Mn^{2+}, and activated the catalytic activity of DNAzyme. The cleavage of mRNA contributed to the down-regulation of Egr-1, which achieved gene silencing and caused apoptosis of tumor cells. The *in vitro* therapeutic effect of DP-PM was evaluated. After treatment, the tumor weight was counted. DP-PM-hNIR group showed completely ablated tumors, indicating the best tumor-suppressive effect (Figure 20.6f). This work provided an effective strategy for the *in vivo* regulation of the catalytic activity of DNAzyme for tumor therapy.

Photodynamic therapy (PDT) is a promising method for cancer treatment. However, relying on external light irradiation, it was difficult for PDT to exert a therapeutic effect in deep tissue areas. The strategy of constructing a self-luminous material system could effectively solve the problem of tissue illumination. As a traditional self-illuminant, the persistent luminescence nanoparticles (PLNPs) exhibited great potential in PDT system, which could steadily emit luminescence for hours or days after energy charging. Based on RCA technology, a PLNP-integrating DNA nanocomplex to realize the activation of PDT under glutathione (GSH) without exogenous laser excitation was explored [18]. The charged PLNP with MnO_2

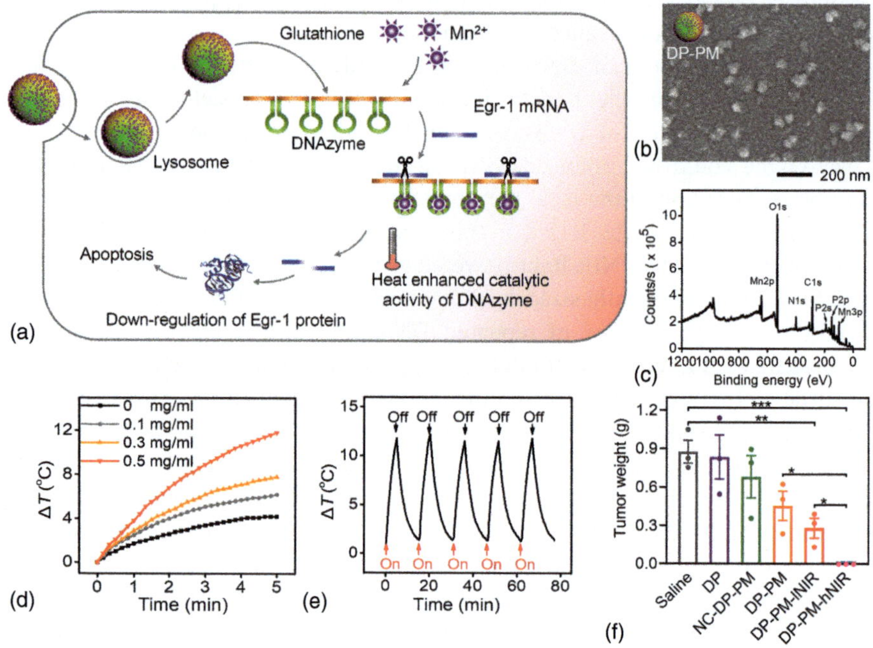

Figure 20.6 RCA-based DNA-polydopamine-MnO_2 nanocomplex for cancer therapy. (a) NIR-Light-Controlled DP-PM photothermal conversion produced a local temperature increase, thereby enhancing the catalytic performance of DNA enzymes in tumor cells and achieving a synergistic tumor suppression effect. (b) SEM image of DP-PM. (c) X-ray photoelectron spectra (XPS) spectrum of DP-PM. (d) Photothermal conversion curves for different concentrations of DP-PM solutions. (e) DP-PM temperature recording over five laser "on"/"off" cycles. (f) The tumor weight in groups treated with different drug formulations. Data represented as mean ± SD, $n = 3$, $^{*}P < 0.05$, $^{**}P < 0.01$, $^{***}P < 0.001$. Source: Huaixin Zhao et al. [78]/Reproduced with permission from American Chemical Society.

coating served as the core of the nanocomplex, and the ultra-long ssDNA chains containing AS1411 aptamer were synthesized *via* RCA to perform as the scaffold. The AS1411 aptamer-loaded photosensitizer silicon phthalocyanine dichloride ($SiPcCl_2$) *via* intercalation interaction and recognized nucleolin that overexpressed on cancer cells for enhanced cell internalization. In the presence of MnO_2, the luminescence of PLNPs was blocked and preserved, while the energy-storage PLNPs were released for self-luminescence in the absence of MnO_2 shell. In the cytoplasm, the MnO_2 shell outside the PLNPs was decomposed by the high concentration of GSH. The product Mn^{2+} was released to catalyze H_2O_2 to form O_2, and it could also be used for magnetic resonance imaging (MRI). PLNPs were released to activate $SiPcCl_2$ to generate $^1O_2$ for oxidative toxicity (Figure 20.7a). The T1-weighted magnetic resonance showed that the tumor site signal was able to persist for 48 hours, demonstrating that this laser-free pre-charged system could maintain continuous self-luminescence without an excitation light source. Multispectral optoacoustic tomography (MSOT) imaging was performed to verify the capability

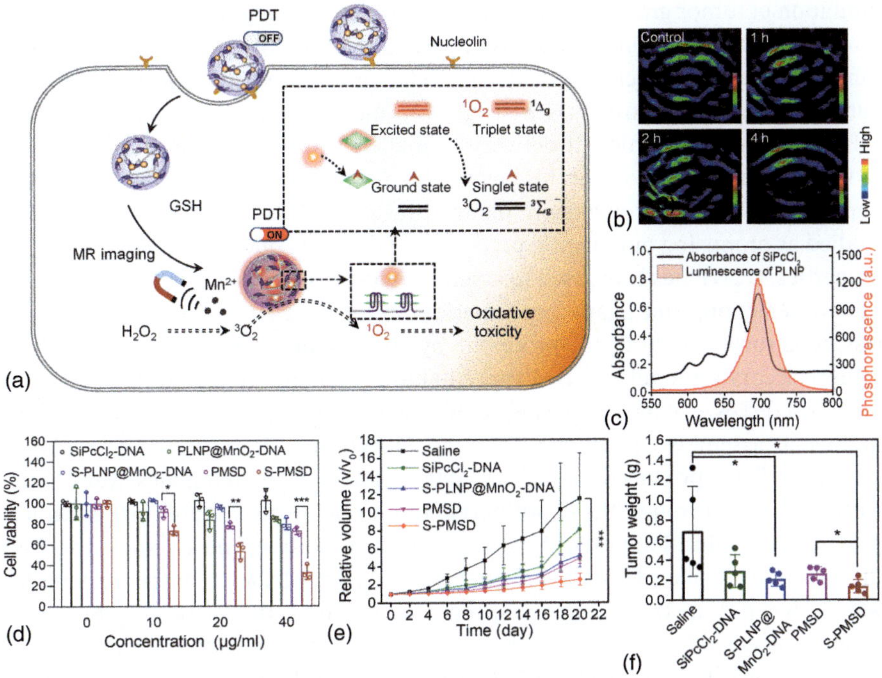

Figure 20.7 RCA-based energy-storing DNA-based nanocomplex for laser-free photodynamic therapy. (a) Laser-free photodynamic therapy of tumors. (b) Representative multispectral optoacoustic tomography (MSOT) images of MCF-7 solid tumors in mice treated with S-PMSD. (c) The emission spectrum of PLNPs and absorbance spectrum of $SiPcCl_2$. (d) Cell viability of MCF-7 cells after different drug formulations treatments. Data represented as mean ± SD, $n = 3$, $^*P < 0.05$, $^{**}P < 0.01$, $^{***}P < 0.001$. (e) The mean growth profile of tumors in different groups treated with different drug formulations. Data represented as mean ± SD, $n = 5$, $^{***}P < 0.001$. (f) The tumor weight in groups treated with different drug formulations. Data represented as mean ± SD, $n = 5$, $^*P < 0.05$. Source: Huaixin Zhao et al. [18]/Reproduced with permission from John Wiley & Sons, Inc.

of oxygen production of S-PMSD (S-PLNP@MnO_2– and $SiPcCl_2$-coloaded DNA nanocomplex), by measuring the oxyhemoglobin saturation at tumor sites. The results of MSOT demonstrated the increased photoacoustic signal intensities at tumor sites, verifying the elevation of blood oxygen. The maximum capability of oxygen production occurred at two hours, demonstrating an improvement in the tumor-depleted environment of S-PMSD (Figure 20.7b). To evaluate the probability of activating $SiPcCl_2$ using the sustained luminescence of PLNPs, the emission spectrum of PLNPs and the absorbance spectrum of $SiPcCl_2$ were characterized. The results indicated that the emission peaks of PLNPs significantly overlapped with the absorption peaks of $SiPcCl_2$ at 697 nm (Figure 20.7c). The cellular apoptosis of MCF-7 cells treated with different drug formulations was further analyzed, which demonstrated an enhanced therapeutic effect in the S-PMSD group (Figure 20.7d). The body weights and tumor sizes of mice were recorded every two days. The tumor growth curves showed that the S-PMSD-treated mice had the strongest inhibition of tumor growth and the smallest tumor size (Figure 20.7e). The average tumor weight in different groups also revealed the S-PMSD group had the greatest tumor suppression effect (Figure 20.7f), indicating this energy-storing DNA-based nanocomplex for laser-free PDT had valuable application prospects.

In addition to the efficient delivery of nucleic acid drugs into the living body, it was extremely important to detect the location and release of the drugs. Lanthanide-doped upconversion nanoparticles (UCNPs) have been extensively studied as promising bioimaging probes in recent years due to their unique optical properties. UCNP has been used to assemble with DNA molecules through electrostatic interactions and ligand coordination. Recently, a NIR-excited nanocomplex was constructed as a signal processor by the assembly of DNA and UCNP to achieve accurate and reliable real-time quantitative monitoring of drugs *in vivo* [79]. The UCNP/DNA nanocomplex (UCDC) signal processor was synthesized *via* interfacial assembly of circular DNA on UCNP and the epitaxial assembly of ultra-long DNA produced by RCA. Firstly, circular DNA templates were bound with the UCNP through the coordination of phosphate groups and rare earth ions (Ln^{3+}). Then, ultra-long ssDNA was synthesized by RCA and entangled to form the epitaxial device of the signal processor (Figure 20.8a). The circle template was designed with three functional regions, including the AS1411 aptamer (red region), the switchable i-motif structure (blue region), and the hairpin structure (green region) (Figure 20.8b). UCDC was subsequently used to detect the *in vivo* release of the model drug mitoxantrone (MTX), which was a widely used antitumor anthracycline, able to effectively intercalate in the double helix region of the hairpin structure. AS1411 aptamer could target nucleolin overexpressed on the plasma membranes of tumor cells for enhanced cellular uptake. The switchable i-motif structure and the hairpin structure responded to the acidic environment in tumors for the release of MTX. The optical properties of NIR-excited UCDC were based on the emission of UCNP and the filtering effect of MTX, which meant the absorption by MTX molecules of light emitted from UCNP. When the distance between UCNP and MTX was within the effective distance, the filtering effect appeared; when the distance increased, the filtering effect became noneffective (Figure 20.8c). The UCL spectrum of UCDC, especially the green emission (UCL_G, 530–570 nm) and the red emission (UCL_R, 645–665 nm) overlapped with the absorption spectrum of MTX, but the NIR emission (UCL_{NIR}, 790–810 nm) hardly did (Figure 20.8d). When MTX was loaded to UCDC, the intensity of UCL_R was reduced due to the filtering effect; when MTX was released, the filtering effect diminished and the intensity of UCL_R recovered as the distance from UCDC to MTX increased. Moreover, the intensity of UCL_{NIR} was not affected by MTX and could be used as an internal reference to improve the accuracy of monitoring. Radiometric imaging for real-time monitoring of drug release was established for monitoring the cumulative release of MTX *in vivo* (Figure 20.8e). Quantitative accuracy was assessed by verifying the linear relationship between $-\log(UCL_R/UCL_{NIR})$ and cumulative MTX release *in vivo* (Figure 20.8f). UCL_R/UCL_{NIR} images of the tumor region were obtained after tail vein injection of UCDC loaded with MTX for different times in a tumor-bearing mouse model, and the $-\log(UCL_R/UCL_{NIR})$ values were calculated for these time

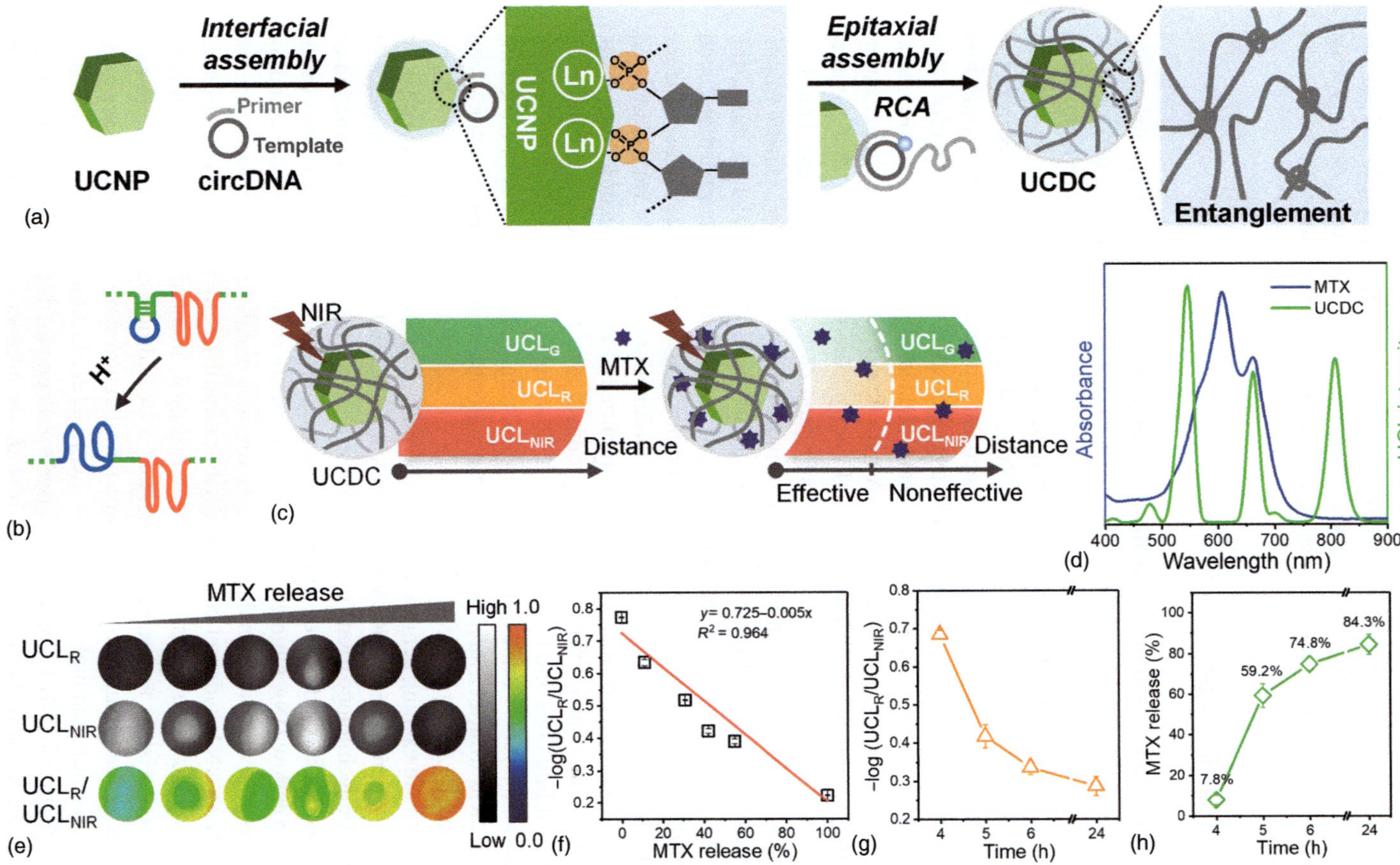

Figure 20.8 DNA signal processor based on UCNP and RCA epitaxial assembly. (a) Construction of UCDC. (b) The DNA transformation based on UCDC in acidic microenvironment. (c) Explanation of the distance dependence of the filtering effect between UCNP and MTX. (d) UCL spectrum of UCDC (green) and absorption spectrum of MTX (blue). (e) Bioimaging subcutaneous UCDC as MTX releases. (f) Corresponding curves of linearly fitted ratio-log(UCL_R/UCL_{NIR}) values to cumulative MTX releases in (e). (g) The scatter plots of the $-\log(UCL_R/UCL_{NIR})$. (h) Cumulative release of MTX at the corresponding time points. Data represented as mean ± SD, $n = 3$. Source: Adapted with permission from Yao et al. [79]. © Elsevier 2022.

points (Figure 20.8g). The cumulative release of MTX at different time points was calculated from the linear relationship and showed an increasing trend with time, reaching 84.3% at 24 hours (Figure 20.8h). This signal processor allowed for the control of drug delivery and the visualization of drug release, enabling robust monitoring and semiquantitative analysis of targeted drug release.

20.5 Conclusion

In this chapter, we summarized the construction of RCA-based DNA nanotechnology and the corresponding applications in cell research. In the part of DNA hydrogel, the synthesis method was double RCA. The ultra-long ssDNA produced *via* RCA formed DNA hydrogel through physical entanglement and non-covalent cross-linking. The supersoft nature of DNA hydrogel allowed it to maximize the protection of cell viability, minimizing mechanical damage to cells. The porous fiber structure of hydrogel facilitated the material exchange of cells. In the part of DNA nanoparticles, the synthesis methods were divided into biomineralization, metal ion compression, and co-assembly. DNA nanoflower formed by biomineralization had the advantages of uniform size and controllable morphology. After forming DNA nanoflower, extra materials such as PM and ZMF, which were coated on their surfaces, endowed DNA nanoflower with more biological functions. Furthermore, different from physical entanglement of ultra-long ssDNA mentioned above, the ssDNA was compressed by metal ions and co-assembled with other particles, such as PLNP, to form multifunctional nanocomplex. In particular, due to the sequence programmability of DNA, different DNA biofunctional elements were integrated into circle template, which were greatly amplified in the subsequent RCA process. The different characteristics of RCA products also made them have different biological functions and effects. Therefore, DNA hydrogel was widely used in cell separation, capture, and culture [49]. And DNA nanoparticles were more used for nucleic acid drug delivery, therapy, and drug release monitoring to achieve cell intervention [23–25].

Despite the exciting achievements in the field of biomedicine, there is still a lot of development space for RCA-based DNA nanotechnology. (i) The intelligence of DNA nanotechnology needs to be further explored. The structure and function of RCA needed to be designed more accurately, and other components were introduced to endow DNA nanotechnology with more diversified functions, responding to different stimuli such as light, heat, redox, and metal. (ii) The physicochemical property of DNA hydrogel could be further optimized. By adjusting the mechanical strength and viscoelasticity of DNA hydrogel, it could adapt to more scenes more appropriately. For example, supersoft DNA hydrogel could pass through narrow gaps, superelastic DNA hydrogel could quickly restore its original appearance after high strain, and high-viscous DNA hydrogel could adhere to the surface of biological tissue. (iii) DNA nanoparticles used for diagnosis, drug delivery, and therapy are needed to meet the requirement for stability in blood circulation, specific targeting of cancer cells, and efficient cellular uptake. Through responding to cellular components, DNA nanoparticles could realize signal amplification to improve their sensitivity in diagnosis. By combining with other active molecules, DNA nanoparticles could achieve high-capability loading and controllable release of drugs and realize combined therapy with other treatments. (iv) Scale and batch production of RCA-based DNA nanotechnology were needed to meet the requirements of wider applications. At present, the cost of RCA-based DNA nanotechnology is still high, which limits its applications in the real world.

With further investigation of biofunctions and properties of DNA, RCA-based DNA nanotechnology will be more intelligent and diversified. We envision that this chapter will attract readers to be more interested in RCA-based DNA nanotechnology and learn more about its applications in cell research.

Acknowledgment

This work was supported in part by National Natural Science Foundation of China (Grant Nos.: 22225505).

References

1 Watson, J.D. and Crick, F.H. (1953). *Nature* 171: 737–738.
2 Dong, Y., Yao, C., Zhu, Y. et al. (2020). *Chem. Rev.* 120: 9420–9481.
3 Seeman, N.C. (1982). *J. Theor. Biol.* 99: 237–247.
4 Krishnan, Y. and Seeman, N.C. (2019). *Chem. Rev.* 119: 6271–6272.
5 Seeman, N.C. and Sleiman, H.F. (2018). *Nat. Rev. Mater.* 3: 17068.
6 Seeman, N.C. (1998). *Angew. Chem. Int. Ed.* 37: 3220–3238.
7 Yang, D., Campolongo, M.J., Tran, T.N.N. et al. (2010). *WIREs Nanomed. Nanobiotechnol.* 2: 648–669.
8 Guo, W., Lu, C.-H., Qi, X.-J. et al. (2014). *Angew. Chem. Int. Ed.* 53: 10134–10138.
9 Li, M., Wang, C., Di, Z. et al. (2019). *Angew. Chem. Int. Ed.* 58: 1350–1354.
10 Guo, X., Li, F., Liu, C. et al. (2020). *Angew. Chem. Int. Ed.* 59: 20651–20658.
11 Liu, C., Chen, Y., Zhao, J. et al. (2021). *Angew. Chem. Int. Ed.* 60: 14324–14328.
12 Wu, Y., Li, C., Boldt, F. et al. (2014). *Chem. Commun.* 50: 14620–14622.
13 Zhao, J., Guo, Y., Tong, Z. et al. (2022). *ACS Appl. Bio Mater.* 5: 3795–3805.
14 Hu, Q., Li, H., Wang, L. et al. (2019). *Chem. Rev.* 119: 6459–6506.
15 Hu, P., Dong, Y., Yao, C., and Yang, D. (2022). *Chem. Asian J.* 17: e202200310.
16 Li, F., Lv, Z., Zhang, X. et al. (2021). *Angew. Chem. Int. Ed.* 60: 25557–25566.
17 Xu, Y., Lv, Z., Yao, C., and Yang, D. (2022). *Biomater. Sci.* 10: 3054–3061.
18 Zhao, H., Li, L., Li, F. et al. (2022). *Adv. Mater.* 34: 2109920.
19 Qi, H., Xu, Y., Hu, P. et al. (2022). *Chin. Chem. Lett.* 33: 1131–1140.
20 Hu, Y. and Niemeyer, C.M. (2019). *Adv. Mater.* 31: 1806294.
21 Yao, C., Xu, Y., Hu, P. et al. (2022). *Acc. Mater. Res.* 3: 42–53.
22 Li, F., Yu, W., Zhang, J. et al. (2021). *Nat. Commun.* 12: 1138.

23 Prinzen, A.L., Saliba, D., Hennecker, C. et al. (2020). *Angew. Chem. Int. Ed.* 59: 12900–12908.

24 Liu, S., Jiang, Q., Zhao, X. et al. (2021). *Nat. Mater.* 20: 421–430.

25 Wang, D., Li, S., Zhao, Z. et al. (2021). *Angew. Chem. Int. Ed.* 60: 15816–15820.

26 Xie, S., Du, Y., Zhang, Y. et al. (2020). *Nat. Commun.* 11: 1347.

27 Qin, W., Chen, L., Wang, Z. et al. (2020). *Adv. Sci.* 7: 2000647.

28 Zhang, S., Chen, C., Xue, C. et al. (2020). *Angew. Chem. Int. Ed.* 59: 14584–14592.

29 Yi, D., Zhao, J., and Li, L. (2021). *Angew. Chem. Int. Ed.* 60: 6300–6304.

30 Han, Z., Wan, F., Deng, J. et al. (2021). *Nano Today* 38: 101203.

31 Zhao, D., Chang, D., Zhang, Q. et al. (2021). *JACS* 143: 15084–15090.

32 Jin, J., Xing, Y., Xi, Y. et al. (2013). *Adv. Mater.* 25: 4714–4717.

33 Li, C., Faulkner-Jones, A., Dun, A.R. et al. (2015). *Angew. Chem. Int. Ed.* 54: 3957–3961.

34 Yang, B., Zhou, B., Li, C. et al. (2022). *Angew. Chem. Int. Ed.* 61: e202202520.

35 Ge, Z., Liu, J., Guo, L. et al. (2020). *JACS* 142: 8800–8808.

36 Li, J., Xun, K., Zheng, L. et al. (2021). *JACS* 143: 4585–4592.

37 Xiao, M., Lai, W., Yu, H. et al. (2021). *JACS* 143: 3448–3454.

38 Li, F., Liu, Y., Dong, Y. et al. (2022). *JACS* 144: 4667–4677.

39 Guo, Z., Zhang, L., Yang, Q. et al. (2022). *Angew. Chem. Int. Ed.* 61: e202111151.

40 Lizardi, P.M., Huang, X.H., Zhu, Z.R. et al. (1998). *Nat. Genet.* 19: 225–232.

41 Ali, M.M., Li, F., Zhang, Z. et al. (2014). *Chem. Soc. Rev.* 43: 3324–3341.

42 Wang, J., Yu, S., Wu, Q. et al. (2021). *Angew. Chem. Int. Ed.* 60: 10766–10774.

43 Zhang, L., Wang, L., Yuan, X. et al. (2021). *Angew. Chem. Int. Ed.* 60: 10745–10755.

44 Kim, M.-G., Park, J.Y., Shim, G. et al. (2015). *Biomaterials* 62: 155–163.

45 Zhan, P., Jiang, Q., Wang, Z.-g. et al. (2014). *ChemMedChem* 9: 2013–2020.

46 Lee, J.B., Peng, S., Yang, D. et al. (2012). *Nat. Nanotechnol.* 7: 816–820.

47 Mohsen, M.G. and Kool, E.T. (2016). *Acc. Chem. Res.* 49: 2540–2550.

48 Yao, C., Zhang, R., Tang, J., and Yang, D. (2021). *Nat. Protoc.* 16: 5460–5483.

49 Hoffman, A.S. (2002). *Adv. Drug Delivery Rev.* 54: 3–12.

50 Kim, E., Agarwal, S., Kim, N. et al. (2019). *ACS Nano* 13: 2888–2900.

51 Li, J., Jie, H.B., Lei, Y. et al. (2015). *Cancer Res.* 75: 508–518.

52 Kansy, B.A., Concha-Benavente, F., Srivastava, R.M. et al. (2017). *Cancer Res.* 77: 6353–6364.

53 Liu, L., Cheung, T.H., Charville, G.W., and Rando, T.A. (2015). *Nat. Protoc.* 10: 1612–1624.

54 Miltenyi, S., Müller, W., Weichel, W., and Radbruch, A. (1990). *Cytometry* 11: 231–238.

55 Oh, S., Jung, S.H., Seo, H. et al. (2018). *Sens. Actuators, B* 272: 324–330.

56 Yao, C., Zhu, C., Tang, J. et al. (2021). *JACS* 143: 19330–19340.

57 Yao, C., Tang, H., Wu, W. et al. (2020). *JACS* 142: 3422–3429.

58 Wang, J., Chao, J., Liu, H. et al. (2017). *Angew. Chem. Int. Ed.* 56: 2171–2175.

59 Hu, Y., Guo, W., Kahn, J.S. et al. (2016). *Angew. Chem. Int. Ed.* 55: 4210–4214.

60 Kacherovsky, N., Cardle, I.I., Cheng, E.L. et al. (2019). *Nat. Biomed. Eng.* 3: 783–795.

61 Hu, X., Wang, Y., Tan, Y. et al. (2017). *Adv. Mater.* 29: 1605235.

62 Li, L., Xu, S., Yan, H. et al. (2021). *Angew. Chem. Int. Ed.* 60: 2221–2231.

63 Mao, X., Liu, M., Yan, L. et al. (2020). *ACS Nano* 14: 8776–8783.

64 Tian, T., Xiao, D., Zhang, T. et al. (2020). *Adv. Funct. Mater.* 31: 2007342.

65 Li, M., Ding, H., Lin, M. et al. (2019). *JACS* 141: 18910–18915.

66 Yin, F., Li, M., Mao, X. et al. (2020). *Angew. Chem. Int. Ed.* 59: 10406–10410.

67 Tang, J., Yao, C., Gu, Z. et al. (2020). *Angew. Chem. Int. Ed.* 59: 2490–2495.

68 Doudna, J.A. and Charpentier, E. (2014). *Science* 346: 1258096.

69 Cox, D.B.T., Platt, R.J., and Zhang, F. (2015). *Nat. Med.* 21: 121–131.

70 Hilton, I.B. and Gersbach, C.A. (2015). *Genome Res.* 25: 1442–1455.

71 Wang, L., Zheng, W., Liu, S. et al. (2019). *ChemBioChem* 20: 634–643.

72 Jiang, Z. and Thayumanavan, S. (2020). *Adv. Ther.* 3: 1900206.

73 Vaughan, H.J., Green, J.J., and Tzeng, S.Y. (2020). *Adv. Mater.* 32: 1901081.

74 Staahl, B.T., Benekareddy, M., Coulon-Bainier, C. et al. (2017). *Nat. Biotechnol.* 35: 431–434.

75 Cong, L., Ran, F.A., Cox, D. et al. (2013). *Science* 339: 819–823.

76 Li, F., Song, N., Dong, Y. et al. (2022). *Angew. Chem. Int. Ed.* 61: e202116569.

77 Yao, C., Qi, H., Jia, X. et al. (2022). *Angew. Chem. Int. Ed.* 61: e202113619.

78 Zhao, H., Zhang, Z., Zuo, D. et al. (2021). *Nano Lett.* 21: 5377–5385.

79 Yao, C., Tang, J., Zhu, C. et al. (2022). *Nano Today* 42: 101352.

21

Precise Integration of Therapeutics in DNA-Based Nanomaterials for Cancer Treatments

Yimeng Li[1], Lijuan Zhu[2], and Chuan Zhang[1]

[1] *School of Chemistry and Chemical Engineering, Frontiers Science Center for Transformative Molecules, Shanghai Key Laboratory for Molecular Engineering of Chiral Drugs, Shanghai Jiao Tong University, 800 Dongchuan Road, Shanghai 200240, China*
[2] *Institute of Molecular Medicine, Renji Hospital, School of Medicine, Shanghai Jiao Tong University, 160 Pujian Road, Shanghai 200127, China*

Cancer has threatened global health as one of the leading lethal diseases. It was estimated that 19.3 million new cancer cases and 10 million cancer deaths occurred globally in 2020 [1]. The future burden of cancer was predicted to be 28.4 million new cases in 2040. Lung cancer is the leading cause of cancer death, followed by colorectal, liver, stomach, and female breast cancers. However, current clinical available treatments, such as surgery and radiotherapy, perform insufficient to combat diverse malignant tumors, particularly in the presence of metastasis. Efficient anti-cancer treatments with tolerant side effects remain a challenge to advanced cancers, which has attracted extensive interest in developing new potent therapies.

Pharmaceutical therapy, including but not limited to small-molecule drugs, proteins, and nucleic acid therapeutical agents, serves as a valuable regimen to treat cancer. While multiple issues could hamper the drug's efficacy, typically including: (i) the physical properties of drug, such as hydrophobicity; (ii) systematic toxicity; and (iii) low bioavailability. For example, the small-molecule drug, doxorubicin (DOX), is a wide-spectrum anti-cancer agent, while its intrinsic hydrophobicity and cytotoxicity impair its clinical performance. Therapeutic antibodies and oligonucleotides are also confronted with the challenge of rapid degradation after administration. Owing to the boom of nanotechnology in the past few decades, nanomaterial offers a powerful platform to deliver pharmaceuticals. With benefits of controllable architecture and imparted functionalities, drug delivery systems (DDSs) at nanoscale could enhance pharmaceutical efficacy. DDSs engineered with different carrier materials have been extensively investigated, including inorganic materials, liposomes, dendrimers, and polymeric micelles [2, 3]. Despite the mass of exploration, the access of varied DDSs to clinical application is still confined due to latent toxicity, inadequate *in vivo* study, defective reproducibility, and high-cost production.

Based on the development of DNA nanotechnology, DNA-based nanomaterials have gained considerable attention in the application of biomedicine. In addition

DNA Nanotechnology for Cell Research: From Bioanalysis to Biomedicine, First Edition. Edited by Zhou Nie.

to therapeutical performance, DNA molecules also show potent promise to constitute DDSs. This is due to its remarkable properties, such as programmable architecture, feasible modification, and advantageous biocompatibility. It has therefore emerged as a promising candidate to assist in the delivery of various therapeutics in cancer treatments [4, 5]. Moreover, synergistic treatments can be accomplished by incorporating multiple agents into a DNA-based nanoplatform, which tends to be a preferable strategy to combat cancer in recent years [6]. In this chapter, the application of DNA-based drug carrier with precise drug integration in synergistic cancer treatments is summarized. The properties and architectures of DNA-based nanomaterials in biomedicine are first generally described, with an emphasis on the advantages of DNA-based DDSs. To achieve a better understanding of employing these materials to constitute therapeutical complexes, the interaction between DNA-based architectures and cells is briefly discussed, followed by general strategies to construct DNA-based DDSs. Lastly, one specific category of DNA-based DDSs that we developed and extensively explored in recent years is particularly presented. This new type of DNA-based construct is characterized by precise integration of chemotherapeutics in DNA strands and designed sequences to regulate the target gene, which is termed as chemogene and can exert synergistic effects for better cancer treatment. Its definition, preparation, biological functions, and clinical potentials are also reviewed in the third part of this chapter.

21.1 DNA-Based Nanomaterials in Biomedicine

The role of DNA molecules in biomedicine can be generally divided into two main categories: therapeutic DNA and carrier DNA. It is known that certain sequence of DNA can be therapeutical alone. For example, aptamers are single-strand nucleic acid molecules with tertiary structure that are capable of binding to specific targets. In the case of cancer treatments, DNA aptamer could potentially trigger immunotherapy by activating or blocking an immune-modulatory receptor [7]. Antisense oligonucleotides (ASOs) are single-stranded nucleic acid molecules that have been utilized to regulate the gene expression. The function of ASO is majorly implemented either by hybridizing with target RNA to form RNA–DNA heteroduplex substrates, which would be subsequently degraded by RNase H enzyme, or by acting as a steric blocker to mask certain sequences within target transcripts [8]. Thereby, DNA-based ASO has the potential to achieve silencing or modulating cancer-causing gene expression in cancer treatments.

Another significant role of DNA is to act as nanocarriers for therapeutic agents. DNA-based nanomaterials exhibit several distinguished properties compared with some widely investigated nanomaterial in the application of biomedicines, such as inherent biocompatibility, programmable architecture, feasible modification, and multiple functionalities. These features allow DNA-based DDSs to serve as a promising nanoplatform to construct therapeutics against diverse diseases, including cancer, and result in a rise of utilization of DNA molecules in cancer treatments. Herein, the properties and architectures of DNA-based nanomaterial are highlighted in this section to elucidate the benefits of DNA-based DDSs. Moreover, the interaction between cells and DNA-based architectures is also discussed to describe the mechanisms of DNA-based DDSs.

21.1.1 Properties of DNA-Based Nanomaterials

As a natural material of genetic information carrier in numerous living creatures, inherent biocompatibility of DNA molecules is of particular importance to benefit the biosafety of DNA-based DDSs. A number of studies have demonstrated the low cytotoxicity and low immunogenicity of several DNA nanostructures at the level of cell and animal experiments, as applied for cancer therapy [9–14]. In addition, there are a few enzymes existing in living organisms that can degrade DNA molecules. Unsurprisingly, DNA-based architectures and DDSs exhibit the property of biodegradability *in vivo*, which can also be advantageous for the demand for responsive drug release.

The behavior of bases interactions in DNA molecules has been well understood, particularly the Watson–Crick base paring (A–T, G–C). Double-stranded DNA (dsDNA) formed by hybridization displays defined conformation and geometry. The typical B-DNA, in the form of double-stranded helix, is characterized by 2 nm of diameter and 3.4 nm per helix turn [15]. This programmable nature of DNA allows the self-assembled DNA architecture with precise predictable sizes and shapes so that the fabricated DNA architectures can be pre-designed to achieve certain purposes. The hydrodynamic sizes of self-assembled DNA architectures can be manipulated in the range of 10–200 nm by rational design, which would benefit from enhanced permeability and retention (EPR) effects to enter tumor regions. Notably, self-assembled DNA architectures show enhanced stability against degradation compared with ssDNA and dsDNA. The character of controllable size and hydrophilic surface also play a role in the extension of circulation *in vivo*.

However, the stability and biodistribution of DNA-based constructs *in vivo* remain a problem to achieve satisfied performance. Confronting with this challenge, chemical modifications provide a solution. Thanks to the progress of solid-phase synthesis technique, large-scale synthesis of oligonucleotides can be achieved by automated phosphoramidite chemistry and is commercially available, even for short strands of DNAs up to 150–200 nt. This technique also enables diverse chemical modifications upon the synthesis of oligonucleotides. While large-scale syntheses of long strands of DNA and DNA nanostructure have also been investigated, such as via bacteriophage- [16, 17] and enzyme-based [18] production. Post-modification via conjugation reaction, as a valuable approach, promotes multiple functionalities of oligonucleotides and DNA-based DDSs. A number of functional groups can be introduced to DNA and serve as handles. Such handles subsequently facilitate the conjugation of DNA with a wide range of molecules (dye, anti-cancer drug, polymer, peptide, and proteins). Details of modification on DNA molecules were recently reviewed by Gothelf and coworker [19]. Multiple functionalities of varied DNA-based DDSs, therefore, can be implemented to trigger designed targeting and stimuli-responsive character. Overall, the feasibility of synthesis in a cost-effective manner, ease of modification, and multiple functionalities, along with inherent biocompatibility and structural programmability lead to DNA molecules a promise of ideal constitute of DDSs and therapeutics.

21.1.2 Architectures of DNA-Based Nanomaterials

The programmable nature of DNA molecules enables the formation of well-defined DNA-based nanoarchitecture. In the early 1980s, Seeman and coworkers reported the pioneering work that utilized DNA as budling block to fabricate immobile junctions [20, 21], which paved the way for the afterwards development of DNA nanotechnology. With four decades of efforts in this field, diverse 2D/3D DNA architectures (Figure 21.1), generally classified as DNA nanostructures and DNA-functionalized nanoparticles, have been developed to promote biomedical applications [4–6, 26, 27]. As a constituent of DDSs, the intrinsic nature of DNA architectures significantly contributes to the stability and cellular interaction of DDSs. Thus, four representative architectures for the purpose of cancer therapy are described below to examine the impacts of architecture. Afterwards, Table 21.1 summarized a few studies on the stability of diverse DNA nanoarchitectures.

21.1.2.1 DNA Polyhedra

DNA polyhedra, such as tetrahedron, cube, bipyramid, octahedra, and icosahedra, are 3D cage-like nanostructures that can be fabricated by tile-based hierarchical assembly of DNA. As a result of the rigid structure, DNA polyhedron is advantageous for serving as scaffolds to load therapeutics. The size of the polyhedron can be manipulated at sub-nanometer range by rational design of DNA sequence. In particular, DNA tetrahedron (TDN) is beneficial for its simple fabrication and remarkable stability, which makes it the most studied polyhedron. Preparation of TDNs can depend on single-strand annealing, three-way-way junction hybridization, and scaffold folding [29]. The characteristic nanostructure of TDNs leads to enhanced endocytosis and the ability to scavenge reactive oxygen species (ROS), which could be advantageous for the treatment of degenerative and inflammatory

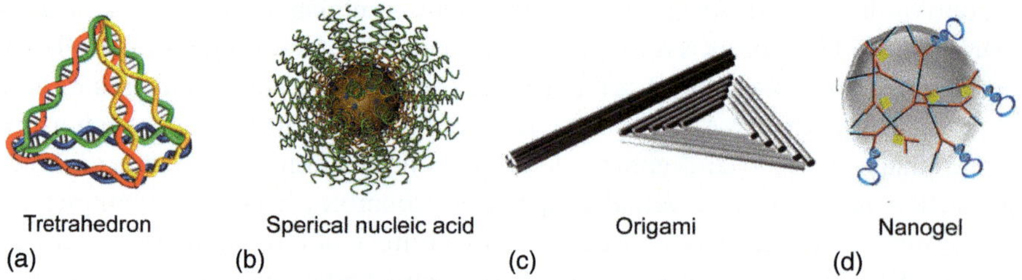

Figure 21.1 Representative DNA architectures used for drug delivery. (a) Tetrahedron. Source: Reproduced with permission from Joaqui-Joaqui et al. [22], © 2022 American Chemical Society. (b) Spherical nucleic acid. Source: Reproduced with permission from Cutler et al. [23], © 2012 American Chemical Society. (c) DNA origami. Source: Reproduced with permission from Jiang et al. [24], © 2012 American Chemical Society. (d) DNA nanogel. Source: Reproduced with permission from Li et al. [25], © 2015 American Chemical Society.

Table 21.1 Stability of DNA nanostructure.

	Size (nm)	Modification	Incubation medium	Stability (h)	Control/Stability (h)
DNA tetrahedron	~ 7		10% FBS	42	dsDNA/0.8
			50% FBS	24	dsDNA/<2
DNA cage	~7		10% FBS	18	ssDNA/<1
		Hexaethylene glycol (HEG)		62	HEG-ssDNA/~28
		T4 ligase		200	ssDNA treated with T4 ligase/ ~ 39
DNA origami	90 × 60		Cell lysates	12	ssDNA, dsDNA/<1
DNA nanotube	80 × 20	Decorated with CpG	FBS-containing medium	>9	ssDNA/~6

Source: Adapted from Jiang and Cai [28].

diseases. Small-sized TDN (~10 nm) displayed a specific pattern of biodistribution, such as kidney accumulation, allowing targeted delivery to specific organs. With its programmable nature, TDN can load chemotherapeutics into the double helices, relying on the porous structure to intercalate and integrate oligonucleotide-based therapeutics by hybridization. For example, Liu et al. developed a dual-targeting DNA TDN (MUC1-Td-AS1411) with MUC1 aptamer and AS1411 aptamer, in which nanocarrier loaded around 44 DOX molecules per TDN via intercalation [30]. Compared with free drug, MUC1-Td-AS1411 showed lower cytotoxicity to MUC1-negative HL-7702 cells, comparative cytotoxicity to sensitive MCF-7 cells, and more effective activity against DOX-resistant MCF-7 cells. Despite the potential of TDNs in drug delivery, further investigations are still necessary to bring new insights into clinical application of TDNs.

21.1.2.2 Spherical Nucleic Acid (SNA)

In 1996, Mirkin et al. reported a synthetic method of polyvalent nucleic acid–nanoparticle conjugates, which is the first well-characterized form of spherical nucleic acid (SNA) constructs and consists of gold core and DNA shell [31]. In 2006, Mirkin and coworkers investigated the role of SNA for intracellular gene regulation [32]. The SNA nanoparticles were less susceptible to enzyme degradation and displayed more effective cellular uptake than linear oligonucleotides, showing a tunable gene knockdown. Following this pioneering work, SNAs whose nanoparticles typically consist of a nanoparticle core densely functionalized with a shell of oriented nucleic acids (typically DNA) have attracted global interest to explore their applications in biomedicine, such as gene regulation and molecular diagnostics [23]. The nucleic acid shell can be fabricated using therapeutic nucleic acids, such as ASOs, siRNAs, and immunostimulatory oligonucleotides, and attached to the core via covalent bond. In addition to a gold core, diverse materials have been employed as SNA core, including quantum dots, SiO_2, polylactic-*co*-glycolic acid (PLGA), proteins, and liposomes [33]. By conjugating the hydrophobic drug paclitaxel onto DNA molecules, Zhang and coworkers synthesized SNA architecture through the self-assembly of the amphiphilic DNA-paclitaxel conjugate, which acted as both drug carrier and therapeutical payload for intracellular gene regulation [34]. Unlike linear nucleic acids, SNAs are able to enter cells in the absence of transfection agent and beneficial from enhanced physiological stability, which allows it serving as a versatile drug delivery system. In particular, Mirkin and coworkers conducted a series of valuable investigations to understand and optimize the performance of SNA as therapeutics [35–39]. In a recent work, Mirkin and coworkers reported a synthetic protocol of ProSNAs (protein core with DNA shell), which encouragingly broadened the application of SNAs as intracellular probes and functional enzyme delivery *in vivo*. SNA represents an extremely promising form of nucleic acid-based therapeutics, and currently two of them (NU-0129 and AST008) are being evaluated in clinical trials for cancer therapy.

21.1.2.3 DNA Origami

In 2006, Rothemund developed the "DNA origami" technique by utilizing "scaffold strand" (long-strand ssDNA) and hundreds of "staple strands" (short stand ssDNA) to obtain several nanoscale shapes and patterns [40], known as a milestone of DNA nanotechnology. Afterwards, numerous efforts have been conducted on the developments of DNA origami to fabricate diverse 3D nano-objects. In particular, the predictable fabrication of DNA origami promotes the size of objects in the range of 50–400 nm, which tends to accumulate in tumor regions due to EPR effects. In the stability study against cell lysates and live cells, naked DNA origami performed stable for 12 hours [41] and gradually degraded during incubation of 72 hours [42] respectively. To enhance the stability of DNA origami, strategies have been conducted by employing protective conjugation onto DNA origami, such as lipid shell and polymer coating [43]. In addition to acting as a 3D container for therapeutics, DNA origami can also display stimuli-triggered reconfiguration to achieve stimuli–response. For example, Tian and coworkers developed assemblies consisting of DNA origami and gold nanorod (AuNR) by the hybridization of the conjugated DNA on AuNR with a complementary handle from DNA origami. This DNA-origami-AuNR assembly can act not only as a probe for cancer theragnostic by accumulating in tumor region but also a promising candidate to achieve photothermal therapy triggered by NIR irradiation [25, 44].

21.1.2.4 DNA Hydrogel

DNA hydrogels are 3D biomaterial with hydrophilic property that can be constructed with either pure DNA or DNA-contained hybrid. DNA hydrogels can be developed via physical self-assembly or chemical crosslinking. Physical interaction-induced hydrogel can be triggered by DNA hybridization, metal–DNA coordination along with ionic interaction between DNA and cationic polymer, while chemical crosslinking can be fulfilled by intermolecular covalent binding of linear DNA–DNA or DNA–polymer [45]. Fabrication of DNA hydrogel can be enforced with flexibility to result highly structured 3D DNA network in different size ranges (bulk hydrogel, microgel, and nanogel) owing to the programmable property of DNA. Similar as peer hydrogel biomaterials, DNA hydrogel is an outstanding carrier candidate to satisfy expected mechanical rigidity and reserve abundant functional sites. Within the crosslinked network, therapeutics are prone to be protected by preventing the permeation of enzymes *in vivo*. As an example, Tan and coworkers synthesized a DNA nanogel through the self-assembly of two Y-shaped monomers along with one DNA linker while the size of it can be controlled by varying the ratio of two monomers [25]. Incorporation of aptamers, disulfide linkages and ASOs were conducted to construct nanogel so that the resulting nanogel successively achieved enzyme-mediated disassembly and targeted gene therapy against cancer cells. Therefore, DNA hydrogels are of potential to act as a smart vehicle to hold diverse therapeutics with a stimuli-responsive character. Nevertheless, *in vivo* investigations of DNA hydrogel-based DDSs are still at extremely preliminary stage, particularly the relative physicochemical properties contributing on aggregation of DNA hydrogel in bloodstream and rapid clearance *in vivo*.

21.1.3 Interactions Between DNA-Based Drug Delivery Systems (DDSs) and Cells

21.1.3.1 Endocytosis of DNA-Based Architectures

Cellular Endocytosis Mechanisms Prior to understand the underlying mechanism of DNA-based DDS, the entry pathway of nano-objects into cells should be addressed. The entry of large molecules, such as nano- and micro-objects, into cells is conducted by endocytosis. Cellular endocytosis can be divided into phagocytosis and pinocytosis, that the former is majorly responsible for the uptake of larger particles (>500 nm) and the latter dominates the uptake of smaller particles [46]. Phagocytosis is commonly possessed by phagocytes in mammals, such as macrophage and neutrophils. With recognition by phagocytes, engulfment of particles can occur through specific ligand-receptor interaction and results to the formation of phagosome (Figure 21.2). Mature phagosome will fuse with lysosome to afford phagolysosome where engulfed particles will be degraded under the acidic condition containing functional enzymes. In contrast, pinocytosis can occur in all types of mammal cells through four main mechanisms: macropinocytosis, clathrin-mediated endocytosis (CME), caveolae-mediated endocytosis, and clathrin- and caveolae-independent endocytosis (Figure 21.2).

Macropinocytosis is an actin-regulated endocytic process in which the internalization proceeds in a non-specific manner. Despite the non-selective uptake, macropinosomes display a micrometer-sized property (0.2–5 μm), which makes them an important entry pathway for macromolecules and large particles. CME is a significant endocytic pathway for cells and is characterized with the involvement of protein clathrin. CME can be classified as receptor-independent uptake and receptor-mediated uptake, which occur via non-specific interaction and specific ligand-receptor binding, respectively. Typically, in mammals, specific binding triggered CME involves the generation of cathrin-coated vesicle (60–120 nm), early

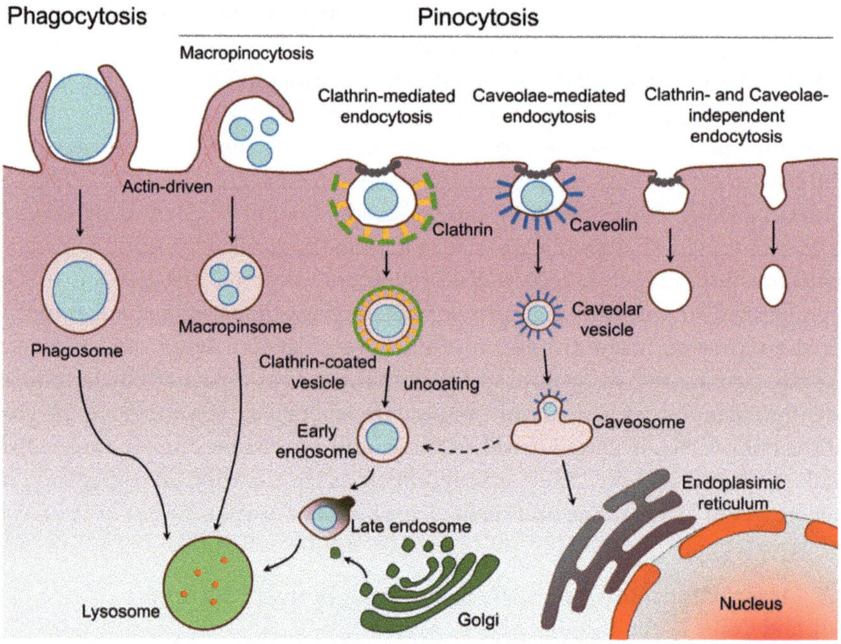

Figure 21.2 Schematic illustration of different mechanisms of cellular endocytosis. Source: Reproduced with permission from Hu et al. [26], © 2019 American Chemical Society.

endosome (pH ~ 6) and late endosome, which ultimately will travel to lysosome. Recruiting CME pathway is a valuable and popular strategy to assist in the design of DDSs, particularly when modified with ligand specific to the receptors in CME pathway. It is also notable that ultimate escape from early endosome should be considered to avoid lysosomal degradation of DDSs so that they can reach the targeted subcellular compartment. Caveolin-dependent endocytosis (CvME) is a receptor-mediated process and involves a flask-shaped caveolae (50–80 nm) after the membrane invagination. Due to caveolae vesicles preferring to transform to either nonenzymatic caveosomes or other non-lysosomal compartments, therapeutics delivered via this pathway is considerably susceptible to escape degradation in lysosome. Therefore, equipped with ligands that can bind the receptors in CvME, such as folic acid and albumin, is particular of values for protecting delivered therapeutics (e.g. genes and proteins) from lysosomal degradation. Nevertheless, the internalization speed via CvME is slower than that via CME, and the vesicle volume for delivered objects in CvME is also smaller than that in CME. In addition to abovementioned pathways, clathrin- and caveolae-independent endocytosis can occur via lipid rafts (40–50 nm), which are cholesterol- and sphingolipid-rich microdomain generated on the cytoplasmic side of any cell surface. However, the mechanism of this pathway is still not well understood and requires further investigations.

Due to the negatively charged property of DNA molecules, internalization of unmodified DNA oligonucleotides by cells can be hindered by electrostatic repulsion. However, constructing DNA-based nanoarchitecture offers a solution to enforce efficient cellular uptake via endocytosis. It is notable that a few factors contribute to the internalization events of DNA-based architecture with cells, as generally described below.

Cellular Entering of DNA Architectures In 2011, the Turberfield group [47] and the Fan group [48] reported the phenomenal internalization of DNA nanostructure (DNA cage and DNA tetrahedron) by several different types of cells in the absence of transfection agents. It was speculated that such internalization occurred via the endocytosis pathway. Cellular uptake of DNA-based nanoarchitecture is remarkable efficient than linear DNA. To investigate the endocytosis pathway employed, Fan and coworkers developed tetrahedral DNA nanostructures (TDNs) (7 nm) and demonstrated that the TDNs interacted with plasma membrane of Hela cells non-specifically and subsequently internalized into cells via CvME [49]. In particular, the size of DNA-based architectures was identified as a significant factor to decide the endocytosis pathway. Typically, caveolar vesicles are able to deliver particles less than 80 nm and thus fabricated particles, such as TDNs and SNAs, in the proper size range can be internalized through CvME. Larger particles over 100 nm will tend to employ CME, micropinocytosis, or multiple endocytosis pathways. In 2013, Mirkin and coworkers also investigated the entry pathway of SNA, whose nanoparticle is usually characterized with a nanoparticle core and a densely packed, highly oriented nucleic acid shell [50]. The fabricated SNAs can efficiently enter more than 50 different cell types. Their work pointed out that the 3D arrangement of nucleotides on SNA made a positive effect on interacting with surface receptors, and the endocytosis employed a lipid-raft-dependent, caveolae-mediated pathway. Meanwhile, sequence composition is also able to affect cellular uptake efficiency; SNAs with high content of G showed higher cellular uptake efficiency than SNAs primarily composed of A, T, and Cs. Furthermore, shapes of DNA-based architecture were also identified as factors to influence cellular uptake. For instance, the architecture characterized by high aspect ratio showed enhanced cellular uptake efficiency [51].

Modifications onto DNA-based architecture can considerably affect its interaction with cells. For instance, it was reported that a DNA nanostructure modified with hydrophobic dendritic chains displayed a rapid cellular uptake, while the cellular uptake of that modified with hydrophilic chains was slow [52]. By surface modification with PEGylated lipid bilayers onto a DNA octahedron, the immune activation was reduced by 2 orders of magnitude below unmodified DNA octahedron [53]. Such modification also facilitated the enhancement of pharmacokinetic bioavailability by about 17-fold.

Modification with targeting ligands has also been widely investigated to enhance therapeutic effects by improving cellular internalization and targeted delivery to specific cells. Example ligands include aptamers, folic acid, cell-penetrating peptides, and iron transport protein transferrin [6]. For instance, AS1411 aptamer is a well-studied aptamer that can improve cellular uptake through an active, temperature-dependent, nucleolin-mediated macropinocytosis. Lin and co-workers developed AS1411-modified tetrahedral DNA nanostructure (Apt-TDN) and confirmed that Apt-TDN displayed higher cellular uptake and nucleus localization in both MCF-7 and L929 cells compared with non-modified TDN [54].

21.1.3.2 Fate of DNA-Based Architectures in Cells

Enzymatic degradation by DNases in cells is typically conducted as the outcome of internalized DNA-based architectures. Whereas the details of degradation are considerably related to the employed endocytosis pathway. For example, the majority of fabricated nanoparticles, including DNA-based architecture, would be recognized and interact with macrophages or other phagocytes as administrated *in vivo*. The internalization occurs via phagocytosis, and the particles ultimately travel to phagolysosomes to be forced to degrade by enzyme. In addition to phagocytosis, DNA-based architectures employing CME will encounter the transportation from vesicles to early endosomes and late endosomes, while eventually travel to lysosomes. Similar as the phagolysosomes, lysosomes are characterized with acidic environment and diverse DNases, which promote the degradation of DNA-based architectures as well as the release of trapped active therapeutics.

However, to satisfy effects of diverse functional therapeutics, targeted sites within cells could be specific subcellular compartments rather than lysosomes. As a result, DNA-based architectures could be designed to either achieve breakdown prior to fusion with endosomes or employ pinocytosis pathways except CME. In the former case, for instance, cationic nanoparticles display the ability to induce osmotic lysis during endosome acidification, based on the proton sponge effect. Accordingly, Gu and coworkers developed a cationic polyethylenimine-modified DNA nanoclew to deliver Cas9/sgRNA, which was identified with successful escape from endosomal transportation and successful delivery from cytoplasm to nuclei [55, 56]. In the latter case, employing CvME is considered to be a popular strategy. Fan and coworkers fabricated TDNs that can be internalized by Hela cells via CvME [49]. Though they were transported to lysosome, functionalization with nucleus-targeting signaling peptides enforced the escape of TDNs from lysosomes and entry to cellular nuclei. Moreover, taking advantage of retrograde endocytic pathway or artificial nanoscale needles also shows potential to facilitate targeted delivery to specific subcellular compartments [26]. As targeted delivery is achieved, remained DNA-based architectures could be degraded within cells and even serve as nutrients for cells.

21.2 Strategies on Constructing DNA-Based DDSs

As discussed above, DNA-based nanomaterial displays several distinctive properties that make them beneficial for serving as drug carriers. Therefore, DNA-based DDSs have been widely investigated as potent platforms in the treatments of cancer. The category of anti-cancer drugs majorly includes small-molecule drugs, nucleic acid drugs, proteins, and peptides (Figure 21.3). This wild category of drugs can correspond to varied cancer therapy methods, such as chemotherapy, phototherapy, gene therapy, and immunotherapy. In general, the strategies to load diverse drugs with

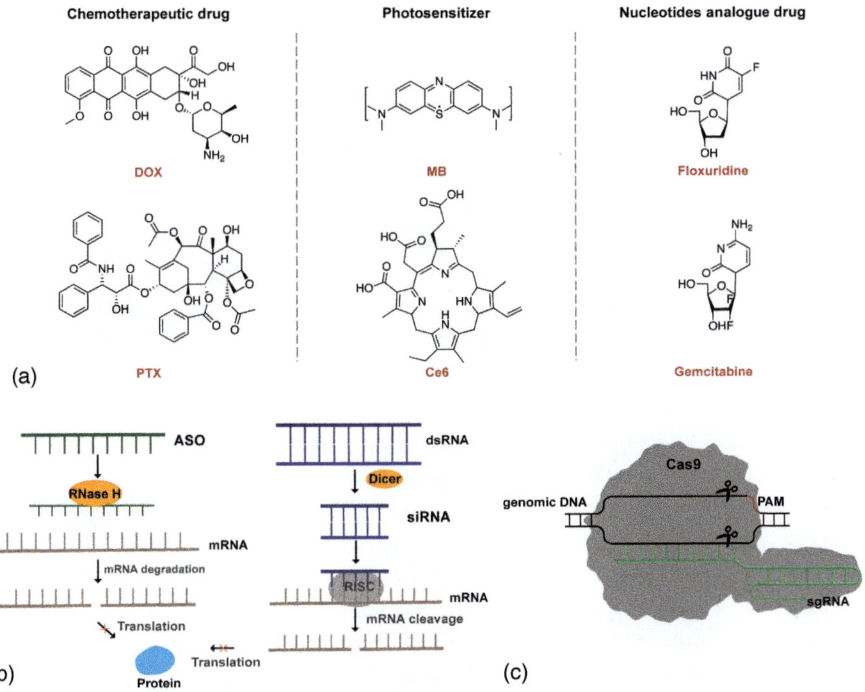

Figure 21.3 Representative anti-cancer therapeutics, including (a) small-molecule drugs; (b) nucleic acid; and (c) protein therapeutics.

DNA-based architectures can be classified as non-covalent interaction and covalent conjugation.

21.2.1 DNA-Based DDSs Engineered Through Non-covalent Interactions

Physical interaction, including electrostatic interaction, hydrophobic interaction, and hydrogen bonding, has been employed as a fundamental drug loading strategy. According to the structure of DNA molecules, intercalation of drugs into polynucleotide bases is a simple approach to load drugs (Figure 21.4a). Successful intercalation requires specific structure of intercalator, typically characterized with intrinsic planar structure or displaying an extended planar component, so that the intercalator would locate between two adjacent pairs of bases. Molecules presenting flat aromatic or hetero-aromatic ring, therefore, are favored to employ intercalation approaches.

DOX is a powerful anti-cancer agent in chemotherapies that can kill tumors by intercalating to DNA base pairs and inhibiting macromolecular biosynthesis. The significant cytotoxicity of DOX against healthy cells makes the desire of proper drug carriers to deliver DOX. Due to the intercalating property of DOX, DNA-based DDSs have been intensively studied to enhance its therapeutic effects. For example, Chang et al. fabricated DNA icosahedra with a five-point-star motif

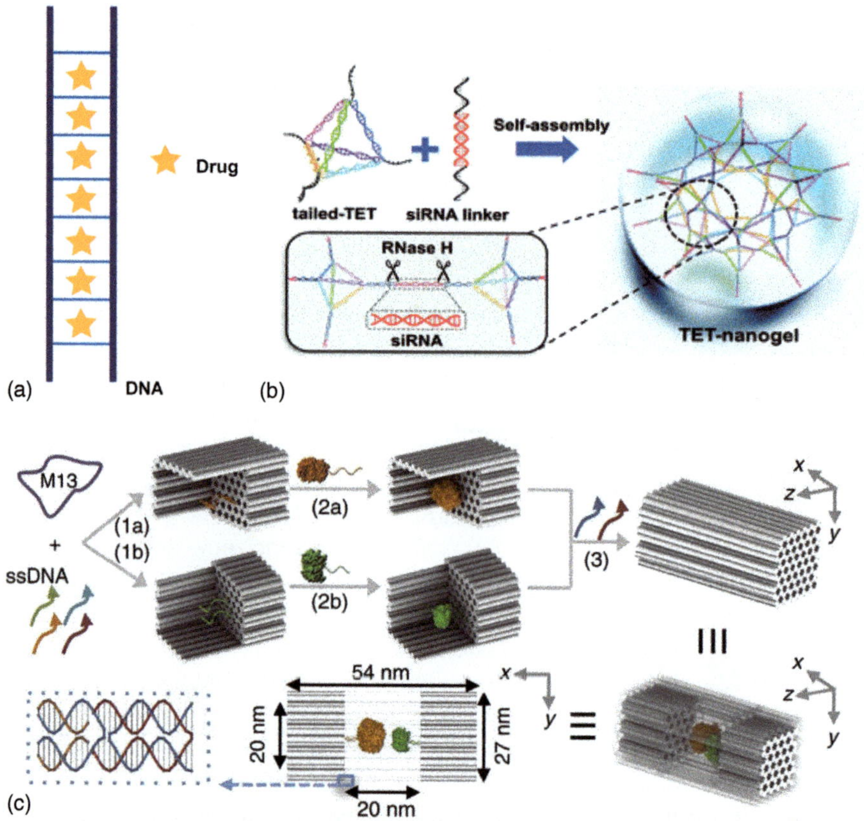

Figure 21.4 Manufacture of DNA-based drug delivery systems utilizing non-covalent interaction. (a) Strategy of intercalation of drugs into duplex nucleic acid helices. (b) Fabrication of DNA nanogel via hybridization-induced self-assembly using tailed DNA tetrahedron and siRNA linker. Source: Reproduced from Xue et al. [57], with permission from the Royal Society of Chemistry. (c) Fabrication of DNA nanocage-enzyme complex via hollow geometry strategy. Source: Zhao et al. [58]/Springer Nature/CC BY 4.0.

and MUC1 aptamer-conjugated six-point-star motif, which loaded DOX into DNA icosahedra via intercalation [59]. Comparing with nude drug and aptamer-excluded complex, the construct (DOX@Apt-DNA-icosa) showed enhanced cellular uptake and cytotoxicity against MUC1-positive human breast cancer cells (MCF-7). Ding and coworkers utilized DNA origami as carriers to intercalate DOX, achieving a high level of DOX loading efficiency (50–60%) [24]. Interestingly, the complex of DOX-origami displayed prominent cytotoxicity to both regular MCF-7 cells and DOX-resistant MCF-7 cells. The performance onto DOX-resistant MCF-7 cells suggested that the DNA nanostructure successfully improved the uptake of DOX by DOX-resistant MCF-7 cells. It is notable that the intercalation of DOX prefers to occur in the CG region of DNA. Hence, DNA architecture with CG-rich property is advantageous for efficient DOX loading and plenty quantity of DOX internalized by targeted cancer cells. This intercalation strategy is important for loading small-molecule drugs, not confined to only DOX. For example, Shieh and coworkers developed a G-quadruplex structure-based AS1411 aptamer and intercalated a photodynamic reagent, porphyrin derivative (TMPyP4), to produce the complex apt-TMP, which showed promising activity against MCF-7 cells by phototherapy [60].

Electrostatic interaction also provides an opportunity to intercalate drugs. As DNA displays negatively charged property, intercalators with positive charge character is capable to interact with DNA backbone. For example, Tian et al. designed a negatively charged and nucleus-targeted DNA tetrahedron and incorporated an anti-cancer positively charged metal complexes into DNA tetrahedron via electrostatic interaction [61]. With assistance of MUC1 and AS1411 aptamer, this functionalized DNA tetrahedron achieved the delivery of anti-cancer agent to nucleus and inhibited the growth and invasion of glioma cells. Overall, intercalation via non-covalent interaction is a simple and low-cost method to achieve rational drug loading, usually exclusive of toxic compounds addition or stringent reaction conditions. Nevertheless, the release of drug could be uncontrollable. Profound optimizations on drug loading sites, drug loading efficiency, and drug release properties are still essential to enhance therapeutic effects.

Moreover, nucleic acid drugs have been emerged as powerful weapons to treat a wide range of diseases. They act as therapeutics by genetically knocking down or inducing protein expression at transcriptional and translational levels. The natural base pairing behavior allows hybridization to act as a robust strategy to deliver nucleic acid drugs with DNA architecture. This low-cost method can afford assemblies with properties of controllable size and shapes and thermodynamic stability. Encouraged by the simplicity and powerfulness, DNA-based DDSs have been widely investigated to deliver nucleic acid drugs by employing hybridization strategy in last few years [8, 62]. For instance, Zhang and coworkers construct a DNA-based nanogel with tailed DNA tetrahedra and siRNA linker [57] (Figure 21.4b). The resulting compact architecture protected the functional siRNA by embedding it inside and successfully knocked down endothelial growth factor protein (EGFP) gene expression for cancer therapy. This study identified the potential of DNA architecture as noncationic vectors to deliver nucleic acid therapeutics. In addition, large particles such as proteins can also employ hollow geometry inclusion strategy to load into DNA architectures. Encapsulation of cargos can take advantage of empty cavities of DNA architectures. Zhao et al. constructed DNA nanocages to encapsulate six enzymes by self-assembly with well-controlled stoichiometry [58] (Figure 21.4c). By investigating the effects of encapsulation and proximal polyanionic surface on enzymatic activity, they demonstrated a significant inverse correlation between the size of enzyme and increased enzymatic activity, as well as positive effects of proximal polyanionic surface of cage on enzymatic stability. This DNA nanocage also increased the stability of enzymes against protease digestion, exhibiting potential as theragnostic nanodevices for future application. Though this strategy provides an opportunity for efficient protection and controlled release of diverse cargos, loading capacity of DNA carriers could be problematic to limit the therapeutic effects.

21.2.2 DNA-Based DDSs Engineered Through Covalent Interactions

For molecules unable to conduct physical interaction, chemical conjugation via a covalent bond has been extensively applied as a drug loading strategy. Covalent modification that enables a mass of therapeutics as cargos in DNA architecture can be generally divided into three strategies: integration of therapeutic nucleoside analogs (NAs); grafting onto DNA backbone; and terminal conjugation.

21.2.2.1 Integration of Therapeutic Nucleoside Analogs (NAs) into Nucleic Acid Strands

A category of anti-cancer drugs is characterized as purine and pyrimidine derivatives and displays similar structures as natural nucleobases. These nucleoside analogue drugs, such as floxuridine, gemcitabine, and decitabine, are prone to be integrated into nucleic acid strands due to their structural similarity. Zhang group used the phosphoramidite form of gemcitabine as monomers to develop a DNA-like polygemcitabine (Ge_{10}) through solid-phase synthesis [63], identifying a feasible method to integrate NAs upon solid-phase synthesis (Figure 21.5a). Based on the property of molecular recognition, the resulting Ge_{10}-generated Ge_{10} nanogel by self-assembling, which provided a precise nanodrug for cancer therapy. Moreover, the integration method was also expanded. Not only integrating floxuridine into DNA strands by solid-phase synthesis, Zhang group also reported the integration of floxuridine into RNA strands via enzyme-mediated transcription [65]. Both strands were conducted to self-assembly to form floxuridine-containing Y-motif DNA and RNA nanogels separately, displaying remarkable inhibitory activity against cancer cells. Furthermore, gemcitabine-containing DNA nanogel was fabricated with a pH-responsive character [66]. As integration of Ge was implemented upon solid-phase synthesis, two specific sticky-end sequences were designed for two Ge-containing Y-motifs, where one sticky end was Ge-rich and the other was partially complementary to the Ge-rich sticky end. Self-assembly of Y-motifs therefore afforded pH-responsive nanogels that were stable at neutral or basic condition and disassembled at acid condition due to the preferred intramolecular quadruplex structure of Ge-rich stick end at acid condition.

Integration of NAs upon solid-phase synthesis can also be applied into other DNA architectures. Zhang and coworkers developed a series of DNA polyhedra with floxuridine-integrated DNA strands (Figure 21.5b) [64]. The resulting DNA polyhedral showed definite drug-loading ratios, tunable size, and morphology while acting as Trojan-horse-like DDSs to hide the anti-cancer drugs. In particular, the buckyball architecture showed the optimum efficiency for drug delivery and anti-cancer activity among several polyhedral architectures. Utilizing this integration method, Zhu et al. developed several aptamers by replacing all natural nucleotides with their NA counterparts, named drugtamer [67]. Based on this strategy, drug loading ratio was extremely enhanced and targeting capability remained. The resulting drugtamers specifically bind to the targeted receptors overexpressed on the surface of certain cancer cells and were qualified for enzyme-mediated release of active therapeutics, which were evaluated as promising NA-based therapeutics against cancer cells.

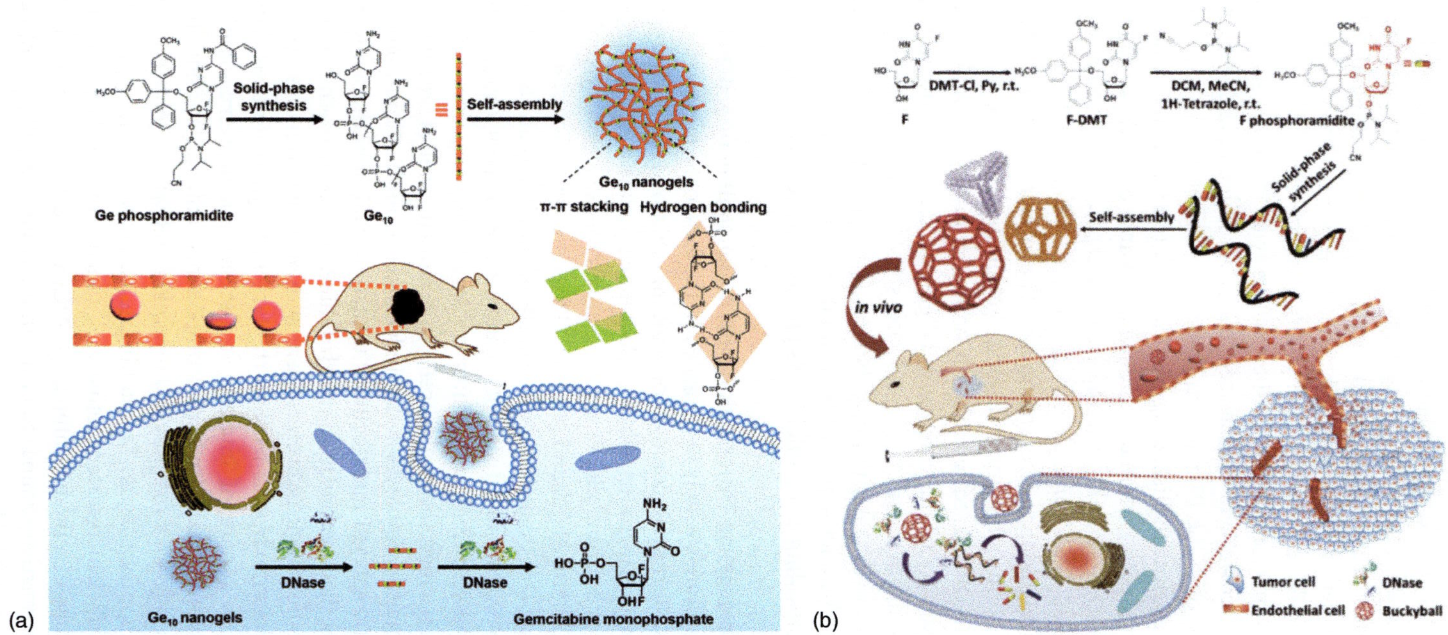

Figure 21.5 Incorporation of nucleoside analogs into DNA-based DDSs. (a) Development of DNA-like polygemcitabine (Ge_{10}) via solid-phase synthesis and Ge_{10}-based nanogel acting as precise nanodrug for cancer therapy. Source: Reproduced from Ma et al. [63], with permission from the Royal Society of Chemistry. (b) Nucleoside analogs-integrated DNA polyhedral acting as Trojan-horse-like DDSs in cancer treatments. Source: Reproduced from Mou et al. [64], with permission from the John Wiley & Sons.

21.2.2.2 Grafting the Chemotherapeutics on DNA Backbone

Phosphorothioate (PS) modification has been explored as a common modification on DNA backbone to improve the pharmacokinetics. It is a feasible modification that can be facilitated at specified position in the DNA sequence and achieved through solid-phase synthesis. Accordingly, therapeutics with functional groups, such as benzyl bromide and carbonyl ethyl bromide, are expected to graft onto PS-modified DNA at precise position by the reaction with PS group. As a proof of concept, Zhang and coworkers synthesized benzyl bromide-modified paclitaxel (PTX) and grafted it onto the PS segment-contained diblock DNA. The resulting conjugate was identified with high drug loading ratio (~53%), demonstrating the feasible conjugation of benzyl bromide-functionalized drug onto PS-modified DNA. Utilizing carbonyl ethyl bromide-modified camptothecin (CPT), Zhang group developed CPT-grafted DNA conjugates to generate CPT-based DNA tetrahedron [68] and DNA nanogel [69] for cancer treatments. The former CPT-DNA conjugates displayed adjustable hydrophilicity by tuning the number and site of PS groups on DNA strands and CPT-based tetrahedron with stimuli-responsive feature was obtained by self-assembly of the conjugates (Figure 21.6a). The latter CPT-DNA conjugates were applied to construct Y motifs and subsequently assembled to form CPT-conjugated DNA hydrogel with injectable feature (Figure 21.6b). This hydrogel achieved promising efficiency to penetrate tumor tissue, kill tumor cells, and inhibit tumor recurrency, providing a potential DDSs for local postsurgical treatments.

21.2.2.3 Directly Conjugating Therapeutic Agents on Terminus of DNA Strands

Therapeutics conjugation is widely investigated in the form of terminal conjugation onto DNA strands. Terminal conjugation can be achieved through reactions between diverse functional groups, while click chemistry plays a significant role in terminal conjugation. For example, Zhang and co-workers successfully conjugated camptothecin [70] (Figure 21.7a) and paclitaxel [34] onto DNA molecules via click chemistry, generating DNA nanostructure by self-assembling. Both functionalized nanostructures showed enhanced stability against nuclease comparing with free DNA. The camptothecin-DNA conjugates achieved light-triggered release of camptothecin, which killed the cancer cells in a similar level of free drug. The paclitaxel-DNA conjugate employed a disulfide linker to fulfill stimuli-responsive release of drug and displayed rapid cellular uptake. In addition, the DNA carrier was developed using therapeutic DNA, which achieved successful gene regulation, providing a platform for single delivery system with multiple functions. Aptamers, known as chemical antibody, can serve as the targeting ligands to assist targeted therapy that there is a raise of interests to develop aptamer–drug conjugates (ApDCs). ApDCs typically contain three components: aptamer ligands, linkers, and drug moieties (Figure 21.7b). Tan and coworkers reported a pioneering work about the construction of a Sgc8c-Dox conjugate based on covalent conjugation and consequently efficient targeted delivery to cancer cells (Figure 21.7c) [72]. Using organic synthesis method to prepare ApDC, aptamers and drugs could be functionalized with reactive groups, such as amino, thiol, and diarylcyclooctyne moiety (DBCO or ADIBO) [73].

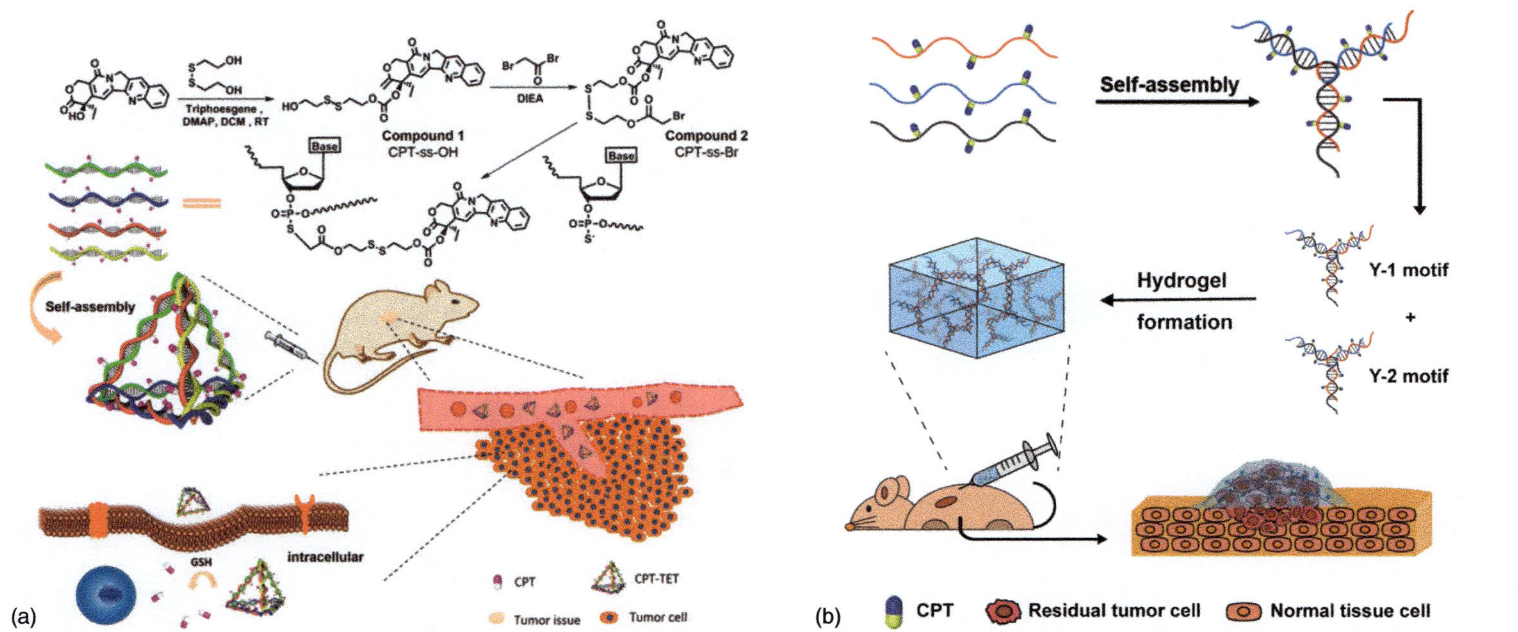

Figure 21.6 Incorporation of drugs into DNA-based DDSs via reaction with PS group on DNA. (a) Synthetic route of camptothecin-DNA conjugates and camptothecin-grafted tetrahedron with stimuli-responsive feature. Source: Reproduced from Zhang et al. [68], with permission from the John Wiley & Sons. (b) Construction of camptothecin-conjugated DNA hydrogel based on Y motifs. Source: Reproduced with permission from Zhang et al. [69], © 2020 American Chemical Society.

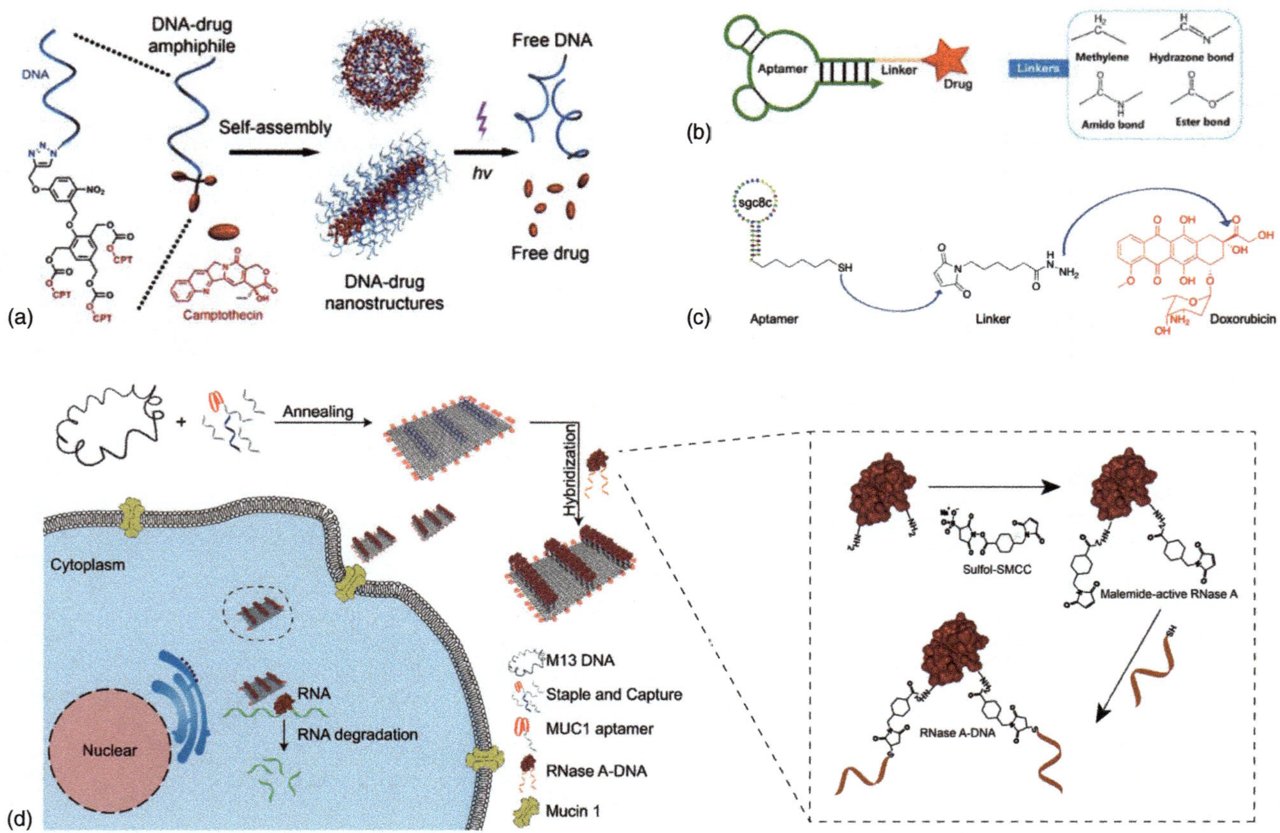

Figure 21.7 Syntheses of drug-DNA conjugates by terminal conjugation on DNA. (a) Camptothecin-grafted DNA nanostructure achieving light-triggered release of drug. Source: Reproduced with permission from Tan et al. [70], © 2015 American Chemical Society. (b) Aptamer–drug conjugates. Source: Reproduced with permission from Aliouat et al. [6], © 2022 Elsevier. (c) ApDC of Sgc8c-Dox conjugate prepared by covalent conjugation. (d) RNase A-loaded DNA origami functionalized with MUC1 aptamer. Source: Reproduced with permission from Zhao et al. [71], © 2019 American Chemical Society.

The strategy of covalent conjugation can be also applied to therapeutics other than small-molecules drugs. For example, cytotoxic protein ribonuclease (RNase) A is a promising protein therapeutics against cancer cells by triggering intracellular RNA degradation. However, the deficiency of targeted intracellular delivery of RNase A hamper the clinical application. Zhao et al. utilized sulfo-SMCC conjugation chemistry to conjugate RNase A onto ssDNA, fabricating an aptamer-conjugated RNase A-loaded DNA origami (Figure 21.7d) [71]. The resulting functional DNA origami construct successfully achieved intracellular delivery of RNase A into tumor cells and potent anti-tumor activity. This DNA origami-protein co-assembling method also provides a valuable platform for the delivery of protein therapeutics.

21.3 Precise Integration of Therapeutics into DNA-Based DDSs to Achieve Synergistic Cancer Treatment

Based on the general strategies described above, numerous therapeutics have been attempted to load into DNA carriers. Efficient therapeutic effects usually require proper stability of the DNA-drug complex in physiological condition as well as controlled release of therapeutics *in vivo*. Consequently, precise integration of therapeutics into DNA carriers that are commonly developed by covalent bonding or hybridization is considered to be a beneficial approach to achieve satisfying therapy. Loading via covalent bonding or hybridization allows for the designable and predictable properties of DDSs. Advantages of DNA-based DDSs have been stressed in this chapter, and developments of those have also been considerably reviewed with recent progress and challenges [4–6, 26, 74, 75]. Notably, therapeutic nucleic acid can also play a role as building block of drug carriers. Utilizing functional nucleic acid to fabricate drug carrier and collaborating with precise integration of drug onto them provides a promising strategy to enhance treatment efficacy and even fulfill synergistic cancer treatments. As a proof of concepts, a novel platform is introduced below to identify the potential of this strategy in future applications.

21.3.1 Chemogenes

Nowadays, chemotherapeutics still serves as the main therapy method in clinics to treat cancer. As DNA is a promising candidate of drug carrier, numerous efforts have been made on the investigation of DNA-chemotherapeutics hybrid nanostructures for anti-cancer therapy [76]. However, resistance of cancer cells to chemotherapeutics remains a huge obstacle in clinic. The chemotherapeutics resistance can occur through different mechanisms, and it is usually associated with abnormal expression of resistance-related proteins. Consequently, a combinational therapy involving chemotherapeutics and gene therapy could be expected to combat drug-resistant cancer cells, in which gene therapy could regulate the expression level of targeted proteins. ASOs have emerged as promising nucleic acid therapeutics that several ASOs are already clinically approved and a few of them are in clinical trials. ASO can achieve gene regulation by binding to specific mRNA and subsequently triggering RNase H-mediated mRNA cleavage. Therefore, the synergistic treatments involving ASOs to collaborate with chemotherapeutics have been explored against drug-resistant tumors [77–79]. Despite the treatment effects, the two elements in such combinational therapy usually make effects separately and require smart DDSs to manipulate drug loading efficiency, drug ratio, and drug release property.

Inspired by the work of facile integration of floxuridine upon solid-phase synthesis of DNA [64], Zhang and coworkers employed this strategy to incorporate drug into functional ASO sequence, developing a concept of two-in-one chemogene [56]. This chemogene platform was expected to achieve down-regulation of targeted drug resistance gene and release nucleoside analogue drug following with intracellular degradation (Figure 21.8a). To examine this concept, floxuridine was incorporated into a well-established ASO sequence that targets the B-cell lymphoma 2 (Bcl-2) family proteins. The functional ASO was synthesized by solid-phase synthesis and conjugated with biodegradable block copolymer, poly(ethylene glycol)-*b*-poly(ε-caprolactone) (PEG-*b*-PCL) to afford self-assembly of a micelle nanostructure. The constructed SNA-like chemogene successfully entered cells in the absence of transfection agents, reversed chemoresistance, and inhibit tumor growth both *in vitro* and *in vivo*. To improve the anti-tumor activity, Zhang and coworkers involved chemotherapeutics paclitaxel (PTX) into chemogene [80]. Two PTX molecules were conjugated to a floxuridine-integrated ASO via click reaction with fluorescent dithiomaleimide (DTM) linker (Figure 21.8b). This PTX-chemogene conjugate successfully formed SNA-like micellar nanoparticles by self-assembly, down regulated the expression of P-glycoprotein, and inhibited tumor growth upon the release of floxuridine and PTX within cells.

To enhance the targeting capacity, aptamers-chemogene has been investigated in a recent work [81]. Liu et al. developed a DNA strand containing a segment of aptamer AS1411 and a segment of ASO targeted to Bcl-2. Chemotherapeutics camptothecin (CPT) was subsequently grafted at predesigned site on the DNA molecule by covalent bonding (Figure 21.8c). As a result, the CPT-DNA conjugate proved the efficiency and controllability of this strategy on drug loading ratio and characterized with precise chemical composition as well as responsive disulfide linkage. The *in vitro* and *in vivo* studies identified the targeted co-delivery of chemo and gene agents, down-regulation of Bcl-2 protein, and efficient apoptosis of cancer cells, demonstrating the enhanced synergistic anti-tumor efficacy. In particular, comparing with the chemogene applying nucleoside substitution strategy, this work employed chemical conjugation to attach drug, which strategy could be practical to a number of chemotherapeutics.

21.3.2 Chemogene-Based DNA Nanomaterials

Chemogene materials are also capable of employing functional genes in complexes by hybridization. Based on the reaction of benzyl bromide and PS group, Zhang and coworkers grafted paclitaxel onto diblock DNA containing one phosphodiester (PO) segment and one PS segment [82] (Figure 21.9a). The resulting amphiphilic

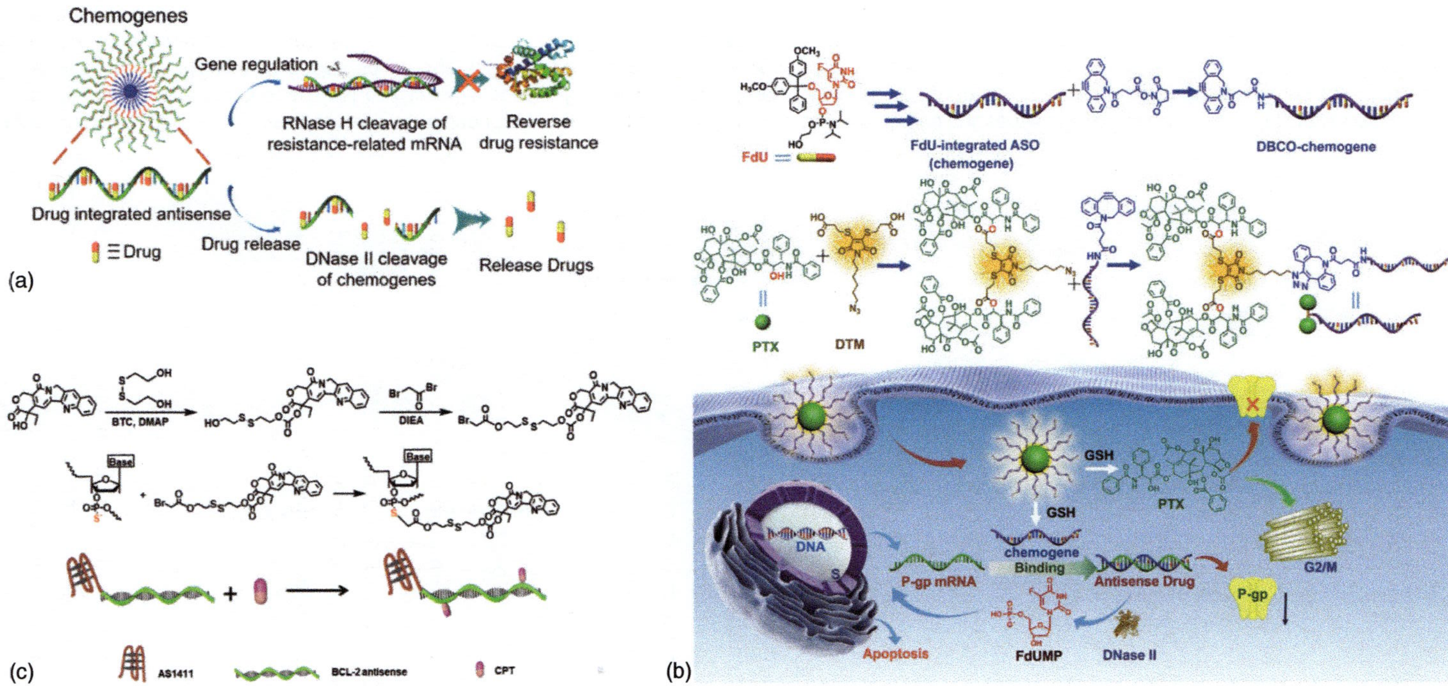

Figure 21.8 (a) Synergistic treatment based on conceptual chemogene. Source: Reproduced with permission from Mou et al. [56], © 2019 American Chemical Society. (b) Floxuridine-integrated chemogene formed SNA-like nanoparticles with stimuli-responsive character. Source: Reproduced with permission from Zhu et al. [80], John Wiley & Sons. (c) Aptamer-functionalized chemogene achieving efficient synergistic anti-cancer treatment. Source: Liu et al. [81]/John Wiley & Sons/CC BY 4.0.

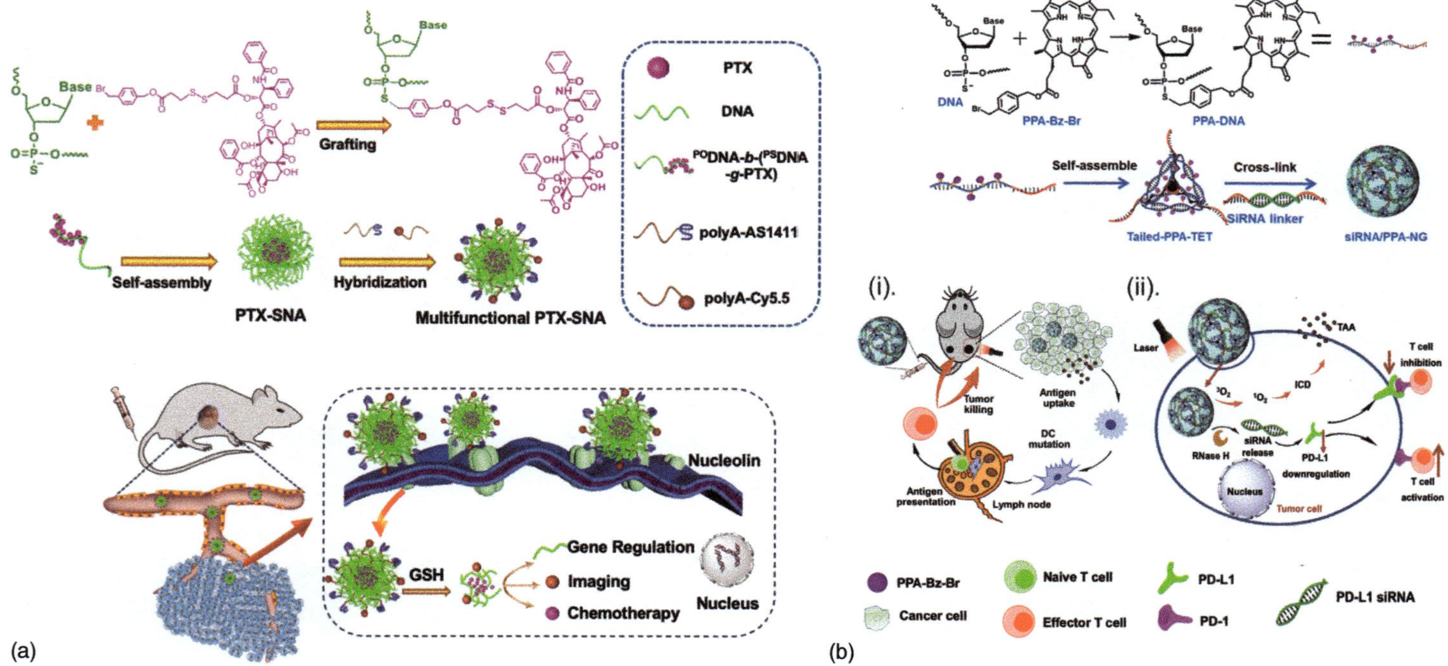

Figure 21.9 Schematic illustrations of chemogene materials. (a) Paclitaxel-loaded SNA employed antisense oligonucleotides by hybridization to generate chemogene material against cancer cells. Source: Reproduced from Guo et al. [82], with permission from the John Wiley & Sons. (b) Synthesis of siRNA and pheophorbide A co-packaged nanogel (siRNA/PPA-NG) by crosslinking of PPA-integrated tetrahedra with siRNA linker, achieving (i) *in vivo* and (ii) intracellular synergistic treatments. Source: Reproduced from Guo et al. [83], with permission from the American Association for the Advancement of Science.

DNA conjugates [^{PO}DNA-*b*-(^{PS}DNA-*g*-PTX)] were conducted to self-assembly to generate PTX-loaded SNA-like micellar nanoparticles (PTX-SNA). This PTX-SNA can be further functionalized with tumor-targeting aptamers, fluorescent probe strands, and ASOs by hybridization. These multifunctional PTX-SNA exhibited efficient specificity to cancer cells along with anti-cancer activity and showed the potential of chemogene materials. In addition to ASO, siRNA is also capable of gene regulation, which therapeutics could be participated into chemogene by hybridization with drug-conjugated DNA. Guo et al. developed photosensitizer pheophorbide A (PPA)-grafted DNA (PPA-DNA) via reactions of benzyl bromide with PS groups (Figure 21.9b) [83]. By self-assembly of PPA-grafted DNA strands, tetrahedra were formed, and the tails on tetrahedra were subsequently functionalized with programmed death ligand-1 (PD-L1) siRNA linker via hybridization, which ultimately led to the generation of siRNA and PPA co-packaged nanogel. The resulting nanogel showed a promising activity against melanoma tumor cells in manner of a synergistic treatment involving photodynamic therapy and immunotherapy, demonstrating a powerful platform for combinational cancer immunotherapy.

Overall, the nanoplatform of chemogene commonly includes small-molecule therapeutics along with gene therapeutics. It is characterized with the precise incorporation of small molecule therapeutics, not only limited to nucleoside analogue drugs. The syntheses of chemogene could be simple, designable, and cost-effective. Based on precise integration of small drug molecules onto functional DNA, it is therefore a promising platform to achieve synergistic cancer treatment. With future investigations, platform of chemogene may provide potent actions to combat drug-resistant cancers in clinic.

References

1 Sung, H., Ferlay, J., Siegel, R.L. et al. (2021). *CA Cancer J. Clin.* 71: 209–249.
2 Tibbitt, M.W., Dahlman, J.E., and Langer, R. (2016). *JACS* 138: 704–717.
3 Mitchell, M.J., Billingsley, M.M., Haley, R.M. et al. (2021). *Nat. Rev. Drug Discovery* 20: 101–124.
4 Jiang, Q., Zhao, S., Liu, J. et al. (2019). *Adv. Drug Delivery Rev.* 147: 2–21.
5 Komiyama, M., Shigi, N., Ariga, K. et al. (2022). *Adv. Funct. Mater.* 32: 2200924.
6 Aliouat, H., Peng, Y., Waseem, Z. et al. (2022). *Biomaterials* 285: 121532.
7 Gilboa, E., McNamara, J., and Pastor, F. (2013). *Clin. Cancer Res.* 19: 1054–1062.
8 Roberts, T.C., Langer, R., and Wood, M.J.A. (2020). *Nat. Rev. Drug Discovery* 19: 673–694.
9 Lee, H., Lytton-Jean, A.K.R., Chen, Y. et al. (2012). *Nat. Nanotechnol.* 7: 389–393.
10 Zhang, Q., Jiang, Q., Li, N. et al. (2014). *ACS Nano* 8: 6633–6643.
11 Chen, Y.J., Groves, B., Muscat, R.A., and Seelig, G. (2015). *Nat. Nanotechnol.* 10: 748–760.
12 Surana, S., Shenoy, A.R., and Krishnan, Y. (2015). *Nat. Nanotechnol.* 10: 741–747.
13 Jiang, D., Sun, Y., Li, J. et al. (2016). *ACS Appl. Mater. Interfaces* 8: 4378–4384.

14 Li, S., Jiang, Q., Liu, S. et al. (2018). *Nat. Biotechnol.* 36: 258–264.

15 Seeman, N.C. (2004). *Sci. Am.* 290: 64–75.

16 Bellot, G., McClintock, M.A., Chou, J.J., and Shih, W.M. (2013). *Nat. Protoc.* 8: 755–770.

17 Kick, B., Praetorius, F., Dietz, H., and Weuster-Botz, D. (2015). *Nano Lett.* 15: 4672–4676.

18 Ducani, C., Kaul, C., Moche, M. et al. (2013). *Nat. Methods* 10: 647–652.

19 Madsen, M. and Gothelf, K.V. (2019). *Chem. Rev.* 119: 6384–6458.

20 Seeman, N.C. (1982). *J. Theor. Biol.* 99: 237–247.

21 Kallenbach, N.R., Ma, R.I., and Seeman, N.C. (1983). *Nature* 305: 829–831.

22 Joaqui-Joaqui, M.A., Maxwell, Z., Ramakrishnam Raju, M.V. et al. (2022). *ACS Nano* 16: 2928–2294.

23 Cutler, J.I., Auyeung, E., and Mirkin, C.A. (2012). *JACS* 134: 1376–1391.

24 Jiang, Q., Song, C., Nangreave, J. et al. (2012). *JACS* 134: 13396–13403.

25 Li, J., Zheng, C., Cansiz, S. et al. (2015). *JACS* 137: 1412–1415.

26 Hu, Q., Li, H., Wang, L. et al. (2019). *Chem. Rev.* 119: 6459–6506.

27 Li, Z., Wang, C., Li, J. et al. (2020). *CCS Chem.* 2: 707–728.

28 Jiang, D., England, C.G., and Cai, W. (2016). *J. Controlled Release* 239: 27–38.

29 Zhang, T., Tian, T., Lin, Y. et al. (2022). *Adv. Mater.* 34: 2107820.

30 Liu, X., Wu, L., Wang, L., and Jiang, W. (2018). *Talanta* 179: 356–363.

31 Mirkin, C.A., Letsinger, R.L., Mucic, R.C., and Storhoff, J.J. (1996). *Nature* 382: 607–609.

32 Rosi, N.L., Giljohann, D.A., Thaxton, C.S. et al. (2006). *Science* 312: 1027–1030.

33 Mokhtarzadeh, A., Vahidnezhad, H., Youssefian, L. et al. (2019). *Trends Mol. Med.* 25: 1066–1079.

34 Tan, X., Lu, X., Jia, F. et al. (2016). *JACS* 138: 10834–10837.

35 Zhang, K., Hao, L., Hurst, S.J., and Mirkin, C.A. (2012). *JACS* 134: 16488–16491.

36 Wu, X.A., Choi, C.H.J., Zhang, C. et al. (2014). *JACS* 136: 7726–7733.

37 Skakuj, K., Wang, S., Qin, L. et al. (2018). *JACS* 140: 1227–1230.

38 Vasher, M.K., Yamankurt, G., and Mirkin, C.A. (2022). *JACS* 144: 3174–3181.

39 Ebrahimi, S.B., Samanta, D., Kusmierz, C.D., and Mirkin, C.A. (2022). *Nat. Protoc.* 17: 327–357.

40 Rothemund, P.W.K. (2006). *Nature* 440: 297–302.

41 Mei, Q., Wei, X., Su, F. et al. (2011). *Nano Lett.* 11: 1477–1482.

42 Shen, X., Jiang, Q., Wang, J. et al. (2012). *Chem. Commun.* 48: 11301–11303.

43 Jiang, Q., Liu, S., Liu, J. et al. (2019). *Adv. Mater.* 31: 1804785.

44 Du, Y., Jiang, Q., Beziere, N. et al. (2016). *Adv. Mater.* 28: 10000–10007.

45 Mo, F., Jiang, K., Zhao, D. et al. (2021). *Adv. Drug Delivery Rev.* 168: 79–98.

46 Zhao, J. and Stenzel, M.H. (2018). *Polym. Chem.* 9: 259–272.

47 Walsh, A.S., Yin, H., Erben, C.M. et al. (2011). *ACS Nano* 5: 5427–5432.

48 Li, J., Pei, H., Zhu, B. et al. (2011). *ACS Nano* 5: 8783–8789.

49 Liang, L., Li, J., Li, Q. et al. (2014). *Angew. Chem. Int. Ed.* 53: 7745–7750.

50 Choi, C.H.J., Hao, L., Narayan, S.P. et al. (2013). *PNAS* 110: 7625–7630.

51 Hamblin, G.D., Carneiro, K.M.M., Fakhoury, J.F. et al. (2012). *JACS* 134: 2888–2891.

52 Bujold, K.E., Fakhoury, J., Edwardson, T.G.W. et al. (2014). *Chem. Sci.* 5: 2449–2455.
53 Perrault, S.D. and Shih, W.M. (2014). *ACS Nano* 8: 5132–5140.
54 Li, Q., Zhao, D., Shao, X. et al. (2017). *ACS Appl. Mater. Interfaces* 9: 36695–36701.
55 Sun, W., Ji, W., Hall, J.M. et al. (2015). *Angew. Chem. Int. Ed.* 54: 12029–12033.
56 Mou, Q., Ma, Y., Ding, F. et al. (2019). *JACS* 141: 6955–6966.
57 Xue, H., Ding, F., Zhang, J. et al. (2019). *Chem. Commun.* 55: 4222–4225.
58 Zhao, Z., Fu, J., Dhakal, S. et al. (2016). *Nat. Commun.* 7: 10619.
59 Chang, M., Yang, C.S., and Huang, D.M. (2011). *ACS Nano* 5: 6156–6163.
60 Shieh, Y.A., Yang, S.J., Wei, M.F., and Shieh, M.J. (2010). *ACS Nano* 4: 1433–1442.
61 Tian, Y., Huang, Y., Gao, P., and Chen, T. (2018). *Chem. Commun.* 54: 9394–9397.
62 Gupta, A., Andresen, J.L., Manan, R.S., and Langer, R. (2021). *Adv. Drug Delivery Rev.* 178: 113834.
63 Ma, Y., Mou, Q., Zhu, L. et al. (2019). *Chem. Commun.* 55: 6603–6606.
64 Mou, Q., Ma, Y., Pan, G. et al. (2017). *Angew. Chem. Int. Ed.* 56: 12528–12532.
65 Ma, Y., Liu, H., Mou, Q. et al. (2018). *Nanoscale* 10: 8367–8371.
66 Pan, G., Mou, Q., Ma, Y. et al. (2019). *ACS Appl. Mater. Interfaces* 11: 41082–41090.
67 Zhu, L., Yang, J., Ma, Y. et al. (2022). *JACS* 144: 1493–1497.
68 Zhang, J., Guo, Y., Ding, F. et al. (2019). *Angew. Chem. Int. Ed.* 58: 13794–13798.
69 Zhang, J., Guo, Y., Pan, G. et al. (2020). *ACS Appl. Mater. Interfaces* 12: 21441–21449.
70 Tan, X., Li, B.B., Lu, X. et al. (2015). *JACS* 137: 6112–6115.
71 Zhao, S., Duan, F., Liu, S. et al. (2019). *ACS Appl. Mater. Interfaces* 11: 11112–11118.
72 Huang, Y.F., Shangguan, D., Liu, H. et al. (2009). *ChemBioChem* 10: 862–868.
73 He, J., Duan, Q., Ran, C. et al. (2023). *Acta Pharm. Sin. B* 13: 1358.
74 Nicolson, F., Ali, A., Kircher, M.F. et al. (2020). *Adv. Sci.* 7: 2001669.
75 Lacroix, A. and Sleiman, H.F. (2021). *ACS Nano* 15: 3631–3645.
76 Huang, X., Blum, N.T., Lin, J. et al. (2021). *Mater. Horiz.* 8: 78–101.
77 MacDiarmid, J.A., Amaro-Mugridge, N.B., Madrid-Weiss, J. et al. (2009). *Nat. Biotechnol.* 27: 643–651.
78 Garbuzenko, O.B., Saad, M., Pozharov, V.P. et al. (2010). *PNAS* 107: 10737–10742.
79 Ling, H., Fabbri, M., and Calin, G.A. (2013). *Nat. Rev. Drug Discovery* 12: 847–865.
80 Zhu, L., Guo, Y., Qian, Q. et al. (2020). *Angew. Chem. Int. Ed.* 59: 17944–17950.
81 Liu, Y., Zhang, J., Guo, Y. et al. (2022). *Exploration* 2: 20210172.
82 Guo, Y., Zhang, J., Ding, F. et al. (2019). *Adv. Mater.* 31: 1807533.
83 Guo, Y., Zhang, Q., Zhu, Q. et al. (2022). *Sci. Adv.* 8: eabn2941.

Index

DNA Nanotechnology for Cell Research: From Bioanalysis to Biomedicine, First Edition. Edited by Zhou Nie.

p

r